Primates of the World

Adult female emperor tamarin (*Saguinus imperator*) in Manu National Park, Peru (photo courtesy of Charles Janson)

Primates of the World

Distribution, Abundance, and Conservation

Jaclyn H. Wolfheim

New York Zoological Society

UNIVERSITY OF WASHINGTON PRESS

Seattle and London

Printed in the United States of America

Library of Congress Cataloging in Publication Data

Wolfheim, Jaclyn H.
Primates of the world.

Bibliography: p.
1. Primates. I. Title.
QL737.P9W64 1982 599.8 82-13464
ISBN 0-295-95899-5

This book was published with the assistance of a grant from the New York Zoological Society.

Dedication

To the memory of Dr. Carl B. Koford,
whose understanding of nature, commitment to conservation,
friendship, and encouragement were a great source of inspiration.

Contents

Introduction

Awareness of the plight of endangered animals and plants and the need for their conservation is increasing, owing to a growing body of research, technical reports, and popular publicity. Yet many of the fundamental questions regarding the status of species, the factors affecting their populations, and methods to protect them have not been answered. Such is the case with the primates, a group represented worldwide by about 150 species and several hundred subspecies. The Order Primates includes prosimians (lemurs and their relatives, galagos, pottos, lorises, and tarsiers), New World primates (cebids, marmosets, and tamarins), Old World monkeys (baboons, macaques, guenons, mangabeys, and leaf eaters), gibbons, great apes, and human beings.

The nonhuman primates are of special interest both intrinsically and because of their close relation to man. In their own right, these animals present a dazzling diversity in morphology, behavior, and ecological adaptations. They are important components of their primarily tropical ecosystems, and their roles in complex tropical forests are only beginning to be studied. Many species are dependent on undisturbed primary forest and can act as indicators revealing the health and viability of the forest habitat.

Nonhuman primates share many anatomical, physiological, and behavioral characteristics with man and are often substitutes for human beings in biological, medical, pharmacological, psychological, and anthropological research. For example, their responses to drugs and vaccines are used as predictors of human reactions, and their social systems provide clues to the origins and evolution of early man. Scientists and laymen alike value the primates for their similarities to human beings and the knowledge of ourselves which they can reveal.

But until now it has been difficult to learn about the situations of populations of primates. Nearly one-third of the species of primates are listed as endangered, rare, or vulnerable by the International Union for the Conservation of Nature and Natural Resources (I.U.C.N. 1972–78). Yet for many of these species, basic information on geographic range, abundance, and natural history is unavailable, incomplete, or scattered throughout the literature. Several recent books (e.g., Bermant and Lindburg 1975, Tattersall and Sussman 1975a, Thorington and Heltne 1976, Clutton-Brock 1977, Rainier and Bourne 1977, Chivers and Lane-Petter 1978, Kleiman 1978) contain detailed essays on the status of single species or groups of species, but the topics and emphases vary greatly and many species have not been treated.

The purpose of this volume is to synthesize the information now available describing the distribution, abundance, habitats, and factors affecting natural populations of each primate species. A uniform format facilitates comparison of species. This information is pertinent to several theoretical and practical questions. How can we decide which species are now in danger of extinction and predict which ones will need protection in the future? How can we quantify this danger and compare the degree of peril of different species? What characteristics of species or their environments are related to survival status? What are the present population levels? What sizes of populations and of habitat reserves are necessary to ensure the viability of a species? What are the effects on various species of different kinds of habitat disturbance, hunting, and collection? Which species are found in national parks or other reserves, and what is the quality of protection in these areas?

Information provided here is offered for use by zoologists, wildlife biologists, ecologists, foresters, governmental agencies, primate researchers, and conservation organizations in designing programs to protect or manage primate populations. These data should be useful in (1) planning the location and size of habitat reserves, (2) devising methods of land use which will minimize damage to primate habitats, (3) formulating regulations for hunting, exploitation, and trade, (4) deciding whether certain species should be cropped, controlled, bred in captivity, used in research, or collected for other purposes, and (5) persuading scientists, governments, industries, conservation groups, educational institutions, funding agencies, and the public of the need for action and support.

But many pieces of the puzzle are still missing, and knowledge of the ecology and status of many species is scanty. By revealing where information is lacking, this compilation should also aid biologists in designing research projects to gather the additional data necessary to plan and implement wise and effective measures for conservation of the primates.

Acknowledgments

I am grateful to the following organizations and individuals, and thank them most sincerely:

The New York Zoological Society, and a grant from the Allied Chemical Foundation to the New York Zoological Society's Charles W. Nichols, Jr. Wildlife Conservation Publications Fund, which made the preparation and publication of this work possible. Dr. F. Wayne King was especially helpful.

The U.S. Fish and Wildlife Service, Department of Interior, which, through its contract No. 14-16-0008-747 to the World Wildlife Fund, supported the early research.

The 111 primate field biologists who contributed unpublished information and opinions. Their contributions were invaluable and I appreciate their help.

Drs. J. Stephen Gartlan and Thelma E. Rowell, for productive discussions in developing some of the ideas presented here.

Drs. R. Bruce Bury, Anne B. Clark, J. Stephen Gartlan, Arthur D. Horn, Carl B. Koford, Russell A. Mittermeier, Lawrence A. Riggs, Thelma E. Rowell, Thomas T. Struhsaker, and Ronald L. Tilson, who reviewed sections of the manuscript and made many constructive suggestions.

Dr. Richard W. Thorington, Jr., and the Smithsonian Institution for supplying office space.

Colorado State University for library research space and excellent services.

Ms. Benella Caminiti, Primate Information Center, University of Washington, and Dr. Sydney Anderson, American Society of Mammalogists, American Museum of Natural History, for useful literature searches.

The Denoyer-Geppert Company for permission to use their copyrighted base maps of Africa and South America.

Times Books for permission to use map projections of Asia and Madagascar based on maps in the *Times Atlas of the World* (1977), copyright John Bartholomew and Son, Ltd.

The H.R.B. Singer Company for the use of their Lasico Autoscaler Planimeter.

Martha Scott, Mary Greenwood, Jill Whelan, Nancy Wilson, and the Natural Resource Ecology Laboratory for assistance; Estelle Duell, Yourlette Hailey, Diane Fox, and Jan Hill for diligent typing; and Pat MacMillan for meticulous inking of the figures.

Dr. R. Bruce Bury for insight, sense, optimism, and, above all, patience during the completion of this volume.

Methods

Data Collection

Letters of inquiry and questionnaires requesting answers to nine specific questions were sent to 226 primate field biologists. The questions concerned the extent of the geographic range, specific study sites, relative and absolute abundance, population parameters (density, group size, home range area), habitats utilized, factors affecting populations, and conservation measures proposed or taken for any species, studied firsthand, in the field.

I received responses via letters, completed questionnaires, and telephone and personal interviews, totaling 285 personal communications containing information on the status of 93 species. These data and opinions are cited in the accounts as personal communication (pers. comm.) with the date of receipt.

I also reviewed more than 1,200 references from the zoological literature, primarily after 1960. Articles were located by computer search by the Primate Information Center and the American Society of Mammalogists, and by personal scanning of books and journals.

No attempt was made to examine every possible relevant report, since this would have been a gargantuan and endless task. Rather, my goal was to locate as much recent, representative data as possible. Information from literature compilations such as those by Hill (1953, 1955, 1957, 1960, 1962, 1966, 1970, 1974), Tappen (1960), Napier and Napier (1967), and Baldwin and Teleki (1972, 1973, 1974, 1977) was only included if recent primary sources of data were unavailable. If information from a personal communication had also been published, only the published version was cited.

Published material was integrated with the personal communications into status accounts for each species under the headings shown below.

Taxonomic Note

In general, the taxonomic arrangements and nomenclature followed are those of Petter and Petter-Rousseaux (1979) for the prosimians (Lorisidae, Tarsiidae, and Lemuroidea [Cheirogaleidae, Lemuridae, Indriidae, Daubentoniidae]), Hershkovitz (1977) for the marmosets and tamarins (Callitrichidae), Cabrera (1958) and Hershkovitz (1972) for the New World monkeys (Cebidae), Thorington and Groves (1970) for the Old World monkeys (Cercopithecidae), Groves (1967b; 1970a,b; 1971a,d; 1972) for the apes (Hylobatidae and Pongidae), and Napier and Napier (1967) for all groups. Other authorities were used in taxonomic decisions for certain species. Major differing opinions on nomenclature or taxonomy are outlined in this section. In controversial cases, the most widely accepted opinion or the most parsimonious grouping for the amount of population data has been accepted. Subspecies are not discussed unless the status of one greatly differs from that of the others.

I treat 151 species and complexes, including 14 Lorisidae and Tarsiidae, 21 Lemuroidea, 16 Callitrichidae, 25 Cebidae, 65 Cercopithecidae, 6 Hylobatidae, and 4 Pongidae.

Geographic Range

The global extent of the natural geographic distribution of the species is described here. The limits of the range were determined from published and unpublished range maps and collecting localities; coordinates and country names are from the *Times Atlas of the World* (1977). Data from introduced populations such as the one on Cayo Santiago Island, Puerto Rico, are not included in this volume. The location of countries and their boundaries are shown in Maps 1 to 4.

Map 1. Countries of Africa and the Arabian peninsula

The range map was drawn as accurately as possible on a continental scale, combining all available general and detailed maps and reports. The sources have not been listed on the map because they are so numerous, but may be found in the geographic range section and distribution table of the account.

In many places inside and outside the accepted range and on its boundaries the presence of species is unknown; some of these areas are indicated by question marks. It is axiomatic that each species occurs only in suitable habitat within the limits of its range and not throughout the whole area delimited by the boundaries.

The particulars of distribution of the species and the places where it has been studied are presented in a large table entitled "Distribution of species X." The countries within the range are listed, usually in clockwise order or from west to east

Map 2. Countries of Central and South America

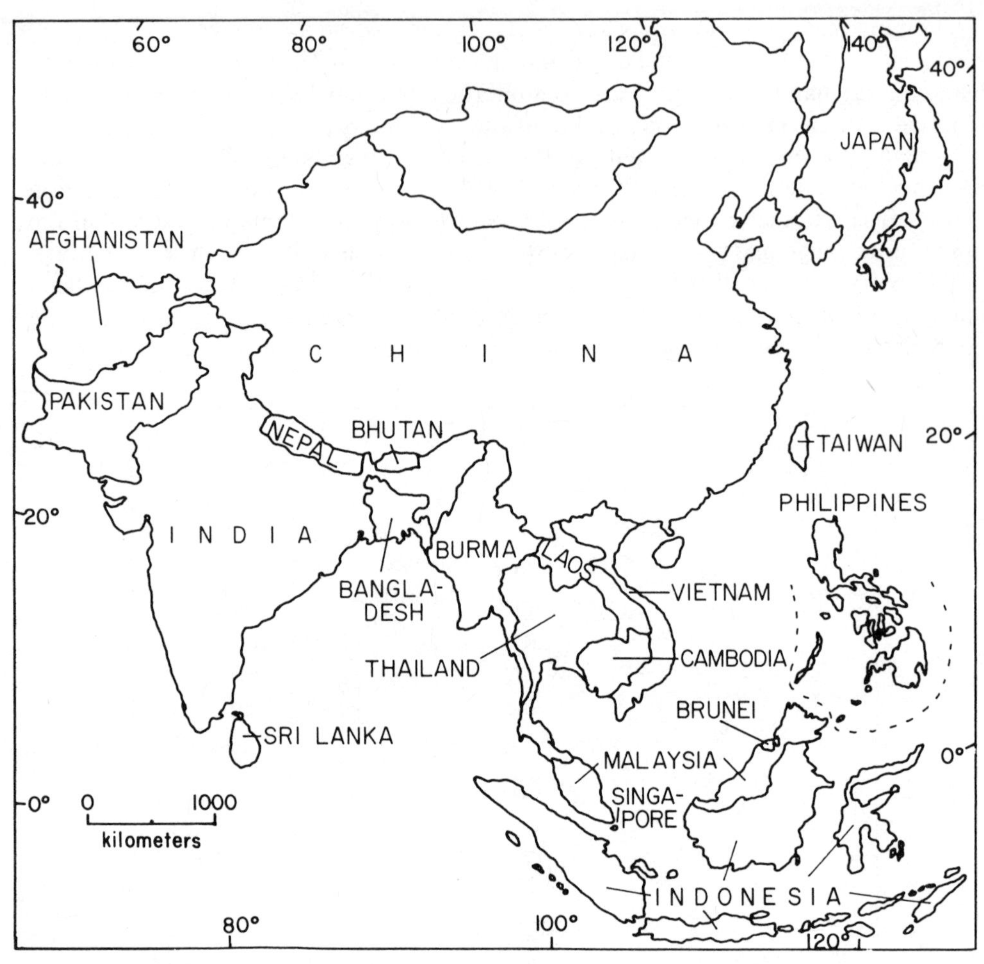

Map 3. Countries of Asia inhabited by primates

or north to south; and within each country, reports of the location, abundance, or status of the species are arranged chronologically.

The data are condensed as much as possible, thus authors reporting the same information are listed together. Square brackets around a reference indicate that its authors mentioned the species at the same site as companion references but did not express the same opinion about abundance. Abbreviations used include "R." for river, "N.P." for national park, and "n.l." denoting that a place mentioned by an author could not be located. Some names are spelled differently by various authors or even on different pages within one paper (for instance in Ethiopia: Shashemenne, Shashamane, or Shashemene, and Webe Shebelle, Webi Shebele, Uebi Shebile, or Webbe Shibeli). It is sometimes impossible to know if one locality is the same as another with a similar name. Thus spellings follow those of the source author or, in cases of controversy, the *Times Atlas of the World* (1977). The precise location of

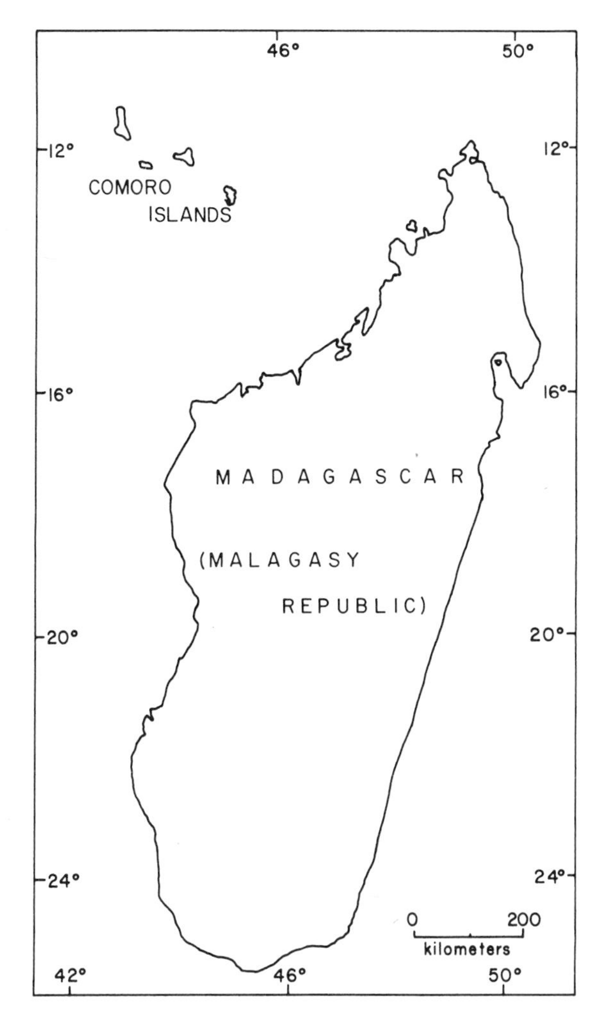

Map 4. Madagascar and the Comoro Islands

many sites can be found in the cited references. If a source reported that a species occurred in a country but did not say where or in what abundance, the reference was not cited here unless no specific evidence was available.

Abundance and Density

Overall abundance and population trends in various countries in recent years (since 1970) are summarized from the table. Estimates of population density, group size, and home range area are presented in tabular form if the data are many. The order in this table follows the order of countries in the previous (distribution) table to facilitate comparison.

Population density is defined as the number of individuals per unit area. Some authors calculate this value using the area of available habitat, others use an entire study plot or reserve, and some do not state their criteria. I have used what they present. The values given here have not been corrected in any way and vary greatly in the number of animals and sizes of areas which they represent.

If an author listed only the number of individuals seen in a region, I did not calculate population density, because I could not know how much of the area was used by the animals. But if the source gave both group density and average group size, I have converted these values to population density. If extremes of variation in population density or home range area were available, these are shown rather than averages. Population density and home range measurements are expressed in metric units (km^2 and ha respectively) throughout.

Group size indicates the number of individuals living together in heterosexual groups. Solitaries were excluded here, since they do not qualify as groups. If a report contained several estimates of mean group size (e.g., in different habitats or at several study sites), I list these as a range of averages under "$\bar{X}$ group size." If the author presented group counts and the total number of groups (N), I have calculated the average group size.

The home range is the area a group of the species occupies and traverses in the course of its normal activities. I have not calculated home range sizes from day ranges or other data. If only the territory size and not the home range was given, I list the former, since home range must be at least as large as territory. Sometimes a report showed the home range area for only one group, or a mean calculated from several groups, not necessarily the same number used to calculate average group size. These values are listed if extremes were not reported.

Habitat

This section includes a general description of the typical vegetation used by the species. Then preferred habitats and other formations which the species can exploit are listed. Where possible, climatic tolerances, elevational distribution, and limiting factors or prerequisites for survival are also given. Complete lists of the tree species in each habitat, plants used, food habits, or dietary requirements are not given but may often be found in the cited references.

Factors Affecting Populations

In the following four subsections, reported occurrences of each factor affecting populations are recounted, and opinions regarding effects on the species are discussed.

Habitat Alteration

This includes all types of habitat changes which are known or suspected to be influencing populations of the species.

Human Predation

Hunting. As defined here, hunting is the killing of members of the species for any of several purposes, excepting those listed below.

Pest Control. This involves efforts to reduce or exterminate populations of primates because of damage to human goods or property.

Collection. The numbers of live individuals gathered and kept alive, exported, or imported into other countries are listed.

Conservation Action

This section contains a summary of the number of parks and reserves in which the species was reported in the distribution table, and of other measures that have been taken or proposed to protect its populations or its natural habitats. Also listed are local, state, national, and international laws or regulations concerning hunting, capture, or trade, as well as pertinent recommendations.

Discussion

The salient information regarding all species of primates is synthesized here from the 151 status acocunts, and some of the implications are discussed.

I measured the sizes of the geographic ranges (as shown on the distribution maps) with a Lasico Autoscaler Planimeter, and obtained the "number of countries" by counting the number of countries or parts of countries included in the distributions of each species. Some of the body weights are averages, some represent one individual only, others are the extremes of a range of variation, and many are from captive animals (Napier and Napier 1967; Rowell, pers. comm. 1978).

The other data in this section are taken from the individual accounts, and the categories are defined accordingly.

References

A few (22) of the references listed were cited in the accounts but not seen. These generally were reports that were old or could not be located, containing information not available elsewhere and recounted in another publication. These were used as rarely as possible, and are identified in the text as, for example, "X (1910) cited by Y (1965)." All other references were examined.

LORISIDAE
Pottos, Galagos, and Lorises

Arctocebus calabarensis Angwantibo or Golden Potto

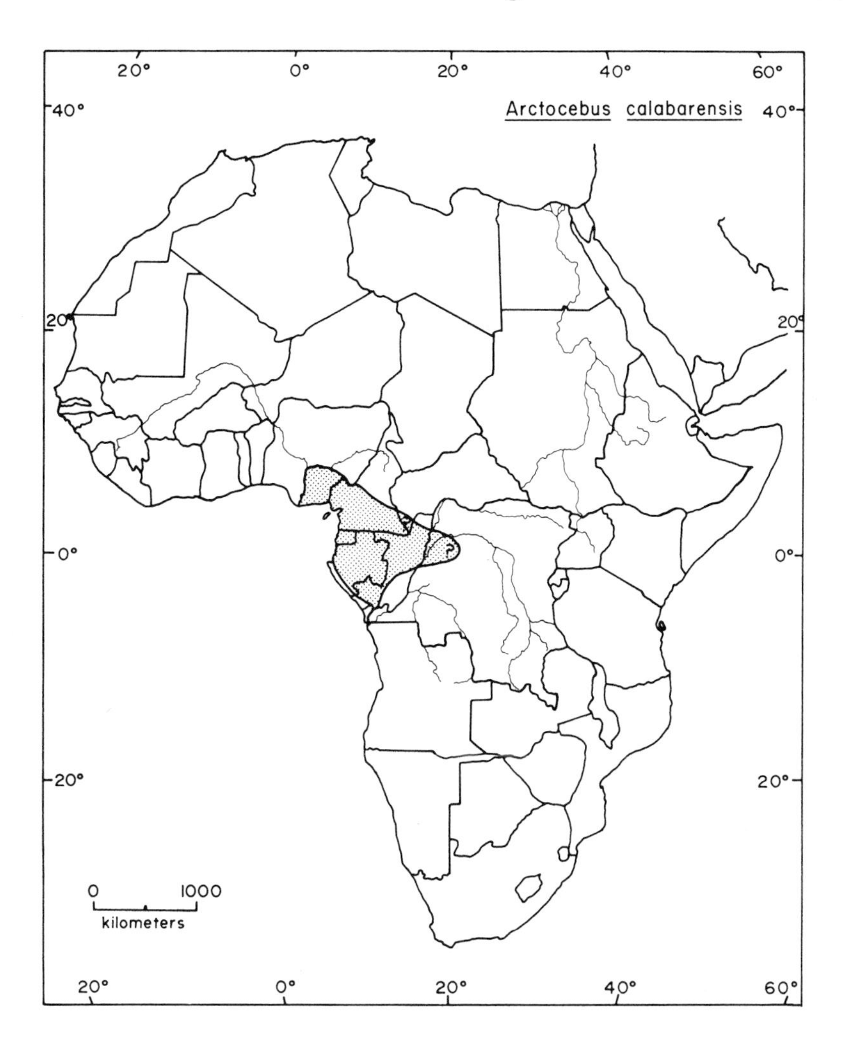

Taxonomic Note

Two subspecies are recognized (Napier and Napier 1967, Petter and Petter-Rousseaux 1979).

Geographic Range

Arctocebus calabarensis is found in western equatorial Africa, from Nigeria to Congo. It occurs as far west as 6°47′E (at 6°11′N) and as far north as 7°41′N, 7°34′E, east and south of the Niger-Benue rivers in southeastern Nigeria (Oates and Jewell 1967). The southern limit of its range is at about 5°S in southern Congo (Vincent 1969) and the eastern limit is either along the Ubangi River at about 18°E in northern Congo (Dorst and Dandelot 1969) or further east to about 20°E in western Zaire

(Vincent 1969, Petter and Petter-Rousseaux 1979). For details of the distribution, see Table 1.

Abundance and Density

The angwantibo has recently been described as abundant but vulnerable in Nigeria, common in Equatorial Guinea, and rare in northeastern Gabon (Table 1). Its present status in Cameroon, Central African Republic, Congo, and Zaire is not known.

Populations of *A. calabarensis* occur at densities of 2 to 7 individuals per km^2 (Charles-Dominique 1971, 1977b), and as high as 18 individuals per km^2 (Charles-Dominique 1978). At night, individuals forage alone 97% of the time and in pairs 2% of the time, and during the day they sleep singly except for mother-infant associations (Charles-Dominique 1977b). No estimates of home range area are available.

Habitat

A. calabarensis is a forest species whose preferred habitat is low, dense growth, often in young secondary forest, farmland edges, and forestry plantations (Jewell and Oates 1969, Oates 1969). These authors found angwantibos also in open habitats such as *Gmelina* plantations, and in wooded savanna, and more abundant in open young plantations than in mature high forest in Nigeria. In Equatorial Guinea this species occurs in secondary forest and commercial plantations (Sabater Pi 1972b).

In northeastern Gabon, *A. calabarensis* is found in primary and secondary forest (Charles-Dominique 1974a), but its distribution in primary forest coincides exactly with tree-fall zones that have been disrupted by strong winds or human activities, and where the regenerating vegetation contains many small lianes and dense undergrowth (Charles-Dominique 1977b, Charles-Dominique and Bearder 1979).

Factors Affecting Populations

Habitat Alteration

Most of the habitats occupied by *A. calabarensis* in Nigeria have been altered by man to some extent, either by cutting for cultivation or wood, tree planting, or other disturbances, and these activities may have increased the forest fringe habitat available to angwantibos (Jewell and Oates 1969, Oates 1969). Nevertheless, deforestation would eliminate this species, and the recent civil war in Nigeria has probably greatly decreased its habitat there (Oates 1969).

In Gabon, limited tree felling has increased the secondary growth available to *A. calabarensis* (Charles-Dominique and Bearder 1979), but no reports were located concerning the overall effects of habitat alteration there or in the other countries of its range.

Human Predation

Hunting. Angwantibos are hunted for food in Nigeria (Jewell and Oates 1969), but are hardly ever hunted in northeastern Gabon (Charles-Dominique 1977b).

Pest Control. No information.

Collection. Three *A. calabarensis* were imported into the United States in 1968–73

(Appendix) from Nigeria and Cameroon. None entered the United States in 1977 (Bittner et al. 1978) or the United Kingdom in 1965–75 (Burton 1978).

Conservation Action

A. calabarensis has been reported in two reserves in Nigeria and one park in Equatorial Guinea (Table 1). Additional preserves or sanctuaries are needed for it in Nigeria and elsewhere (Jewell and Oates 1969, Happold 1971).

This species is included in Class B of the African Convention, giving it legal protection from hunting, killing, capture, or collection except by special authorization (Organization of African Unity 1968).

Table 1. Distribution of *Arctocebus calabarensis* (countries are listed in order from north to south)

Country	Presence and Abundance	Reference
Nigeria	Only E of Cross R. (extreme SE corner) (map)	Booth 1958
	Localities listed and mapped at Akpacha, Nsukka, Akpaka Forest Reserve near Onitsha, Mamu Reserve, Owerri, Umuahia, Elele, Aba, and Calabar, all in SE E of Niger R. and S of Benue R. including W of Cross R.; abundant	Oates and Jewell 1967, Oates 1969
	Localities mapped in SE; not secure; may become scarce if not protected	Jewell and Oates 1969
	In SE, E of Niger R. (maps)	Vincent 1969, Petter and Petter-Rousseaux 1979
	E of Niger R.	Happold 1971
Cameroon	In S half (maps)	Booth 1958, Vincent 1969, Petter and Petter-Rousseaux 1979
Central African Republic	Questionable in SW corner (map)	Vincent 1969
Equatorial Guinea	Throughout (maps)	Booth 1958, Vincent 1969, Petter and Petter-Rousseaux 1979
	Throughout, especially in interior; common	Sabater Pi 1966
	Localities listed and mapped at Egnonayong, Movo, and Benito R. (all W central)	Jones 1969
	In Monte Raices Territorial Park (in NW, E of Bata)	I.U.C.N. 1971
	Localities listed and mapped at Machinda, Achimeláng, Ndjiakom, Alarmitáng, and Adjap (in W); not very rare	Sabater Pi 1972b

Table 1 (continued)

Country	Presence and Abundance	Reference
Gabon	Rare	Malbrant and Maclatchy 1949
	Throughout except along W coast S of about 1°S (maps)	Vincent 1969, Petter and Petter-Rousseaux 1979
	Rare near Makokou (NE)	Charles-Dominique 1971
	At Makokou, Ogooué-Ivindo basin	Charles-Dominique 1974a, 1977b, 1978; Charles-Dominique and Bearder 1979
Congo	Rare	Malbrant and Maclatchy 1949
	In NW, central, and SW (maps)	Vincent 1969, Petter and Petter-Rousseaux 1979
Zaire	Questionable in W between about 2°N and 1°S and about 20°E (map)	Vincent 1969
	In W (map)	Petter and Petter-Rousseaux 1979

Euoticus elegantulus Western Needle-clawed Galago

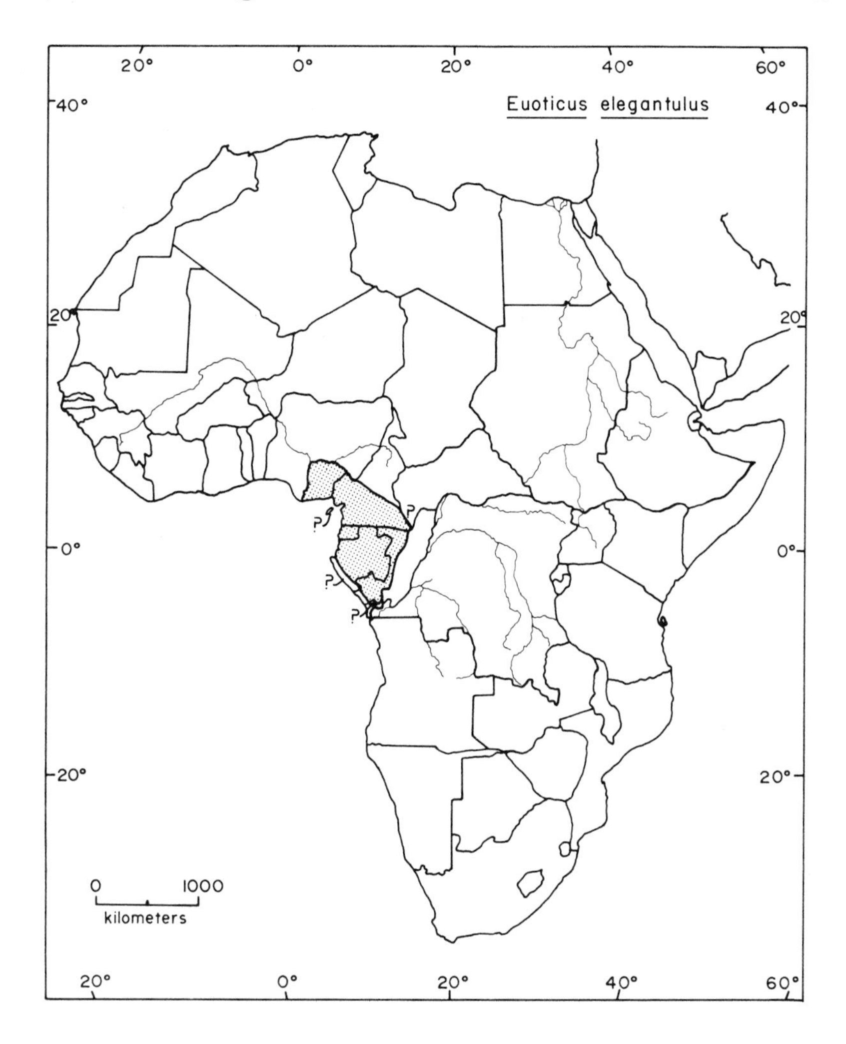

Taxonomic Note

Two subspecies are recognized in this species, which is sometimes placed within the genus *Galago* (Napier and Napier 1967, Kingdon 1971, Petter and Petter-Rousseaux 1979).

Geographic Range

Euoticus elegantulus is found in western Africa from southeastern Nigeria to southern Congo. It has been recorded as far west as 6°48′E (at 5°6′N) near the Niger River delta and as far north as 6°N, 7°30′E in the same area (Oates and Jewell 1967). This species occurs as far south as about 6°S near the mouth of the Congo River, and east to about 16°E at the southeastern corner of Cameroon (Vincent 1969, 1978;

Petter and Petter-Rousseaux 1979) or further east to about 19°E along the Ubangi River (Kingdon 1971). For details of the distribution, see Table 2.

Abundance and Density

The western needle-clawed galago is rare in Nigeria and Equatorial Guinea (Table 2). Its status elsewhere in its range is not known.

In northeastern Gabon, population densities of *E. elegantulus* vary from 3 to 30 ($\bar{X}$ = 15) individuals per km^2 (Charles-Dominique 1971). Individuals forage alone 76% of the time and in groups of 2 to 4 24% of the time, and form daytime sleeping groups of as many as 7 members (Charles-Dominique 1977b). No estimates of home range area were located.

Habitat

E. elegantulus is a forest species which occupies both primary and secondary forest. It has been recorded in young and old secondary forest in Nigeria (Jewell and Oates 1969), and in vegetation at the edges of clearings, and large trees in coffee plantations, in Equatorial Guinea (Oates 1969). In Cameroon it also occupies young secondary forest and plantations (Molez 1976).

At one study site the highest densities of *E. elegantulus* were found in secondary forest and riverine vegetation, with lower densities along paths in primary forest, in the interior of primary forest, and in flooded primary forest (Charles-Dominique 1971). There the species utilized the highest trees (*Albizzia gummifera* and *Pentacletra eetveldeana*) which dominated the bushy undergrowth of the secondary forest, and large trees in primary forest (Charles-Dominique 1977b, Charles-Dominique and Bearder 1979).

Factors Affecting Populations

Habitat Alteration

Since *E. elegantulus* thrives in secondary forest, it is likely to tolerate limited habitat disturbance. But any large-scale deforestation will decrease its available habitat. For example, the recent civil war in Nigeria has probably severely affected its habitats there (Oates 1969). No specific information is available concerning the effects of habitat alteration on this species anywhere in its range.

Human Predation

Hunting. *E. elegantulus* is hardly ever hunted in northeastern Gabon (Charles-Dominique 1977b). No reports from other countries were received.

Pest Control. No information.

Collection. Two *E. elegantulus* were exported from Equatorial Guinea between 1958 and 1968 (Jones, pers. comm. 1974). None were imported into the United States in 1968–1973 (Appendix) or in 1977 (Bittner et al. 1978), or into the United Kingdom in 1965–75 (Burton 1978).

Conservation Action

E. elegantulus has been reported in one reserve in Nigeria and one in Cameroon (Table 2). A forest park or sanctuary is needed for it in Nigeria (Happold 1971). This species is listed in Class B of the African Convention, giving it legal protection from killing or capture except by special authorization (Organization of African Unity 1968).

Table 2. Distribution of *Euoticus elegantulus* (countries are listed in order from north to south)

Country	Presence and Abundance	Reference
Nigeria	In SE corner E of Cross R. only (map)	Booth 1958
	Localities listed and mapped at Elele (5°6′N, 6°48′E) and Mamu Reserve (further N and E) (both W of Cross R.)	Oates and Jewell 1967, Jewell and Oates 1969
	Not abundant	Oates 1969
	In SE, E of Niger R. (maps)	Vincent 1969, 1978; Kingdon 1971; Petter and Petter-Rousseaux 1979
	E of Niger R.	Happold 1971, Monath and Kemp 1973
Cameroon	Throughout S (maps)	Booth 1958, Vincent 1969
	At Bakundu Reserve (W)	Gartlan and Struhsaker 1972
	Near Nkolguem I, 15 km N of Yaoundé (S central)	Molez 1976, Vincent 1978
Central African Republic	In SW (map)	Kingdon 1971
Equatorial Guinea	Throughout mainland (maps)	Booth 1958; Vincent 1969, 1978; Kingdon 1971
	On Macias Nguema (formerly Fernando Po)	Dorst and Dandelot 1969
	A few in interior; very scarce	Sabater Pi 1966
	Localities listed and mapped at Achimeláng (1°41′N, 9°49′E) and Eyamoyong (1°37′N, 9°49′E) (W central)	Jones 1969
	Locality mapped at Eyamoyong	Sabater Pi 1972b
Gabon	Throughout	Malbrant and Maclatchy 1949
	Throughout except W coast S of about 1°S (maps)	Vincent 1969, 1978; Petter and Petter-Rousseaux 1979
	In Makokou area, Ogooué-Ivindo basin (NE)	Charles-Dominique 1971, 1974a, 1977b, 1978; Charles-Dominique and Bearder 1979
	Throughout (map)	Kingdon 1971
Congo	Throughout	Malbrant and Maclatchy 1949, Kingdon 1971
	In W (maps)	Vincent 1969, 1978; Petter and Petter-Rousseaux 1979
Angola	In E Cabinda (maps)	Vincent 1969, 1978; Petter and Petter-Rousseaux 1979
Zaire	In extreme SW, N of Congo R. (maps)	Vincent 1969, 1978; Petter and Petter-Rousseaux 1979

Euoticus inustus Eastern Needle-clawed Galago

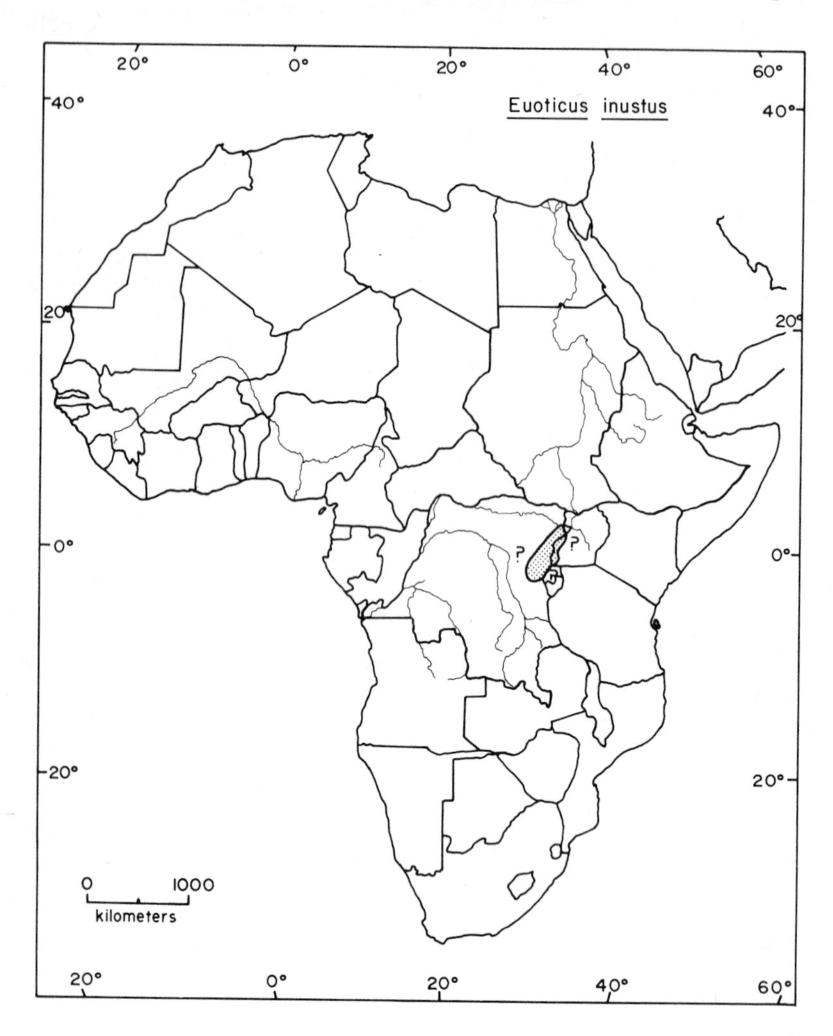

Taxonomic Note

This is a monotypic species which is placed by some authors in the genus *Galago* (Napier and Napier 1967, Groves 1974, Petter and Petter-Rousseaux 1979), and whose specific status is regarded by some as hypothetical (Dorst and Dandelot 1969).

Geographic Range

Euoticus inustus is found in northeastern Zaire and southwestern Uganda. It has been recorded as far north and east as 2°N, 30°45′E, and as far south and west as 2°30′S,27°45′E (Rahm 1965), but the precise limits of its range have not been determined (Rahm 1966, Vincent 1969, Kingdon 1971). For details of the distribution, see Table 3.

Abundance and Density

E. inustus is thought to be fairly rare in northeastern Zaire (Table 3), but no details regarding its status, populations, group sizes, or home ranges could be located.

Habitat

E. inustus is restricted to forest habitats, such as forests in which *Parinari excelsa* is the dominant tree, and forest margins (Kingdon 1971). It has been observed in dense humid forest (Rahm 1965), at medium elevations (Rahm 1966), between 740 and 1,600 m altitude (Vincent 1969, 1972).

Factors Affecting Populations

No information.

Conservation Action

E. inustus may occur in one reserve in Uganda (Table 3), and is legally protected from killing or capture (except by special authorization) by its listing in Class B of the African Convention (Organization of African Unity 1968).

Table 3. Distribution of *Euoticus inustus*

Country	Presence and Abundance	Reference
Zaire	Localities mapped in N Kivu Province (E) and Haut-Zaire Province (NE)	Rahm 1965
	Localities listed include Irangi, Kisanga, Djugu, Nsangani, Kinyawanga, and Lesse (all in E); probably has larger range than is known	Rahm 1966
	Six localities mapped in NE; W extent of range is not known	Vincent 1969
	Localities mapped in NE	Kingdon 1971
	Localities listed and mapped include: Djugu, Irangi, Kinyawanga, Kisanga, Lesse, Lundjulu, Lutunguru, and Nsangani; range probably extends further W	Vincent 1972
	Localities mapped in NE	Groves 1974
	Fairly rare in Haut-Zaire Province (NE)	Heymans 1975
Uganda	In Ankole District (SW)	Rahm 1966
	Localities mapped in SW	Vincent 1969
	Localities mapped in SW; E extent of range not known	Kingdon 1971
	Localities listed and mapped at Kalinzu, Kanyawara, and Kayonza (all in SW); range probably extends further N and S	Vincent 1972
	In Kibale Forest Reserve 0°13′-0°41′N, 30°19′-30°32′E (but identification tentative)	Struhsaker 1975
Rwanda	Range probably extends into N	Vincent 1972

Galago alleni Allen's Galago

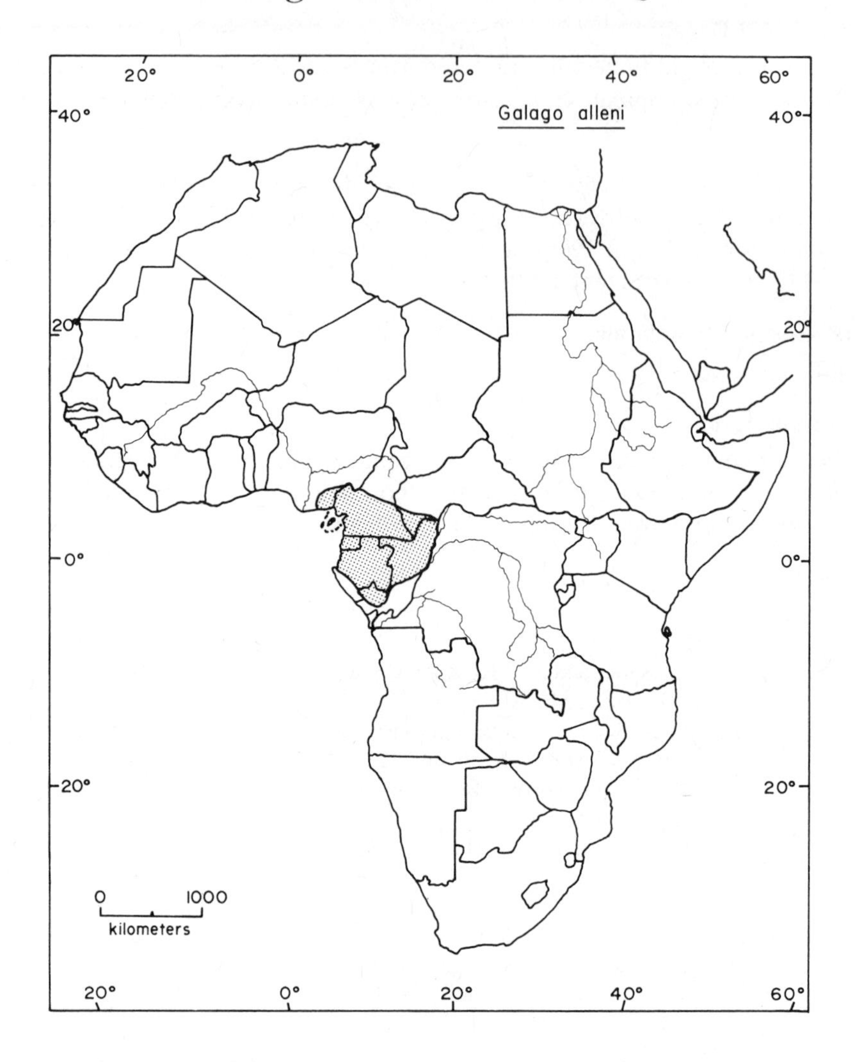

Taxonomic Note

This is a monotypic species (Napier and Napier 1967, Petter and Petter-Rousseaux 1979).

Geographic Range

Allen's galago is found from southeastern Nigeria to southern Congo. It occurs as far west as 6°48′E (at 5°6′N) near the delta of the Niger River (Oates and Jewell 1967). The southern limit of its range is near 4°S in southern Congo, and the eastern limit is near 18°–18°30′E in southwestern Central African Republic and northeastern Congo (Vincent 1969, 1978; Petter and Petter-Rousseaux 1979). For details of the distribution, see Table 4.

Abundance and Density

Galago alleni has recently been described as rare in Nigeria, and as abundant on Macias Nguema, Equatorial Guinea (Table 4). No recent data on its status in the other countries of its range were located.

In northeastern Gabon, population densities of Allen's galago vary from 3 to 25 individuals per km^2 ($\bar{X}$ = 15) (Charles-Dominique 1971, 1974a). At night, individuals forage alone 86% of the time and in groups of 2 or 3 14% of the time; during the day, small sleeping clusters of 1 to 4 are formed (Charles-Dominique 1977b). Females occupy home ranges of 3.9 to 16.6 ha ($\bar{X}$ = 10 ha) and males have ranges of 17 to 50 ha ($\bar{X}$ = 32 ha) (Charles-Dominique 1977a). At one site in Cameroon, the average home range area was 0.82 ha (Molez 1976). Groups of as many as 4 individuals were seen on Macias Nguema (Jewell and Oates 1969).

Habitat

G. alleni is a forest species, most abundant in mature forest, especially in wet areas including montane forest, and in bushes in grassland (Jewell and Oates 1969). It also occupies mist forest, swampy areas, and wet lowland forest, and is uncommon in secondary growth in Nigeria (Oates 1969). In Equatorial Guinea and Cameroon, however, it has been recorded in young secondary forest and in plantations (Sabater Pi 1972b, Molez 1976).

In northeastern Gabon, the highest population densities of *G. alleni* are in flooded primary forest, with lower densities in nonflooded primary forest (Charles-Dominique 1971). There the species is rarely encountered in secondary forest (Charles-Dominique 1977b).

Factors Affecting Populations

Habitat Alteration

G. alleni can utilize disturbed habitats to some extent but is found in reduced densities there, and prefers mature forest (see above). In Nigeria it occupies secondary forest if some large trees remain (Jewell and Oates 1969), but in Gabon it is found in degraded areas only in flooded marshy patches not suitable for cultivation and so not cleared (Charles-Dominique 1977b).

The recent civil war in Nigeria has probably resulted in a decrease in the amount of habitat available to *G. alleni* there (Oates 1969). No information is available concerning the effects of habitat alteration on this species elsewhere in its range.

Human Predation

Hunting. G. alleni is rarely hunted in northeastern Gabon (Charles-Dominique 1977b). No reports from other areas were received.

Pest Control. No information.

Collection. One *G. alleni* was imported into the United States between 1968 and 1973 (Appendix) from Cameroon. None entered the United States in 1977 (Bittner et al. 1978) or the United Kingdom in 1965–75 (Burton 1978).

Conservation Action

G. alleni has been reported in one reserve in Cameroon (Table 4). A park or sanctuary for this species is needed in Nigeria (Happold 1971). No reports of conservation measures in other countries were located.

G. alleni is listed in Class B of the African Convention, giving it protection from killing or capture except with special authorization (Organization of African Unity 1968).

Table 4. Distribution of *Galago alleni* (countries are listed in order from north to south)

Country	Presence and Abundance	Reference
Nigeria	E of Cross R. only (extreme SE corner) (map)	Booth 1958
	Localities listed and mapped at Elele near Niger R. delta, and Oban, E of Cross R.	Oates and Jewell 1967, Jewell and Oates 1969
	Not very abundant	Oates 1969
	In SE (maps)	Vincent 1969, 1978; Petter and Petter-Rousseaux 1979
	E of Niger R.	Happold 1971
	Rare	Monath and Kemp 1973
Cameroon	In S half (maps)	Booth 1958, Vincent 1969, Petter and Petter-Rousseaux 1979
	In Bakundu Reserve (in W)	Gartlan and Struhsaker 1972
	At Nkolguem I, 15 km N of Yaoundé	Molez 1976, Vincent 1978
Central African Republic	In SW (maps)	Vincent 1969, 1978; Petter and Petter-Rousseaux 1979
Equatorial Guinea	Throughout (maps)	Booth 1958; Vincent 1969, 1978; Petter and Petter-Rousseaux 1979
	Abundant on mountains of Macias Nguema (formerly Fernando Po)	Jewell and Oates 1969
	Locality mapped at Ntòbo (W central)	Sabater Pi 1972b
Gabon	Rather rare throughout	Malbrant and Maclatchy 1949
	Throughout except SW (maps)	Vincent 1969, 1978; Petter and Petter-Rousseaux 1979
	Near Makokou, Ogooué-Ivindo basin (NE, 0°34′N, 12°52′E)	Charles-Dominique 1971, 1974a, 1977a, b, 1978; Charles-Dominique and Bearder 1979
Congo	Rather rare throughout	Malbrant and Maclatchy 1949
	Throughout N and W central; absent in SW, S, and E central (maps)	Vincent 1969, 1978; Petter and Petter-Rousseaux 1979

Galago crassicaudatus

Greater or Thick-tailed Galago or Bushbaby

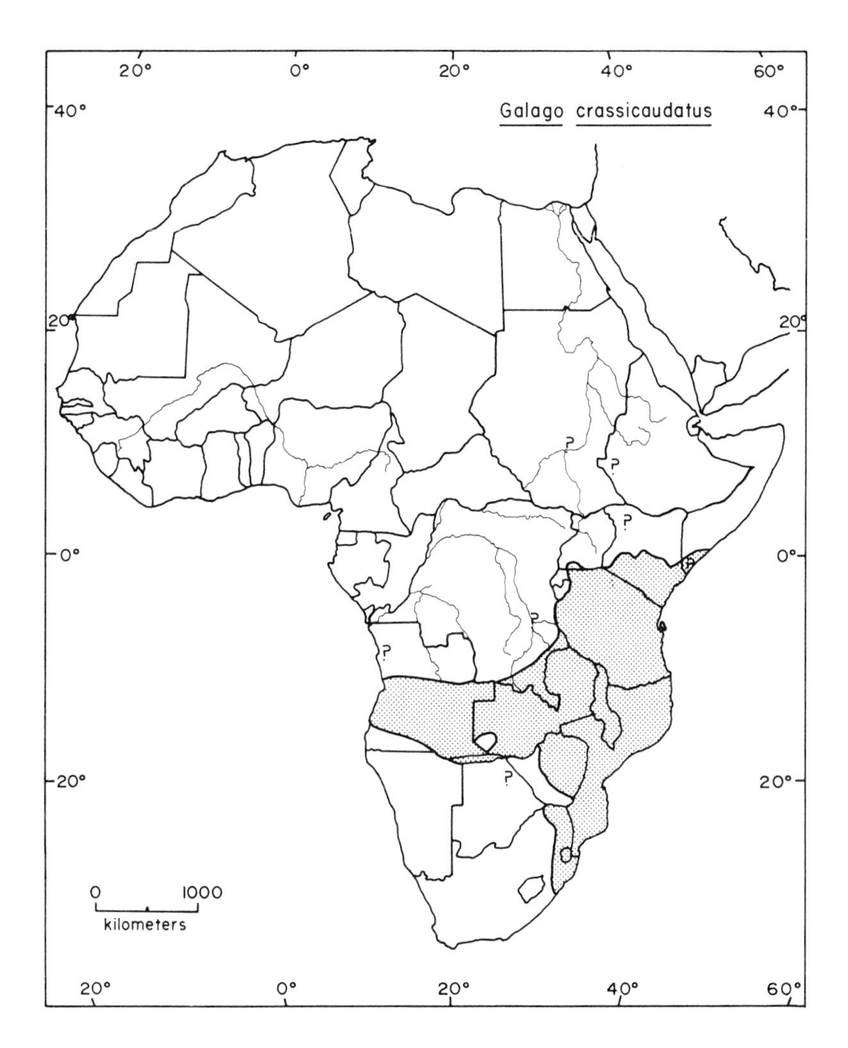

Taxonomic Note

Ten (Napier and Napier 1967) or eleven subspecies (Petter and Petter-Rousseaux 1979) are recognized.

Geographic Range

The greater galago is found in eastern and southern Africa, from southern Somalia to eastern South Africa and west to the Atlantic coast of Angola. Some maps show the range extending north to about 10°N in Sudan and also into southwestern Ethiopia (Vincent 1969, Petter and Petter-Rousseaux 1979), while others show the northern limit at about 0°30′S, 42°30′E in Somalia (Dorst and Dandelot 1969, Kingdon 1971), but no firsthand data are available from any of these countries. *Galago crassicaudatus* occurs as far south as about 29°S on the coast of Natal, South Africa (Doyle and Bearder 1977), and from about 9–11°S to 15°S in Angola (Vincent 1969, Kingdon 1971, Petter and Petter-Rousseaux 1979). For details of the distribution, see Table 5.

Abundance and Density

G. crassicaudatus has recently been described as common with stable populations in Zambia and South Africa, common in Malawi, and stable in Zimbabwe (Table 5). No assessments of its status in the other countries of its range were located.

At one study site in South Africa the density of *G. crassicaudatus* populations was 200 individuals per km^2, with groups of 3 to 10 individuals associating with each other, and individuals occupying home ranges of 5 to 12 ha (Clark, pers. comm. 1978). In other areas, population densities varied from 72 to 125 individuals per km^2, groups contained 2 to 6 members, and home ranges averaged 7 ha in area (Bearder and Doyle 1974a).

Habitat

G. crassicaudatus is found in savanna, bush, woodland, and forest habitats. For example, in Kenya and Tanzania it occupies a variety of moist vegetational types including riverine, bamboo, and montane forests, and plantations and gardens, but is particularly numerous in coastal thicket forest (Kingdon 1971). In Tanzania it has also been reported in bushland of *Acacia, Combretum,* and *Commiphora* (Swynnerton 1958), and in woodland-forest mosaics including *Brachystegia, Mapaca, Combretum, Diplorhynchus,* and *Acacia* (Itani and Nishida, pers. comm. 1974).

In Zambia, Malawi, and Zimbabwe, greater galagos utilize montane forest, bamboo thickets, gum, cypress plantations, and evergreen riverine forest (Smithers 1966) as well as woodland (Ansell 1960a; Sheppe and Osborne 1971; Attwell, pers. comm. 1974). In Malawi this species is most common in rain forest and riparian growth (Long 1973), while in Angola it is typical of savanna (Hayman 1963).

G. crassicaudatus has been most intensively studied in South Africa where it is mainly confined to dense evergreen indigenous forest and riparian bush in well-watered coastal regions or inland montane forest (Bearder and Doyle 1974a). Its populations reach their highest densities in dense coastal dune forest, temperate forest, and woodland mixed with timber plantations, and are less dense in escarpment riparian bush and scrub, and lowveld riparian bush and savanna (Bearder and Doyle 1974a, Doyle and Bearder 1977). It is often found in woodland, forest, and dense bush along rivers (Bourquin et al. 1971, Pringle 1974), and seems to require a continuous canopy and a year-round food supply of fleshy fruits or *Acacia* gums (Clark, pers. comm. 1978). This species has been reported as high as 3,500 m elevation (Haltenorth 1975).

Factors Affecting Populations

Habitat Alteration

Greater galagos can utilize artificial habitats such as suburban gardens and plantations of eucalyptus, mango, cypress, coffee (Kingdon 1971), blue gum, black wattle, and pine (Bearder and Doyle 1974a). But although they may sleep in or move through these exotic trees, their feeding is centered on the adjoining natural vegetation, and if none of this remains they cannot survive (Bearder, pers. comm. 1974).

In South Africa much of the natural habitat on which *G. crassicaudatus* depends has been exploited by man and replaced by cropland or relatively sterile timber plantations (Doyle and Bearder 1977). Although the rate of destruction of bush has now slowed (Bearder, pers. comm. 1974), future clearing for development schemes may be detrimental to *G. crassicaudatus* there (Doyle and Bearder 1977). In Zambia the "clearing of woodlands for agriculture and other purposes are not likely to deplete the habitat [of *G. crassicaudatus*] seriously" (Ansell, pers. comm. 1974). No reports of the effect of habitat alteration are available from the other countries of the range.

Human Predation

Hunting. G. crassicaudatus is eaten in some parts of East Africa (Kingdon 1971) and is occasionally killed for its meat or pelt in South Africa (Bearder, pers. comm. 1974). No information was located concerning hunting in other areas.

Pest Control. Thick-tailed galagos are known to raid beer "factories" of local tribes in South Africa (Pienaar 1967), and to eat cultivated fruits such as papaya there (Clark, pers. comm. 1978) and in Tanzania (Kano 1971a). They are blamed for killing chickens in Zimbabwe, Zambia, and Malawi (Smithers 1966), but no evidence for this behavior was found in South Africa (Bearder, pers. comm. 1974). None of these authors mentioned efforts to control the activities of *G. crassicaudatus*, and in South Africa at least, the species seems to be tolerated by the local people (Bearder, pers. comm. 1974; Clark, pers. comm. 1978).

Collection. G. crassicaudatus is frequently kept as a pet in Zambia (Ansell 1960a), but is considered by many to be an unacceptable pet because of its fierce bite (Smithers 1966). It is seldom kept in captivity in South Africa (Bearder, pers. comm. 1974).

According to U.S. Department of Interior statistics, a total of 59 *G. crassicaudatus* were imported into the United States in 1968–73 (Appendix), primarily from Kenya and Somalia. But the Committee on Conservation of Nonhuman Primates (1975a) reports that 137 *G. crassicaudatus* were imported into the United States in 1972–73 for biomedical research. No greater galagos entered the United States in 1977 (Bittner et al. 1978) or the United Kingdom in 1965–75 (Burton 1978).

Conservation Action

G. crassicaudatus is found in three established national parks, one proposed one, and several more in Zambia and Zimbabwe, as well as at least four habitat reserves (Table 5). In South Africa it is protected by laws which demand heavy penalties for killing, capturing, or keeping it (Doyle and Bearder, pers. comm. 1974). It is not in need of any legal protection in Zambia (Ansell, pers. comm. 1974). No reports were received on conservation efforts in other countries.

This species is listed in Class B of the African Convention, which prohibits their killing and capture except with special authorization (Organization of African Unity 1968).

Table 5. Distribution of *Galago crassicaudatus* (countries are listed in clockwise order)

Country	Presence and Abundance	Reference
Somalia	In SW along coast	Dorst and Dandelot 1969, Kingdon 1971
	In SW corner as far E as mouth of Juba R. (map)	Vincent 1969
Kenya	Throughout except NE corner (map)	Vincent 1969
	Localities mapped throughout S half	Kingdon 1971
	In S and W (map)	Petter and Petter-Rousseaux 1979
Uganda	Throughout except SW corner (map)	Vincent 1969
	One locality mapped in S	Kingdon 1971
	In E (map)	Petter and Petter-Rousseaux 1979
Tanzania	In Serengeti N.P. (N)	Swynnerton 1958, Hendrichs 1970
	Throughout (map)	Vincent 1969
	Localities listed and mapped in W between 5°–7°S and 29°30′–31°E including: Kasoge, upper Issa R., Mienge, upper Kasolo R., Mount Wansisi, Tumbatumba, Yaruvano Mts., Mkomwami, Filabanga, Ikuga, and Busebia	Kano 1971a
	Throughout; localities mapped in N, NW, central, and S	Kingdon 1971
	In Mahali Mountains proposed N.P., E shore Lake Tanganyika	Itani and Nishida, pers. comm. 1974
	On Mount Meru (NE)	Haltenorth 1975
	Throughout except SE (map)	Petter and Petter-Rousseaux 1979
Zambia	Throughout except SW Barotse (Western Province); common in most places	Ansell 1960a
	Throughout except SW corner (map)	Smithers 1966
	On Fort Jameson (Chipata) Plateau (SE)	Wilson 1968
	Throughout (maps)	Vincent 1969, Kingdon 1971
	Common near Kafu Flats of Zambezi R.	Sheppe and Osborne 1971
	Locality listed near Mufulira (N central, 12°30′S, 28°12′E)	Ansell 1973
	Localities listed at Mutundu Camp (8°33′S, 30°14′E) and 8°35′S, 30°37′E (NE)	Ansell 1974
	Common and populations stable; in several national parks	Ansell, pers. comm. 1974
	Throughout except NW corner (map)	Petter and Petter-Rousseaux 1979
Malawi	In Karonga (Northern), Central, and Southern provinces	Ansell et al. 1962

Table 5 (continued)

Country	Presence and Abundance	Reference
Malawi (cont.)	Throughout (maps)	Smithers 1966, Vincent 1969, Kingdon 1971, Petter and Petter-Rousseaux 1979
	In Nsanje District (Port Herald) (extreme S); common	Long 1973
Mozambique	At Zinave R. and at Panzila	Dalquest 1968
	Throughout (maps)	Vincent 1969, Kingdon 1971, Petter and Petter-Rousseaux 1979
	Between Zambezi R. (central) to Limpopo R. (S)	Bearder and Doyle 1974a
Zimbabwe	In NE, central, and E (map); sparse in W	Smithers 1966
	Throughout (maps)	Vincent 1969, Kingdon 1971, Petter and Petter-Rousseaux 1979
	In central and E as far W as mouth of Umguza R. NW of Bulawayo (in SW); to Essexvale S of Bulawayo but not in rest of W and SW; population stable; in several national parks	Attwell, pers. comm. 1974
	Near Umtali (E central)	Bearder and Doyle 1974a
Botswana	In NE (maps)	Vincent 1969, Petter and Petter-Rousseaux 1979
South Africa	In Ndumu Game Reserve, NE border of Natal	Paterson et al. 1964
	Sparsely distributed in Kruger N.P., NE border of Transvaal (localities mapped)	Pienaar 1967
	In NE and E (maps)	Vincent 1969, Kingdon 1971, Petter and Petter-Rousseaux 1979
	In Hluhluwe and Umfolozi game reserves and corridor between them (E Natal, NE Zululand) (map)	Bourquin et al. 1971
	In several protected areas	Bearder and Doyle 1972
	Locally abundant and remaining stable or increasing in some areas; as far S as Zululand in Natal (extreme E)	Bearder, pers. comm. 1974
	In Drakensberg foothills, E Transvaal; in Natal and Zululand	Bearder and Doyle 1974a
	Localities mapped in E Natal between Louwsburg and Donnybrook	Pringle 1974
	In NE and E, as far S as about 30°S (map); not threatened with extinction	Doyle and Bearder 1977
	In N Transvaal, N of Louis Trichardt (23°01′S, 29°43′E); stable or increasing there	Clark, pers. comm. 1978

Table 5 (continued)

Country	Presence and Abundance	Reference
Namibia	In Caprivi Strip (NE) (maps)	Vincent 1969, Petter and Petter-Rousseaux 1979
Angola	Localities listed at Alto Chicapa (10°52′S, 19°17′E), S of Cazombo (11°54′S, 22°56′E) (E central), and Camissombo	Hayman 1963
	In Luando Reserve	Cabral 1967
	Throughout except in N and SW (map)	Vincent 1969
	Throughout except SE (map)	Kingdon 1971
	In S half except SE corner (map)	Petter and Petter-Rousseaux 1979
Zaire	In SE (maps)	Vincent 1969, Kingdon 1971, Petter and Petter-Rousseaux 1979
	In Kundelungu N.P. (in SE, about 10°S, 27°E)	I.U.C.N. 1975

Galago demidovii Demidoff's or Dwarf Galago

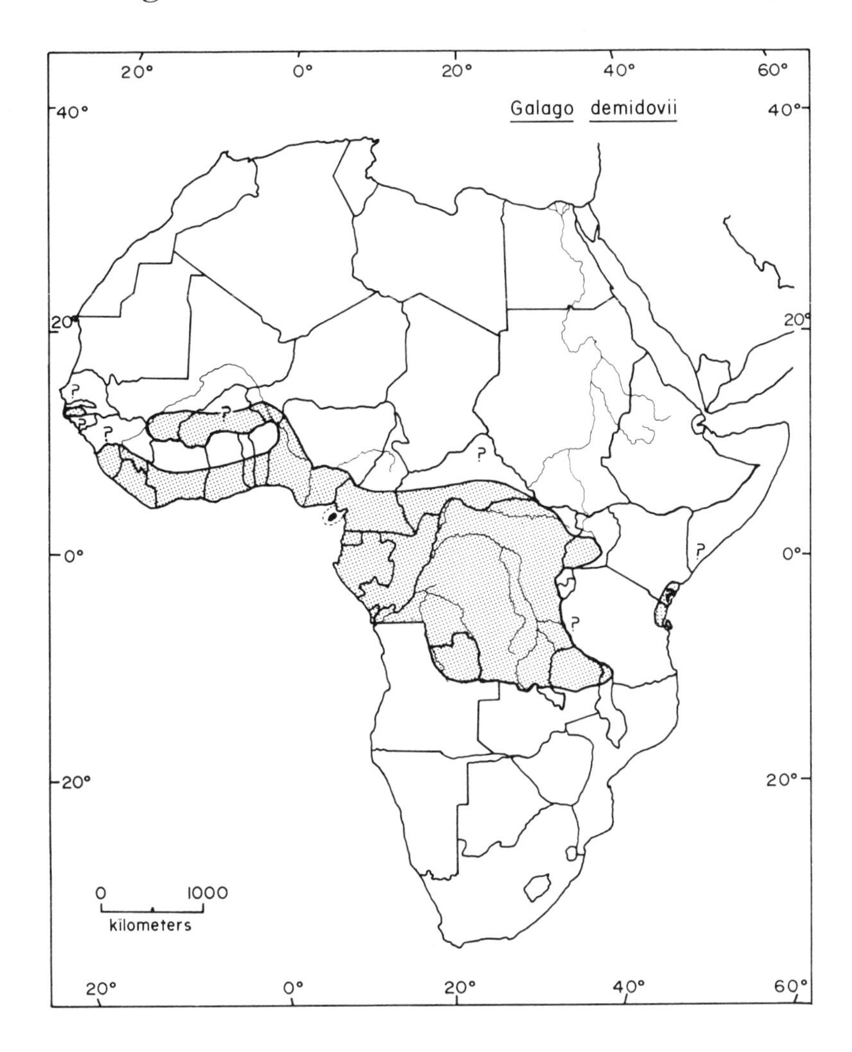

Taxonomic Note

Seven subspecies are recognized (Napier and Napier 1967, Petter and Petter-Rousseaux 1979).

Geographic Range

The dwarf galago is found throughout central Africa from Senegal to Zambia. It has been recorded as far north and west as about 13°N, 16°40′W in southwestern Senegal (Dupuy 1971b) and as far south as about 11°S, 30°E in Zambia (Vincent 1969). Some authors show the range extending east into eastern Kenya and Tanzania (Dorst and Dandelot 1969, Vincent 1969) and even into Somalia (Petter and Petter-Rousseaux 1979), while others place its eastern limit in Uganda (Pitman 1954, Kingdon 1971). For details of the distribution, see Table 6.

Abundance and Density

G. demidovii has been described as common in Equatorial Guinea, Uganda, and Nigeria, and locally common at specific sites in Liberia, Ivory Coast, Gabon, and Congo (Table 6). Its status throughout most of its range is unknown.

In northeastern Gabon, *G. demidovii* is found at densities of 32 to 300 individuals per km^2 ($\bar{X}$ = 50 to 80) (Charles-Dominique 1971, 1972, 1978). This species forms small sleeping groups of 2 to 5 (Vincent 1968), as large as 10 individuals (Kingdon 1971, Charles-Dominique 1977b). During nocturnal foraging, home ranges of females are 0.6 to 1.4 ha ($\bar{X}$ = 0.8 ha) and males 0.5 to 2.7 ha ($\bar{X}$ = 1.8 ha) (Charles-Dominique 1974a, 1977b).

Habitat

G. demidovii is a forest species which can utilize many types of dense vegetation. It has been recorded in primary, secondary, gallery, and relict humid forest, savanna-forest edge and mosaic, woodland, and, rarely, plantations and open habitats (Vincent 1969). It occupies lowland as well as montane, swamp, and gallery forest, bamboo, and *Cynometra* (Kingdon 1971).

In Liberia, dwarf galagos are most often found in narrow strips of secondary *Musanga-Harungana* scrub alongside paths and clearings (Coe 1975). In Ghana their habitat consists of low, tangled bush in high forest, secondary forest, coastal scrub (Booth 1956a), riparian forest, and thicket (Booth 1958), and the species tends to favor dense low secondary vegetation (Jeffrey 1974). Nigerian *G. demidovii* are most common in thick secondary or epiphytic growth (Jewell and Oates 1969) and also thrive in wooded savanna (Oates 1969).

In Cameroon, *G. demidovii* has been observed in young secondary forest (Struhsaker 1970, Molez 1976). In Gabon it lives in dense vegetation of fine branches and foliage invaded by small lianes in primary and secondary forest and inundated forest along rivers, and is frequently encountered along paths and roads (Charles-Dominique 1972, 1974a, 1977b). In Congo it occupies secondary forest and gallery forest (Vincent 1968), in Zaire it has been recorded in montane forest (Rahm 1965), and in Angola it occupies rain forest (Hayman 1963).

G. demidovii occurs between 0 and 2,000 m elevation (Vincent 1969). Most of the authors consulted agreed that its preferred habitats are dense, bushy, secondary growths of thick, intertwined vegetation (Struhsaker 1970, Charles-Dominique 1971, 1974c).

Factors Affecting Populations

Habitat Alteration

Since *G. demidovii* prefers secondary vegetation, some types of habitat disturbance, such as selective logging, may increase the habitat available to it. For example, in Cameroon this species reinvades secondary bush derived from plantations abandoned 10 to 40 years earlier (Struhsaker 1970). And in Liberia the intrusion of man has probably increased the numbers of *G. demidovii* by increasing the amount of secondary growth (Coe 1975).

But extensive deforestation would surely eliminate dwarf galagos, and has occurred, for example, over large portions of southern Ghana (Jeffrey 1975b) and southern Nigeria (Oates 1969). No direct information is available concerning the effects of human activities on populations of *G. demidovii* in any of the countries of its range.

Human Predation

Hunting. G. demidovii is hunted for food in Ghana (Asibey 1974), but is rarely hunted in Gabon (Charles-Dominique 1977b). No reports were located from other countries.

Pest Control. No information.

Collection. Between 1968 and 1973, a total of 215 *G. demidovii* were imported into the United States (Appendix), primarily from Benin. None entered the United States in 1977 (Bittner et al. 1978) or the United Kingdom in 1965–75 (Burton 1978). In 1975, 60 were exported from Liberia for German biomedical research (Jeffrey 1977).

Conservation Action

G. demidovii is found in one national park, one territorial park, and five reserves (Table 6). It is listed in Class B of the African Convention, which prohibits its hunting, killing, capture, or collection except by special authorization (Organization of African Unity 1968). No other provisions or recommendations for its conservation were located.

Table 6. Distribution of *Galago demidovii* (countries are listed in order from west to east and then clockwise)

Country	Presence and Abundance	Reference
Senegal	In S central, S of Gambia border (map)	Vincent 1969
	Locality mapped in SW along S border of Gambia; not seen since 1808	Dupuy 1971b
	Throughout to N border (map)	Petter and Petter-Rousseaux 1979
Guinea-Bissau	Present (maps)	Booth 1958, Petter and Petter-Rousseaux 1979
Guinea	Present (maps)	Booth 1958, Petter and Petter-Rousseaux 1979
	Localities mapped in SE	Vincent 1969
Sierra Leone	Present (maps)	Booth 1958, Petter and Petter-Rousseaux 1979
	In N and SE (maps)	Vincent 1969
Liberia	Throughout (maps)	Booth 1958, Petter and Petter-Rousseaux 1979
	Localities mapped in SW, central, and N	Vincent 1969
	Common in Mount Nimba area (NE)	Coe 1975
Ivory Coast	In S (map)	Booth 1958
	Localities mapped in W and SE	Vincent 1969
	Abundant near Lamto (in S near 6°N on Bandama R.)	Bourlière et al. 1974
	Throughout (map)	Petter and Petter-Rousseaux 1979

Table 6 (continued)

Country	Presence and Abundance	Reference
Mali	In SE (maps)	Vincent 1969, Petter and Petter-Rousseaux 1979
Upper Volta	Localities mapped in SW	Vincent 1969
	Throughout (map)	Petter and Petter-Rousseaux 1979
Ghana	Localities mapped throughout S	Booth 1956a, Vincent 1969
	At Sefwi Wiawso and in Bia Tributaries North Forest Reserve (SW)	Jeffrey 1975b
Togo	In S (maps)	Booth 1958, Vincent 1969
Benin	In S (map)	Booth 1958
Niger	In SW (maps)	Vincent 1969, Petter and Petter-Rousseaux 1979
Nigeria	Throughout S (map)	Booth 1958, Vincent 1969, Petter and Petter-Rousseaux 1979
	Localities listed and mapped at Mamu, Obudu, Akpaka, Elele, and Oban (in SE); common	Jewell and Oates 1969, Oates 1969
	In Upper Ogun and adjacent Old Oyo reserves (W)	Adekunle 1971
Cameroon	In S, S of about 5°N (map)	Booth 1958
	Localities mapped in W and S	Vincent 1969
	Near Idenau (4°15′N, 9°E)	Struhsaker 1970
	At Idenau and in Bakundu Reserve (both in W)	Gartlan and Struhsaker 1972
	Near Nkolguem I, 15 km N of Yaoundé	Molez 1976
Equatorial Guinea	Throughout (map)	Booth 1958
	Abundant throughout	Sabater Pi 1966
	Common in mountains of Macias Nguema (formerly Fernando Po)	Jewell and Oates 1969
	Localities listed and mapped in W at Ndyiakom and Benito R., and central at Ayena and Mount Alén	Jones 1969
	Localities mapped in W and S central	Vincent 1969
	In Monte Raices Territorial Park (in NE, E of Bata)	I.U.C.N. 1971
	Locality mapped at Mount Alén (central)	Sabater Pi 1972b
Gabon	Throughout	Malbrant and Maclatchy 1949
	Localities mapped in N	Vincent 1969
	Very common in Makokou area in NE, Ogooué-Ivindo basin (0°34′N, 12°52′E)	Charles-Dominique 1971, [1972, 1974a, 1974c, 1977b, 1978]
Congo	Throughout	Malbrant and Maclatchy 1949
	Localities listed near Brazzaville (SE),	Vincent 1968

Table 6 (continued)

Country	Presence and Abundance	Reference
Congo (cont.)	Gangalingolo, Makanna; common near Patte d'Oie	Vincent 1968 (cont.)
	Localities mapped in N and S	Vincent 1969
Central African Republic	In SW, W, and central (map)	Vincent 1969
Zaire	On Idjwi Island, Lake Kivu (E)	Rahm 1965
	Localities listed at Irangi, Tshabondo, Kisanga, Kabunga, Niamirwa, and Bunyakiri (all in E, E of Lualaba R., between 2°N and 5°S)	Rahm 1966
	Localities mapped throughout	Vincent 1969
	In NE (map)	Rahm 1972
	Localities mapped in E	Groves 1974
	In Haut-Zaire Province (in N): common N and S of Aruwimi R.	Heymans 1975
	In Kundelungu N.P. (in SE, about 10°S, 27°E)	I.U.C.N. 1975
	In mountains W of Lake Kivu	Hendrichs 1977
Uganda	S and W of Victoria Nile; E limit is in W	Pitman 1954
	In Bwamba Forest (W)	de Vos 1969
	In W and SW (map)	Vincent 1969
	Localities mapped in W and SW; common	Kingdon 1971
	In Budongo Forest (W) and Mabira Forest (central)	Rahm 1972
	Localities mapped in SW	Groves 1974
	In Kibale Forest Reserve (0°13′–0°41′N, 30°19′–30°32′E) (in W)	Struhsaker 1975
Kenya	At mouth of Tana R., E coast (maps)	Dorst and Dandelot 1969, Vincent 1969
	Along E coast (map)	Petter and Petter-Rousseaux 1979
Somalia	In extreme SW (map)	Petter and Petter-Rousseaux 1979
Tanzania	In NE (map)	Vincent 1969
	In W and NE (map)	Petter and Petter-Rousseaux 1979
Malawi	Locality mapped at N end of Lake Malawi	Vincent 1969
	In Misuku Mts. in N	Ansell and Ansell 1973
Zambia	In N near Lake Bangweulu (about 11°S, 30°E) (map)	Vincent 1969
Angola	Locality listed at R. Camuanza, tributary of R. Luachimo, near Dundo	Hayman 1963
	Localities mapped in NE, N of 11°S and E of 18°E	Barros Machado 1969

Galago senegalensis Lesser Galago or Lesser Bushbaby

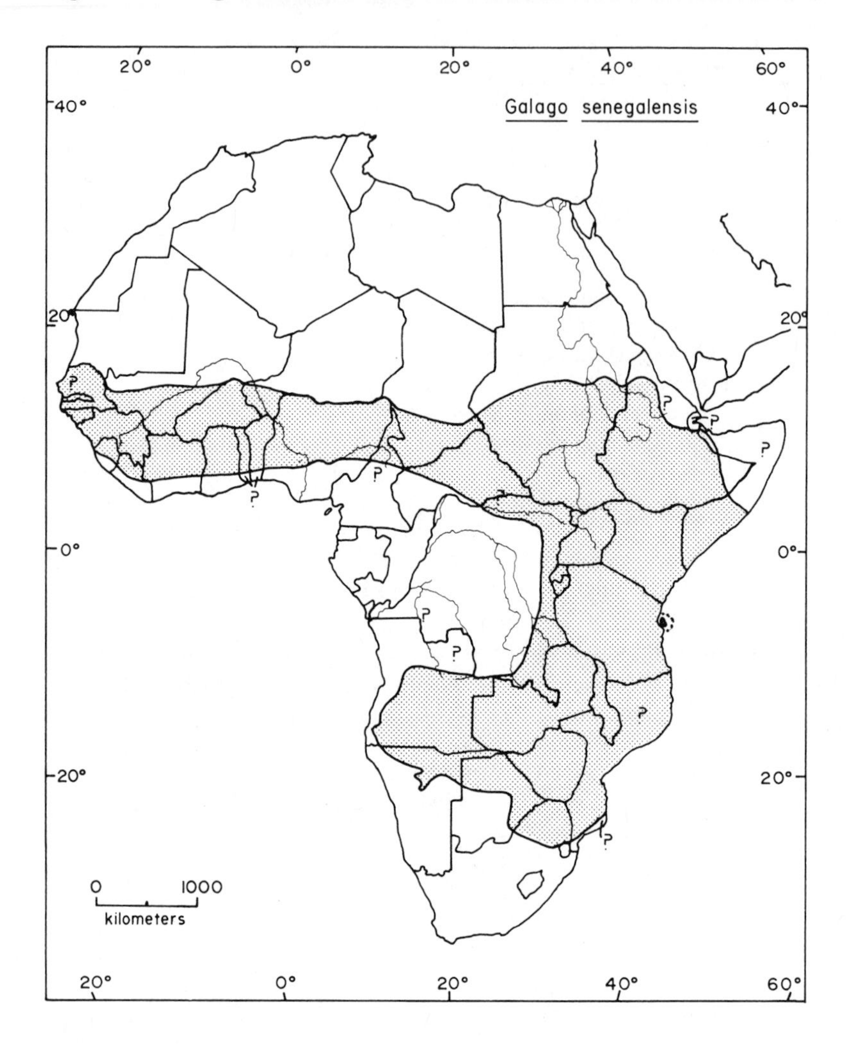

Taxonomic Note

Nine subspecies are recognized (Napier and Napier 1967, Petter and Petter-Rousseaux 1979) including the form *zanzibaricus*, considered to be a separate species by some authors (Kingdon 1971, Groves 1974).

Geographic Range

Galago senegalensis is found throughout subsaharan Africa except in the Congo-Zaire River basin, the Upper Guinea coast, and the southern tip of the continent. It occurs as far north as about 16°N in Senegal (Vincent 1969, Petter and Petter-Rousseaux 1979) and 15°N in Ethiopia (Kock 1969), and as far south as about 26°S in South Africa (Bearder and Doyle 1974a, Doyle and Bearder 1977) and 21°30′S in Namibia (Coetzee, pers. comm. 1974). For details of the distribution, see Table 7.

Abundance and Density

Within the last ten years *G. senegalensis* has been described as locally common or abundant in parts of Senegal, Niger, Nigeria, Sudan, Zambia, Malawi, South Africa, and Angola, scarce in Namibia but stable there and in Zimbabwe, Zambia, and South Africa (Table 7). In the 1960s it was thought to be locally increasing in Nigeria and Mozambique, but no recent information is available from these countries or from most of the countries within the range of this species (Table 7).

Population densities of *G. senegalensis* vary from 0.77 to 500 individuals per km^2 and home ranges from 2.8 to 10 ha (Table 8). Individuals usually forage alone but sleep in small family groups of as many as 9 members (Table 8) that also engage in some social behavior during their waking time (Bearder and Doyle 1974b, Doyle and Bearder 1977).

Habitat

Lesser galagos are found primarily in savanna and woodland habitats including semiarid areas, open woodland, orchard bush and scrub, isolated thickets, and primary forest (Bearder and Doyle 1974a). For example, in Senegal they occupy Sudan and guinea savanna (Dupuy 1971b), and in Ghana they utilize grass-covered orchard bush in guinea savanna and thorn scrub in Sudan savanna (Booth 1956a, 1958). In Niger they are found in homogeneous wooded grassland in guinea savanna (Poché 1973), and near the northern limit of their range in Sudan they occupy semidesert areas as well as dense forests of *Acacia seyal* and *Balanites aegyptica* (Butler 1966).

In East Africa, *G. senegalensis* is typical of savanna habitat especially where *Acacia* is present (Pitman 1954), in bushland consisting of *Acacia*, *Combretum*, and *Commiphora*, and in riverine forest (Swynnerton 1958). In Zambia it has been found in *Brachystegia* woodland (Wilson 1968), and in Zimbabwe it is common in *Acacia* and mopane woodland and in *Baikiaea plurijuga* on Kalihari sand (Wilson 1975).

In Botswana this species is predominantly associated with *Acacia* and also with mopane, *Terminalia*, *Combretum*, and *Baikiaea* woodlands (Smithers 1968). In Namibia its preferred habitats are woodlands of *Phyllogeiton discolor*, *Sclerocarpus africana*, and *Combretum* (Coetzee, pers. comm. 1974), and it also utilizes thornbush veld (Sauer 1974). In South Africa, *G. senegalensis* has been studied in woodland, woodland-savanna, and *Acacia delagoensis* communities (Pienaar 1967), dry thornveld savanna (Bearder and Doyle 1974b), *Acacia karoo* thickets, wooded valleys, riparian bush, and savanna in lowveld and on escarpments (Doyle and Bearder 1977).

This species can occupy relatively open deciduous vegetation but reaches its highest population densities in compact *Acacia* thickets (Charles-Dominique and Bearder 1979) and does not occur in completely open treeless areas (Starck and Frick 1958, Sheppe and Osborne 1971, Bearder and Doyle 1974a). It has been reported as high as 2,000 m elevation in Kenya (Haltenorth 1975), and up to 1,500 in South Africa (Bearder and Doyle 1974a), as well as in lowland areas (Kingdon 1971).

Factors Affecting Populations

Habitat Alteration

Destruction of natural vegetation is not threatening populations of *G. senegalensis* in some of the countries of its range. In Zambia suitable woodland is abundant, and clearing for agriculture and other purposes is not likely to deplete the habitat seriously (Ansell, pers. comm. 1974). And in Namibia more than 95% of the habitat remains intact despite some felling of trees for building and firewood (Coetzee, pers. comm. 1974).

In South Africa the rate of bush destruction has decreased recently because of cattle ranching, game farming, and the establishment of sanctuaries. Since *G. senegalensis* can adapt to limited habitat alteration and thrives in secondary forest and man-made habitats, it does not suffer from limited disturbance (Bearder, pers. comm. 1974). But in the future the destruction of natural vegetation may prove harmful to *G. senegalensis* habitats in South Africa, and no plans have been made for the conservation of populations in areas earmarked for housing or agricultural developments (Doyle and Bearder 1977).

In some regions, habitat alteration has already decreased the vegetation available to *G. senegalensis*. In southwest Niger, for example, the establishment of farms and the destruction of brush by cattle, fire, and drought have eliminated much suitable habitat (Poché 1973). These processes are also occurring in many other countries, but no direct evidence of their effects on *G. senegalensis* populations is available.

Human Predation

Hunting. Lesser galagos are hunted for food in Ghana (Asibey 1974) and in parts of east Africa (Kingdon 1971).

Pest Control. This species is not a pest in Zambia (Ansell, pers. comm. 1974) or in South Africa (Bearder 1974). No reports from other countries were located.

Collection. *G. senegalensis* is popular as a pet in East Africa (Kingdon 1971) and is caught as a pet in Zambia but "this is not on such a scale as to affect its status" (Ansell, pers. comm. 1974). In South Africa the species has disappeared in some areas because of heavy collecting to sell to tourists as pets (Bearder, pers. comm. 1974).

The collection of *G. senegalensis* for research or zoos is negligible in Zambia (Ansell, pers. comm. 1974). None are exported from Namibia (Coetzee, pers. comm. 1974). The species is used to a small extent for research in South Africa (Bearder, pers. comm. 1974).

Between 1968 and 1973 a total of 680 *G. senegalensis* were imported into the United States (Appendix), primarily from Botswana, Somalia, and Kenya. In 1977 three shipments of unspecified sizes entered this country from Kenya (Bittner et al. 1978). This species is sometimes used in research in the United States (Comm. Cons. Nonhuman Prim. 1975a). None were imported into the United Kingdom in 1965–75 (Burton 1978).

Conservation Action

G. senegalensis has been reported in eight national parks and several unspecified ones in Zambia and Zimbabwe, and three reserves (Table 7). It is legally protected from hunting, export, and keeping in captivity in Namibia (Coetzee, pers. comm. 1974), Zimbabwe (Attwell, pers. comm. 1974), and South Africa, where capture permits are granted to only a few research organizations (Bearder, pers. comm. 1974). In Zambia, lesser galagos have no legal protection and are in need of none (Ansell, pers. comm. 1974).

In southeast Zimbabwe an area near the Haroni-Lusitu confluence should be added to the Chimanimani National Park to protect one race of *G. senegalensis* (*granti*) which occurs there (Attwell, pers. comm. 1974). And in Niger the protection of the "W" National Park must be strengthened by additional personnel, equipment, money, and enforcement of regulations if the deterioration of its habitats and animal populations is to be halted (Poché 1973). No other recommendations for the conservation of *G. senegalensis* were located.

This species is listed in Class B of the African Convention, protecting it from killing and capture except by special authorization (Organization of African Unity 1968).

Table 7. Distribution of *Galago senegalensis* (countries are listed in order from west to east, then clockwise)

Country	Presence and Abundance	Reference
Senegal	Throughout (maps)	Booth 1958, Vincent 1969, Kingdon 1971, Petter and Petter-Rousseaux 1979
	Scattered throughout; very abundant in Niokolo-Koba N.P. (SE)	Dupuy 1971b
	In Basse Casamance N.P. (SW)	Dupuy 1973
	Along lower Senegal R. upstream from Matam (15°40′N, 13°18′W)	Bourlière et al. 1976
	Very common in Niokolo-Koba N.P.	Dupuy and Verschuren 1977
Guinea-Bissau	Throughout (maps)	Booth 1958, Vincent 1969, Kingdon 1971, Petter and Petter-Rousseaux 1979
Guinea	Throughout (maps)	Booth 1958, Vincent 1969, Kingdon 1971, Petter and Petter-Rousseaux 1979
Sierra Leone	In N (maps)	Booth 1958, Vincent 1969, Kingdon 1971, Petter and Petter-Rousseaux 1979
Liberia	In N (map)	Petter and Petter-Rousseaux 1979
Ivory Coast	In N (maps)	Booth 1958, Vincent 1969, Kingdon 1971, Petter and Petter-Rousseaux 1979
Mali	In S (maps)	Booth 1958, Vincent 1969, Kingdon 1971, Petter and Petter-Rousseaux 1979

Table 7 (continued)

Country	Presence and Abundance	Reference
Upper Volta	Throughout (maps)	Booth 1958, Vincent 1969, Kingdon 1971, Petter and Petter-Rousseaux 1979
Ghana	Localities mapped in N	Booth 1956a
Togo	In N (maps)	Booth 1958, Petter and Petter-Rousseaux 1979
	Throughout (maps)	Vincent 1969, Kingdon 1971
Benin	In N (maps)	Booth 1958, Petter and Petter-Rousseaux 1979
	Throughout (maps)	Vincent 1969, Kingdon 1971
Niger	In SW (maps)	Booth 1958, Vincent 1969, Petter and Petter-Rousseaux 1979
	Common in "W" N.P. (SW corner)	Poché 1973
	In S including "W" N.P.	Poché 1976
Nigeria	In N (maps)	Booth 1958, Vincent 1969, Kingdon 1971, Petter and Petter-Rousseaux 1979
	Locally abundant in Yankari Game Reserve (9°34′–10°N, 10°17′–10°47′E)	Sikes 1964a
	Estimated population in Yankari Game Reserve: 1,800–2,000 individuals and increasing	Sikes 1964b
	Abundant in Borgu Game Reserve (9°45′–10°23′N, 3°40′–4°32′E); estimated population: 3,000; not in danger there	Howell 1968
	Increasing slowly in Borgu Game Reserve	Howell 1969
	Recorded from W of Niger R. and N of Benue R. but not E of Niger R.	Jewell and Oates 1969
Cameroon	In N half (maps)	Booth 1958, Vincent 1969, Kingdon 1971
	Common in Bouba-Ndjida N.P. (in NE at 8°25′–9′N, 14°25′–14°55′E)	Lavieren and Bosch 1977
	In extreme N only (map)	Petter and Petter-Rousseaux 1979
Chad	In S (maps)	Booth 1958, Vincent 1969, Kingdon 1971, Petter and Petter-Rousseaux 1979
Central African Republic	In N half (maps)	Vincent 1969, Petter and Petter-Rousseaux 1979
Zaire	In SE and SW (map)	Vincent 1969
	In E and SE (map)	Kingdon 1971
	Localities mapped in SE	Groves 1974

Table 7 (continued)

Country	Presence and Abundance	Reference
Zaire (cont.)	In Kundelungu N.P. (in SE near 10°S, 27°E)	I.U.C.N. 1975
	In NE, E, and SE (map)	Petter and Petter-Rousseaux 1979
Sudan	S of 13°N (map) including localities listed: Rashad, Um Berembeita, Kadero, Habila, Abu Gabeiha, Abu Kershola, Talodi, Delami, Agur, Dilling (common), Kadugli, El Muglad, and Sitra; extremely common in Nuba Mts., N of Bahr el Arab R.	Butler 1966
	Localities mapped as far N as 14°N	Kock 1969
	In S (maps)	Vincent 1969, Kingdon 1971, Petter and Petter-Rousseaux 1979
	Two localities mapped in SE	Groves 1974
Ethiopia	In S and E to E Harar	Starck and Frick 1958
	In Danakil Depression (E)	Glass 1965
	Locality mapped in NW at 15°N, 37°30′E	Kock 1969
	In S (maps)	Vincent 1969, Kingdon 1971
	In Bole Valley (9°25′N, 38°00′E) (central)	Dunbar and Dunbar 1974a
	Localities mapped in NW, E central, and S	Groves 1974
Afars-Issas	Throughout (maps)	Vincent 1969, Kingdon 1971
Somalia	Throughout (maps)	Vincent 1969, Kingdon 1971
	In NW and S (map)	Petter and Petter-Rousseaux 1979
Kenya	Localities mapped in S half	Kingdon 1971
	Localities mapped in N, central, SE, and SW	Groves 1974
	In Mnazini Forest, lower Tana R. (SE)	Andrews et al. 1975
	On Mount Elgon (W central)	Haltenorth 1975
Uganda	Throughout	Pitman 1954
	In Bwamba Forest (W central)	de Vos 1969
	Localities mapped in E and W	Kingdon 1971
	Localities mapped throughout	Groves 1974
Tanzania	In Serengeti N.P. (N)	Swynnerton 1958, Hendrichs 1970
	At Sibwesa (6°30′S, 30°45′E) near Kasoge (E central)	Kano 1971a
	Localities mapped throughout and on Zanzibar Island	Kingdon 1971, Groves 1974
	On Mahali Peninsula, E shore Lake Tanganyika (6°S, 30°E)	Itani and Nishida, pers. comm. 1974
Zambia	Throughout; common in most areas	Ansell 1960a
	Throughout but localized	Smithers 1966

Table 7 (continued)

Country	Presence and Abundance	Reference
Zambia (cont.)	In Chipangali area, Fort Jameson (Chipata) Plateau (SE)	Wilson 1968
	At Chaanga (16°27′S, 28°24′E) (in S) and reported at other localities listed	Ansell 1969
	Common on Kafu Flats of Zambezi R.	Sheppe and Osborne 1971
	Near Mufulira	Ansell 1973
	Common and populations stable; in several N.P.s	Ansell, pers. comm. 1974
	Localities mapped in E and central	Groves 1974
Malawi	At Ncheu (14°50′S, 34°45′E) and Cape Maclear (SW shore of Lake Malawi)	Ansell et al. 1962
	Throughout but localized	Smithers 1966
	Fairly common at Chididi, Nsanje District (extreme S)	Long 1973
Mozambique	Along Zinave R. and at Panzila; more common in 1965 than in 1963	Dalquest 1968
	In N and E (maps)	Vincent 1969, Kingdon 1971
	Localities mapped in NE and central	Groves 1974
	Throughout (map)	Petter and Petter-Rousseaux 1979
Zimbabwe	Throughout but localized	Smithers 1966
	Throughout; population stable; in several N.P.s	Attwell, pers. comm. 1974
	In E highlands	Bearder and Doyle 1974a
	Localities mapped in E	Groves 1974
	Locally common in Wankie N.P. (extreme W) (localities listed)	Wilson 1975
Botswana	Localities mapped in N and E; not in central or SW Kalahari except near Damara Pan, SE of Ghanzi	Smithers 1968
South Africa	Localities mapped in N and S Kruger N.P. (NE); not common there	Pienaar 1967
	In Transvaal NE and E of Pretoria; absent from Zululand; abundant and populations stable	Bearder, pers. comm. 1974
	In Transvaal: in N including Springbok Flats and Waterberg Mts.; in NE in Drakensberg escarpment and E lowlands; in S	Bearder and Doyle 1974a
	In N Transvaal	Bearder and Doyle 1974b, Martin and Bearder 1978
	In Transvaal as far S as about 26°S (map); not threatened	Doyle and Bearder 1977

Table 7 (continued)

Country	Presence and Abundance	Reference
Namibia	In N and NE as far S as 20°30′S and in Caprivi Strip (map); scarce but population stable	Coetzee, pers. comm. 1974
	In Otavi and Waterberg Mts. (N central) and along Okavango R. (NE)	Sauer 1974
Angola	Localities listed at Cazombo (11°54′S, 22°56′E), Alto Cuílo (10°04′S, 19°36′E), Alto Chicapa (10°52′S, 19°17′E), and Sandano (11°40′S, 20°48′E) (all E central)	Hayman 1963
	Common in Luando Reserve	Cabral 1967
	Throughout except NW and SW corners (map)	Vincent 1969
	Throughout except N and SW (map)	Petter and Petter-Rousseaux 1979

Table 8. Population parameters of *Galago senegalensis*

Population Density (individuals/km^2)	Group Size $\bar{X}$	Group Size Range	Group Size No. of Groups	Home Range Area (ha)	Study Site	Reference
0.96					Nigeria	Sikes 1964a
0.77					Nigeria	Howell 1968
	2.5				South Africa	Bearder, pers. comm. 1974
87–500		2–6		2.8	South Africa	Bearder and Doyle 1974a
				10	South Africa	Martin and Bearder 1978
		≤9			Namibia	Sauer 1974

Loris tardigradus Slender Loris

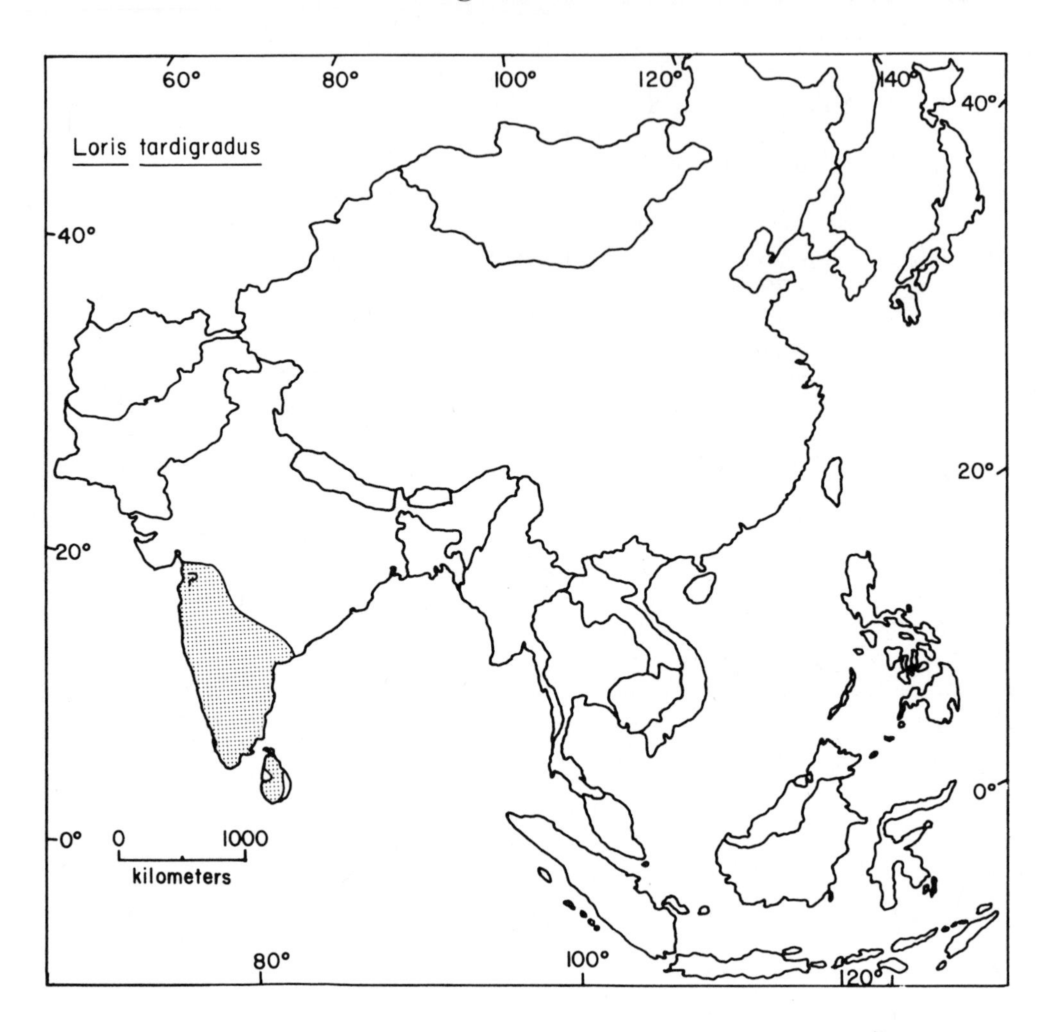

Taxonomic Note

Six subspecies are recognized (Napier and Napier 1967, Petter and Petter-Rousseaux 1979).

Geographic Range

Loris tardigradus is found in southern India and in Sri Lanka (Ceylon). In India it occurs throughout the south, probably as far north as the Tapti River (about 21°N) (Roonwal and Mohnot 1977), and in Sri Lanka it is found in all regions except along the southeastern coast and in the central west (Petter and Petter-Rousseaux 1979). For details of the distribution, see Table 9.

Abundance and Density

L. tardigradus is considered abundant in India but its status in Sri Lanka is not known (Table 9). Population densities of 0.01 individuals per km^2 have been observed (Petter and Hladik 1970). Groups of 2 to 4 sleep together during the day (Rahaman and Parthasarathy 1970), and at night adults forage singly or in pairs (Prater 1971) in home ranges of about one hectare per individual (Hladik 1975, 1979).

Habitat

The slender loris is a forest and woodland species which occupies tropical rain forest, open woodland, swampy coastal forest (Mohnot 1978), scrub jungles, and casurina groves (Rahaman and Parthasarathy 1970). In India it has been seen in *Acacia planifrons* in scrub jungle at sea level, and in evergreen forest at 762 m elevation (Webb-Peploe 1947), and in one montane area is confined to deciduous forest up to 762 m (Hutton 1949). Elsewhere in India it occurs as high as 1,850 m altitude (Roonwal and Mohnot 1977).

In Sri Lanka, *L. tardigradus* is found in lowland, medium, and highland wet zones from 0 m to more than 1,524 m elevation, and in monsoon scrub jungle, monsoon forest, and grassland habitats in dry zones (Eisenberg and McKay 1970). It utilizes secondary forest where *Cassia roxburghii* or *Chloroxylon*, *Berrya*, *Vitex*, and *Schleichera* are the dominant plants (Petter and Hladik 1970), dense deciduous forest, and montane forest (Hladik 1979). Its population densities are higher in dry forest than in humid forest (Petter and Hladik 1970).

Factors Affecting Populations

Habitat Alteration

No information.

Human Predation

Hunting. No information.

Pest Control. No information.

Collection. The eyes of *L. tardigradus* are said to be potent love charms and are also used as cures for certain eye diseases, hence many individuals of this species are captured and sold annually in India (Prater 1971), some to research laboratories for the study of eye problems (Mohnot 1978). *L. tardigradus* is also collected by snake charmers, who use it as an object of amusement to attract crowds (Rahaman and Parthasarathy 1970).

In 1968–73 a total of 24 *L. tardigradus* were imported into the United States (Appendix). None entered the United States in 1977 (Bittner et al. 1978) or the United Kingdom in 1965–75 (Burton 1978). This species is not used in research in the United States (Comm. Cons. Nonhuman Prim. 1975a).

Conservation Action

L. tardigradus is found in the Wilpattu National Park in Sri Lanka (Table 9). No other information was located regarding its conservation.

Table 9. Distribution of *Loris tardigradus*

Country	Presence and Abundance	Reference
India	Common in S Tinnevelly, Madras State (now Tamil Nadu) near Dohnavur	Webb-Peploe 1947
	Rare in High Wavy Mts., Madura District	Hutton 1949
	In S in Eastern Ghats (E), Mysore, Malabar, Travancore, and Coorg (W)	Ellerman and Morrison-Scott 1966
	In S in Eastern Ghats, W to Mangalore, Malabar, Wynaad, S Coorg, and Kerala (map)	Roonwal and Mohnot 1977
	Abundant in S	Mohnot 1978
	Throughout S as far N as Tapti R. in W and Godavari R. in E (map)	Petter and Petter-Rousseaux 1979
Sri Lanka	In S, SW, NE, and NW	Eisenberg and McKay 1970
	At Polonnaruwa (in E at 7°56′N, 81°02′E) and Kandy (central, at 7°17′N, 80°40′E)	Petter and Hladik 1970
	In Wilpattu Reserve (in NW)	Dubost and Hladik 1971
	Throughout Wilpattu N.P.	Eisenberg and Lockhart 1972
	At Polonnaruwa	Hladik 1975
	Throughout except SE and W central (map)	Petter and Petter-Rousseaux 1979

Nycticebus coucang Slow Loris

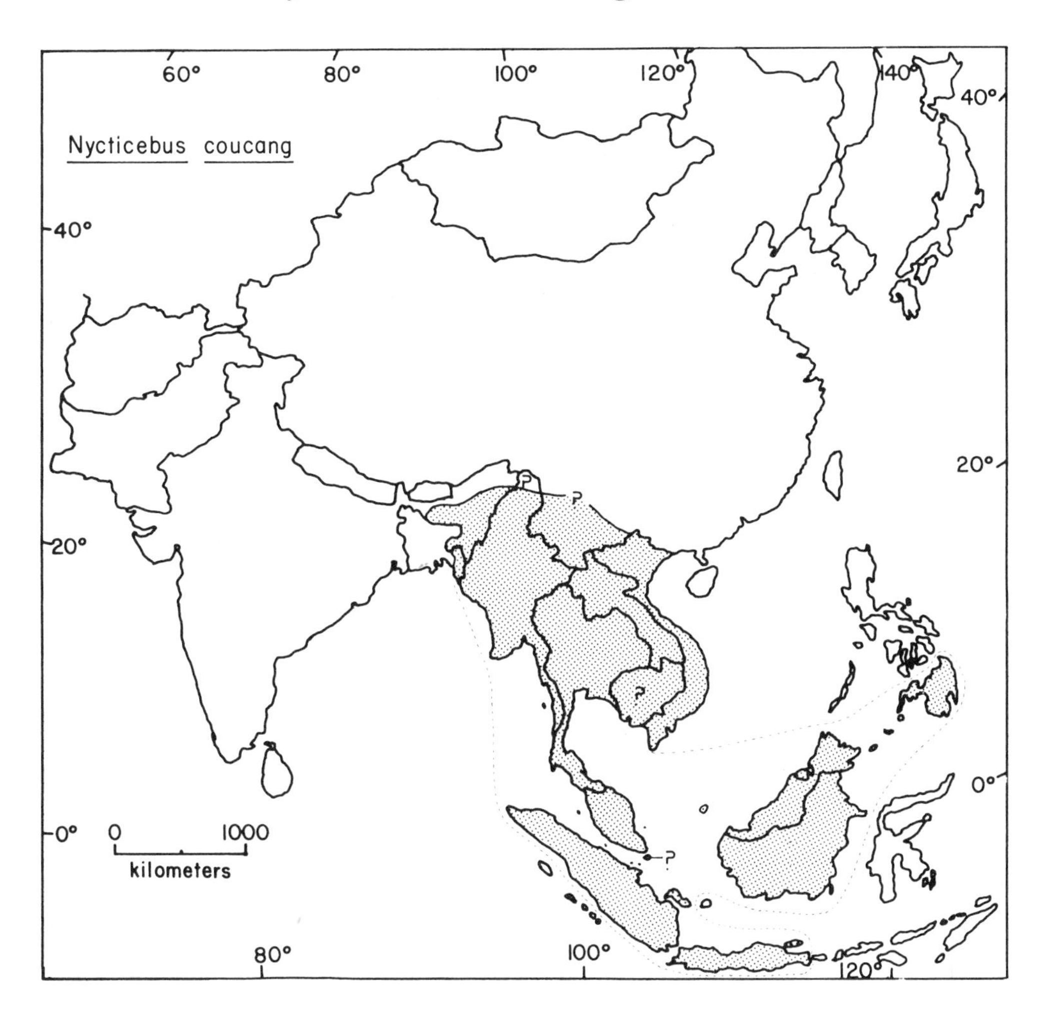

Taxonomic Note

Four (Groves 1971c) to nine subspecies (Napier and Napier 1967) are recognized excluding *pygmaeus*, which is listed by some authors as a race of *N. coucang* (Petter and Petter-Rousseaux 1979) but is considered by others to be a distinct species, and is treated separately here.

Geographic Range

Nycticebus coucang is found in southeastern Asia from Assam, India, and Bangladesh to the southern Philippines and Java, Indonesia. The northwestern limit of its range is the southward bend of the Brahmaputra River at about 26°N, 90°E (Groves 1971c). This species occurs as far east as Mindanao Island, Philippines (about 126°E), and as far south as Java (about 8°S) (Groves 1971c). For details of distribution, see Table 10.

Abundance and Density

N. coucang has been described as abundant on the Malaysian Peninsula but is declining in one area there and in the Philippines (Table 10). Its status throughout the rest of its range is not known.

No estimates of population density or home range area were located. Slow lorises are found singly, in couples, or with young (Medway 1969a).

Habitat

N. coucang is a forest species which is found in many types of vegetation including bamboo and evergreen forest (Fooden 1971a), lowland primary forest (Harrison 1969), scrub, and suburban gardens (Krishnamurti 1968). In India it occupies dense tropical rain forest (Roonwal and Mohnot 1977), and in Sabah it has been observed in low montane forest between 1,280 and 1,646 m elevation (Lim and Heyneman 1968).

In several areas, *N. coucang* is found in secondary growth (Dào Van Tien 1961, Harrison 1966, Elliot and Elliot 1967, Lekagul and McNeely 1977), but in peninsular Malaysia it is reported to be more abundant in primary forest than in secondary (Lim 1969), while in Sumatra and Borneo the reverse is true (MacKinnon 1973, Wilson and Wilson 1975a).

Factors Affecting Populations

Habitat Alteration

The slow loris is able to exploit some types of disturbed habitats including secondary forest (see above), plantations, farms, gardens (Dào Van Tien 1961, Krishnamurti 1968, Medway 1969a), and logged forest (Davis 1962, Wilson and Wilson 1975a). But this species is dependent on forest and is eliminated from deforested areas. For example, it is being depleted by the destruction of forest in the Philippines (Rabor 1968).

No direct evidence regarding the effect of habitat alteration elsewhere in *N. coucang*'s range is available, but the prolonged war in Indochina has certainly been detrimental to habitats there (Orians and Pfeiffer 1970).

Human Predation

Hunting. In Indonesia, *N. coucang* is frequently killed and used to make traditional medicines (MacKinnon 1973, Brotoisworo 1978). No other reports of hunting were located.

Pest Control. Slow lorises do utilize food from plantations (Muul, pers. comm. 1974), but none of the sources consulted mentioned efforts to control these activities.

Collection. N. coucang is captured for sale as a pet in Assam (Mohnot 1978), and is also collected for sale in Indonesia (Brotoisworo 1978). Between 1968 and 1973 a total of 421 *N. coucang* were imported into the United States (Appendix), primarily from Thailand. None entered the United States in 1977 (Bittner et al. 1978) or the United Kingdom in 1965–75 (Burton 1978). This species is used in biomedical research in the United States to a small extent (Comm. Cons. Nonhuman Prim. 1975a).

Conservation Action

N. coucang has been reported in one national park in Sabah, one reserve in peninsular Malaysia, and two sanctuaries in Thailand (Table 10). It is protected by law in Indonesia (Brotoisworo 1978), Malaysia (Krishnamurti 1968, Medway 1969a, Chin 1971, Yong 1973), Thailand (Lekagul and McNeely 1977), and Singapore (Burkill 1961).

Table 10. Distribution of *Nycticebus coucang* (countries are listed in order from west to east, then clockwise)

Country	Presence and Abundance	Reference
Bangladesh	In Chittagong (SE)	Ellerman and Morrison-Scott 1966, Groves 1971c
India Assam	Localities listed and mapped in Goalpara, Naga Hills, Changpang, and Tipperah (NW, E, and SW)	Groves 1971c
	Throughout central and S (map)	Roonwal and Mohnot 1977
Burma	In Chaukan Pass area (in N)	Milton and Estes 1963
	Localities listed and mapped in N at Bhamo and Sumprabum; in W at Kindat, W of Chindwin R., Chin Hills; central at Taungyi, Kyeikpadein; S at Bassein, N of Pegu, Toungoo East, Thaundaung; and Tenasserim at Mergui town, King's Island, Mergui Archipelago, Hinlaem, and Amherst	Groves 1971c
China	In Hsi-Shuan-Pan-Na area, S Yunnan	Kao et al. 1962
	Locality mapped at Hsi-Shuan-Pan-Na at about 22°N, 101°E	Groves 1971c
Thailand	Localities listed at Ban Mae Lamao (Changwat Kamphaeng Phet) at about 16°45′N, 98°45′E, and Ban Mae Na Rhee at about 16°30′N, 99°30′E (both in SW)	Fooden 1971a
	Localities listed and mapped at Chiengmai and Rahang (NW), Khao Soi Dow (S), and Trang and Khaw Song (peninsular)	Groves 1971c
	At Ban Thap Plik and Ban Krung Khayan (both S peninsular)	Fooden 1976
	Throughout (map)	Lekagul and McNeely 1977
	In Khao Soi Dao Wildlife Sanctuary, Chanthaburi Province (SE), and Huay Kha Khaeng Game Sanctuary (in W, near 15°29′N)	Mittermeier 1977b
Laos	Localities listed in N central at Xieng Khouang (19°21′N, 103°23′E), Vientiane (17°59′N, 102°38′E), and S at Pakse (15°N, 105°50′E) and Thateng	Delacour 1940

Table 10 (continued)

Country	Presence and Abundance	Reference
Laos (cont.)	Localities listed and mapped at Xieng Khouang, Vientiane, Pakse, and Thateng	Groves 1971c
Cambodia	Throughout (maps)	Lekagul and McNeely 1977, Petter and Petter-Rousseaux 1979
Vietnam	At Hue (central)	Delacour 1940
	In Thai Nguyen region (21°35′–21°47′N, 105°38′–106°02′E), including at Dinh Ca (N central)	Dào Van Tien 1961
	Localities listed and mapped at Dinh Ca (N), Lao Bao, Hue (central), and Nha Trang (SE)	Groves 1971c
Philippines	Being depleted	Rabor 1968
	Localities listed and mapped on Mindanao at Catagan (may be introduced), and on Bongao in Tawitawi Islands (SE of Mindanao)	Groves 1971c
Indonesia	Localities listed and mapped in Kalimantan (Borneo) at Samarinda (E central), Puruk Tjahu, Barito R. (central), Sungei Landak, Sungei Kapuas, Sungei Sakaiam (W); on Bangka at Klabat Bay; on North Natuna at Sibintang and Bunguran; on Sumatra at Padang Highlands, Batu Sangkar, Tarussan Bay, Tapanuli Bay, Padang Bahrang, Tandjong Morawa, and Siantar (W); on Pulao Tebingtinggi (off E central coast); on Riau Archipelago at Batam (off E central coast); on Java at Lodojo Kediri, Tjaruban, and Djakarta	Groves 1971c
	In N Sumatra	MacKinnon 1973
	In E Kalimantan, W of Renggang at about 1°S, 117°E	Wilson and Wilson 1975a
Malaysia: North Borneo	Not uncommon in Sabah near Kalabakan (SE)	Davis 1962
	Near Mount Kinabalu, Sabah (N)	Lim and Heyneman 1968
	In Sarawak	Chin 1971
	Localities listed and mapped at Menggatal, Ranau, Mount Kinabalu (W Sabah), and Mount Dulit (NE Sarawak)	Groves 1971c
	In Kinabalu N.P., N Sabah	Jenkins 1971, Phillipps 1973
Peninsular	Common throughout	Harrison 1966
	In Ulu Langat region and S of Raub, Pahang State	Elliot and Elliot 1967

Table 10 (continued)

Country	Presence and Abundance	Reference
Malaysia: Peninsular (cont.)	Throughout lowlands	Krishnamurti 1968
	In disjunct distribution throughout (map)	Lim 1969
	Widespread; also on islands of Penang, Pangkor, and Tioman	Medway 1969a
	Localities listed and mapped at Johore Lamu, Jambu, Biserat (S), Trengganu (E), Kepong, Ulu Gombak, Kuala Lumpur, Rumpin R., Batu Gantong, Changkat Mentri, Pulao Pangkor, Dindings (W), Penang Island (NW), and Tioman Island (SE)	Groves 1971c
	At Gunong Benom (just S of 4°N)	Medway 1972
	Common and abundant throughout	Muul, pers. comm. 1974
	Decreasing in Bukit Lanjan Forest Reserve near Kepong, Selangor (3°11′N, 101°37′E)	Lim et al. 1977
Singapore	May remain in Bukit Timah and Water Catchment Area	Burkill 1961
	Present	Medway 1969a

Nycticebus pygmaeus Lesser Slow Loris

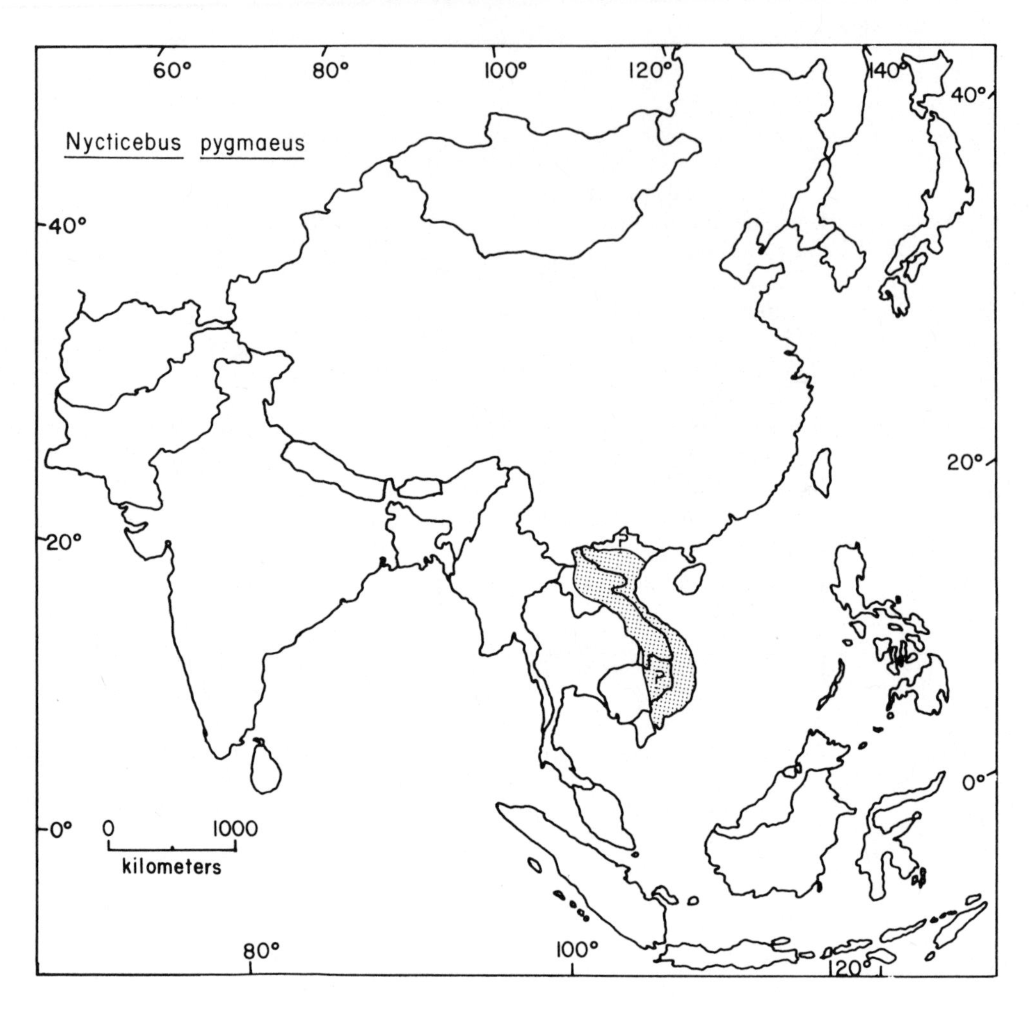

Taxonomic Note

This is a monotypic species (Napier and Napier 1967) which some authors consider to be a race of *Nycticebus coucang* (Petter and Petter-Rousseaux 1979).

Geographic Range

Nycticebus pygmaeus is found in Vietnam and Laos. It has been recorded as far north as 22°04′N, 103°10′E and as far south as 10°46′N, 106°43′E in Vietnam, and as far west as 102°06′E (at 21°04′N) in northern Laos (Groves 1971c). For details of the distribution, see Table 11.

Abundance and Density

No information.

Habitat

No information.

Factors Affecting Populations

Habitat Alteration

The prolonged war in Indochina damaged many habitats (Orians and Pfeiffer 1970) and was probably detrimental to *N. pygmaeus* populations.

Human Predation

No information.

Conservation Action

N. pygmaeus has not been reported in any habitat reserves (Table 11). It is listed as threatened under the U.S. Endangered Species Act (U.S. Department of Interior 1977b).

Table 11. Distribution of *Nycticebus pygmaeus*

Country	Presence and Abundance	Reference
Vietnam	In Phu Quy region (19°20′N, 105°26′E) (NE)	Dào Van Tien 1963
	In Yen Bai Province at Luc Yen (21°40′N, 104°48′E) (N)	Dào Van Tien 1967
	Localities listed at Trang Bom, Bien Hoa Province (SE); near Saigon, Gia Dinh Province (S); Blao, Lam Dong Province (S central); Ban Me Thuot, Darlac Province, and Kontum, Kontum Province (both W central); Nha Trang, Khanh Hoa Province, and Hue, Thua Thien Province (both E central)	Van Peenen et al. 1969
	Localities listed and mapped in N at Lai Chau, Phu Qui, Hoi Xuan, Thai Nien, Hoa Binh, and in S at Saigon, Blao, Nha Trang, Pleiku, Kontum, and Hue	Groves 1971c
Laos	Localities listed and mapped in N at Phong Saly and S at Plateau des Bolovens and Thateng	Groves 1971c

Perodicticus potto Bosman's Potto

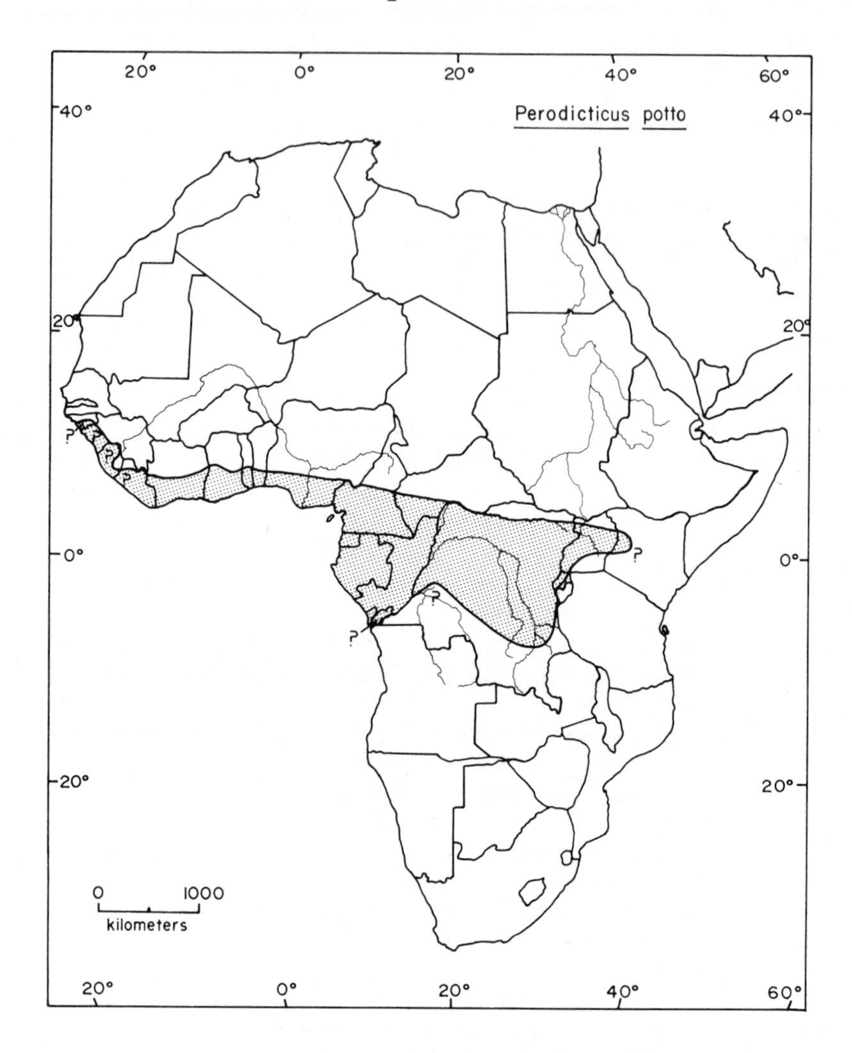

Taxonomic Note

Five subspecies are recognized (Napier and Napier 1967, Petter and Petter-Rousseaux 1979).

Geographic Range

Perodicticus potto is found in western and central Africa, from the Upper Guinea region to western Kenya. It has been reported as far north and west as about 12°N, 15°W in Guinea-Bissau (Booth 1958), but no recent records are available from west of eastern Liberia. It occurs as far east as at least 35°40′E in Kenya (Kingdon 1971), and as far south as about 8°S in Zaire (Vincent 1969, Petter and Petter-Rousseaux 1979). For details of the distribution, see Table 12.

Abundance and Density

P. potto has been described as common in one region of Ivory Coast, in Equatorial Guinea, and in northern Zaire, as rare in Nigeria and Angola, and as decreasing in northeastern Liberia (Table 12). Its status throughout the rest of its range is not known.

In northeastern Gabon, *P. potto* population densities vary from 5 to 28 ($\bar{X} = 8$) individuals per km^2 (Charles-Dominique 1971). Females occupy home ranges of 6 to 9 ha ($\bar{X} = 7.5$) and males range over 9 to 13 ha ($\bar{X} = 12$) (Charles-Dominique 1974b). Individuals forage alone 96% of the time and in pairs 4% of the time, and sleep singly except for mothers with infants (Charles-Dominique 1977b). In Nigeria and in East Africa, Bosman's potto is also solitary (Jewell and Oates 1969, Kingdon 1971).

Habitat

P. potto is a forest species which utilizes high forest, coastal scrub, forest outliers in savanna (Booth 1956a), riparian forest, and woodland (Booth 1958). It has been recorded in gallery forest in Ivory Coast (Bourlière et al. 1974), low dense growth (Jeffrey 1974), farmland, and cultivated trees in Ghana (Jeffrey 1975b), old secondary forest (Jewell and Oates 1969) and wooded savanna (Oates 1969) in Nigeria, and young secondary forest and plantations in Cameroon (Molez 1976).

In Equatoria Guinea, *P. potto* is most often encountered in secondary forest, agricultural forest, farms, and plantations (Sabater Pi 1972b). In Gabon its populations are densest in flooded primary forest and riverine forest, and less dense in secondary forest (Charles-Dominique 1971). And in Zaire this species occupies low, medium, and montane forests as well as secondary growth of parasoliers (*Musanga*) (Rahm 1966).

In East Africa, *P. potto* is typical of secondary forest and clearings along forest margins, and also is found in isolated trees in savanna, in lower montane forest, and in swamp forest, but is not common in climax forest with a high, closed canopy (Kingdon 1971). It is sometimes associated with the date palm (*Phoenix reclinata*), and occurs up to at least 1,829 m in Uganda (Pitman 1954) and 3,500 m in Rwanda (Vincent 1969).

Factors Affecting Populations

Habitat Alteration

P. potto is able to exploit several types of disturbed habitats and in some areas, at least, is commoner in farmland and secondary forest than in mature high forest (Booth 1956a, Kingdon 1971). It can occupy degraded forests, farms, and plantations (Sabater Pi 1966, 1972b; Jeffrey 1970) and is even found in the city of Accra (Booth 1956a).

But it cannot survive in completely deforested areas devoid of trees (Oates 1969), and its habitat is known to be decreasing in some regions. For example, in southwestern Ghana large tracts of vegetation are being cleared and burned every dry season, and extensive road building, logging, and erosion are destroying available habitat (Jeffrey 1975b). In southeastern Nigeria the recent civil war is suspected to have greatly decreased the habitat available to all wildlife, including *P. potto* (Oates 1969). And in Kenya the Kakamega Forest is retreating, owing to charcoal burning and agriculture, and much of the remaining forest in the western part of the country will probably be destroyed by the expanding human population (Zimmerman 1972). No specific information is available concerning the effect of habitat alteration on *P. potto* throughout most of its range.

Human Predation

Hunting. P. potto is hunted in northeastern Liberia (Coe 1975), and is eaten in western Ghana, although its meat is renowned for its toughness (Jeffrey 1970). It is rarely hunted in Gabon (Charles-Dominique 1977b).

Pest Control. No information.

Collection. Bosman's pottos are sometimes collected for export. For example, 21 individuals were exported from Equatorial Guinea between 1958 and 1968 (Jones, pers. comm. 1974).

In 1968–73, a total of 73 *P. potto* were imported into the United States (Appendix), primarily from Benin. None entered the United States in 1977 (Bittner et al. 1978) or the United Kingdom in 1965–75 (Burton 1978).

Conservation Action

Bosman's pottos have been reported in one reserve in Nigeria, another in Uganda, and one park in Equatorial Guinea (Table 12). They are listed in Class B of the African Convention, giving them legal protection from killing or capture except by special authorization (Organization of African Unity 1968).

Table 12. Distribution of *Perodicticus potto* (countries are listed in order from west to east)

Country	Presence and Abundance	Reference
Guinea-Bissau	In central (map)	Booth 1958
Guinea	In W (map)	Booth 1958
Sierra Leone	In central and S (map)	Booth 1958
Liberia	Throughout (map)	Booth 1958
	In E half (map)	Vincent 1969
	At Mount Nimba (NE); disappearing from area	Coe 1975
Ivory Coast	In S half (maps)	Booth 1958, Vincent 1969
	Common near Lamto (in S near 6°N on Bandama R.)	Bourlière et al. 1974
Ghana	Localities mapped throughout S	Booth 1956a
	In S half (maps)	Booth 1958, Vincent 1969
	In W	Jeffrey 1970
	Localities listed in SW at Sefwi Wiawso (6°10′N, 2°30′W), Pampramase, and Papase	Jeffrey 1975b

Table 12 (continued)

Country	Presence and Abundance	Reference
Togo	In S (maps)	Booth 1958, Vincent 1969
Benin	In S (maps)	Booth 1958, Vincent 1969
Nigeria	In S (maps)	Booth 1958, Vincent 1969
	Localities mapped at Nsukka (about 7°N, 7°30′E), and Mamu Reserve (about 6°N, 7°20′E) (in SE)	Jewell and Oates 1969
	Rare in E	Oates 1969
	Very rare	Monath and Kemp 1973
Cameroon	In S half (maps)	Booth 1958, Vincent 1969
	Near Nkolguem I, 15 km N of Yaoundé (S central)	Molez 1976
Equatorial Guinea	Throughout (maps)	Booth 1958, Vincent 1969
	Common throughout	Sabater Pi 1966
	Localities listed and mapped in NW at Nkgomakak, Ngozok, Minkan, San Cristobal, Eyamoyong, and Benito R.	Jones 1969
	In Monte Raices Territorial Park (in NW, E of Bata)	I.U.C.N. 1971
	Localities listed and mapped in NW and central including at Alep, Akòra, Ntòbo, Ntuba, Mbongeté, Mfàman, Achimeláng, Nseín, and Mbongeté	Sabater Pi 1972b
Gabon	Throughout (map)	Vincent 1969
	Near Makokou, Ogooué-Ivindo basin (in NE)	Charles-Dominique 1971, 1974a,b,c, 1977b; 1978; Charles-Dominique and Bearder 1979
	Throughout except along W coast (map)	Petter and Petter-Rousseaux 1979
Congo	Throughout (maps)	Vincent 1969, Petter and Petter-Rousseaux 1979
Angola	In Cabinda (N of mouth of Congo R.) (maps)	Vincent 1969, Petter and Petter-Rousseaux 1979
	Rare	Bothma 1975
Central African Republic	In SW (maps)	Vincent 1969, Petter and Petter-Rousseaux 1979
Zaire	In NE in Ituri forest near Epulu R.	Curry-Lindahl 1956
	In E between Lualaba R. and Rift, and between 5°S and N bank of Congo (Zaire R.); localities listed at Irangi, Tshabondo, Kabunga, Kalimbi, Kisanga, Kabongola, and Shabunda-Kalima	Rahm 1966
	Throughout N, central, and E (map)	Vincent 1969

Table 12 (continued)

Country	Presence and Abundance	Reference
Zaire (cont.)	Common throughout Haut-Zaire Province (N) on right side of Zaire R.	Heymans 1975
Rwanda	In Bugesera forest	Curry-Lindahl 1956
	In Rugege forest, E of Lake Kivu	Rahm 1965, Elbl et al. 1966
Uganda	Throughout most of S, S of 2°30′N	Pitman 1954
	In Bwamba Forest (W)	de Vos 1969
	Throughout (map)	Vincent 1969
	Localities mapped in W and S, S of about 2°N	Kingdon 1971
	In Zika Forest on Entebbe Peninsula, NW shore of Lake Victoria	McCrae et al. 1971
	In Budongo Forest (W) and Mabira Forest (S central) (map)	Rahm 1972
	In Kibale Forest Reserve (0°13′–0°41′N, 30°19′–30°32′E) (in W)	Struhsaker 1975
Kenya	E limit is Kakamega Forest, NE of Kisumu (in SW)	Pitman 1954
	In Kakamega Forest (0°09′–0°22′N, 34°50′–34°58′E)	Elbl et al. 1966, Rahm 1972, Zimmerman 1972
	In central (map)	Vincent 1969
	As far E as Mau Forest (about 35°40′E) (SW) (map)	Kingdon 1971
	At Muguga (1°14′S, 36°40′E) (but may have been transported to there)	Peirce 1975

TARSIIDAE

Tarsiers

Tarsius bancanus Horsfield's Tarsier

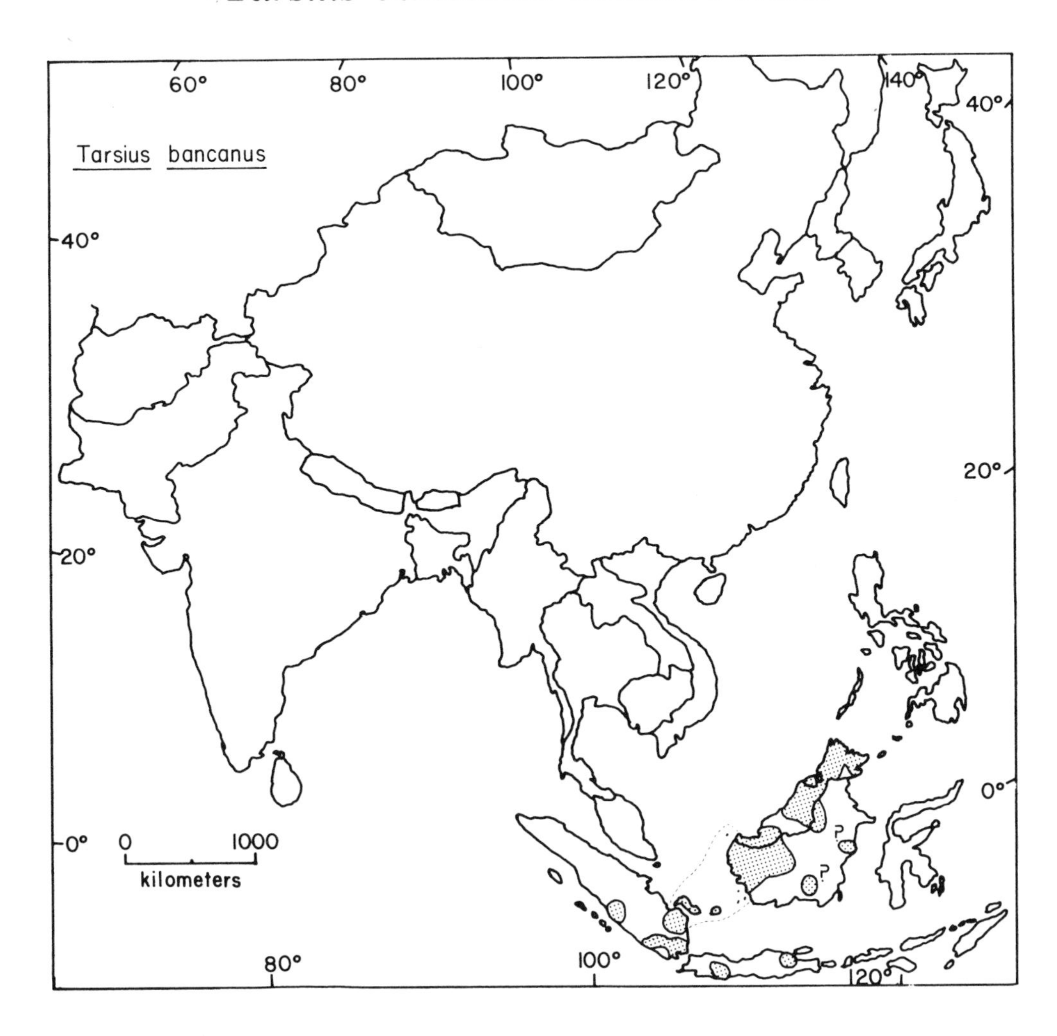

Taxonomic Note

Four subspecies are recognized (Napier and Napier 1967).

Geographic Range

Tarsius bancanus is found in Indonesia, in the North Bornean states (Sabah and Sarawak) of Malaysia, and in Brunei. It occurs from about 101°E in southern Sumatra to about 119°E in eastern Sabah, as far north as about 7°N in northern Sabah and as far south as about 8°S on Java (Petter and Petter-Rousseaux 1979). For details of the distribution, see Table 13.

Abundance and Density

T. bancanus is considered to be an endangered species (Niemitz 1973), although it is not rare in Sarawak and no specific information is available concerning its status elsewhere in its range (Table 13). Population densities vary from 80 to 116 individuals per km^2; and pairs, possibly accompanied by an infant or juvenile, inhabit areas of 0.7 to 1.6 ha (Niemitz 1973, 1979), or 2 to 3 ha (Fogden 1974).

Habitat

Horsfield's tarsiers inhabit primary and secondary forest, coastal shrubs, and plantation edges (Niemitz 1973, 1979). They are more numerous in secondary growth than in mature primary forest (Davis 1962, Niemitz 1979) and seem to prefer edges and secondary vegetation with an abundance of saplings to other types of habitats (Fogden 1974).

Factors Affecting Populations

Habitat Alteration

T. bancanus can utilize some types of disturbed habitats, such as old logged forest (Davis 1962), and may benefit from the replacement of primary forest with secondary growth that would increase the habitat suitable for it (Phillipps 1973). But deforestation of large areas would eliminate this species, since it does not often migrate over long distances (Niemitz 1973). Also, the use of certain persistent pesticides, such as DDT in plantations, can be harmful to this insectivorous species. If large numbers of sprayed insects are ingested, the insecticide may accumulate to be released during subsequent metabolism or lactation (Niemitz 1973).

Human Predation

Hunting. No information.

Pest Control. No information.

Collection. In Sarawak, *T. bancanus* is often captured and kept in captivity, where it usually dies within three days (Niemitz 1973).

Fifty *T. bancanus* were imported into the United Kingdom between 1965 and 1975 (Burton 1978).

Conservation Action

T. bancanus occurs in one national park and two reserves in Sabah and several parks and reserves in Sarawak (Table 13). Mining concessions and tourist development could decrease the value of Kinabalu National Park, but administrators are trying to restrict this damage (Phillipps 1973). No reports were located of habitat reserves for *T. bancanus* in Indonesia. This species is legally protected in Sarawak (Chin 1971), Sabah (de Silva 1971), and Indonesia (Brotoisworo 1978).

To improve its status in Sarawak, better planning of forest clearance and a change to nonaccumulating pesticides are necessary (Niemitz 1973).

Table 13. Distribution of *Tarsius bancanus*

Country	Presence and Abundance	Reference
Indonesia	N of Mahakam R., E central Kalimantan (map)	Wilson and Wilson 1975a
	In S and E Sumatra, E and W Java, Bangka, Belitung, Serasan, Karimata Islands, and SE and W Kalimantan (map)	Petter and Petter-Rousseaux 1979
Malaysia	Localities listed in Sabah near Sandakan (NE), Sepilok Forest Reserve, Sapagaya Forest Reserve, and Bukit Kretam (NE)	Davis 1962
	In N Borneo	Kuntz 1969
	Rare in Kinabalu N.P., N Sabah	Jenkins 1971
	Not rare in Sarawak; localities listed: Ulu Baram, Ulu Tinjar, Ulu Niah, on Rejang, in Sedong area, near Kuching (SW); in several parks and reserves	Niemitz 1973
	Common around Poring; in Kinabalu N.P. (N Sabah)	Phillipps 1973
	In Semengo Forest Reserve, S of Kuching, SW Sarawak	Fogden 1974
	In Sarawak	Niemitz 1979
	In SW, W, E, and N Sarawak and SW and NE Sabah (map)	Petter and Petter-Rousseaux 1979
Brunei	Present (map)	Petter and Petter-Rousseaux 1979

Tarsius spectrum Spectral, Celebesian, or Eastern Tarsier

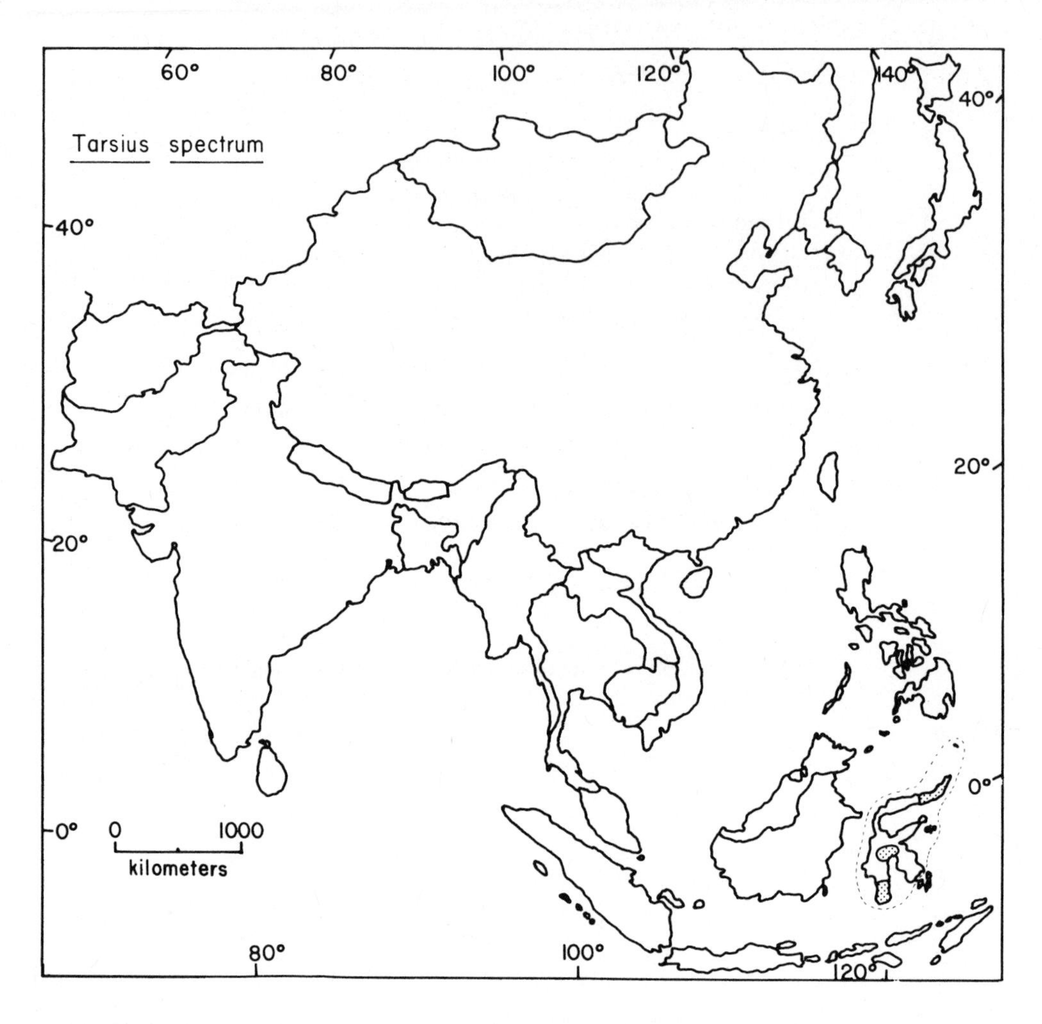

Taxonomic Note

Five subspecies are recognized (Napier and Napier 1967), but some have been described on slender grounds (Hill 1955).

Geographic Range

Tarsius spectrum is found in southern, central and northern Sulawesi (Celebes), Indonesia, and on Peleng Island to the east and the Sangihe Islands to the north of Sulawesi (Petter and Petter-Rousseaux 1979, map). It has recently been studied in the Tangkoko Batuangus Reserve in northern Sulawesi (Anon. 1978). No specific information is available concerning the details of its distribution.

Abundance and Density

T. spectrum apparently lives in small territorial family groups (Anon. 1978). No other data concerning its populations were located.

Habitat

No information.

Factors Affecting Populations

Habitat Alteration

No information.

Human Predation

Hunting. No information.
Pest Control. No information.
Collection. Four *T. spectrum* were imported into the United Kingdom between 1965 and 1975 (Burton 1978).

Conservation Action

In the Tangkoko Batuangus Reserve, J. and K. MacKinnon and I. Tarmujia have studied the resident wildlife and have begun developing the reserve, training local people, and surveying other areas suitable for the establishment of more reserves for *T. spectrum* and other species (Anon. 1978).

T. spectrum is protected by law in Indonesia (Brotoisworo 1978).

Tarsius syrichta Philippine Tarsier

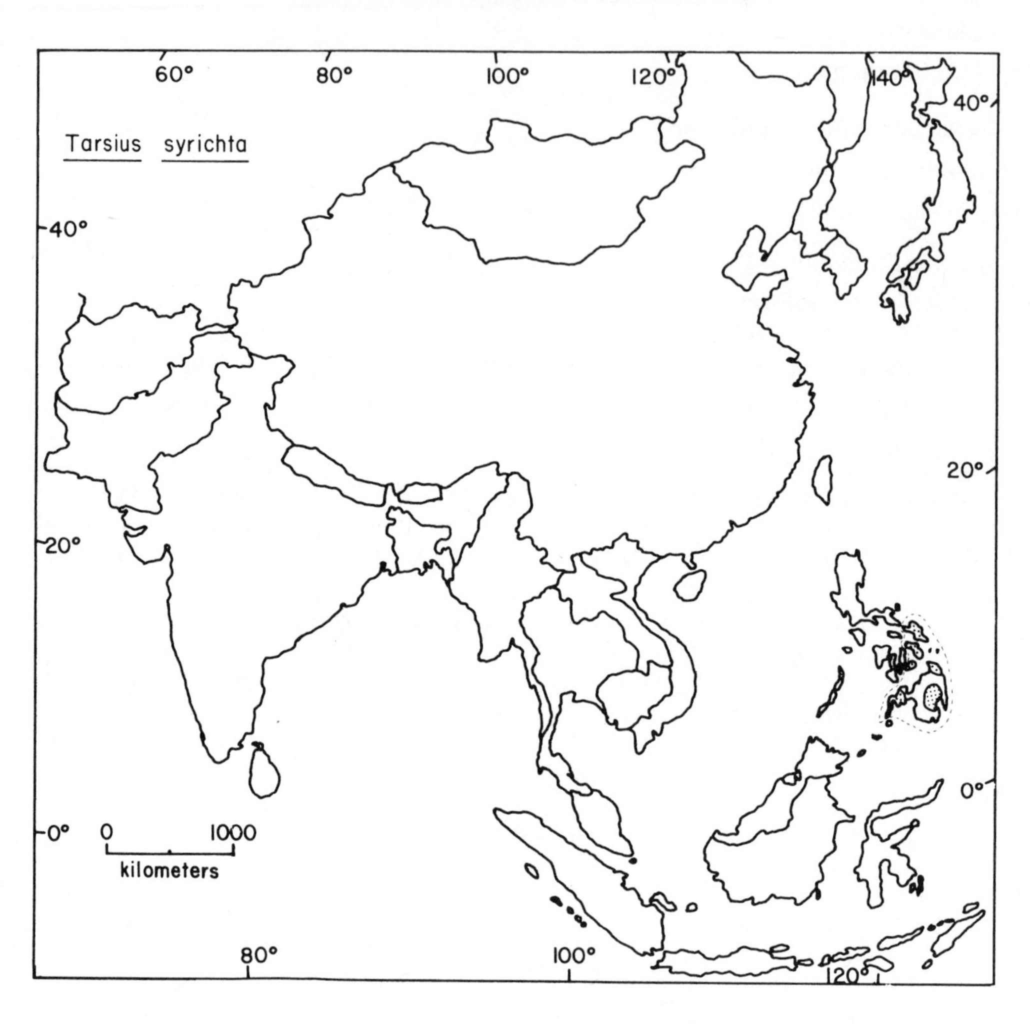

Taxonomic Note

Three subspecies are recognized (Napier and Napier 1967), but are questionably distinct (Hill 1955).

Geographic Range

Tarsius syrichta is found on the southern islands of the Philippine archipelago, on Samar, Leyte, Dinagat, Siargao, Bohol, and northern, central, and southwestern Mindanao (Petter and Petter-Rousseaux 1979). The details of its distribution are not known.

Abundance and Density

No information.

Habitat

No information.

Factors Affecting Populations

Habitat Alteration

Populations of the Philippine tarsier are being reduced, owing to destruction of their forest habitats (Rabor 1968).

Human Predation

Hunting. This species is not usually hunted (Rabor 1968).

Pest Control. No information.

Collection. One unidentified *Tarsius* was imported into the United States in 1968, and one entered from the Philippines in 1973 (Appendix). None were imported into the United States in 1977 (Bittner et al. 1978). Ten *T. syrichta* entered the United Kingdom between 1965 and 1975 (Burton 1978).

Conservation Action

No parks or reserves occur within the range of *T. syrichta*, and a study to determine suitable protected areas is needed (Mittermeier 1977b).

CHEIROGALEIDAE
Dwarf Lemurs and Mouse Lemurs

Allocebus trichotis Hairy-eared Dwarf Lemur

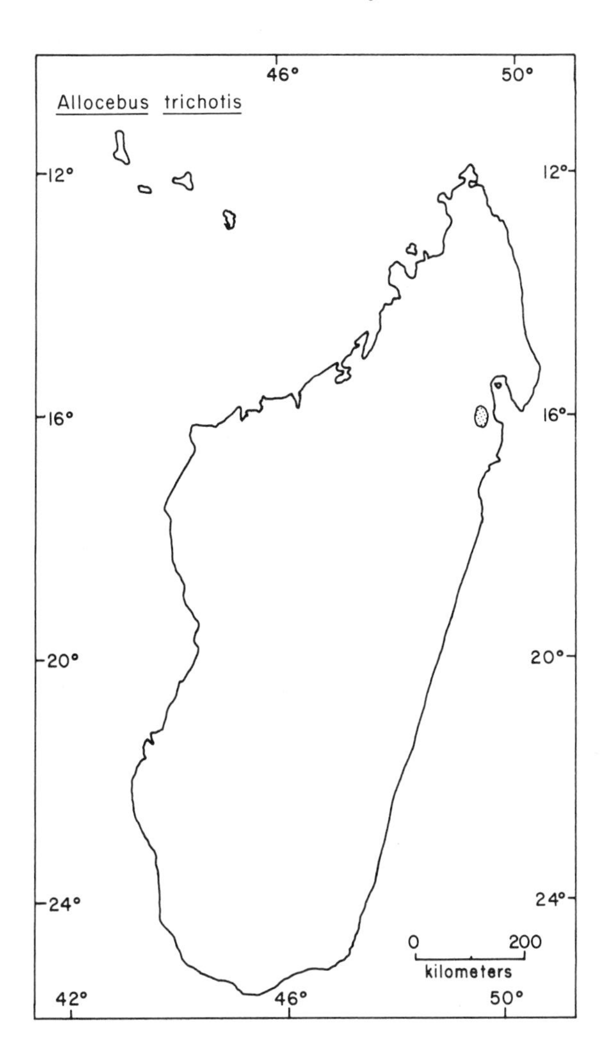

Taxonomic Note

This is a monotypic species (Petter and Petter-Rousseaux 1979).

Geographic Range

Allocebus trichotis is found in eastern Madagascar, near the east coast at about 16°S, 49°30′E. Its range includes a small area north of the Mananara River, west of the town of Mananara (16°10′S, 49°46′E) (Tattersall 1977c, Petter and Petter-Rousseaux 1979, locality mapped). Further details of its distribution are not available.

Abundance and Density

This species is very rare (Petter 1965, I.U.C.N. 1972) and was considered extinct until recently (Petter 1972). It is suspected to be declining and on the brink of extinction (Richard and Sussman 1975), but no specific information is available.

Habitat

A. trichotis is found in high rain forest (I.U.C.N. 1972).

Factors Affecting Populations

Habitat Alteration

The forested area within the range of *A. trichotis* is decreasing because of timbering, agriculture, and industry (Richard and Sussman 1975).

Human Predation

No information.

Conservation Action

A. trichotis does not occur within any habitat reserves, but is legally protected from hunting and export (Richard and Sussman 1975) and is included in Class A of the African Convention (Organization of African Unity 1968). It is listed in Appendix I of the Convention on International Trade in Endangered Species of Wild Fauna and Flora, and as endangered under the U.S. Endangered Species Act (U.S. Department of Interior 1977a,b).

Cheirogaleus major Greater Dwarf Lemur

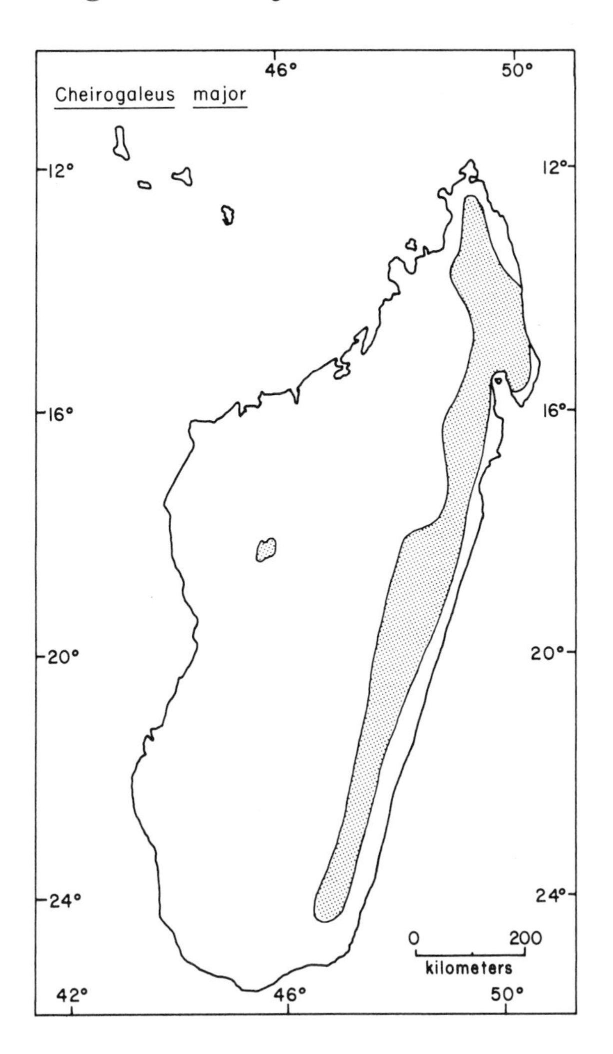

Taxonomic Note

Two subspecies are recognized (Petter and Petter-Rousseaux 1979).

Geographic Range

Cheirogaleus major is found in eastern Madagascar and in one small area north of the Manambolo River in western Madagascar. In the east its range extends from about 12°30′S to 24°40′S in a strip slightly inland from the east coast, and in the west it occurs between about 18°15′–18°30′S and 45°30′–45°45′E (Petter and Petter-Rousseaux 1979). For details of the distribution, see Table 14.

Abundance and Density

C. major is suspected to be declining (Richard and Sussman 1975) and solitary (Petter 1962a), but no direct information on its populations is available.

Habitat

C. major is a forest species which occupies both rain forest and dry forest (Petter 1962a, 1972).

Factors Affecting Populations

Habitat Alteration

The forests in which *C. major* occurs are diminishing because of industry, timbering, and agriculture (Richard and Sussman 1975).

Human Predation

No information.

Conservation Action

C. major is found in five reserves (Table 14), is legally protected from hunting and export (Richard and Sussman 1975), and is included in Class A of the African Convention (Organization of African Unity 1968). It is listed in Appendix I of the Convention on International Trade in Endangered Species of Wild Fauna and Flora, and as endangered under the U.S. Endangered Species Act (U.S. Department of Interior 1977a,b).

Table 14. Distribution of *Cheirogaleus major*

Country	Presence and Abundance	Reference
Malagasy Republic	In E and W	Petter 1962a
	Localities mapped in E	Petter 1962b
	In E	Petter 1965, 1972
	Very numerous at Mahambo on E coast between Tamatave and Fénérive	Petter and Petter 1967
	On Nosy Mangabe Island in Bay of Antongil (in NE)	Petter and Peyriéras 1970a
	In 4 strict nature reserves of Zahamena, Betampona (both in E), Marojejy (in NE), and Tsaratanana (in N)	I.U.C.N. 1971
	In NE and SE, questionable in central plateau	Martin 1972b
	In 4 reserves in E, N, and NE including Nosy Mangabe; declining; habitat being reduced	Richard and Sussman 1975
	N and E of Antsohihy in W; on Massif du Tsaratanana in N; throughout E except extreme N	Tattersall 1977c
	Throughout E; in W, N of Manambolo R. (map)	Petter and Petter-Rousseaux 1979

Cheirogaleus medius Fat-tailed Dwarf Lemur

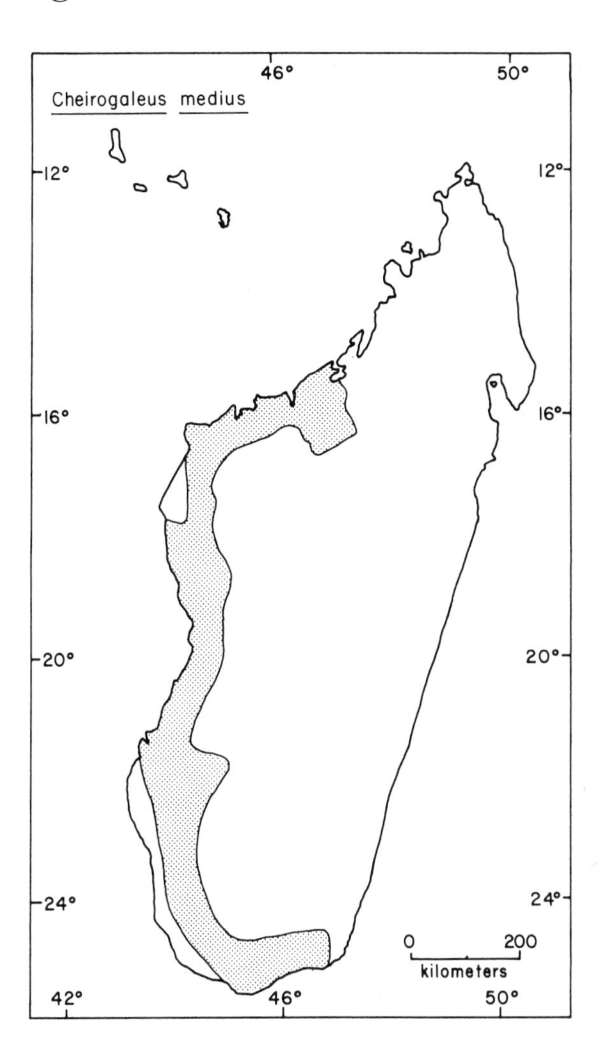

Taxonomic Note

This is a monotypic species (Petter and Petter-Rousseaux 1979).

Geographic Range

Cheirogaleus medius is found in western and southern Madagascar. In the west it occurs along the coast as far north as about 15°20′S and east to the Mahajamba River at about 47°E, and in the south its range extends to the southern tip of Madagascar and as far east as about 46°40′E (Petter and Petter-Rousseaux 1979). For details of the distribution, see Table 15.

Abundance and Density

C. medius has recently been described as nowhere common, vulnerable, and scarce and declining (Table 15). Its population density has been estimated as 300 to 400 individuals per km^2 (Petter 1978) and 400 individuals per km^2 (Hladik 1979), and it is thought to be solitary (Petter 1962a). No data concerning home range size were located.

Habitat

The fat-tailed dwarf lemur occupies dry forest (Petter 1962a) including deciduous forest (Petter 1978) and closed canopy forest dominated by *Tamarindus indica* trees (Sussman 1977a).

Factors Affecting Populations

Habitat Alteration

The range of *C. medius* is contracting rapidly as a result of forest and scrub clearance and degradation (I.U.C.N. 1972), industry, timbering, and agriculture (Richard and Sussman 1975). Even in one legally protected area, clandestine tree felling is threatening its survival (Martin 1973).

Human Predation

Hunting. No information.

Pest Control. No information.

Collection. In 1968–73 a total of three *C. medius* were imported into the United States (Appendix). In 1977 none entered the United States (Bittner et al. 1978). Ten were imported into the United Kingdom in 1965–75 (Burton 1978).

Conservation Action

C. medius has been reported in eight reserves and one national park, although one of the reserves and the park are outside of its range limits (Table 15). It is legally protected from hunting and export, but more meaningful fines and education are necessary to improve its protection (Richard and Sussman 1975). This species is listed in Class A of the African Convention (Organization of African Unity 1968), in Appendix I of the Convention on International Trade in Endangered Species of Wild Fauna and Flora, and as endangered under the U.S. Endangered Species Act (U.S. Department of Interior 1977a,b).

Table 15. Distribution of *Cheirogaleus medius*

Country	Presence and Abundance	Reference
Malagasy Republic	In W	Petter 1962a
	Localities mapped in SW and W	Petter 1962b
	In W and S	Petter 1965
	At Lambomakandro (in SW)	Jolly 1966
	Near Betsiboka R. in NW and near Mangoky R. in SW; in Ankarafantsika Reserve and Montagne d'Ambre N.P. (outside its range)	Fisher et al. 1969
	In 6 strict nature reserves: Tsingy du Bemaraha (W. central), Ankarafantsika (NW), Andohahela (SE), Tsingy de Namoroka (W), Tsimanampetsotsa Lake (SW), and Betampona (E) (outside its range)	I.U.C.N. 1971
	Not abundant, 60 km N of Morondava (on W coast at about 20°S)	Petter et al. 1971, 1975; Petter 1978
	Vulnerable	I.U.C.N. 1972
	In N, NW, and throughout W	Martin 1972b
	Throughout W but nowhere common	Petter 1972
	In Mandena forestry reserve, near Fort Dauphin (in SE)	Martin 1973
	At Antserananomby (N of Mangoky R. in SW) and Berenty (along Mandrare R. in S)	Sussman 1974, 1977
	Also at Ampijoroa forestry station in Ankarafantsika Reserve (in NW at 16°35′S, 46°82′E)	Sussman and Richard 1974, Tattersall and Sussman 1975a, Richard 1978
	In Berenty Reserve in S and another in NW; scarce and declining; habitat being reduced	Richard and Sussman 1975
	Throughout lowlands S and W of Antsohihy	Tattersall 1977c
	In W and S along coast (map)	Petter and Petter-Rousseaux 1979

Microcebus coquereli Coquerel's Mouse Lemur

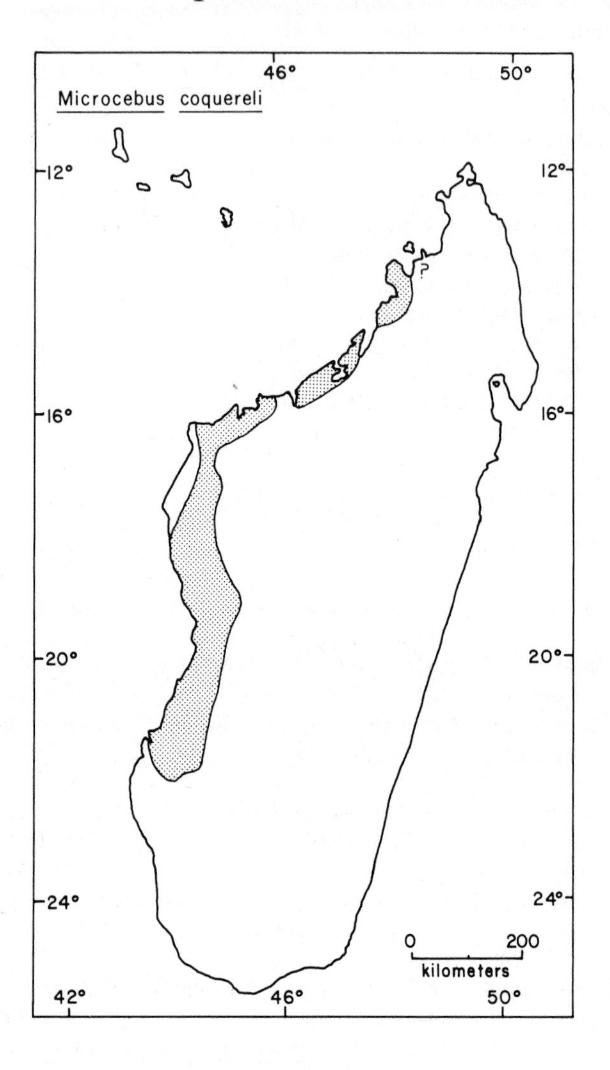

Taxonomic Note

This is a monotypic species (Petter and Petter-Rousseaux 1979).

Geographic Range

Microcebus coquereli is found along the coast of western Madagascar. It occurs as far north as Ampasimena (13°34′S, 48°03′E) and as far south as about 21°50′S, 44°E along the Mangoky River (Petter and Petter-Rousseaux 1979). For details of the distribution, see Table 16.

Abundance and Density

Although populations of *M. coquereli* have been described as locally abundant in some areas, in general they are considered to be very rare, vulnerable, declining, and probably on the brink of extinction (Table 16). Density estimates vary from 50 to 210 individuals per km^2 (Petter et al. 1975) to 250 to 300 individuals per km^2 (Petter et al. 1971). This species is solitary (Petter 1962a), and females have home ranges of 10 ha while males have home ranges of 8 ha (Pages 1978).

Habitat

M. coquereli is found in tropical dry deciduous forest (Pages 1978, Petter 1978). At one study site this forest contained many baobab trees and a dense underbrush of lianes and bushes (Petter et al. 1971).

Factors Affecting Populations

Habitat Alteration

The forest in which *M. coquereli* lives is being reduced by timbering, cultivation, and industry (I.U.C.N. 1972, Richard and Sussman 1975). No further details are available.

Human Predation

No information.

Conservation Action

M. coquereli is found in at least three reserves (Table 16). It is legally protected from hunting and export (Richard and Sussman 1975), and listed in Class A of the African Convention (Organization of African Unity 1968), in Appendix I of the Convention on International Trade in Endangered Species of Wild Fauna and Flora, and as endangered under the U.S. Endangered Species Act (U.S. Department of Interior 1977a, b).

Table 16. Distribution of *Microcebus coquereli*

Country	Presence and Abundance	Reference
Malagasy Republic	Locality mapped in SW	Petter 1962b
	Rare	Petter 1965
	In Tsingy du Bemaraha (W central) and Tsimanampetsotsa Lake (SW) Strict Nature Reserves	I.U.C.N. 1971
	Limited to W coast between Onilahy R. (about 23°50′S), Fierenana R., and Mahavavy R. (about 46°E); localities listed and mapped at Ankazoabo (SW), near Morondava (W. central) and Ampasindava (NW); very rare	Petter et al. 1971
	Vulnerable; limited to W; locally fairly abundant but threatened	I.U.C.N. 1972
	Near Analabe, 60 km N of Morondava (near coast at about 20°S)	Petter et al. 1975, Petter 1978

Table 16 (continued)

Country	Presence and Abundance	Reference
Malagasy Republic (cont.)	In Analabe Reserve and another reserve in W; extremely rare, declining, and probably on the brink of extinction; habitat being reduced	Richard and Sussman 1975
	In N in Ambanja region (mapped); in W in Belo-sur-Tsiribihina–Morondava region; in SW in Ankazoaba region	Tattersall 1977c
	Near Morondava	Pages 1978
	Along W coast from Ampasimena to Mangoky R. (map)	Petter and Petter-Rousseaux 1979

Microcebus murinus Lesser Mouse Lemur

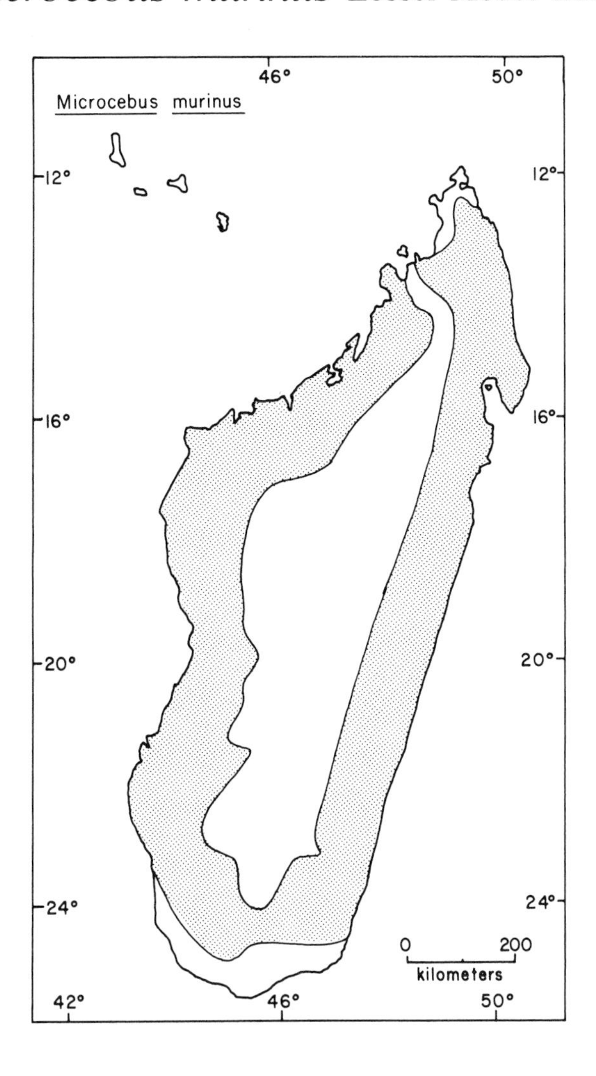

Taxonomic Note

Recently Petter and Petter-Rousseaux (1979) have treated the brown mouse lemur as a separate species (*Microcebus rufus*). But since previous authors considered this form to be a subspecies of *M. murinus* and no specific information is available concerning the status of the brown mouse lemur, it is treated here as part of *M. murinus*, of which there is one other race (Petter 1962b, 1972; Martin 1973).

Geographic Range

Microcebus murinus is found along the western and eastern coasts of Madagascar, and in the north and south of the island inland from the coasts. It occurs as far north as about 12°30′S and as far south as 25°S, and is found throughout most of the island within these limits except for the central plateau and the northern and southern ends (Petter and Petter-Rousseaux 1979). For details of the distribution, see Table 17.

Abundance and Density

M. murinus has recently been described as relatively abundant and is not considered threatened at the present time (Table 17). This species has been studied at local densities of 250 to 360 individuals per km^2 (Charles-Dominique and Hladik 1971), 300 to 400 per km^2 (Petter 1978), and 360 to 400 per km^2 (Hladik 1979). Individuals move solitarily but sleep in groups (Petter 1962a) consisting of 2 to 15 adults (Martin 1973), males often alone or in pairs and females in groups of an average of 4 (Martin 1972a). The area of individual home ranges is not known.

Habitat

M. murinus is a forest species which utilizes humid forest (Charles-Dominique and Hladik 1971), closed canopy forest dominated by *Tamarindus indica* (Sussman 1977a), and deciduous forest (Petter 1978). It also occupies Didiereaceae bush (Charles-Dominique and Hladik 1971) and is able to colonize xerophytic bush areas and lush gallery forest with equal ease (Martin 1972b). Lesser mouse lemurs are also found in gardens and patches of wasteland in towns, and at one study site were much more common in secondary vegetation bordering roads than in nearby primary littoral forest (Martin 1973). All of their habitats contain relatively dense foliage with large numbers of small branches and lianes (Martin 1972a).

Factors Affecting Populations

Habitat Alteration

The forest habitat of *M. murinus* is being reduced by timbering, agriculture, and industry, and the low bush habitat is being destroyed by heavy grazing by cattle and goats (Richard and Sussman 1975). In one legally protected area clandestine tree felling is suspected to have been responsible for a decline in the size of female nesting groups between 1968 and 1970, and is considered a serious threat to the resident population (Martin 1973).

Human Predation

Hunting. No information.

Pest Control. No information.

Collection. Between 1968 and 1973 a total of 15 *M. murinus* were imported into the United States (Appendix). None entered the United States in 1977 (Bittner et al. 1978). Twenty-six were imported into the United Kingdom in 1965–75 (Burton 1978).

Conservation Action

M. murinus is found in one national park, ten strict nature reserves, and four other reserves (Table 17). It is legally protected from hunting and export but more meaningful fines and education are necessary for effective enforcement (Richard and Sussman 1975). The species is included in Class A of the African Convention (Organization of African Unity 1968), Appendix I of the Convention on International Trade in Endangered Species of Wild Fauna and Flora, and the endangered list of the U.S. Endangered Species Act (U.S. Department of Interior 1977a,b).

Table 17. Distribution of *Microcebus murinus*

Country	Presence and Abundance	Reference
Malagasy Republic	In E, W, Sambirano area (NW), and S	Petter 1962a
	Localities mapped in SW, S, SE, E, NE, and NW	Petter 1962b
	In W including Ankarafantsika and on Nosy Bé Island (NW)	Petter 1965
	At Morondava and Lambomakandro (in W) and Ifotaka (in S)	Jolly 1966
	On Nosy Mangabe Island, Bay of Antongil (NE)	Petter and Peyriéras 1970a, Davis 1975
	In S at Périnet and N of Bebarimo	Charles-Dominique and Hladik 1971
	In Montagne d'Ambre N.P. (in N) and 10 strict nature reserves: Tsingy du Bemaraha, Tsingy de Namoroka (both in W), Ankarafantsika (NW), Tsimanampetsotsa Lake (SW), Andohahela (SE), Zahamena, Betampona (both in E), Marojejy (NE), Tsaratanana (N), and Lokobe (on Nosy Bé Island off NW)	I.U.C.N. 1971
	Not abundant 60 km N of Morondava near Lake Andranolava (about 20°N)	Petter et al. 1971, 1975 [Petter 1978, Hladik 1979]
	In E, W, N, and S (map); common in Mandena forestry reserve (SE)	Martin 1972a
	In E, W, N, and NW	Martin 1972b
	In Mandena forestry reserve, near Fort Dauphin (SE); overall not threatened but locally vulnerable	Martin 1973
	At Antserananomby and Tongobato (N of Mangoky R. in SE) and Berenty (on Mandrare R. in S)	Sussman 1974, 1977a
	Also at Ampijoroa forestry station (16°35′S, 46°82′E, in NW), and near Hazafotsy (24°84′S, 46°50′E, in S)	Sussman and Richard 1974, Richard 1978
	Relatively abundant; habitat being reduced; in 8 reserves including Analabe (in W), Berenty (in S), Nosy Bé, and Nosy Mangabe as well as 3 in E and 1 in NW	Richard and Sussman 1975
	At Ampijoroa station in Ankarafantsika Forest Reserve (NW)	Tattersall and Sussman 1975a

Table 17 (continued)

Country	Presence and Abundance	Reference
Malagasy Republic (cont.)	In Natural Reserve No. 11 (Andohahela) near Hazofotsy (in SE at 24°49′S, 46°37′W)	Russell and McGeorge 1977
	In N throughout W as far N as Sambirano R., also N, E, and SE of this river and throughout NE except on Cap d'Ambre	Tattersall 1977c
	Throughout coastal areas except on N and S ends (map)	Petter and Petter-Rousseaux 1979

Phaner furcifer

Fork-marked Dwarf Lemur or Fork-crowned Lemur

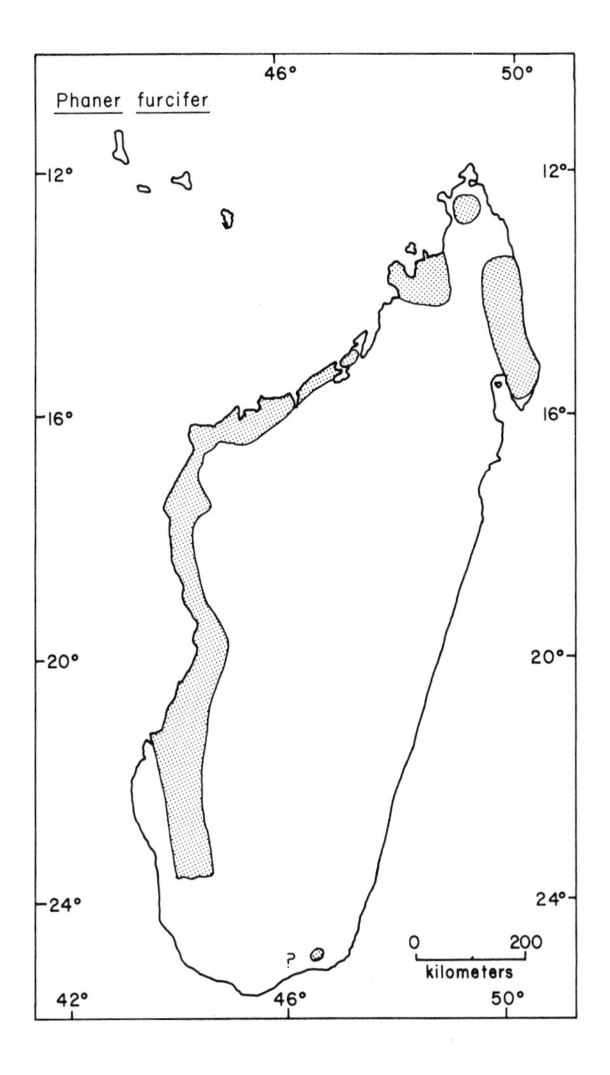

Taxonomic Note

This is a monotypic species (Petter and Petter-Rousseaux 1979).

Geographic Range

Phaner furcifer is found in western, northwestern, northern, and northeastern Madagascar. In the west it occurs along the coast as far south as the mouth of the Mangoky River (about 21°20′S) and inland as far south as the Onilahy River (about 23°30′S). Its range extends as far north as about 12°30′S, and in the northeast as far south as about 15°45′S on Cap Masoala (Petter and Petter- Rousseaux 1979). Recently, a population of *Phaner* that may represent a distinct species has been reported from southeastern Madagascar at 24°49′S, 46°37′E (Russell and McGeorge 1977). For details of the distribution, see Table 18.

Abundance and Density

The range of *P. furcifer* is described as contracting, and its populations are thought to be declining (Table 18), but no specific information is available.

Population densities at two study sites were estimated at 550 and 850 individuals per km^2 (Petter et al. 1975), but in a later report the maximum density of this species was listed as 100 to 200 individuals per km^2 (Petter 1978). *P. furcifer* is assumed to be solitary (Richard and Sussman 1975), and home ranges of about 1 ha are shared by 12 individuals (Petter et al. 1971).

Habitat

P. furcifer is found in deciduous forest characterized by numerous baobabs (*Adansonia* sp.) and bushes of *Terminalia* and *Rhophalocarpus* (Petter et al. 1971, 1975). It also utilizes closed canopy forest dominated by *Tamarindus indica* (Sussman 1977a), and arid *Didierea* bush forest dominated by *Alluaudia procera*, and adjacent gallery forest (Russell and McGeorge 1977).

This species occurs more often in secondary than in primary forest, but not in zones in which a continuous canopy is lacking, and has been recorded as high as 1,000 m elevation (Petter et al. 1975).

Factors Affecting Populations

Habitat Alteration

The forest habitat of *P. furcifer* is being destroyed by cutting and burning (I.U.C.N. 1972), for timber, industry, and agriculture (Richard and Sussman 1975).

Human Predation

No information.

Conservation Action

P. furcifer is found in one national park and four habitat reserves (Table 18). It is listed in Class A of the African Convention (Organization of African Unity 1968), in Appendix I of the Convention on International Trade in Endangered Species of Wild Fauna and Flora, and as endangered under the U.S. Endangered Species Act (U.S. Department of Interior 1977a,b). It is legally protected from hunting and export, but measures needed to improve its prospects for survival include conservation education (Richard and Sussman 1975) and protection of remaining forest habitats in the west (I.U.C.N. 1972, Petter et al. 1975).

Table 18. Distribution of *Phaner furcifer*

Country	Presence and Abundance	Reference
Malagasy Republic	In W and in Sambirano region (NW)	Petter 1962a
	Localities mapped in SW and NW	Petter 1962b
	In N and W including near coast in Bay of Ampasindava and in SW near Tuléar	Petter 1965
	At Morondava (in W) and Lambomakandro (in SW)	Jolly 1966
	At Montagne d'Ambre N.P. (extreme N) and 2 strict nature reserves: Tsingy du Bemaraha (W central) and Tsimanampetsotsa Lake (SW)	I.U.C.N. 1971
	Localities listed and mapped at Morondava, Beroboka, Tabiky, Ampasindava, Ambanja, Montagne d'Ambre, Manja, and Masoala Peninsula; throughout W, N of Onilahy R. and Fiherenana R., in NW and NE	Petter et al. 1971
	Range contracting	I.U.C.N. 1972
	In NE and W	Martin 1972b
	In W and on Montagne d'Ambre	Petter 1972
	At Analabe N of Morondava (20°S, 44°33′E)	Pariente 1974, Petter 1978
	At Antserananomby and Tongobato (in SW, N of Mangoky R. at 21°46′S, 44°07′E)	Sussman 1974; Sussman and Richard 1974; Sussman 1975, 1977a
	Throughout W, N of Onilahy R. and Fiherenana R.; in E near Hiaraka on Masoala Peninsula; in N at Roussettes on Montagne d'Ambre; study sites: Analabe and Antserananomby	Petter et al. 1975
	In Analabe reserve and another reserve in NW; declining; habitat being reduced	Richard and Sussman 1975
	In Natural Reserve No. 11 near Hazofotsy (24°49′S, 46°37′E) in Mananara R. basin (in SE)	Russell and McGeorge 1977
	In N on Montagne d'Ambre and Massif du Tsaratanana; in NE in Hiaraka region of Masoala Peninsula (map)	Tattersall 1977c
	In W, NW, N, and NE (map)	Petter and Petter-Rousseaux 1979

LEMURIDAE
Lemurs

Hapalemur griseus Gray Gentle Lemur

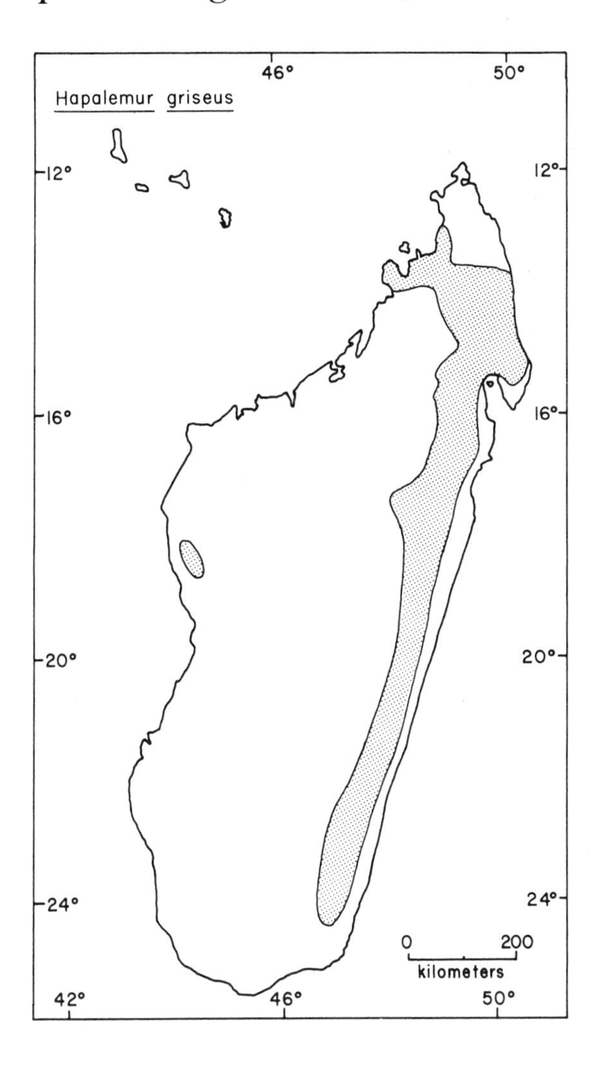

Taxonomic Note

Three subspecies are recognized (Rumpler 1975, Petter and Petter-Rousseaux 1979).

Geographic Range

Hapalemur griseus is found in eastern, west central, and northwestern Madagascar. In the east it occurs from about 13°40′S to about 24°30′S in a strip parallel to the coast and adjacent to it in the north; in the west it is found in a small area between about 18°10′–18°50′S and 44°20′–44°40′E north of the Manambolo River and along the northwestern coast between about 13°S and 14°S (Petter and Petter-Rousseaux 1979). For details of the distribution, see Table 19.

Abundance and Density

H. griseus has been described as numerous in the northwest and exceptionally abundant in several areas in the northeastern and eastern portions of its range, but is also described as vulnerable, and as very sparse along the northeastern coast and declining in some regions of the east, for example, near Lake Alaotra (Table 19). Its status in the west is not known.

At one study site, population density of *H. griseus* was 47 to 62 individuals per km^2 and groups consisted of 3 to 6 members (minimum $\bar{X} = 2.8$) (Pollock 1979). Other workers have reported groups of 2 to 5 (Petter 1962a), 3 to 5 with congregations of 12 and 30 to 40 (Petter and Peyriéras 1970b, 1975), 5 to 15 (Martin 1972b), and 2 to 10 (Petter 1972).

Habitat

H. griseus is found in humid forests and marshy vegetation. Most of its populations live in swampy areas or rain forests dominated by bamboo trees (Petter 1972, Hladik 1975, Pollock 1979). For example, at one study site this species inhabited clumps of bamboo among periodically inundated *Afromomum* trees and bushes (Petter and Peyriéras 1970b, 1975). These authors added that the two widely distributed subspecies of *H. griseus* occur at all elevations up to 1,000 m, and are rare in dense primary forest.

The gray gentle lemur race that is found only at Lake Alaotra (*H. g. alaotrensis* Rumpler 1975) occupies semiaquatic vegetation and marshes, including papyrus swamps and reeds (Petter and Peyriéras 1970b).

Factors Affecting Populations

Habitat Alteration

Bamboo often invades humid areas in which the primary forest has been cut and burned. This improves the habitat for *H. griseus*, and population densities of the species in several disturbed areas in eastern Madagascar are higher than those in undisturbed habitats (Petter and Peyriéras 1970b, 1975). But where the human population settles the cleared land, bamboo cannot invade and *H. griseus* habitats are eliminated (I.U.C.N. 1972). In general, the forest habitats of this species are decreasing, owing to timbering, industry, and agriculture (Richard and Sussman 1975).

Near Lake Alaotra large areas of marshland are regularly burned, directly killing *H. griseus*, destroying its habitat, and threatening its future (Petter and Peyriéras 1970b, 1975). Marsh drainage for rice cultivation is also reducing its habitat in this area (I.U.C.N. 1972).

Human Predation

Hunting. *H. griseus* is heavily hunted for food in human populated areas (Petter and Peyriéras 1970b). For example, its populations in the eastern interior between Tsarabaria and Cap Est are sparse, owing to intensive hunting by local inhabitants (Tattersall 1977c). This species is also hunted near Lake Alaotra, where large numbers are cornered and killed or captured for later consumption during the annual reed burning (Petter and Peyriéras 1970b, 1975).

Pest Control. No information.

Collection. Between 1968 and 1973, 3 *H. griseus* were imported into the United States (Appendix). None entered the United States in 1977 (Bittner et al. 1978) or the United Kingdom in 1965–75 (Burton 1978).

Conservation Action

H. griseus is found in one national park and six reserves (Table 19). It is legally protected from hunting and export in Madagascar, but more meaningful fines and education are necessary to improve enforcement (Richard and Sussman 1975). Also needed is a reserve for the form found near Lake Alaotra (I.U.C.N. 1972).

This species is included in Class A of the African Convention (Organization of African Unity 1968), is in Appendix I of the Convention on International Trade in Endangered Species of Wild Fauna and Flora, and is listed as endangered under the U.S. Endangered Species Act (U.S. Department of Interior 1977a,b).

Table 19. Distribution of *Hapalemur griseus*

Country	Presence and Abundance	Reference
Malagasy Republic	In E	Petter 1962a
	Localities mapped throughout E, N, and W	Petter 1962b
	At Bemangidy (SE) and Périnet (E central)	Jolly 1966
	Throughout, including near Ampasindava, on NW coast, at Tsaratanana (N central), Hiaraka (Masoala Peninsula, NE), near Maroantsetra (NE), and Périnet (E central); exceptionally abundant near Sambava and Vohémar (NE), near Tamatave, between Fénérive and Vavatena, and near Mahanoro-Marolambo (all in E); isolated population at Lake Alaotra (about 17°30′S, 48°30′E)	Petter and Peyriéras 1970b
	In Montagne d'Ambre N.P. (in N) and 4 strict nature reserves: Tsingy du Bemaraha and Tsingy de Namoroka (both in W), Zahamena and Betampona (both in E)	I.U.C.N. 1971
	Locally abundant but declining, especially near Lake Alaotra; vulnerable	I.U.C.N. 1972
	Throughout E, W, and N	Martin 1972b
	Numerous throughout NE coast and on Ampasindava Peninsula on NW coast; occasional near Périnet (E), W of Maroantsetra and on Masoala Peninsula (NE); exceptionally abundant near Mahanoro-Marolambo (in E)	Petter and Peyriéras 1975

Table 19 (continued)

Country	Presence and Abundance	Reference
Malagasy Republic (cont.)	In 4 reserves in E, NE, and NW (map); declining; habitat being reduced	Richard and Sussman 1975
	In N throughout E, S of Andapa-Sambava area; very sparse between Sambava and Bay of Antongil; in Ambanja-Ampasindava Peninsula region of NW (map)	Tattersall 1977c
	Throughout E, locality mapped at Lake Alaotra; in small area of W central and along NW coast (map)	Petter and Petter-Rousseaux 1979
	At Analamazoatra (E central, locality mapped)	Pollock 1979

Hapalemur simus Broad-nosed Gentle Lemur

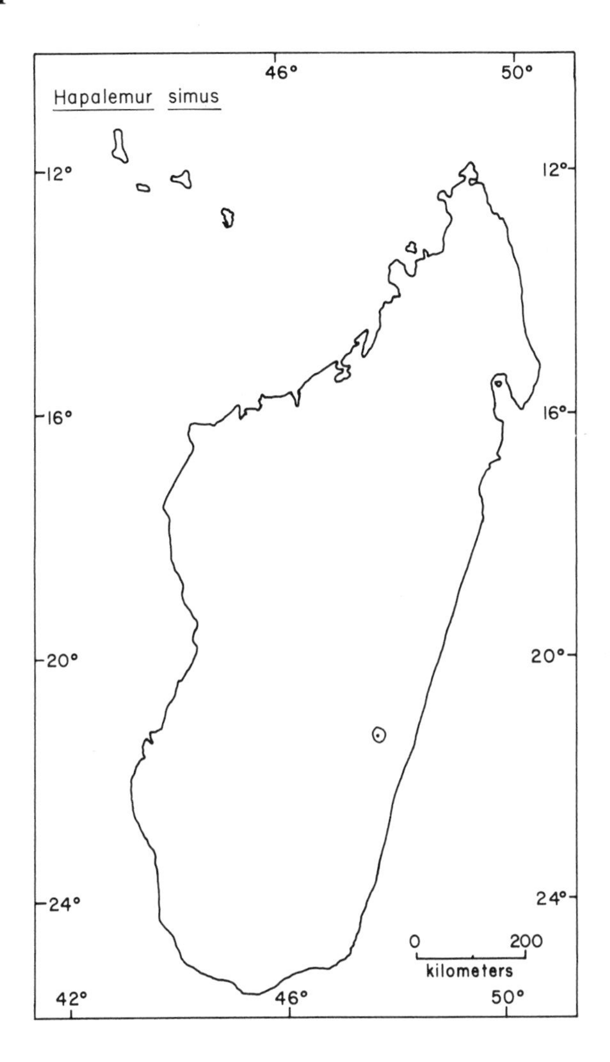

Taxonomic Note

This is a monotypic species (Rumpler 1975, Petter and Petter-Rousseaux 1979).

Geographic Range

Hapalemur simus occurs in eastern Madagascar. It was previously reported near Lake Alaotra (Petter 1962b), but the form there was subsequently found to be a race of *H. griseus* (Petter and Peyriéras 1970b). For several years *H. simus* was believed to be extinct, but recently a population was discovered (Petter 1975) east of Fianarantsoa at about 21°20′S, 47°40′E (Petter and Petter-Rousseaux 1979), about 50 km west of Ifanadiana in East Betsiléo, southeastern Madagascar (I.U.C.N. 1972). Further details of its distribution are not available.

Abundance and Density

H. simus is considered to be very rare (I.U.C.N. 1972; Petter 1972, 1975), declining, and probably on the brink of extinction (Richard and Sussman 1975). No specific information regarding its populations is available.

Habitat

The region in which *H. simus* is found is dense bamboo forest bordered by small remnants of the eastern tropical rain forest (I.U.C.N. 1972). No details of the habitat requirements of this species are known.

Factors Affecting Populations

Habitat Alteration

The habitat of *H. simus* has been greatly altered by human activity (I.U.C.N. 1972), including timbering, agriculture, and industry (Richard and Sussman 1975).

Human Predation

No information.

Conservation Action

H. simus does not occur in any habitat reserves but is legally protected from hunting and export (Richard and Sussman 1975). An ecological survey and the establishment of a special reserve are needed for this species (I.U.C.N. 1972).

H. simus is listed in Class A of the African Convention (Organization of African Unity 1968), in Appendix I of the Convention on International Trade in Endangered Species of Wild Fauna and Flora, and as endangered under the U.S. Endangered Species Act (U.S. Department of Interior 1977a,b).

Lemur catta Ring-tailed Lemur

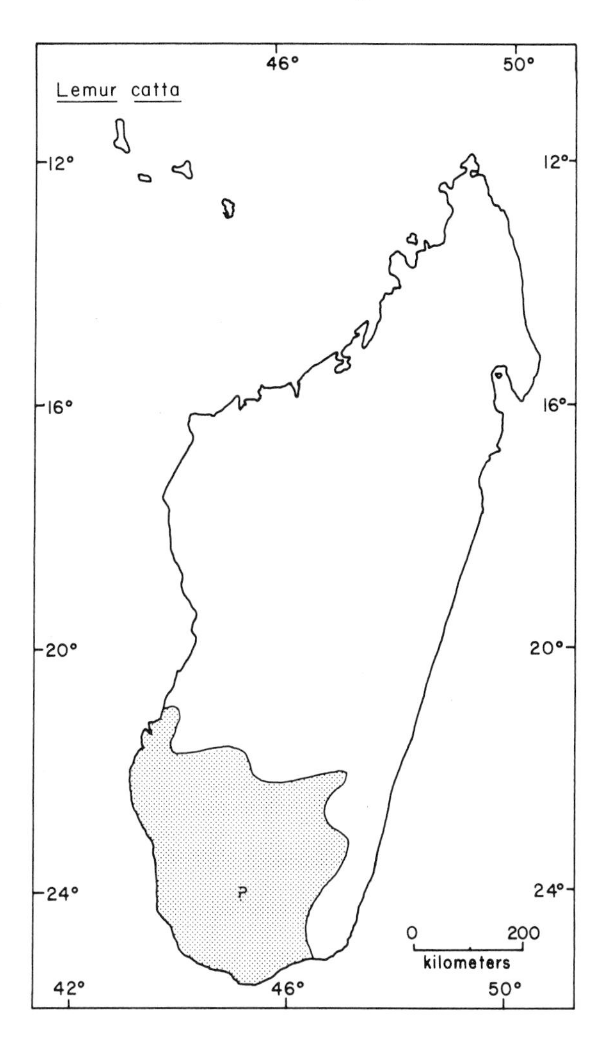

Taxonomic Note

This is a monotypic species (Petter and Petter-Rousseaux 1979).

Geographic Range

Lemur catta is found in southern and southwestern Madagascar. Its range extends as far north as about 21°S along the west coast, southward and eastward around the southern end of the island as far east as about 47°E, and northward to about 22°S in the east (Sussman 1977c, Petter and Petter-Rousseaux 1979). For details of the distribution, see Table 20.

Abundance and Density

Populations of *L. catta* are abundant compared with populations of other Malagasy primates but are scarce in absolute terms, and are declining rapidly (Sussman, pers. comm. 1974; Richard and Sussman 1975). Population densities vary from 74 to 300 individuals per km^2, with groups of 5 to 30 ($\bar{X}$ = 13 to 18.8) occupying home ranges of 5.7 to 23.1 ha (Table 21).

Habitat

L. catta is a forest species which occupies both wet and dry zones covered by deciduous canopy forest, low-profile brush and scrub forest, and desertlike *Didierea* forest (Sussman, pers. comm. 1974, 1977c; Sussman and Richard 1974). It has been studied in one closed canopy forest dominated by *Tamarindus indica*, *Acacia*, and *Terminalia*, and in another containing large *Albizzia polyphylla* and *Celtis bifida* (Sussman 1977a), as well as in gallery forest (Jolly 1966, 1972; Sussman and Richard 1974; Budnitz and Dainis 1975; Sussman 1977c).

In dry areas, *L. catta* is able to live in xerophytic bush (Martin 1972b), in rocky areas where only patches of forest remain (Petter 1972, Sussman and Richard 1974), and in the scrubby transition zone between gallery forest and subdesert vegetation (Budnitz and Dainis, pers. comm. 1974). It can also utilize the very arid *Didierea* forests dominated by *Alluaudia* trees (Sussman and Richard 1974) if these are adjacent to wetter areas where it can obtain water (Mertl, pers. comm. 1975).

Factors Affecting Populations

Habitat Alteration

The most important reason for the decline of *L. catta* populations is continuous and rapid deforestation via clearing for industry and plantations, timbering for local use and export, and slash and burn agriculture (Sussman, pers. comm. 1974; Richard and Sussman 1975). This species can only survive in primary vegetation and does not thrive in secondary growth (Sussman, pers. comm. 1974), so even rotational cultivation destroys its habitats. Clearing for agriculture is proceeding at a fast rate in southern Madagascar, and is the main threat to *L. catta*'s habitat (Budnitz and Dainis, pers. comm. 1974) along with burning and reforestation with *Eucalyptus* trees, which it cannot utilize (Mertl, pers. comm. 1975).

Human Predation

Hunting. Ring-tailed lemurs are hunted with dogs (Sussman 1977a) for food by the local people, but the number killed each year is thought to be small (Sussman, pers. comm. 1974).

Pest Control. One worker has reported that *L. catta* can be a pest to crops and is shot or stoned by farmers to prevent this (Mertl, pers. comm. 1975), but other researchers state that this species does not utilize cultivated foods (Sussman, pers. comm. 1974; Budnitz and Dainis, pers. comm. 1974).

Collection. The collection of *L. catta* for zoos and museums was a problem in the past, but has now been effectively curtailed (Sussman, pers. comm. 1974). Between 1968 and 1973, 74 *L. catta* were imported into the United States (Appendix). In 1977 none entered the United States (Bittner et al. 1978). Twenty-five were imported into the United Kingdom in 1965–75 (Burton 1978).

Conservation Action

L. catta is found in three strict nature reserves and three other reserved areas (Table 20). It is protected by some local taboos against killing it, but these are breaking down with the acceptance of foreign religions (Mertl, pers. comm. 1975). The species has also been granted legal protection against hunting and against timbering and burning in reserves, but the laws are not well publicized or enforced (Budnitz and Dainis, pers. comm. 1974; Sussman, pers. comm. 1974). More meaningful fines and conservation education are needed to improve this situation (Richard and Sussman 1975).

L. catta is listed in Class A of the African Convention (Organization of African Unity 1968), in Appendix I of the Convention on International Trade in Endangered Species of Wild Fauna and Flora, and as endangered under the U.S. Endangered Species Act (U.S. Department of Interior 1977a,b).

Table 20. Distribution of *Lemur catta*

Country	Presence and Abundance	Reference
Malagasy Republic	In S	Petter 1962a
	Localities mapped in SW and SE	Petter 1962b
	Study sites: Berenty Reserve (on Mandrary R.), Bevala, and Ifotaka (all in SE) and Lambomakandro (in SW)	Jolly 1966
	In Berenty Reserve	Klopfer and Jolly 1970, Jolly 1972, Budnitz and Dainis 1975
	In 3 strict nature reserves: Andohahela, Andringitra (both in SE), and Tsimanampetsotsa Lake (in SW)	I.U.C.N. 1971
	In SW	Martin 1972b
	In S and SW as far N as Morondava	Petter 1972
	Declining except in reserves	Budnitz and Dainis, pers. comm. 1974
	Study sites: Antserananomby (in SW, N of Mangoky R. at 21°46′S, 44°07′E) and Berenty (in S, on Mandrare R.)	Sussman 1974, 1977a,b
	Northernmost locality is Kirindy (E of Belo-sur-Mer), E to borders of Central Plateau (map); populations discontinuous; relatively abundant but declining rapidly	Sussman, pers. comm. 1974
	Throughout S and SW as far N as Morondava R. (map)	Sussman and Richard 1974
	Moderately scarce and declining	Mertl, pers. comm. 1975
	In Berenty Reserve and 3 other reserves in SW and SE; relatively abundant but declining; habitat being reduced	Richard and Sussman 1975

Table 20 (continued)

Country	Presence and Abundance	Reference
Malagasy Republic (cont.)	At Antserananomby and in forest S of Mangoky R.	Sussman 1975
	In Natural Reserve No. 11 (Andohahela) near Hazofotsy (24°49′S, 46°37′W) (in SE)	Russell and McGeorge 1977
	Localities listed and mapped as far NE as borders of Central Plateau include: N of Ranohira, Isalo National Reserve, W and SE of Ihosy, S of Ambalavao, Beomby, Berenty, Ejeda, Fiherenana I and II, Lambomakandro, Ankerandrere, Ambohy Menafifi, Antserananomby, Ianadranto, Mangoky, N Bengily, and Soaserana; northernmost record: Kirindy reserve, E of Belo-sur-Mer (20°44′S, 44°00′E, 45 km S of Morondava R., just S of Maharivo R.); easternmost record: S of Manamboro (25°02′S, 46°48′E)	Sussman 1977c
	Along SW coast S of about 21°S; along S coast to nearly 47°E and N to about 22°S in a U shaped distribution (map)	Petter and Petter-Rousseaux 1979

Table 21. Population parameters of *Lemur catta*

Population Density (individuals/km^2)	Group Size $\bar{X}$	Range	No. of Groups	Home Range Area (ha)	Study Site	Reference
		20–24	1	5.7	Madagascar	Jolly 1966
		22	1		Madagascar	Klopfer and Jolly 1970
		8–24			Madagascar	Jolly 1972
		≤30			Madagascar	Martin 1972b
215–250	18	15–20	3	6.0–8.8	Madagascar	Sussman 1974, 1977a
167	13–17	5–22		6.0–23.1	Madagascar	Budnitz and Dainis 1975
74–220					Madagascar	Mertl, pers. comm. 1975
200–300	15–18			6–9	Madagascar	Richard and Sussman 1975
215–250	18.8	15–20	8	6–9	Madagascar	Sussman 1977b

Lemur coronatus Crowned Lemur

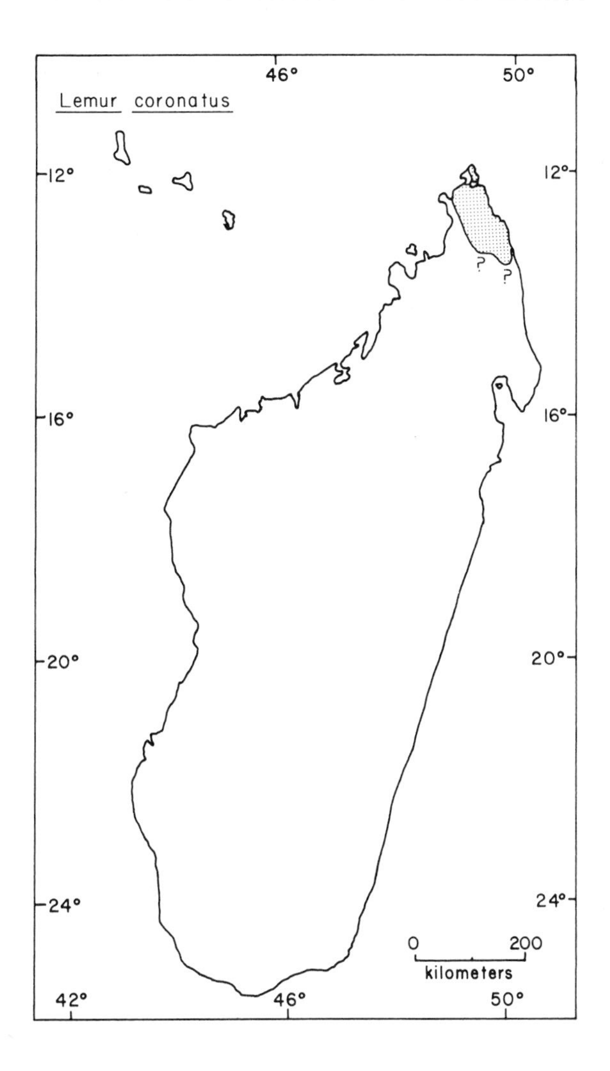

Taxonomic Note

This form was previously considered to be a race of the mongoose lemur, *Lemur mongoz* (Petter 1962b, 1972) but recently has been treated as a distinctive, monotypic species (Rumpler 1975; Tattersall 1976b, 1977c; Petter and Petter-Rousseaux 1979).

Geographic Range

Lemur coronatus is found in northern and northeastern Madagascar. It occurs on the Cap d'Ambre on the northern tip of the island and south along the northeastern coast, at least to the Fanambana River at about 13°25′–13°30′S (Tattersall 1977c, Petter and Petter-Rousseaux 1979). The southern and southwestern limits of its range are not known (Tattersall 1977c). For details of the distribution, see Table 22.

Abundance and Density

L. coronatus has been described as very rare, vulnerable, and declining (Table 22). No details concerning its population density, group size, or home range area are available.

Habitat

L. coronatus is typical of dry seasonal forest, such as the very arid forest of the Cap d'Ambre and the dry forest on the lower slopes of the Montagne d'Ambre but not the humid forests at higher elevations (Tattersall 1977c). It also occupies savanna, dry bush, and the edges of humid forest (I.U.C.N. 1972).

Factors Affecting Populations

Habitat Alteration

The habitat of *L. coronatus* is being destroyed and degraded (I.U.C.N. 1972). No further details are available.

Human Predation

Hunting. *L. coronatus* is hunted even in the Montagne d'Ambre National Park, and this hunting is increasing (I.U.C.N. 1972).

Pest control. This species frequently causes damage to crops and plantations and is regularly killed by farmers (I.U.C.N. 1972).

Collection. No crowned lemurs were imported into the United States in 1968–73 (Appendix) or in 1977 (Bittner et al. 1978), nor into the United Kingdom in 1965–75 (Burton 1978).

Conservation Action

The crowned lemur is found in one national park (Table 22) but has not been reported in other habitat reserves. It is legally protected from hunting and export (Richard and Sussman 1975), is listed in Class A of the African Convention (Organization of African Unity 1968), in Appendix I of the Convention on International Trade in Endangered Species of Wild Fauna and Flora, and as endangered under the U.S. Endangered Species Act (U.S. Department of Interior 1977a,b).

Table 22. Distribution of *Lemur coronatus*

Country	Presence and Abundance	Reference
Malagasy Republic	Locality mapped in NE	Petter 1962b
	Confined to a small area near Montagne d'Ambre; very rare	Fisher et al. 1969
	In Montagne d'Ambre N.P., extreme N	I.U.C.N. 1971
	Vulnerable; N of Bay of Bombetoka in W and Bay of Antongil in E, N and W of Montagne d'Ambre; population and habitat decreasing	I.U.C.N. 1972
	In extreme N and on Montagne d'Ambre	Petter 1972

Table 22 (continued)

Country	Presence and Abundance	Reference
Malagasy Republic (cont.)	On Cap d'Ambre (peninsula N of Diégo-Suarez), on Montagne d'Ambre, in Ankarana region (SW of Anivorano Nord); in E as far S as Fanambana R. (S of Vohémar) (map); S and SW limits uncertain	Tattersall 1977c
	In N from about 12°15′S to about 13°15′S (map)	Petter and Petter-Rousseaux 1979

Lemur fulvus Brown Lemur

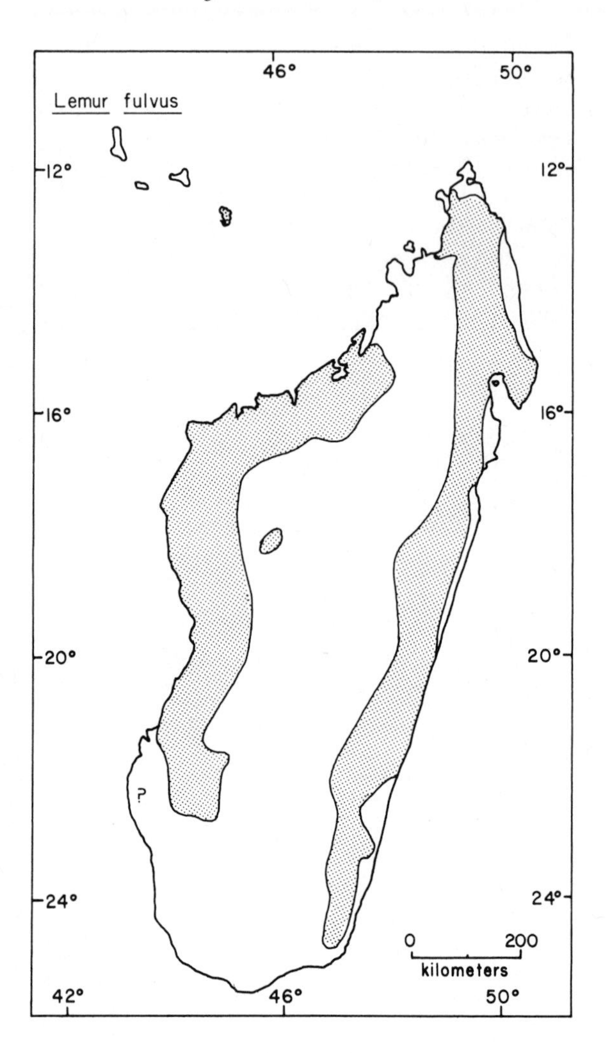

Taxonomic Note

The members of this group have sometimes been placed within *Lemur macaco* (Petter 1962b, 1972; Jolly 1966; I.U.C.N. 1972), and the taxonomic questions — including the correct name for this species, whether it is conspecific with *L. macaco*, and the number and names of the races it includes — have not been resolved (Groves 1974, 1978). But most recent authors (Tattersall and Schwartz 1974, Rumpler 1975, Sussman 1975, Petter and Petter-Rousseaux 1979) treat *Lemur fulvus* as a distinct species, and these will be followed here. Seven subspecies are recognized, excluding *flavifrons* listed by older authors, and including *albocollaris* (Rumpler 1975, Petter and Petter-Rousseaux 1979), which should be called *cinereiceps* according to Groves (1974, 1978).

Geographic Range

Lemur fulvus is found in eastern, northern, and western Madagascar. It occurs in the east along or near the coast from about 12°30′S to about 24°45′S (Petter and Petter-Rousseaux 1979) and in the north as far west as about 13°22′S, 48°53′E near Beramanja on the Ifasy River (Tattersall 1976c). In the west its range extends from about 14°45′S (Tattersall 1977c) to Lambomakandro at 22°45′S, 44°40′E on the south bank of the Fiherenana River (Sussman 1977c). This species also occurs on Mayotte in the Comoro Islands (Tattersall 1976a, 1977a,b,c). For details of the distribution, see Table 23.

Abundance and Density

Two of the races of *L. fulvus* are considered endangered and decreasing, and although the species as a whole is locally abundant in some areas, it is declining and its habitat is being reduced throughout its range (Table 23).

Population densities in this species can reach high levels: 900 to 1,222 individuals per km^2. Groups vary in size from 2 to 29 ($\bar{X}$ = 9.1 to 20) and occupy ranges of 0.75 to 7 ha (Table 24).

Habitat

The preferred habitat of *L. fulvus* is primary, continuous, closed canopy forest (Sussman, pers. comm. 1974, 1977c). At several study sites it occupies deciduous forest composed of *Tamarindus indica* in consociation with baobabs (*Adansonia madagascariensis*), *Acacia rovumae*, *Ficus soroceoides*, and other tall trees, with an underlayer of lower trees or a thick underbrush of lianes and bushes (Sussman and Richard 1974; Sussman 1975, 1977a,b). It can also utilize continuous canopy forests where they are mixed with brush and scrub forest, but is not found in pure stands of this latter habitat type (Sussman 1974, 1977c). On Madagascar, *L. fulvus* does not thrive in secondary growth vegetation (Sussman, pers. comm. 1974).

On Mayotte, almost all the forest is secondary, and *L. fulvus* occupies it; the species is found in lower numbers above 400 m than at lower elevations (Tattersall 1977a,b).

Factors Affecting Populations

Habitat Alteration

L. fulvus populations are declining primarily because of continuous and rapid deforestation, including clearing for industry and plantations, timbering for local use and export, and slash and burn agriculture (Sussman, pers. comm. 1974; Richard and Sussman 1975). In eastern and western Madagascar the habitat is also being burned and degraded by woodcutters and cattle, and in the north near Montagne d'Ambre, extensive timbering is restricting the range (I.U.C.N. 1972).

On Mayotte, large-scale clearance of vegetation has resulted in denuded land covered only with stunted grass, especially in the south; the spread of the introduced tree *Litsea tersa*, which *L. fulvus* can use, is ameliorating the situation somewhat in the north (Tattersall 1977b).

Human Predation

Hunting. L. fulvus is intensively hunted in northeastern Madagascar (Tattersall 1977c), and hunting has also contributed to its decline in the northern, western, and eastern regions of the island (I.U.C.N. 1972). It is hunted for food in the southwestern portion of its range, but the number killed is small and remaining stable, and this factor is minor compared with habitat alteration (Sussman, pers. comm. 1974). *L. fulvus* is also hunted for food and sport on Mayotte, and although hunting is increasing, it is not yet seriously threatening *L. fulvus* there (Tattersall 1977a).

Pest Control. L. fulvus does feed in plantations, for example, near Joffreville (I.U.C.N. 1972) and on Nosy Komba (Bomford 1976b), but no reports of efforts to control this activity were located.

Collection. The collection of *L. fulvus* in Madagascar for foreign zoos and museums was a problem in the past, but strict enforcement of export laws has now curtailed this (Sussman, pers. comm. 1974). On Mayotte, children collect infant *L. fulvus* by killing their mothers (Tattersall 1977a).

No *L. fulvus* were imported into the United States in 1968–73 (Appendix) or in 1977 (Bittner et al. 1978). Six entered the United Kingdom in 1965–75 (Burton 1978).

Conservation Action

L. fulvus is found in one national park and nine reserves (Table 23). In some areas local taboos or traditions effectively prevent it from being killed (Bomford 1976b). The species is legally protected in Madagascar, but more meaningful fines and education are needed to improve the effectiveness of the laws (Richard and Sussman 1975). Legal restrictions on timbering are not adequately publicized or enforced (Sussman, pers. comm. 1974). Similarly, on Mayotte, legal restraints on hunting and keeping *L. fulvus* should be more widely advertised and enforced, and prohibitions against cutting streamside vegetation should be extended from 15 m to 30 m from watercourses (Tattersall 1977a). Better management and protection of the Montagne d'Ambre National Park, and the elimination of poaching inside the park and reserves, are also necessary (I.U.C.N. 1972).

L. fulvus is listed in Class A of the African Convention (Organization of African Unity 1968), in Appendix I of the Convention on International Trade in Endangered Species of Wild Fauna and Flora, and as endangered under the U.S. Endangered Species Act (U.S. Department of Interior 1977a,b).

Table 23. Distribution of *Lemur fulvus*

Country	Presence and Abundance	Reference
Malagasy Republic	In E and W including Lake Ampijoroa in Ankarafantsika forest	Petter 1962a
	Localities mapped in E, N, NW, and SW	Petter 1962b
	At Morondava (W), Lambomakandro (SW), Ankarafantsika (NW), Périnet (E), and Bemangidy (SE)	Jolly 1966
	In Montagne d'Ambre N.P. (extreme N) and 6 strict nature reserves:	I.U.C.N. 1971

Table 23 (continued)

Country	Presence and Abundance	Reference
Malagasy Republic (cont.)	Ankarafantsika (NW), Zahamena (E), Marojejy (NE), Betampona (E), Tsingy du Bemaraha, and Tsingy de Namoroka (both in W)	I.U.C.N. 1971 (cont.)
	Abundant 60 km N of Morondava near Analabe and Lake Andranolava at about 20°S	Petter et al. 1971, 1975 [Petter 1978]
	Race in W central and in E between 20° and 22°S (*L. f. rufus*): endangered, locally common in a few small areas including Antsalova Reserve but numbers and habitat declining; race in extreme N in and around Montagne d'Ambre N.P. (*L. f. sanfordi*): endangered, numbers and range decreasing, habitat deteriorating	I.U.C.N. 1972
	In SE as far as Farafangana; in NE between Andapa and Mananara; in N at Montagne d'Ambre; in E central and in W near Ankarafantsika	Petter 1972
	Localities mapped in E, N, NW, W, and SW	Groves 1974
	At Antserananomby and Tongobato (in SW, N of Mangoky R.)	Sussman 1974, 1977a,b
	In W, as far N as forest of Katsepy across Betsiboka R. from Majunga (15°48′S, 46°11′E) and as far S as forest of Lambomakandro, N of Sakaraha along Fiherenana R. (22°44′S, 44°43′E) (map); populations discontinuous; relatively abundant but rapidly declining; habitat being reduced	Sussman, pers. comm. 1974
	Throughout W from near Majunga to Fiherenana R. (map) including Antserananomby and Tongobato at 21°46′S, 44°07′E and Ampijoroa forestry station (16°35′S, 46°82′E) in Ankarafantsika Reserve	Sussman and Richard 1974, Richard 1978
	At Ankarafantsika Forest Reserve	Harrington 1975, Tattersall and Sussman 1975a
	In 11 reserves including Analabe (W), Nosy Mangabe (NE), and others in SE, E, NW, and W; relatively abundant but declining; habitat being reduced	Richard and Sussman 1975
	Throughout E, N, and W (map) including 3 study sites near Mangoky R. (21°46′S, 44°07′E)	Sussman 1975
	On Nosy Komba Island (off NW coast)	Bomford 1976b

Table 23 (continued)

Country	Presence and Abundance	Reference
Malagasy Republic (cont.)	In NW near Beramanja and between Analalava and Betsiboka R.; in N (map); in E near Andasibé (Périnet), Ankoby, Ambarakaraka, and Marovatokely	Tattersall 1976c
	Localities listed and mapped in S at Fiherenana, Lambomakandro, Antanimangotroky, Antserananomby, Ianadranto, Malindira, Mangoky, Tarata, Tongobato, and Belo-sur-Mer ; in W at Ambatovoamba, Andranomena, Angansiva, Ankilimare, Befasy, Beroboka, Manamby, Maroary, Tanambao, Ampijoroa, Katsepy, Madirovalo, Marohogo, and Mitsinjo; as far E as Mandabe (23°03′S, 44°56′E)	Sussman 1977c
	In N on Montagne d'Ambre S and W of Diégo-Suarez and S to Anivorano Nord; in E from Fanambana R. S of Vohémar to N of Tamatave; in W, W and S of Antsohihy to Ambato-Boeni and Tsaramandroso; isolates near Beramanja and Andasibé (Périnet) (map)	Tattersall 1977c
	Throughout E from N of Fort Dauphin to 12°30′S; in W from Mangoky R. N to about 15°N (map)	Petter and Petter-Rousseaux 1979
Comoro Islands	On Mayotte Island (SE one of group)	Groves 1974
	Minimum population about 50,000 individuals on Mayotte	Tattersall 1976a, 1977a
	98 localities listed and mapped in S, central, E, and W Mayotte; may have been introduced by man; habitat being reduced	Tattersall 1977b
	On Mayotte (map)	Tattersall 1977c

Table 24. Population parameters of *Lemur fulvus*

Population Density (individuals/km^2)	Group Size			Home Range Area (ha)	Study Site	Reference
	$\bar{X}$	Range	No. of Groups			
	~20				Madagascar	Petter 1972
900–1,222	9.5	4–17	17	0.75–1.0	Madagascar	Sussman 1974, 1975, 1977a,b
		12	2	7	Madagascar	Harrington 1975
		6–8			Madagascar	Pollock 1975
1,000	9.1	2–29	100		Mayotte Island	Tattersall 1977b

Lemur macaco Black Lemur

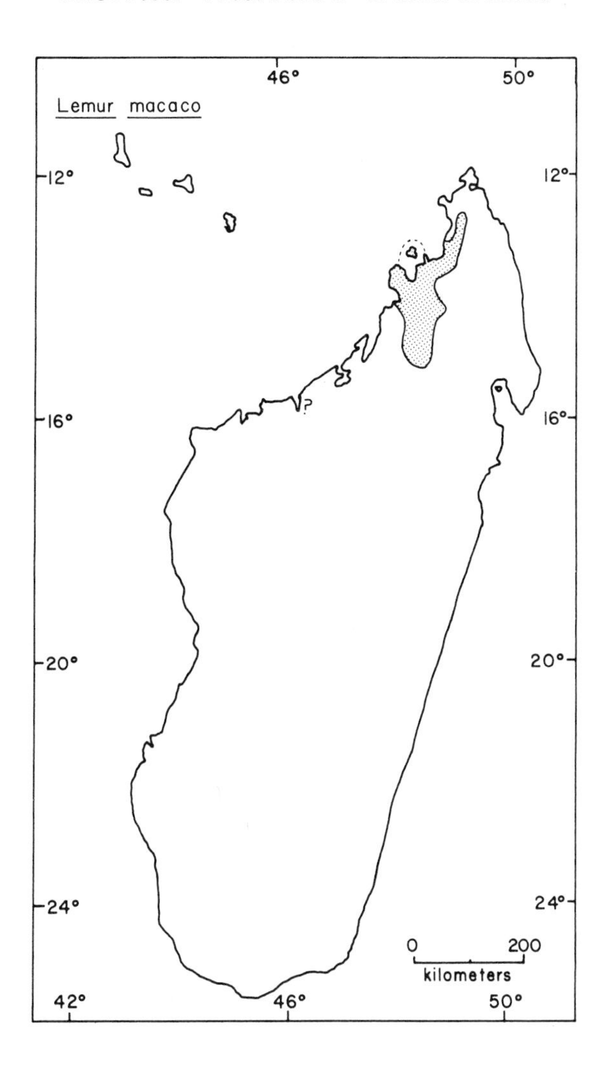

Taxonomic Note

Two subspecies are recognized (Rumpler 1975, Petter and Petter-Rousseaux 1979).

Geographic Range

Lemur macaco is found in northwestern Madagascar and on Nosy Bé and Nosy Komba islands off the northwestern coast (Petter 1962a). On the main island it occurs on the Ampasindava Peninsula and along the coast north of this (Petter and Petter-Rousseaux 1979). It has been reported as far north as Anivorano Nord (at 12°42′S, 49°12′E) and as far south as Befandriana Nord (15°14′S, 48°33′E) (Tattersall 1976c, 1977c). For details of the distribution, see Table 25.

Abundance and Density

L. macaco populations are declining, and one race (*L. m. flavifrons*) is very rare and may be extinct (Table 25). The average group size is about 9.5 individuals (Richard and Sussman 1975). No data are available concerning population density or home range area.

Habitat

L. macaco inhabits very humid forest (I.U.C.N. 1972). No details of its requirements have been reported.

Factors Affecting Populations

Habitat Alteration

The forests in which *L. macaco* lives are being reduced because of clearing for timber, industry, and agriculture (Davis 1975, Richard and Sussman 1975). Large areas have been burned or converted to plantations; in a few areas *L. macaco* populations have increased in these plantations, but in many regions habitat deterioration has eliminated this species (I.U.C.N. 1972).

Human Predation

Hunting. *L. macaco* is not usually hunted (Tattersall 1977c).

Pest Control. Black lemurs do enter plantations and eat cultivated foods, including bananas, coffee beans, and cacao pods (Davis 1975), and recently the protection previously provided by local taboos has been waning and people have been poisoning and shooting the animals (I.U.C.N. 1972, Davis 1975).

Collection. Eleven *L. macaco* were imported into the United States in 1968–73 (Appendix). None entered the United States in 1977 (Bittner et al. 1978). One was imported into the United Kingdom in 1965–75 (Burton 1978).

Conservation Action

L. macaco is found in two reserves (Table 25), but the Lokobé Reserve is in need of better protection and expansion (I.U.C.N. 1972). The species is legally protected, but more meaningful fines and education are necessary (Richard and Sussman 1975). *L. macaco* is listed in Class A of the African Convention (Organization of African Unity 1968), in Appendix I of the Convention on International Trade in Endangered Species of Wild Fauna and Flora, and as endangered under the U.S. Endangered Species Act (U.S. Department of Interior 1977a,b).

Table 25. Distribution of *Lemur macaco*

Country	Presence and Abundance	Reference
Malagasy Republic	In Sambirano region (NW), on Nosy Komba, and in Lokobé Reserve of Nosy Bé	Petter 1962a
	Localities mapped in NW	Petter 1962b
	At Ambanja (NW) and on Nosy Bé	Jolly 1966
	In 2 strict nature reserves of Tsaratanana (N central) and Lokobé (on Nosy Bé)	I.U.C.N. 1971
	Endangered; declining, habitat reduced; one race (*L. m. flavifrons*) may be extinct	I.U.C.N. 1972
	In NW near Ambanja, at Bay of Ampasindava, on Nosy Bé; form near Maromandia (*L. m. flavifrons*) is very rare	Petter 1972
	Localities mapped in NW N of Bay of Ampasindava and as far S as Majunga (this is outside the range limits)	Groves 1974
	In Lokobé Reserve (Nosy Bé) and on Nosy Komba	Davis 1975
	In Lokobé Reserve; overall: declining, habitat being reduced	Richard and Sussman 1975
	Common in Beramanja region (about 13°22′S, 48°53′E); from Anivorano Nord to S of Maromandia along coast and to Befandriana Nord in interior (map)	Tattersall 1976c
	Range includes W part of Massif du Tsaratanana, Ampasindava Peninsula, Nosy Bé, and Nosy Komba (map)	Tattersall 1977c
	In NW on Nosy Bé (map)	Petter and Petter-Rousseaux 1979

Lemur mongoz Mongoose Lemur

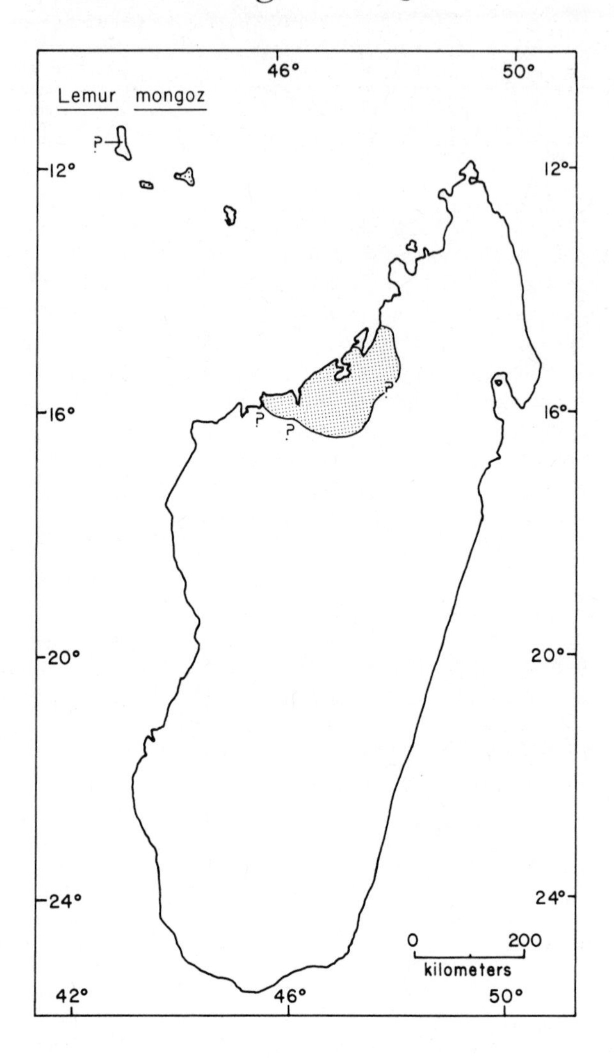

Taxonomic Note

This is a monotypic species (Rumpler 1975, Petter and Petter-Rousseaux 1979).

Geographic Range

Lemur mongoz is found in northwestern Madagascar and on the Comoro Islands. On Madagascar it occurs north of about 16°30′S and west of about 48°E (Petter and Petter-Rousseaux 1979). Along the coast it has been reported as far north as Analalava (14°38′S, 47°46′E) and as far south as Lake Kinkony (16°09′S, 45°50′E) (Tattersall 1977c, 1978). And on the Comoro Islands the species is found on Mohéli and Anjouan, the central islands of the group (Tattersall 1976b, 1977a). It has also been reported on Grand Comoro (Petter and Petter-Rousseaux 1979). For details of the distribution, see Table 26.

Abundance and Density

L. mongoz is considered scarce, vulnerable, and declining with a shrinking habitat on Madagascar, and relatively abundant on Mohéli (Table 26).

In one area, population density of *L. mongoz* was about 350 individuals per km^2 (Richard and Sussman 1975). On Madagascar this species lives in groups of 6 to 8 (Jolly 1966), 2 to 4 (Harrington 1975, Sussman and Tattersall 1976), and 2 to 6 (Harrington 1978). On the Comoro Islands its groups number 2 to 7 ($\bar{X}$ = 3.7, N = 22) on Mohéli, and 2 to 4 ($\bar{X}$ = 3.1, N = 26) on Anjouan (Tattersall 1976b). Home range area is about 0.5 to 1.0 ha (Richard and Sussman 1975).

Habitat

L. mongoz is a forest species which has been reported in dry forest (Petter 1962a) and native deciduous tropical forest containing some introduced trees (Harrington 1978). It utilizes continuous canopy forest, canopy forest mixed with brush and scrub, riverine forest, and cliff edge or hilly forest (Sussman 1977c).

On the Comoro Islands *L. mongoz* is as successful in the secondary vegetation which covers most of Mohéli as in the primary forest, and also occurs in secondary forest and in highland rain forest above 600 to 700 m elevation on Anjouan (Tattersall 1976b, 1977a).

Factors Affecting Populations

Habitat Alteration

The forest areas in which *L. mongoz* lives are diminishing in size because of clearing for timber, industry, and agriculture (I.U.C.N. 1972, Richard and Sussman 1975). No specific information is available from its range in Madagascar, but in its habitat on Anjouan island, extensive areas have been cleared, including parts of the central rain forest, the cultivation of steep slopes is causing serious erosion which prevents secondary succession, and the rate of destruction of vegetation is accelerating (Tattersall 1977a).

Human Predation

Hunting. Although it has been reported that *L. mongoz* is trapped intensively (I.U.C.N. 1972), other workers consider that hunting is not a serious problem for this species on Madagascar (Richard and Sussman 1975) or on the Comoro Islands (Tattersall 1977a).

Pest Control. No information.

Collection. On the Comoro Islands, local children commonly catch *L. mongoz* infants by killing the mothers (Tattersall 1977a).

Between 1968 and 1973, 25 *L. mongoz* were imported into the United States (Appendix). None entered the United States in 1977 (Bittner et al. 1978). One was imported into the United Kingdom in 1965–75 (Burton 1978).

Conservation Action

On Madagascar, *L. mongoz* occurs within one reserve (Table 26) and is legally protected, but more meaningful fines and education are necessary for its conservation (Richard and Sussman 1975). On the Comoro Islands it is illegal to kill or keep *L. mongoz* without a license, and exports are controlled, but these laws should be more restrictive, more widely publicized, and more strictly enforced (Tattersall 1977a). In addition, legal restraints preventing the destruction of vegetation within 15 m of a stream should be extended to 30 m and enforced, and cutting at high elevations should be prohibited (Tattersall 1977a).

L. mongoz is listed in Class A of the African Convention (Organization of African Unity 1968), in Appendix I of the Convention on International Trade in Endangered Species of Wild Fauna and Flora, and as endangered under the U.S. Endangered Species Act (U.S. Department of Interior 1977a,b).

Table 26. Distribution of *Lemur mongoz*

Country	Presence and Abundance	Reference
Malagasy Republic	In W	Petter 1962a
	Locality mapped in NW	Petter 1962b
	In Ankarafantsika (NW)	Jolly 1966, Petter 1972
	Rare; habitat reduced	Fisher et al. 1969
	In 2 strict nature reserves of Ankarafantsika (NW) and Tsingy de Namoroka (in W outside range limits)	I.U.C.N. 1971
	Vulnerable; decreasing; range contracting	I.U.C.N. 1972
	At Ampijoroa forestry station (16°35′S, 46°82′E), Ankarafantsika Reserve	Sussman and Richard 1974; Harrington 1975, 1978; Richard 1978
	Populations scarce and declining; habitat decreasing	Richard and Sussman 1975
	Abundant at Ampijoroa, 100 km S of Majunga	Tattersall and Sussman 1975c, Sussman and Tattersall 1976
	Localities listed and mapped at Ampijoroa, Lake Kinkony, and Madirovalo (16°33′S, 46°24′E)	Sussman 1977c
	In NW, N, and E of Betsiboka R. along coast as far N as Analalava; S and W of Betsiboka R. at least as far as Lake Kinkony and E of it at least to Ambato-Boéni (16°26′S, 46°43′E) (maps); S and SW limits not known	Tattersall 1977c, 1978
	In NW from about 15°20′–16°30′S and 46°40′–48°E but not along coast (map)	Petter and Petter-Rousseaux 1979
Comoro Islands	On Mohéli and Anjouan but not on Grand Comoro	Tattersall and Sussman 1975c
	Localities listed and mapped throughout Mohéli and in central and E Anjouan; probably introduced by man; relatively abundant on Mohéli	Tattersall 1976b
	On Mohéli and Anjouan	Tattersall 1977a, 1978
	On Mohéli, Anjouan, and Grand Comoro (map)	Petter and Petter-Rousseaux 1979

Lemur rubriventer Red-bellied Lemur

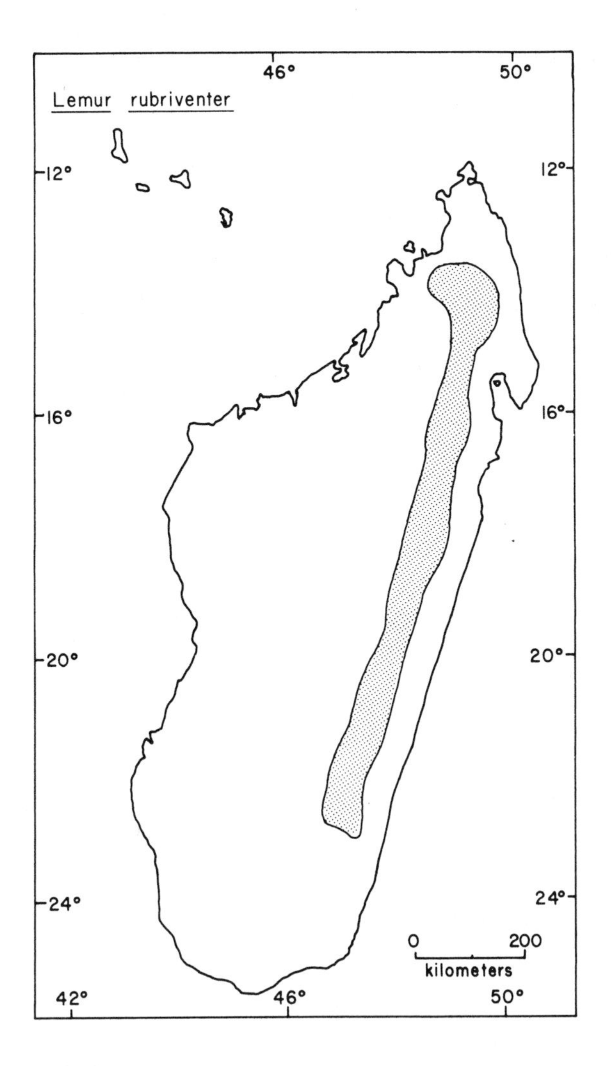

Taxonomic Note

This is a monotypic species (Rumpler 1975, Petter and Petter-Rousseaux 1979).

Geographic Range

Lemur rubriventer is found in eastern Madagascar, in a long narrow area about 50 km inland from and parallel to the east coast. The range extends from about 13°40′S in the north to the Mananara River at about 23°S in the south, and as far west as about 48°30′E (including the Massif du Tsaratanana) in the north and 46°40′E in the south (Petter and Petter-Rousseaux 1979). For details of the distribution, see Table 27.

Abundance and Density

L. rubriventer is rare and declining and is suspected of being on the verge of extinction (Table 27). Very little information is available concerning its populations.

No estimates of population density, average group size, or home range area were located. In one study three groups were seen which included 2 to 4 members each (Pollock 1979), and elsewhere groups of 4 to 5 have been reported (Jolly 1966).

Habitat

L. rubriventer occupies rain forest (Petter 1962a) and is more abundant at high altitudes than at lower ones on the Massif du Tsaratanana (Tattersall 1977c). No details of its requirements have been reported.

Factors Affecting Populations

Habitat Alteration

The forests in which *L. rubriventer* lives are being cleared for timber, industry, and cultivation (Richard and Sussman 1975).

Human Predation

No information.

Conservation Action

L. rubriventer is not known to occur in any habitat reserves (Richard and Sussman 1975). It is listed in Class A of the African Convention (Organization of African Unity 1968), in Appendix I of the Convention on International Trade in Endangered Species of Wild Fauna and Flora, and as endangered under the U.S. Endangered Species Act (U.S. Department of Interior 1977a,b).

Table 27. Distribution of *Lemur rubriventer*

Country	Presence and Abundance	Reference
Malagasy Republic	In E	Petter 1962a, Martin 1972b
	Localities mapped throughout E	Petter 1962b
	Rare; throughout E including Tsaratanana Mts.	Petter 1972
	Extremely rare, declining, probably on the brink of extinction; habitat decreasing	Richard and Sussman 1975
	Rare in N; on Massif du Tsaratanana (map); S limit unknown	Tattersall 1977c
	Throughout E (map)	Petter and Petter-Rousseaux 1979
	At Analamazoatra and Vohidrazana (both E central, map); locally rare	Pollock 1979

Lepilemur Group   Sportive Lemur

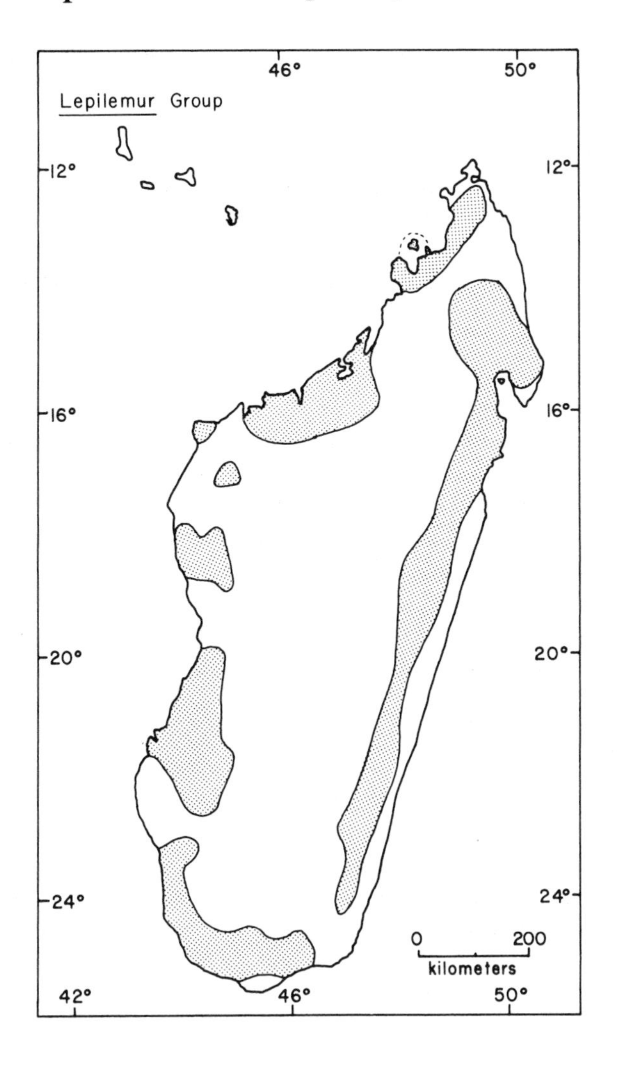

Taxonomic Note

All sportive lemurs were previously considered to belong to one species, *Lepilemur mustelinus* (Petter and Petter-Rousseaux 1960; Petter 1962b, 1965, 1972; Napier and Napier 1967), but recently several authors have split this into six (Rumpler and Albignac 1978) or seven species (Rumpler 1975, Petter and Petter-Rousseaux 1979). Because there is very little information available concerning most of these forms, the whole complex will be treated here together as *Lepilemur*.

Geographic Range

Lepilemur is found throughout most of the periphery of Madagascar, along or near the northern, northwestern, western, southern, and eastern coasts of the island. It occurs as far north as about 12°20′S, south of Diégo Suarez, throughout most of the northwest, western, and southern regions in a disjunct distribution, and parallel to the eastern coast north as far as about 14°S (Petter and Petter-Rousseaux 1979). For details of the distribution, see Table 28.

Abundance and Density

The western and southern forms of *Lepilemur* are considered endangered, the one on Nosy Bé is rare, and all populations of the genus are thought to be declining (Table 28).

Sportive lemur population densities vary from 100 to 810 individuals per km^2, group sizes are 2 to 6, and home ranges 0.15 to 0.46 ha (Table 29).

Habitat

Members of the *Lepilemur* group occupy several different types of habitat including rain forest in the east and north, dry forest in the west, and arid thornbush in the south (Petter 1962a). The form along the east coast (*mustelinus*) for example, is found in dense evergreen rain forest, while the one in the south (*leucopus*) inhabits arid Didiereaceae bush dominated by *Alluaudia procera* and *A. ascendens* with a dense shrubby underbrush, and gallery forest dominated by *Tamarindus* with a scanty undergrowth (Hladik and Charles-Dominique 1974, Pariente 1974, Sussman 1977a). The form found in the west, south of 20°N (*ruficaudatus*), occupies primary deciduous forest (Petter et al. 1975, Petter 1978).

Factors Affecting Populations

Habitat Alteration

The habitats of *Lepilemur* are diminishing, owing to clearing for timber, industry, and cultivation (Richard and Sussman 1975). For instance, the southern form (*leucopus*) is being threatened because of the destruction of the Didiereaceae bush by cutting, burning, grazing, cultivation, and erosion (Hladik and Charles-Dominique 1971, I.U.C.N. 1972). The forests of the west are rapidly disappearing, owing to repeated fires and excessive exploitation, and on Nosy Bé forest destruction and degradation are also reducing available habitat for *Lepilemur* (I.U.C.N. 1972).

Human Predation

Hunting. In western Madagascar, *Lepilemur* is hunted for its flesh (I.U.C.N. 1972). No other reports of hunting were located.

Pest Control. No information.

Collection. On Nosy Bé, sportive lemurs are caught for sale (I.U.C.N. 1972). The magnitude of this trade is not known.

One *Lepilemur* was imported into the United States in 1968–73 (Appendix). None entered the United States in 1977 (Bittner et al. 1978) or the United Kingdom in 1965–75 (Burton 1978).

Conservation Action

Lepilemur is found in one national park and ten habitat reserves (Table 28). It is legally protected in Madagascar, but more meaningful punishments and conservation education are needed (Richard and Sussman 1975), as are additional reserves and stronger protection for existing ones (I.U.C.N. 1972). All forms in this group are listed in Class A of the African Convention (Organization of African Unity 1968), Appendix I of the Convention on International Trade in Endangered Species of Wild Fauna and Flora, and as endangered under the U.S. Endangered Species Act (U.S. Department of Interior 1977a,b).

Table 28. Distribution of *Lepilemur*

Country	Presence and Abundance	Reference
Malagasy Republic	Localities listed throughout E, N, and W (map)	Petter and Petter-Rousseaux 1960
	In E, Sambirano (NW), W, and S	Petter 1962a
	Localities mapped throughout E, NW, and W	Petter 1962b
	In E, SE, S, W, and on Nosy Bé Island (NW)	Petter 1965, 1972
	At Ifotaka, Bevala, and Berenty (in SE) and Nosy Bé Island (NW)	Jolly 1966
	Near Berenty, N of Bebarimo, and in Andohahela Reserve (all in SE)	Charles-Dominique and Hladik 1971
	In Montagne d'Ambre N.P. extreme N, and 8 strict nature reserves: Tsingy du Bemaraha (W central), Ankarafantsika (NW), Lokobé (on Nosy Bé, off NW), Tsaratanana (N), Betampona and Zahamena (both in E), Andohahela (SE), and Tsimanampetsotsa Lake (SW)	I.U.C.N. 1971
	Near Analabe and Lake Andranolava, N of Morondava (in W)	Petter et al. 1971, Petter 1978
	Form in W (*ruficaudatus*, includes *edwardsi* of Petter and Petter-Rousseaux [1979]): endangered, declining, threatened with extinction; one in S (S of Onilahy R.) (*leucopus*): extremely endangered, locally abundant but declining; one on Nosy Bé Island (*dorsalis*): rare; all 3 ranges being reduced	I.U.C.N. 1972
	In E, N, NW, and W	Martin 1972b
	At Berenty (25°S, 46°24′E) NW of Amboasary	Pariente 1974
	At Antserananomby and Tongobato (in SW) and Berenty (in S)	Sussman 1974, 1977a
	At Antserananomby and Tongobato (N of Mangoky R.), Berenty reserve, Hazafotsy reserve (24°84′S, 46°50′E), and Ampijoroa foresty station, Ankarafantsika Reserve (16°35′S, 46°82′E)	Sussman and Richard 1974

Table 28 (continued)

Country	Presence and Abundance	Reference
Malagasy Republic (cont.)	In Lokobé Reserve, Nosy Bé Island	Davis 1975
	At Mandrare R. (in S)	Hladik 1975
	At Antserananomby on tributary of Mangoky R. and 60 km N of Morondava near Analabe and Lake Andranovala (both in W)	Petter et al. 1975
	In Analabe and Berenty reserves and 6 other reserves in E, NW, W, and SW; declining	Richard and Sussman 1975
	At Ampijoroa in Ankarafantsika Forest Reserve	Tattersall and Sussman 1975c
	At Andohahela Reserve near Hazofotsy (24°49′S, 46°37′W)	Russell and McGeorge 1977
	At Morondava, Antsalova, and Sakaraha (W), Majunga-Ampijoro, Ambanja, and Nosy-Bé (NW); Forêt de Sahafary, Montagne d'Ambre, Ankarana, and Chaîne de l'Andrafiamena (N); Périnet, Foulpointe, and Tamatave (in E); and Berenty (in S)	Rumpler and Albignac 1978
	Throughout N, NW, much of W, S, and E S of 14°N (map)	Petter and Petter-Rousseaux 1979

Table 29. Population parameters of *Lepilemur*

Population Density (individuals/km²)	Group Size $\bar{X}$	Group Size Range	Group Size No. of Groups	Home Range Area (ha)	Study Site	Reference
200–810					Madagascar	Charles-Dominique and Hladik 1971
250–300					Madagascar	Petter et al. 1971
300				0.18–0.30	Madagascar	Charles-Dominique 1974c
200–450		2–6		0.15–0.46	Madagascar	Hladik and Charles-Dominique 1974
100–260					Madagascar	Petter et al. 1975
200–300					Madagascar	Petter 1978
300					Madagascar	Hladik 1979

Varecia variegata Ruffed Lemur

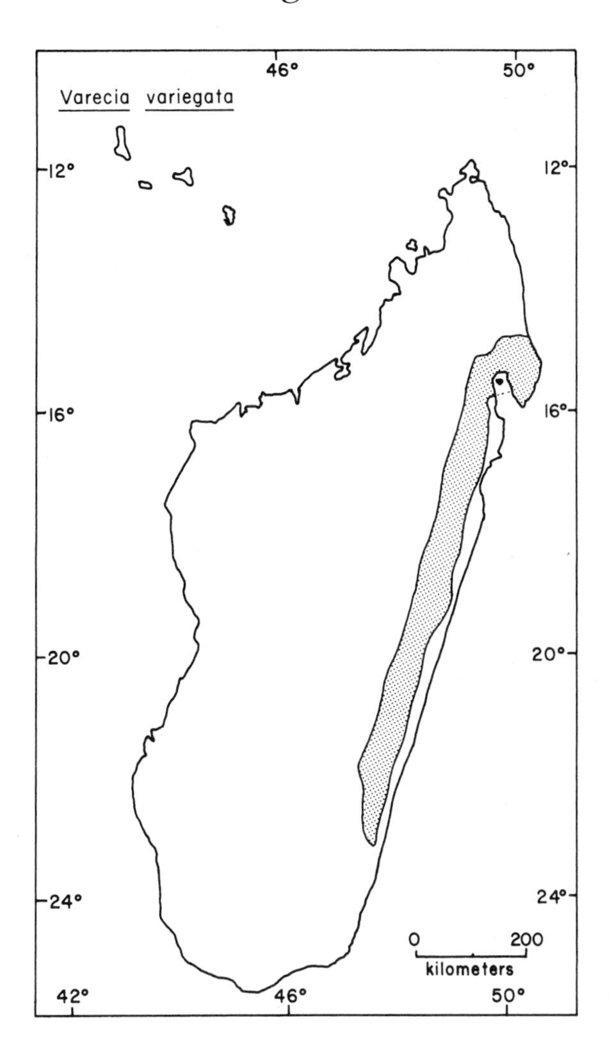

Taxonomic Note

Two subspecies are recognized (Rumpler 1975, Petter and Petter-Rousseaux 1979) in this species, which was formerly placed in the genus *Lemur* (Napier and Napier 1967).

Geographic Range

Varecia variegata is found in eastern Madagascar, in a narrow strip parallel to the east coast from the Masoala Peninsula to the Mananara River. Its northern limit is at about 14°45′S near Antalaha, and its southern limit at about 23°15′S near the mouth of the Mananara River; the range extends west as far as about 47°30′E at 22°S (Petter and Petter-Rousseaux 1979). For details of distribution, see Table 30.

Abundance and Density

V. variegata populations are sparse in the northern portion of the range and generally declining (Table 30). No estimates of population density, average group size, or home range area are available. Groups of this species vary in size from 2 to 5 (Petter 1962a, Martin 1972b), 4 to 10 (Petter 1972), and 2 to 4 (Pollock 1979).

Habitat

V. variegata is a forest species found in humid rain forest (Petter 1962a, Tattersall 1977c). No details of its requirements have been reported.

Factors Affecting Populations

Habitat Alteration

The forests inhabited by *V. variegata* are declining in area because of cutting for timber, industry, and agricultural land (Richard and Sussman 1975).

Human Predation

Hunting. In the northern part of its range, *V. variegata* populations have been decimated by hunting (Tattersall 1977c). No reports are available from the southern portion of the range.

Pest Control. No information.

Collection. Collection of ruffed lemurs for export has been a major reason for the decline of this species in some areas (Richard and Sussman 1975).

Between 1968 and 1973, eight *V. variegata* were imported into the United States (Appendix). None entered the United States in 1977 (Bittner et al. 1978). One was imported into the United Kingdom in 1965–75 (Burton 1978).

Conservation Action

V. variegata is found in four reserves (Table 30) and is legally protected from hunting and export, but more meaningful fines and education are necessary to insure its conservation (Richard and Sussman 1975). The species is listed in Class A of the African Convention (Organization of African Unity 1968), in Appendix I of the Convention on International Trade in Endangered Species of Wild Fauna and Flora, and as endangered under the U.S. Endangered Species Act (U.S. Department of Interior 1977a,b).

Table 30. Distribution of *Varecia variegata*

Country	Presence and Abundance	Reference
Malagasy Republic	In E	Petter 1962a, Martin 1972b
	Localities mapped in E and NE	Petter 1962b
	In 2 strict nature reserves: Zahamena and Betampona (both in E, NW of Tamatave)	I.U.C.N. 1971
	N of Bay of Antongil and between Maroantsetra and Manakara	Petter 1972
	In 4 reserves including on Nosy Mangabe, 2 in E, NW of Tamatave, and Andringitra (in E near Ambalavao); declining; habitat being reduced	Richard and Sussman 1975
	W of Bay of Antongil and Antainambalana R. including Nosy Mangabe (map); S as far as S of Farafangana near Manombo; sparse in N; previously N at least to Antalaha	Tattersall 1977c
	Throughout E along coast between Antalaha and Mananara R. (map)	Petter and Petter-Rousseau 1979
	At Betampona and Fierenana (both in E, NW of Tamatave)	Pollock 1979

INDRIIDAE

Avahi, Indris, and Sifakas

Avahi laniger Avahi

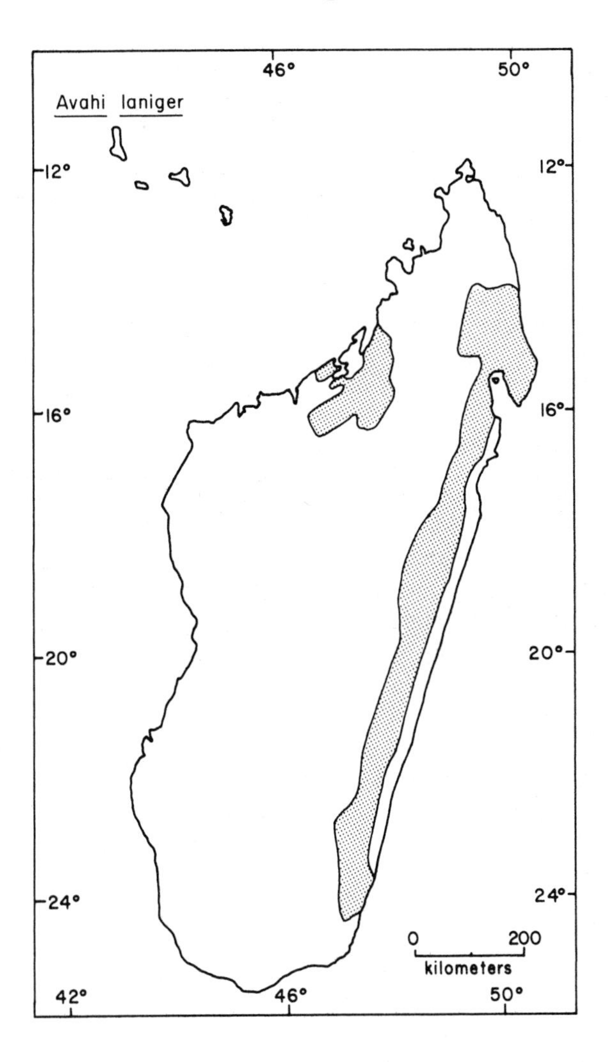

Taxonomic Note

Two subspecies are recognized (Petter and Petter-Rousseaux 1979).

Geographic Range

Avahi laniger is found in eastern and northwestern Madagascar. In the east its range extends in a narrow area parallel to the east coast from about 14°S to 24°30′S, and reaches the coast at the northern and southern ends of this strip (Petter and Petter-Rousseaux 1979). In the west *A. laniger* occurs from near Antsohihy (14°50′S, 47°58′E) in the north, to the east bank of the Betsiboka River near Ambato-Boéni (16°26′S, 46°43′E) in the south (Tattersall 1977c). For details of the distribution, see Table 31.

Abundance and Density

The western form of *A. laniger (A. l. occidentalis)* has been described as threatened and vulnerable with a shrinking range, and the species as a whole is declining (Table 31).

No data concerning population density or home range area were located. Group sizes are 2 or 3 (Petter 1962a, 1975; Petter and Pariente 1971), 2 to 5 (Martin 1972b), 2 to 6 (Petter 1972), and 2 to 4 ($\bar{X}$ = 2.5, N = 6) (Pollock 1975).

Habitat

Avahis are found in forest habitats, including humid coastal rain forest in the east and drier seasonal forest in the west (Petter 1962a, 1972, 1975; Pollock 1977a). No details of their requirements are available.

Factors Affecting Populations

Habitat Alteration

The major threat to the survival of *A. laniger* is the destruction of forest via clearing for timber, industry, and cultivation (Richard and Sussman 1975). For example, in southeastern Madagascar near Fort Dauphin, clandestine tree-felling is decreasing its available habitat (Martin 1973), and in the northwest, uncontrolled clearing and burning are threatening the race found there (Fisher et al. 1969, I.U.C.N. 1972).

Human Predation

Hunting. A. laniger is intensively hunted in northeastern Madagascar (Tattersall 1977c), and is also killed for its meat in the southeast (Martin 1973) and northwest (Fisher et al. 1969).

Pest Control. No information.

Collection. No *A. laniger* were imported into the United States in 1968–73 (Appendix) or 1977 (Bittner et al. 1978), nor into the United Kingdom in 1965–75 (Burton 1978).

Conservation Action

A. laniger is found in three habitat reserves (Table 31). It is legally protected against hunting and export, but more meaningful fines and education are needed to insure its conservation (Richard and Sussman 1975).

This species is listed in Class A of the African Convention (Organization of African Unity 1968), in Appendix I of the Convention on International Trade in Endangered Species of Wild Fauna and Flora, and as endangered under the U.S. Endangered Species Act (U.S. Department of Interior 1977a,b).

Table 31. Distribution of *Avahi laniger*

Country	Presence and Abundance	Reference
Malagasy Republic	In E and W	Petter 1962a
	At Bemangidy in SE	Jolly 1966
	Abundant in E; W form (*A. l. occidentalis*) in NW from Ampasindava Bay to Bombétoka: threatened	Fisher et al. 1969
	In 4 strict nature reserves: Ankarafantsika and Zahamena (NW), Betampona (E), and Tsingy de Namoroka (in W, outside range limits)	I.U.C.N. 1971
	W form: (*A. l. occidentalis*): vulnerable; range shrinking	I.U.C.N. 1972
	In E, N, NW, and W	Martin 1972b
	In E, W, and SW	Petter 1972
	In Mandena forestry reserve near Fort Dauphin (SE) in 1968 but not sighted in 1970	Martin 1973
	In NW at Ampijoroa forest station (16°35′S, 46°82′E), Ankarafantsika Reserve	Sussman and Richard 1974, Tattersall and Sussman 1975a, Richard 1978
	In 3 reserves in NE and NW; declining; habitat being reduced	Richard and Sussman 1975
	In E and NW	Pollock 1977a
	In E as far N as Marojezy-Sambava region (14°16′S, 50°10′E); sparse in E interior between Tsarabaria and Cap Est; in W from Antsohihy to Ambato-Boeni on Betsiboka R., including along coast S of Analalava and W of Andranoboka (map)	Tattersall 1977c
	In E along coast in N and S, in inland strip parallel to coast in E central; in NW from slightly N of Sofia R. to E bank of Betsiboka R. but not along coast (map)	Petter and Petter-Rousseaux 1979
	In E at Analamazoatra, Fierenana, and Vohidrazana	Pollock 1979

Indri indri Indris

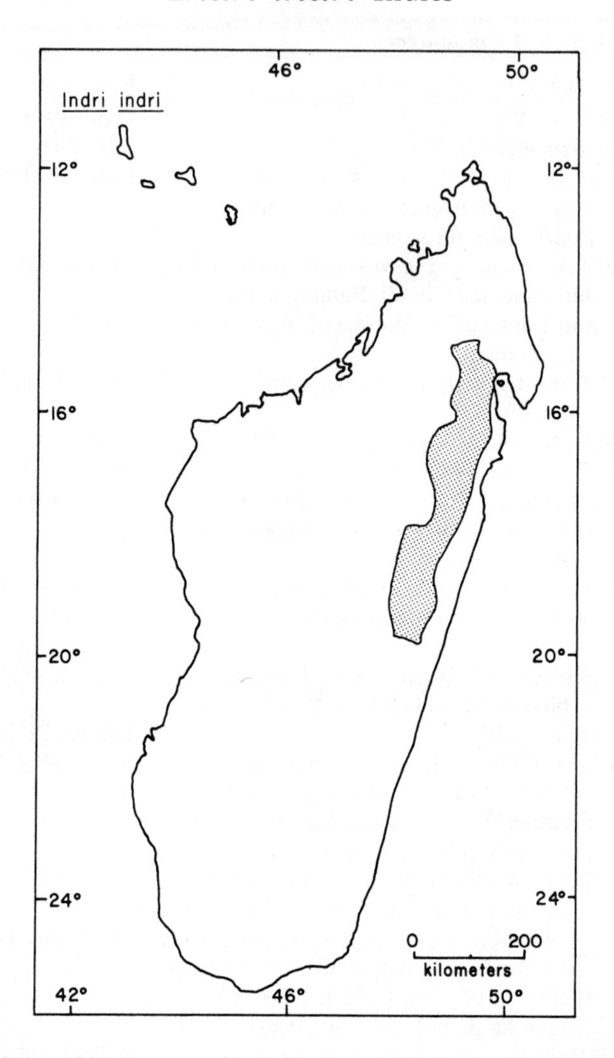

Taxonomic Note

This is a monotypic species (Petter-Rousseaux 1979).

Geographic Range

Indri indri is found in eastern Madagascar south of the Tangbale River, which flows into the Bay of Antongil, and north of the Mangoro River. It occurs as far north as about 15°S and as far south as about 19°45′S, in a strip running parallel to the coast which extends farthest west (to about 48°E) at its southern end (Petter and Petter-Rousseaux 1979). For details of the distribution, see Table 32.

Abundance and Density

I. indri has recently been described as endangered, becoming rare, and generally declining (Table 32).

Population densities of *I. indri* vary from 3 (I.U.C.N. 1972) to 8 to 16 individuals per km^2 (Pollock 1975, 1977a). Group sizes are 2 to 4 (Petter 1962a, Petter and Peyriéras 1974), 3 to 5 (Fisher et al. 1969, Bomford 1976b), 2 to 6 (Petter 1972), and 2 to 5 ($\bar{X}$ = 3.1, N = 18 groups) (Pollock 1975). Each group occupies a home range of about 100 ha (Petter and Peyriéras 1974).

Habitat

I. indri is a forest species, found in humid montane rain forest (Petter 1962a, Petter and Pariente 1971, Petter and Peyriéras 1974, Pollock 1977a). It has been observed as high as 1,300 m elevation (Pollock 1975).

Factors Affecting Populations

Habitat Alteration

The forest habitats of *I. indri* are being destroyed by abusive exploitation and slash and burn agriculture (Petter and Pariente 1971), and those forests which remain are now isolated patches in an expanding wave of agriculture and timbering (Pollock 1977b). For example, the forest near Périnet has been heavily exploited and altered (Petter and Peyriéras 1974), and large-scale clearing and burning are affecting many areas (Fisher et al. 1969, I.U.C.N. 1972).

Indrises are especially sensitive to disturbance, and loud noises and activity (such as approaches by people, helicopters, airplanes, and other machinery) increase infant mortality due to falling (Petter and Pariente 1971) and cause resident groups to flee, producing abnormally high densities on the fringes of disturbed forests (Petter and Peyriéras 1974).

Human Predation

Hunting. In the past *I. indri* was protected from hunting by several local taboos and beliefs (I.U.C.N. 1972, Tattersall 1972). Recently, however, hunting of this species has increased (Petter and Pariente 1971, I.U.C.N. 1972, Petter and Peyriéras 1974).

Pest Control. No information.

Collection. No *I. indri* were imported into the United States in 1968–73 (Appendix) or 1977 (Bittner et al. 1978), nor into the United Kingdom in 1965–75 (Burton 1978).

Conservation Action

I. indri occurs in three reserves (Table 32). The protection of the special reserve of Périnet is considered crucial for the survival of this species (Bomford 1976b). Heavy pressure is being exerted on the government to relinquish land inside reserves (Fisher et al. 1969), and the assistance from international conservation groups is needed to combat this pressure and maintain effective protection (Pollock 1977b). Existing reserves should receive improved protection, and additional ones should be established (I.U.C.N. 1972).

Indrises are legally protected in Madagascar, but more meaningful fines and education are necessary for their conservation (Richard and Sussman 1975). They are listed in Class A of the African Convention (Organization of African Unity 1968), in Appendix I of the Convention on International Trade in Endangered Species of

Wild Fauna and Flora, and as endangered under the U.S. Endangered Species Act (U.S. Department of Interior 1977a,b).

Table 32. Distribution of *Indri indri*

Country	Presence and Abundance	Reference
Malagasy Republic	In N half of E	Petter 1962a, Martin 1972b
	Localities mapped in E	Petter 1962b
	At Périnet (E central)	Jolly 1966
	In E between Bay of Antongil and Masora R.; declining sharply; in Massif de Betampona, Masoala forest, and Périnet reserves	Fisher et al. 1969
	In Zahamena Strict Nature Reserve (E central)	I.U.C.N. 1971
	Between Cap Masoala and about 18°S, including site E of Maroantsetra; becoming rare	Petter and Pariente 1971
	Endangered; declining rapidly; in 3 reserves	I.U.C.N. 1972
	In N half of E coast, including forest of Fampanambo (NE of Maroantsetra,), Fierenana (N of Lake Alaotra), Périnet reserve at Analamazoatra, and Lakato concession	Petter and Peyriéras 1974
	At Analamazoatra near Périnet (18°56′S, 48°24′E), Vohidrazana near Maromiza (S of Périnet), and Sahamanga near Fierenana (N of Périnet); throughout E, N of about 20°S, including on Masoala Peninsula (maps)	Pollock 1975, 1977a
	In 2 reserves in E; scarce, declining; habitat being reduced	Richard and Sussman 1975
	In Périnet reserve	Bomford 1976b
	As far N as near Maroantsetra; not on Masoala Peninsula (map)	Tattersall 1977c
	In E between 15°S and 19°45′S (map)	Petter and Petter-Rousseaux 1979

Propithecus diadema Diademed Sifaka

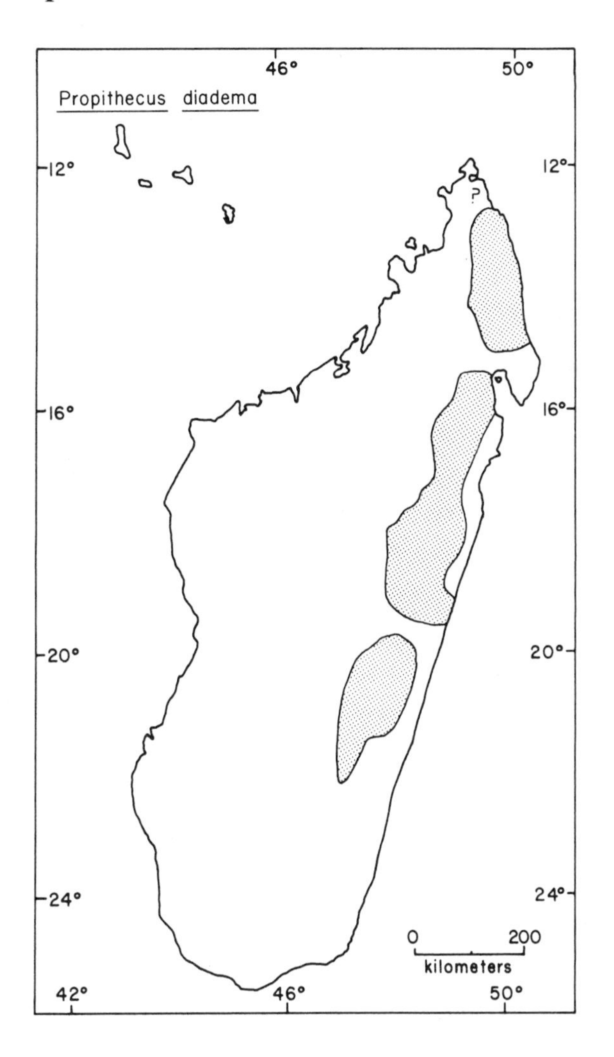

Taxonomic Note

Five subspecies are recognized (Petter and Petter-Rousseaux 1979).

Geographic Range

Propithecus diadema is found in northeastern and eastern Madagascar, in several discontinuous populations along and parallel to the east coast. It occurs as far north as about 12°45′S, as far south as about 22°15′S, and as far west as about 47°E near the southern limit of its range (Petter and Petter-Rousseaux 1979). For details of the distribution, see Table 33.

Abundance and Density

The northernmost race of *P. diadema (P. d. perrieri)* is known to be very rare, and the species as a whole is thought to be scarce, declining, and probably on the brink of extinction (Table 33).

The population density of the northernmost race is about 3 individuals per km^2 (I.U.C.N. 1972). No other estimates of density or of home range area are available. Group sizes are 3 or 4 (Petter and Pariente 1971), 2 to 6 (Petter 1972), and 2 to 5 (Pollock 1979).

Habitat

P. diadema is found in humid rain forest (Petter 1962a) and in drier, more seasonal vegetation (Tattersall 1977c). No details of its requirements have been reported.

Factors Affecting Populations

Habitat Alteration

The forest habitat of *P. diadema* is being reduced by timbering, industry, and agriculture (Richard and Sussman 1975). In the extreme northeast, savannah fires are gradually reducing the forested area available to it (I.U.C.N. 1972).

Human Predation

Hunting. P. diadema is actively hunted throughout its range by local people who appreciate its meat and by Europeans who seek its thick, silky fur (Petter and Pariente 1971). For example, it is intensively hunted in the northeastern interior between Tsarabaria and Cap Est (Tattersall 1977c). It is reported not to be hunted farther north in the Analamera forest (I.U.C.N. 1972).

Pest Control. No information.

Collection. No *P. diadema* were imported into the United States in 1968–73 (Appendix) or in 1977 (Bittner et al. 1978), nor into the United Kingdom in 1965–75 (Burton 1978).

Conservation Action

P. diadema occurs in two to four reserves (Table 33), and a well-protected area is needed in the northern portion of the range (I.U.C.N. 1972). The species is legally protected from hunting and export, but more meaningful fines and conservation education are necessary (Richard and Sussman 1975). *P. diadema* is listed in Class A of the African Convention (Organization of African Unity 1968), in Appendix I of the Convention on International Trade in Endangered Species of Wild Fauna and Flora, and as endangered under the U.S. Endangered Species Act (U.S. Department of Interior 1977a,b).

Table 33. Distribution of *Propithecus diadema*

Country	Presence and Abundance	Reference
Malagasy Republic	In E	Petter 1962a
	Localities mapped in E and NE	Petter 1962b
	At Périnet (E central)	Jolly 1966
	In 2 strict nature reserves: Zahamena (E central) and Marojejy (NE); assumed to be also in Andringitra (SE) and Betampona (E central)	I.U.C.N. 1971
	Rare; in E, NE, SE	Petter and Pariente 1971
	Race in extreme NE (*P. d. perrieri*): rare, restricted to Analamera forest, Andrafiamena range, right bank Irofo R.; range contracting, population about 500 individuals	I.U.C.N. 1972
	In extreme N, in Sambava-Andapa region (farther S), between Mananara and Tamatave, in E central region, and near Idanadiana (Ifanadiana?)	Petter 1972
	In 3 reserves (but 1 listed is outside range limits) and possibly in 1 other reserve; scarce, declining, extremely rare and probably on the brink of extinction; habitat being reduced	Richard and Sussman 1975
	Northernmost form (*P. d. perrieri*): very rare, known from Andrafiamena forest SE of Anivorano Nord (12°42′S, 49°12′E); form on Cap Est (*P. d. candidus*): sparse, as far N as at least Daraina, along coast as far S as Cap Est (map)	Tattersall 1977c
	Along NE and E coast from 12°45′S to 22°15′S (map)	Petter and Petter-Rousseaux 1979
	At Analamazoatra, Fierenana, and Vohidrazana (all E central) (map)	Pollock 1979

Propithecus verreauxi Verreaux's Sifaka

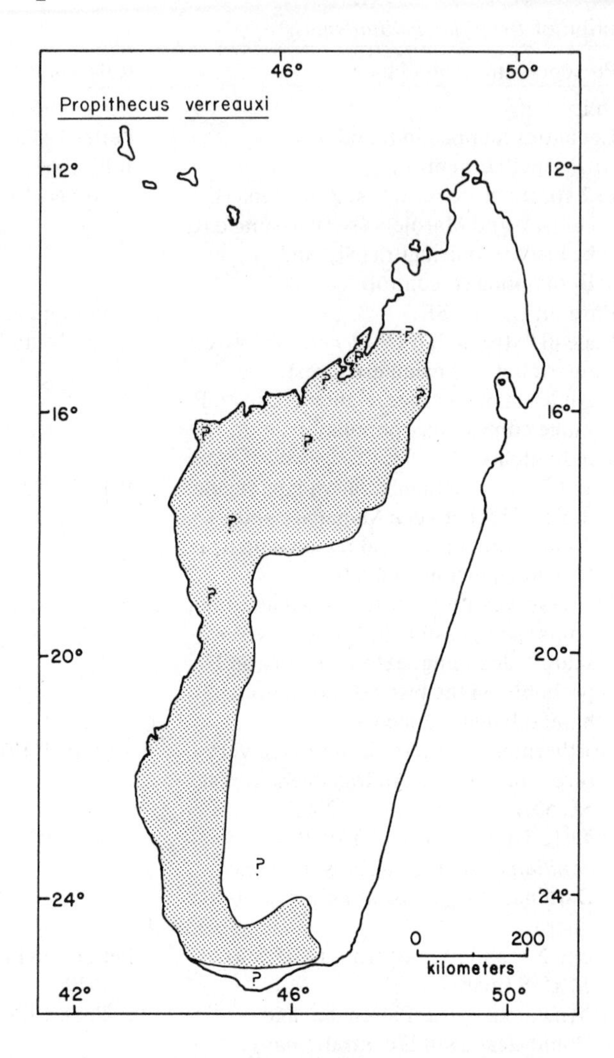

Taxonomic Note

Five subspecies are recognized (Petter and Petter-Rousseaux 1979).

Geographic Range

Propithecus verreauxi is found in western, southwestern, and southern Madagascar. Its range extends as far north as about 14°40′S, as far east as about 48°30′E in the northern part of the range, south almost to the southern tip of the island, and east to about 46°30′ in the south (Sussman and Richard 1974, Sussman 1977c, Tattersall 1977c, Richard 1978, Petter and Petter-Rousseaux 1979). For details of the distribution, see Table 34.

Abundance and Density

P. verreauxi is an endangered species whose populations are locally abundant compared with those of other Madagascan primates but are generally scarce and declining rapidly (Table 34).

Population densities at various study sites have been estimated at 38 to 530 individuals per km^2 (Table 35). Groups include 2 to 15 members ($\bar{X}$ = 3.0 to 7.8) and occupy home ranges of 1 to 20 ha (Table 35).

Habitat

Verreaux's sifakas are found in a wide variety of forest habitats, from lush wet areas to parched dry ones. They are not restricted to primary vegetation but thrive in secondary forest, and their flexibility in choice of habitats is one of their prominent characteristics (Richard, pers. comm. 1973).

This species occupies rich, mixed deciduous and evergreen forests, and gallery and closed canopy forest dominated by tamarind trees (Jolly 1972, Sussman 1977a, Richard 1978). It also utilizes dry deciduous forests and very dry Didiereaceae forests dominated by *Alluaudia ascendens* and *A. procera*, with *Acacia* thorns, aloes, and fleshy euphorb trees (Jolly 1966; Richard 1974a, 1978). *P. verreauxi* can occupy continuous canopy forest alone or mixed with brush and scrub, but is not found in pure brush and scrub forest (Sussman 1974, 1977c).

Factors Affecting Populations

Habitat Alteration

Throughout its range *P. verreauxi* is being threatened by destruction of its habitats by clearing, burning, and cultivation (Fisher et al. 1969, I.U.C.N. 1972, Richard and Sussman 1975). For example, in western Madagascar, clearcutting for timber is proceeding, followed by replacement of forests with sisal, coffee, and cotton plantations, and in the south, *Didierea* forest is being felled for firewood and building materials (Richard, pers. comm. 1973). At Berenty, the reserved forest is surrounded by sisal fields that virtually isolate the resident *P. verreauxi* population (Jolly 1966), and in nearby areas the destruction of natural vegetation is threatening the species (Hladik and Charles-Dominique 1971).

Most of the accessible forests of Madagascar have already been disturbed and transformed by man through extensive timbering, slash and burn cultivation, cutting for firewood, and cattle grazing, and these processes are continuing to reduce the habitat of *P. verreauxi* (Richard 1978). The habitats within the ranges of two of the races of this species (*P. v. coronatus* and *P. v. majori*) have been almost completely destroyed, and even the most widespread subspecies (*P. v. verreauxi*) is being threatened by the rapid disappearance of the southern and western forests (I.U.C.N. 1972).

Human Predation

Hunting. P. verreauxi is persistently hunted for food and sport in many regions (Petter 1972; Gustafson, pers. comm. 1975; Richard 1978). In some areas it is still protected by local beliefs and taboos (Hladik and Charles-Dominique 1971, Tattersall 1972), but in others, hunting, especially by Europeans and other nonindigenous people using guns, is common (Richard, pers. comm. 1973).

Pest Control. No information.

Collection. Very few *P. verreauxi* are exported from Madagascar (Richard, pers. comm. 1973; Gustafson, pers. comm. 1975). Between 1968 and 1973 a total of 11 were imported into the United States (Appendix). None entered the United States in 1977 (Bittner et al. 1978) or the United Kingdom in 1965–75 (Burton 1978).

Conservation Action

P. verreauxi occurs in seven habitat reserves (Table 34), but in some of these (e.g., Ampijoroa and Hazafotsy) livestock grazing is permitted (Richard 1978), and in many, people with a subsistence economy are resident (Richard, pers. comm. 1973). More effective habitat protection is essential for these reserves, and additional reserves have been recommended in the Mahafaly Tombs area south of Ampanihy, in the Bongolava forest, and in the Tsiroanomandidy region (I.U.C.N. 1972).

P. verreauxi is legally protected from hunting and export, but more meaningful fines and conservation education are necessary if it is to survive (Richard and Sussman 1975). The species is listed in Class A of the African Convention (Organization of African Unity 1968), in Appendix I of the Convention on International Trade in Endangered Species of Wild Fauna and Flora, and as endangered under the U.S. Endangered Species Act (U.S. Department of Interior 1977a,b).

Table 34. Distribution of *Propithecus verreauxi*

Country	Presence and Abundance	Reference
Malagasy Republic	In S and W including Ankarafantsika (100 km E of Majunga)	Petter 1962a
	Localities mapped in S and W	Petter 1962b
	At Ankarafantsika (NW), Morondava (W central), Lambomakandro (SW), Ifotaka, Bevala, and Berenty (in SE)	Jolly 1966
	Endangered; in NW, SW, and S from Fort Dauphin to Bay of St.-Augustin	Fisher et al. 1969
	Near Berenty (S)	Hladik and Charles-Dominique 1971, Jolly 1972
	In 5 strict nature reserves: Ankarafantsika (NW), Tsingy du Bemaraha and Tsingy de Namoroka (W), Tsimanampetsotsa Lake (SW), and Andohahela (SE)	I.U.C.N. 1971
	Endangered; range contracting; originally in SW as far N as Tsiribihina R., N of Betsiboka R., between Manambolo R. and Mahavavy R., Sakaraha forest near Lambomakandro, and N of Amboasary W of Fort Dauphin; all races threatened with extinction	I.U.C.N. 1972

Table 34 (continued)

Country	Presence and Abundance	Reference
Malagasy Republic (cont.)	In Ankarafantsika Reserve and Bongolava forest (W)	Petter and Pariente 1971
	Abundant N of Morondava near Analabe and Lake Andranolava (W)	Petter et al. 1971, 1975
	In N, NW, W, and SW	Martin 1972b
	Disappearing very rapidly; at Ankarafantsika, Antsalova-Soalala region, between Majunga and Soalala; throughout W, S, and SW; form in SW near Sakaraha (*P. v. majori*): on verge of extinction	Petter 1972
	Declining; populations discontinuous in W, S, and SW (map)	Richard, pers. comm. 1973
	Abundant at Ampijoroa (16°35′S, 46°82′E) (W); near Hazafotsy (24°85′S, 46°50′E) (S); at Evasy, Ejeda, Antserananomby, and Berenty (S); throughout W, SW, and S (map)	Richard 1974a, 1978; Sussman and Richard 1974
	At Antserananomby, Tongobato, and Berenty (SW and S)	Sussman 1974, 1975, 1977a
	In S at Berenty: scarce, declining, range being reduced	Gustafson, pers. comm. 1975
	At Ampijoroa in Ankarafantsika Forest Reserve (W)	Harrington 1975, Tattersall and Sussman 1975a
	In 7 reserves including Analabe (W), Berenty (S), 3 others in W, 1 in SW, 1 in SE; relatively abundant but absolutely scarce and declining; habitat being reduced	Richard and Sussman 1975
	In Andohahela Reserve near Hazofotsy (24°49′S, 46°37′W)	Russell and McGeorge 1977
	Localities listed and mapped in S at Beomby, Berenty, and Ejeda; in SW at Lambomakandro, Antserananomby, Ianadranto, Malindira, Mangoky, Tarata, Tongobato, and Belo-sur-Mer; and W at Ambatovoamba, Andranomena, Angansiva, Ankilimare, Befasy, Beroboka, Manamby, Maroary, Tanambao, Ampijoroa, Katsepy, Madirovalo, and Mitsinjo	Sussman 1977c
	In W as far N as N and E of Antsohihy (14°50′S, 47°58′E), E at least to Antetemazy, W of Befandriana Nord (15°14′S, 48°33′E) (map)	Tattersall 1977c
	In W, SW, and S (map)	Petter and Petter-Rousseaux 1979

Table 35. Population parameters of *Propithecus verreauxi*

Population Density (individuals/km²)	Group Size $\bar{X}$	Range	No. of Groups	Home Range Area (ha)	Study Site	Reference
	4.6–5	2–10	25	2.2–2.6	Madagascar	Jolly 1966
		6–8			Madagascar	Fisher et al. 1969
		3–5			Madagascar	Hladik and Charles-Dominique 1971
		3–6		1	Madagascar	Petter and Pariente 1971
		5–15			Madagascar	Martin 1972b
		2–6			Madagascar	Petter 1972
		4–8	4	6.8–8.5	Madagascar	Richard 1974b
		5–8	4	5–6	Madagascar	Sussman and Richard 1974
38–203					Madagascar	Gustafson, pers. comm. 1975
	3	2–5	5	~20	Madagascar	Pollock 1975
140–240	~6	3–13		2–6	Madagascar	Richard and Sussman 1975
362–530	5–7.8	3–12		6.8–8.5	Madagascar	Richard 1978

DAUBENTONIIDAE

Aye-aye

Daubentonia madagascariensis Aye-aye

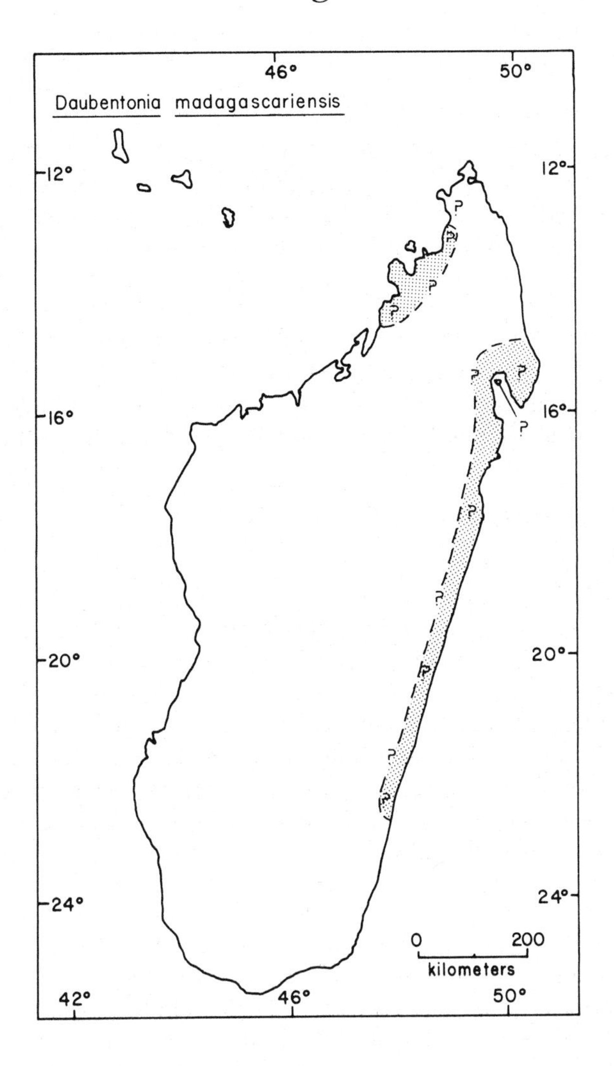

Taxonomic Note

This is a monotypic species (Petter and Petter-Rousseaux 1979).

Geographic Range

Daubentonia madagascariensis was originally found in eastern, northern, and northwestern Madagascar. Old reports have placed it as far south as Farafangana (22°50′S, 47°50′E) and as far north as Antalaha (14°53′S, 50°16′E) along the east coast, east of the Montagne d'Ambre in the extreme north, and on the northwest coast between Ambilobe (13°10′S, 49°03′E) and Analalava (14°38′S, 47°46′E) (Petter 1977). But recent field surveys have failed to reveal populations of *D. madagascariensis* in any of these areas, and the latest range map available (Petter and Petter-Rousseaux 1979) shows this species present only on Nosy Mangabe Island, where it was introduced in 1966–67 (Petter and Peyriéras 1970a, Petter 1977). For details of its distribution, see Table 36.

Abundance and Density

D. madagascariensis is an endangered species which is probably on the verge of extinction (Table 36). It seems to be strictly solitary except for groupings of mother and young (Petter 1962a, 1972), and has a home range of about 4.86 ha per individual (Petter and Petter 1967). No estimates of typical population densities were located.

Habitat

Aye-ayes are found in rain forest (Petter 1962a) at low elevations (Petter and Peyriéras 1970a). They are limited to coastal areas and have rarely been encountered above 200 m, although they have been reported from high altitudes (Petter 1977). They are able to utilize plantations and large trees near villages (Petter and Petter 1967).

Factors Affecting Populations

Habitat Alteration

The main reason for the decline of *D. madagascariensis* is the destruction of its forest habitat by cutting and burning (Petter and Peyriéras 1970a, Petter 1972). Along the east coast north of Tamatave, the coastal forest has disappeared almost entirely, because of exploitation, burning, and replacement with imported tree species; this process is continuing and the remaining pockets of forest are being destroyed (Petter 1977). South of Tamatave the forests are also rapidly being reduced, leaving little hope for the future of *D. madagascariensis* (Petter 1977). Road building and the construction of new villages are further decreasing forested areas within the aye-aye's range (Petter and Petter 1967).

Human Predation

Hunting. In the past, *D. madagascariensis* was protected by a local belief that anyone who harmed one would die within a year (Tattersall 1972). Today, villagers kill this species whenever possible because it is thought to be evil and bring bad luck or death (Fisher et al. 1969, Tattersall 1972, Fitter 1974, Petter 1977). This killing has resulted in its extermination near many villages and in the last remainders of the coastal forest (Petter and Peyriéras 1970a).

Pest Control. *D. madagascariensis* does enter fruit, clove, and coffee plantations, and maize patches. It eats mangoes, coconuts, litchis, and other cultivated fruits, and in many areas is killed to prevent this consumption of human food (Petter and Petter 1967, Petter 1977).

Collection. No aye-ayes were imported into the United States in 1968–73 (Appendix) or in 1977 (Bittner et al. 1978), nor into the United Kingdom in 1965–75 (Burton 1978).

Conservation Action

In 1957 a small reserve was established for *D. madagascariensis* at Mahambo, but by 1963 the reserve had been cut in half, was without resident aye-ayes, and had become very difficult to protect (Petter and Peyriéras 1970a). In 1966–67, nine *D. madagascariensis* were captured on the mainland and released into a special reserve on Nosy Mangabe Island, where mango and coconut trees had been planted as food sources for them (Petter and Peyriéras 1970a, Petter 1977). At least one individual was still alive on the island in 1975, but protection and study of the aye-ayes there have been hampered by a lack of money and transportation (Bomford 1976a).

D. madagascariensis is legally protected against hunting and export, but more meaningful fines and education are necessary for its conservation (Richard and Sussman 1975). Also needed are population surveys in remote areas and the establishment of securely guarded reserves (I.U.C.N. 1972, Fitter 1974).

D. madagascariensis is listed in Class A of the African Convention (Organization of African Unity 1968), in Appendix I of the Convention on International Trade in Endangered Species of Wild Fauna and Flora, and as endangered under the U.S. Endangered Species Act (U.S. Department of Interior 1977a,b).

Table 36. Distribution of *Daubentonia madagascariensis*

Country	Presence and Abundance	Reference
Malagasy Republic	In E and Sambirano region (NW)	Petter 1962a
	Localities mapped in E, NE, and NW	Petter 1962b
	Never common; believed extinct until rediscovered in 1957 near Mahambo (13 km S of Fénérive, on E coast near 17°21′S); in 1964 several remained in area but habitat was being reduced	Petter and Petter 1967
	Rare and gravely endangered; in NE between Antalaha and Mananjary; population about 50 individuals; habitat decreasing	Fisher et al. 1969
	Rare and endangered; in 1966 transferred 9 to special reserve on Nosy Mangabe near Maroantsetra in Bay of Antongil	Petter and Peyriéras 1970a
	In Nosy Mangabe special reserve	I.U.C.N. 1971, Harroy 1972
	Endangered; previously throughout E and NW coasts; now scattered on NE and NW coasts; severely depleted, threatened with extinction	I.U.C.N. 1972
	Habitat being reduced	Kuhn 1972
	Very rare	Petter 1972
	Reduced to near extinction	Tattersall 1972
	Population probably less than 50 individuals	Fitter 1974
	Extremely rare, declining, probably on the brink of extinction; habitat being reduced; in Nosy Mangabe Reserve	Richard and Sussman 1975
	Sighted one on Nosy Mangabe in 1975	Bomford 1976a
	Originally on E coast between Antalaha and Mananjary, and on NW coast between Ambilobe and Analalava; old reports	Petter 1977

Table 36 (continued)

Country	Presence and Abundance	Reference
Malagasy Republic (cont.)	from SE and NW; localities listed where searched for but not found; seen near Sahafany; reported rare near Ambaliha, Bay of Ampasindava	Petter 1977 (cont.)
	Evidence from Sambava (NE coast), Montagne d'Ambre (N), Ankobakobaka W of Befandriana Nord, and Mahambo; gravely menaced but distribution may be wider than believed	Tattersall 1977c
	On Nosy Mangabe Island off NE coast near Maroantsetra (map)	Petter and Petter-Rousseaux 1979

CALLITRICHIDAE
Marmosets and Tamarins

Callimico goeldii Goeldi's Monkey or Goeldi's Marmoset

Taxonomic Note

This is a monotypic species, included by some authors in the family Callitrichidae (Napier and Napier 1967) and placed by others in its own family Callimiconidae (Hershkovitz 1977).

Geographic Range

Callimico goeldii is found in western South America from southern Colombia to northern Bolivia. It occurs as far north as 1°07′N, 76°38′W in Colombia (Hernández-Camacho and Cooper 1976) and as far south as about 13°S in Peru (Hershkovitz 1977). The range extends west to about 77°30′W in Ecuador and northern Peru, and as far east as 64°45′W and perhaps farther east south of the Rio Solimões, Brazil (Hershkovitz 1977). For details of the distribution, see Table 37.

Abundance and Density

C. goeldii is considered rare in Colombia, Peru, and Brazil (Table 37), but very little is known about its distribution or abundance. No estimates of population density or home range area are available. Group sizes of 20 to 30 (Walker 1968), 2 to 5 (Whittemore 1972), and 3 or 4 (Janson and Terborgh, in press) have been reported.

Habitat

Goeldi's monkeys occupy bamboo forest along rivers (Whittemore 1972) and non-flooding forest in level areas and on rolling hills (Hernández-Camacho and Cooper 1976). They have also been observed in mixed forest and scrub, primarily of low and secondary growth, on a periodically flooded island (Moynihan 1976a). No further details of their habitat requirements are known.

Factors Affecting Populations

Habitat Alteration

In western Putumayo, Colombia, slash and burn agriculture has destroyed portions of *C. goeldii* habitat (Whittemore 1972). No data regarding habitat disturbance in the rest of the range could be located.

Human Predation

Hunting. Hunting as game is listed as one of the uses of *C. goeldii* in Peru (Castro 1976). This species is seldom hunted in Bolivia (Freese et al., in prep.).

Pest Control. No information.

Collection. C. goeldii has been collected heavily in the past for the pet trade (Fisher et al. 1969, Simon 1969, Castro 1976). Between 1962 and 1972, 82 *C. goeldii* were listed as being exported from Peru, but this total may have included other species (Soini 1972). From 1968 to 1973, 179 *C. goeldii* were imported into the United States (Appendix). None were imported into the United States in 1977 (Bittner et al. 1978) or into the United Kingdom between 1965 and 1975 (Burton 1978).

Conservation Action

C. goeldii is rare in the Manú National Park in Peru, but has not been reported in any other protected areas (Table 37). It is protected under the Brazilian Endangered Species Act (Mittermeier and Coimbra-Filho 1977) and Supreme Decree No. 934-73-AG in Peru, but these laws are difficult to enforce (I.U.C.N. 1978).

C. goeldii is listed in Appendix I of the Convention on International Trade in Endangered Species of Wild Fauna and Flora, and as endangered under the U.S. Endangered Species Act (U.S. Department of Interior 1977a,b). Several authors have recommended status surveys and the establishment of habitat reserves for this species (Fisher et al. 1969; Simon 1969; Coimbra-Filho 1972a; I.U.C.N. 1972, 1978; Freese 1975). A study of its status and of reserve sites in northern Bolivia has now received support (Mittermeier 1977b).

Table 37. Distribution of *Callimico goeldii* (countries are listed in counterclockwise order)

Country	Presence and Abundance	Reference
Colombia	In Amazonian drainage	Hernández-Camacho and Barriga-Bonilla 1966
	In Puerto Umbría region and near R. Uchapayaco, R. Guineo, and upper R. Putumayo	Whittemore 1972
	Between R. Putumayo and R. Caquetá at R. Igara-Paraná near La Chorrera and Quebrada del Hacha, and on lower R. Guamués (map); questionable N of Mocoa and in E Amazonas	Hernández-Camacho and Cooper 1976
	In Putumayo region	Moynihan 1976a
	Localities listed and mapped in S in basins of R. Caquetá and R. Putumayo	Hershkovitz 1977
	Rare	I.U.C.N. 1978
Ecuador	In E (map); no records	Hershkovitz 1977
Peru	In E to basin of upper R. Ucayali	Cabrera 1958
	Apparently rare; recorded only from Cerro Azul near Contamana, R. Ucayali	Grimwood 1969
	In Madre de Dios, R. Ucayali basin, and at Mishana, S bank R. Nanay	Sioni 1972
	In Manú N.P., Madre de Dios and Cuzco	Castro 1976
	Locality listed in Contamana, Loreto	Napier 1976
	Localities listed and mapped in N and E including basin of R. Madre de Dios, upper R. Juruá and R. Purús, and lower R. Marañón	Hershkovitz 1977
	Rare	Castro 1978
	Rare	I.U.C.N. 1978
	Rare at Cocha Cashu, Manú N.P.	Janson and Terborgh, in press
Bolivia	In N in basins of upper R. Madeira and R. Beni	Cabrera 1958
	In basin of R. Madre de Dios (in NW corner); may be throughout NW (map)	Hershkovitz 1977
	At Cobija (11°01′S, 68°45′W) (in NW)	Freese et al., in prep.
Brazil	At Seringal Oriente, W Acre	Carvalho 1957
	In extreme W in basin of upper R. Madeira	Cabrera 1958
	Endangered	Padua et al. 1974
	At R. Iaco, R. Xapuri, and upper R. Juruá in Acre, and at R. Mu (7–8°S, 72°W)	Napier 1976
	In W Amazonas between R. Japurá and R. Solimões as far E as mouth of R. Japurá N of Tefé; S of R. Solimões between R. Juruá and R. Jurupari, and R. Purus and R. Xapuri; throughout Acre; questionable as far E as R. Madeira (map)	Hershkovitz 1977

Table 37 (continued)

Country	Presence and Abundance	Reference
Brazil (cont.)	Naturally rare and sparsely distributed	Mittermeier and Coimbra-Filho 1977
	Rare	Mittermeier et al., in press

Callithrix argentata Bare-ear or Silvery Marmoset

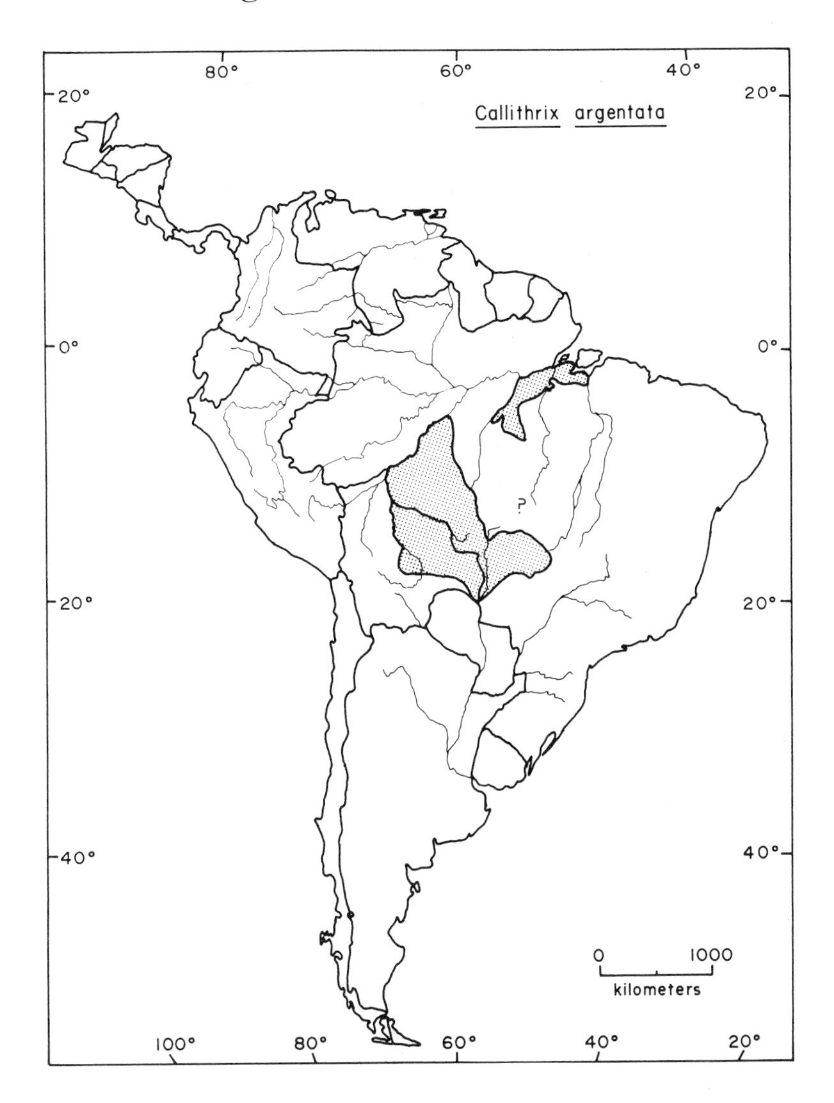

Taxonomic Note

Three subspecies are recognized (Cabrera 1958; Hershkovitz 1968, 1977).

Geographic Range

Callithrix argentata is found in central Brazil and eastern Bolivia. It occurs as far north as about 1°S at the mouth of the Rio Amazonas, as far east as about 49°20′W at the mouth of the Rio Tocantins, as far south as 20°S along the Rio Tacuarí, and as far west as about 65°W along the Río Mamoré in Bolivia (Hershkovitz 1977). For details of the distribution, see Table 38.

Abundance and Density

One race of *C. argentata* is vulnerable, another is common and not endangered, and the third is abundant at one site in southeastern Bolivia but its overall status is not known (Table 38).

At the study site in Bolivia, the population density of *C. argentata* was 5.5 individuals per km^2, and groups included 5 to 8 members ($\bar{X}$ = 5, N = 2), (Heltne et al. 1976). In Brazil, *C. argentata* groups consist of 2 to 4 individuals (Mittermeier and Coimbra-Filho 1977). No estimates of home range size are available.

Habitat

In Brazil, *C. argentata* is found in primary and secondary *várzea* forest (seasonally inundated by white-water rivers) and primary and secondary *terra firme* forest (nonflooding) (Mittermeier and Coimbra-Filho 1977). It has been recorded between 700 and 900 m elevation (Napier 1976). In Bolivia, *C. argentata* has been observed in seasonally dry, short deciduous broadleaf forest in a white-water drainage (Freese et al., in prep.).

Factors Affecting Populations

Habitat Alteration

In eastern Pará, extensive areas of forest are being cleared to make way for agriculture and cattle pasture (Ayres 1978). In western Pará, between the Rio Jamanxim and Rio Cupari, clearcutting and disturbance caused by the Transamazonian Highway are threatening the range of one race of *C. argentata* (*C. a. leucippe*) (Mittermeier, Bailey, and Coimbra-Filho 1978). Elsewhere the effect of habitat alteration on this species has not been studied.

Human Predation

Hunting. C. argentata is seldom hunted for food in Brazil (Mittermeier and Coimbra-Filho 1977) or in Bolivia (Freese et al., in prep.).

Pest Control. No information.

Collection. C. argentata is occasionally captured and kept as a pet in Brazil, and stuffed specimens are also offered for sale (Mittermeier, Bailey, and Coimbra-Filho 1978). This species is trapped for export from Bolivia (Heltne et al. 1976). The northernmost population in Brazil may be large enough to withstand controlled cropping of a few hundred individuals per year (Mittermeier, Coimbra-Filho, and Van Roosmalen 1978).

Between 1968 and 1970 a total of 119 *C. argentata* were imported into the United States (Appendix). None entered the United States in 1971–73. Five were imported into United States from Bolivia in 1977 (Bittner et al. 1978), and 107 were imported into the United Kingdom between 1965 and 1975 (Burton 1978).

Conservation Action

C. argentata is found in one biological reserve in Brazil (Table 38), and is protected from commercial exploitation in Brazil (Mittermeier and Coimbra-Filho 1977). One race (*C. a. leucippe*) is listed as vulnerable and is in need of a habitat reserve

(I.U.C.N. 1978). If cropping programs are initiated, population studies should be done first in order to determine the potentials for long-term yield (Heltne et al. 1976; Mittermeier, Coimbra-Filho, and Van Roosmalen 1978).

Table 38. Distribution of *Callithrix argentata*

Country	Presence and Abundance	Reference
Brazil	In Pará at least as far W as R. Tapajós and into N and W Mato Grosso	Cabrera 1958
	In Pará between R. Tocantins, R. Iriri, and R. Tapajós; in Mato Grosso from about 12°S to R. Tacuarí; as far W as R. Madeira; questionable in central Rondônia (map)	Hershkovitz 1968
	Localities listed on R. Tocantins, R. Curuá, R. Tapajós (near Santarém) in Mato Grosso, and Serra da Chapada	Napier 1976
	Localities listed and mapped in Pará between R. Tocantins and R. Tapajós; in SE Amazonas between R. Madeira and R. Aripuanã; throughout Rondônia; and in W Mato Grosso as far E as about 53°W; questionable in N Mato Grosso and S Pará	Hershkovitz 1977
	In Pará E of R. Xingu	Ayres 1978
	Common and not endangered in N of range between lower R. Tapajós and R. Tocantins and R. Iriri (*C. a. argentata*); vulnerable between R. Jamanxim and R. Cuparí, E of R. Tapajós (*C. a. leucippe*); status of southern form (*C. a. melanura*): not known	Mittermeier, Bailey, and Coimbra-Filho 1978
	In Caracara Biological Reserve	Mittermeier et al., in press
Bolivia	In E	Cabrera 1958
	In E throughout most of Santa Cruz and E El Beni as far W as R. Mamoré (map)	Hershkovitz 1968
	As far W as R. Beni	Avila-Pires 1969b
	Abundant at San José de Chiquitos (in SE at 17°50′S, 60°43′W) (locality mapped)	Heltne et al. 1976
	Localities listed and mapped in SE, central, and SW Santa Cruz; W as far as R. Mamoré (map)	Hershkovitz 1977

Callithrix humeralifer Tassel-ear Marmoset

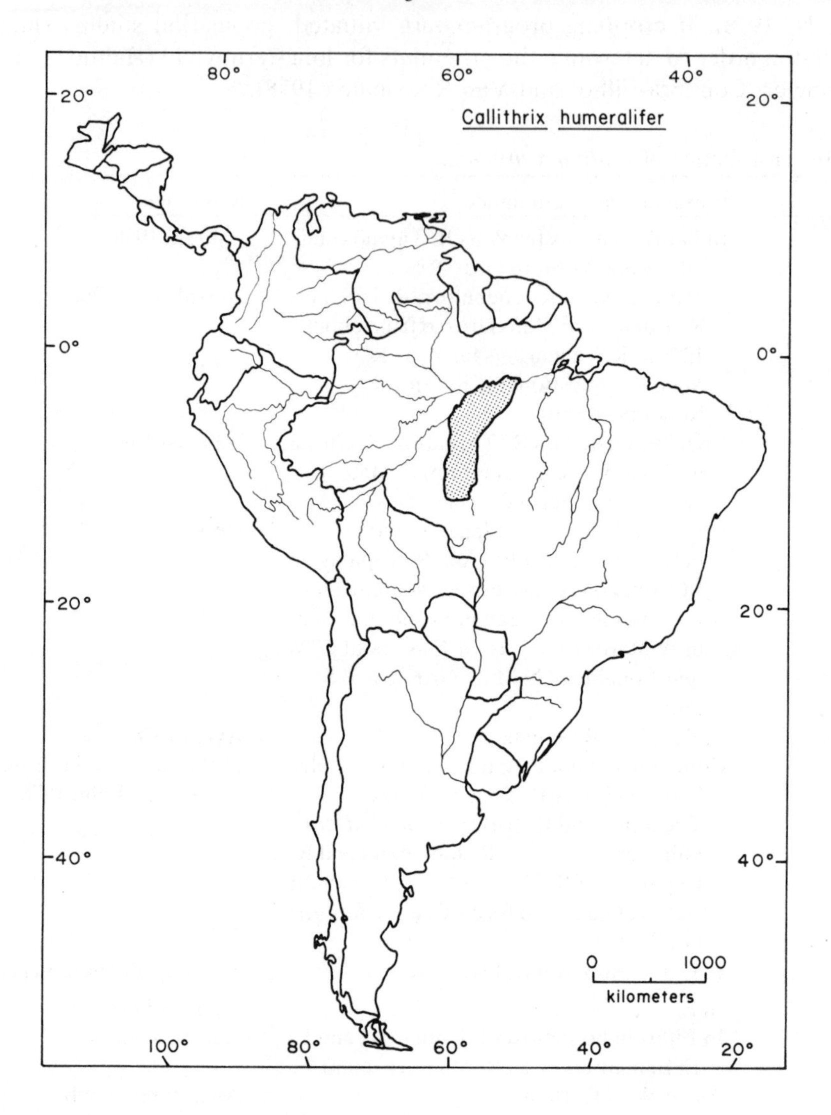

Taxonomic Note

Three subspecies are recognized (Hershkovitz 1977). Some earlier workers treat one of these, the silky marmoset (*C. h. chrysoleuca*) as a separate species (Cabrera 1958, Napier and Napier 1967, Avila-Pires 1969b).

Geographic Range

Callithrix humeralifer is found in central Brazil south of the Rio Amazonas between the Riọ Madeira and the Rio Tapajós. It occurs as far west as the mouth of the Rio Aripuanã (5°05′S, 60°20′W) and southward east of this river and the Rio

Roosevelt to about 11°S (Hershkovitz 1977). Its eastern limit is formed by the Rio Juruena and Rio Tapajós, as far east as the mouth of the Rio Tapajós at about 2°12′S, 54°54′W (Hershkovitz 1977). For details of the distribution, see Table 39.

Abundance and Density

C. humeralifer is thought to be threatened with extinction because of its very small range (Hershkovitz 1972), but nothing is known about the actual status of its populations. No estimates of population density, group size, or home range area are available.

Habitat

No information.

Factors Affecting Populations

Habitat Alteration

The Transamazonian Highway runs through the range of *C. humeralifer*, but the effect on its populations is not known.

Human Predation

Hunting. This species is usually not hunted for food (Mittermeier and Coimbra-Filho 1977).

Pest Control. No information.

Collection. A total of 27 *C. humeralifer* were imported into the United States between 1968 and 1973 (Appendix), primarily from Colombia, where this species does not occur. None entered the United States in 1977 (Bittner et al. 1978), nor were any imported into the United Kingdom between 1965 and 1975 (Burton 1978).

Conservation Action

C. humeralifer does not occur in any national parks or reserves (I.U.C.N. 1978) but is legally protected from commercial exploitation (Mittermeier and Coimbra-Filho 1977). Breeding colonies in Rio de Janeiro and a survey to determine status of this species are being planned (I.U.C.N. 1978).

Table 39. Distribution of *Callithrix humeralifer*

Country	Presence and Abundance	Reference
Brazil	S of R. Amazonas between R. Tapajós and R. Madeira	Cabrera 1958, Avila-Pires 1969b
	S of R. Amazonas in E Amazonas and W Pará between R. Madeira and R. Canumã to about 5°S, and in triangular area W of R. Tapajós	Hershkovitz 1968
	Range precariously small; threatened with extinction	Hershkovitz 1972

Table 39 (continued)

Country	Presence and Abundance	Reference
Brazil (cont.)	Localities listed include Boim, R. Tapajós; Villa Braga (4°25′S, 56°17′W); Borba, R. Madeira; and Itaituba (4°17′S, 55°59′W)	Napier 1976
	Localities listed and mapped in E Amazonas, W Pará, NW Mato Grosso, and tiny corner of E Rondônia	Hershkovitz 1977

Callithrix jacchus Tufted-ear or Common Marmoset

Taxonomic Note

Five subspecies are recognized (Hershkovitz 1972, 1977; Napier 1976). These forms are considered to be full species (*C. jacchus*, *C. penicillata*, *C. geoffroyi*, *C. flaviceps*, and *C. aurita*) by some authors (Coimbra-Filho and Mittermeier 1973b, Mittermeier et al., in press).

Geographic Range

Callithrix jacchus is found in eastern Brazil from northern Ceará to southern São Paulo. It occurs as far north as 3°03′S, 39°38′W along the northern coast of Ceará, as far south as about 24°30′S along the Rio Ribeira, as far west as 53°W along the Rio

Araguaia in Goiás, and as far east as about 35°W along the coast of Pernambuco (Hershkovitz 1975, 1977). For details of the distribution, see Table 40.

Abundance and Density

In the northern portion of its range *C. jacchus* is common, although one form is in a vulnerable position; in the southern portion of the range this species is vulnerable or endangered (Table 40). No estimates of population density, average group size, or home range area are available.

Habitat

C. jacchus is found in old forest, young forest, swamp forest, and plantations (Laemmert et al. 1946). It also occupies wooded savanna, dry caatinga forest (Avila-Pires 1969b), Atlantic coastal evergreen tropical seasonal rain forest (Cerqueira, pers. comm. 1974), and plateau *cerrados* vegetation (Vanzolini 1978). It has been recorded from sea level to 800 m elevation (Coimbra-Filho and Mittermeier 1973b).

Factors Affecting Populations

Habitat Alteration

The fragmentation of previously extensive forests into scattered isolated lots, and the cultivation and burning of wooded savannas and scrublands, have resulted in many semi-isolated populations of *C. jacchus* (Hershkovitz 1968). This habitat disturbance is continuing and is threatening several forms of *C. jacchus* (Mittermeier, Coimbra-Filho, and Van Roosmalen 1978). Deforestation is especially extensive in Rio de Janeiro, eastern São Paulo, Espírito Santo, southeastern Minas Gerais, and southeastern Bahia, where the habitats of the forms *aurita, flaviceps*, and one population of *penicillata* have been greatly reduced (Mittermeier et al., in press). The northernmost form (*jacchus*) is highly adaptable, can live in many forest types including highly degraded forest remnants, and is not being threatened by habitat alteration (Mittermeier et al., in press).

Human Predation

Hunting. No information.

Pest Control. C. jacchus occupies plantations of cacao, bananas, and coconut (Laemmert et al. 1946), and was once abundant in orchards and gardens (Hershkovitz 1977). No efforts to control this species were mentioned.

Collection. C. jacchus has been collected for yellow fever and malarial research in Brazil (Laemmert et al. 1946, Waddell and Taylor 1946, Kumm and Laemmert 1950, Deane et al. 1969). The northern forms of this species are thought to be abundant enough to withstand sustained yield cropping of several hundred per year (Mittermeier, Coimbra-Filho, and Van Roosmalen 1978).

Between 1968 and 1972 a total of 611 *C. jacchus* were imported into the United States (Appendix). In 1977 one shipment of an unspecified size was imported into the United States from Bolivia (Bittner et al. 1978). In 1965–75 a total of 6,196 *C. jacchus* were imported into the United Kingdom (Burton 1978). *C. jacchus* is sometimes used in biomedical research in the United States (Comm. Cons. Nonhuman Prim. 1975a).

Conservation Action

C. jacchus occurs in eight national parks and five reserves, and may also occur in five other national parks, one reserve, and one state park (Table 40). It is protected against commercial exploitation in Brazil (Mittermeier and Coimbra-Filho 1977) but is not listed as endangered there (I.U.C.N. 1978).

Recommendations for the conservation of the threatened southern forms include a total ban on export, shooting, or interference, as well as management of home environments by reforestation and supply of known limiting factors, census, and observation (Kleiman 1976), and also inclusion in Brazil's endangered species list, creation of reserves, and captive breeding (I.U.C.N. 1978). The northernmost form is also in need of breeding colonies, rigorous management of genetic stocks, control of export quotas, and control of conditions of shipment (Kleiman 1976). Before cropping programs are initiated, studies should be performed to determine long-term yield potentials (Mittermeier, Coimbra-Filho, and Van Roosmalen 1978).

The two southern forms of *C. jacchus* (*aurita* and *flaviceps*) are listed in Appendix I of the Convention on International Trade in Endangered Species of Wild Fauna and Flora (U.S. Department of Interior 1977a).

Table 40. Distribution of *Callithrix jacchus*

Country	Presence and Abundance	Reference
Brazil	Near Ilhéus, SE coast of Bahia	Laemmert et al. 1946, Waddell and Taylor 1946
	Localities mapped in Ceará, Bahia, Espírito Santo, Rio de Janeiro, Minas Gerais, and Goiás	Kumm and Laemmert 1950
	From Ceará to Bahia, in Minas Gerais, Espírito Santo, Guanabara, São Paulo, and Goiás	Cabrera 1958
	Localities mapped in E Piauí; W, S, and E Ceará (questionable in N and in Rio Grande do Norte); E Paraiba, Pernambuco, Alagoas, and Sergipe; E and S Bahia; most of Minas Gerais, Espírito Santo, Rio de Janeiro, and Guanabara; S São Paulo (questionable in central); and S and E Goiás	Hershkovitz 1968
	Abundant	Avila-Pires 1969a
	In Maranhão, Bahia, Minas Gerais, São Paulo, Goiás, and Espírito Santo; introduced into Tijuca, Guanabara, about 1900 and spread into Rio de Janeiro	Avila-Pires 1969b
	Localities mapped in Ceará, Alagoas, Bahia, Espírito Santo, Minas Gerais, São Paulo, and Goiás	Deane et al. 1969
	Relatively abundant between R. Doce and R. São Mateus in Espírito Santo (between	Coimbra-Filho 1971

Table 40 (continued)

Country	Presence and Abundance	Reference
Brazil (cont.)	R. Jucu and R. Itaunas); also from R. São Francisco, Bahia to Maranhão	Coimbra-Filho 1971 (cont.)
	In Caparaó N.P., Espírito Santo and Minas Gerais; and Tijuca N.P., Guanabara	I.U.C.N. 1971
	Population in SE Minas Gerais, SE and central Espírito Santo, and NE Rio de Janeiro (*flaviceps*): reduced, disjunct, declining, endangered	Coimbra-Filho 1972a
	Localities listed in Tijuca N.P. and Federal District	Cerqueira, pers. comm. 1974
	One form (*flaviceps*): rare; has disappeared from some areas, e.g., Serra de Macaé, Rio de Janeiro	I.U.C.N. 1974
	In Caparaó N.P. and Nova Lombardia Federal Biological Reserve (on E coast of Espírito Santo)	Padua et al. 1974
	Localities listed and mapped as in 1968 except adding S Piauí, all of Paraíba and Pernambuco, N and W Bahia N and W of R. São Francisco, and W Goiás to R. Araguaia and NE to R. Manuel Alves Grande; questionable in central Maranhão, central Ceará, NE Bahia, and NE Minas Gerais	Hershkovitz 1975, 1977
	Two forms (*flaviceps* and *aurita*, both in S of range) threatened so severely that they are near extinction; northern form (*jacchus*) not in danger	Kleiman 1976
	N form (*jacchus*) not in danger	Mittermeier et al. 1976
	Localities listed in Espírito Santo, Ceará, Pernambuco, Bahia, Goiás, and Minas Gerais	Napier 1976
	S form (*aurita*): vulnerable, probably in Serra dos Orgãos N.P., Rio de Janeiro; Itatiaia N.P., Rio de Janeiro and Minas Gerais; and Serra da Bocaina N.P., Rio de Janeiro and Sao Paulo; *flaviceps*: endangered; in Caparaó N.P. and Nova Lombardia Federal Biological Reserve	I.U.C.N. 1978
	N 3 forms common, other 2 endangered or vulnerable	Mittermeier, Coimbra-Filho, and Van Roosmalen 1978
	In need of protection	Vanzolini 1978
	From N to S: (*jacchus*): common; in Sete Cidades N.P., Ubajara N.P., Tijuca N.P. and 2 biological reserves of Serra Negra and Maceio; *penicillata*: one form	Mittermeier et al., in press

Table 40 (continued)

Country	Presence and Abundance	Reference
Brazil (cont.)	common another vulnerable; in Brasília N.P., Emas N.P., Chapado dos Veadeiros N.P., Monte Pascoal N.P., biological reserves of Pau Brasil and Una, and may be in Araguaia N.P. and Serra da Canastra N.P.; *geoffroyi*: common; may occur in Nova Lombardia and Sooretama biological reserves and Rio Doce State Park; *flaviceps*: endangered; may be in Caparaó N.P. and Nova Lombardia Federal Biological Reserve; *aurita*: one form endangered, one vulnerable; probably in Serra dos Orgãos N.P., Itatiaia N.P., and Serra da Bocaina N.P.	Mittermeier et al., in press (cont.)

Cebuella pygmaea Pygmy Marmoset

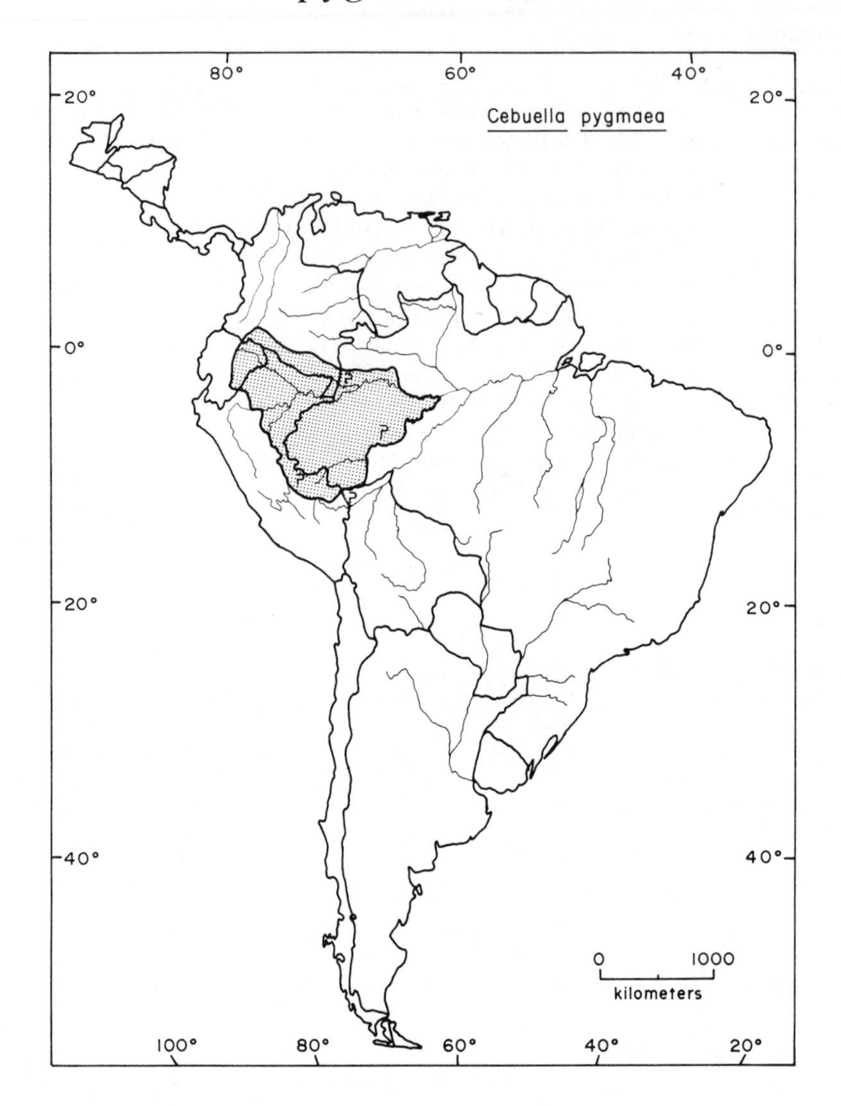

Taxonomic Note

Two subspecies have been described (Cabrera 1958).

Geographic Range

Cebuella pygmaea is found in western South America from southern Colombia to southeastern Peru. The northern limit of its range is near Mocoa, Colombia (1°07′N, 76°38′W) (Hernández-Camacho and Cooper 1976), and the southernmost record is at Cocha Cashu, Peru (11°50′S, 71°23′W) (Janson and Terborgh, in press). *C. pygmaea* occurs as far west as 78°19′W (at 2°43′S) in Ecuador, and as far east as the mouth of the Rio Purus (about 61°30′W) in Brazil (Hershkovitz 1977). For details of the distribution, see Table 41.

Abundance and Density

C. pygmaea is locally abundant in parts of Colombia and Peru, but is considered vulnerable in Peru; it is common and not endangered in Brazil, and its status in Colombia, Ecuador, and Bolivia is not known (Table 41).

Group size in this species varies from 3 to 20, with average sizes of 6.4 to 10 (Table 42). Population densities of 34 to 560 individuals per km^2 and home ranges of 0.1 to 1.3 ha have been reported (Table 42).

Habitat

Colombian pygmy marmosets are typical of mature, nonflooding forest, where they are associated with the guarango tree (*Parkia* spp.) (Hernández-Camacho and Cooper 1976). They also occupy the edges of forest, patches of secondary growth, and low and degraded or unnatural woodland (Moynihan 1976a,b), are most abundant in thick brush and second growth, and prefer thick cover along streams (Hershkovitz 1977).

In Peru, *C. pygmaea* utilizes both inundatable and noninundatable forests as well as disturbed habitats, but is restricted to areas with concentrations of exudate-source trees (Freese et al. 1978). In one study it was found most frequently (92%) near present or old water courses, populations were the densest in heterogeneous forest inundatable to a maximum depth of 2 m, and the most important sap food sources were *Vochysia lomatophylla* trees (Soini 1978). Other workers have recorded *C. pygmaea* in evergreen, mesophytic, broadleaf forest on flood plains and high ground in black- and white-water drainages (Freese et al., in prep.), and in edge vegetation along rivers, clearings, and old banana fields (Castro and Soini 1978).

In western Brazil, *C. pygmaea* is found in bamboo thickets in rain forest (Vanzolini, pers. comm. 1974). And in the Amazonian region it occupies primary and secondary *terra firme* (nonflooding) forests (Mittermeier and Coimbra-Filho 1977).

Factors Affecting Populations

Habitat Alteration

C. pygmaea has a remarkable ability to live as a near commensal of man in highly degraded habitats. In Colombia, for example, the partial clearing of forests has provided new patches of suitable edge and secondary growth, and the planting of hedges and exotic food plants has improved the habitat for *C. pygmaea* (Moynihan 1976a,b). The guarango tree is not a desirable timber species and is often left standing (Hernández-Camacho and Cooper 1976). Nevertheless, these last authors cite habitat destruction as the most serious threat to *C. pygmaea* in Colombia, and it is also a threat to this species in Peru (Freese 1975).

In eastern Acre, Brazil, clearing of forest for pasture is proceeding at a rapid rate, and 50% of the habitat originally available to *C. pygmaea* will probably be destroyed by 1984 (Vanzolini, pers. comm. 1974). Elsewhere in Brazil, *C. pygmaea* has still been able to adapt well to man-altered environments (Mittermeier and Coimbra-Filho 1977).

Human Predation

Hunting. C. pygmaea is not usually hunted for food because of its small size. This is true in Colombia (Hernández-Camacho and Cooper 1976), Peru (Encarnación et al. 1978, Freese et al. 1978), and Brazil (Mittermeier and Coimbra-Filho 1977).

Pest Control. No information.

Collection. Pygmy marmosets are sometimes trapped by local people in Colombia, and in one area are used to pick lice from people's hair (Hernández-Camacho and Cooper 1976). In some areas collection pressure for the local pet trade and for export is intense and has seriously affected populations (Moynihan 1976b). *C. pygmaea* is captured infrequently for the local pet trade in Peru (Ramírez et al. 1978) and Brazil (Mittermeier and Coimbra-Filho 1977).

C. pygmaea has been collected fairly heavily for the export trade. For example, collection for biomedical research has affected populations in Peru (Castro 1976), and future intensive trapping programs could severely reduce remaining stocks (Freese et al. 1978).

Between 1962 and 1968, 5,297 *C. pygmaea* were exported from Peru; and in 1970 Peru banned further export (Soini 1972). In 1970, 165 *C. pygmaea* were exported from Colombia (Green 1976), and in 1971, 110 were exported from Colombia (Cooper, pers. comm. 1974). But United States import statistics show 172 and 127 *C. pygmaea* entering the United States from Colombia in 1970 and 1971 respectively (Paradiso and Fisher 1972, Clapp and Paradiso 1973).

Between 1968 and 1973 a total of 1,396 *C. pygmaea* were imported into the United States (Appendix). None entered the United States in 1977 (Bittner et al. 1978). Two hundred fifty were imported into the United Kingdom between 1965 and 1975 (Burton 1978).

Conservation Action

C. pygmaea is found in two national parks and two national reserves in Peru, but not in any protected areas in the other countries of its range (Table 41). It has been protected by Peruvian law since 1970 (Castro 1976), and is also legally protected from commercial exploitation in Brazil (Mittermeier and Coimbra-Filho 1977). In Colombia its capture was limited to 60 per license holder between 1 June and 30 September 1972 (Cooper and Hernández-Camacho 1975), but no information concerning more recent regulations is available.

Recommendations for actions to conserve this species include strict regulation of trade, the establishment of breeding colonies (Moynihan 1976b), limitation of export from Peru to less than 50 to 100 per year for breeding purposes only (Freese 1975), and protection of its Peruvian food tree (*Vochysia lomatophylla*) by logging restrictions (Soini 1978).

C. pygmaea was listed in Appendix I of the Convention of International Trade in Endangered Species of Wild Fauna and Flora (U.S. Department of Interior 1977a), but has been removed from that appendix (Cook 1979).

Table 41. Distribution of *Cebuella pygmaea* (countries are listed in counterclockwise order)

Country	Presence and Abundance	Reference
Colombia	In Putumayo and Amazonas S of R. Caquetá (map); may be farther N in basin of upper R. Guaviare	Hernández-Camacho and Cooper 1976
	Dense population in basin of R. La Tagua, middle basin R. Caquetá	Izawa 1976
	Near El Pepino and Rumiyaco, Putumayo	Moynihan 1976b
	Localities listed and mapped S of R. Caquetá	Hershkovitz 1977
Ecuador	In E	Cabrera 1958
	Along R. Pastaza near E border	Grimwood 1969
	One locality at Santa Cecilia (0°03′S, 76°58′W), Napo-Pastaza	Baker 1974
	Locality listed on R. Copataza	Napier 1976
	Localities listed and mapped in E to R. Pastaza	Hershkovitz 1977
Peru	In N, E of Andes	Cabrera 1958
	Along R. Amazonas and lower R. Napo including Iquitos and Chimbote E and S to at least Santa Cruz on R. Huallaga (5°20′S, 75°30′W); not rare	Grimwood 1969
	Not rare	Soini 1972
	Vulnerable; in Manú N.P., Madre de Dios and Cuzco; Tingo Maria N.P., Huánuco; Pacaya and Samiria national reserves, Loreto	Castro 1976
	Localities listed in Loreto and Amazonas	Napier 1976
	Localities listed and mapped throughout N and E; questionable near 5°S and 10°S	Hershkovitz 1977
	At R. Aucayo and R. Tahuayo, Loreto	Castro and Soini 1978
	On R. Yavineto (R. Putumayo), N Loreto	Encarnación et al. 1978
	Locally abundant along R. Nanay; common along R. Ampiyacu (in NE)	Freese et al. 1978
	Along R. Nanay	Ramírez et al. 1978
	In Loreto at R. Manití, R. Tahuayo, R. Orosa, R. Aucayo, and R. Nanay	Soini 1978
	Rare at Cocha Cashu, Manú N.P.	Janson and Terborgh, in press
Bolivia	At Cobija (in NW at 11°01′S, 68°45′W)	Heltne et al. 1976; Freese et al., in prep.
Brazil	At Seringal Oriente, W Acre	Carvalho 1957
	Along R. Solimões to basin of R. Juruá	Cabrera 1958
	In W Amazonas from R. Içá to R. Tefé (map)	Avila-Pires 1974
	Not rare in basins of R. Purus and R. Acre, E Acre	Vanzolini pers. comm., 1974
	Localities listed at Santarita and Tabatinga (4°14′S, 69°44′W)	Napier 1976

Table 41 (continued)

Country	Presence and Abundance	Reference
Brazil (cont.)	Localities listed and mapped in W Amazonas; questionable between R. Japurá and R. Içá, in central Acre, and near 7°S, W of R. Purus	Hershkovitz 1977
	Not endangered	Mittermeier and Coimbra-Filho 1977; Mittermeier, Bailey, and Coimbra-Filho 1978
	Common	Mittermeier, Coimbra-Filho, and Van Roosmalen 1978; Mittermeier et al., in press

Table 42. Population parameters of *Cebuella pygmaea*

Population Density (individuals/km²)	Group Size $\bar{X}$	Group Size Range	Group Size No. of Groups	Home Range Area (ha)	Study Site	Reference
		≤10–15			S Colombia	Hernández-Camacho and Cooper 1976
	10	4–20			S Colombia	Izawa 1976
		3–6	6		S Colombia	Moynihan 1976b
560		5–10	4	0.8–1.3	N Peru	Castro and Soini 1978
		3	1		N Peru	Encarnación et al. 1978
		7–9	1		NE Peru	Freese et al. 1978
		7	1	0.3	NE Peru	Ramírez et al. 1978
34–247	6.4	4–9	12	0.1–0.4	NE Peru	Soini 1978
		4–8			SE Peru	Janson and Terborgh, in press

Leontopithecus rosalia Lion Tamarin

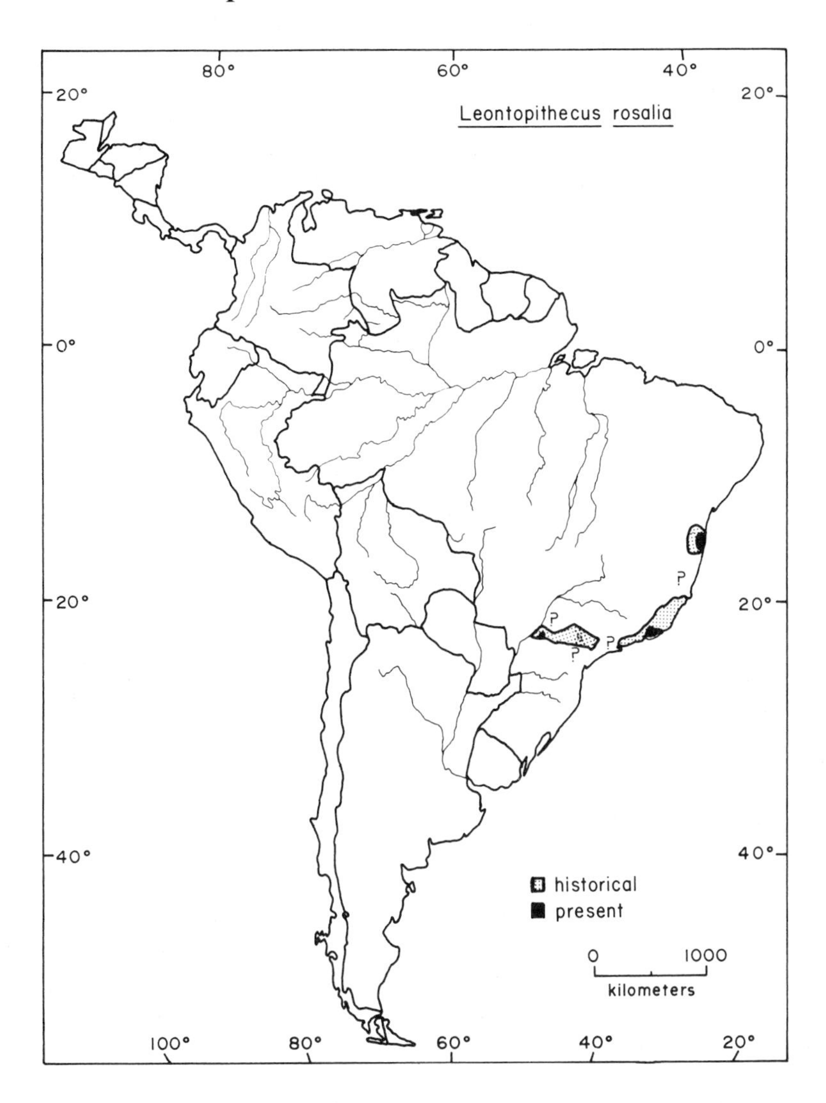

Taxonomic Note

Three subspecies are recognized: *L. r. rosalia* (golden lion tamarin), *L. r. chrysopygus* (golden-rump lion tamarin), and *L. r. chrysomelas* (golden-head or gold and black lion tamarin) (Coimbra-Filho and Magnanini 1972; Coimbra-Filho and Mittermeier 1972, 1973a; Hershkovitz 1972, 1977).

Geographic Range

Leontopithecus rosalia is found in southeastern Brazil from coastal Bahia to São Paulo. Its range originally extended from the Rio das Contas at about 14°S in southeastern Bahia to about 24°S along the Rio Paranapanema, and west as far as the Rio

Paraná (about 53°W) in São Paulo (Coimbra-Filho and Mittermeier 1973a, 1977; Hershkovitz 1977). The range has been reduced and fragmented, and the species now occurs in only three areas: between Ilhéus (14°50′S, 39°06′W) and Una (15°18′S, 39°06′W), coastal Bahia (*chrysomelas*); in the basin of the Rio São João (near 23°30′S), Rio de Janeiro (*rosalia*); and in the Morro do Diabo State Park (22°31′S, 52°10′W), southwestern São Paulo (*chrysopygus*) (Mittermeier et al. 1976, Coimbra-Filho and Mittermeier 1977). For details of the distribution, see Table 43.

Abundance and Density

L. rosalia is imminently threatened with extinction. Its total population was estimated at 1,000 to 1,300 in 1972, 700 to 1,200 in 1973, and 500 to 600 in 1976, 1977, and 1978 (Table 43). All three races are declining and endangered.

No estimates of population density or home range area are available. Groups usually include 2 to 8 members ($\bar{X}$ = 3 or 4, *rosalia*), with sightings of groups of 15 or 16 (*rosalia*), 4 to 12 (*chrysomelas*), and 2 to 5 ($\bar{X}$ = 3.3, N = 3, *chrysopygus*) also reported (Coimbra-Filho and Mittermeier 1973a, 1977).

Habitat

L. rosalia is a tropical humid forest species which prefers mature primary forest but can also utilize various kinds of secondary forest (Coimbra-Filho and Mittermeier 1977). *L. r. rosalia* is found in seasonal tropical forest (classification of Veloso 1966) in which the dominant tree species are *Ficus, Inga, Platymenia*, and *Tapirira guianensis*; *L. r. chrysomelas* occupies Atlantic rain forest and shaded cacao plantations; and *L. r. chrysopygus* is found in dense, undisturbed riparian extensions of the coastal forest, characterized by *Cedrella*, *Aspidosperma*, *Gallesia*, *Ficus*, and *Arecastrum romanzoffianum* (Coimbra-Filho and Mittermeier 1973a, 1977). *L. r. chrysomelas* has also been recorded in old and young type forests, and in swamp forest (Laemmert et al. 1946).

Lion marmosets are typical of lowland forest from sea level to 300 m (*rosalia*), sea level to 112 m (*chrysomelas*), and as high as 700 m (*chrysopygus*) (Coimbra-Filho and Mittermeier 1973a). Within the forest they tend to select areas with heavy vine and epiphytic growth which provide cover, and with tree holes for sleeping sites (Coimbra-Filho and Mittermeier 1977, Coimbra-Filho 1978).

Factors Affecting Populations

Habitat Alteration

Deforestation followed by the replacement of forest with agriculture is the primary cause of the decline of *L. rosalia*. The range of the species is in the most densely populated part of Brazil, and that rapidly developing country views forests as unwanted barriers to its progress which should be stripped of their valuable timber and cleared for agriculture or pasture to feed its growing population and aid its economic progress and expansion (Coimbra-Filho and Mittermeier 1977).

Climatic changes may have been involved in the fragmentation of the originally continuous forests which *L. rosalia* occupied (Coimbra-Filho and Mittermeier 1972). Beginning in the seventeenth century, the large-scale cutting of forests for sugar cane plantations in Rio de Janeiro and coffee plantations in Guanabara eliminated most suitable habitat for *L. r. rosalia* (Coimbra-Filho and Mittermeier 1973a). Deforestation for banana plantations and housing projects (I.U.C.N. 1972), bridge and highway building (Perry 1971, Fitter 1974), and the alteration of the watershed of the Rio São João (Coimbra-Filho and Magnanini 1972) have further reduced *L. r. rosalia*'s habitat. The original 2 million ha range was reduced to less than 20,000 ha by 1968, and to isolated forest plots by 1971 (Magnanini 1978).

Habitat destruction within the range of *L. r. rosalia* is so widespread that this race is now restricted to small, widely scattered forest patches between which gene flow is impossible, and which are continuing to shrink because of burning for charcoal production and clearing for farmland (Coimbra-Filho and Mittermeier 1977). Most of the remaining forested areas are being degraded by fire and cutting and are in various stages of secondary growth, and the *L. r. rosalia* which inhabit them seem to be smaller and less brilliant in color and shine, and may be adversely affected socially by their disturbed habitat conditions (Coimbra-Filho 1978). "The entire valley of the Rio São João, which contains the last populations of *L. r. rosalia*, has been slated for an enormous agricultural project which will ensure the destruction of all remaining natural habitat" (Mittermeier et al. 1976).

In southeastern Bahia, the forests have been destroyed for lumber and agricultural land, and most of the former habitat of *L. r. chrysomelas* now consists of cacao plantations (Coimbra-Filho and Mittermeier 1973a). This subspecies has been seen in plantations in which some large forest trees remain, but generally lion tamarins are "unable to adapt to altered conditions," and do not utilize disturbed forests (Coimbra-Filho and Mittermeier 1973a). The remaining forests within the range of *L. r. chrysomelas* contain valuable timber trees (such as Brazilian rosewood, *Dalbergia nigra*, Leguminosae), and are being cleared to make way for rubber and cacao plantations (Mittermeier et al. 1976).

The habitat of *L. r. chrysopygus* has been even more thoroughly destroyed than that of the other two races. By 1966 the forest cover of São Paulo had been reduced from 64.7% (in 1911) to 13.7%, by logging for lumber, agriculture, and pasture (Coimbra-Filho and Mittermeier 1973a). About 90 ha of suitable habitat are thought to remain in the Morro do Diabo State Park (I.U.C.N. 1974), but frequent fires, "some intentionally set," have destroyed more than one-third of the park (Coimbra-Filho and Magnanini 1972). In addition, a railroad and a highway divide the park in two, and the park is isolated and surrounded by some detrimental activities, such as the recent use of the herbicides 2,4,5-T and 2,4-D by local farmers (Coimbra-Filho and Mittermeier 1977, Coimbra-Filho 1978).

Human Predation

Hunting. No information.

Pest Control. No information.

Collection. L. rosalia has long been popular as a zoo exhibit in the United States and Europe, has been sold in the pet trade, and has occasionally been used in research, and these demands have led to considerable trapping pressure on *L. r. rosalia* for export to foreign markets (Coimbra-Filho and Mittermeier 1977). For example, one collector obtained 200 lion tamarins in five years near Estanislau and Presidente, and another captured more than 300 in six years near Poco das Antas and Rio Iguape (Coimbra-Filho and Mittermeier 1973a). Overcollection for commercial trade has contributed strongly to the decline of *L. r. rosalia* (I.U.C.N. 1972, Coimbra-Filho and Mittermeier 1973a).

Between 1968 and 1970, 349 *L. rosalia* were imported into the United States (Appendix). In 1970, 122 of the 150 imported came from Colombia and Peru (Paradiso and Fisher 1972) where *L. rosalia* does not occur. No *L. rosalia* were imported into the United States in 1971–73 (Appendix) or in 1977 (Bittner et al. 1978). Eighteen were imported into the United Kingdom between 1965 and 1975 (Burton 1978).

Conservation Action

L. rosalia occurs in one state park and two biological reserves (Table 43), but the park is vulnerable to the effects of nearby habitat alteration (see above), and the reserves are being established with a great deal of difficulty and delay (Coimbra-Filho and Mittermeier 1977, Magnanini 1978). The Poco das Antas Reserve is the last hope for the survival of *L. r. rosalia* in its natural habitat, but a large portion of it must be reforested, and even then it may not be able to survive in the midst of hundreds of thousands of hectares of cultivated land (Coimbra-Filho and Mittermeier 1977). The government of São Paulo is interested in establishing a reserve near Galia for the recently discovered small population of *L. r. chrysopygus* there (Mittermeier et al., in press).

In spite of international concern for the fate of *L. rosalia*, the governmental agencies responsible for the preservation of Brazil's fauna have done very little to insure its survival in its natural habitat, and a few dedicated Brazilian conservationists are waging a difficult battle to save even the smallest remnants of southeastern Brazil's fauna and flora from destruction (Coimbra-Filho and Mittermeier 1977). For example, the National Research Council of Brazil and the Brazilian Foundation for the Conservation of Nature financed a study of the ecology and status of *L. rosalia* by A. F. Coimbra-Filho (de Melo Carvalho 1971b).

Individual lion tamarins are being collected from threatened areas and taken to the Tijuca Bank in Tijuca National Park, Rio de Janeiro, for participation in a captive breeding program there (Magnanini and Coimbra-Filho 1972, Fitter 1974, Magnanini et al. 1975). Captive breeding is also being attempted in the United States, but many difficulties remain to be overcome before breeding colonies produce a significant number of surplus animals (Hill 1970, Perry 1971, Avila-Pires 1972, Hampton 1972, Kleiman and Jones 1978).

A conference was held in 1972 to discuss the status of the lion tamarin (Bridgewater 1972). The recommendations for research priorities which were offered included (1) field and laboratory studies of nonendangered tamarins and marmosets to gain information applicable to captive breeding of *L. rosalia*, and (2) experimental breeding focused on nonendangered species in the United States and *L. rosalia* at the Tijuca Bank (Kleiman 1972). A second conference in 1975 recommended (1) a complete ban on export, shooting, or interference with *L. rosalia*, (2) urgent study by census and observations, and (3) close management of remaining home environments by reforestation and the supply of known limiting factors (Kleiman 1976). One of these factors is shelter holes for sleeping sites, which are not abundant in young secondary forest with trees of small diameter. These could be supplied by hanging shelter boxes, which would improve the quality of degraded habitats for *L. rosalia* (Coimbra-Filho 1978).

Brazilian laws forbid the capture, hunting, purchase, and sale of *L. rosalia*, but do not prohibit destruction of its habitat (Coimbra-Filho and Mittermeier 1977). *L. rosalia* is listed in Appendix I of the Convention on International Trade in Endangered Species of Wild Fauna and Flora, and as endangered under the U.S. Endangered Species Act (U.S. Department of Interior 1977a,b). It may not be imported into the United Kingdom except for scientific, educational, or propagation purposes (I.U.C.N. 1972).

Table 43. Distribution of *Leontopithecus rosalia*

Country	Presence and Abundance	Reference
Brazil	Near Ilhéus (*chrysomelas*)	Laemmert et al. 1946, Waddell and Taylor 1946, Kumm and Laemmert 1950
	In coastal Bahia (*chrysomelas*), Rio de Janeiro (*rosalia*), and São Paulo (*chrysopygus*)	Hershkovitz 1949, Cabrera 1958
	rosalia: previously from S Espírito Santo through coastal Rio de Janeiro to W of Itaguai and throughout Guanabara; now only near Rio São João (map); estimated population: 600	Coimbra-Filho 1969
	Estimated population of 100 *chrysopygus* in Morro do Diabo State Forest near Teodoro Sampaio, counties of Presidente Wenceslau and Presidente Epitácio near confluence of R. Paraná and R. Paranapanema, SW São Paulo	Coimbra-Filho 1970
	chrysomelas: in Una area, Bahia; *rosalia:* only in counties of Rio Bonito and Silva Jardim, both sides of R. São João, Rio de Janeiro; estimated population less than 1,000; *chrysopygus:* thought to be extinct but recently rediscovered in W São Paulo; all endangered	de Melo Carvalho 1971b
	chrysomelas: in small area of S Bahia; the most endangered race; *rosalia:* mostly confined to municipalities of Cabo Frio, Casimiro de Abreu, and Silva Jardim, along tributaries of R. São João; extinct in Espírito Santo; declining; *chrysopygus:* population about 100; range about 100 km^2	Coimbra-Filho and Magnanini 1972

Table 43 (continued)

Country	Presence and Abundance	Reference
Brazil (cont.)	Range fragmented and reduced (map)	Coimbra-Filho and Mittermeier 1972
	chrysomelas: in great danger of extinction; *rosalia:* previous range 50,000 km^2 in Rio de Janeiro and Espírito Santo, now extinct in Espírito Santo, range in Rio de Janeiro 900 km^2, population about 500; *chrysopygus:* considered extinct until rediscovery in 1970; population about 100	Hampton 1972
	Precariously small and rapidly contracting range; all races threatened with extinction	Hershkovitz 1972
	chrysomelas: endangered, population 400–500; *rosalia:* endangered, population 400–600; *chrysopygus:* rare, population in Morro do Diabo: about 200	I.U.C.N. 1972
	In 1968 *rosalia* in significant numbers only in R. São João basin; small remnant populations in Poco das Antas region	Magnanini and Coimbra-Filho 1972
	chrysomelas: in isolated localities in municipalities of Una, Ilhéus, Buerarema, and Itabuna, Bahia, and in at least 1 tract S of R. Mucuri, N Espírito Santo; population 200–300; *rosalia:* extinct in Espírito Santo and Guanabara, now only in basin of R. São João in municipalities of Araruama, São Pedro da Aldeia, Cabo Frio, Casimiro de Abreu, and Silva Jardim near Sobrara, S R. Aldeia Velha, Gaviões, Bananeiras, Correntezas, and Poco das Antas; range about 900 km^2 and decreasing rapidly; population in 1970: about 400; *chrysopygus:* in Morro do Diabo State Forest, São Paulo; population: 100–500 (map)	Coimbra-Filho and Mittermeier 1973a
	chrysomelas: endangered; population in 1971: 200–300; *rosalia:* endangered; population in 1970: 400, range 900 km^2; *chrysopygus:* endangered; population: 100–500 in about 90 km^2 of suitable habitat	I.U.C.N. 1974
	Threatened so severely that they are near extinction; total population about 600; range now 2% of original	Kleiman 1976
	chrysomelas: population about 200; *rosalia:*	Mittermeier et al. 1976

Table 43 (continued)

Country	Presence and Abundance	Reference
Brazil (cont.)	population 100–200; *chrysopygus:* population about 200; all are decreasing (map)	Mittermeier et al. 1976 (cont.)
	Localities listed and mapped; originally uncommon and restricted in range; declining in numbers and habitat; *chrysomelas:* near Una, Buerarema, Itabuna, and perhaps Ilhéus; not in N Espírito Santo; population about 200; *rosalia:* entirely restricted to R. São João basin in municipalities of Silva Jardim, Casimiro de Abreu, Cabo Frio, Araruama, São Pedro da Aldeia, and possibly Rio Bonito and Saquarema; population 100–200, range less than 900 km^2, in the greatest danger; *chrysopygus:* restricted to Morro do Diabo State Park and private land in Galia, W of Baurú, central São Paulo; population about 200; the least endangered	Coimbra-Filho and Mittermeier 1977
	Localities listed and mapped; questionable between R. Doce and R. Jequitinhonha, E Minas Gerais, and in SE São Paulo	Hershkovitz 1977
	rosalia: in Gaupi, Araruama County, and Silva Jardim County, Rio de Janeiro; *chrysopygus:* rare in Morro do Diabo Park, small population near Galia, mainly along R. Paranapanema	Coimbra-Filho 1978
	Endangered; population about 500–600 and declining; *chrysomelas:* about 200 in SE Bahia between R. das Contas and R. Pardo, to 100 km W of coast along R. Gongogi; *rosalia:* 100–200, all in R. São João basin, Rio de Janeiro, range less than 900 km^2; *chrysopygus:* about 200 in Morro do Diabo State Park and Galia, São Paulo	I.U.C.N. 1978
	rosalia vanishing; little cause for optimism; population in 1975: 100–200 (map)	Magnanini 1978
	All races endangered	Mittermeier, Coimbra-Filho, and Van Roosmalen 1978
	chrysomelas: in Una Biological Reserve; *rosalia:* in Poco das Antas Biological Reserve; *chrysopygus:* in Morro do Diabo State Park; all endangered; on the brink of extinction	Mittermeier et al., in press

Saguinus bicolor Brazilian Bare-face or Pied Tamarin

Taxonomic Note

Three subspecies are recognized (Hershkovitz 1966a, 1977).

Geographic Range

Saguinus bicolor is found in northern Brazil north of the Rio Amazonas in northeastern Amazonas and northwestern Pará. It occurs as far west as about 62°W along the Rio Negro, as far east as about 56°W at the mouth of the Rio Erepecurú, and as far south as about 3°S along the Rio Amazonas (Hershkovitz 1966a, 1972, 1977). The northern limit of the range is not known, but may lie as far north as 0° in southern Roraima, or even farther north (Hershkovitz 1977). This species was previously believed to occur also in northern Peru near Pebas and the mouth of the Río Napo (Hershkovitz 1949, Cabrera 1958), but recent evidence does not support this belief. For details of the distribution, see Table 44.

Abundance and Density

S. bicolor is suspected to be threatened with extinction because of its small range, but the westernmost race (*S. b. bicolor*) is thought to be common, and no information is available regarding the other two forms (Table 44). No estimates of population density, group size, or home range area were located.

Habitat

S. bicolor has been observed in swamp forest (Deane et al. 1971) and secondary growth and edge habitats (Thorington 1975). It seems to be quite adaptable and even occurs within the city limits of Manaus (Mittermeier and Coimbra-Filho 1977).

Factors Affecting Populations

Habitat Alteration

Since *S. bicolor* is able to utilize disturbed habitats, its abundance may be increasing owing to the activities of man (Thorington 1975). But its proximity to the rapidly growing Manaus area (Mittermeier, Bailey, and Coimbra-Filho 1978), and its tiny range in the midst of steady human pressure (Hershkovitz 1972), suggest that its populations may be suffering from the effects of habitat alteration.

Human Predation

Hunting. This species "is apparently not persecuted" (Mittermeier, Bailey, and Coimbra-Filho 1978).

Pest Control. No information.

Collection. *S. bicolor* was not imported into the United States between 1968 and 1973 (Appendix) or in 1977 (Bittner et al. 1978). None were imported into the United Kingdom in 1965–75 (Burton 1978).

Conservation Action

S. bicolor does not occur in any habitat reserves (Table 44). It is protected from commercial exploitation by Brazilian law (Mittermeier and Coimbra-Filho 1977).

This species is listed in Appendix I of the Convention on International Trade in Endangered Species of Wild Fauna and Flora, and as endangered under the U.S. Endangered Species Act (U.S. Department of Interior 1977a,b). A status survey, reserves, and a captive breeding program have been recommended for it (I.U.C.N. 1978).

Table 44. Distribution of *Saguinus bicolor*

Country	Presence and Abundance	Reference
Brazil	N of R. Amazonas between R. Erepecurú (Óbidos) and R. Yamundú (Faro) in NW Pará, and at Manaus, E bank of R. Negro	Hershkovitz 1949, Cabrera 1958
	N of R. Amazonas between lower R. Erepecurú and lower R. Negro (map)	Hershkovitz 1966a
	In Amazonas	Deane et al. 1969
	At Pôrto Mauá, near Manaus	Deane et al. 1971
	Original maximum range from R. Erepecurú (= Cuminá) to R. Negro (map) now fragmented and precariously small; threatened with extinction	Hershkovitz 1972
	From R. Paru de Oeste to R. Jamundá, W to Uatumã and to lower R. Negro	Avila-Pires 1974
	Not endangered (map)	Thorington 1975
	Localities listed at R. Negro and R. Nhamundá	Napier 1976
	Localities listed and mapped include R. Nhamundá, Castanhal (1°50′S, 56°58′W), R. Piratucu (1°52′S, 56°55′W), San José, R. Erepecurú, tributary of R. Paru de Oeste (1°30′S, 56°W), and near Manaus	Hershkovitz 1977
	W race (*S. b. bicolor*) not endangered; no data about others	Mittermeier, Bailey, and Coimbra-Filho 1978
	W race common but may be vulnerable; no data about others	Mittermeier, Coimbra-Filho, and Van Roosmalen 1978

Saguinus fuscicollis Saddle-back Tamarin

Taxonomic Note

Thirteen subspecies are recognized (Hershkovitz 1966a, 1968, 1977).

Geographic Range

Saguinus fuscicollis is found in western South America from southern Colombia to central Bolivia. It occurs as far north as at least the south bank of the Río Guaviare (at 2°34′N, 72°38′W) in northwestern Vaupés, Colombia (Hernández-Camacho and Cooper 1976), and as far south as about 17°S in Bolivia (Hershkovitz 1968, 1977). Its western limit is at about 78°W in Ecuador, and its eastern limit is at about 61°30′W at the mouth of the Rio Purus or perhaps farther east along the Rio Madeira (Hershkovitz 1968, 1977). For details of the distribution, see Table 45.

Abundance and Density

S. fuscicollis is locally common in some parts but uncommon in other parts of Peru and Bolivia, some of the races in Brazil are common, and the species is considered not threatened in Colombia but vulnerable in Peru (Table 45). Its status in Ecuador is not known.

Group size in this species varies from 2 to 40, with average sizes of 4 to 7; population densities are 2.4 to 30 individuals per km^2, and home ranges are 60 to 100 ha per group (Table 46).

Habitat

S. fuscicollis is a forest species which can utilize secondary as well as primary forest (Hernández-Camacho and Cooper 1976). In Putumayo, Colombia, it prefers dense tangles and crowded clusters of trees and secondary vegetation produced near windfalls to more mature, open forest (Moynihan 1976a,b). In Peru it thrives in both heavily disturbed, largely secondary forest and virgin primary forest (Castro and Soini 1978, Freese et al. 1978).

In Peru and Bolivia, saddle-back tamarins have been recorded in evergreen, mesophytic, broadleaf forests on high ground and on flood plains of black and white-water drainages (Freese et al., in prep.). In Brazil they have been observed in primary and secondary *várzea* forests (seasonally flooded by white-water rivers) (Mittermeier and Coimbra-Filho 1977). This species occurs from lowland areas to as high as about 1,200 m (Hershkovitz 1977) and even 1,370 m elevation (Grimwood 1969).

Factors Affecting Populations

Habitat Alteration

The range of *S. fuscicollis* has been described as "virtually unbroken" over more than one-half million square miles (Hershkovitz 1968), and as uninterrupted in Colombia (Hernández-Camacho and Cooper 1976). In Peru this species is able to maintain its populations well in areas of heavy forest degradation, and is relatively unaffected by human disturbance (Freese et al. 1976, 1978). In northern Bolivia, however, extensive land clearing has contributed to the sharp decline of *S. fuscicollis* near Riberalta (Heltne et al. 1976). And in general the accelerated rate of deforestation is thought to be a potential threat to this species (Kleiman 1976).

Human Predation

Hunting. *S. fuscicollis* is sometimes killed for food in Peru and Bolivia (Freese 1975, Castro 1976, Heltne et al. 1976), but this hunting is so infrequent that it has not reduced population densities (Freese et al., in prep.).

Pest Control. Saddle-back tamarins raid plantations in Colombia (Moynihan 1976a), and feed in orchards in Peru (Castro and Soini 1978), but none of the authors consulted mentioned efforts to control these activities.

Collection. S. fuscicollis has not been collected in large numbers in Colombia (Hernández-Camacho and Cooper 1976). It has been collected frequently in Peru, and once made up the majority of the 1,000 to 3,000 tamarins exported annually from Iquitos (Soini 1972). In 1976–77, 100 *S. fuscicollis* were exported from Peru (Castro 1978), and its populations there are considered abundant enough to withstand exploitation of about 1,000 individuals per year (Freese 1975, Neville 1976b).

This species has also been trapped for export from Bolivia, and 250 to 500 could probably be removed yearly from the northern part of that country without significantly reducing populations (Heltne et al. 1976). It is one of the most frequently exported callitrichids in Brazil, where its populations could probably withstand cropping of several hundred per year (Mittermeier, Coimbra-Filho, and Van Roosmalen 1978). *S. fuscicollis* suffers an especially high mortality rate during capture and transport (Grimwood 1969, Mittermeier et al. 1976).

Between 1968 and 1973 a total of 893 *S. fuscicollis* were imported into the United States (Appendix), all from Peru in 1971–73. In 1977, 73 entered the United States (Bittner et al. 1978). Between 1965 and 1975 a total of 96 were imported into the United Kingdom (Burton 1978). This species is frequently used in biomedical research in the United States (Gengozian 1969, Thorington 1972, Comm. Cons. Nonhuman Prim. 1975a).

Conservation Action

S. fuscicollis occurs in one national park in Colombia and two national parks and two national reserves in Peru (Table 45). It is legally protected against hunting, capture, or export in Peru (Castro 1976) and against commercial exploitation in Brazil (Mittermeier and Coimbra-Filho 1977).

Suggestions that have been made for the conservation of this species include field surveys to determine possible levels of exploitation in northern Bolivia, the removal of only one or two juveniles per troop in trapping operations (Heltne et al. 1976), studies to determine the potential for long-term yields from Brazilian populations (Mittermeier, Coimbra-Filho, and Van Roosmalen 1978), captive breeding colonies with genetic management of stocks in home countries, and control of export quotas and conditions during shipment to minimize losses (Kleiman 1976).

Table 45. Distribution of *Saguinus fuscicollis* (countries are listed in counterclockwise order)

Country	Presence and Abundance	Reference
Colombia	Along R. Caquetá and R. Putumayo	Cabrera 1958
	In S, W of R. Caguán to E base of Cordillera Oriental (map)	Hershkovitz 1966a
	Throughout S Amazonas and Putumayo S of R. Caquetá-Caguán and E of Andean foothills, except questionable S of R. Putumayo (S triangle of Amazonas) (map)	Hershkovitz 1968
	In La Macarena N.P., S Meta	Green 1972
	In most of Caquetá, in S Meta S of R. Guayabero, in Vaupés as far E as San José del Guaviare, in Amazonas S of R. Yarí except extreme S, and most of Putumayo; questionable E of R. Yarí, N of R. Caquetá and N of R. Guaviare (map); probably not threatened	Hernández-Camacho and Cooper 1976
	Along R. Peneya, middle basin of R. Caquetá and R. Putumayo	Izawa 1976
	On island in R. Guineo, at El Pepino and Rumiyaco (between Mocoa and Puerto Asís), NW Putumayo, and near Valparaíso, Caquetá	Moynihan 1976a,b
	In S between R. Caguán-Caquetá and R. Putumayo (map)	Hershkovitz 1977
Ecuador	In E (maps)	Cabrera 1958; Hershkovitz 1966a, 1968
	Localities listed include R. Napo, R. Copataza, R. Coca and R. Napo confluence, R. Pastaza (1°28′S, 78°07′W), and Oriente near Aguarico (0°, 76°20′W)	Napier 1976
	In E from R. Napo W to R. Santiago and foothills of Cordillera Oriental (map)	Hershkovitz 1977
Peru	In N and NE in basin of R. Amazonas and R. Marañón, in E in basin of R. Ucayali and R. Huallaga, and in SE	Cabrera 1958
	Throughout E (map)	Hershkovitz 1966a
	Throughout E except questionable in S Loreto E of R. Ucayali (map)	Hershkovitz 1968
	In N in basins of R. Napo, R. Samiria, R. Yavarí; in E in basins of R. Ucayali (in Loreto, San Martín, and NE Huánuco), R. Marañón, R. Amazonas, R. Pacaya; in SE in Cuzco and Puno	Grimwood 1969
	At R. Llullapichis, branch of R. Pachitea (central)	Koepcke 1972
	Common in R. Nanay basin	Soini 1972

Table 45 (continued)

Country	Presence and Abundance	Reference
Peru (cont.)	Along R. Ucayali, R. Javarí, and R. Huallaga	Avila-Pires 1974
	Abundant in NE	Neville 1975
	In Manú N.P., in Madre de Dios and Cuzco; Tingo María N.P., Huánuco; and 2 national reserves of Pacaya and Samiria, Loreto; vulnerable	Castro 1976
	Common near Iquitos (R. Nanay), along R. Orosa, and R. Samiria (localities mapped in N and NE)	Freese et al. 1976
	Localities listed along upper and lower R. Ucayali, at Pebas, Contamana, and Iquitos, Loreto; and Andoas, Oriente; Chinchavita, R. Pachitea, and Tingo María, Huánuco; R. Perené, Junín; Yurac Yacu and R. Mayo, San Martín; and R. San Miguel, Cuzco	Napier 1976
	Along R. Samiria and R. Pacaya	Neville et al. 1976
	Throughout E in basin of R. Napo, from Amazonas and N San Martín through Huánuco and Pasco to R. Perené in N Junín, E to R. Ucayali and R. Javari S to Cerro Azul in Loreto; in SE in Madre de Dios, Puno, and Cuzco (map)	Hershkovitz 1977
	In fair numbers along R. Aucayo, R. Tahuayo, and R. Nanay, Loreto	Castro and Soini 1978
	In Madre de Dios near Iberia and Iñapari between R. Acre and R. Madre de Dios (in S near E border)	Encarnación and Castro 1978
	In N at R. Yavineto (R. Putumayo)	Encarnación et al. 1978
	Fairly abundant along R. Samiria; low abundance along R. Orosa and at Pucallpa; also near Iquitos and at Cocha Cashu, Manú N.P.	Freese et al. 1978
	Uncommon at Cocha Cashu, Manú N.P.	Janson and Terborgh, in press
Bolivia	In N	Cabrera 1958
	In NW as far S as 16°S and as far E as R. Beni (map)	Hershkovitz 1966a
	In NW as far S as 17°S, as far W as foothills of Andes, as far E as R. Mamoré (map)	Hershkovitz 1968
	Common at Cobija–R. Acre (NW corner); not common at Riberalta (in N) or at Ixiamas (in W at 13°46′S, 68°08′W) (localities listed and mapped)	Heltne et al. 1976
	In N as far S as 17°S in Pando, N La Paz, and Beni, as far E as R. Mamoré (map)	Hershkovitz 1977

Table 45 (continued)

Country	Presence and Abundance	Reference
Brazil	At Cruzeiro do Sul, Pedra Preta, and Seringal Oriente, W Acre	Carvalho 1957
	In W in basins of R. Solimões, R. Purus, R. Acre, and R. Juruá	Cabrera 1958
	In W, S of R. Solimões and W of R. Tarauacá, S of R. Purus and W of R. Madeira; questionable NW of R. Purus and between mouths of R. Purus and R. Madeira (map)	Hershkovitz 1966a
	In W, S of R. Japurá and R. Solimões, W of R. Pixuna and R. Purus; questionable W of mouth of R. Madeira, N of R. Purus, N of upper R. Tapauá, and W of R. Juruá (map)	Hershkovitz 1968
	N of R. Solimões to R. Içá; S of R. Solimões between R. Juruá, R. Purus, and R. Madeira; S of R. Pixuna into Bolivia; in W Acre on right bank of R. Juruá	Avila-Pires 1974
	Localities listed on upper R. Purus, R. Xapuri, R. Acre, Lago de Mapixi (5°43′S, 63°54′W), and R. Tonantins, Amazonas	Napier 1976
	In W, S of R. Japurá and R. Solimões, and W of R. Pixuna and R. Madeira; questionable between R. Juruá and R. Jutaí, N of upper R. Tapauá, W of mouth of R. Madeira, N and W of R. Purus, and W of R. Tarauacá (map)	Hershkovitz 1977
	Two races between R. Javari and R. Jutaí (*S. f. fuscicollis* and *S. f. melanoleucus*) common; no data about other races	Mittermeier, Bailey, and Coimbra-Filho 1978; Mittermeier, Coimbra-Filho, and Van Roosmalen 1978
	All races common except 4 of unknown status: in W Acre (*S. f. acrensis*), in central Acre (*S. f. cruzlimai*), between R. Juruá and R. Pixuna (*S. f. avilapiresi*), and between R. Pixuna and R. Madeira (*S. f. crandalli*)	Mittermeier, Coimbra-Filho, and Van Roosmalen, in press

Table 46. Population parameters of *Saguinus fuscicollis*

Population Density (individuals/km²)	Group Size X̄	Range	No. of Groups	Home Range Area (ha)	Study Site	Reference
		5–20			S Colombia	Hernández-Camacho and Cooper 1976
		≤40	5	100	S Colombia	Izawa 1976
	4.25	2–7	20		S Colombia	Moynihan 1976a
12	4				NE Peru	Neville 1976b
		2–10	25		NE Peru	Castro and Soini 1978
13.6	6	2–8	20		SE Peru	Encarnación and Castro 1978
8.4–30	6	2–9			Peru	Freese et al. 1978
10–18	7	3–13	7	60–70	SE Peru	Janson and Terborgh, in press
2.4–29.4	6	2–9	53		Peru and Bolivia	Freese et al., in prep.
3.6–27.6	6		19		Bolivia	Heltne et al. 1976
		8	1		W Brazil	Carvalho 1957

Saguinus imperator Emperor Tamarin

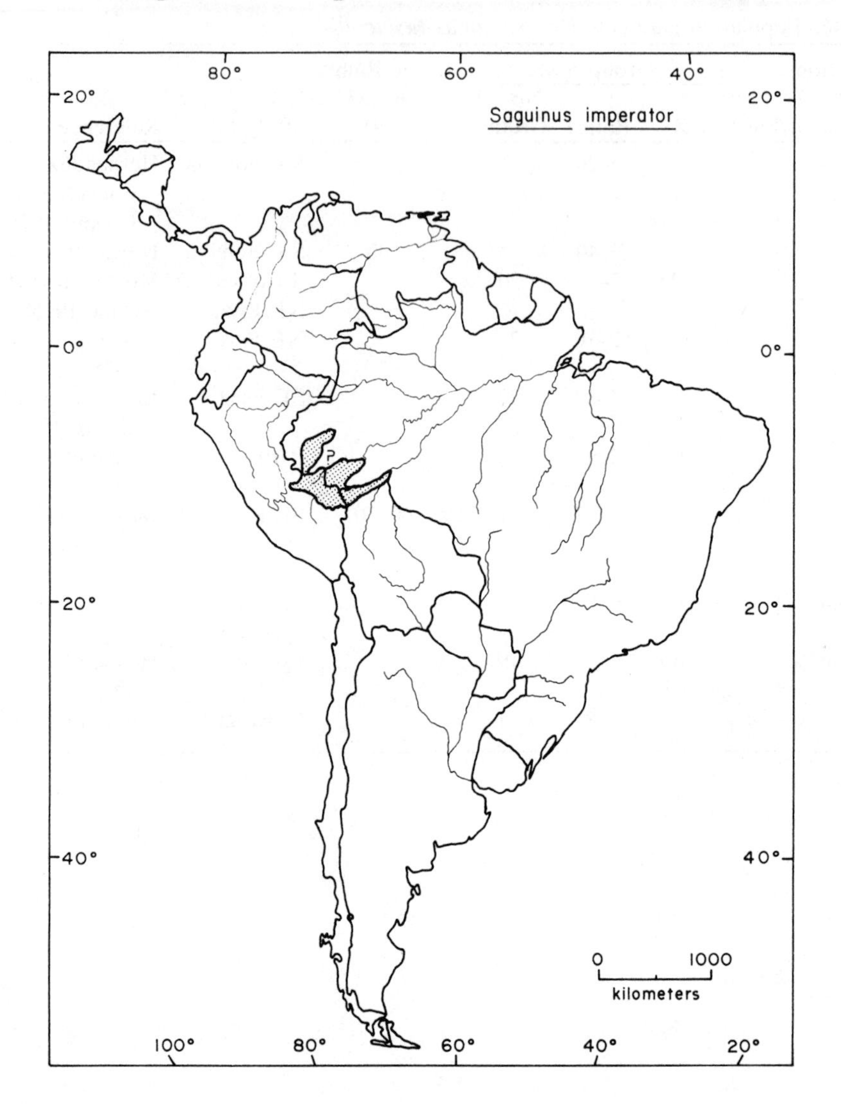

Taxonomic Note

Two subspecies have been described (Cabrera 1958), but the species is now considered to be monotypic (Hershkovitz 1968, 1977).

Geographic Range

Saguinus imperator is found in west central South America, in southeastern Peru, northwestern Bolivia, and southwestern Brazil. It occurs as far west as the Río Tambo, 10°42′S, 73°47′W, in northeastern Junín; as far south as 12°36′S, 70°06′W, Madre de Dios, Peru; as far north as about 6°30′S along the Rio Juruá, Amazonas, Brazil; and as far east as about 65°20′W at the confluence of the Río Abuna and Río

Madre de Dios at the northern tip of Bolivia (Hershkovitz 1977). For details of the distribution, see Table 47.

Abundance and Density

S. imperator is common in one area but uncommon at another site in Peru, is considered vulnerable in that country (Table 47), and is suspected of being scarce throughout its range (I.U.C.N. 1978). Its status in Bolivia and Brazil is not known.

Population density estimates vary from 2.02 to 12 individuals per km^2, group sizes of 2 to 40 with averages of 3 to 15 per group have been reported, and, in one area, home ranges were 30 to 40 ha in size (Table 48).

Habitat

S. imperator has been observed in the vine-tangled canopy of Amazonian lowland rain forest at 300 m elevation (Gardner, pers. comm. 1974; 1976), in high, noninundated forest with scarce underbrush (Encarnación and Castro 1978), and in evergreen, mesophytic, broadleaf forest in a seldom flooded, white-water drainage (Freese et al., in prep.).

Factors Affecting Populations

Habitat Alteration

No information.

Human Predation

Hunting. S. imperator is shot for food in the Río Curanja basin, where its teeth are also used as ornaments (Gardner, pers. comm. 1974). It is not hunted in the Cocha Cashu area (Freese et al., in prep.). No reports are available from Bolivia or Brazil.

Pest Control. No information.

Collection. Captured juvenile *S. imperator* are sometimes sold to animal traders in Peru (Gardner, pers. comm. 1974), and this species may be collected for the pet trade (Castro 1976). None were imported into the United States in 1968–73 (Appendix) or into the United Kingdom in 1965–75 (Burton 1978). Five were imported into the United States from Bolivia in 1977 (Bittner et al. 1978).

Conservation Action

S. imperator occurs in one national park in Peru (Table 47), and is legally protected against commercial exploitation in Brazil (Mittermeier and Coimbra-Filho 1977).

Export of this species from Peru should be restricted to institutions concerned with breeding them (Freese 1975), and a status survey is needed (I.U.C.N. 1978).

Table 47. Distribution of *Saguinus imperator*

Country	Presence and Abundance	Reference
Peru	In basin of upper R. Purús and tributaries	Cabrera 1958
	In SE Loreto as far W as 73°W and in most of Madre de Dios as far S as R. Madre de Dios (about 12°30′S) (map)	Hershkovitz 1968
	Along R. Curanja, SE Loreto, and in Manú N.P., Madre de Dios	Grimwood 1969
	Common along R. Curanja, SE Loreto; population stable 1966–71	Gardner, pers. comm. 1974
	In Manú N.P.; vulnerable	Castro 1976
	Common near Balta, R. Curanja	Gardner 1976
	Localities listed and mapped from R. Urubamba, in Loreto, and R. Juárez and R. Manú, in Madre de Dios	Hershkovitz 1977
	In E Madre de Dios between R. Acre and R. Madre de Dios near E border; decreasing near Iberia, R. Tahuamanu	Encarnación and Castro 1978
	At Cocha Cashu, Manú N.P.	Freese et al. 1978
	Uncommon at Cocha Cashu, Manú N.P.	Janson and Terborgh, in press
Bolivia	In NW between R. Abuna and R. Madre de Dios (map)	Hershkovitz 1968, 1977
Brazil	In W Acre at Pedra Preta and Seringal Oriente	Carvalho 1957
	In W along upper R. Juruá and upper R. Purus and tributaries	Cabrera 1958
	In SW Amazonas and in Acre, between R. Juruá and R. Tarauacá, and R. Purus and R. Acre (map)	Hershkovitz 1968
	Locality listed at R. Purus, Amazonas	Napier 1976
	Localities listed and mapped between R. Juruá and R. Tarauacá (questionable E of R. Tarauacá) and R. Purus and R. Acre	Hershkovitz 1977

Table 48. Population parameters of *Saguinus imperator*

Population Density (individuals/km²)	Group Size			Home Range Area (ha)	Study Site	Reference
	X̄	Range	No. of Groups			
		30–40			Peru	Grimwood 1969
	10–15				Peru	Gardner, pers. comm. 1974
2.02	8		4		Peru	Encarnación and Castro 1978
5.4	3	2–3	4		Peru	Freese et al. 1978
6–12	3	3–5	6	30–40	Peru	Janson and Terborgh, in press

Saguinus inustus Mottle-face Tamarin

Taxonomic Note

This is a monotypic species (Hershkovitz 1966a, 1977).

Geographic Range

Saguinus inustus is found in eastern Colombia and northwestern Brazil. Its northern limit is the Río Guaviare, as far north as 4°N at its confluence with the Río Atabapo and as far west as about 73°30′W (Hernández-Camacho and Cooper 1976). Its range extends south to almost 2°S along the Rio Japurá and east to 64°49′W along the Rio Negro (Hershkovitz 1977). For details of the distribution, see Table 49.

Abundance and Density

S. inustus has never been studied in Colombia but is believed not to be threatened and to live in small groups (Hernández-Camacho and Cooper 1976). No data are available regarding its status in Brazil, its population densities, average group size, or home range area.

Habitat

S. inustus inhabits rain forest in Colombia (Hernández-Camacho and Cooper 1976). No further details were located.

Factors Affecting Populations

No information.

Conservation Action

No information.

Table 49. Distribution of *Saguinus inustus*

Country	Presence and Abundance	Reference
Colombia	Throughout Vaupés and Guainía (map)	Hershkovitz 1966a
	Throughout Vaupés and Guainía (map) including Mitú area on both sides of R. Vaupés; as far N as San José del Guaviare and possibly farther NW along R. Ariari; as far S as R. Apaporis; probably not threatened	Hernández-Camacho and Cooper 1976
	Localities listed and mapped in Vaupés; from base of the Macarena Mts. probably as far E as R. Atabapo-Guainía at E border; between R. Guaviare and R. Apaporis (map)	Hershkovitz 1977
Brazil	Along lower R. Solimões between R. Negro and R. Japurá	Cabrera 1958
	In NW Amazonas N of R. Japurá and W of about 65°W, S of R. Negro (map)	Hershkovitz 1966a, 1969
	Localities listed and mapped at mouth of R. Papurí (0°36′N, 69°13′W), opposite Tahuapunta, R. Uaupés (0°36′N, 64°11′W) and Jauanari, R. Negro (0°32′S, 64°49′W)	Hershkovitz 1977
	Status unknown	Mittermeier, Bailey, and Coimbra-Filho 1978

Saguinus labiatus
Red-chested Moustached or Red-bellied Tamarin

Taxonomic Note

Two subspecies are recognized (Hershkovitz 1968, 1977).

Geographic Range

Saguinus labiatus is found in midwestern South America in western Brazil, northwestern Bolivia, and southeastern Peru. An isolated record exists from the Rio Tonantins, 2°48′S, 67°46′W, but otherwise the species occurs south of the Rio Solimões as far north as about 3°S, and as far east as about 59°W at the mouth of the Rio

Madeira (Hershkovitz 1977). The southernmost and easternmost sighting of *S. labiatus* is near Iberia, Peru (11°22′S, 69°35′W) (Encarnación and Castro 1978). There is a possibility that this species also occurs in southern Colombia (Hernández-Camacho and Cooper 1976). For details of the distribution, see Table 50.

Abundance and Density

S. labiatus is common in one area of northern Bolivia but its status in Brazil and Peru is not known (Table 50). Near Cobija, Bolivia, population density was 27.6 individuals per km^2, and 15 groups had 1 to 13 members ($\bar{X}$ = 6) (Heltne et al. 1976). In eastern Madre de Dios, Peru, population density was 3.78 individuals per km^2, and 5 groups had 3 to 6 members ($\bar{X}$ = 6) (Encarnación and Castro 1978). No estimates of home range area are available.

Habitat

In Bolivia, *S. labiatus* has been recorded in evergreen, mesophytic, broadleaf forest on high ground in a white-water drainage (Freese et al., in prep.). In Peru this species prefers high, nonflooding forest with scarce underbrush (Encarnación and Castro 1978). No reports were located regarding its habitats in Brazil.

Factors Affecting Populations

Habitat Alteration

No information.

Human Predation

Hunting. In some areas of southeastern Peru, *S. labiatus* is heavily hunted (Encarnación and Castro 1978), but in northern Bolivia it is seldom hunted (Freese et al., in prep.).

Pest Control. No information.

Collection. S. labiatus has recently been collected for export in fairly large numbers from Bolivia, and populations are abundant enough there to supply 250 to 500 per year without significant reduction (Heltne et al. 1976).

A total of 103 *S. labiatus* were imported into the United States in 1968–73 (Appendix), and 60 were imported (from Bolivia) in 1977 (Bittner et al. 1978). One-hundred thirty were imported into the United Kingdom in 1965–75 (Burton 1978).

Conservation Action

S. labiatus is not found in any reserved habitats or parks. It is legally protected from commercial exploitation in Brazil (Mittermeier and Coimbra-Filho 1977).

Recently, 270 *S. labiatus* were trapped for export from Bolivia, but the government rescinded the export permit and the animals were released onto a peninsula on the Río Yapacani in central Bolivia, far outside their natural range. These tamarins should be carefully monitored to determine their breeding rate and possible escape into adjacent areas, and any trapping in Bolivia should be professionally supervised and preceded by field surveys to determine potential yields (Heltne et al. 1976).

Table 50. Distribution of *Saguinus labiatus*

Country	Presence and Abundance	Reference
Brazil	In basin of R. Solimões and upper R. Purus and their affluents	Cabrera 1958
	Localities listed include Lago de Mapixi (E of R. Purus), Humaitá (R. Madeira), Bom Lugaz (R. Purus), Paraná do Jacaré (R. Solimões), and R. Tonantins	Napier 1976
	Localities listed and mapped between R. Purus and R. Madeira as far SW as 10°11′S, 67°33′W between R. Iriquiri and R. Acre; 1 record on R. Tonantins between R. Içá and R. Japurá	Hershkovitz 1977
	Status unknown	Mittermeier, Coimbra-Filho, and Van Roosmalen 1978
Bolivia	Common near Cobija in R. Acre area (NW corner of Pando) (map)	Heltne et al. 1976
Peru	In basin of R. Amazonas	Cabrera 1958
	Near Iberia, Miraflores, Yaverija, Primavera, Iñapari, and Noaya, all between R. Acre and R. Tahuamanu in E Madre de Dios near E border	Encarnación and Castro 1978

Saguinus leucopus

White-footed or Silvery Brown Bare-face Tamarin

Taxonomic Note

This is a monotypic species (Hershkovitz 1977).

Geographic Range

Saguinus leucopus is found in northern Colombia between the Río Cauca and the Río Magdalena, in southern Bolívar, northeastern and eastern Antioquia, northeastern Caldas, and northern Tolima (Hernández-Camacho and Cooper 1976). It occurs as far west as about 76°W, as far east as about 74°W, as far north as about 9°N near

Mompós, and south at least to Mariquita (5°11′N, 74°54′W) (Hernández-Camacho and Cooper 1976). For details of distribution, see Table 51.

Abundance and Density

No opinions are available regarding the status of this species. Population densities of 2.5 to 15.2 individuals per km^2 were observed at one study site in Bolívar (Green 1978b). Group sizes of 2 to 6 ($\bar{X}$ = 4.3, N = 3) (Bernstein, pers. comm. 1974), 3 to 12 or more (Hernández-Camacho and Cooper 1976), and 10 to 15 (Hershkovitz 1977) have been reported.

Habitat

S. leucopus is a forest dweller, most frequently seen at forest fringes, near streams, or in secondary vegetation (Hernández-Camacho and Cooper 1976). It utilizes remnant primary forest and recent secondary growth (Bernstein et al. 1976a), young (5–15 years old), and mature (more than 15 years old) secondary forest in the humid tropical forest zone (Green 1978b), and in one area at least, spends a high proportion of its time in primary forest (Neyman, pers. comm. 1978). It is found from near sea level to about 1,500 m elevation (Hershkovitz 1977).

Factors Affecting Populations

Habitat Alteration

The habitat of *S. leucopus* has been greatly reduced owing to the clearing of forests, especially during the last 20 years (Hernández-Camacho and Cooper 1976). Although this species is able to thrive in secondary forest and the small forest remnants left after agricultural clearing, widespread deforestation within its range has fragmented and destroyed much of the habitat once available to it (Bernstein et al. 1976a, Green 1978b).

Human Predation

Hunting. S. leucopus is not hunted for food (Hershkovitz 1977).

Pest Control. S. leucopus is reported to raid crops and may sometimes be killed as an agricultural pest (Green 1976).

Collection. S. leucopus has been collected for export. Thirty-three were imported to the United States in 1968, but none entered the United States in 1969–73 (Appendix) or in 1977 (Bittner et al. 1978). Two were imported into the United Kingdom in 1965–75 (Burton 1978).

Populations of *S. leucopus* in central Bolívar are probably not abundant enough to withstand even modest trapping pressure (Bernstein et al. 1976a).

Conservation Action

S. leucopus was not reported as occurring in any parks or reserves. It is listed in Appendix I of the Convention on International Trade in Endangered Species of Wild Fauna and Flora, and as threatened under the U.S. Endangered Species Act (U.S. Department of Interior 1977a,b).

Table 51. Distribution of *Saguinus leucopus*

Country	Presence and Abundance	Reference
Colombia	In N from confluence of R. Cauca and R. Magdalena at 9°N through S Bolívar to 6°N in E and W Antioquia (map)	Hershkovitz 1949
	In N, S of confluence of R. Cauca and R. Magdalena	Cabrera 1958
	In N (maps)	Hershkovitz 1966a, 1969
	Near Ventura, Bolívar (8°22′N, 74°22′W)	Bernstein et al. 1976a
	Near Puerto Rico, Tequisio region, Bolívar (8°34′N, 74°14′W)	Green 1976, 1978b
	E of R. Cauca as far S as about 7°N; W of R. Magdalena as far S as 5°11′N, questionable farther S (map); N between rivers to Mompós; on all of larger river islands and foothills of central Andes	Hernández-Camacho and Cooper 1976
	Localities listed include Medellín, Antioquia, and Valdivia, R. Cauca	Napier 1976
	In N between R. Cauca and R. Magdalena from their confluence S almost to Medellín, and S in E into W Caldas to about 5°30′N (map); localities listed in Antioquia and Bolívar	Hershkovitz 1977

Saguinus midas Midas or Black Tamarin

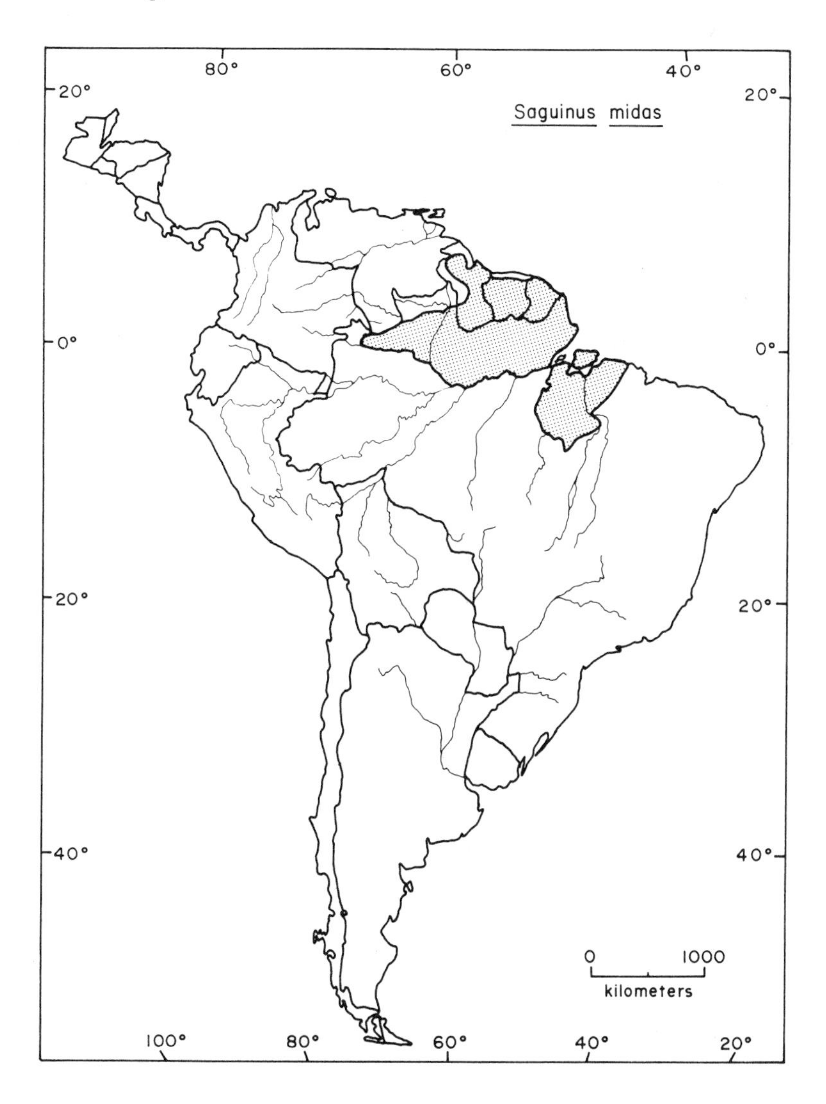

Taxonomic Note

Two subspecies are recognized (Hershkovitz 1969, 1977), including the negro tamarin (*S. m. niger*), treated as a distinct species (*S. tamarin*) by older authors (Avila-Pires 1958, 1969a; Cabrera 1958; Napier and Napier 1967).

Geographic Range

Saguinus midas is found in northeastern South America in the Guianas and northern Brazil. It occurs as far north as about 7°N, 60°W in Guyana, and in Brazil is found as far east as about 46°W (at 1°S) at the mouth of the Rio Gurupi, as far south as about 8°S, 51°W along the Rio Fresco (an east bank tributary of the Rio Xingu), and as far west as about 67°W at the equator along the Rio Negro (Hershkovitz 1977). For details of the distribution, see Table 52.

Abundance and Density

S. midas is common in Surinam, French Guyana, and Brazil, but is considered vulnerable in Guyana and possibly also in eastern Pará, Brazil, south of the Rio Amazonas (Table 52).

In Guyana, population densities varied from 2.1 to 13.9 individuals per km^2 (Muckenhirn et al. 1975); at one site in Surinam, density was 23.5 per km^2 (Mittermeier 1977a); and in Amapá, Brazil, density was 15 to 32 per km^2 (Thorington 1968b). Reported group sizes for *S. midas* include: 5 to 20 (Husson 1957); 2 to 6 ($\bar{X}$ = 3.77, N = 10) (Thorington 1968b); 2 to 6 ($\bar{X}$ = 4.8 to 5.2, N = 5) (Muckenhirn et al. 1975); 4 to 10 (Hershkovitz 1977); and 2 to 12 ($\bar{X}$ = 5.87, N = 13), one with a home range of about 9 ha (Mittermeier 1977a).

Habitat

S. midas is a forest species which can utilize a wide variety of forest habitats. For example, in Surinam it is found in high and low rain forest, ridge forest, lowland and mountain savanna forest, liane forest, riverbank marsh forest, creek forest, and swamp forest, and does well in edge habitats and secondary vegetation (Mittermeier 1977a). In Guyana this species thrives in "mixed habitats" (Muckenhirn and Eisenberg 1978).

In Amapá, Brazil, *S. midas* occupies tall forest, low secondary growth, and *Cecropia* trees, and probably survives best in areas that include secondary or edge vegetation for feeding and deep forest for shade (Thorington 1968b). It has been observed in primary and secondary *terra firme* (nonflooding) forest in the Amazonian region (Mittermeier and Coimbra-Filho 1977), and has been collected from 5 to 50 m elevation (Napier 1976).

Factors Affecting Populations

Habitat Alteration

S. midas is an adaptable species which benefits from certain kinds of limited habitat alteration. For instance, in Guyana it seems to increase initially in areas of expanding agriculture (Muckenhirn and Eisenberg 1978), and in Surinam is more abundant in disturbed areas than in unbroken high forest (Mittermeier 1977a). In several parts of its range limited disturbance, such as that produced by slash and burn agriculture, is increasing the habitat available to *S. midas* (Mittermeier, Coimbra-Filho, and Van Roosmalen 1978). But in eastern Pará, Brazil, widespread deforestation for agriculture and cattle ranching is decreasing its habitat (Ayres 1978).

Human Predation

Hunting. *S. midas* is occasionally hunted for its meat, but is not a high priority target (Mittermeier, Bailey, and Coimbra-Filho 1978).

Pest Control. No information.

Collection. This species is sometimes kept as a pet in Surinam (Husson 1957,

Mittermeier 1977a) and in Brazil (Ayres 1978). It has also been collected for export, and populations are considered abundant enough to withstand cropping of (1) moderate numbers from agricultural areas in Guyana (Muckenhirn et al. 1975), and (2) probably several hundred per year in Surinam and Brazil (Mittermeier, Coimbra-Filho, and Van Roosmalen 1978). But in these latter countries, "if commercial exploitation continues to be prohibited, it (*S. midas*) should remain abundant for a long time to come" (Mittermeier, Bailey, and Coimbra-Filho 1978).

Fifty-four *S. midas* were imported into the United States in 1968–73 (Appendix), and none entered the United States in 1977 (Bittner et al. 1978). A total of 216 were imported into the United Kingdom between 1965 and 1975 (Burton 1978).

Conservation Action

S. midas is found in one forest reserve in Guyana and one nature park and five nature reserves in Surinam (Table 52). It is legally protected from commercial exploitation in Brazil (Mittermeier and Coimbra-Filho 1977), and from hunting in northern Surinam and transport in the rest of that country (Mittermeier 1977a). Before any large-scale trapping programs are initiated, population surveys should be done to determine long-term yield potentials (Mittermeier, Coimbra-Filho, and Van Roosmalen 1978).

Table 52. Distribution of *Saguinus midas* (countries are listed in clockwise order)

Country	Presence and Abundance	Reference
Guyana	At Lua and R. Mararuni	Avila-Pires 1964
	Throughout except in N along coast and in W central (map)	Hershkovitz 1969
	Localities listed and mapped in E at Berbice, Moraballi Forest Reserve, and Apoteri; only E of R. Essequibo and S of R. Rupununi	Muckenhirn et al. 1975
	Localities listed at Bonasica, R. Demerara, R. Essequibo, R. Supinaam, coast region, and Moon Mts. in S	Napier 1976
	Localities listed and mapped in N central; not in NW and W central	Hershkovitz 1977
	Restricted distribution; somewhat vulnerable	Muckenhirn and Eisenberg 1978
Surinam	Common; in NE, N of Moengo Tapoe	Husson 1957
	In NE, S of Affobakka Dam	Walsh and Gannon 1967
	Throughout except along coast in N (map)	Hershkovitz 1969
	Abundant in E along R. Maroni	Durham, pers. comm. 1974
	Localities listed at Paramaribo, R. Pará, and Zanderij	Napier 1976
	Localities listed and mapped in NE and central; distribution disjunct, absent between lower R. Saramacca and R. Coesewijne	Hershkovitz 1977

Table 52 (continued)

Country	Presence and Abundance	Reference
Surinam (cont.)	Localities listed and mapped in N and in S central; common in interior but absent in young coastal plain; not in danger; in Brownsberg Nature Park and nature reserves of Raleighvallen-Voltzberg, Tafelberg, Eilerts de Haan Gebergte, and Sipaliwini Savanna	Mittermeier 1977a
	Very common	Mittermeier, Bailey, and Coimbra-Filho 1978
	Also in Brinckheuvel Nature Reserve	Mittermeier, Coimbra-Filho, and Van Roosmalen, in press
French Guiana	Throughout (map)	Hershkovitz 1969
	Abundant in W along R. Maroni	Durham, pers. comm. 1974
	At Ipousin, R. Approuague (in E)	Napier 1976
	Localities listed and mapped in N and NE	Hershkovitz 1977
	Very common; near Saül in interior S of Cayenne	Mittermeier, Bailey, and Coimbra-Filho 1978
Brazil	Locality mapped at Nova Timboteua, E Pará	Kumm and Laemmert 1950
	Common near Belém, Pará	Avila-Pires 1958
	N of R. Amazonas from Manaus near R. Negro to delta	Avila-Pires 1964
	In Amapá, W of Porto Platon	Thorington 1968b
	Abundant	Avila-Pires 1969a
	In Utinga reserved forest near Belém, and S of Belém, E Pará	Carvalho and Toccheton 1969
	In Pará	Deane et al. 1969
	N of R. Amazonas throughout Amapá, N Pará, S and E Roraima, and N Amazonas as far W as R. Negro; S of R. Amazonas in E Pará between R. Xingu and R. Gurupi (map)	Hershkovitz 1969
	Near Belém, E Pará	Pine 1973
	N of R. Amazonas, E of R. Negro, and S of R. Amazonas, between R. Tocantins and R. Xingu, including islands S of Marajó	Avila-Pires 1974
	In Amapá near Macapá, Pôrto Grande, and R. Amapari	Thorington, pers. comm. 1974
	Localities listed at Óbidos, Igarapé Assu (Açu), R. Tocantins, and R. Xingu, Pará	Napier 1976
	Localities listed and mapped in Amapá, N Pará, S Roraima, N Amazonas E of R. Negro, and E Pará between R. Xingu and R. Gurupi, including the island of Marajó	Hershkovitz 1977

Table 52 (continued)

Country	Presence and Abundance	Reference
Brazil (cont.)	Not in danger; frequent in NE; near Belém	Mittermeier and Coimbra-Filho 1977
	Frequent near Belém	Ayres 1978
	Common	Mittermeier, Bailey, and Coimbra-Filho 1978
	S race (*S. m. niger*) may be vulnerable	Mittermeier, Coimbra-Filho, and Van Roosmalen, in press

Saguinus mystax Black-chested Moustached Tamarin

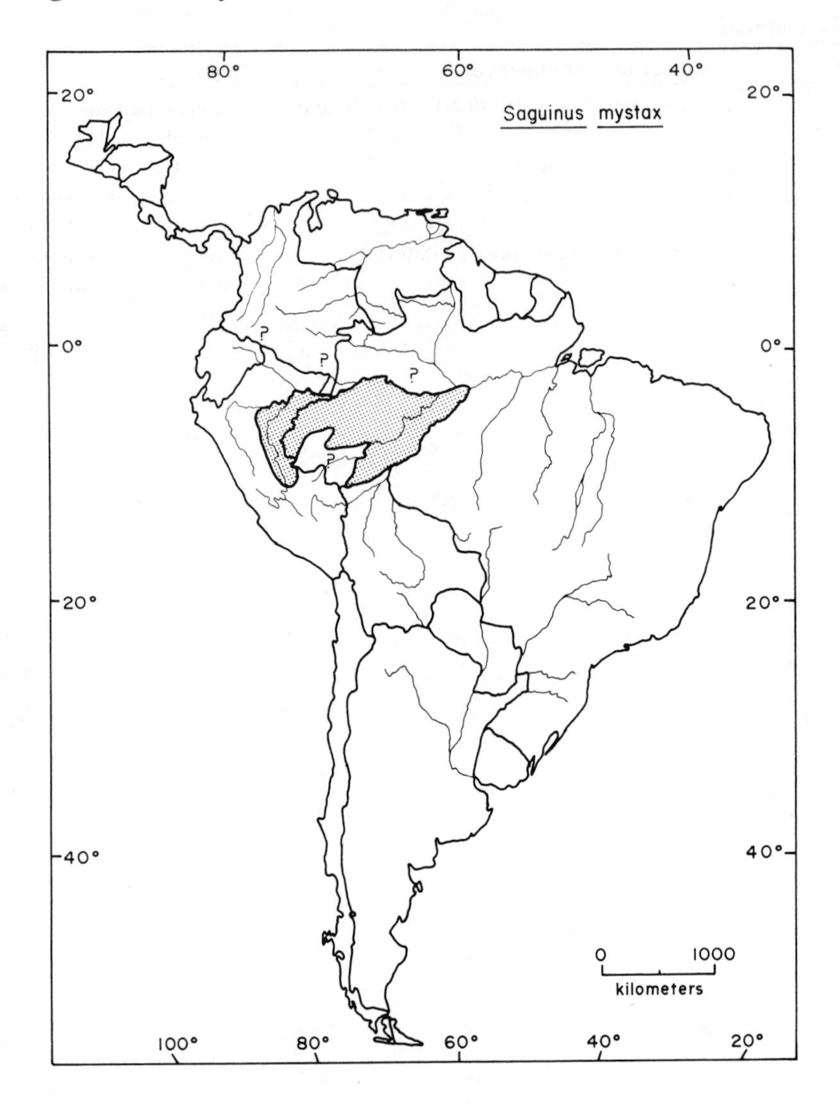

Taxonomic Note

Three subspecies are recognized (Hershkovitz 1968, 1977).

Geographic Range

Saguinus mystax is found in northeastern Peru and northwestern Brazil. Its northern limit is the Rio Solimões as far north as about 2°30′S, 66°W, and as far east as the mouth of the Rio Madeira at about 59°W (Hershkovitz 1977). It occurs as far west as about 76°W along the Río Huallaga, and as far south as about 11°S in eastern Junín, Peru and eastern Acre, Brazil (Hershkovitz 1977). This species may also occur in southern Colombia, but documentation is lacking. For details of the distribution, see Table 53.

Abundance and Density

S. mystax is considered vulnerable in Peru and common west of the Rio Juruá in Brazil; its status east of the Rio Juruá is not known (Table 53).

Groups of 2 to 6 have been observed in Loreto, Peru (Castro and Soini 1978). Otherwise no information is available regarding group size, population density, or home range area of this species.

Habitat

In Loreto, Peru, *S. mystax* is usually found in fairly undisturbed primary forest but sometimes also occupies old secondary forest (Castro and Soini 1978). In Brazil it has been seen in primary *várzea* forest (periodically flooded by a white-water river) (Mittermeier and Coimbra-Filho 1977). It has been recorded as high as 610 m elevation (Napier 1976).

Factors Affecting Populations

Habitat Alteration

S. mystax habitats are suffering from an accelerated rate of deforestation (Kleiman 1976). No further details are available.

Human Predation

Hunting. S. mystax is sometimes hunted for food in Peru (Castro 1976, Castro and Soini 1978).

Pest Control. These tamarins do feed in fruit orchards, especially when the domesticated "uvilla" tree (*Pourouma* sp.) is in fruit (Castro and Soini 1978). No efforts at control were mentioned.

Collection. S. mystax has been collected for the pet trade and for biomedical research (Castro 1976), and has been exported from Peru (Soini 1972, Freese 1975), Colombia (Cooper, pers. comm. 1974), and Brazil, where its populations could probably withstand cropping of several hundred per year (Mittermeier, Coimbra-Filho, and Van Roosmalen 1978). It is difficult to ship and suffers high mortality rates during capture, transport, and holding (Mittermeier et al. 1976). A total of 24,077 *S. mystax* were exported from Peru between 1964 and 1974 (Castro 1976), and 445 left that country in 1976–77 (Castro 1978).

In 1968–73, 4,950 *S. mystax* were imported into the United States (Appendix), primarily from Colombia and Peru. In 1977, 90 entered the United States from Peru (Bittner et al. 1978). Between 1965 and 1975, 30 *S. mystax* were imported into the United Kingdom (Burton 1978). This species is used in medical research in the United States (Thorington 1972, Comm. Cons. Nonhuman Prim. 1975a).

Conservation Action

S. mystax is found within two national reserves in Peru (Table 53), and is protected from hunting, capture, and export in Peru (Castro 1976), and commercial exploitation in Brazil (Mittermeier and Coimbra-Filho 1977). Suggestions for its conservation include the control of export quotas and conditions, the development of breeding centers (with management of genetic stocks) in the countries of origin (Kleiman 1976), and population studies to determine long-term yield potentials before cropping programs are initiated (Mittermeier, Coimbra-Filho, and Van Roosmalen 1978).

Table 53. Distribution of *Saguinus mystax*

Country	Presence and Abundance	Reference
Peru	In Loreto in basins of R. Amazonas and affluents, R. Napo, R. Ucayali, and upper R. Purús	Cabrera 1958
	In NE, S of R. Amazonas and R. Marañón, E of R. Huallaga (map)	Hershkovitz 1968
	In NE near Pebas and in basins of R. Ucayali and R. Napo	Grimwood 1969
	In national reserves of Pacaya and Samiria, Loreto; vulnerable	Castro 1976
	Localities listed at R. Pachitea, Huánuco, and Contamana (7°19′S, 75°04′W), Loreto	Napier 1976
	Localities listed and mapped throughout E, S of R. Amazonas-Marañón, along base of Andes to junction of R. Urubamba and R. Ucayali	Hershkovitz 1977
	In Loreto in basins of R. Aucayo and R. Tahuayo	Castro and Soini 1978
Brazil	Locality mapped at Benjamin Constant, S of R. Solimões in W Amazonas	Kumm and Laemmert 1950
	At Seringal Oriente, W Acre	Carvalho 1957
	S of R. Amazonas between R. Purus and R. Juruá and W of R. Juruá	Cabrera 1958
	S of R. Solimões and W of R. Madeira-Abuna throughout SW Amazonas and Acre, except absent between upper R. Juruá and R. Tarauacá and between upper R. Purus and R. Acre (map)	Hershkovitz 1968
	From S of R. Içá to lower R. Purus and into Acre	Avila-Pires 1974
	Localities listed in Amazonas on R. Juruá between 68°–71°W, at Tefé, on lower R. Purus at Lago de Ayapuá, at Fonte Boa, and on the R. Juruá at 6°36′S	Napier 1976
	Localities listed and mapped throughout SW Amazonas W of R. Madeira, except between lower R. Juruá and R. Acre; in W Acre and E Acre between R. Abuna	Hershkovitz 1977

Table 53 (continued)

Country	Presence and Abundance	Reference
Brazil (cont.)	and R. Acre; questionable N of R. Solimões and in central Acre between R. Tarauacá and lower R. Purus	Hershkovitz 1977 (cont.)
	Westernmost race (*S. m. mystax*) common; status of populations E of R. Juruá not known	Mittermeier and Coimbra-Filho 1977; Mittermeier, Bailey, and Coimbra-Filho 1978; Mittermeier, Coimbra-Filho, and Van Roosmalen 1978
Colombia	In Puerto Umbría region, Putumayo (R. Guineo area)	Whittemore 1972
	No documented records	Hernández-Camacho and Cooper 1976
	Questionable in S Amazonas (map)	Hershkovitz 1977

Saguinus nigricollis Black-mantle or Red and Black Tamarin

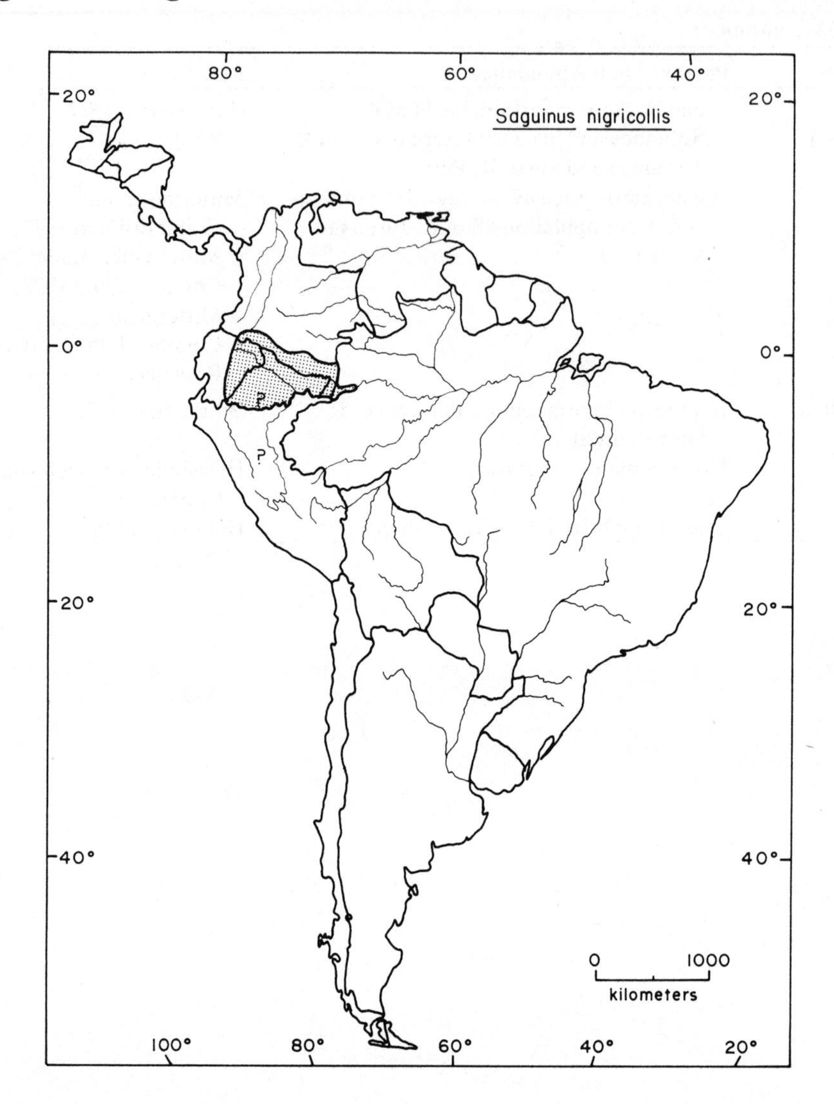

Taxonomic Note

Two subspecies are recognized (Hershkovitz 1977), including the Río Napo tamarin regarded as a separate species (*S. graellsi*) by some authors (Cabrera 1958, Hershkovitz 1966a, Napier and Napier 1967, Hernández-Camacho and Cooper 1976).

Geographic Range

Saguinus nigricollis is found in southern Colombia, eastern Ecuador, northern Peru, and a small part of western Brazil. The northern limit of its range is the Río Caquetá, as far north as about 1°N, 77°W (Hernández-Camacho and Cooper 1976).

The species occurs as far west as about 78°20′W in Ecuador and as far east as about 68°W between the Rio Içá and the Rio Solimões in Brazil (Hershkovitz 1977). In Peru the range may extend as far south as 5°S, 75°W along the Río Marañón (Hershkovitz 1977) or even farther south along the Río Huallaga, Río Ucayali, and Río Pachitea (Grimwood 1969). For details of the distribution, see Table 54.

Abundance and Density

S. nigricollis is abundant near Leticia, Colombia, along the Río Ampiyacu in northeastern Peru, and in its small range in Brazil, and is rare along the Río Nanay in Peru (Table 54). Its status throughout most of Colombia, Peru, and Ecuador is not known.

Population densities of 19.2 individuals per km^2 (Freese et al. 1978) and 10 to 13 per km^2 (Izawa 1978) have been reported. Group size in this species varies from an average of 5 to 10 (Hernández-Camacho and Cooper 1976), 4 to 12 ($\bar{X}$ = 6.2, N = 10) (Moynihan 1976a), 2 to 7 ($\bar{X}$ = 6, N = 7) (Freese et al., in prep.), and 4 to 8 ($\bar{X}$ = 6.3, N = 10), with home ranges of 30 to 50 ha (Izawa 1978).

Habitat

S. nigricollis is a forest species which utilizes both primary and secondary forest. In Colombia it is found in undisturbed rain forest and near human habitations, and seems to be adaptable (Hernández-Camacho and Cooper 1976). In Peru it has been observed in riverbank and hillside forest (Encarnación et al. 1978), and in evergreen, broadleaf, mesophytic forest in a black and white-water drainage (Freese et al., in prep.).

S. nigricollis appears to thrive in second growth and edge habitats (Thorington 1975), and is as much at home in thick secondary vegetation as in high virgin or reestablished forest (Hershkovitz 1977). It has been recorded as high as 914 m elevation (Napier 1976).

Factors Affecting Populations

Habitat Alteration

S. nigricollis is able to maintain near natural numbers in areas of forest disturbance (Freese et al. 1976). Because of its ability to utilize secondary growth, it may be increasing in abundance owing to the activities of man (Thorington 1975).

Human Predation

Hunting. Although *S. nigricollis* is hunted in Peru (Castro 1976), this pressure is not severe (Freese et al., in prep.).

Pest Control. No information.

Collection. *S. nigricollis* has been collected for export from Peru and Colombia for biomedical research and the pet trade (Soini 1972, Hershkovitz 1966a, Grimwood 1969, Hernández-Camacho and Cooper 1976). It has also been exported frequently from Brazil, where its populations may be abundant enough to withstand cropping of several hundred per year (Mittermeier, Coimbra-Filho, and Van Roosmalen 1978).

In 1968–73 a total of 10,274 *S. nigricollis* were imported into the United States

(Appendix), primarily from Colombia and Peru. In 1977, 50 entered the United States from Bolivia (Bittner et al. 1978), where this species does not occur. Between 1965 and 1975, 473 *S. nigricollis* were imported into the United Kingdom (Burton 1978). *S. nigricollis* is used in biomedical research in the United States (Comm. Cons. Nonhuman Prim. 1975a).

Conservation Action

S. nigricollis does not occur within any habitat reserves or parks (Table 54). It is legally protected from hunting, capture, and export in Peru (Castro 1976), and from commercial exploitation in Brazil (Mittermeier and Coimbra-Filho 1977). Before any cropping programs are initiated, populations should be studied to determine their potentials for long-term yield (Mittermeier, Coimbra-Filho, and Van Roosmalen 1978).

Table 54. Distribution of *Saguinus nigricollis*

Country	Presence and Abundance	Reference
Colombia	In Leticia panhandle S of R. Putumayo (map)	Hershkovitz 1966a, Thorington 1975
	S of R. Caquetá throughout Putumayo and S two-thirds of Amazonas (map); frequent near Leticia	Hernández-Camacho and Cooper 1976
	Near Santa Rosa, Putumayo	Moynihan 1976a
	S of R. Putumayo in SW Putumayo (map)	Hershkovitz 1977
	In basin of R. Peneya, near R. Caquetá (0°2′N, 74°05′W)	Izawa 1978
Ecuador	In E	Cabrera 1958
	In E as far N as R. Aguarico and W to E slopes of Cordillera Oriental (map)	Hershkovitz 1966a
	Localities listed include R. Pastaza, R. Napo, R. Suno, and Oriente	Napier 1976
	Localities listed and mapped throughout E	Hershkovitz 1977
Peru	In basins of R. Napo and R. Ucayali (in N and NE)	Cabrera 1958
	In N, N of R. Marañón and R. Amazonas, E of R. Santiago, S of R. Napo in W and R. Putumayo in E (map)	Hershkovitz 1966a
	In basins of R. Napo, R. Pachitea, R. Huallaga, and at Yarina Cocha, Ganzo Azul, and Pucallpa on R. Ucayali; probably in reserves of R. Pacaya and R. Samiria	Grimwood 1969
	Rare in R. Nanay basin	Soini 1972
	Not in any protected areas; vulnerable	Castro 1976
	Frequent along R. Ampiyacu (in NE at 3°10′S, 71°50′W)	Freese et al. 1976, 1978

Table 54 (continued)

Country	Presence and Abundance	Reference
Peru (cont.)	In Loreto at Tarapoto (2°10′S, 74°W), Pebas, and R. Napo	Napier 1976
	In N Loreto as far E as R. Ucayali; questionable NE of R. Napo W of its mouth and S of 3°S; in NE corner between R. Amazonas and R. Putumayo E of mouth of R. Napo (map)	Hershkovitz 1977
	On R. Yavineto (R. Putumayo), N border	Encarnación et al. 1978
Brazil	Between R. Solimões and R. Içá in W Amazonas (map)	Hershkovitz 1966a
	At Santarita, upper Solimões, Amazonas	Napier 1976
	Localities listed and mapped at Santa Rita and São Paulo de Olivença between R. Içá and R. Solimões	Hershkovitz 1977
	Common; abundant	Mittermeier, Coimbra-Filho, and Van Roosmalen 1978

Saguinus oedipus Crested Tamarin

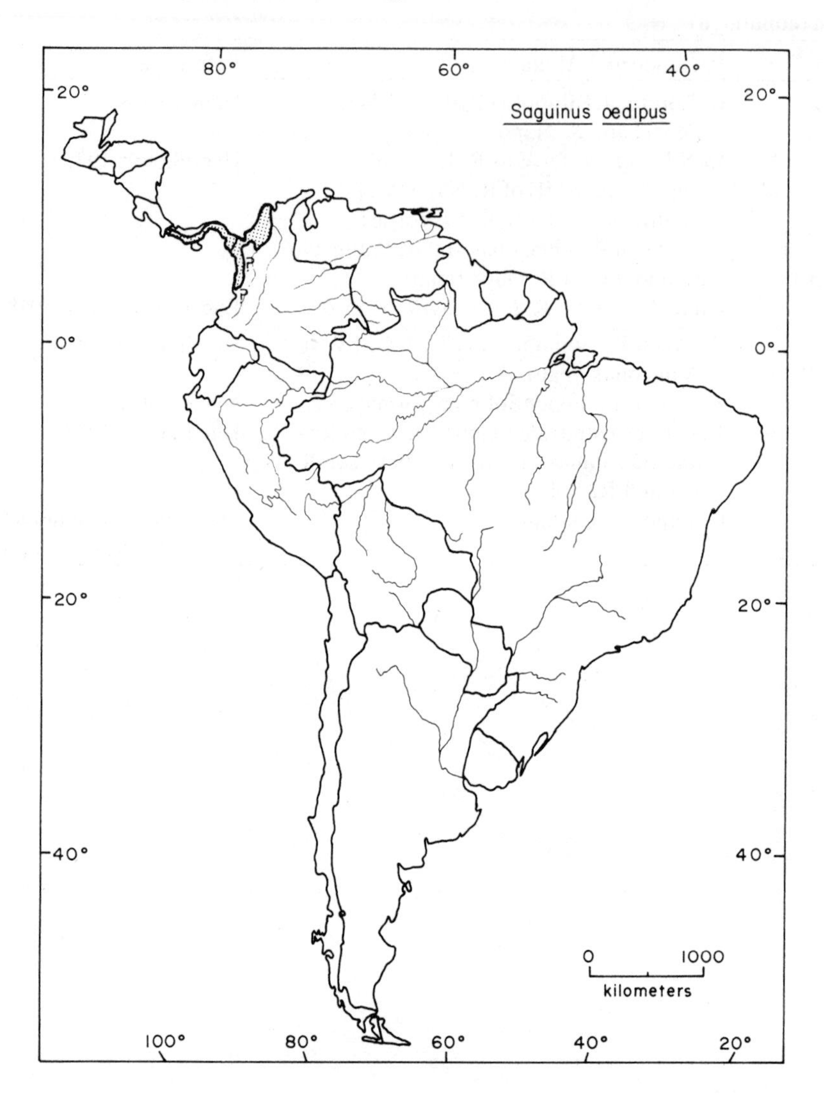

Taxonomic Note

Two subspecies are recognized (Hershkovitz 1977). The westernmost form (*geoffroyi*) is treated as a separate species by many authors (Hall and Kelson 1959; Moynihan 1970, 1976a, Hernández-Camacho and Cooper 1976)

Geographic Range

Saguinus oedipus is found in Central America and northwestern South America, from southeastern Costa Rica to northern Colombia. The western limits of its range is at about 83°10′W, just west of the Costa Rican border (Hall and Kelson 1959, Hershkovitz 1977), and the eastern limit is at about 74°30′W along the Río Cauca,

Bolívar, Colombia (Hernández-Camacho and Cooper 1976). *S. oedipus* occurs as far north as Barranquilla (11°10′N, 74°50′W) and as far south as 4°20′S, and perhaps farther south, along the Pacific coast of Colombia (Hernández-Camacho and Cooper 1976, Hershkovitz 1977). For details of the distribution, see Table 55.

Abundance and Density

S. oedipus is locally abundant and considered common in Panama, and the eastern Colombian form (*S. o. oedipus*) is endangered (Table 55). The status of populations in Costa Rica and western Colombia is not known.

Population densities of 20 to 30 per km^2 in the Canal Zone, Panama (Dawson 1978), and 30 to 180 per km^2 in northwestern Colombia (Neyman 1978) have been observed. Group size in *S. oedipus* varies from 2 to 9 ($\bar{X}$ = 3.9, N = 23) (Moynihan 1970), 3 to 12 or more (Hernández-Camacho and Cooper 1976), 2 to 6 ($\bar{X}$ = 4) (Hershkovitz 1977), 2 to 19 ($\bar{X}$ = 6.39±0.31, N = 71), with home ranges of 26 to 32 ha (Dawson 1978), and 3 to 13, with home ranges of 7.8 to 10 ha (Neyman 1978).

Habitat

S. oedipus is a forest species most frequently found along forest edges and in secondary growth. In Panama it utilizes scrub forest and forest edges (Handley 1966), is most common in low bushy secondary vegetation, tangles, and vines, and does not flourish in high rain forest or mature, humid seasonal forest with a sparse understory (Moynihan 1967, 1970, 1976a). Also in Panama, *S. oedipus* has been recorded in the secondary growth of former agricultural plots (Telford et al. 1972), and in second-growth dry tropical forest (Fleming 1973, Dawson 1978). Ideal habitat for it there appears to be low brush and vine-covered edge vegetation interspersed with taller, nondeciduous trees (Dawson 1976).

On the Pacific coast of Colombia, *S. oedipus* is also most abundant in secondary forest or early secondary growth on forest fringes or near streams, and seldom inhabits deep climax forest; in northern Colombia it is found in mature rain forest and deciduous forest but is also most frequent at forest fringes and in secondary growth (Hernández-Camacho and Cooper 1976). At one study site it was more abundant in a large block of wet forest than in nearby similar gallery forest (Struhsaker et al. 1975), and at another it occupied very dry tropical forest (Neyman 1978).

S. oedipus is found from sea level to nearly 1,500 m elevation, and is "as much at home in tall forest as in low dense second growth," although it may achieve higher population densities or be more conspicuous in secondary than in primary forest (Hershkovitz 1977).

Factors Affecting Populations

Habitat Alteration

Limited habitat disturbance such as selective logging or shifting agriculture can benefit populations of *S. oedipus* by increasing the extent of secondary growth vegetation (Moynihan 1970). Along the Atlantic coast of Panama, for example, this species has been able to invade and inhabit abandoned fields and has become a commensal of man (Moynihan 1976a). Where a forest is allowed to mature, and habitat disturbance is prevented, crested tamarins decline markedly in abundance (Moynihan 1970).

Populations of *S. oedipus* may have increased since the arrival of the Spaniards in Panama, owing to the decline of the Amerinds and their extensive agriculture, and the consequent regeneration of forest unavailable to primates in pre-Columbian times (Bennett 1968). This author also postulated that *S. oedipus* may have invaded Panama from Colombia during the last few centuries, through a newly established corridor of secondary growth.

But extensive permanent clearing of forest is causing the decline of *S. oedipus* populations in Panama and Colombia. The amount of land being cleared in Panama is greater than the amount reverting to forest (owing to the rapidly increasing human population), and has resulted in a net habitat loss for *S. oedipus* (Dawson, pers. comm. 1974). Extensive forested areas still remain in the Darién region (Thorington 1975).

In northern Colombia most forest has been cleared and converted to cattle pasture, and although *S. oedipus* can survive in small forest remnants, these too are being destroyed (Hernández-Camacho and Cooper 1976). Only about 5% of the original forest cover remains in the northern three-quarters of the eastern region of *S. oedipus*'s range, and this forest, which consists of more than 270 isolated tiny patches, is continually being cut for fence posts and building lumber (Neyman 1978). The destruction of these patches is expected to continue at a rapid pace, since unexploited private land may be legally invaded and expropriated by colonists (Neyman 1978). Along the Pacific coast of Colombia, habitat alteration has probably not been extensive (Hernández-Camacho and Cooper 1976).

Human Predation

Hunting. *S. oedipus* is intensively hunted in many parts of Panama, and this is thought to have reduced population densities and average group sizes in some localities (Moynihan 1970). But its small size and the high cost of ammunition make this species an economically marginal target, and some authors believe that hunting for meat and sport is probably a negligible factor in *S. oedipus*'s mortality rate (Dawson 1976, Hershkovitz 1977).

Pest Control. *S. oedipus* feeds on bananas and avocados in Panama (Dawson, pers. comm. 1974) and plantains and corn in Colombia (Neyman, pers. comm. 1974). No mention of efforts to control this species was located.

Collection. In Panama, *S. oedipus* is commonly found as a pet in rural homes (Sousa et al. 1974), and is hunted intensively to obtain the young for sale as pets (Moynihan 1976a). It is also kept as a pet in Colombia (Neyman 1978).

The eastern race in Colombia (*S. o. oedipus*) has been heavily collected for export for at least 10 to 15 years, and about 30,000 to 40,000 were exported between 1960 and 1972 (Hernández-Camacho and Cooper 1976). As the remaining forest habitats of *S. oedipus* in northern Colombia are destroyed, the resident crested tamarins are being collected, and the population is being progressively depleted (Neyman 1978).

Although the original demand for this species was for the pet trade, much of the current demand is for biomedical research, and Colombia has received many requests for export permits from laboratories in the United States and West Germany (Cooper and Hernández-Camacho 1975). *S. oedipus* is used frequently for biomedical research in the United States (Comm. Cons. Nonhuman Prim. 1975a).

Between 1968 and 1973 a total of 13,879 *S. oedipus* were imported into the United States (Appendix). Many came from Colombia but were listed as originating from Brazil, Peru, and Paraguay (Paradiso and Fisher 1972, Clapp and Paradiso 1973, Clapp 1974), where this species does not occur. No *S. oedipus* were imported into the United States in 1977 (Bittner et al. 1978). In 1965–75, 1,042 *S. oedipus* were imported into the United Kingdom (Burton 1978).

Conservation Action

Except for a small population on Barro Colorado Island, *S. oedipus* is not found in any protected reserves or parks (Table 55). No special laws protect it in Panama or the Canal Zone, although commercial trapping and sale are regulated by the Departments of Treasury and of Natural Resources of Panama (Dawson, pers. comm. 1974).

In 1969 Colombia forbade all hunting of *S. oedipus*, but in 1972 it amended the law to allow commercial hunting (capture) in the northern region at the rate of 25 *S. oedipus* per month per special permit holder; in 1973 the quota for *S. oedipus* was set at 500, with no more than 100 to be captured per month (Cooper and Hernández-Camacho 1975). These authors appealed to investigators to "avoid using this species for research whenever possible and to refuse to buy illegally exported animals." In return they promised that "Colombia will continue to give serious consideration to permit requests for studies requiring *S. oedipus* for which other species cannot easily be substituted."

Other recommendations for the conservation of *S. o. oedipus* include the establishment of a reserve in the Serranía de San Jacinto (Neyman 1977) and several other reserves in different representative habitats (Neyman 1978), the establishment of breeding colonies and study of basic husbandry and laboratory and field biology (Neyman 1978), a total ban on export, shooting, or interference, a census, and the management of home environments by reforestation and the supply of limiting factors (Kleiman 1976). A study has now been funded which will investigate the status of this species in two newly created reserves, establish additional parks or reserves, develop an educational park at Monteria, Córdoba, and initiate a conservation education campaign to increase public awareness of *S. oedipus*'s situation in Colombia (Mittermeier 1977b).

S. oedipus is listed in Appendix I of the Convention on International Trade in Endangered Species of Wild Fauna and Flora, and as endangered under the U.S. Endangered Species Act (U.S. Department of Interior 1977a,b).

Table 55. Distribution of *Saguinus oedipus*

Country	Presence and Abundance	Reference
Costa Rica	Along SE border (map)	Hall and Kelson 1959
	In Coto region of SE (map)	Hershkovitz 1977
Panama	Scarce in Coto region of extreme W	Carpenter 1935
	Throughout portion E of canal (map)	Hershkovitz 1949
	Localities listed E of canal in Darién, R. Bayano valley, and Madden Lake area; W of canal in Panamá, Colón, and Chiriquí	de Rodaniche 1957
	Throughout (map)	Hall and Kelson 1959
	Abundant in Canal Zone and E; locally distributed in W	Handley 1966
	Throughout except in NW (map)	Hershkovitz 1966a
	In E from 77°W–80°30′W and perhaps farther W; absent on Azuero Peninsula	Bennett 1968
	Throughout except NW and Azuero Peninsula (map)	Hershkovitz 1969
	Locally distributed in E; in SW on Burica Peninsula near R. Coto	Méndez 1970
	In E to central; locally abundant along Pacific coast; on Ancon Hill (E of canal between Ancon and Balboa), Rodman Naval Base (W of Canal), N of Balboa, Cerro Campana, and Barro Colorado Island	Moynihan 1970
	In San Blas, E of Mulatupo Island	Telford et al. 1972
	Probably fewer than 50 on Barro Colorado Island, Canal Zone	Eisenberg and Thorington 1973
	At Balboa, Canal Zone	Fleming 1973
	Abundant in W Canal Zone	Dawson, pers. comm. 1974
	E of canal in N Panamá and SE Colón (map)	Sousa et al. 1974
	Localities mapped in central and E	Deane 1976
	Localities listed at Chepo and Tocoumé	Napier 1976
	Localities listed and mapped E of 80°W in W and central Panamá, Canal Zone, E and W San Blas, and throughout Darién	Hershkovitz 1977
	At Rodman Ammunition Depot (8°57′N, 79°37′W) in Canal Zone W of canal	Dawson 1978
	Common	Mittermeier, Coimbra-Filho, and Van Roosmalen, in press
Colombia	In Chocó along W coast, in basins of R. Atrato, R. Cauca, and R. Magdalena in N from Gulf of Darién into Antioquia and E to R. San Jorge, Bolívar; N along W bank R. Magdalena to Cartagena and S. Atlántico (map)	Hershkovitz 1949

Table 55 (continued)

Country	Presence and Abundance	Reference
Colombia (cont.)	In Chocó and between Gulf of Darién and R. Magdalena	Cabrera 1958
	In W as far S as 4°N and in NW as far N as 11°N (maps)	Hershkovitz 1966a, 1969
	E form (*S. o. oedipus*) threatened with extinction; in NW between R. Atrato and R. Magdalena, in Bolívar, Sucre, Antioquia, Atlántico, and Córdoba	Hershkovitz 1972
	Moderately scarce near Tolú, SE Sucre; declining	Neyman, pers. comm. 1974
	At Estanzuela (9°32′N, 75°32′W) (in NW near coast)	Struhsaker et al. 1975
	In W and NW (map)	Thorington 1975
	Near Guaranada, Majagual, and Sucre	Green 1976
	Along Pacific coast probably as far S as R. San Juan (N of 4°N in S Chocó); E to R. Atrato, questionable farther E; in NW from Urabá region in NW Antioquia S at least to R. León, and in Córdoba, Sucre, N Bolívar, and Atlántico; E limit is W bank of lower R. Magdalena and R. Cauca (map); population size unknown but concern for survival is increasing	Hernández-Camacho and Cooper 1976
	E form (*S.o. oedipus*) endangered	I.U.C.N. 1976
	E form threatened so severely that is near extinction	Kleiman 1976
	Localities listed and mapped in N Chocó and W Córdoba as far E as R. Sinú; extends E to lower R. Cauca-Magdalena through Atlántico, Bolívar, and NW Antioquia	Hershkovitz 1977
	Becoming scarce near Tocurá, R. Sinú; near Colosó, S Serranía de San Jacinto	Neyman 1977
	NE of Tolú, Sucre (9°34′N, 75°27′W); threatened (map)	Neyman 1978
	W form (*S.o. geoffroyi*) common; E form (*S.o. oedipus*) endangered	Mittermeier, Coimbra-Filho, and Van Roosmalen, in press

CEBIDAE
Cebid Monkeys

Alouatta belzebul Red-handed Howler Monkey

Taxonomic Note

Five subspecies are recognized (Cabrera 1958, Napier and Napier 1967).

Geographic Range

Alouatta belzebul is found in northeastern Brazil south of the Rio Amazonas in the states of Maranhão, Pará, and eastern Amazonas. It occurs as far west as the Rio Madeira and as far east as the islands of Marajó and Mexiana at the Mouths of the Amazon (Cabrera 1958). The limits of its range have not been determined. For details of the distribution, see Table 56.

Abundance and Density

No information.

Habitat

A. belzebul has been recorded in *terra firme* primary forest (Mittermeier and Coimbra-Filho 1977). No further details are available.

Factors Affecting Populations

Habitat Alteration

The construction of the Transamazonian and Belém-Brasília highways is opening and altering natural habitats within the range of *A. belzebul* (Wozniewicz 1974, Vanzolini 1978). The habitat of this species is also being destroyed by clearing for agriculture and cattle ranching in eastern Pará (Ayres 1978).

Human Predation

Hunting. A. belzebul is hunted for its flesh in Pará along the Transamazonian Highway (Smith 1976), along the Rio Tapajós (Mittermeier and Coimbra-Filho 1977), and east of the Rio Xingu (Ayres 1978).

Pest Control. No information.

Collection. No *A. belzebul* were imported into the United States between 1968 and 1973 (Appendix) or in 1977 (Bittner et al. 1978). A total of 37 were imported into the United Kingdom in 1965–75 (Burton 1978).

Conservation Action

A. belzebul is protected from commercial exploitation by the Brazilian Faunal Protection Law (Mittermeier and Coimbra-Filho 1977).

Table 56. Distribution of *Alouatta belzebul*

Country	Presence and Abundance	Reference
Brazil	In Maranhão and Pará	Kumm and Laemmert 1950
	On both sides of R. Pará, islands of Marajó and Mexiana, between R. Tocantins and mountains of NE Maranhão, S of R. Amazonas between R. Tocantins and R. Tapajós, in E Amazonas between R. Tapajós and R. Madeira; may also be N of R. Amazonas	Cabrera 1958
	S of Belém, E Pará	Carvalho and Toccheton 1969
	In Pará and Amazonas	Deane et al. 1969
	Four localities mapped in E Amazonas and Pará	Deane 1976
	In Pará: Igarapé Assu (Açu) (1°07′S, 47°36′W), Marajó, R. Jamanxim (S of ~5°S, 56°W), Santarém, Oriximiná (1°47′S, 55°48′E); in Maranhão: Humberto de Campos (2°34′S, 43°30′W) (localities listed)	Napier 1976
	Between Altamira, Itaituba, and Marabá, along Transamazonian Highway, Pará	Smith 1976
	In E Pará E of R. Xingu, including S of Ourém near Igarapé Pedral	Ayres 1978

Alouatta caraya Black Howler Monkey

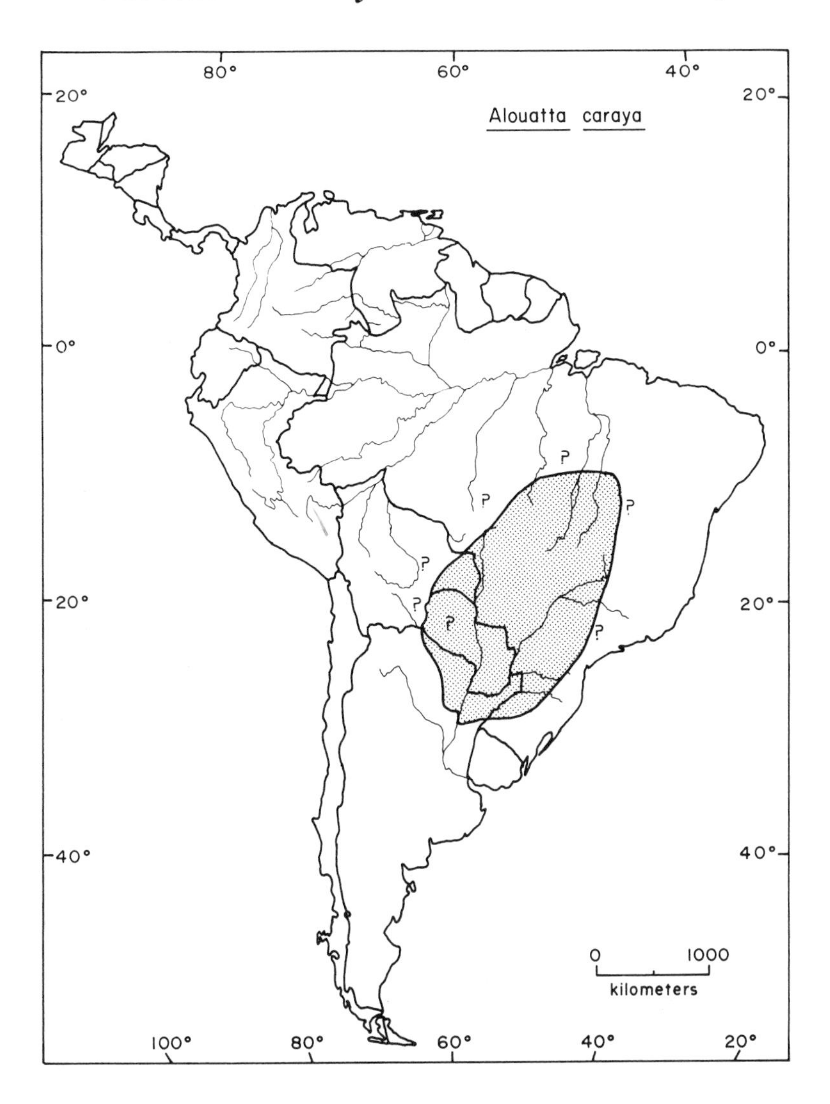

Taxonomic Note

This is a monotypic species (Cabrera 1958, Napier and Napier 1967).

Geographic Range

Alouatta caraya is found in eastern Bolivia, northern Argentina, Paraguay, and southern Brazil (Cabrera 1958). It has been recorded as far north as about 10°S in northern Goiás, Brazil (Deane 1976), and as far south as 29°S in Corrientes, Argentina (Colillas and Coppo 1978). It occurs as far east as eastern Goiás and western Minas Gerais in southeast Brazil (Cabrera 1958), but the limits of its range there and its westward extent in Bolivia are not known. For details of the distribution, see Table 57.

Abundance and Density

The status of the black howler monkey is not known for any of the countries in which it occurs. At one study site in Bolivia, population density was 0.8 individuals per km^2, consisting of two troops of two members each (Heltne et al. 1976). In Argentina, group sizes ranged from 3 to 10 ($\bar{X}$ = about 5, N = 10; Krieg 1928 cited by Pope 1968), 4 to 14 ($\bar{X} = 7.9$, N = 17; Pope 1968), and up to 20 ($\bar{X} = 5.5$ to 12.6, N = 40), with home ranges of about 1 ha per group (Colillas and Coppo 1978).

Habitat

In Brazil, *A. caraya* has been observed in high primary forest with swamps and watercourses (Kühlhorn 1955), and in semiarid caatinga and plateau *cerrado* formations (Vanzolini 1978). In Argentina it is found in tall streamside forests and, during dry seasons, in lower vegetation near large rivers (Colillas and Coppo 1978). At one site in eastern Bolivia it occupies seasonally very dry, short, deciduous, broadleaf forest on high ground in a white-water drainage (Freese et al., in prep.).

Factors Affecting Populations

Habitat Alteration

A. caraya adapts well to the proximity of humans and to deforestation, and is able to occupy small, isolated *capões* (Avila-Pires 1966). In Argentina it often lives near villages and is fed by the local people (Colillas and Coppo 1978).

But in 1970 thousands of black howler monkeys were destroyed in southern Mato Grosso, Brazil, during the large burnings carried out to create pastureland (Coimbra-Filho 1974). No information was located regarding the impact of other habitat alteration on this species.

Human Predation

Hunting. A. caraya has been eliminated by hunting near Bella Vista, Argentina (Malinow 1968), and is also hunted in eastern Bolivia (Heltne et al. 1976).

Pest Control. No information.

Collection. A total of 55 *A. caraya* were imported into the United States between 1968 and 1973, primarily from Paraguay (Appendix). None entered the United States in 1977 (Bittner et al. 1978). Thirteen were imported into the United Kingdom in 1965–75 (Burton 1978).

Conservation Action

A. caraya is found in two national parks in Brazil (Table 57), and is protected from commercial exploitation in that country (Mittermeier and Coimbra-Filho 1977), but is in need of study and further protection (Vanzolini 1978). No reports of conservation efforts in Bolivia, Argentina, or Paraguay were located.

Table 57. Distribution of *Alouatta caraya* (countries are listed in counterclockwise order)

Country	Presence and Abundance	Reference
Bolivia	In E	Cabrera 1958
	Not common near San José de Chiquitos (17°50′S, 60°43′W) (in SE)	Heltne et al. 1976, [Freese et al., in prep.]
Argentina	In N in Formosa, E Salta, Misiones, and N Corrientes	Cabrera 1958
	In NE on islands in R. Paraná (28°30′S, 59°W, NW Corrientes); range extends S about 100 km including Bella Vista; reported from R. Pilcomayo (in N)	Malinow 1968
	On Tragadero Sur, island in R. Paraná	Pope 1968
	In NE	Ringuelet 1970
	Localities listed: R. Bermejo, Chaco; Itati, Corrientes (both in NE)	Napier 1976
	In N central and NE (map), S to 29°S in Corrientes and 27°S in Chaco, W to 60°W	Colillas and Coppo 1978
Paraguay	Present	Cabrera 1958
	One locality: Tayru (about 27°S, 58°33′W) (SW corner)	Napier 1976
Brazil	In Goiás, Mato Grosso, Minas Gerais, and Rio Grande do Sul	Kumm and Laemmert 1950
	In S Mato Grosso near Alto Paraná and R. Ivinheima	Kühlhorn 1955
	From Mato Grosso to S Goiás, W Minas Gerais, São Paulo, Paraná, and Santa Catarina	Cabrera 1958
	In Paraná, Mato Grosso, Goiás, Minas Gerais	Deane et al. 1969
	Ín Iguaçu N.P. (W Paraná) and Brasília N.P. (Federal District)	I.U.C.N. 1971
	In NE Mato Grosso and central Goiás (localities mapped)	Deane 1976
	Localities listed in Goiás and Mato Grosso	Napier 1976
	From coast of NE to N Argentina	Vanzolini 1978

Alouatta fusca Brown Howler Monkey

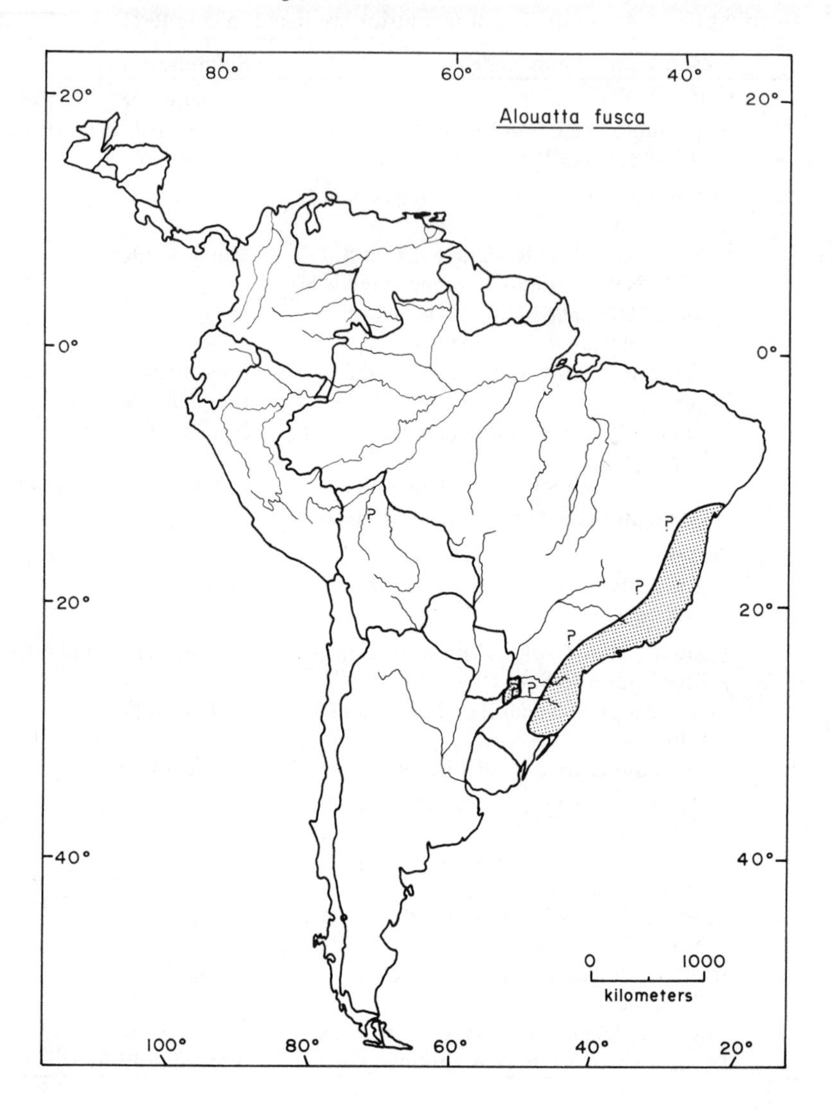

Taxonomic Note

This species is also referred to as *Alouatta guariba* (Crespo 1954, Cabrera 1958). Three subspecies are recognized (Cabrera 1958, Napier and Napier 1967).

Geographic Range

Alouatta fusca is found in southeastern Brazil. Old references also report it from northern Bolivia and northeastern Argentina (Crespo 1954, Cabrera 1958), but no recent confirmation of its presence in these two countries is available. In Brazil, *A. fusca* occurs along the eastern coast from about 12°S to 30°S (Napier 1976). The western limits of the range are not known. For details of the distribution, see Table 58.

Abundance and Density

A. fusca has recently been described as locally numerous in several parts of southeastern Brazil, but is near extinction in Rio de Janeiro State, and is considered by at least one authority to be endangered (Table 58). No information is available regarding its population densities, group sizes, or home range areas.

Habitat

Brown howler monkeys have been found in "old-type forest" (Laemmert et al. 1946), in forested, coastal lowlands, and in plains with small remnants of forest along rivers or creeks (Deane et al. 1971). They have been collected from 400 to 1,150 m elevation (Napier 1976). No further details of their habitat preferences were located.

Factors Affecting Populations

Habitat Alteration

The forests of southeastern Brazil have been extensively cut and replaced with agriculture (Coimbra-Filho and Mittermeier 1973a). The forests which remain are rapidly being cleared for plantations, charcoal production, and development, and this deforestation is the main factor responsible for the decline of *A. fusca* populations (I.U.C.N. 1978).

Human Predation

Hunting. A. fusca is hunted in Brazil and is prized for the quality of its meat and its large size (I.U.C.N. 1978).

Pest Control. No information.

Collection. No *A. fusca* were imported into the United States between 1968 and 1973 (Appendix) or in 1977 (Bittner et al. 1978). Twenty-four were imported into the United Kingdom in 1965–75 (Burton 1978).

Conservation Action

A. fusca is found in four or five national parks, three state parks, and two federal biological reserves in Brazil (Table 58), and is protected by Brazilian law against commercial exploitation (Mittermeier and Coimbra-Filho 1977). Nevertheless, "study leading to its protection is urgent" (Vanzolini 1978).

Table 58. Distribution of *Alouatta fusca*

Country	Presence and Abundance	Reference
Brazil	Near Ilhéus (14°50′S, 39°06′W), S Bahia	Laemmert et al. 1946
	In Espírito Santo and Santa Catarina	Kumm and Laemmert 1950
	Along Atlantic coast from Bahia to Rio Grande do Sul	Crespo 1954
	In E Bahia, Minas Gerais, Espírito Santo, Rio de Janeiro, E São Paulo, Paraná, Santa Catarina, Rio Grande do Sul; possibly also in W near Bolivian border	Cabrera 1958

Table 58 (continued)

Country	Presence and Abundance	Reference
Brazil (cont.)	In E Espírito Santo, São Paulo, Santa Catarina, and Rio Grande do Sul (localities mapped)	Deane et al. 1969
	Very numerous near São Paulo city, in Santa Catarina, and locally in Rio Grande do Sul; rare near Sooretama (N Espírito Santo)	Deane et al. 1971
	In Monte Pascoal N.P., Bahia, and Sooretama Biological Reserve, Espírito Santo	I.U.C.N. 1971
	Localities mapped along coast from 20°S–30°S	Deane 1976
	From about 12°S in Bahia to about 30°S in Rio Grande do Sul; localities also listed from Espírito Santo, Minas Gerais, Paraná, Santa Catarina, and São Paulo	Napier 1976
	Extremely reduced in numbers, bordering on extinction in Rio de Janeiro State	Coimbra-Filho 1978
	In Itatiaia N.P. (Rio de Janeiro and Minas Gerais), Serra dos Orgãos N.P. (Rio de Janeiro), Serra da Bocaina N.P., Nova Lombardia Federal Biological Reserve, Rio Doce State Park, Campos do Jordão State Park (22°45′S, 45°33′W), Jacupiranga State Park (22°38′S, 47°58′W), and perhaps Caparaó N.P.	I.U.C.N. 1978
	Endemic to Atlantic coast; endangered	Vanzolini 1978
Bolivia	In N	Cabrera 1958
	From Puerto Salinas, Beni (in N)	Hill 1962
Argentina	In Misiones (in NE); westernmost record	Crespo 1954
	In Misiones	Cabrera 1958

Alouatta palliata Mantled Howler Monkey

Taxonomic Note

Many authors have referred to this species as *Alouatta villosa*, but *A. palliata* is now more generally accepted. Seven subspecies are recognized (Hall and Kelson 1959, Napier and Napier 1967), excluding *A. pigra*, which is treated as a separate species following Smith (1970).

Geographic Range

Alouatta palliata is found in Central America and northern South America from southern Mexico to northwestern Peru. Its range extends north as far as about 18°N in Tabasco, Mexico, and west to about 95°30′W in Oaxaca (Hall and Kelson 1959,

Smith 1970). It occurs as far south as the Department of Tumbes (near 4°S) in Peru (Castro 1976) and as far east as about 75°W in northeastern Córdoba, Colombia (Hernández-Camacho and Cooper 1976). For details of the distribution, see Table 59.

Abundance and Density

In the past, *A. palliata* populations were described as locally abundant in parts of Honduras and Panama (Table 59). Recently they have been judged to be decreasing but not endangered in Costa Rica, endangered in Peru, and declining in Mexico, El Salvador, and Panama (Table 59). Their status in Guatemala, Honduras, Nicaragua, Colombia, and Ecuador is not known.

This species has been most intensively studied on Barro Colorado Island (B.C.I., Table 60), where the factors regulating population density and dispersion have been investigated and reviewed by several authors (Chivers 1969; Baldwin and Baldwin 1972a; Heltne, Turner, and Scott 1976; Mittermeier 1977a). The various studies of B.C.I. mantled howler monkeys have revealed population densities of 15 to 80 individuals per km^2 and troop home ranges of 7.9 to 76 ha (Heltne and Thorington 1976). In mainland Panama one *A. palliata* population achieved a density of at least twelve times the maximum described for B.C.I., with "no conspicuous behavioral or physiological abnormalities" observed (Baldwin and Baldwin 1976).

Average group sizes for *A. palliata* are 5.2 to 18.9 (Table 60), although groups of up to 50 have been reported (Hall and Kelson 1959). Mainland Central American home ranges are 3.2 to 10 ha in area (Table 60).

Habitat

A. palliata is a forest species, found in several kinds of forest vegetation. In some areas it prefers dense primary forest and does not cross open country (Carpenter 1934). For example, in Mexico, it is found in tall virgin stands of heavy tropical rain forest and cloud forest (Leopold 1959; Alvarez del Toro, pers. comm. 1974).

At one study site in northwestern Costa Rica, *A. palliata* utilized mature or nearly mature evergreen forest, including riparian, almost exclusively during the dry season, although deciduous forest was available (Freese 1976). In another study area the most favorable habitat for *A. palliata* was moist forest on alluvial woodland, but the species also occupied more marginal habitats such as dry, deciduous, hillside forest and narrow riparian forests (Heltne, Turner, and Scott 1976). These last authors believe that the Costa Rican tropical dry forest in general may represent marginal habitat for *A. palliata*, but another worker found this species to be more common in tropical deciduous forest than in the wetter rain forests of the Caribbean and Pacific lowlands (Wilson, pers. comm. 1974). In the Panama Canal Zone, *A. palliata* inhabits lowland moist tropical deciduous forest (Fleming 1973), and in Colombia it is found in most nonflooding forest, from rain forest to semievergreen (Hernández-Camacho and Cooper 1976).

Mantled howler monkeys occupy coastal marshy areas as well as inland lowland forest in Panama (Baldwin and Baldwin 1976). On Coiba Island they were observed in one nearly pure stand of mangroves, as well as hillside forest (Milton and Mittermeier 1977). They also inhabit the broken swampy forest of the coastal plain in Honduras (Trapido and Galindo 1955), but are absent from the coastal mangrove zone and swampy forests of Colombia (Hernández-Camacho and Cooper 1976).

In Costa Rica, *A. palliata* does not thrive in secondary forest (Freese, pers. comm. 1974), but in southern Mexico it prefers somewhat xeric subclimax or secondary forest (Smith 1970). In Chiapas it occupies low thicketlike vegetation when no tall trees are available (Alvarez del Toro 1977). It can utilize fairly small patches of forest and strips between cultivated fields and pastures and near villages (Trapido and Galindo 1955; Glander 1971; Heltne, Turner, and Scott 1976).

A. palliata has been reported as high as 800 m in Mexico (Alvarez del Toro, pers. comm. 1974), 914 m in Panama (Galindo and Srihongse 1967), and 2,000 m in Guatemala (Haltenorth 1975), and in many lowland areas.

In summary, mantled howler monkeys are capable of utilizing a variety of forest habitats, but when a choice is available, and perhaps depending on the area and the season, they tend to prefer evergreen over deciduous, dry over wet, primary over secondary, and lowland over highland forests.

Factors Affecting Populations

Habitat Alteration

The forest habitat of *A. palliata* is being destroyed in Mexico, El Salvador, Costa Rica, Panama, and Colombia. Deforestation is probably also occurring in the other countries of the range, but no direct information is available.

In Mexico, clearing of the lowland forest is the principal factor responsible for the decline of *A. palliata* populations (Leopold 1959, Alvarez del Toro, pers. comm. 1974). Throughout southern Mexico, large areas of forest are being cleared annually for agriculture, and Leopold (1959) predicted that "lumber interests will sooner or later find means of logging the mahogany and red cedar" of Mexico's remaining tropical forests.

In El Salvador the destruction of the tropical evergreen forest was the major factor responsible for the extermination of *A. palliata* along the coast (Daugherty 1972). More than half of the original forest in the northwest has also been cut, leaving only a small patch (12 km^2) of remaining suitable habitat (Daugherty 1973).

In Costa Rica, clearcut logging for timber and cultivation are destroying about 50,000 ha or 1% of the forest per year and this rate is accelerating (Freese, pers. comm. 1974). This clearcutting not only reduces the resources available to *A. palliata* but leaves some areas without monkeys (because the animals cannot reach the patches via arboreal pathways and cannot cross open areas), and creates abnormally high population densities in some remaining patches of forest; when the populations readjust to these smaller areas, significantly smaller troop sizes may result in response to the reduced food supply (Heltne, Turner, and Scott 1976).

Extensive deforestation has also drastically reduced the habitat of *A. palliata* in Panama. The absence of this species over most of central and southern Panama is due primarily to the "destruction of forests and woodlands . . . , the rapid creation of permanent pastures," and tree cutting in the last refuges along rivers (Bennett 1968). The clearing of forest is being promoted by the implementation of an old land reform law that makes unexploited forest land subject to government seizure (Baldwin and Baldwin 1976). Spurred by the rapidly (31% per decade) growing human population, this clearing has relegated *A. palliata* to a few scattered and remote areas of Panama (Baldwin and Baldwin 1976).

In Colombia, deforestation is threatening *A. palliata* in the Río Atrato and Río Sinú basins, which are suffering extensive forest exploitation and destruction (Hernández-Camacho and Cooper 1976). In general, although *A. palliata* is able to coexist with man and utilize disturbed habitats to some extent (Smith 1970; Freese, pers. comm. 1974; Wilson, pers. comm. 1974), large-scale deforestation is the principal threat to its survival in every area for which information is available.

In addition to forest clearing, three other kinds of habitat changes have affected *A. palliata* populations in northwestern Costa Rica (Heltne, Turner, and Scott 1976). Fires (both planned and accidental) have destroyed forested areas; the course of the Río Higuerón was shifted to facilitate drainage and clearing of forest; and heavy doses of pesticide were sprayed over the area to protect crops. Following the spraying, several dead mantled howler monkeys were found with no evidence of disease or injury, but with "gastric and associated glands swollen, puffy, inflamed and loaded with insecticides" (Heltne, Turner, and Scott 1976). Pesticides were also cited as detrimental to *A. palliata* population in southern Panama (Baldwin and Baldwin 1976).

In Tabasco, Mexico, recent habitat disturbances, as a result of road construction and increased agriculture, have allowed *A. palliata* to cross an area of forest which it previously did not traverse, and may be responsible for the present sympatry of this species with *A. pigra* (Smith 1970).

Human Predation

Hunting. A. palliata is heavily hunted for its meat and pelt in Mexico (Leopold 1959, Alvarez del Toro, pers. comm. 1974). It is also killed for use as bait by big cat hunters, and hunting is "slowly annihilating" the populations in Chiapas (Alvarez del Toro 1977). *A. palliata* is intensively hunted everywhere in Guatemala (Hendrichs 1977), but not in Costa Rica (Freese, pers. comm. 1974).

Hunting has eliminated the remnant populations of *A. palliata* left after destruction of most of the forest in much of south central Panama (Bennett 1968). And in Colombia, hunting exerts a significant pressure on populations of mantled howler monkeys (Hernández-Camacho and Cooper 1976).

Pest Control. No information.

Collection. A total of 372 *A. palliata* were imported into the United States between 1968 and 1973 (Appendix), primarily from Colombia and Nicaragua. None entered the United States in 1977 (Bittner et al. 1978). One hundred were imported into the United Kingdom in 1965–75 (Burton 1978). Howler monkeys do not usually survive well in captivity and are not often used in biomedical research (Thorington 1972).

Conservation Action

A. palliata is found in one national park in Nicaragua, two in Colombia, two parks and two reserves in Costa Rica, and one national forest in Peru (Table 59).

Although *A. palliata* is legally protected in Mexico, stricter enforcement as well as habitat protection are necessary for its survival there (Leopold 1959, Alvarez del Toro 1977). Efforts are under way to preserve some of its natural habitat in Chiapas (Alvarez del Toro, pers. comm. 1974).

A national park or wildlife refuge should be established in Chiriquí, Panama, to preserve the fauna of this region, including *A. palliata* (Baldwin and Baldwin 1976). This species does not occur in any national parks in Peru, and is not protected by Peruvian laws (Castro 1976). In order to conserve its populations in Tumbes, a guard post should be established in the Tumbes National Forest to prevent poaching and forest destruction there, and a campaign for conservation education should be initiated (Mittermeier 1977b).

A. palliata is listed in Appendix I of the Convention on International Trade in Endangered Species of Wild Fauna and Flora, and as endangered under the U.S. Endangered Species Act (U.S. Department of Interior 1977a,b).

Table 59. Distribution of *Alouatta palliata* (countries are listed in order from north to south)

Country	Presence and Abundance	Reference
Mexico	In SE in S Veracruz, NE Oaxaca, W Tabasco, N Chiapas, absent on Yucatán Peninsula (localities mapped; others *A. pigra* according to Smith 1970)	Hall and Kelson 1959
	In S Veracruz, N and SE Chiapas (localities mapped); fairly common but range being reduced and numbers shrinking steadily	Leopold 1959
	In NE Oaxaca in Isthmus of Tehuantepec, Juchitán district near Veracruz state line, near 95°04′W (locality mapped)	Goodwin 1969
	In S Veracruz, NE Oaxaca, W Tabasco, and NW and central Chiapas (localities mapped)	Smith 1970
	Declining in Chiapas	Alvarez del Toro, pers. comm. 1974
	In NW Chiapas (localities listed)	Alvarez del Toro 1977
Guatemala	In S (map) (other localities are *A. pigra* according to Smith 1970)	Hall and Kelson 1959
	In narrow strip across S, questionable in extreme S (map)	Smith 1970
	Near Chilasco (may be *A. pigra*)	Haltenorth 1975
	In Sierra de Chamá, W of Cobán (15°28′N, 90°20′W) (in S)	Hendrichs 1977
El Salvador	Throughout (map)	Hall and Kelson 1959
	Throughout except SW corner (map)	Smith 1970
	Became uncommon by 1880; nearly exterminated along coast; in NW into twentieth century; now infrequent stragglers in Montecristo area of extreme NW	Daugherty 1972
	Extinct in all but NW	Daugherty 1973

Table 59 (continued)

Country	Presence and Abundance	Reference
Honduras	Abundant in N between La Ceiba and Tela	Trapido and Galindo 1955
	Throughout (map)	Hall and Kelson 1959
	Throughout except NW (map)	Smith 1970
Nicaragua	Throughout (maps)	Hall and Kelson 1959, Smith 1970
	In Saslaya N.P., Zelaya (in NE)	Ryan 1978
Costa Rica	At Limón, Alajuela; in Guanacaste, at Santa Cruz, La Cruz, Cañas Dulces; rare near Puntarenas; formerly abundant NW of Puerto Cortes and Puntarenas, now absent there	Vargas-Mendez and Elton 1953
	Throughout (map)	Hall and Kelson 1959
	Less common in lowlands of Caribbean and Pacific than in N; seen in Guanacaste and Heredia; range has decreased but population not sharply declining; not endangered	Wilson, pers. comm. 1974
	In Santa Rosa N.P. (in NW), Cahuita N.P. (in SE), and Cabo Blanco National Reserve (S tip Nicoya Peninsula)	I.U.C.N. 1975
	Throughout except central plateau; abundance variable; declining; in Santa Rosa N.P. (in NW)	Freese 1976
	Populations declining and distressed NW and SW of Las Cañas, Guanacaste (in NW)	Heltne, Turner, and Scott 1976
	Near Las Cañas, Guanacaste	Glander 1977, 1978
	In Monteverde Reserve	C.H.F. Rowell, pers. comm. 1978
Panama	Small population in Coto region (near W border)	Carpenter 1935
	In Darién (E), from Bayano R. valley W to Canal Zone, Colón W of Canal, and Chiriquí	de Rodaniche 1957
	Throughout and on Coiba Island (map)	Hall and Kelson 1959
	Locally abundant throughout below 1,554 m	Handley 1966
	In S and N Darién and central Panamá (localities listed)	Galindo and Srihongse 1967
	Formerly widespread in lowland Chiriquí, range there now greatly reduced, only near coast; in 1920s common in S central (Coclé, Herrera, Los Santos), now absent in most of region; frequent in R. Tonosí valley, S Los Santos	Bennett 1968
	High density in Chiriquí (at 8°19′N, 82°38′W)	Baldwin and Baldwin 1972a, 1973

Table 59 (continued)

Country	Presence and Abundance	Reference
Panama (cont.)	At Cristobal (N central near 9°N, 80°W)	Fleming 1973
	In Darién, Los Santos and Panamá Provinces	Sousa et al. 1974
	Common in early 1950s; now scarce and restricted to scattered remote areas; localities mapped in S Chiriquí; endangered if current rate of development continues	Baldwin and Baldwin 1976
	Localities mapped in central and E	Deane 1976
	On Coiba Island off SW	Milton and Mittermeier 1977
Colombia	From Cartagena S to R. Sinú, R. Atrato, Chocó, and along Pacific coast	Hershkovitz 1949
	In Farallones de Cali N.P., Valle, and Las Orquideas N.P., Antioquia	Green 1972
	Throughout Pacific lowlands (in W) including basins of R. Atrato and R. Sinú (map); may extend farther NE toward Cartagena; status unknown	Hernández-Camacho and Cooper 1976
Ecuador	Along W coast	Hershkovitz 1949
	W of Andes	Cabrera 1958
	Localities listed in Esmeraldas (NW), Manabí (W), and Los Ríos (W)	Baker 1974
	Extends S to 3–4°S in W; localities listed include: Mindo (0°2′S, 78°49′W), R. Blanco, and NW of Quito	Napier 1976
Peru	May be in Zarumilla Province, Tumbes (extreme NW, on coast)	Grimwood 1969
	In Tumbes; endangered	Castro 1976
	S limit is on Pacific coast of Tumbes including Tumbes National Forest	Mittermeier 1977b

Table 60. Population parameters of *Alouatta palliata*

Population Density (individuals /km²)	Group Size $\bar{X}$	Range	No. of Groups	Home Range Area (ha)	Study Site	Reference
		5–25			Mexico	Leopold 1959
	20				Mexico	Alvarez del Toro, pers. comm. 1974
5					Guatemala	Hendrichs 1977
	8.1	3–24	8		NW Costa Rica	Freese 1976
	9–15.4		117		NW Costa Rica	Heltne, Turner, and Scott 1976
		13	1	10	NW Costa Rica	Glander 1977
1,030	18.9	7–28	8	3.2–6.9	SW Panama	Baldwin and Baldwin 1972a
	5.2	2–9	5		Coiba Island, Panama	Milton and Mittermeier 1977
	17.3	4–35	23		B.C.I.	Carpenter 1934
	8.1	3–17	29		B.C.I.	Collias and Southwick 1952
	14.7	11–18	12		B.C.I.	Chivers 1969
	16.2	13–23	6		B.C.I.	Mittermeier 1973b
	13.8	7–31	27		B.C.I.	Smith 1977

Alouatta pigra Guatemalan Howler Monkey

Taxonomic Note

Older authors regarded this form as a subspecies of *Alouatta palliata* (Hall and Kelson 1959, Leopold 1959). But it has been shown to have characteristics which are distinct and nonintergrading with those of *A. palliata* (Smith 1970), and many workers now consider it to be a separate species (Hershkovitz 1972, Jones et al. 1974).

Geographic Range

Alouatta pigra is found in southeastern Mexico, central and northern Guatemala, Belize, and perhaps northern Honduras (Smith 1970). It has been recorded as far north and east as about 21°N, 88°W in Yucatán, as far west as about 92°30′W in

Tabasco, and as far south as about 15°N in southern Guatemala (Smith 1970). For details of the distribution, see Table 61.

Abundance and Density

A. pigra is reported to be declining in Mexico; no evidence is available concerning its status elsewhere (Table 61). In the Tikal N.P. in northeastern Guatemala, population density of *A. pigra* was 5 individuals per km^2; 4 groups consisted of 6 or 7 members each, with an average of 6.25 per group (Coelho et al. 1976).

Habitat

A. pigra occupies one of the few relatively undisturbed and extensive rain forests remaining in middle America and seems to prefer extensive, undisturbed, mesic tropical forest (Smith 1970). In Mexico it occupies "quasi-rainforest" (Jones et al. 1974). In Guatemala it has been observed in thick, semideciduous forest from sea level to 250 m elevation (Coelho et al. 1977).

Factors Affecting Populations

Habitat Alteration

Forest clearing was responsible for the early disappearance of *A. pigra* from most of the Yucatán Peninsula (Leopold 1959), and habitat destruction is continuing to contribute to its decline in Mexico (I.U.C.N. 1978). In Guatemala the primary forest in areas inhabited by *A. pigra* has been reduced to patches by fruit cultivation firms (Boshell and Bevier 1958) and shifting agriculture (Coelho et al. 1977).

Habitat alteration due to road construction and spreading agriculture may be responsible for the present sympatry between *A. pigra* and *A. palliata* in Tabasco, Mexico (Smith 1970). Since *A. palliata* can occupy disturbed and secondary forest while *A. pigra* cannot, habitat disturbance may allow *A. palliata* to replace the more specialized *A. pigra* (Smith 1970).

Human Predation

Hunting. *A. pigra* is reported to be hunted in Mexico (I.U.C.N. 1978), but is protected by its reputed foul taste in Guatemala (Boshell and Bevier 1958).

Pest Control. No information.

Collection. No information.

Conservation Action

A. pigra occurs within one national park in Guatemala (Table 61). It is listed as threatened under the U.S. Endangered Species Act (U.S. Department of Interior 1977b), and may be included with *A. palliata* in Appendix I of the Convention on International Trade in Endangered Species of Wild Fauna and Flora (I.U.C.N. 1978).

Table 61. Distribution of *Alouatta pigra* (countries are listed in order from north to south)

Country	Presence and Abundance	Reference
Mexico	In E Chiapas, E Tabasco, S Campeche, and SW Quintana Roo (map)	Hall and Kelson 1959
	Localities mapped in Yucatán and W Chiapas; had disappeared from Yucatán by 1917	Leopold 1959
	Localities mapped in E Chiapas, E Tabasco, central Campeche, and E Yucatán	Smith 1970
	Localities listed from N and N central Campeche, S, central, E, and N Quintana Roo; probably occurs throughout Yucatán Peninsula	Jones et al. 1974
	In Palenque and Lacandona forests of Chiapas	Alavarez del Toro 1977
	Declining	I.U.C.N. 1978
Guatemala	At low densities in lower Motagua valley (in SE)	Boshell and Bevier 1958
	In N, central, and SE (localities mapped)	Smith 1970
	In Tikal N.P. (in NE near 17°13′N, 89°24′W)	I.U.C.N. 1975, Coelho et al. 1976
Belize	Present	Disney 1968
	3 localities mapped	Smith 1970
Honduras	Questionable in NE (should be NW?)	Smith 1970

Alouatta seniculus Red Howler Monkey

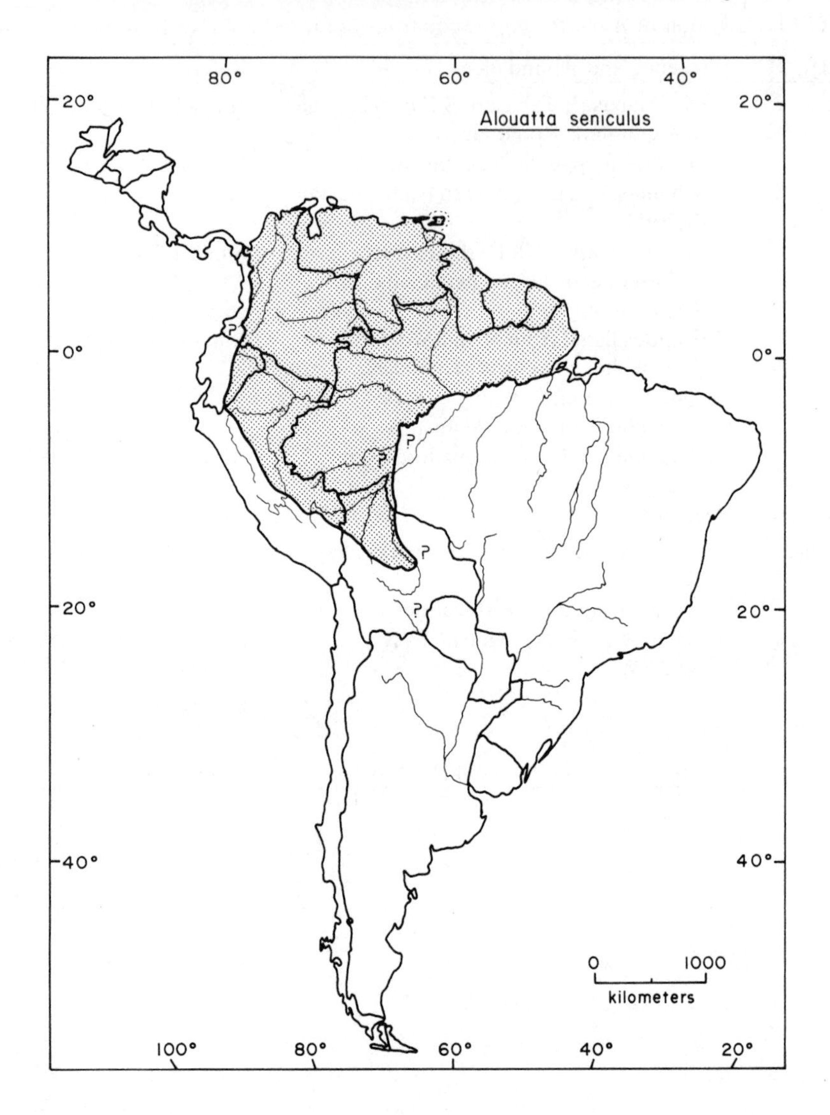

Taxonomic Note

Four (Cabrera 1958) or five (Napier and Napier 1967) subspecies are recognized.

Geographic Range

Alouatta seniculus is found in northwestern South America, from northern Colombia to central Bolivia and east into northern Brazil. The northern limit of its range is near Riohacha (11°34′N, 75°58′W), and the western limit is near Sautata (7°52′N, 77°06′W), Colombia (Hernández-Camacho and Cooper 1976). It has been recorded as far south as Sandia, Peru (14°18′S, 69°25′W) (Grimwood 1969), and Buena Vista,

Bolivia (17°28′S, 63°37′W) (Napier 1976). It is found as far east as the Atlantic coast north of the mouth of the Amazon in Amapá, Brazil (Cabrera 1958). For details of the distribution, see Table 62.

Abundance and Density

A. seniculus has recently been described as locally common in some parts of Peru and Bolivia, as abundant in Brazil and Surinam, and as perhaps increasing in some areas of Venezuela (Table 62). But its populations are also judged to be decreasing locally and vulnerable in Peru, exterminated in many parts of Trinidad, and possibly threatened in Guyana, Surinam, and French Guiana (Table 62). They are not common in some portions of Colombia, but their status there and in Ecuador is not known.

Population density estimates vary from 1.6 to 120 individuals per km^2 (Table 63). The average group size is between 3 and 11, with some groups as large as 25 and home ranges of 0.7 to 400 ha (Table 63).

Habitat

A. seniculus is a forest species which is found in a wide variety of habitats. In Colombia it occupies mangrove swamp, gallery forest, deciduous tropical forest, rain forest, oak and other cloud forest, small isolated forest patches, and secondary growth (Hernández-Camacho and Cooper 1976). It has been seen in lakeside and riverine vegetation (Klein and Klein 1976), and in northern Colombia is more abundant in riparian forest than in nearby dry upland forest (Struhsaker et al. 1975). It is able to live in old secondary forest (Neyman 1977), and inhabits young (5 to 15 years old) and mature (more than 15 years old) secondary forest, but not forest less than 5 years old (Green 1978b).

In Peru, *A. seniculus* occurs throughout the low selva zone and the lowest portion of the high selva zone (Grimwood 1969). It has been recorded in lowland rain forest (Gardner, pers. comm. 1974), and on flood plains and high ground in black-water and white-water drainages in evergreen, broadleaf forests (Freese et al., in prep.). Brazilian red howler monkeys utilize primary *terra firme* (nonflooding) forest, primary *várzea* forest (flooded by white-water rivers), and *igapó* forests (flooded by black and clear-water rivers) (Mittermeier and Coimbra-Filho 1977).

In Surinam, *A. seniculus* is common in inland high tropical rain forest, riverbank high forest, riverbank marsh forest, swamp forest, and swamp wood, and is also present in inland low tropical rain forest, ridge forest, savanna forest, mountain savanna forest, liane forest, *Mora* forest, "scholbos" (island marsh forest in swampy coastal plain), *Pina* marsh forest, swamp scrub, *Avicennia* mangrove forest, and cultivated plantations (Mittermeier 1977a). In Guyana, population densities were relatively high in mid- and late-successional forests, but lower in younger regenerating areas (Muckenhirn and Eisenberg 1978).

Venezuelan *A. seniculus* has been observed in gallery forest (Pirlot 1963), tall deciduous forest, thorn forest, cloud forest, savanna, and palm tree savanna (Neville 1972). In one study, 69% of the *A. seniculus* seen were near streams or in other moist areas; 80% were in the canopy of evergreen forest, and the remainder were in deciduous forest or in scattered trees in savanna or orchards (Handley, pers. comm. 1974). Fifty-four percent were in tropical humid forest, 25% in tropical dry forest, and the rest in tropical very humid, and premontane humid and very humid forests (life zones of Ewel and Madriz 1968).

On Trinidad, *A. seniculus* has been recorded in evergreen seasonal forest (Neville 1972), including one variety named the Crappo-Guatecare type, characterized by *Ischnosiphon arouma* and a variety of other trees (Downs et al. 1968).

A. seniculus is found at a wide range of elevations, from sea level to higher than 3,200 m in the central Andes of Colombia (Hernández-Camacho and Cooper 1976). In Venezuela it occurs up to 1,000 m elevation (Neville 1972), but the majority are seen below 500 m (Handley, pers. comm. 1974).

Factors Affecting Populations

Habitat Alteration

"In areas where all other species of monkeys have been driven out or caused to disappear because of deforestation . . . the red howler monkey may still be found occupying the last isolated stands of timber" (Hershkovitz 1949). But red howler monkeys so isolated in forest patches with limited food supplies "eat less, travel less, their growth becomes stunted, their resistance to disease and parasites is reduced," they are easier prey for predators, show characteristics of "localized degeneration," and eventually will be exterminated (Hershkovitz 1949).

Thus, although this species has a "great ability to survive environmental fluctuations" (Hershkovitz 1949) and is "highly adaptable" and able to utilize small forest patches, secondary growth, and other disturbed habitats, forest destruction nevertheless poses the greatest threat to its survival, at least in Colombia (Hernández-Camacho and Cooper 1976). Much of the forest of northern Colombia has been cleared for cultivation and pastureland, and in many areas only isolated forest patches remain (Comm. Cons. Nonhuman Prim. 1975b, Bernstein et al. 1976a, Green 1978b, Neyman 1978).

In Peru many of the areas in which *A. seniculus* was sighted have undergone light to moderate timbering (Freese et al., in prep.), but no data are available on the effects of this on *A. seniculus* populations. At one site in northern Bolivia extensive land clearing has contributed to the decline of monkey populations (Heltne et al. 1976).

Throughout the Guianas, forests are being cleared for settlements and roads, and the wilderness is rapidly being opened (Poonai 1973). In northeastern Surinam the Affobakka Dam on the Surinam River flooded about 2,330 km^2 of forest habitat; 479 individual *A. seniculus* were transported to higher ground (Walsh and Gannon 1967), but the species cannot reinvade this portion of its former range. Surinam's *A. seniculus* usually disappears in areas of habitat disturbance, but is considered adaptable to human encroachment (Mittermeier 1977a). In Guyana, *A. seniculus* can survive in areas of low intensity, highly selective logging (Muckenhirn et al. 1975), but in general is not considered adaptable to human encroachment (Muckenhirn and Eisenberg 1978).

In parts of the northern llanos of Venezuela, recent control of dry season fires has resulted in an increase of forest and would be expected to lead to an increase in *A. seniculus* populations (Neville 1972). But in other parts of the country the cloud forest habitats of this species are being destroyed (Mondolfi 1976). On Trinidad, forest cutting has restricted *A. seniculus* to remnant patches (Neville 1976a).

In summary, *A. seniculus* populations are known to have declined because of deforestation in Colombia, Peru, Bolivia, and Trinidad. Forest clearing is also occurring in the other countries of the range, but its effects have not been measured.

Human Predation

Hunting. Red howler monkey flesh is highly prized as a food item in central Colombia, and the species has been eliminated from many suitable relict forests by hunting (Hernández-Camacho and Cooper 1976). *A. seniculus* is also hunted heavily in several regions of Peru (Gardner, pers. comm. 1974; Neville et al. 1976; Encarnación and Castro 1978; Freese et al., in prep.), and populations are much larger in areas in which hunting is effectively prohibited (Freese 1975). The meat of *A. seniculus* is not highly esteemed in Peru (Grimwood 1969), but consumption of it in the Iquitos area probably does represent a significant drain on the population (Freese et al. 1976, Neville et al. 1976).

A. seniculus is also hunted for food in Bolivia, where some populations have been greatly reduced (Heltne et al. 1976), in Brazil (Mittermeier and Coimbra-Filho 1977), in Surinam (Husson 1957, Mittermeier 1977a), and in Trinidad (Neville 1976a), and is shot for sport (Neville 1972) and meat (Mondolfi 1976) in Venezuela.

This species is killed for use as bait in jaguar and ocelot traps in Peru (Encarnación and Castro 1978), Bolivia (Heltne et al. 1976), and Brazil (Mittermeier and Coimbra-Filho 1977). Its hyoid bone is used as a drinking and medicinal device in Colombia (Hernández-Camacho and Cooper 1976), Peru (Castro 1976), and Surinam (Mittermeier 1977a).

Pest Control. Red howler monkeys do not normally supplement their food supply with items from cultivated fields (Hershkovitz 1949). Thus they would not be expected to be subject to control efforts.

Collection. A. seniculus is collected for export from Colombia (Green 1976), but it brings a low price on the export market and consequently few are exported (Cooper and Hernández-Camacho 1975). Only 55 were exported from Peru between 1962 and 1971, owing to the lack of demand, scarcity, and difficulty of trapping this species (Soini 1972).

Red howler monkeys are occasionally kept as pets in Brazil (Mittermeier and Coimbra-Filho 1977) and in Surinam (Mittermeier 1977a), but they do not survive well in captivity and usually die within a few months (Husson 1957).

Between 1968 and 1973 a total of 74 *A. seniculus* were imported into the United States (Appendix), primarily from Colombia but also from Nicaragua, Honduras, and Paraguay, where the species does not occur. None entered the United States in 1977 (Bittner et al. 1978). In 1965–75, 139 *A. seniculus* were imported into the United Kingdom (Burton 1978).

Conservation Action

A. seniculus is found in eleven established national parks, one projected national park, and a total of fifteen reserves, wildlife sanctuaries, and nature parks (Table 62). But some of these areas are not fully protected. For example, La Macarena National Park in Colombia has shrunk to less than half its original size since 1971, owing to illegal encroachment by agriculturists, and will have totally disappeared within a few years unless immediate and effective protective measures are taken (Struhsaker 1976b). Similarly, the two national reserves in Peru are not effectively patrolled against poaching (Castro 1976).

Peruvian law (Supreme Decree 934-73-AG) prohibits hunting, capture, and commerce in Amazonian monkeys, including *A. seniculus* (Castro 1976), but the hunting prohibition is not adequately enforced (Neville et al. 1976), and should be if populations are to be conserved (Freese 1975). Brazilian law protects *A. seniculus* from commercial exploitation (Mittermeier and Coimbra-Filho 1977). In Surinam, protective laws prohibit hunting in the northern part of the country and transporting monkey meat to northern population centers (Mittermeier 1977a). Legal restrictions prevent the export of *A. seniculus* from Venezuela (Mondolfi 1976).

Table 62. Distribution of *Alouatta seniculus* (countries are listed in counterclockwise order)

Country	Presence and Abundance	Reference
Colombia	Throughout except in SW; from N Chocó (R. Atrato) and Darién, Cartagena, and Santa Marta region S	Hershkovitz 1949
	E of San Martín (3°40′N, 73°20′E)	Thorington 1968a
	Near R. Magdalena, Santander (near 6°38′N, 73°55′W)	Sturm et al. 1970
	In Puracé N.P., Cauca-Huila; Cueva de los Guácharos N.P., Huila; Isla de Salamanca N.P., Magdalena; Tayrona N.P., Magdalena; Farallones de Cali N.P., Valle; Las Orquideas N.P., Antioquia; La Macarena N.P., Meta; Los Nevados N.P. (projected, Caldas-Quindío-Tolima); and El Tuparro Faunistic Territory, Vichada	Green 1972
	In La Macarena N.P.	Hunsaker 1972
	In Isla de Salamanca N.P., in N on Caribbean coast, near mouth of R. Magdalena	I.U.C.N. 1975
	In La Macarena N.P. (3°N, 73–74°W), N bank R. Guayabero	Klein and Klein 1975
	At sites in N near coast, and NE near Venezuelan border (localities listed)	Struhsaker et al. 1975, Scott et al. 1976
	Near Ventura, Bolívar, between R. Cauca and R. Magdalena (8°22′N, 74°22′W)	Bernstein et al. 1976a
	Near Puerto Rico, Tequisio region, N Bolívar (8°34′N, 74°14′W)	Green 1976, 1978b

Table 62 (continued)

Country	Presence and Abundance	Reference
Colombia (cont.)	Throughout except on Pacific coast and desert of Guajira Peninsula (in N), questionable in SW (map); in most lowland areas; not common in Amazonian region and parts of R. Magdalena and R. Cauca valleys	Hernández-Camacho and Cooper 1976
	At R. Peneya, tributary of R. Caquetá (in S)	Izawa 1976
	In Caquetá region and in Meta	Moynihan 1976a
	In good numbers in La Macarena N.P.	Struhsaker 1976b
	E and W of R. Magdalena in Bolívar, Cordoba, César, Meta, and Santander (localities listed)	Yunis et al. 1976
	Right bank of R. Duda, W boundary of La Macarena N.P.	Izawa and Mizuno 1977
	In N, NE of Tolú, Sucre (about 9°34′N, 75°27′W); also near Colosó, Serranía de San Jacinto, and along R. Sinú near Montería	Neyman 1977, 1978
Ecuador	In E	Hershkovitz 1949
	At Macas (central, 2°22′S, 78°08′W) and R. Copataza	Napier 1976
Peru	In NE	Hershkovitz 1949
	In N, E of Andes	Cabrera 1958
	Along R. Napo, R. Marañón, R. Huallaga, R. Ucayali, and R. Curanja, Loreto, and in San Martín, Huánuco, Pasco, Junín, Cuzco, Puno, and Madre de Dios (localities listed); has disappeared near major rivers and settlements; fairly common elsewhere	Grimwood 1969
	Common at R. Curanja, SE Loreto; decreased 1966–71	Gardner, pers. comm. 1974
	Along R. Samiria, R. Pucallpa, and R. Orosa (in N); in Manú N.P. (in E); uncommon along R. Ampiyacu (in NE)	Freese 1975
	In Manú N.P., Madre de Dios and Cuzco; Tingo Maria N.P., Huánuco; and Pacaya and Samiria National Reserves, Loreto; vulnerable	Castro 1976
	Abundant in R. Samira basin; common in R. Pacaya basin (both in N)	Freese et al. 1976, Neville et al. 1976
	At headwaters of R. Acre and R. Tahuamanu (in E, near 11°S, 70–71°W, in Madre de Dios)	Encarnación and Castro 1978
	At R. Yavineto (R. Putumayo), Loreto	Encarnación et al. 1978
	Common at Cocha Cashu, Manú N.P.	Janson and Terborgh, in press

Table 62 (continued)

Country	Presence and Abundance	Reference
Peru (cont.)	Along upper R. Nanay (3°45′S, 74°10′W); R. Orosa (3°22′S, 72°07′W); R. Samiria (4°45′S, 74°20′W); at Panguana (9°35′S, 74°57′W); and Cocha Cashu (11°50′S, 71°23′W)	Freese et al., in prep
Bolivia	In central	Cabrera 1958
	Abundant at El Triunfo (central, 15°15′S, 64°15′W); uncommon at Cobija (in NW at 11°01′S, 68°45′W); at Ixiamas (13°46′S, 68°08′W)	Heltne et al. 1976; Freese et al., in prep
	Localities listed: Buena Vista and Sara Province	Napier 1976
Brazil	At Seringal Oriente, W Acre	Carvalho 1957
	In W as far S as R. Purus; in Amazonian region from R. Negro E to coast	Cabrera 1958
	In R. Negro region near Manaus (N central)	Avila-Pires 1964
	In Amazonas near Manaus and in Roraima (in N; localities mapped)	Deane et al. 1969
	Along Manaus-Itacoatiara Road, Amazonas	Deane et al. 1971
	Localities mapped in NW Acre, Amazonas, Roraima, and Amapá	Deane 1976
	Localities listed: R. Solimões, R. Juruá, R. Negro, and R. Mamoré, W Rondônia	Napier 1976
	In N, NW, and NE in Amazonas, Acre, Roraima, N Pará, and Amapá	Neville 1976a
	In Amazonia N of mainstream and E to R. Purus S of mainstream	Mittermeier 1977a
	Near Oriximiná (1°45′S, 55°50′W); at R. Jacurapá (in W near Leticia); and at R. Panauá (in W); abundant along mainstream R. Amazonas; not endangered	Mittermeier and Coimbra-Filho 1977
French Guiana	Present	Cabrera 1958
	Will be threatened if present pressure continues	Poonai 1973
	Near Saül (185 km SW of Cayenne)	Mittermeier, Bailey, and Coimbra-Filho 1978
Surinam	Quite common throughout; localities listed in NE, N, and E	Husson 1957
	Present	Cabrera 1958, Fooden 1964
	In NE near Affobakka	Walsh and Gannon 1967
	Will be threatened if present pressure continues	Poonai 1973
	Localities listed include Paramaribo and R. Coppename	Napier 1976

Table 62 (continued)

Country	Presence and Abundance	Reference
Surinam (cont.)	Common throughout interior and almost all coastal region (localities mapped); in Brownsberg Nature Park (in NE) and 5 nature reserves: Raleighvallen-Voltzberg (N central), Tafelberg (central), Eilerts de Haan Gebergte, Sipaliwini Savanna (S central), and Wia-Wia (NE); in little danger	Mittermeier 1977a
Guyana	Present	Cabrera 1958
	Will be threatened if present pressure continues	Poonai 1973
	In central and N; localities mapped include R. Berbice, Apoteri, Matthews Ridge, Pomeroon, and 3 forest reserves: 24 Mile, Moraballi, and H.M.P.S.	Muckenhirn et al. 1975
	Localities listed include R. Essequibo, R. Mazaruni (in W), R. Rupununi (in W), and R. Supinaam	Napier 1976
	Widespread; moderately vulnerable	Muckenhirn and Eisenberg 1978
Venezuela	At Parmana, Guárico (left bank R. Orinoco)	Racenis 1952
	Throughout including W of L. Maracaibo, from Cordillera de Mérida through basin of R. Orinoco into S	Cabrera 1958
	At Palmar, Zulia (in W)	Pirlot 1963
	In E Guárico near Calabozo (8°58′N, 67°28′W), and Henry Pittier (Rancho Grande) N.P., N of Maracay, Aragua (N central); may be increasing locally	Neville 1972, 1976a
	Throughout in states of Falcón, Apure, Amazonas, Bolívar, Zulia, Monagas, Trujillo, and Guárico (localities mapped)	Handley, pers. comm. 1974
	One locality mapped (central)	Deane 1976
	Amply distributed throughout including Guatopo N.P. (N central); locally abundant	Mondolfi 1976
	Localities listed in Mérida (in SW)	Napier 1976
Trinidad	On E central coast	Downs et al. 1968
	In Bush Bush Forest (E central); exterminated in many areas and range is contracting	Neville 1972
	One locality on S coast	Napier 1976
	Remnant populations in 3 wildlife sanctuaries: Bush Bush, Central Range, and Trinity Hills (in SE); and in NW, SW, and E (localities mapped)	Neville 1976a

Table 63. Population parameters of *Alouatta seniculus*

Population Density (individuals/km²)	Group Size $\bar{X}$	Range	No. of Groups	Home Range Area (ha)	Study Site	Reference
		2–6			N Colombia	Bernstein et al. 1976a
	6–8	2–15			Colombia	Hernández-Camacho and Cooper 1976
	5.4	3–11	29	300–400	S Colombia	Izawa 1976
11.6–29		3–6	6–15		Central Colombia	Klein and Klein 1976
20.4–21.9					N Colombia	Green 1978b
		≤25			Peru	Grimwood 1969
24–29.5	5	2–7			NE and E Peru	Freese 1975
15	5	4–6	2		NE Peru	Neville et al. 1976
18–48	6	4–7	22	10–20	SE Peru	Janson and Terborgh, in press
8–120	3–7.4	1–14	37		Peru and Bolivia	Freese et al., in prep.
8–120	8	1–14	13		Bolivia	Heltne et al. 1976
	4.3	2–7	4		Brazil	Mittermeier and Coimbra-Filho 1977
		5–6			Surinam	Husson 1957
17	4.3	2–8	24	6–11	Surinam	Mittermeier 1977a
1.6–63.8	5.3	2–11	20		Guyana	Muckenhirn et al. 1975
61–108	8.5	4–16	30	0.7–7.1	Venezuela	Neville 1972
	5.4–11	4–15	19		Venezuela	Neville 1976a
114	8	7–9	2	6.6	Trinidad	Neville 1972

Aotus trivirgatus Night Monkey, Owl Monkey, or Douroucouli

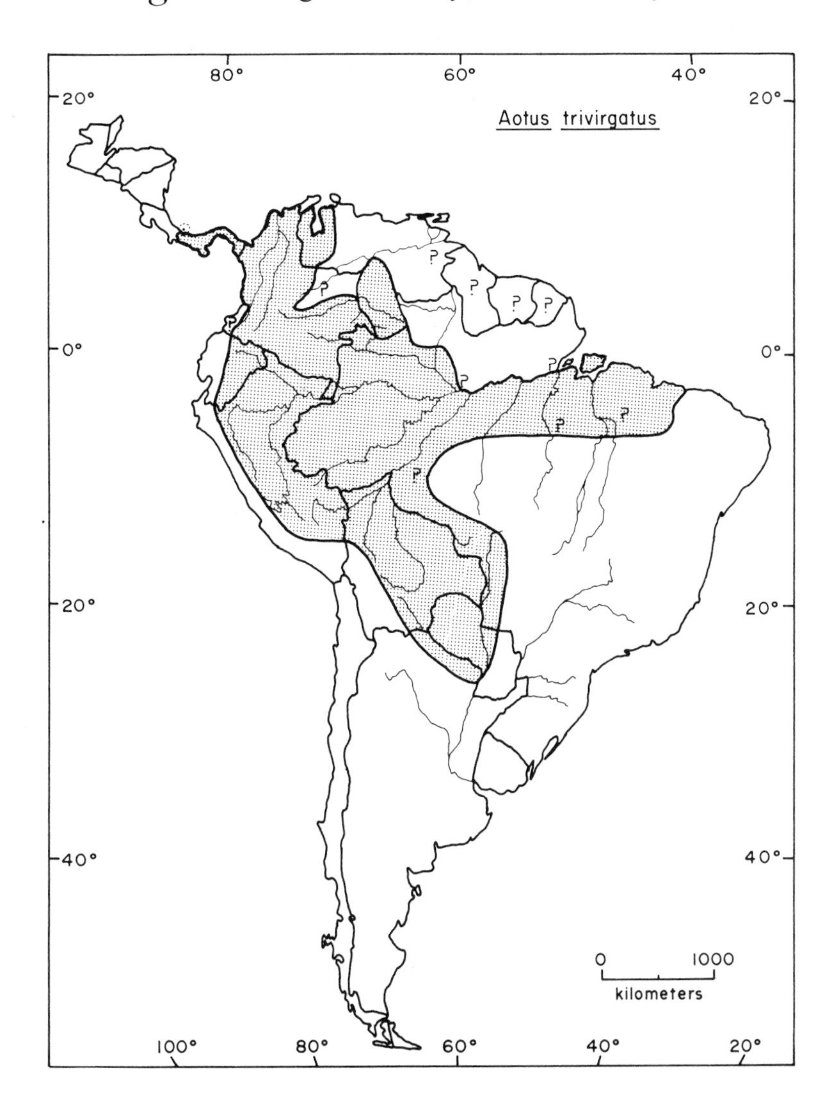

Taxonomic Note

Some workers consider several forms of night monkeys to be distinct species (Brumback et al. 1971; Brumback 1973, 1974, 1976). But most authors treat all *Aotus* as one species with eight (Hershkovitz 1949) or nine races (Napier and Napier 1967), and it is treated as such here. The recognized subspecies may not be valid genetic entities (Thorington 1976, Thorington and Vorek 1976).

Geographic Range

Aotus trivirgatus is found from Panama southward through northern and central South America to northern Argentina. It has been recorded as far west as Chiriquí,

Panama (de Rodaniche 1957), and Isla Bastimentos (west of 82°W), off the northwest coast of Panama (Handley 1966), and as far north as about 11°30′N in northeastern Colombia (Hernández-Camacho and Cooper 1976). The range extends as far south as about 25°–27°S in northern Argentina (Thorington 1975, Rathbun and Gache 1977, Colillas and Coppo 1978), and as far east as about 43°W in eastern Maranhão, northeast Brazil (Kumm and Laemmert 1950). For details of the distribution, see Table 64.

Abundance and Density

A. trivirgatus has recently been described as abundant in Peru and parts of Panama; as "successful" in Colombia but with its range shrinking in the north; as vulnerable in Peru; as locally common but nowhere abundant and with a decreasing range in Argentina; and as not endangered in Brazil (Table 64). Its present status in Ecuador, Bolivia, Paraguay, and most of Panama is not known, but the species as a whole is thought to be in little danger of extinction (Thorington and Vorek 1976).

Night monkeys live in small family groups consisting of one pair and their young (Grimwood 1969, Hernández-Camacho and Cooper 1976). Group sizes are usually 2 to 7 (Table 65), but large feeding congregations of up to 35 individuals have been observed (Durham 1972). Population density estimates vary from 1.5 to 150 per km^2, and one group had a home range of 3.1 ha (Table 65).

Habitat

A. trivirgatus is a forest species found in many types of forest vegetation. In Panama it lives in evergreen forest (Handley 1966), usually of tall mature stature (Moynihan 1967), but has also been observed in secondary growth (Telford et al. 1972, Fleming 1973, Moynihan 1976a). In Colombia it occupies every major forest zone except mangrove swamp and montane savanna-scrub (Hernández-Camacho and Cooper 1976). These authors found that it can also occupy secondary growth and disturbed habitats, and believe that it is successful because of its wide range of habitats. Others working in Colombia have concluded that *A. trivirgatus* occupies only older diverse forests, avoiding young, undiversified ones (Heltne 1978), and that it is more abundant in large blocks of wet forest than in similar, nearby gallery forest (Struhsaker et al. 1975).

Peruvian night monkeys are adaptable to a variety of habitats throughout the low selva and lower levels of the high selva zones (Grimwood 1969). At specific study sites they have been observed in lowland and highland wet forest (Durham 1972); riverine and clearing-edge vegetation, tangled vines, and bamboo stands (Gardner, pers. comm. 1974); continuous primary tropical forest (Wright 1978); and flood plain and high ground evergreen, mesophytic, broadleaf forests in black-water river drainages (Freese et al., in prep.). In Bolivia they were found in dryer forest on the edge of the *chaco* (Heltne et al. 1976), and in high ground evergreen forest in a white-water drainage (Freese et al., in prep.).

Argentine *A. trivirgatus* occupies both low and tall forests with dense canopies and high species diversities, which occur in a disjunct mosaic of forest islands surrounded by nonforest vegetation and in gallery forests along rivers (Rathbun and

Gache 1977). In Paraguay it has been recorded in dry thorn forest with dense and spiny undergrowth (Wetzel and Lovett 1974). In Brazil it has been sighted in *igapó* forest, periodically flooded by a black-water river (Mittermeier and Coimbra-Filho 1977).

During one study in Venezuela, 96% of all *A. trivirgatus* were seen in evergreen forest; 56% of these were in moist areas and 44% in dry habitats (Handley, pers. comm. 1974). Seventy-nine percent of all sightings were in tropical humid forest, 14% in premontane very humid, and the remainder in tropical very humid and tropical dry forest (life zones of Ewel and Madriz 1968). Ninety-nine percent were below 500 m elevation, and 1% were between 500 and 1,200 m (Handley, pers. comm. 1974). Elsewhere, *A. trivirgatus* has been recorded as high as 1,500 m (Durham 1975) and 3,200 m (Hernández-Camacho and Cooper 1976).

Factors Affecting Populations

Habitat Alteration

A. trivirgatus can survive in close association with man, and "seems to be exterminated only in areas in which the forest is completely destroyed" (Thorington and Vorek 1976). This is occurring in several portions of its range. For example, in northern Colombia its habitat is being rapidly reduced by deforestation (Thorington 1975). Forests suitable for night monkeys now cover only 1 to 5% of the total area there, and forest cutting is proceeding at a fast pace, for instance, in the Sierra Nevada de Santa Marta (northeast Magdalena) and in the Serranía de San Lucas (southeast Bolívar) (Heltne 1978). In many areas of northern Colombia the remaining forest areas are threatened by intensive colonization and will probably disappear within the next five years; much of the deforested land is already showing signs of serious soil desiccation and erosion (Heltne 1978).

At one study site in northern Bolivia, extensive land clearing has contributed to the sharp decline of monkey populations (Heltne et al. 1976). In northern Argentina the forests have been greatly fragmented and reduced because of agricultural clearing, and deforestation for timber and land clearance is proceeding steadily (Rathbun and Gache 1977). These authors found no evidence of long-term planning of forest exploitation or of efforts toward reforestation, and concluded that the greatest threat to *A. trivirgatus* in Argentina was extensive habitat alteration. In eastern Pará, Brazil, large areas of forest have also been destroyed for agriculture and cattle raising (Ayres 1978).

No information is available regarding the effect of habitat disturbance on *A. trivirgatus* throughout most of its range.

Human Predation

Hunting. *A. trivirgatus* is sometimes killed for food in Peru (Gardner, pers. comm. 1974; Freese 1975; Neville et al. 1976) and in Brazil (Mittermeier and Coimbra-Filho 1977). Its fur is also used to decorate bridles in Colombia (Hernández-Camacho and Cooper 1976).

Pest Control. No information.

Collection. Night monkeys are popular as pets (Hershkovitz 1972), and are kept as pets, for example, in northern Colombia (Heltne 1978), northern Argentina (Rathbun and Gache 1977), and Brazil (Mittermeier and Coimbra-Filho 1977).

A. trivirgatus has been heavily collected for export for medical research, especially from Colombia and Peru. In the late 1960s and early 1970s, several thousand were collected annually from the region of the lower Río Cauca and neighboring middle Río Magdalena valleys of Colombia; the effect of this on the local status of the species is not yet known (Hernández-Camacho and Cooper 1976). In 1970 and 1971 totals of 2,825 and 1,526 *A. trivirgatus* respectively were listed as being exported legally from Colombia (Cooper and Hernández-Camacho 1975). Some of the 5,133 unidentified monkeys that left that country in 1970 were probably also night monkeys, since 3,683 individuals of this species were reported to have entered the United States from Colombia in 1970 (Paradiso and Fisher 1972).

From 1964 to 1974, 3,798 *A. trivirgatus* were exported from Peru (Castro 1976). Before 1974 the species had not been heavily exploited in Bolivia, but in the future collection will probably occur there (Thorington and Vorek 1976), and 500 to 1,000 per year could be obtained from forests destined for clearcutting in the central part of that country (Heltne et al. 1976).

Between 1968 and 1973 a total of 23,505 *A. trivirgatus* were imported into the United States (Appendix), primarily from Colombia and Peru. In 1977, 16 night monkeys entered the United States (Bittner et al. 1978). From 1965 to 1975, 1,182 *A. trivirgatus* were imported into the United Kingdom (Burton 1978).

This species is used extensively in biomedical research in the United States (Thorington 1972, Comm. Cons. Nonhuman Prim. 1975a). One laboratory alone used more than 4,600 imported *A. trivirgatus* in 1969–73 (Schmidt 1973).

Conservation Action

A. trivirgatus occurs in four existing national parks, one projected national park, and one faunistic territory in Colombia, and one national park and two reserves in Peru (Table 64). But at least one of the parks in Colombia is not adequately protected against illegal encroachment by agriculturists and is consequently shrinking in size (Struhsaker 1976b), and in the two reserves in Peru, hunting is not effectively controlled (Castro 1976). No information was located regarding protected habitat areas in the other countries of the range.

In 1971 and 1972 Colombia passed legislation restricting capture of *A. trivirgatus* to 100 individuals per month for each permit holder, to a total of 700 between 1 March and 1 November, and limiting permits to collectors presenting a certificate of need from an established and approved biomedical research institution (Cooper and Hernández-Camacho 1975, Green 1976). The number of permits granted and the enforceability of this regulation are not known, but "until reliable data become available regarding trapping methods used, specific areas of exploitation, biological cycles of *A. trivirgatus* and the size of the regional population, licenses for trapping and export are being carefully screened" (Hernández-Camacho and Cooper 1976).

Detailed ecological and distributional information and the establishment of forest reserves are crucial for the conservation of this species in northern Colombia (Struhsaker et al. 1975). More specifically, no private trapping should be permitted in the Sierra Nevada de Santa Marta region, no more than one *A. trivirgatus* per year should be removed from any forest patch of less than 20 ha, yearly inventories should be made to determine population trends in remnant forests, and new arrangements should be made with the agrarian reform agency to reduce pressure on the remaining forested areas there (Heltne 1978).

In Peru, Supreme Decree No. 934-73-AG prohibits the hunting, capture, or commerce in Amazonian wildlife (Castro 1976). This has stopped the export of *A. trivirgatus* from Peru, but enforcement of hunting regulations is difficult (Freese 1975). The trapping of Bolivian populations should be done only in areas where clearcutting of forest is more than 50% complete and continuing, and by a professional collecting team to minimize losses due to stress, disease, and poor nutrition during transport (Heltne et al. 1976).

Recommendations for conserving *A. trivirgatus* in Argentina include: (1) increase enforcement of laws protecting it; (2) establish completely protected reserves; (3) grant permission to capture only in areas destined for deforestation; and (4) make a long-term study of social structure and population dynamics to determine sustained yield potential and breeding parameters (Rathbun and Gache 1977).

Brazilian law prohibits commercial exploitation of *A. trivirgatus* (Mittermeier and Coimbra-Filho 1977), and legal restrictions prevent the export of this species from Venezuela (Mondolfi 1976).

No recommendations or reports of conservation action were received for *A. trivirgatus* in Panama, Ecuador, or Paraguay.

Table 64. Distribution of *Aotus trivirgatus* (countries are listed in counterclockwise order)

Country	Presence and Abundance	Reference
Panama	Throughout E (map)	Hershkovitz 1949
	E of canal in Bayano R. valley, Madden Lake area, Darién, Panamá E and W of canal, and Chiriquí (SW)	de Rodaniche 1957
	Throughout E of 81°W (map)	Hall and Kelson 1959
	Common in E and possibly on Caribbean coast of W; rare on Azuero Peninsula; localities listed include: Darién, San Blas, Panamá, Canal Zone, and Bocas del Toro (Isla Bastimentos)	Handley 1966
	At Cerro Azul (E of canal)	Galindo and Srihongse 1967
	Between 77°W and 80°–81°W	Bennett 1968
	Throughout E of about 81°20′W and on island in archipelago of Bocas del Toro (map)	Mendez 1970
	In San Blas	Telford et al. 1972
	At Balboa and Cristobal	Fleming 1973
	In Panamá, Colón, and Darién	Sousa et al. 1974
	Common at Rodman Ammunition Depot (8°57′N, 79°37′W)	Dawson 1976

Table 64 (continued)

Country	Presence and Abundance	Reference
Colombia	In Bolívar, Magdalena, Atlántico, NW Antioquia, Chocó, and in Andes; questionable along W coast S of about 4°N; absent on Guajira Peninsula (map)	Hershkovitz 1949
	E of San Martín (central, 3°40′N, 73°20′E)	Thorington 1968a
	In Tayrona N.P., Magdalena; Farallones de Cali N.P., Valle; Las Orquideas N.P., Antioquia; La Macarena N.P., Meta; Los Nevados N.P. (projected, Caldas-Quindío-Tolima); and El Tuparro Faunistic Territory, Vichada	Green 1972
	At El Diamante (in NE) and Estanzuela (NW near coast)	Struhsaker et al. 1975
	In Serranía de San Lucas, S Bolívar; range in N greatly reduced	Thorington 1975
	Near Puerto Rico, Tequisio, Bolívar	Green 1976, 1978b
	Throughout except Guajira Desert in NE, NE plains, and mountains; questionable in SW along coast and in E Meta, Arauca, Boyacá, and S and E Vichada (all in E central) (map); very successful	Hernández-Camacho and Cooper 1976
	Along R. Peneya, middle basin of R. Caquetá and R. Putumayo (in S)	Izawa 1976
	Right bank of R. Duda, W boundary of La Macarena N.P.	Izawa and Mizuno 1977
	Localities listed in R. Magdalena basin and Sierra Nevada de Santa Marta in Magdalena, Bolívar, Boyacá, Antioquia; habitat rapidly disappearing in N	Heltne 1978
	NE of Tolú, Sucre near Caribbean coast (9°34′N, 75°27′W)	Neyman 1978
Ecuador	In Andes	Hershkovitz 1949
	In Andes and E	Cabrera 1958
	In E half (map)	Thorington 1975
	Localities listed include Napo (E central), Quito, R. Napo (0°5′S, 77°W, in NE), Gualaquiza (3°31′S, 78°W), and Macas (2°22′S, 78°08′W) (both in SE)	Napier 1976
	Localities mapped in E	Thorington and Vorek 1976
Peru	In Andes and E	Cabrera 1958
	Near Iquitos, R. Yavarí Mirim, R. Ucayali, R. Curanja, in Loreto; Amazonas, Huánuco, Junín, Ayacucho, Cuzco, and Puno; reasonably abundant	Grimwood 1969
	At R. Curanja, SE Loreto, population stable 1966–71	Gardner, pers. comm. 1974

Table 64 (continued)

Country	Presence and Abundance	Reference
Peru (cont.)	In SE at R. Piachaca, San Gaban–R. San Juan, Karos-Pilcopata, and Salvación-Madre de Dios	Durham 1975
	In Pacaya National Reserve (in NE between R. Marañón and R. Ucayali)	I.U.C.N. 1975
	Throughout central and E (maps)	Thorington 1975, Thorington and Vorek 1976
	In Manú N.P., Madre de Dios and Cuzco, and Pacaya and Samiria national reserves, Loreto; vulnerable	Castro 1976
	Localities listed include Amazonas, Chanchamayo (n.l.), Huánuco, Loreto, R. Pacaya, R. Ucayali, Pozuzo	Napier 1976
	In SE near Iberia and Iñapari between R. Acre and R. Madre de Dios	Encarnación and Castro 1978
	In NE near Iquitos	Moro 1978
	On R. Pichis near 10°S, 75°W, Pasco (central)	Wright 1978
	Along R. Nanay and R. Orosa (in NE)	Freese et al., in prep
	Uncommon at Cocha Cashu, Manú N.P.	Janson and Terborgh, in press
Bolivia	In Chaco (E) and Andes (central to W)	Hershkovitz 1949
	In central	Cabrera 1958
	At R. Surutu and Buena Vista, Santa Cruz de la Sierra (in SE) (locality mapped)	Soria 1959
	Throughout except SW (map)	Thorington 1975
	Localities listed include Cochabamba and Santa Cruz (both central)	Napier 1976
	Localities mapped in central and N central	Thorington and Vorek 1976
	Near Santa Cruz de la Sierra (near 17°27′S, 63°34′W)	Cambefort and Moro 1978
	At Cobija (NW corner)	Freese et al., in prep
Argentina	In Chaco (N central)	Hershkovitz 1949
	In Formosa (N central)	Cabrera 1958
	In N (map)	Thorington 1975
	In N in E Formosa and NE Chaco as far W as about 60°08′W (at 26°S) and S to 27°20′S, N of Corrientes (map); along many rivers (listed), common along several; nowhere abundant; in Formosa, total available range about 645 km^2, total population about 11,600 individuals; distribution disjunct, fragmented, and shrinking	Rathbun and Gache 1977
	In Chaco and Formosa (map)	Colillas and Coppo 1978
Paraguay	In Chaco (in W)	Hershkovitz 1949
	From Chaco to R. Paraguay basin (W half)	Cabrera 1958

Table 64 (continued)

Country	Presence and Abundance	Reference
Paraguay (cont.)	At Puerto Casado (central) (map)	Soria 1959
	Localities listed in W and S Boquerón (in W)	Wetzel and Lovett 1974
	Throughout W and E of R. Paraguay (map)	Thorington 1975
	One locality listed: central Chaco, N of Concepción	Napier 1976
	Two localities mapped in central	Thorington and Vorek 1976
Brazil	In N and upper (E) Amazon region, Chaco area, and Serra da Chapada, Mato Grosso	Hershkovitz 1949
	In SW Mato Grosso, E Pará, and E Maranhão (localities mapped)	Kumm and Laemmert 1950
	In S central, in Amazonian region to Marajó Island, and in basin of R. Solimões (in NW)	Cabrera 1958
	Locality S of Belém, E Pará	Carvalho and Toccheton 1969
	In Amazonas and Pará	Deane et al. 1969
	Throughout W; in central, S of R. Amazonas to mouth (map)	Thorington 1975
	Localities listed include R. Solimões, R. Purus, Mato Grosso	Napier 1976
	Localities mapped along R. Amazonas (questionable N of it in E) and in W and central	Thorington and Vorek 1976
	Not endangered	Mittermeier and Coimbra-Filho 1977
	In E Pará E of R. Xingu including near Igarapé Pedral, S of Ourém	Ayres 1978
French Guiana	In Guianas	Cabrera 1958
Surinam	In Guianas	Cabrera 1958
Guyana	In Guianas	Cabrera 1958
	Questionable in NW (map)	Thorington 1975
Venezuela	In S	Cabrera 1958
	In basins of R. Yasa and R. Tucuco, Sierra de Perijá, Zulia (in W)	Pirlot 1963
	Localities listed and mapped in Amazonas (in S), Trujillo, Táchira, and Zulia (all in W)	Handley, pers. comm. 1974
	In S and W (map)	Thorington 1975
	Locally distributed S of R. Orinoco in Amazonas from Brazo Casiquiare and R. Negro to Puerto Ayacucho and R. Manapiare; in NE Bolívar; N of R. Orinoco only around Lake Maracaibo in Táchira, Trujillo, and region of Perijá	Mondolfi 1976

Table 64 (continued)

Country	Presence and Abundance	Reference
Venezuela (cont.)	Localities listed include Mérida (in W) and San Carlos	Napier 1976
	Localities mapped in S and W	Thorington and Vorek 1976

Table 65. Population parameters of *Aotus trivirgatus*

Population Density (individuals/km^2)	Group Size X̄	Range	No. of Groups	Home Range Area (ha)	Study Site	Reference
		2–3			Panama	Moynihan 1967
		3–5			Colombia	Hernández-Camacho and Cooper 1976
1.5					N Colombia	Green 1978b
18–150	2–3	2–7	5		N Colombia	Heltne 1978
25–50	15.6	2–35	9		SE Peru	Durham 1972
4.5	3	2–3	4		SE Peru	Encarnación and Castro 1978
20–30		2–3			NE Peru	Moro 1978
		2–4		3.1	Central Peru	Wright 1978
		2–4			SE Peru	Janson and Terborgh, in press
18	2.9	2–4	25		N Argentina	Rathbun and Gache 1977

Ateles paniscus Spider Monkey

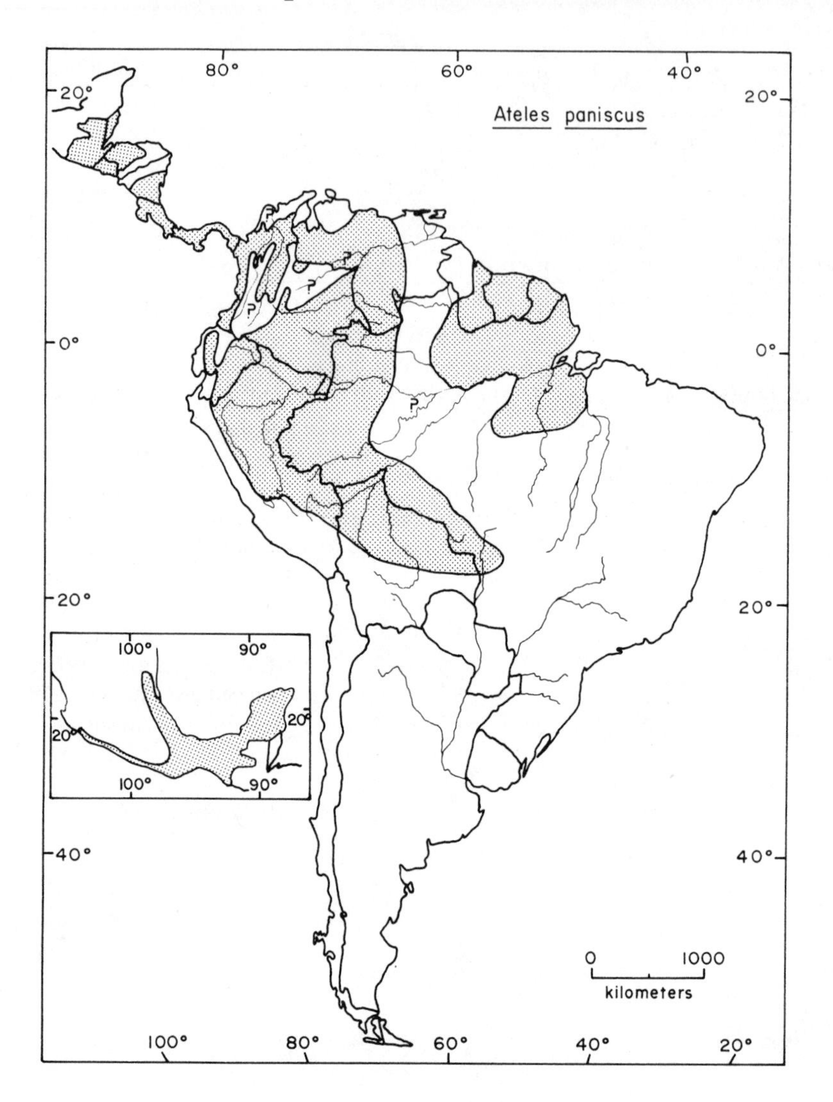

Taxonomic Note

Older authorities recognized four species of spider monkeys (Kellogg and Goldman 1944, Cabrera 1958). Several recent workers have regarded all *Ateles* as conspecific (Moynihan 1967, Hershkovitz 1972, Hernández-Camacho and Cooper 1976). Other authors believe that the older species classification is incorrect, but that more than one species probably exist (Thorington 1976). Since the taxonomic issues are still unresolved, all *Ateles* are treated together as *Ateles paniscus* here. Sixteen races have been described (Kellogg and Goldman 1944, Napier and Napier 1967).

Geographic Range

Spider monkeys are found in Central America and South America, from southern Mexico to central Bolivia and Brazil. Their northernmost record is at Escandón, at about 23°N, 99°W in southern Tamaulipas, Mexico (Kellogg and Goldman 1944), and the southernmost record is at Buena Vista, Bolivia, at 17°28′S, 63°37′W (Napier 1976). The range extends as far west as about 104°W (near 19°N) on the west coast of Mexico (Hall and Kelson 1959, Leopold 1959), and as far east as Cametá, Brazil, at about 2°12′S, 49°30′W (Kellogg and Goldman 1944). For details of the distribution, see Table 66.

Abundance and Density

A. paniscus has recently been described as locally abundant in parts of Mexico, Costa Rica, Panama, Peru, Surinam, and French Guiana, but is also known to be decreasing in numbers in all of these countries except the last, as well as in Guatemala, El Salvador, northern Colombia, and northwestern Bolivia (Table 66). Its populations are considered endangered in Costa Rica and Peru, and vulnerable in Brazil, Surinam, and Guyana (Table 66). No information is available regarding its present status in Belize, Honduras, Nicaragua, Panama, most of Colombia, Ecuador, most of Bolivia, or Venezuela.

Spider monkeys have been observed at densities of 1 to 35 individuals per km^2 and in groups of up to 59 members (Table 67). Average group sizes are between 2.3 and 18.5 members, and home ranges are 157 to 750 ha in area (Table 67).

Habitat

A. paniscus is a forest monkey, most typical of evergreen tropical rain forest. It is more abundant in wet than dry forest and in evergreen than deciduous forest in Costa Rica (Freese, pers. comm. 1974) and in Venezuela (Handley, pers. comm. 1974).

In Mexico this species has been observed in "quasi-rainforest" (Jones et al. 1974) and mangrove swamp, but the latter is not its typical habitat (Eisenberg and Kuehn 1966, Alvarez del Toro 1977). In Guatemala it has been studied in dry forest (Quick 1975) and thick semideciduous rain forest (Coelho, Coelho, Bramblett, Bramblett, and Quick 1977).

In Colombia, *A. paniscus* occupies "hygrotropophytic," rain, and cloud forests (Hernández-Camacho and Cooper 1976), seasonally inundated heterogeneous forest and swamp forest (Klein and Klein 1976). In Peru it is found in wet forest, gallery forest, cloud forest, and bamboo thickets (Durham 1971), and in Peru and Bolivia utilizes evergreen, mesophytic, and broadleaf forests in white-water drainages (Freese et al., in prep.). In southeast Peru it seems to be limited by the presence of large hardwoods to provide canopy height, and is typical of forests containing *Sabal, Ficus,* and *Inga laurina* (Durham, pers. comm. 1974).

Brazilian spider monkeys have been seen in *terra firme* (nonflooding) primary forest (Mittermeier and Coimbra-Filho 1977). In Surinam this species is largely restricted to undisturbed high rain forest, is sometimes found in montane savanna forest, marsh forest, and *Epurua falcata* savanna forest, and seems to avoid edge habitats and riverbank forest (Mittermeier 1977a). In one study in Venezuela, 96% of all spider monkeys were seen in tropical humid forest, with the remainder in tropical very humid forest and tropical dry forest (Handley, pers. comm. 1974; life zones of Ewel and Madriz 1968).

A. paniscus is found at a wide range of elevations, from sea level and lowland areas to 2,500 m (Hernández-Camacho and Cooper 1976). It occurs in montane areas in Mexico (Leopold 1959), Guatemala (Boshell and Bevier 1958, Haltenorth 1975), Panama (Handley 1966, Galindo and Srihongse 1967), Peru (Durham 1971), and Surinam (Husson 1957). But, it is more often seen at lower (under 800 m) elevations than at higher ones in Colombia (Hernández-Camacho and Cooper 1976). It was found by one researcher only below 500 m in Venezuela (Handley, pers. comm. 1974), and achieves higher densities and larger group sizes in lowland areas than in highland ones in Peru (Durham 1971, 1975). Thus although *A. paniscus* can occupy high altitudes, it seems to be better adapted to lowland habitats.

This species can utilize secondary forest but prefers mature primary growth (Leopold 1959; Moynihan 1967, 1976a; Freese pers. comm. 1974). In Colombia it is most common in mature forest but can occupy remnant or degraded forest (Hernández-Camacho and Cooper 1976) and secondary growth more than five years old (Green 1978b). In Surinam it is largely restricted to undisturbed primary forest (Mittermeier 1977a).

Thus, in summary, the preferred habitat of *A. paniscus* throughout its range seems to be primary, lowland, humid evergreen rain forest, but under some circumstances the species also occupies old secondary, highland, dry, or deciduous forests. It has not been reported in any nonforest habitats.

Factors Affecting Populations

Habitat Alteration

Cutting of the tropical forest for lumber and agricultural land is destroying the habitat of *A. paniscus* in Mexico (Leopold 1959, Alvarez del Toro 1977), Guatemala (Boshell and Bevier 1958), El Salvador (Daugherty 1972), Costa Rica (Freese, pers. comm. 1974; Wilson, pers. comm. 1974), Panama (Mendez 1970), Colombia (Hernández-Camacho and Cooper 1976, Klein and Klein 1976), northern Bolivia (Heltne et al. 1976), and eastern Pará, Brazil (Ayres 1978). This species can tolerate a limited amount of logging, but requires a continuous forest for survival (Leopold 1959). Tall trees seem to be a vital component of its niche (Freese, pers. comm. 1974), and the tall trees it prefers (*Mora*, *Cedrus*, mahogany, etc.) are also preferred by lumbermen (Durham, pers. comm. 1974). Even selective logging is detrimental to it, and the cutting of small forest patches can interfere with travel routes and prevent its movements (Klein and Klein 1976).

Much of the land is already deforested in northern Colombia (Green 1978b). There, *A. paniscus* has been restricted to small remaining isolated forest pockets and is unlikely to be able to migrate elsewhere, although it may survive in the patches if some large trees remain (Bernstein et al. 1976a). Habitat destruction has not been as severe in Amazonian Peru or Brazil or in the Guianas, but since *A. paniscus* is not adaptable to human encroachment (Muckenhirn and Eisenberg 1978), cutting the forest in these areas is expected to be detrimental to it (Mittermeier 1977a).

Human Predation

Hunting. Spider monkey meat is considered particularly delicious by many groups of people, and the species is heavily hunted in Mexico (Alvarez del Toro 1977), Guatemala (Hendrichs 1977), Costa Rica (Freese, pers. comm. 1974; Wilson, pers. comm. 1974), Panama (Bennett 1968, Mendez 1970), Colombia (Hernández-Camacho and Cooper 1976, Klein and Klein 1976), Peru (Grimwood 1969, Soini 1972), Bolivia (Heltne et al. 1976), Brazil (Mittermeier and Coimbra-Filho 1977), Venezuela (Mondolfi 1976), and Surinam (Husson 1957, Mittermeier 1977a).

As early as 1944, *A. paniscus* was considered to be in danger of extermination in many areas because of this hunting (Kellogg and Goldman 1944). In Mexico it was "almost universally overhunted" in the past (Leopold 1959), and is still being ruthlessly persecuted for its flesh (Alvarez del Toro 1977). In El Salvador this species has become extremely scarce (Daugherty 1972).

Since its meat is so desirable, *A. paniscus* is often the first species to be exterminated when a new area is colonized (Gardner, pers. comm. 1974; Klein and Klein 1976). Even light to moderate hunting can eliminate *A. paniscus* from large areas (Freese 1975). This species may be especially vulnerable to extinction by hunting because of its rather slow reproductive rate. Females first breed in their fourth year and bear young only once every two years, making population recruitment relatively slow (Klein and Klein 1976).

Hunting of *A. paniscus* for food is especially severe along the major rivers of northern Loreto, Peru, where meat is collected for shipment to Iquitos (Durham 1975, Freese et al. 1976, Neville et al. 1976). Hunting is also blamed for the absence of this species in areas of suitable forest elsewhere in Peru (Grimwood 1969; Freese et al., in prep.), in northern Bolivia (Heltne et al. 1976), in Colombia (Hernández-Camacho and Cooper 1976), and in Guatemala (Boshell and Bevier 1958).

A. paniscus is also killed for use as bait in large cat traps in Peru (Encarnación and Castro 1978), Bolivia (Heltne et al. 1976), and Brazil (Mittermeier and Coimbra-Filho 1977).

Pest Control. Spider monkeys raid orchards and cornfields in Mexico (Leopold 1959), but none of the sources consulted mentioned efforts to control this species.

Collection. Spider monkey juveniles are considered desirable pets (Kellogg and Goldman 1944, Hershkovitz 1972), and young ones obtained by killing the mother are kept as pets in Mexico (Alvarez del Toro 1977), Colombia (Bernstein et al. 1976b, Green 1976), Peru (Durham, pers. comm. 1974), Brazil (Mittermeier and Coimbra-Filho 1977), and Surinam (Husson 1957, Mittermeier 1977a).

A. paniscus is also captured for export. For example, in 1964–74 a total of 4,036 were exported from Peru (Castro 1976). Between 1968 and 1973, 12,075 spider monkeys were imported into the United States (Appendix), primarily from Nicaragua, Colombia, and Peru. None entered the United States in 1977 (Bittner et al. 1978). In 1965–75, 1,588 spider monkeys were imported into the United Kingdom (Burton

1978). This species has been used in research in the United States (Thorington 1972, Comm. Cons. Nonhuman Prim. 1975a).

Conservation Action

A. paniscus is found within ten national parks, eight reserves, and one nature park (Table 66). But none of these protected areas are in Mexico, Honduras, Panama, Ecuador, Bolivia, Brazil, French Guiana, or Guyana. At least one of the national parks in Colombia is shrinking in size and needs better protection from encroaching agriculturists if it is to survive (Struhsaker 1976b), and the two reserves in Peru are not adequately patrolled against hunting (Castro 1976).

In Mexico, hunting regulations protect *A. paniscus* in writing but are not enforced (Alvarez del Toro 1977). Efforts have been made to establish tropical forest preserves in Chiapas (Alvarez del Toro, pers. comm. 1974), but no information is available on their success. In Costa Rica, strict protection of the remaining populations and their habitat is necessary if the species is to survive (Freese, pers. comm. 1974; Wilson, pers. comm. 1974). In Colombia a capture quota of 250 *A. paniscus* (and 50 in any one month) was imposed in 1973 (Cooper and Hernández-Camacho 1975). No more recent information was located.

Peruvian law prohibits the hunting, capture, and export of *A. paniscus*, but the hunting provision is not enforced, and strict enforcement and special conservation efforts are needed in that country (Neville 1976b, Neville et al. 1976). For example, export should be limited to an average of less than 50 to 100 per year over the next 10 years, and only to institutions concerned with breeding the species (Freese 1975). In northern Colombia, population levels are so low that even modest hunting or trapping probably could not be tolerated (Bernstein et al. 1976a).

In Surinam, hunting regulations protect *A. paniscus* in the north and northeast, and prohibit shipment of monkey meat from the interior to northern population centers (Mittermeier 1977a). Brazilian law protects *A. paniscus* from commercial exploitation there (Mittermeier and Coimbra-Filho 1977), and legal restrictions prevent export from Venezuela (Mondolfi 1976).

The races of spider monkeys that are found in Nicaragua, Costa Rica, and Panama are listed in Appendix I of the Convention on International Trade in Endangered Species of Wild Fauna and Flora, and as endangered under the U.S. Endangered Species Act (U.S. Department of Interior 1977a,b).

Table 66. Distribution of *Ateles paniscus* (countries are listed from north to south and counterclockwise within South America)

Country	Presence and Abundance	Reference
Mexico	In E, S of 23°N and throughout S, E of about 96°W; localities listed and mapped in Tamaulipas, San Luis Potosí, Veracruz, Oaxaca, Tabasco, Campeche, and Quintana Roo	Kellogg and Goldman 1944
	Along W coast S of about 20°N; in E, S of 24°N; throughout S (map)	Hall and Kelson 1959
	From S Tamaulipas to SE, including Veracruz, Oaxaca, and Yucatán; isolated population on W coast of Jalisco (map); numbers shrinking steadily	Leopold 1959
	30 km W of Acapetahua, Chiapas	Eisenberg and Kuehn 1966
	Localities mapped in N and E Oaxaca	Goodwin 1969
	Localities mapped in Campeche, Quintana Roo, and Yucatán; locally common	Jones et al. 1974
	Declining in Chiapas; becoming scarce; will soon also disappear elsewhere in Mexico	Alvarez del Toro 1977
Guatemala	Localities mapped in S and NE	Kellogg and Goldman 1944
	In Motagua valley (in SE); not in high densities; rare or extinct in many areas	Boshell and Bevier 1958
	In Rio Dulce N.P. (in SE, near 15°30′N, 88°40′W)	I.U.C.N. 1971
	On Orizala Volcano	Haltenorth 1975
	In Tikal N.P. (in NE, near 17°06′N, 89°30′W)	I.U.C.N. 1975; Quick 1975; Coelho et al. 1976, 1977; Cant 1978
	Localities listed on Pacific coast, Vera Paz, and Atitlán Volcano	Napier 1976
Belize	Present (maps)	Kellogg and Goldman 1944, Hall and Kelson 1959
El Salvador	Throughout; one locality in S (maps)	Kellogg and Goldman 1944, Hall and Kelson 1959
	Abundant in late nineteenth century, now seriously depleted along coast	Daugherty 1972
Honduras	Throughout except S and E (localities listed and mapped)	Kellogg and Goldman 1944
	In NW along coast between La Ceiba and Tela	Trapido and Galindo 1955
	Throughout (map)	Hall and Kelson 1959
Nicaragua	Throughout except in N (localities listed and mapped)	Kellogg and Goldman 1944
	Throughout (map)	Hall and Kelson 1959
	In Saslaya N.P., Zelaya (in NE)	Ryan 1978

Table 66 (continued)

Country	Presence and Abundance	Reference
Costa Rica	Throughout; localities listed and mapped	Kellogg and Goldman 1944
	At Puntarenas, Montañas de Verdún, Quepos, Las Mellejones, Miravalles, Tierras Morenas, and in Guanacaste	Vargas-Mendez and Elton 1953
	Throughout (map)	Hall and Kelson 1959
	Rare in Cabo Blanco Nature Reserve, S tip of Nicoya Peninsula	Harroy 1972
	Throughout except central plateau and much of NW; scarce and declining in most areas; abundant in some parts of NE (e.g., Tortuguero N.P.)	Freese, pers. comm. 1974
	Extinct or rare in many areas; sparsely distributed along Pacific coast N of Osa Peninsula and throughout Caribbean lowlands; endangered	Wilson, pers. comm. 1974
	In Santa Rosa N.P., in NW (in Guanacaste, NW of Liberia)	Freese 1976
	On Osa Peninsula and in Monteverde Reserve	C.H.F. Rowell, pers. comm. 1978
Panama	Very plentiful in Coto region (in W)	Carpenter 1935
	Localities listed and mapped throughout except in N	Kellogg and Goldman 1944
	In Darién, Panamá E of canal, Colón W of canal, Chiriquí (in SW), and Bocas del Toro (in NW)	de Rodaniche 1957
	Throughout (map)	Hall and Kelson 1959
	Common in basins of R. Bayano, R. Chucunaque, and R. Tuira (in E); localities listed in Darién, San Blas, and Panamá	Handley 1966
	In S and N Darién and at Cerro Azul, W Panamá	Galindo and Srihongse 1967
	Overall fairly abundant	Bennett 1968
	Throughout except SW and S central (map); not common; probably has declined appreciably in many places	Mendez 1970
	In Darién and Panamá	Sousa et al. 1974
Colombia	Throughout except strip of W coast, Magdalena valley, extreme N, and SE (localities listed and mapped)	Kellogg and Goldman 1944
	In N in Magdalena, Atlántico, Bolívar, the Santanders, and extreme SE Guajira	Hershkovitz 1949
	In W from Bolívar to Nariño (in SW); in E to foot of Cordillera Oriental; in N as far as Santander; in NE E of R. Atrato	Cabrera 1958
	Along R. Lebrija	Tinoco 1969

Table 66 (continued)

Country	Presence and Abundance	Reference
Colombia (cont.)	In Santander, R. Magdalena valley	Sturm et al. 1970
	In Farallones de Cali N.P., Valle; Orquideas N.P., Antioquia; and La Macarena N.P., Meta	Green 1972
	In La Macarena N.P. (central)	Hunsaker 1972, Struhsaker 1976b
	On N bank R. Guayabero, La Macarena N.P.	Klein and Klein 1975, 1976, 1977
	Near Ventura, between R. Cauca and R. Magdalena in Bolívar (in N); numbers reduced	Bernstein et al. 1976a
	Near Puerto Rico, Tequisio region, Bolívar; on W bank of R. Magdalena near Sucre, Guaranada, and Majagual (in N)	Green 1976
	Throughout W, S, and N except in NE and Guajira, along Caribbean coast and on N and W slopes of Sierra Nevada de Santa Marta, upper R. Cauca and R. Magdalena valleys (map); questionable in several areas of E central, W central, and N	Hernández-Camacho and Cooper 1976
	Along R. Peneya, tributary of R. Caquetá (in S)	Izawa 1976
	Along right bank of lower R. Duda, W boundary of La Macarena N.P.	Izawa and Mizuno 1977
	Near Colosó, S Serranía de San Jacinto	Neyman 1977
Ecuador	Localities mapped throughout except along W coast	Kellogg and Goldman 1944
	W and E of the Andes	Cabrera 1958
	Localities listed in Esmeraldas (in NW), Guayas, Los Ríos, Manabí, and Pichincha (all in W)	Baker 1974
Peru	Localities mapped throughout E and N	Kellogg and Goldman 1944
	In E and NE in R. Ucayali, R. Urubamba, and R. Marañón basins	Cabrera 1958
	In Loreto, San Martín, Huánuco, Pasco, Ayacucho, Puno, Madre de Dios, and Cuzco; along Ríos Ucayali, Curanja, Napo, Tigre, Amazonas (at Iquitos), Marañón, and Samiria; in Manú N.P. and Samiria and Pacaya reserves; has disappeared from many parts of its range and is declining everywhere else	Grimwood 1969
	In Cuzco, Puno, and Madre de Dios; 4 localities listed between 12°S–13°30′S, 70°W–71°06′W	Durham 1971

Table 66 (continued)

Country	Presence and Abundance	Reference
Peru (cont.)	Scarce near Iquitos and elsewhere; have disappeared from many areas; in urgent need of protection	Soini 1972
	At R. Curanja, SE Loreto (in E); decreased 1966–71	Gardner, pers. comm. 1974
	Moderately abundant in SE	Durham 1975
	Extinct along middle and lower R. Nanay; rare in R. Samiria basin; absent or rare in all regions of N but most remote ones; in Manú N.P.; populations in S more plentiful than in N	Freese 1975
	In Manú N.P., Madre de Dios and Cuzco, and Pacaya and Samiria reserves, Loreto; endangered	Castro 1976
	Seriously threatened along R. Samiria and R. Pacaya (between lower R. Ucayali and R. Huallaga/Marañón) (in N); generally threatened throughout	Neville et al. 1976
	Vulnerable	Castro 1978
	At headwaters of R. Acre and R. Tahuamanu (in SE)	Encarnación and Castro 1978
	Near R. Yavineto (R. Putumayo), Loreto	Encarnación et al. 1978
	Common at Cocha Cashu, Manú N.P.	Janson and Terborgh, in press
Bolivia	Localities mapped in N	Kellogg and Goldman 1944
	In N and central, as far S as N Santa Cruz	Cabrera 1958
	Uncommon at Ixiamas (in NW) and El Triunfo (central); apparently extinct over large areas in extreme NW	Heltne et al. 1976
	Localities listed: R. Yapacani and Buena Vista, Santa Cruz	Napier 1976
Brazil	Localities mapped in W, NW, and NE, N and S of lower R. Amazonas	Kellogg and Goldman 1944
	In Pará and Amazonas; abundant in parts of Mato Grosso	Kumm and Laemmert 1950
	In NW in basins of R. Negro and R. Solimões; in N, E of R. Branco and lower R. Negro; S of R. Amazonas between R. Tocantins and R. Tapajós in Pará and in Amazonas, Mato Grosso, and Acre	Cabrera 1958
	In central Amazonas and Pará (localities mapped)	Deane et al. 1969
	Along Manaus-Itacoatiara Road, Amazonas	Deane et al. 1971
	Localities mapped in E Amazonas and W Pará	Deane 1976

Table 66 (continued)

Country	Presence and Abundance	Reference
Brazil (cont.)	Vulnerable	Mittermeier and Coimbra-Filho 1977
	In E Pará E of R. Xingu	Ayres 1978
French Guiana	Localities mapped in N and E	Kellogg and Goldman 1944
	Very abundant in remote areas	Durham, pers. comm. 1974
Surinam	Locality mapped in N	Kellogg and Goldman 1944
	Quite common in interior; localities listed: Tibiti (in N) and Nassau Mts. (in E)	Husson 1957
	Very abundant in remote areas	Durham, pers. comm. 1974
	Rapidly declining	Mittermeier and Milton 1976
	Restricted to interior except in W where reaches coastal plain; localities mapped in S and central as far N as about 5°15′N; in Brownsberg Nature Park and 4 nature reserves; Raleighvallen-Voltzberg, Tafelberg, Eilerts de Haan Gebergte, and Sipaliwini Savanna; locally common but very vulnerable	Mittermeier 1977a
Guyana	Localities mapped in N and S	Kellogg and Goldman 1944
	Localities listed: Bonasica and R. Supinaam	Napier 1976
	Localities listed and mapped include: R. Berbice (in NE), Apoteri (central), and R. Rupununi; in need of conservation	Muckenhirn et al. 1975
	Restricted range; absent from NW; extremely vulnerable	Muckenhirn and Eisenberg 1978
Venezuela	Localities mapped in central and S	Kellogg and Goldman 1944
	In NW from Sierra de Perijá to Lake Maracaibo	Hershkovitz 1949
	In NW, W of Lake Maracaibo, and central and S between R. Orinoco and R. Caura	Cabrera 1958
	Localities mapped in Amazonas, Apure, Zulia, and Trujillo (in S and W)	Handley, pers. comm. 1974
	In Amazonas from R. Negro, upper R. Orinoco, and R. Ventuari to San Juan de Manapiare; in Bolívar along R. Caura; in basin of Lake Maracaibo in Sierra de Perijá, from R. Guasare to plains of Trujillo, Mérida, and Táchira; in Barinas, Portuguesa, and Apure; in Guatopo N.P. (N central); rare in Cordillera de la Costa (N central coast)	Mondolfi 1976
	Localities listed: left bank of Brazo Casiquiare (Amazonas), and in Mérida	Napier 1976

Table 67. Population parameters of *Ateles paniscus*

Population Density (individuals /km²)	Group Size X̄	Group Size Range	Group Size No. of Groups	Home Range Area (ha)	Study Site	Reference
		2–50			Mexico	Hall and Kelson 1959
		10–50			Mexico	Leopold 1959
	4.9	3–6	9		Mexico	Eisenberg and Kuehn 1966
	12				Mexico	Alvarez del Toro, pers. comm. 1974
		2–59			Guatemala	Quick 1975
45					Guatemala	Coelho et al. 1976
27.8	4.3	2–28			Guatemala	Cant 1978
		2–20			Costa Rica	Freese 1976
		12–15			Colombia	Tinoco 1969
12–15	3.5	2–22		259–389	Colombia	Klein and Klein 1975
14					N Colombia	Bernstein et al. 1976a
		≤15–30			Colombia	Hernández-Camacho and Cooper 1976
	8.4	3–30	19		S Colombia	Izawa 1976
20–33					N Colombia	Green 1978b
6–24	4.5–18.5	3–25	19	300–750	SE Peru	Durham 1971, 1972, 1975
22.4	7	2–24			E Peru	Freese 1975
14–35	3.8	2–17	46		SE Peru	Janson and Terborgh, in press
2–8.5	5		5		Bolivia	Heltne et al. 1976
9.5	9.5	7–12	4	157	Surinam	Mittermeier 1977a
1–8	2.3	2–8	19		Guyana	Muckenhirn et al. 1975

Brachyteles arachnoides Wooly Spider Monkey

Taxonomic Note

This is a monotypic species (Cabrera 1958, Napier and Napier 1967).

Geographic Range

Brachyteles arachnoides is found along the coast of southeastern Brazil from Bahia to Paraná. Its range previously extended north to 13°30′S but has receded to 15°S, once extended west to almost 52°W along the Rio Paraná, but now occurs only to 48°30′W, and includes localities as far south as 25°S (Aguirre 1971). Recent reports indicate that the range has now shrunk further and includes only eastern São Paulo and parts of Rio de Janeiro and Minas Gerais (I.U.C.N. 1978). For details of the distribution, see Table 68.

Abundance and Density

B. arachnoides is an endangered species whose numbers and range are both small and decreasing (Table 68). Its population is estimated to have declined from 400,000 before 1500 to about 3,000 in 1971, 2,000 in 1972, and less than 1,000 in 1977; its range encompasses less than 300,000 km^2, and only a few centers of local abundance have been reported (Table 68).

In the 1940s, groups of 30 to 40 *B. arachnoides* were common, but in the 1970s group sizes were 6 to 20 (Aguirre 1971). An average group of 8 to 10 members occupies an area of about 500 ha, and population densities vary from 4 to 52 individuals per km^2 (Aguirre 1971).

Habitat

B. arachnoides is found in primary rain forest from 40 m to 1,500 m elevation (Aguirre 1971). No further details are available.

Factors Affecting Populations

Habitat Alteration

The forest habitats of *B. arachnoides* are being rapidly eliminated (Aguirre 1971) by woodcutting for fuel, and clearing for agriculture and settlement (I.U.C.N. 1971, 1974, 1978).

Human Predation

Hunting. This species has been heavily hunted for its meat and its skin (Aguirre 1971), and is still being hunted for food (Coimbra-Filho, pers. comm. 1977 cited by I.U.C.N. 1978).

Pest Control. No information.

Collection. A total of 64 *B. arachnoides* were imported into the United States between 1968 and 1973 (Appendix). None entered the United States in 1977 (Bittner et al. 1978). Twenty-four were imported into the United Kingdom in 1965–75 (Burton 1978). This species is not used in research in the United States (Thorington 1972).

Conservation Action

B. arachnoides occurs in five national parks, one state park, seven federal and state reserves, and four private reserves (Table 68). But hunting occurs even within the national parks, and more effective protection — through greater mobility of guards, educational campaigns, adequate enforcement budgets, and additional habitat reserves — is urgently needed (Aguirre 1971).

Since 1967, wooly spider monkeys have been protected by Brazilian law from hunting, capture, purchase, and export (Coimbra-Filho and Mittermeier 1977), but enforcement of the hunting provision is difficult and inadequate (I.U.C.N. 1972). The Brazilian National Research Council and the Brazilian Foundation for Conservation of Nature sponsored Aguirre's (1971) study of the status of this species and methods for its conservation (de Melo Carvalho 1971a). The species is classified as endangered in Brazil, and plans have been made to reintroduce it into the Tijuca National Park in Rio de Janeiro city (Padua et al. 1974).

B. arachnoides is listed in Appendix I of the Convention on International Trade in Endangered Species of Wild Fauna and Flora and as endangered under the U.S. Endangered Species Act (U.S. Department of Interior 1977a,b). It is illegal to import into the United Kingdom except for scientific, educational, or propagation purposes (I.U.C.N. 1972).

Table 68. Distribution of *Brachyteles arachnoides*

Country	Presence and Abundance	Reference
Brazil	Localities mapped in Minas Gerais and Rio de Janeiro	Kumm and Laemmert 1950
	In Bahia, Minas Gerais, Espírito Santo, Rio de Janeiro, Guanabara, and São Paulo	Cabrera 1958
	In Caparaó N.P. and Brigadeiro Reserve	Avila-Pires 1969a
	Locality mapped in Espírito Santo	Deane et al. 1969, Deane 1976
	On the verge of extinction	Hershkovitz 1969
	Before 1500: from R. Paraguaçu (about 13°13′N on coast) to R. São Francisco (near 25°S, 54°W), estimated population of 400,000. Now from 15°S–25°S and W to 48°30′W, including S Bahia, parts of Espírito Santo, Rio de Janeiro, Guanabara, Minas Gerais; Região Sul, São Paulo, and N Paraná; range area about 300,000 km^2, population about 3,000; in Monte Pascoal N.P., S Bahia; Serra dos Orgãos N.P., Rio de Janeiro; Itatiaia N.P., Rio de Janeiro; Nova Lombardia Federal Biological Reserve, Espírito Santo; Coronel Fabriciano Florestal Park, Minas Gerais; 4 other federal and state reserves, and 4 private reserves (these and other localities listed and mapped)	Aguirre 1971
	Confined to narrow strip of coastal mountains (Serra de Mar)	de Melo Carvalho 1971a
	Numerous near Santa Leopoldina, Espírito Santo	Deane et al. 1971
	In Caparaó N.P., Espírito Santo and Minas Gerais	I.U.C.N. 1971
	In Serra da Bocaina N.P. (Serra do Mar)	Harroy 1972
	Range within last 200 years from about 17°S to 25°S from Bahia to Paraná (map); now fragmented and precariously small; threatened with extinction	Hershkovitz 1972

Table 68 (continued)

Country	Presence and Abundance	Reference
Brazil (cont.)	Endangered; S Bahia to W São Paulo; in 8 national or state and 3 private reserves	I.U.C.N. 1972
	Endangered; localities listed from Vieira (1944, 1955); Coimbra-Filho (1972b) estimates population at 2,000; range shrinking; small populations in 3 national parks and reserves and Rio Doce State Park	I.U.C.N. 1974
	Endangered; estimated population less than 1,000; sparsely distributed in São Paulo, Rio de Janeiro, and Minas Gerais; Vanzolini reports it restricted to E São Paulo; largest populations in Serra da Bocaina N.P. and in central São Paulo	I.U.C.N. 1978

Cacajao calvus Red or Red-faced and White or Bald Uakari

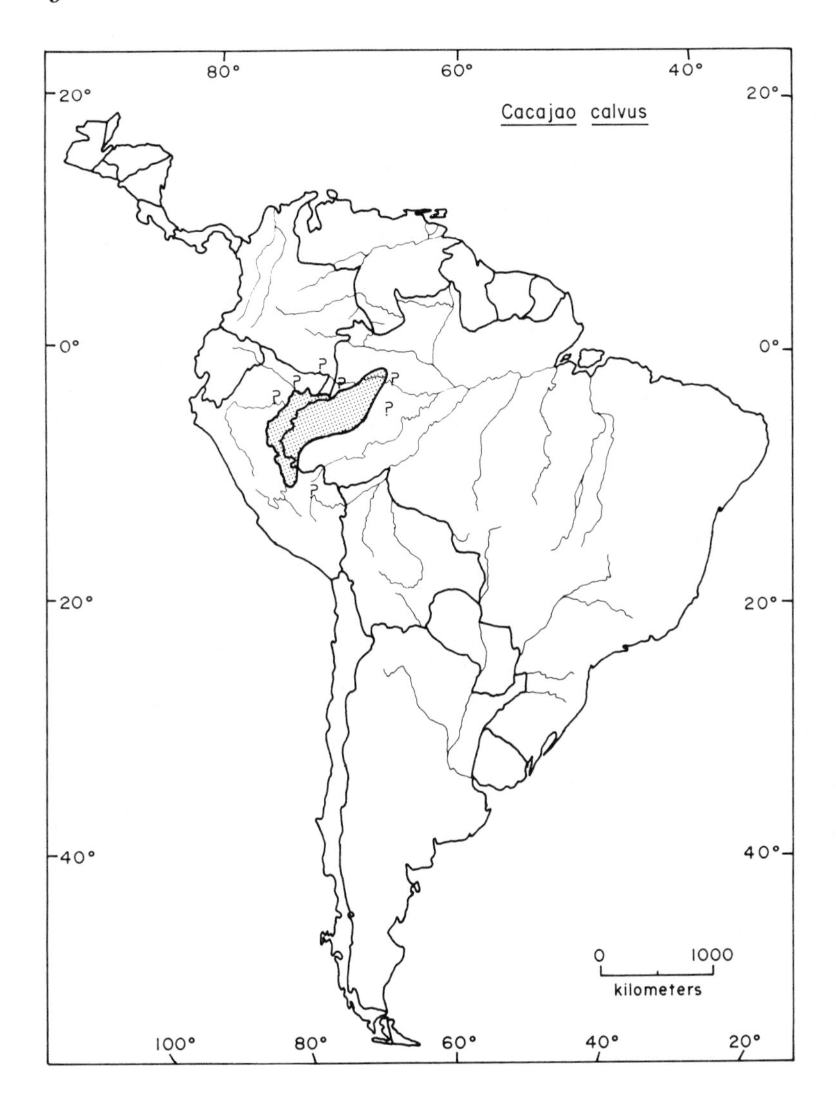

Taxonomic Note

Some authors consider the red uakari (*C. c. rubicundus*) and the white uakari (*C. c. calvus*) to be separate species (Cabrera 1958, Napier and Napier 1967), but they are treated here as conspecifics following Hershkovitz (1972). They are distinguished as red or white, where necessary, to avoid confusion.

Geographic Range

Cacajao calvus is found in the upper Amazon basin, in northwestern Brazil, eastern Peru, and possibly southern Colombia. The range limits have not been precisely delimited. The white form occurs north of the Rio Solimões on the fluvial island formed by that river, the Rio Japurá, and the Rio Auati-Paraná, and perhaps farther west (I.U.C.N. 1974). The red form is found south of the Rio Solimões-Amazonas as far west as 75°W along the Río Ucayali, as far north and east as about 2°33′S, 65°30′W at the mouth of the Rio Juruá, and as far south as about 11°S (Hershkovitz 1972). It has also been reported farther west along the Ríos Napo, Putumayo, and Marañón (Grimwood 1969), farther east as far as the Rio Purus (Mittermeier and Coimbra-Filho 1977), and farther south to about 13°30′S in southern Madre de Dios and northern Puno, Peru (Durham 1972, I.U.C.N. 1972). For details of the distribution, see Table 69.

Abundance and Density

C. calvus is rare, declining, and endangered in Peru, and the white form is thought to be endangered in Brazil (Table 69). Previously both forms were considered endangered throughout their ranges, but now their official status has been changed to "indeterminate" (I.U.C.N. 1978).

No estimates of population density, average group size, or home range area are available. Groups of 10 to 18 (Durham 1972) and up to more than 30 (Mittermeier and Coimbra-Filho 1977) have been reported.

Habitat

C. calvus inhabits tropical rain forest subject to periodic flooding (Napier and Napier 1967). The white form occupies an island of *várzea* forest, seasonally flooded by a white-water river, and the red form has been seen in *igapó* forest, flooded by a black-water river (Mittermeier and Coimbra-Filho 1977). In Peru this species was sighted in heavy canopy forest in *Cecropia* trees (Durham, pers. comm. 1974).

Factors Affecting Populations

Habitat Alteration

The area in which the white uakari is found is remote from centers of human population, and the habitat there has been disturbed only to a small extent (Mittermeier, in prep.). Human settlement has been blamed for the contraction of the ranges of both races of *C. calvus* (I.U.C.N. 1972), but the extent of habitat alteration is not known.

Human Predation

Hunting. The decline of the red uakari in Peru is due primarily to hunting, because this species is easy to kill and its flesh is highly esteemed (Grimwood 1969). In the past, heavy hunting was also reported for Brazil, but more recent studies have found *C. calvus* to be hunted only infrequently for fish bait or for human consumption (Mittermeier and Coimbra-Filho 1977).

Pest Control. No information.

Collection. Capture for export has exerted pressure on *C. calvus* populations, especially in Peru (Grimwood 1969). For example, between 1962 and 1968 a total of 433 red uakaris were exported from Iquitos (Soini 1972), and in 1970, 20 were exported from Colombia (Green 1976).

C. calvus is also captured for the pet trade in Peru (Castro 1976, I.U.C.N. 1978), but is no longer in demand as a pet in Brazil (Mittermeier and Coimbra-Filho 1977).

Between 1968 and 1973 a total of 211 *C. calvus* were imported into the United States (Appendix). None entered the United States in 1977 (Bittner et al. 1978). Uakaris are not used in research in the United States (Thorington 1972). From 1965 to 1975, 7 were imported into the United Kingdom (Burton 1978).

Conservation Action

C. calvus is classified as an endangered species in Brazil (Padua et al. 1974). It is legally protected from all forms of commerce there, but enforcement in the Amazonian region is quite difficult (Mittermeier and Coimbra-Filho 1977). This species is also completely protected by law in Peru, but effective control of hunting is also difficult there (Castro 1976).

C. calvus does not occur in any protected areas in Peru or Brazil. The fluvial island between the Rios Japurá-Solimões and Auati-Paraná should be set aside as a reserve for the white uakari (I.U.C.N. 1974), and other large reserves for this species in natural ecosystems are needed in Brazil (Coimbra-Filho 1972a) and Peru. Export from Peru should be limited to less than 50 to 100 per year and only to institutions concerned with breeding them (Freese 1975). Research on the status of *C. calvus* populations is badly needed in both Peru and Brazil (Freese 1975, I.U.C.N. 1978).

C. calvus is listed in Appendix I of the Convention on International Trade in Endangered Species of Wild Fauna and Flora, and as endangered under the U.S. Endangered Species Act (U.S. Department of Interior 1977a,b).

Table 69. Distribution of *Cacajao calvus*

Country	Presence and Abundance	Reference
Colombia	Possible but unlikely in Amazonas between R. Caquetá and R. Putumayo (white form)	Hernández-Camacho and Cooper 1976
Brazil	In W Amazonas between R. Japurá and R. Juruá	Cabrera 1958
	White form from only 1 locality: opposite Fonte Boa between R. Japurá and N bank R. Solimões (map); rare and in precarious position; red form throughout SW Amazonas between R. Solimões and R. Juruá; threatened with extinction	Hershkovitz 1972
	Both forms endangered; white form from Tefé to mouth of R. Içá (citing older authors)	I.U.C.N. 1972
	White form endangered; on fluvial island formed by R. Solimões, R. Japurá, R. Auati-Paraná, possibly to W near São Paulo de Olivença (Avila-Pires)	I.U.C.N. 1974

Table 69 (continued)

Country	Presence and Abundance	Reference
Brazil (cont.)	Localities mapped in W Acre and Amazonas	Deane 1976
	Localities listed at Paraná do Manhana (white form); Paraná do Jacaré and Auati-Paraná; S to R. Eiru, right bank R. Juruá, 7°S, 70°W (red form)	Napier 1976
	Red form S of R. Solimões as far E as R. Juruá and perhaps to W bank of R. Purus; also to N between R. Solimões and R. Içá, possibly to R. Auati-Paraná; white form between R. Auati-Paraná, R. Solimões and R. Japurá, endangered	Mittermeier and Coimbra-Filho 1977
	Red form seen along R. Jacurapá, small W bank tributary of R. Içá; white form endangered: range small	Mittermeier, in prep.
Peru	In N along R. Amazonas to lower R. Napo and in E basin of R. Ucayali	Cabrera 1958
	Confined to NE Amazon region, N of R. Amazonas in basins of R. Napo and R. Putumayo and S in basins of R. Yavarí and R. Ucayali to about 7°S; localities listed include Cerro Azul near Contamana, and Maynas on R. Marañón; everywhere rare and declining	Grimwood 1969
	Possible sighting of white form near R. Inambari, N Puno–S Madre de Dios (about 13°30′S, 70°W)	Durham 1972
	Only S of R. Amazonas and E of R. Ucayali, S to 11°S (map)	Hershkovitz 1972
	Confirmed in R. Madre de Dios (about 12°40′S) (citing others)	I.U.C.N. 1972
	More abundant S of R. Amazonas than N	Soini 1972
	Sighted in Puno (white form)	Durham, pers. comm. 1974
	Endangered; not in protected areas (red form)	Castro 1976
	Appears to be rare (red form)	Mittermeier, in prep.

Cacajao melanocephalus Black-headed Uakari

Taxonomic Note

This is a monotypic species (Cabrera 1958, Napier and Napier 1967).

Geographic Range

Cacajao melanocephalus is found in northwestern South America from central Colombia and southern Venezuela into northern Brazil. It occurs as far west as the northern Macarena Mountains (about 3°20′N, 74°W) and as far north as about 4°N along the Río Guaviare in Colombia (Hernández-Camacho and Cooper 1976). The southern limit of its range is the Rio Solimões, as far south as about 4°S, as far east as the mouth of the Rio Negro (about 60°W), and north of the Rio Negro as far east as the Rio Araca and perhaps farther east (Mittermeier and Coimbra-Filho 1977). For details of the distribution, see Table 70.

Abundance and Density

C. melanocephalus is considered to be in a precarious position in Colombia, and vulnerable in Brazil, although it is abundant in some areas (Table 70). Its status in Venezuela is not known. It was previously listed as endangered (I.U.C.N. 1972), but is now classified as vulnerable (I.U.C.N. 1978).

No estimates of population density, average group size, or home range area are available. Groups of up to more than 20 (Mittermeier and Coimbra-Filho 1977) and up to 30 (Hernández-Camacho and Cooper 1976) have been reported.

Habitat

Black-headed uakaris have been observed in *igapó* (seasonally or permanently flooded) forests of small black-water rivers and lakes in Brazil (Mittermeier and Coimbra-Filho 1977). In Venezuela they were recorded in evergreen forest near streams between sea level and 500 m elevation (Handley, pers. comm. 1974), in the tropical humid forest zone (life zones of Ewel and Madriz 1968). No further details of their habitat preferences are available.

Factors Affecting Populations

Habitat Alteration

Previously the range of *C. melanocephalus* was thought to be contracting because of human settlement (I.U.C.N. 1972). In Colombia this species does not coexist well with humans in areas of settlement and degraded forest (Hernández-Camacho and Cooper 1976), but no details are available regarding the extent of alteration of its habitat.

The range of *C. melanocephalus* in Brazil is remote from centers of human activity (Mittermeier and Coimbra-Filho 1977). Because the human population there is small and the *igapó* forests have little commercial value, habitat destruction has not been significant (Mittermeier, in prep.). No information is available regarding the Venezuelan portion of the range.

Human Predation

Hunting. The meat of *C. melanocephalus* is highly prized in parts of Colombia, and hunting may have contributed to its scarcity there (Hernández-Camacho and Cooper 1976). In Brazil this species is not usually hunted for food but is shot for fish, turtle, and cat bait (Mittermeier and Coimbra-Filho 1977).

Pest Control. No information.

Collection. Four *C. melanocephalus* were imported into the United States in 1968, and none from 1969 to 1973 (Appendix) or in 1977 (Bittner et al. 1978). None were imported into the United Kingdom between 1965 and 1975 (Burton 1978).

Conservation Action

C. melanocephalus is found in one national park in Colombia (Table 70), but this park is shrinking because of agricultural encroachment and must have better protection if it is to survive (Struhsaker 1976b). *C. melanocephalus* is in need of a habitat reserve in the Rio Negro area of Brazil (I.U.C.N. 1978). It is classified as an endangered species in Brazil and thus is protected from commercial exploitation there (Padua et al. 1974, Mittermeier and Coimbra-Filho 1977).

This species is also listed in Appendix I of the Convention on International Trade in Endangered Species of Wild Fauna and Flora, and as endangered under the U.S. Endangered Species Act (U.S. Department of Interior 1977a,b).

Table 70. Distribution of *Cacajao melanocephalus*

Country	Presence and Abundance	Reference
Colombia	In La Macarena N.P., Meta	Green 1972, I.U.C.N. 1975
	In SW, Guainía and Vaupés (map); questionable N of R. Guaviare and S of R. Apaporis; distribution poorly known (localities listed); situation very precarious	Hernández-Camacho and Cooper 1976
Venezuela	In extreme SW in basin of Brazo Casiquiare	Cabrera 1958
	Localities listed and mapped in S Amazonas at R. Mavaca and Brazo Casiquiare, SE and SW of Esmeralda	Handley, pers. comm. 1974
	Locality listed on R. Yatua at 1°43′N, 66°30′W	Napier 1976
Brazil	In basin of R. Negro	Cabrera 1958
	In Amazonas	Deane et al. 1969
	Localities listed on R. Solimões and at Manacaquiru (Manacapuru?), Amazonas	Napier 1976
	Abundant N of R. Solimões between R. Negro and R. Japurá, rare elsewhere; N of R. Negro E to R. Araçá and perhaps to or beyond R. Branco; vulnerable	Mittermeier and Coimbra-Filho 1977
	Common along several small W bank tributaries of R. Negro (e.g., R. Cuiuni, R. Uneiuxi); not in danger between R. Japurá, R. Negro, and R. Solimões	Mittermeier, in prep.

Callicebus moloch Dusky Titi

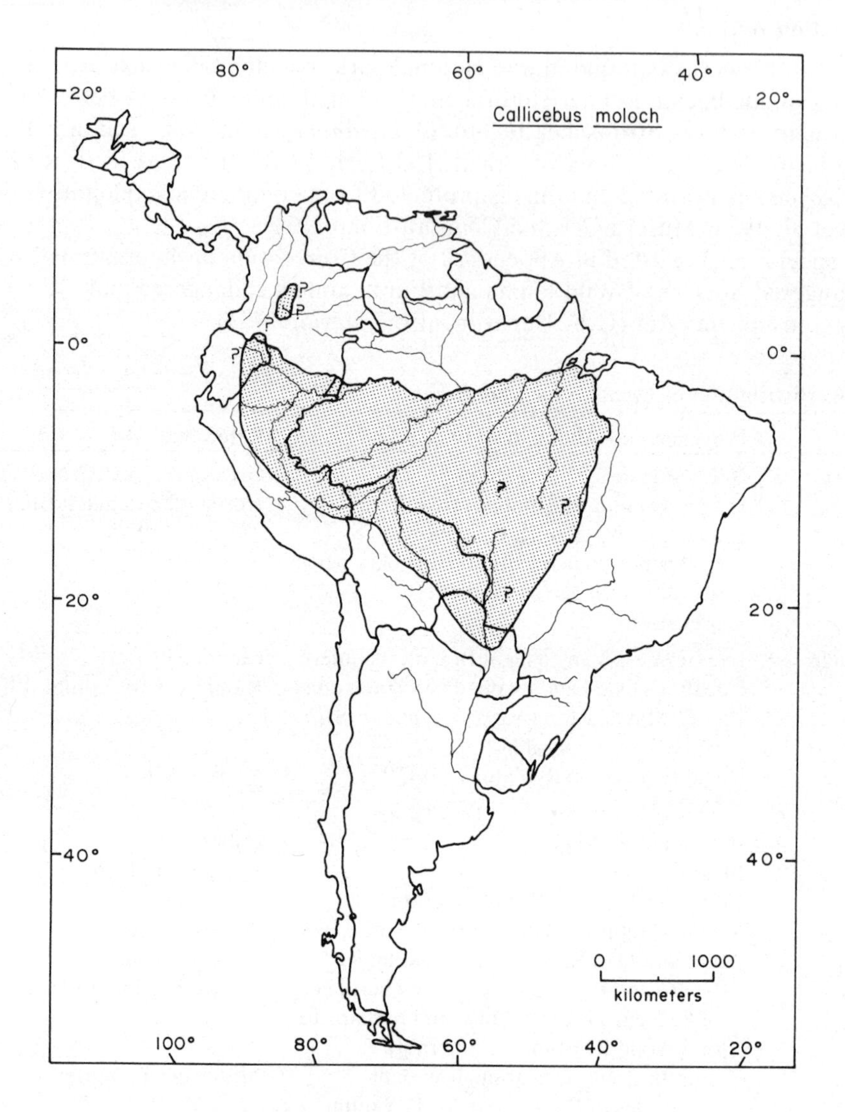

Taxonomic Note

Seven subspecies are recognized (Hershkovitz 1963, Napier and Napier 1967).

Geographic Range

Dusky titis are found in northwestern and central South America, from central Colombia to Paraguay. They occur as far north as Medina (4°31′N, 73°21′W) in southeastern Cundinamarca, Colombia, and as far south as about 23°S, 57°W, 48 km north of Concepción, Paraguay (Hershkovitz 1963). Their range extends west to at least 78°W (Hershkovitz 1963) and perhaps to 79°37′W (Baker 1974) in Ecuador, and their easternmost record is at São João de Araguaya (5°24′S, 48°41′W), in eastern Pará, Brazil (Hershkovitz 1963). For details of the distribution, see Table 71.

Abundance and Density

The geographic range of *Callicebus moloch* is shrinking in Colombia, and populations are considered vulnerable in Peru, but the species is common in one portion of northern Bolivia and is not endangered in Brazil (Table 71). Its status in Ecuador, Paraguay, and most of Bolivia is not known.

Population densities at several sites in Peru varied from 1.8 to 16.2 individuals per km^2 (Freese 1975; Freese et al., in prep.), and in Bolivia a density of 14.7 per km^2 was found (Heltne et al. 1976). These authors saw groups of 1 to 4 and estimate the average group size as 3. In the Manú National Park, population density was 14 to 31 individuals per km^2, groups included 3 to 4 members ($\bar{X} = 3$, $N = 17$), and home range areas were 10 to 12 ha (Janson and Terborgh, in press).

In Colombia, groups of *C. moloch* consist of 2 to 7 individuals (Hernández-Camacho and Cooper 1976). In one study, 9 groups included 2 to 4 members ($\bar{X} = 3.2$), and each group occupied a home range of about 400 ha (Mason 1968).

Habitat

C. moloch is a forest species which seems to prefer dense thickets and humid forests. It favors areas with thick underbrush and is often found in swampy, wet, and gallery forests (Moynihan 1976a). In Colombia it occupies lowland gallery and piedmont forest, Amazonian rain forest (Hernández-Camacho and Cooper 1976), and both inundated and noninundated forest (Klein and Klein 1976).

In Peru this species utilizes evergreen, mesophytic, broadleaf forest on high ground in white-water drainages and on flood plains of black and white-water rivers (Freese et al., in prep.). It is also found in secondary growth consisting of cane, bamboo, and *Cecropia* (Gardner, pers. comm. 1974). In Brazil, *C. moloch* has been recorded in *terra firme* (nonflooding) primary forest (Mittermeier and Coimbra-Filho 1977).

Dusky titis are found from sea level to 850 m elevation (Hershkovitz 1963), including at 500 m in Colombia (Hernández-Camacho and Cooper 1976) and 823 m in Peru (Napier 1976).

Factors Affecting Populations

Habitat Alteration

C. moloch has been described as an adaptable species, able to exist in close proximity to human habitations (Mittermeier and Coimbra-Filho 1977), and is able to survive in small patches of isolated gallery forest (Mason 1968). Nevertheless, in Meta, Colombia, north of the Río Ariari, rapid deforestation due to clearing of land for agriculture by the rapidly expanding human population has reduced the suitable habitat for *C. moloch* by more than 50% during the last 30 years (Hernández-Camacho and Cooper 1976). The range there is continuing to shrink, and these authors predict that in southwestern Colombia, rapid settlement due to the development of a major oilfield will further reduce the available habitat for *C. moloch*.

In eastern Pará, Brazil, the forest habitat of *C. moloch* is being cleared for agriculture and cattle ranching (Ayres 1978). No information was located regarding the effect of habitat alteration on this species in the rest of Brazil or in Ecuador, Peru, Bolivia, or Paraguay.

Human Predation

Hunting. C. moloch is not hunted for food in central Colombia, but is hunted in the southern trapezium (Hernández-Camacho and Cooper 1976). It is hunted heavily in some regions of Peru (Freese 1975, Neville 1975, Castro 1976), and is occasionally hunted in Brazil (Mittermeier and Coimbra-Filho 1977).

Pest Control. No information.

Collection. Dusky titis are kept as pets in Peru (Castro 1976) and Brazil (Mittermeier and Coimbra-Filho 1977). They are also captured for export from Amazonian Colombia (Hernández-Camacho and Cooper 1976), and from Peru, which exported a total of 768 between 1964 and 1974 (Castro 1976).

From 1968 to 1973 a total of 212 *C. moloch* were imported into the United States (Appendix), primarily from Colombia and Peru. One *C. moloch* entered the United States in 1977 (Bittner et al. 1978). From 1965 to 1975, 11 of this species were imported into the United Kingdom (Burton 1978). *C. moloch* are not frequently used in biomedical research in the United States (Thorington 1972, Comm. Cons. Nonhuman Prim. 1975a).

Conservation Action

C. moloch is found in two national parks and two reserves (Table 71). But one of these, La Macarena National Park in Colombia, is steadily shrinking because of agricultural encroachment, and must receive improved protection if it is to survive (Struhsaker 1976b). Also, the two reserves in Peru are not effectively patrolled against poaching (Castro 1976). Throughout Peru, in fact, *C. moloch* is legally protected from hunting, capture, or commerce (Supreme Decree 934-73-AG, 1973), but the hunting prohibition is not enforced and should be, where feasible (Freese 1975, Neville 1975).

Brazilian law also protects *C. moloch* against commercial exploitation (Mittermeier and Coimbra-Filho 1977). No reports were located regarding conservation action in Ecuador, Bolivia, or Paraguay.

Table 71. Distribution of *Callicebus moloch* (countries are listed in counterclockwise order)

Country	Presence and Abundance	Reference
Colombia	In Cordillera Oriental	Cabrera 1958
	Localities listed and mapped in central along E base of Andes and Macarena Mts. from R. Guayabero to upper R. Meta, in states of Cundinamarca and Meta	Hershkovitz 1963
	Near San Martín, NW Meta	Mason 1966, 1968, 1971; Thorington 1968a

Table 71 (continued)

Country	Presence and Abundance	Reference
Colombia (cont.)	In La Macarena N.P., Meta	Green 1972, Klein and Klein 1976
	In W Meta between R. Upía and R. Guayabero, including La Macarena N.P.; questionable to N, E, and W; in SW Putumayo between R. Guamués and R. Sucumbios; in trapezium of extreme S Amazonas between R. Putumayo and R. Amazonas (map); range in Meta and Putumayo shrinking; status of Amazonas population unknown	Hernández-Camacho and Cooper 1976
	Near Valparaíso, Caquetá, and Santa Rosa, Putumayo	Moynihan 1976a
	Localities listed on upper R. Meta and at New Grenada	Napier 1976
	Along lower R. Duda, W boundary of La Macarena N.P.	Izawa and Mizuno 1977
Ecuador	In extreme N	Cabrera 1958
	Localities listed and mapped in N and NE along R. Napo, R. Aguarico, R. Bobonaza, R. Copataza, and at Payamino	Hershkovitz 1963
	Localities listed in Napo-Pastaza at Santa Cecilia and on R. Napo near Limón Cocha at 0°25′S, 79°37′W (but this is W of province border)	Baker 1974
	Localities listed on R. Napo and R. Copataza	Napier 1976
Peru	In N in basin of R. Napo, in E in basin of upper R. Ucayali, and in SE	Cabrera 1958
	Localities listed and mapped throughout E in Loreto, Amazonas, San Martín, Huánuco, Junín, Madre de Dios, Cuzco, and Puno	Hershkovitz 1963
	In N except between R. Napo and R. Putumayo; throughout E; not common but widespread away from settlements; not endangered; in Manú N.P.	Grimwood 1969
	At R. Llullapichis, branch of R. Pachitea (central)	Koepcke 1972
	Decreased 1966–71 near villages along R. Curanja, SE Loreto	Gardner, pers. comm. 1974
	In NE along R. Nanay, R. Orosa, and R. Pucallpa; in SE at Manú N.P.	Freese 1975
	In Manú N.P. and Pacaya and Samiria national reserves (in NE); vulnerable	Castro 1976

Table 71 (continued)

Country	Presence and Abundance	Reference
Peru (cont.)	Localities listed in Huánuco, Loreto, San Martín, at Ríos Ucayali, Tahuamanu, Pachitea, Mayo, Pastaza, and Huallaga	Napier 1976
	Between R. Acre and R. Madre de Dios, E Madre de Dios	Encarnación and Castro 1978
	Along R. Yavineto (R. Putumayo), N Loreto	Encarnación et al. 1978
	Common at Cocha Cashu, Manú N.P.	Janson and Terborgh, in press
Bolivia	In N and central	Cabrera 1958
	Localities listed and mapped in N, central, and E, in Beni along R. Mamoré and R. Beni, and at Mojos; in Cochabamba (R. Chaparé) and Santa Cruz	Hershkovitz 1963
	Common at Cobija, R. Acre (on NW border); low densities at Ixiamas (in W, 13°46′S, 68°08′W)	Heltne et al. 1976
	Localities listed at Buena Vista (17°28′S, 63°37′W) and Sara Province (central)	Napier 1976
Paraguay	Present	Cabrera 1958
	At Puerto Casado, 22°15′S, 57°55′W (locality mapped)	Soria 1959
	Localities listed and mapped in Chaco at Puerto Casado and N of Concepción (both central)	Hershkovitz 1963
	Localities listed in Chaco: central and near San Salvador	Napier 1976
Brazil	At Santarém, Pará (S of R. Amazonas)	Kumm and Laemmert 1950
	In Amazonas in basins of upper R. Solimões, R. Juruá, R. Purus, and R. Madeira; in lake zone W of R. Tapajós; along lower R. Amazonas E of R. Tapajós in Acre; in W and NW Mato Grosso	Cabrera 1958
	Localities listed and mapped in Amazonas, Pará, Rondônia, Acre, and Mato Grosso (i.e., throughout W central)	Hershkovitz 1963
	In Amazonas	Deane et al. 1969
	2 localities mapped in Acre	Deane 1976
	Localities listed in Amazonas and Pará	Napier 1976
	Not in danger	Mittermeier and Coimbra-Filho 1977
	In E Pará, E of R. Xingu	Ayres 1978

Callicebus personatus Masked Titi

Taxonomic Note

Four subspecies are recognized (Napier and Napier 1967).

Geographic Range

Callicebus personatus is found in eastern Brazil in the states of Bahia, Espírito Santo, Minas Gerais, Rio de Janeiro, and São Paulo (Cabrera 1958). The range extends as far north as about 12°S in Bahia and as far south as about 23°S in Rio de Janeiro (Napier 1976). The species occurs as far east as the Atlantic coast, and at least as far west as 48°W near 19°S in southwestern Minas Gerais (Kumm and

Laemmert 1950), but the western limit of the range has not been determined. For details of the distribution, see Table 72.

Abundance and Density

C. personatus is considered to be endangered or possibly vulnerable, and its populations have been described as at low densities, rare in one locality, and numerous in another (Table 72). Very little is known about the abundance of this species, and no estimates of population density, group size, or home range area are available.

Habitat

C. personatus occupies Atlantic coastal forest (Vanzolini 1978) and swamp forest (Laemmert et al. 1946). It has been recorded at 300 to 900 m elevation (Napier 1976).

Factors Affecting Populations

Habitat Alteration

Habitat destruction is widespread and accelerating within the range of *C. personatus*, and remaining forests are rapidly being cut to clear land for agriculture and pasture and for charcoal production (I.U.C.N. 1978).

Human Predation

Hunting. Hunting for food is a serious threat to *C. personatus* (I.U.C.N. 1978).

Pest Control. No information.

Collection. No *C. personatus* were imported into the United States between 1968 and 1973 (Appendix) or in 1977 (Bittner et al. 1978), nor is this species used in research in the United States (Thorington 1972). A total of 14 were imported into the United Kingdom between 1965 and 1975 (Burton 1978).

Conservation Action

C. personatus is found in five national parks, one state park, and two federal biological reserves (Table 72). Nevertheless, study leading to its protection is thought to be urgent (Vanzolini 1978).

Commerce in *C. personatus* has been forbidden by Brazilian law since 1967 (Mittermeier and Coimbra-Filho 1977).

Table 72. Distribution of *Callicebus personatus*

Country	Presence and Abundance	Reference
Brazil	Near Ilhéus, coast of Bahia (14°50′S, 39°06′W)	Laemmert et al. 1946
	At São Domingos, Espírito Santo; localities mapped in N and S Espírito Santo and central and W Minas Gerais	Kumm and Laemmert 1950
	From Espírito Santo and São Paulo	Deane et al. 1969
	At Itarana, Espírito Santo	Aguirre 1971
	Rare near São Paulo; numerous near Sooretama, N Espírito Santo	Deane et al. 1971
	Localities listed in Lamarão, Bahia; Engenheiro Reeve, Espírito Santo; Piquete, São Paulo; Formosa	Napier 1976
	At low densities; along R. Itabapoana, R. Doce, R. São Mateus, Serra da Mantiqueira; in Monte Pascoal N.P., Serra dos Orgãos N.P., Itatiaia N.P., Serra da Bocaina N.P., Caparaó N.P., Sooretama Federal Biological Reserve, Nova Lombardia F.B.R., Rio Doce State Park; status indeterminate, possibly vulnerable	I.U.C.N. 1978
	Endangered	Vanzolini 1978

Callicebus torquatus Collared Titi or Widow Monkey

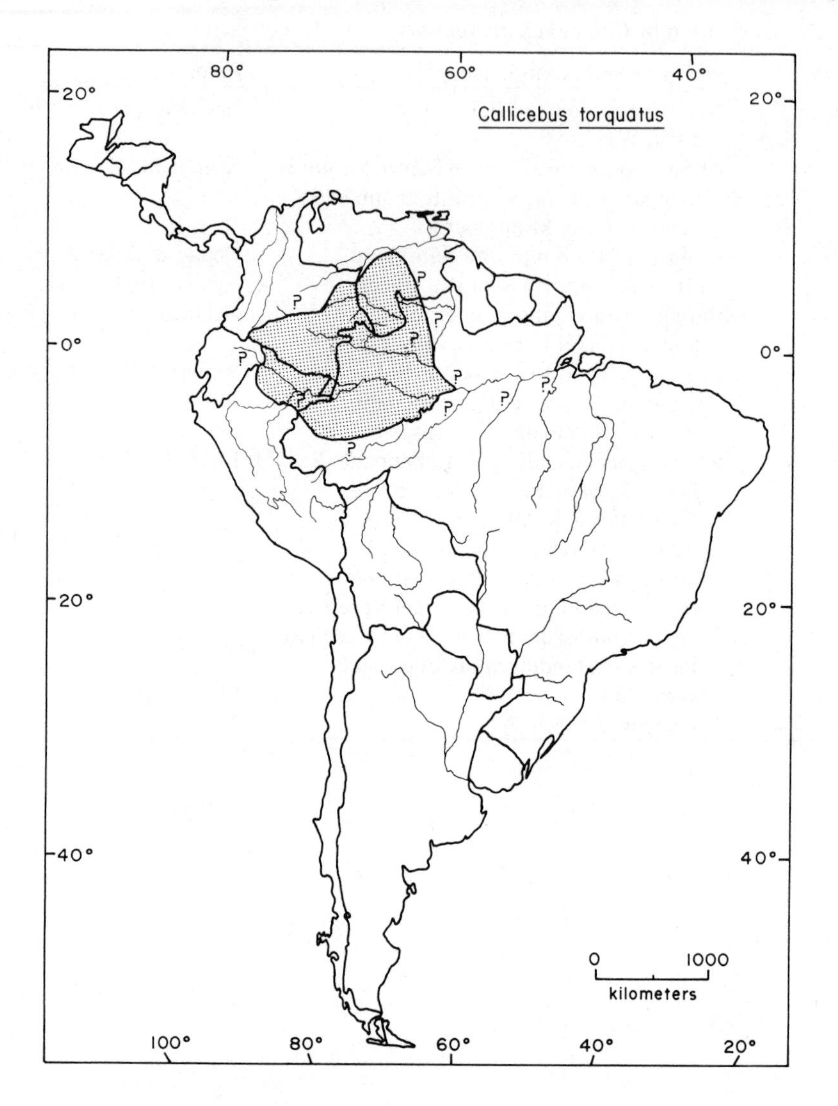

Taxonomic Note

Three (Hershkovitz 1963, Napier and Napier 1967) or four (Cabrera 1958) subspecies are recognized.

Geographic Range

Collared titis are found in northwestern South America from southern Colombia and Venezuela into northeastern Peru and northwestern Brazil. They occur as far north as Maripa, Venezuela (7°24′N, 65°10′W) (Mondolfi 1976), and as far west as about 77°W in southwestern Putumayo, Colombia (Hernández-Camacho and Cooper 1976). The range extends as far south as 7°S along the Rio Juruá and as far east as at least Codajás at 3°55′S, 62°W on the Rio Solimões (Hershkovitz 1963). This species has also been reported farther east near Manaus at 60°W on the Rio Negro (Avila-Pires 1964) and near Altamira at 3°13′S, 52°15′W on the Rio Xingu (Mittermeier and Coimbra-Filho 1977). For details of the distribution, see Table 73.

Abundance and Density

Callicebus torquatus is not thought to be severely threatened in Colombia except in the western portion of its range there, and is not in danger in Brazil, but is in a vulnerable position in Peru and is locally rare in Venezuela (Table 73).

Population densities in northeastern Peru vary from 2 to 15.2 individuals per km^2 (Freese et al., in prep.). Group sizes of 2 to 5 (Hernández-Camacho and Cooper 1976) and 2 to 4 ($\bar{X}$ = 3, N = 8) (Izawa 1976) have been observed in Colombia, and groups of 3 or 4 ($\bar{X}$ = 3.5, N = 6) were reported in Peru (Freese 1975), where one group had a home range of about 20 ha (Kinzey et al. 1977).

Habitat

C. torquatus is a lowland rain forest species, found in mature or mixed forest on well-drained ground (Moynihan 1976a) and from narrow gallery forest to continuous rain forest in Colombia (Hernández-Camacho and Cooper 1976). In Peru it is said to prefer higher terrain away from rivers (Soini 1972), but has also been observed in palm forests along streams and the bases of slopes, in noninundated evergreen seasonal formations (Kinzey 1977a), and in evergreen, mesophytic, broadleaf forests on flood plains and high ground of black-water drainages (Freese et al., in prep.).

In Brazil, *C. torquatus* has been found in *igapó* forest (seasonally flooded with black water) and both primary and secondary *várzea* forest (seasonally flooded with white water) (Mittermeier and Coimbra-Filho 1977). In one study in Venezuela, 87% of *C. torquatus* were in tropical humid forest and 13% in tropical very humid forest (life zones of Ewel and Madriz 1968), all in evergreen forest near streams or other moist areas between sea level and 500 m elevation (Handley, pers. comm. 1974).

Factors Affecting Populations

Habitat Alteration

In the western portion of its range in Colombia, heavy settlement and clearing of forest for agriculture have decreased *C. torquatus*'s available habitat (Hernández-Camacho and Cooper 1976). Elsewhere this species seems to be adaptable and able to survive in proximity to human habitations (Freese 1975, Mittermeier and Coimbra-Filho 1977), but no information is available regarding the overall impact of habitat disturbance on its populations.

Human Predation

Hunting. Collared titis are not heavily hunted in Colombia (Hernández-Camacho and Cooper 1976), but are hunted for their meat in Peru (Neville 1975, Castro 1976) and Brazil (Mittermeier and Coimbra-Filho 1977).

Pest Control. No information.

Collection. C. torquatus are short lived under ordinary conditions of captivity and are not commonly kept as pets (Hershkovitz 1963). But they are occasionally captured for use as pets in Peru (Castro 1976) and Brazil (Mittermeier and Coimbra-Filho 1977).

C. torquatus is rarely exported from Peru (Soini 1972). It was not imported into the United States between 1968 and 1973 (Appendix) or in 1977 (Bittner et al. 1978), nor into the United Kingdom between 1965 and 1975 (Burton 1978). It is not used in research in the United States (Thorington 1972).

Conservation Action

C. torquatus is found in one national park in Colombia (Table 73), but this park is shrinking in size and is in need of better protection (Struhsaker 1976b). It does not occur within any protected areas in Peru (Castro 1976) and no reports were located of protected habitat for it in Brazil or Venezuela.

This species is legally protected against hunting, capture, and commerce in Peru (Neville et al. 1976), but the hunting prohibition is not enforced, and should be, where feasible (Freese 1975). *C. torquatus* is also protected by law against commerce in Brazil (Mittermeier and Coimbra-Filho 1977), and against export from Venezuela (Mondolfi 1976).

Table 73. Distribution of *Callicebus torquatus* (countries are listed in counterclockwise order)

Country	Presence and Abundance	Reference
Colombia	In Putumayo basin	Cabrera 1958
	Between R. Guaviare-Guayabero and R. Putumayo; localities listed and mapped in Vichada, Vaupés, Putumayo, and Caquetá	Hershkovitz 1963
	In La Macarena N.P., Meta	Green 1972
	Throughout S half, S of R. Tomo and R. Guaviare-Guayabero and E of Cordillera Oriental in S Vichada, Guainía, Vaupés, Amazonas, extreme S Meta, Caquetá, and Putumayo (map); questionable further N; probably not severely threatened except in W	Hernández-Camacho and Cooper 1976
	In basins of R. Putumayo, R. Caquetá, and R. Yali (Yarí?) (all in S)	Izawa 1976
	On S bank of R. Guayabero, La Macarena N.P.	Klein and Klein 1976
	Near El Pepino, Rumiyaco, and Santa Rosa, Putumayo	Moynihan 1976a
	At El Refugio, near La Macarena N.P.	Struhsaker 1976b
Ecuador	No information	
Peru	In Putumayo basin, N Loreto	Cabrera 1958
	In N as far S as N bank of R. Marañón in basins of R. Napo and R. Tigre; localities	Hershkovitz 1963

Table 73 (continued)

Country	Presence and Abundance	Reference
Peru (cont.)	listed and mapped on R. Nanay, and R. Putumayo	Hershkovitz 1963 (cont.)
	On left bank of R. Amazonas between R. Tigre and R. Putumayo	Grimwood 1969
	In R. Nanay basin	Soini 1972
	Along R. Nanay and R. Ampiyacu; status probably good	Freese 1975
	Vulnerable; not in any protected areas	Castro 1976
	Localities mapped in NE	Freese et al. 1976
	Locality listed in Loreto at 2°40′S, 70°30′W	Napier 1976
	Near R. Nanay at 73°30′W, 4°10′S	Kinzey et al. 1977
Brazil	In Amazonian region N and S of R. Solimões	Cabrera 1958
	In W Amazonas S of R. Solimões to R. Purus and N of R. Solimões to R. Negro; localities listed and mapped	Hershkovitz 1963
	Near Manaus, R. Negro	Avila-Pires 1964
	Locality mapped in W Amazonas S of Leticia	Deane et al. 1969, Deane 1976
	Localities listed in Amazonas on Rios Solimões, Negro, Purus, and Iça	Napier 1976
	Common in Amazonian region; also reported common near Altamira, R. Xingu (by Nigel Smith); not in danger	Mittermeier and Coimbra-Filho 1977
Venezuela	In Amazonas (S)	Cabrera 1958
	In basins of upper R. Orinoco and Brazo Casiquiare, Amazonas; localities listed and mapped include Santa Bárbara (in W on R. Orinoco) and Mount Duida, R. Cunucunuma	Hershkovitz 1963
	Localities listed and mapped in S Amazonas, SE, SW, and NW of Esmeralda on R. Cunucunuma, R. Orinoco, R. Mavaca, and Brazo Casiquiare	Handley, pers. comm. 1974
	Locally distributed in Amazonas from R. Negro, Brazo Casiquiare, and upper R. Orinoco (R. Cunucunuma) to R. Ventuari (near 5°N); rare in W Bolivar, on R. Caura at Maripa and on R. Antabari, an affluent of R. Paragua	Mondolfi 1976
	Localities listed on R. Yatua (S Amazonas) and at about 5°N along R. Orinoco	Napier 1976

Cebus albifrons White-fronted Capuchin

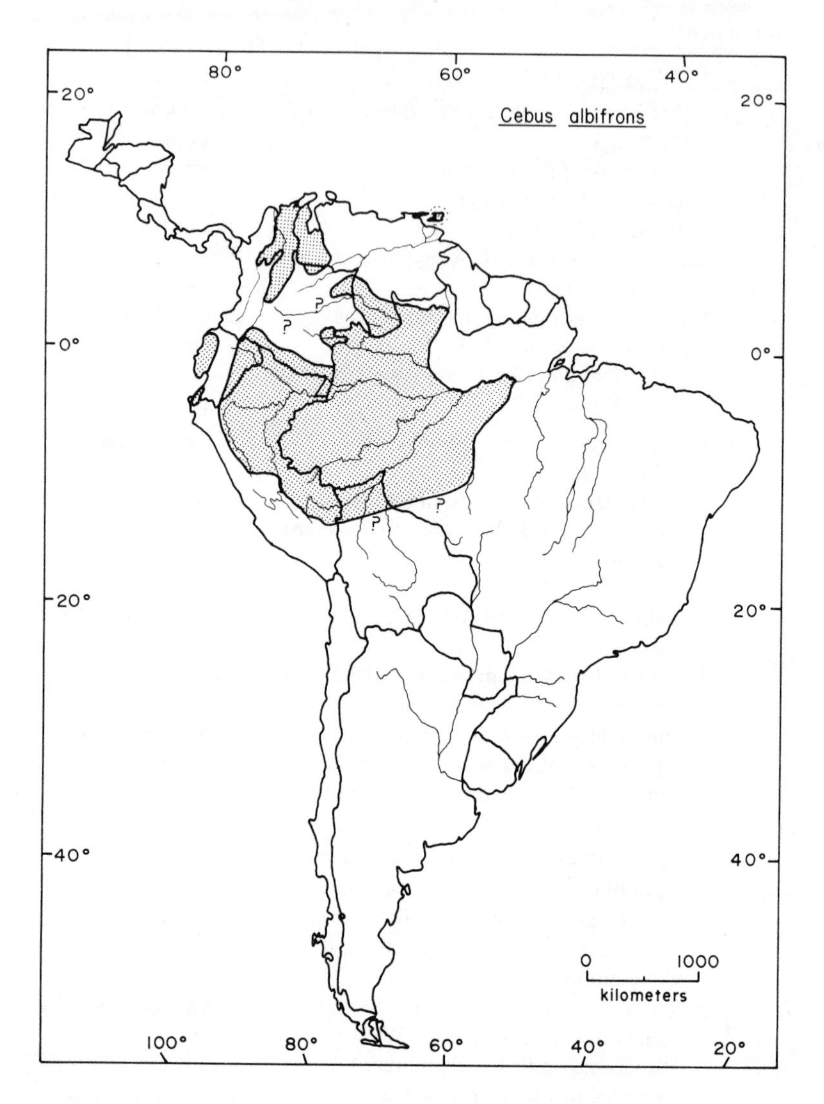

Taxonomic Note

Various authors list 11 (Cabrera 1958), 12 (Napier and Napier 1967), or 13 subspecies (Hershkovitz 1949).

Geographic Range

Cebus albifrons is found in northern and central South America, from northern Colombia and Trinidad to southern Peru and central Brazil. Its northern limit is at Ríohacha (11°34′N, 72°58′W), on the Caribbean coast of Guajira, Colombia (Hernández-Camacho and Cooper 1976), and its southernmost locality is on the Río Piachaca (13°42′S, 70°24′W) in southern Madre de Dios, Peru (Durham 1975). This species occurs as far west as about 81°W on the Pacific coast of Ecuador and as far east as about 55°W at the mouth of the Rio Tapajós in Pará, Brazil (Hershkovitz 1949). For details of the distribution, see Table 74.

Abundance and Density

C. albifrons was not described as abundant in any of the countries of its range. Its populations are declining in parts of Peru and Brazil, are vulnerable in Peru but not endangered in Brazil, and have been reduced to remnants in Trinidad (Table 74). Its status in Colombia, Ecuador, Bolivia, and Venezuela is not known.

At one study site in Colombia, the population density of *C. albifrons* was 3.8 to 15.3 individuals per km^2 (Green 1978b). In Peru this species has been observed at densities of 2 to 24 per km^2 (Freese 1975) and 10 to 34 per km^2 (Janson and Terborgh, in press).

Large groups of 15 to 20 (Green 1976) and 20 to 30 (Hernández-Camacho and Cooper 1976) are common. In southern Colombia group size varied from 10 to 50 ($\bar{X}$ = 19, N = 7 groups), and one group had a home range of about 300 ha (Izawa 1976). In Peru, groups of 8 to 34 (Durham 1975), 1 to 5 ($\bar{X}$ = 2.4, N = 10) (Freese et al., in prep.), and 8 to 16 ($\bar{X}$ = 7.45, N = 20) with home ranges of more than 200 ha (Janson and Terborgh, in press) have been reported.

Habitat

C. albifrons is found in deciduous, humid, gallery, and brackish-water mangrove forests, and prefers primary or advanced secondary forest to younger vegetation (Hernández-Camacho and Cooper 1976). In Colombia it occupies isolated forest patches as well as continuous forest (Bernstein et al. 1976a), and waterlogged and well-drained forest (Moynihan 1976a), and is more abundant in wet gallery forest than nearby dry forest (Struhsaker et al. 1975).

White-fronted capuchins have been observed in evergreen, mesophytic, broadleaf forests on high ground and on flood plains of white and black-water rivers in Peru, and on high ground in white-water drainages in Bolivia (Freese et al., in prep.). Brazilian populations have been recorded in seasonally or permanently flooded back-water *igapó* forest and primary white-water *várzea* forest (Mittermeier and Coimbra-Filho 1977).

In one study in Venezuela, all *C. albifrons* were seen in mature upland evergreen forest: 60% in tropical humid forest, 23% in tropical very humid forest, and 17% in premontane very humid forest (Handley, pers. comm. 1974, life zones of Ewel and Madriz 1968). At one site on Trinidad this species occupied an evergreen seasonal hardwood forest on a sandy island in a freshwater swamp (Downs et al. 1968).

C. albifrons occurs in habitats from sea level to 2,000 m elevation (Hershkovitz 1949). It seems to be more common in humid than dry forests and in primary than secondary growth.

Factors Affecting Populations

Habitat Alteration

C. albifrons is an adaptable species that can survive in very reduced and degraded habitats, yet the greatest threat to its survival in northern Colombia is the complete destruction of the habitat for cattle grazing and farming, which has occurred in many areas (Hernández-Camacho and Cooper 1976). The cutting of forests has also reduced its available habitat in Trinidad (Neville 1976a), but is not yet a serious threat in Brazil (Mittermeier and Coimbra-Filho 1977). No information was located regarding the effect of habitat alteration on this species in Ecuador, Peru, Bolivia, or Venezuela.

Human Predation

Hunting. White-fronted capuchins are hunted for food in the middle Magdalena valley and the Amazon region of Colombia (Hernández-Camacho and Cooper 1976), in Venezuela (Mondolfi 1976), and on Trinidad (Neville 1976a). In Peru, hunting is heavy in some areas (Neville et al. 1976), and where *C. albifrons* populations are low, an increase in hunting could lead to local extinction (Freese 1975). Hunting of *C. albifrons* for food and as bait for large cat traps occurs in northern Peru (Encarnación et al. 1978) and Brazil (Mittermeier and Coimbra-Filho 1977).

Pest Control. C. albifrons is an infamous crop pest in northern Colombia (Green 1976). It often forages in cornfields, and is killed as a pest when possible (Hernández-Camacho and Cooper 1976). In Peru this species feeds on sugar cane, limes, and bananas (Durham, pers. comm. 1974), but no information on control efforts is available. In Brazil it is not persecuted as a pest to crops (Mittermeier and Coimbra-Filho 1977).

Collection. White-fronted capuchins are frequently kept as pets in Brazil, but are not exported from that country (Mittermeier and Coimbra-Filho 1977). In the past they have been collected for export from several countries, especially Peru and Colombia. For example, 1,592 *C. albifrons* were exported from Peru in 1964–74 (Castro 1976), and 1,852 *C. albifrons* were imported into the United States from Colombia in 1970 (Paradiso and Fisher 1972).

Between 1968 and 1973 a total of 18,793 *C. albifrons* entered the United States (Appendix), primarily from Colombia and Peru and also from several countries where this species does not occur. None entered the United States in 1977 (Bittner et al. 1978). In 1965–75, 553 *C. albifrons* were imported into the United Kingdom (Burton 1978).

C. albifrons has been used in research in the United States (Thorington 1972, Comm. Cons. Nonhuman Prim. 1975a).

Conservation Action

C. albifrons occurs in three national parks, three wildlife sanctuaries, two reserves, one faunistic territory, and one national forest (Table 74). But in the two reserves in Peru, enforcement of antipoaching regulations is not adequate (Castro 1976). In fact, throughout Peru the law (Supreme Decree 934-73-AG, 1973) prohibits killing, capture, or commerce in *C. albifrons*, but the hunting portion of the law is not effectively enforced (Freese 1975, Neville et al. 1976). One proposal for conservation of this species suggests the establishment of a guard post to prevent poaching and forest destruction in the Tumbes National Forest and a campaign for public education in this area (Mittermeier 1977b).

Legal restrictions prevent the export of *C. albifrons* from Venezuela (Mondolfi 1976) and Brazil (Mittermeier and Coimbra-Filho 1977). Colombian law permits the capture of 10 *C. albifrons* per month by each license holder (Cooper and Hernández-Camacho 1975). These permits are issued only if trappers can present a certificate of need from an approved, established, research institution (Green 1976).

Table 74. Distribution of *Cebus albifrons* (countries are listed in counterclockwise order)

Country	Presence and Abundance	Reference
Colombia	In N in lower R. Sinú, R. Rancheria, and R. Cauca-Magdalena areas; in R. Orinoco drainage and Amazonian area (map)	Hershkovitz 1949
	Not W of R. Magdalena-Cauca	Hershkovitz 1955
	In NE, N coast between R. Magdalena and Sierra Nevada de Santa Marta; Bolívar between R. Magdalena and R. Cauca; in valley of R. César; on W slope of Cordillera Oriental; in central Magdalena valley from S Magdalena to Cundinamarca and Tolima; in SE	Cabrera 1958
	In Santander at 6°38′N, 73°55′W	Sturm et al. 1970
	In Isla de Salamanca N.P. and Tayrona N.P., Magdalena; in El Tuparro Faunistic Territory, Vichada	Green 1972
	Localities listed in N and NE in NE Magdalena, NW and E Guajira, and central César	Struhsaker et al. 1975
	Near Ventura, Bolívar	Bernstein et al. 1976a
	Near Puerto Rico, Bolívar (8°34′N, 74°14′W)	Green 1976, 1978b
	In Magdalena, César, S Guajira, Bolívar, E Sucre, NE and E Antioquia, Santander, W Boyacá, E Caldas, W Cundinamarca, N Tolima, Norte de Santander, NW Arauca, E Vichada, E Vaupés, Putumayo, and S Amazonas (map); questionable elsewhere in E and central	Hernández-Camacho and Cooper 1976
	Along R. Peneya, middle basin of R. Caquetá and R. Putumayo	Izawa 1976
Ecuador	In W along coast N of 2°S and in E (map)	Hershkovitz 1949
	W of Andes	Cabrera 1958
	May be introduced in W	Hershkovitz 1972
	Localities listed in Esmeraldas (in NW) and Los Ríos at 0°13′S, 79°31′W (but this is N of Los Ríos)	Baker 1974
	Localities listed include Mindo (0°2′S, 78°48′W), Gualea, Quito, and Archidona	Napier 1976

Table 74 (continued)

Country	Presence and Abundance	Reference
Peru	Throughout E (map)	Hershkovitz 1949
	In NE in basins of R. Ucayali and R. Huallaga between R. Marañón and R. Napo; in E in basins of R. Urubamba and upper R. Madre de Dios	Cabrera 1958
	In Amazonian region and probably in coastal Tumbes (in NW); distribution patchy	Grimwood 1969
	Population decreased 1966–71 at R. Curanja, SE Loreto	Gardner, pers. comm. 1974
	In SE at R. Piachaca, San Gaban–R. San Juan, Karos-Pilcopata, and Manú N.P., Madre de Dios; abundant	Durham 1975
	Low population levels along R. Nanay near Iquitos and Pucallpa, R. Ucayali; also in Manú N.P.	Freese 1975
	In Manú N.P. and Pacaya and Samiria national reserves (in N); vulnerable	Castro 1976
	Localities listed include Cuzco, Loreto, R. Ucayali, R. Tahuamanu, and R. Cosireni	Napier 1976
	Not abundant along R. Samiria and R. Pacaya between R. Ucayali and R. Huallaga-Marañón	Neville et al. 1976
	On Pacific coast in Tumbes including Tumbes National Forest	Mittermeier 1977b
	In N along R. Yavineto (R. Putumayo) near Bellavista	Encarnación et al. 1978
	Common at Cocha Cashu, Manú N.P.	Janson and Terborgh, in press
Bolivia	In N, N of about 12°N (map)	Hershkovitz 1949
	Uncommon near Cobija, R. Acre (extreme NW)	Heltne et al. 1976
	At Cobija and Riberalta (in N at 12°05′S, 65°50′W)	Freese et al., in prep.
Brazil	Throughout NW, W of R. Negro-Branco and R. Tapajós (map)	Hershkovitz 1949
	At São Salvador, extreme W Acre	Carvalho 1957
	In Amazonas	Deane et al. 1969
	Localities listed include Rios Juruá, Negro, Tapajós, and Tonantins, and near Santarém	Napier 1976
	Diminishing but not endangered in Amazonian region	Mittermeier and Coimbra-Filho 1977
Venezuela	In S in Amazonian region and upper Orinoco drainage, and in W, W and S of Lake Maracaibo (map)	Hershkovitz 1949

Table 74 (continued)

Country	Presence and Abundance	Reference
Venezuela (cont.)	In W from Sierra de Perijá to Lake Maracaibo and S Zulia; in S in upper Orinoco and Brazo Casiquiare basins	Cabrera 1958
	In Zulia along R. Yasa, Sierra de Perijá	Pirlot 1963
	Localities listed and mapped in extreme W (S Zulia, Táchira, W Apure), and in S central Amazonas (R. Orinoco and R. Mavaca)	Handley, pers. comm. 1974
	In W, in N central Sierra de Perijá between R. Palmar and R. Tucuco, N of Montes de Oca near R. Guasare, S of Lake Maracaibo; in S from El Duida in Alto Orinoco to Brazo Casiquiare and R. Negro	Mondolfi 1976
Trinidad	On E central coast	Downs et al. 1968
	Almost certainly introduced there	Hershkovitz 1972
	Remnant populations in SE and central (localities mapped) including in wildlife sanctuaries of Bush Bush, Trinity Hills, and Central Range	Neville 1976a

Cebus apella Black-capped or Tufted Capuchin

Taxonomic Note

Eleven subspecies are recognized (Cabrera 1958, Napier and Napier 1967).

Geographic Range

Cebus apella is found in northern and central South America, from central Colombia to southern Brazil. It occurs as far west as about 77°W in southwestern Colombia (Hernández-Camacho and Cooper 1976) and as far east as 39°E (Kumm and Laemmert 1950) and perhaps even farther east, along the coast of Brazil. Its northernmost record is on Isla de Margarita, Venezuela, at about 11°N, 64°W (Handley, pers. comm. 1974), and its southernmost locality is at Trés Passos, in northern Rio

Grande do Sul, Brazil (27°33′S, 53°55′W) (Kumm and Laemmert 1950). For details of the distribution, see Table 75.

Abundance and Density

C. apella is known to be abundant in some parts of Bolivia, Peru, and Surinam, and has been described as "not in danger" in the latter two countries (Table 75). But other authors consider it to be vulnerable in Peru, and somewhat vulnerable in Guyana. It is known to be declining in some parts of Colombia, Peru, and Brazil, and although it is almost extinct in coastal Rio de Janeiro, it is thought to be in no danger in Amazonian Brazil (Table 75). No recent information is available regarding its status in Ecuador, Argentina, or Paraguay.

Population density in *C. apella* varies from 2 to 111 individuals per km^2 (Table 76). Groups include 2 to more than 30 members, with average groups numbering 5 to 20, and home ranges of 80 to 150 ha (Table 76).

Habitat

C. apella is a forest monkey which, in Colombia, occupies virtually every type of humid forest from gallery to palm to rain forest, including both seasonally flooded and nonflooded forest, cloud forest, isolated forest, and secondary growth (Hernández-Camacho and Cooper 1976), creek levee bank heterogeneous forest, and swamp forest (Klein and Klein 1976). It is a flexible species, and in Surinam is common in high and low rain forest, riverine forest, liane forest, marsh forest, swamp forest, and various secondary formations (Mittermeier 1977a).

In Peru, *C. apella* has been observed in evergreen, mesophytic, broadleaf forests on high ground, and on black-water flood plain (Freese et al., in prep.). In Bolivia it occupies dry forest on the edge of the *chaco* (Heltne et al. 1976) and evergreen and deciduous forests on high ground in white-water drainages (Freese et al., in prep.). And in Argentina, *C. apella* utilizes fairly dry to fairly humid forests (Crespo 1950).

In Brazil this species has been recorded in remnants of climax and young forests on steep hills (Causey et al. 1948), old forests (Laemmert et al. 1946), and high Amazonian and coastal forests (Deane et al. 1971). In the Amazonian region it has been found in *terra firme* (nonflooding) primary forest, primary *várzea* forest (seasonally flooded by white-water rivers), and *igapó* forests (flooded by black or clear-water rivers) (Mittermeier and Coimbra-Filho 1977).

In one study in Venezuela, 73% of all *C. apella* occurred near streams in evergreen forest, 66% in the humid tropical forest, 25% in premontane dry forest, and 9% in premontane humid forest (Handley, pers. comm. 1974; life zones of Ewel and Madriz 1968). All these habitats were lower than 500 m elevation, but *C. apella* has been recorded up to 1,500 m in Peru (Grimwood 1969) and 2,700 in Colombia (Hernández-Camacho and Cooper 1976).

Factors Affecting Populations

Habitat Alteration

C. apella is an adaptable species which may benefit fro[illegible]ertain kinds of habitat disturbances (Mittermeier 1977a), such as the early stage[illegible]gricultural expansion

(Muckenhirn and Eisenberg 1978). Nevertheless, habitat destruction is the principal threat to its populations in Colombia (Hernández-Camacho and Cooper 1976), and has contributed to the decline of primate populations in northern Bolivia (Heltne et al. 1976).

In southern Mato Grosso, Brazil, thousands of tufted capuchins were destroyed in 1970 during extensive burning to create pasture land (Coimbra-Filho 1974). In Amazonian Brazil, deforestation is not yet widespread enough to threaten *C. apella* (Mittermeier and Coimbra-Filho 1977), but the habitat of this species is being destroyed by clearing for agriculture and cattle ranching in eastern Pará, at least (Ayres 1978). No information was located regarding the extent of habitat alteration elsewhere in the range.

Human Predation

Hunting. C. apella is hunted for its flesh in Colombia (Hernández-Camacho and Cooper 1976) and Peru (Grimwood 1969; Gardner, pers. comm. 1974; Castro 1976; Neville et al. 1976). In several regions of Peru this hunting is especially intense and has greatly reduced *C. apella* populations (Freese 1975, Encarnación and Castro 1978). The same is true in Bolivia (Heltne et al. 1976).

Tufted capuchins are also killed for food in Brazil (Smith 1976, Mittermeier and Coimbra-Filho 1977), Surinam (Husson 1957, Mittermeier 1977a), and Venezuela (Mondolfi 1976). They are also used as bait for large cat traps in Peru (Encarnación et al. 1978), Bolivia (Heltne et al. 1976), and Brazil (Mittermeier and Coimbra-Filho 1977).

Pest Control. Tufted capuchins feed on crops, especially immature corn, and are thus killed as pests in Colombia (Hernández-Camacho and Cooper 1976) and Guyana (Muckenhirn and Eisenberg 1978). They are also considered destructive to cacao, citrus, palm, and corn cultivation in Surinam (Husson 1957, Mittermeier 1977a), are shot to protect crops in southeastern Bolivia (Heltne et al. 1976), and are commonly accused of raiding crops and gardens in Peru (Grimwood 1969). They are not persecuted as crop pests in Brazilian Amazonia (Mittermeier and Coimbra-Filho 1977).

Collection. C. apella is frequently kept as a pet in Surinam (Husson 1957, Mittermeier 1977a), Brazil (Mittermeier and Coimbra-Filho 1977, Ayres 1978), and Peru (Castro 1976, Encarnación et al. 1978). It has also been captured for export in large numbers, especially in Peru. Between 1964 and 1974 a total of 18,986 *C. apella* were exported from Peru (Castro 1976). In Guyana this species is abundant enough in some areas to "support localized trapping of moderate numbers" (Muckenkirn et al. 1975).

Between 1968 and 1973, 7,261 *C. apella* were imported into the United States (Appendix), primarily from Peru and Paraguay. In 1977, 104 entered the United States from Bolivia (Bittner et al. 1978). *C. apella* is used in biomedical research in the United States (Thorington 1972, Comm. Cons. Nonhuman Prim. 1975a). In 1965–75 a total of 942 *C. apella* were imported into the United Kingdom (Burton 1978).

Conservation Action

C. apella is found in eight national parks, one nature park, and ten reserves (Table 75). But not all of these reserves are adequately protected from habitat destruction or hunting. For example, La Macarena National Park in Colombia is rapidly shrinking, owing to encroachment by agriculturists, and will disappear if protection is not enforced (Struhsaker 1976b). Also, the Pacaya and Samiria national reserves in northern Peru are not patrolled effectively against poaching (Castro 1976).

Hunting, capture, and commercialization of *C. apella* are illegal in Peru (Supreme Decree 934-73-AG, 1973) (Castro 1976), but the hunting prohibition has not been well enforced and should be, in order to protect this species (Freese 1975). In Surinam, *C. apella* had previously been classified as game (Schulz et al. 1977), but new legislation now protects it from hunting in the north and from transport to centers of population (Mittermeier 1977a). Legal restrictions prohibit the export of *C. apella* from Venezuela (Mondolfi 1976) and commercial exploitation in Brazil (Mittermeier and Coimbra-Filho 1977).

Table 75. Distribution of *Cebus apella* (countries are listed in counterclockwise order)

Country	Presence and Abundance	Reference
Colombia	E of San Martín, Meta	Thorington 1967
	In Puracé N.P., Cauca-Huila; Cueva de los Guácharos N.P., Huila; El Tuparro Faunistic Territory, Vichada	Green 1972
	In La Macarena N.P., Meta	Hunsaker 1972
	On N bank R. Guayabero, La Macarena N.P., Meta	Klein and Klein 1975, 1976
	Throughout Amazonian region and E (map); in Tierradentro region, E Cauca; declining in upper Magdalena valley (Huila) and in areas of settlement in E plains	Hernández-Camacho and Cooper 1976
	Along R. Peneya, middle basin of R. Caquetá and R. Putumayo (in S)	Izawa 1976
	At Tolima (2°20′N)	Napier 1976
	Common in La Macarena N.P.	Struhsaker 1976b
	On R. Duda, W boundary of La Macarena N.P.	Izawa and Mizuno 1977
Ecuador	E of Andes	Cabrera 1958
Peru	In Amazon region throughout Loreto and in San Martín, Amazonas, Huánuco, Pasco, Junín, Cuzco, Puno, and Madre de Dios; widespread and common; abundant in most areas; not in danger of extinction	Grimwood 1969
	Along R. Curanja, SE Loreto	Gardner, pers. comm. 1974
	Along R. Nanay near Iquitos, R. Orosa, R. Samiria (all in NE) and at Cocha Cashu, Manú N.P. (in SE)	Freese 1975

Table 75 (continued)

Country	Presence and Abundance	Reference
Peru (cont.)	In Manú N.P., Tingo María N.P., and Pacaya and Samiria national reserves; vulnerable	Castro 1976
	Abundant in R. Samiria basin, frequent in R. Orosa basin	Freese et al. 1976
	Localities listed in Cuzco, Huallaga, Huánuco, Loreto, San Martín, and on R. Tahuamanu	Napier 1976
	Not abundant along R. Samiria and R. Pacaya between lower R. Ucayali and R. Huallaga-Marañón	Neville et al. 1976
	In E Madre de Dios between R. Acre and R. Madre de Dios; disappearing in Iberia-Miraflores and R. Acre areas	Encarnación and Castro 1978
	Common at Cocha Cashu, Manú N.P.	Janson and Terborgh, in press
Bolivia	In N, central, and NE	Cabrera 1958
	At R. Surutu and Buena Vista, Santa Cruz de la Sierra, in SE (locality mapped)	Soria 1959
	At Cobija (in NW on R. Acre) and Riberalta (in N); abundant at El Triunfo (central); not common at San José de Chiquitos (SE) (localities mapped)	Heltne et al. 1976; Freese et al., in prep.
	Localities listed include Buena Vista, in Santa Cruz, R. Yapacani, Sara, Tarija, Chimate, and Yungas de Cochabamba	Napier 1976
	Near Santa Cruz de la Sierra (17°27′S, 63°34′W)	Cambefort and Moro 1978
Argentina	Fairly abundant near Salta (in NW); throughout NE to 28°S (map)	Crespo 1950
	In N in extreme SE Jujuy, Salta, Formosa, Chaco, and Misiones from E bank of R. Paraná to E border	Cabrera 1958
	At R. Santa María, NW Salta, and in Chaco (localities mapped)	Soria 1959
	In Iguazu N.P., Misiones (in NE)	I.U.C.N. 1975
	In Misiones	Colillas and Coppo 1978
Paraguay	Throughout (map)	Crespo 1950
	At Capitán Meza (in S); locality mapped	Soria 1959
	At Villarrica, Guiara (Guaíra) (24°45′S, 56°28′W) (in S)	Napier 1976
Brazil	Near Ilhéus, E Bahia	Laemmert et al. 1946
	Near Passos, S Minas Gerais	Causey et al. 1948
	Localities mapped in Amazonas, Bahia, Espírito Santo, Federal District, Goiás, Maranhão, Mato Grosso, Minas Gerais,	Kumm and Laemmert 1950

Table 75 (continued)

Country	Presence and Abundance	Reference
Brazil (cont.)	Pará, Paraná, Pernambuco, Rio de Janeiro, Rio Grande do Sul, Santa Catarina, and São Paulo; very common and widespread	Kumm and Laemmert 1950 (cont.)
	Localities listed in S Mato Grosso	Kühlhorn 1955
	Also in Piauí (in NE)	Cabrera 1958
	Locality mapped at Bahú, São Paulo	Soria 1959
	Abundant near Manaus, Amazonas, along R. Negro	Avila-Pires 1964
	In E Pará near Belém	Carvalho and Toccheton 1969
	In Amazonas, Rondônia, Pará, Espírito Santo, São Paulo, Paraná, Santa Catarina, Mato Grosso, Goiás, and Minas Gerais	Deane et al. 1969
	In São Paulo, Santa Catarina, Amazonas along Manaus-Itacoatiara Road; numerous near Sooretama in Espírito Santo	Deane et al. 1971
	In Caparáo N.P., Espírito Santo and Minas Gerais, and Tijuca N.P., Guanabara	I.U.C.N. 1971
	In S Mato Grosso	Coimbra-Filho 1974
	Localities mapped in Amazonas, Pará, and W Acre	Deane 1976
	Localities listed	Napier 1976
	Diminishing but not endangered in Amazon region	Mittermeier and Coimbra-Filho 1977
	In E Pará, E of R. Xingu	Ayres 1978
	Quite common in Morro do Diabo State Park, SW São Paulo; extremely reduced numbers, bordering on extinction in coastal Rio de Janeiro	Coimbra-Filho 1978
French Guiana	Present	Cabrera 1958
	At Approuagua	Napier 1976
Surinam	Common; in NE near Moengo Tapoe and in N at Tibiti	Husson 1957
	Localities listed include Paramaribo and R. Coppename	Napier 1976
	Widespread and abundant throughout interior and most of coastal area; localities listed and mapped include Brownsberg Nature Park and 6 nature reserves: Raleighvallen-Voltzberg, Tafelberg, Eilerts de Haan Gebergte, Sipaliwini Savanna, Wia-Wia, and Galibi; common and in little danger	Mittermeier 1977a

Table 75 (continued)

Country	Presence and Abundance	Reference
Guyana	Localities mapped in NE and central including R. Rupununi, R. Essequibo, and R. Berbice	Muckenhirn et al. 1975
	Localities listed include Bonasica, Essequibo, and Moon Mts. (in S)	Napier 1976
	Restricted to E and S; somewhat vulnerable	Muckenhirn and Eisenberg 1978
Venezuela	Localities mapped in central Amazonas (in S) and Isla de Margarita (off N coast)	Handley, pers. comm. 1974
	In Amazonas locally distributed along upper R. Orinoco from San Fernando de Atabapo (4°03′N, 67°45′W) to Brazo Casiquiare and R. Negro; on Isla de Margarita in Cerro el Copey	Mondolfi 1976

Table 76. Population parameters of *Cebus apella*

Population Density (individuals/km²)	Group Size $\bar{X}$	Group Size Range	Group Size No. of Groups	Home Range Area (ha)	Study Site	Reference
		5–7	2		Central Colombia	Thorington 1967
6–10		6–12			Central Colombia	Klein and Klein 1975
		3–30			Colombia	Hernández-Camacho and Cooper 1976
		18–30			Colombia	Izawa 1976
		≥30			Peru	Grimwood 1969
2–36	10				E Peru	Freese 1975
15–46	5–10		2–3		NE Peru	Neville et al. 1976
7	8	2–8	11		SE Peru	Encarnación and Castro 1978
20–35	5.6	4–12	46	80–90	SE Peru	Janson and Terborgh, in press
2–55	10	6–10	24		Bolivia	Heltne et al. 1976
		4–20			Argentina	Crespo 1950
	10–20	≥30			Surinam	Husson 1957
13	10–20			100–150	Surinam	Mittermeier 1977a
6–111	8–12	2–20	19		Guyana	Muckenhirn et al. 1975

Cebus capucinus White-throated or White-faced Capuchin

Taxonomic Note

Five subspecies are recognized (Hershkovitz 1949, Napier and Napier 1967).

Geographic Range

Cebus capucinus is found in Central America and northwestern South America, from Honduras to Ecuador. It occurs as far north as the northern coast of Honduras (Hall and Kelson 1959), and as far west as the area west of La Ceiba (15°45′N, 86°45′W) (Trapido and Galindo 1955). In Colombia its range extends as far north and east as Barranquilla (11°10′N, 74°50′W), and as far south as the Ecuador border

(Hernández-Camacho and Cooper 1976). No information was located regarding the limits of its range in Ecuador. For details of the distribution, see Table 77.

Abundance and Density

White-throated capuchins are locally abundant in parts of Costa Rica and Panama, but they are also declining in both countries and in one region of Colombia, and are in danger of extinction in southwestern Panama. No information is available concerning their status in Honduras, Nicaragua, or Ecuador.

The only population density estimate available is 50 individuals per km^2 at one site in the Santa Rosa National Park, Costa Rica (Freese, pers. comm. 1974). There, 15 to 20 *C. capucinus* troops included 15 to 20 members each (Freese 1976), and one troop had a home range of about 50 ha (Freese 1978).

In southwestern Panama, *C. capucinus* groups numbered 2 to 30, and one group occupied a home range of 32 to 40 ha (Baldwin and Baldwin 1976, 1977). On Barro Colorado Island, one group included 15 members and defended an area of 85.5 ha (Oppenheimer 1969). And on Coiba Island, 13 groups of 5 to 20 members were recorded (Milton and Mittermeier 1977). In Colombia, group size varies from 6 to 15 individuals (Hernández-Camacho and Cooper 1976).

Habitat

C. capucinus is a forest monkey found in several types of wet, dry, primary, and secondary forest. In Costa Rica it utilizes tall evergreen and deciduous dry forest, tall primary lowland wet forest, secondary growth, and mangrove forest (Freese, pers. comm. 1974). In seasonally or chronically dry areas the availability of standing water may be an important determinant of its local density and distribution (Freese 1978).

In Panama, *C. capucinus* is most common in humid areas on well-drained ground, but also extends into mangrove and a variety of second growth habitats (Moynihan 1976a). In southern Chiriquí, *C. capucinus* troops spent at least 50% of their time in mangrove and marshy forest (Baldwin and Baldwin 1976). At other study sites this species occupied moist tropical forest (Fleming 1973), young secondary forest, ecotone between secondary forest and old forest, and ecotone between secondary forest and open areas (Milton and Mittermeier 1977).

Colombian *C. capucinus* prefers primary or old secondary forest, is also found in highly degraded remnant forests and freshwater seasonally flooded forests, but is absent in xerophytic, mangrove, natal, and lowland coastal mixed forests (Hernández-Camacho and Cooper 1976). In one study it was found to be more abundant in a large block of wet forest than in a similar gallery forest (Struhsaker et al. 1975).

C. capucinus has been recorded from sea level to 1,800 m elevation in Panama (Handley 1966, Sousa et al. 1974), and up to 2,100 m elevation in Colombia (Hernández-Camacho and Cooper 1976).

Factors Affecting Populations

Habitat Alteration

Clearing of forests is reducing available habitat for *C. capucinus* in the three countries for which information is available: Costa Rica, Panama, and Colombia.

In Costa Rica about 1% of the remaining forest is being cut annually, and this rate is accelerating, so that in 14 to 20 years all forest accessible to logging will be destroyed (Freese, pers. comm. 1974). Northwestern Costa Rica's forests have been extensively cleared for cattle grazing, and almost all primates are absent from the densely populated central plateau, where the most intensive habitat alteration has occurred. The primary threats to *C. capucinus* populations in Costa Rica are "clear-cut logging for timber and clearing of forests for agriculture" (Freese, pers. comm. 1974).

In Panama, tree removal, rapid and widespread creation of new and permanent pastures, and destruction of forests and woodlands have greatly reduced the range of *C. capucinus* (Bennett 1968). The rapidly growing human population and the implementation of an old land reform law allowing expropriation of unexploited property have accelerated the rate of deforestation (Baldwin and Baldwin 1976). For example, from 1950 to 1960 the forest cover of Chiriquí decreased from 61% to 50%, and the application of insecticides has probably also been detrimental to primate populations there (Baldwin and Baldwin 1976).

White-throated capuchins in Colombia are adaptable. They are able to survive in reduced numbers even in badly degraded habitats such as cattle pastures, as long as a few trees such as *Scheelea magdalenica* or other palms survive (Hernández-Camacho and Cooper 1976). Nevertheless, the greatest threat to *C. capucinus* in Colombia is the complete destruction of forest such as has already occurred in the Cauca valley and in northern Bolívar (Hernández-Camacho and Cooper 1976).

No data are available regarding the effect of habitat alteration on this species in Honduras, Nicaragua, or Ecuador.

Human Predation

Hunting. C. capucinus is hunted for food in some parts of Costa Rica (Wilson, pers. comm. 1974), but this hunting probably exerts an insignificant pressure on the population (Freese, pers. comm. 1974). No other reports of hunting were located.

Pest Control. C. capucinus does feed on crops, including corn and fruit, but in Costa Rica this activity seems to be rare and farmers generally tolerate it (Freese, pers. comm. 1974). In Chiriquí, Panama, this species is considered a pest to agriculture and is harassed, trapped, and shot in defense of crops (Baldwin and Baldwin 1976). And in Colombia, *C. capucinus* is also killed as a pest in cultivated areas (Hernández-Camacho and Cooper 1976).

Collection. Trapping for export, probably primarily to zoos and pet dealers, has had a significant effect on *C. capucinus* populations in Panama (Baldwin and Baldwin 1976). One dealer in Chiriquí exported about 600 *C. capucinus* between 1952 and 1960, but then the species became too scarce for exploitation (Baldwin and Baldwin 1976).

C. capucinus have also been collected in northern Colombia (Hernández-Camacho and Cooper 1976). In 1970, 264 *Cebus* (of three species) were legally exported from Colombia (Cooper and Hernández-Camacho 1975), but 719 *C. capucinus* were imported from Colombia into the United States (Paradiso and Fisher 1972).

Between 1968 and 1973 a total of 8,248 *C. capucinus* were imported into the United States (Appendix), primarily from Colombia but also from Peru and several other countries where the species does not occur. In 1965–75, 2,051 *C. capucinus* were imported into the United Kingdom (Burton 1978). None entered the United States in 1977 (Bittner et al. 1978).

Conservation Action

C. capucinus is found in at least four national parks and one reserve, and may also occur in one other park and one other reserve (Table 77). It was not reported in any protected areas in Honduras, Panama, or Ecuador.

In Costa Rica the National Park Service faces many problems, such as a lack of funds to pay guards and buy equipment, and growing demand for land to be used for agriculture. If it can survive these pressures, *C. capucinus* will be protected in several existing and proposed national parks (Freese, pers. comm. 1974). A national park or wildlife refuge is needed to preserve the monkey populations of Chiriquí, Panama (Baldwin and Baldwin 1976).

Colombian legislation imposed a capture quota of 150 *C. capucinus* and a maximum of 25 in any one month, between April and December 1973 (Cooper and Hernández-Camacho 1975). No later information was located.

Table 77. Distribution of *Cebus capucinus* (countries are listed in order from north to south)

Country	Presence and Abundance	Reference
Honduras	In N between La Ceiba and Tela	Trapido and Galindo 1955
	In E and SE (localities mapped)	Hall and Kelson 1959
Nicaragua	Throughout E and central (localities mapped)	Hall and Kelson 1959
	At Matagalpa (12°52′N, 85°58′W)	Napier 1976
	In Saslaya N.P., Zelaya (in NE)	Ryan 1978
Costa Rica	Throughout except in NW (localities mapped)	Hall and Kelson 1959
	Rare in Cabo Blanco Nature Reserve (S tip Nicoya Peninsula, Puntarenas)	Harroy 1972
	Throughout including NW (Nicoya Peninsula, Santa Rosa N.P., near Liberia, Volcan Rincon de la Vieja); absent on central plateau; overall abundant but declining; density and rate of decline locally variable	Freese, pers. comm. 1974
	In Guanacaste and Pacific and Caribbean lowlands; still common in many areas	Wilson, pers. comm. 1974
	In Santa Rosa N.P. (in NW); not confirmed in Cabo Blanco Nature Reserve	I.U.C.N. 1975
	Population of 250 to 350 in Santa Rosa N.P.	Freese 1976
	At Nicoya (in NW)	Napier 1976
	On Osa Peninsula (in SW), and Monteverde Reserve	C. H. F. Rowell, pers. comm. 1978

Table 77 (continued)

Country	Presence and Abundance	Reference
Panama	In Coto region, extreme W	Carpenter 1935
	In Chiriquí (SW), Los Santos (SE Azuero Peninsula), Panamá E and W of canal including Bayano R. area, and Darién (in SE)	de Rodaniche 1957
	Throughout including Coiba Island (off SW coast); localities mapped	Hall and Kelson 1959
	Locally abundant	Handley 1966
	In S and N Darién and central Panamá E of canal (localities listed)	Galindo and Srihongse 1967
	Throughout, including Coiba Island (may be introduced there); in 1920 was common in S central (in Coclé, Herrera, and Los Santos), now absent over most of this area; range greatly reduced in Chiriquí	Bennett 1968
	At Cristobal (NW Canal Zone)	Fleming 1973
	In Alanje and Baru districts, W Chiriquí, in Panamá, and Darién	Sousa et al. 1974
	In S Chiriquí (localities mapped); only in scattered and remote areas; common in early 1950s, now scarce; endangered	Baldwin and Baldwin 1976, 1977
	Locally rare in Canal Zone	Dawson 1976
	Localities mapped in E and central	Deane 1976
	Localities listed in Chiriquí, Coiba Island, Insoleta Island, Brava Island, Cébaco Island, and Sevilla Island (all to W)	Napier 1976
	On Coiba Island (off SW coast)	Milton and Mittermeier 1977
Colombia	In W and on Gorgona Island off SW coast	Hershkovitz 1949
	Extends E across R. Atrato to left banks of R. Cauca and lower R. Magdalena	Hershkovitz 1969
	In Farallones de Cali N.P., Valle; Orquideas N.P., Antioquia; La Macarena N.P., Meta (outside range)	Green 1972
	In NW near Estanzuela, Pispiche, R. Esmeralda, and R. Sinú	Struhsaker et al. 1975
	On W bank of R. Magdalena near Sucre, Guaranada, and Majagual	Green 1976
	Along Caribbean and Pacific coasts from Barranquilla to S border (map); in SW Atlántico E to W bank of lower R. Magdalena and middle and lower R. San Jorge; in NW Antioquia, Córdoba, Sucre, N Bolívar, Chocó, Valle, W Quindío, N and W Cauca, W Nariño, and Gorgona Island; populations very reduced in upper	Hernández-Camacho and Cooper 1976

Table 77 (continued)

Country	Presence and Abundance	Reference
Colombia (cont.)	Cauca valley; questionable in N and S Antioquia and E Nariño	Hernández-Camacho and Cooper 1976 (cont.)
	Localities listed include Gorgona Island, R. San Juan, Chocó, Jiminez (in W), Medellín	Napier 1976
	Near Colosó, S Serranía de San Jacinto	Neyman 1977
	Near Tolú, Sucre (9°34′N, 75°27′W)	Neyman 1978
Ecuador	In W	Hershkovitz 1949
	In W on W slopes of Cordillera Occidental	Cabrera 1958
	In NW	Hershkovitz 1969

Cebus nigrivittatus Weeper Capuchin

Taxonomic Note

Five subspecies are recognized (Hershkovitz 1949, Cabrera 1958, Napier and Napier 1967).

Geographic Range

Cebus nigrivittatus is found in northern South America, in Venezuela, the Guianas, and northern Brazil. It has been recorded as far northwest as almost 11°N, 69°W, in Falcón, Venezuela (Handley, pers. comm. 1974), and may occur even farther west (Mondolfi 1976). Its range extends east as far as eastern French Guiana (Hershkovitz 1949), perhaps even to 50°W in northeastern Brazil, and south to

Manaus at 3°S, 60°W, at the mouth of the Rio Negro on the Rio Amazonas (Avila-Pires 1964). For details of the distribution, see Table 78.

Abundance and Density

C. nigrivittatus populations are somewhat vulnerable in Guyana and Surinam (Table 78). Their status in Venezuela, French Guiana, and Brazil is not known.

In Guyana, population densities varied from 1.2 to 73.8 individuals per km^2 (Muckenhirn et al. 1975), and in Surinam, density was estimated at 4.33 individuals per km^2 (Mittermeier 1977a). Group sizes of 10 to 20 (Husson 1957, Mittermeier 1977a), 19 to 33 (N = 2) (Oppenheimer and Oppenheimer 1973), and 4 to 15 ($\bar{X}$ = 6.2 to 9.0, N = 11) (Muckenhirn et al. 1975) have been reported. No estimates of home range area are available.

Habitat

C. nigrivittatus is a tropical forest species found from lowland to cloud forest habitats (Mondolfi 1976). In one study in Venezuela, 88% of all *C. nigrivittatus* were observed in evergreen forest, with 4.5% in gallery forest in savanna and 0.9% each in deciduous forest and cloud forest (Handley, pers. comm. 1974). Fifty-four percent of these were in tropical humid forest, 16% in premontane humid forest, 14% in tropical dry forest, 7% in tropical very humid forest, and 3% each in premontane very humid and premontane rain forest (life zones of Ewel and Madriz 1968). These weeper capuchins were found predominantly (89%) at low elevations (0 to 500 m), with fewer sightings made up to 2,000 m altitude (Handley, pers. comm. 1974).

In Surinam, *C. nigrivittatus* is primarily restricted to undisturbed, inland tropical, high rain forest, but is occasionally seen in liane forest and montane savanna forest (Mittermeier 1977a). In Guyana it thrives in mixed habitats (Muckenhirn and Eisenberg 1978). No information is available regarding its habitats in French Guiana or Brazil.

Factors Affecting Populations

Habitat Alteration

C. nigrivittatus is "moderately adaptable" to human encroachment, and its populations seem to increase initially in areas of expanding agriculture (Muckenhirn and Eisenberg 1978). In one area of Venezuela, pesticide runoff from nearby rice fields was suspected of being harmful to *C. nigrivittatus* populations (Oppenheimer and Oppenheimer 1973). Otherwise no information was located concerning the impact of habitat alteration on this species.

Human Predation

Hunting. Weeper capuchins are hunted for their meat in Venezuela (Mondolfi 1976) and Surinam (Husson 1957, Mittermeier 1977a).

Pest Control. No information.

Collection. C. nigrivittatus is kept as a pet in Surinam (Husson 1957), and is abundant enough in some parts of Guyana to support moderate localized trapping (Muckenhirn et al. 1975).

A total of 253 were imported into the United States between 1968 and 1973 (Appendix). None entered the United States in 1977 (Bittner et al. 1978). One hundred fifty were imported into the United Kingdom between 1965 and 1975 (Burton 1978).

Conservation Action

C. nigrivittatus is found in two national parks in Venezuela, three forest reserves in Guyana, and one nature park and four nature reserves in Surinam (Table 78). It is legally protected from export in Venezuela (Mondolfi 1976), from hunting in northern Surinam (Mittermeier 1977a), and from all commerce in Brazil (Mittermeier and Coimbra-Filho 1977).

Table 78. Distribution of *Cebus nigrivittatus* (countries are listed in clockwise order)

Country	Presence and Abundance	Reference
Venezuela	Localities mapped in N, central, and SE	Hershkovitz 1949
	In N and S and in E in basin of lower R. Orinoco	Cabrera 1958
	Along Caracol Stream and R. Guárico in N	Neville 1972
	S of Calabozo, Guárico (N central)	Oppenheimer and Oppenheimer 1973
	Localities listed and mapped in Falcón and Carabobo (in N), Bolívar and Monagas (in E), and Amazonas and Apure (in S and central)	Handley, pers. comm. 1974
	Throughout except in basin of Lake Maracaibo (in NW); including in Cordillera de la Costa and Cordillera de Cerro Jaua (Bolívar); in Guatopo N.P. and Península de Paria (Sucre)	Mondolfi 1976
	Localities listed in lower R. Orinoco (in E) and R. Yatua (extreme S)	Napier 1976
	At Hato Masaguaral, Camatagua Reservoir, and Henry Pittier N.P. (all in N central)	Neville 1976a
Guyana	Present	Hershkovitz 1949, Cabrera 1958
	Localities listed and mapped in N and central at 3 forest reserves (H.M.P.S., 24 Mile, and Moraballi), at Apoteri, Matthews Ridge, and Pomeroon	Muckenhirn et al. 1975
	Localities listed include Bonasica, R. Essequibo, Potaro Road, R. Mazaruni, R. Supinaam, and coast region	Napier 1976
	Somewhat vulnerable	Muckenhirn and Eisenberg 1978
Surinam	Present	Hershkovitz 1949, Cabrera 1958, Fooden 1964

Table 78 (continued)

Country	Presence and Abundance	Reference
Surinam (cont.)	Not common; one locality in Nassau Mts. (in E)	Husson 1957
	Rare; somewhat vulnerable; restricted to interior; localities listed and mapped throughout S of about 5°20′N; in Brownsberg Nature Park and 4 nature reserves of Raleighvallen-Voltzberg, Tafelberg, Eilerts de Haan Gebergte, and Sipaliwini	Mittermeier 1977a
French Guiana	Locality mapped in N	Hershkovitz 1949
	Present	Cabrera 1958
Brazil	In lower R. Amazonas to R. Negro	Hershkovitz 1949
	In basin of R. Amazonas and tributaries near N central border	Cabrera 1958
	In R. Negro region near Manaus	Avila-Pires 1964
	In Amazonas	Deane et al. 1969
	In N	Deane 1976
	Localities listed on upper R. Catrimani (W Roraima) and NW of Manaus	Napier 1976

Chiropotes albinasus White-nosed Saki

Taxonomic Note

This is a monotypic species (Cabrera 1958, Napier and Napier 1967).

Geographic Range

Chiropotes albinasus is found in central Brazil south of the Rio Amazonas in eastern Amazonas, western Pará, northern Mato Grosso, and northern Rondônia. Its range extends from the Rio Madeira as far west as about 63°W, to the Rio Xingu at about 52°W, as far north as the Rio Amazonas (about 2°S), and as far south as at least the Rio Jiparaná at about 8°–10°S (Napier 1976, Mittermeier and Coimbra-Filho 1977). For details of the distribution, see Table 79.

Abundance and Density

C. albinasus is sparsely distributed and locally decreasing, and its range is shrinking (Table 79). It is considered vulnerable by the I.U.C.N. (1978) and endangered under Brazilian law (Padua et al. 1974). This species has been seen in groups of 15 to 25 (Ayres 1978). No data are available concerning population density, average group size, or home range area.

Habitat

White-nosed sakis have been observed in *igapó* forest seasonally or permanently flooded by clear-water rivers (Mittermeier and Coimbra-Filho 1977), in *terra firme* (nonflooding) forest, and occasionally in *várzea* (flooded by white-water rivers) forest (Ayres 1978).

Factors Affecting Populations

Habitat Alteration

The habitat of *C. albinasus* is being deforested to clear land for the Transamazonian Highway and for cattle pasture (I.U.C.N. 1974, Mittermeier and Coimbra-Filho 1977). The highway runs directly through the center of the range of *C. albinasus* at about 4°S, then along the west bank of the Rio Tapajós and westward at about 7°S to the Rio Madeira (Vanzolini 1978).

Human Predation

Hunting. *C. albinasus* is heavily hunted for food in the Rio Tapajós area, and is also shot to bait cat traps and to obtain its tail for a duster (Mittermeier and Coimbra-Filho 1977).

Pest Control. No information.

Collection. No *C. albinasus* were imported into the United States in 1968–73 (Appendix) or in 1977 (Bittner et al. 1978), nor into the United Kingdom in 1965–75 (Burton 1978).

Conservation Action

C. albinasus occurs in one new national park (Table 79), and more parks or reserves are needed (Coimbra-Filho 1972a, I.U.C.N. 1978). It is protected from capture, collection, sale, or export under the Brazilian Endangered Species Act (I.U.C.N. 1972), but enforcement in the Amazonian region is very difficult (Mittermeier and Coimbra-Filho 1977).

This species is listed in Appendix I of the Convention on International Trade in Endangered Species of Wild Fauna and Flora, and as endangered under the U.S. Endangered Species Act (U.S. Department of Interior 1977a,b).

Table 79. Distribution of *Chiropotes albinasus*

Country	Presence and Abundance	Reference
Brazil	S of R. Amazonas to N Mato Grosso	Cabrera 1958
	S of R. Amazonas between R. Madeira and R. Xingu	Napier and Napier 1967
	Rare	I.U.C.N. 1972
	Vulnerable; between R. Madeira and R. Xingu S of R. Amazonas to Mato Grosso; sparsely distributed, locally decreasing; in 1 new N.P. near Itaituba, W bank of R. Tapajós	I.U.C.N. 1974
	Endangered	Padua et al. 1974
	Locality mapped in W Pará	Deane 1976
	From R. Xingu to W of R. Tapajós near mouth at Arapiuns and S to R. Jiparaná and R. Theodore Roosevelt in Rondônia and Mato Grosso; locality at R. Iriri (left tributary of R. Xingu)	Napier 1976
	Between R. Xingu and R. Madeira, S at least to R. Jiparaná; locality at R. Tapajós near Monte Cristo; range shrinking; locally disappearing (e.g., near R. Tapajós)	Mittermeier and Coimbra-Filho 1977
	At Aripuanã, 9°10′S, 60°40′W, NW Mato Grosso	Ayres 1978
	Vulnerable; sparsely distributed; range decreasing	I.U.C.N. 1978

Chiropotes satanas Black or Bearded Saki

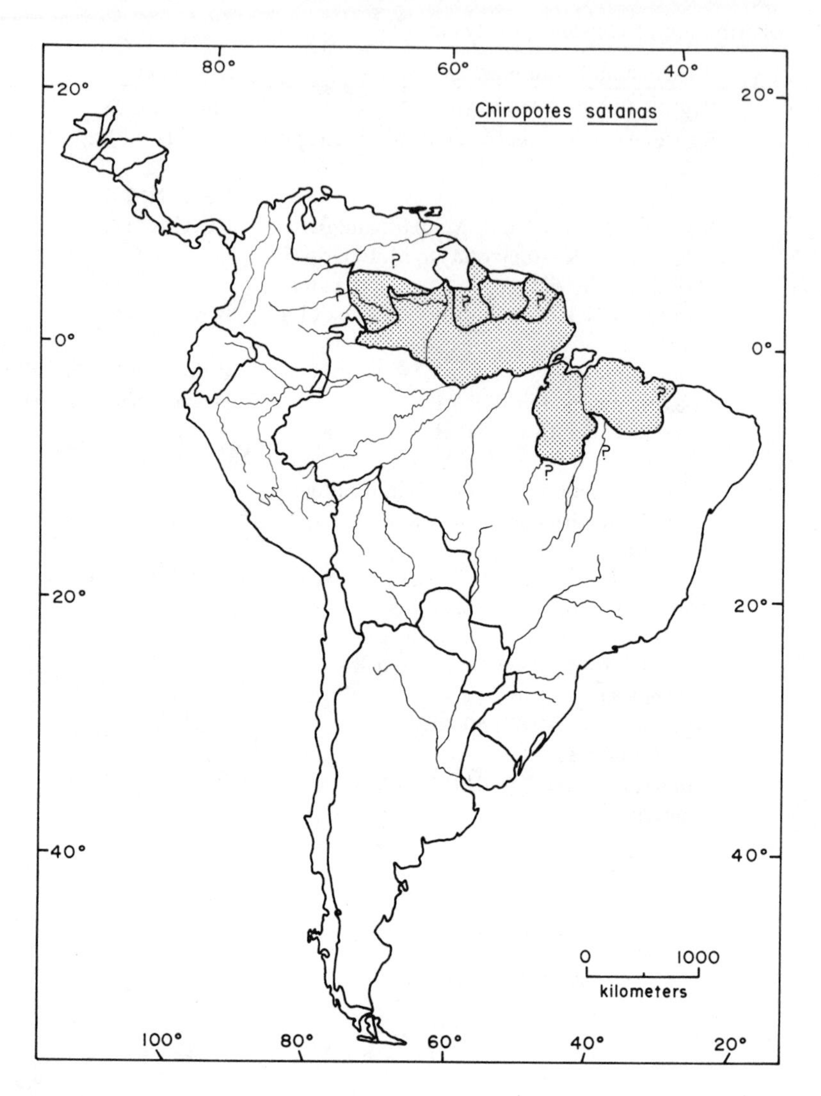

Taxonomic Note

Two subspecies are recognized (Cabrera 1958, Napier and Napier 1967).

Geographic Range

Chiropotes satanas is found in northern South America, in Venezuela, the Guianas, and northern Brazil. It occurs as far northwest as Puerto Ayacucho, Venezuela (5°39′N, 67°32′W), and perhaps farther north in that country (Mondolfi 1976) and westward into eastern Colombia (Hernández-Camacho and Cooper 1976), and as far north as about 6°N in Guyana (Muckenhirn et al. 1975). In Brazil the range of this species extends east to at least 43°W, 6°S in eastern Maranhão (Deane et al.

1969), and south at least to Gradaús, Pará, at 7°43′S, 51°10′W (Ayres 1978), but the exact limits have not been determined. For details of the distribution, see Table 80.

Abundance and Density

C. satanas is uncommon and vulnerable in Guyana and Surinam, and is declining in the southeastern portion of its range in Brazil (Table 80). Its status in Venezuela, northern Brazil, and French Guiana is not known.

In Guyana, population densities of 3.9 to 82.4 individuals per km^2 have been recorded (Muckenhirn et al. 1975), while in Surinam, density was estimated to be less than 7.53 individuals per km^2 and home range area at least 200 ha per group (Mittermeier 1977a). Group size in this species varies from 6 to 10 (Husson 1957), 9 to 11 (N = 2, Ayres 1978), 4 to 20 ($\bar{X}$ = 10.3 to 13.1, N = 10, Muckenhirn et al. 1975), and 16 to 27 ($\bar{X}$ = 22.7, N = 3, Mittermeier 1977a).

Habitat

C. satanas is a high forest species found primarily in humid tropical rain forest. In Venezuela it was recorded most frequently (80% of localities) in evergreen forest near streams, and less often (16%) in deciduous forest, all between 135 and 161 m elevation (Handley, pers. comm. 1974). Of these sites, 80% were in tropical humid forest and 20% in tropical very humid forest (life zones of Ewel and Madriz 1968).

In Surinam, *C. satanas* utilizes high tropical rain forest, riverbank high forest, and mountain savanna forest (Mittermeier 1977a). In Brazil this species has been observed in "high forest" (Deane et al. 1971) and *igapó* swamp forest (seasonally or permanently flooded with black or clear-water rivers) (Mittermeier 1977a).

Factors Affecting Populations

Habitat Alteration

The habitat of *C. satanas* east of the Rio Xingu in Brazil has been extensively deforested for agriculture, cattle ranching, and colonization, and *C. satanas* has disappeared from several areas where forest is no longer present (Ayres 1978). More than 60% of the forest cover has been destroyed. The area has a large human population, and it will probably grow because the Transamazonian Highway intersects it, thus habitat alteration is expected to increase (Ayres 1978).

C. satanas in Surinam is not adaptable to human intrusion and disappears in the face of habitat alteration (Mittermeier 1977a). In Guyana, however, it can survive in areas of low-intensity selective logging (Muckenhirn et al. 1975) and is considered "slightly adaptable" to human encroachment (Muckenhirn and Eisenberg 1978).

Human Predation

Hunting. The black saki is hunted for its meat in Venezuela (Mondolfi 1976), Surinam (Husson 1957, Mittermeier 1977a), and Brazil (Mittermeier and Coimbra-Filho 1977). It is also shot as bait for cat traps (Ayres 1978) and its tail is used as an ornament by Venezuelan Indians (Mondolfi 1976).

Pest Control. No information.

Collection. Young *C. satanas* are sometimes kept as pets in Surinam (Husson 1957). Between 1968 and 1973 a total of 44 *C. satanas* were imported into the United States (Appendix). None were imported into the United States in 1977 (Bittner et al. 1978) or into the United Kingdom in 1965–75 (Burton 1978).

Conservation Action

C. satanas is found in one nature park and four nature reserves in Surinam, but no reserves were listed for it in other countries (Table 80). One proposed reserve which would protect a sizable resident population is south of Ourém, near the Rio Sujo, east of the Rio Tocantins in eastern Pará, Brazil (Ayres 1978).

This species is legally protected from export in Venezuela (Mondolfi 1976), from hunting in northern Surinam (Mittermeier 1977a), and from commerce in Brazil (Mittermeier and Coimbra-Filho 1977).

Table 80. Distribution of *Chiropotes satanas* (countries are listed in clockwise order)

Country	Presence and Abundance	Reference
Colombia	May occur in E Guainía or SE Vichada on left bank of R. Orinoco or R. Atabapo, or on right bank of upper R. Negro or R. Guainía	Hernández-Camacho and Cooper 1976
Venezuela	In S	Cabrera 1958
	Localities listed and mapped in S Amazonas, SE, NW, and W of Esmeralda, SE and SW of Puerto Ayacucho, along R. Cunucunuma, R. Mavaca, R. Orinoco, and R. Manapiare	Handley, pers. comm. 1974
	Locally distributed S of R. Orinoco in Amazonas along R. Negro, Brazo Casiquiare, R. Ventuari, and R. Manapiare, and from upper Orinoco to Puerto Ayacucho; in Bolívar at Caño Maniapure and near R. Caura	Mondolfi 1976
Guyana	Present	Cabrera 1958
	Localities listed and mapped in E including R. Berbice (NE) and Apoteri (4°S, 58°35′W), R. Demerara, and R. Abary	Muckenhirn et al. 1975
	Localities listed from interior Demerary (Demerara), Macusi District, R. Essequibo, and R. Supinaam	Napier 1976
	Generally uncommon, moderately vulnerable; only E of R. Essequibo, absent in NW	Muckenhirn and Eisenberg 1978
Surinam	Not common; 1 locality in Nassau Mts. (in E)	Husson 1957
	Present	Cabrera 1958, Fooden 1964

Table 80 (continued)

Country	Presence and Abundance	Reference
Surinam (cont.)	Rare, vulnerable; restricted to interior; localities listed and mapped throughout central including Brownsberg Nature Park and 4 nature reserves: Raleighvallen-Voltzberg, Tafelberg, Eilerts de Haan Gebergte, and Sipaliwini Savanna	Mittermeier 1977a
French Guyana	Present	Cabrera 1958
Brazil	N and S of R. Amazonas	Cabrera 1958
	Near Manaus, R. Negro	Avila-Pires 1964
	In E Pará near Belém	Carvalho and Toccheton 1969
	Localities listed in Amazonas near Manaus, in Pará, and E Maranhão (mapped)	Deane et al. 1969
	Along Manaus-Itacoatiara Road, E Amazonas	Deane et al. 1971
	Near Manaus (locality mapped)	Deane 1976
	Localities listed at Igarapé Assu (Açu), in E Pará, and upper R. Catrimani, in W Roraima; E to Bragança (1°02′S, 46°46′W) and W at least to R. Tocantins	Napier 1976
	Originally throughout E Pará E of R. Xingu; probably to Zona dos Cocais de Sampaio (Meio Norte); localities listed include Anhanga, Benevides, 300 km S of Belém, Itupiranga, Ourém, and vicinity, Castanhal, Cametá, and Portel, in E Pará; Gradaús, R. Fresco, in S Pará; and Piroculina, in Maranhão; has disappeared from areas (e.g., near Belém and Brigantina)	Ayres 1978

Lagothrix flavicauda

Yellow-tailed or Hendee's Wooly Monkey

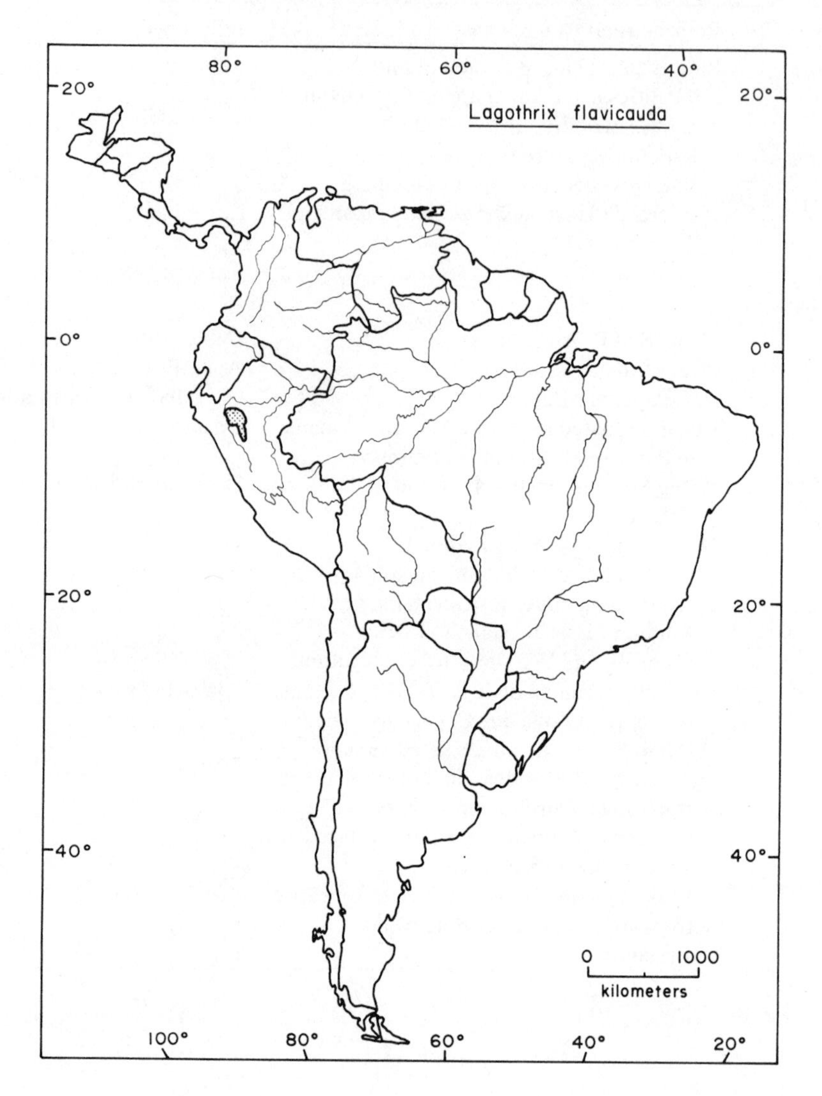

Taxonomic Note

This is a monotypic species (Cabrera 1958, Fooden 1963, Napier and Napier 1967).

Geographic Range

Lagothrix flavicauda is found in northern Peru, in southern Amazonas and western San Martin. The range extends from about 5°40′S to 7°45′S, and from about 78°40′W along the Río Marañón eastward to about 77°10′W (Mittermeier, Macedo-Ruíz, Luscombe, and Cassidy 1977). But these limits are only approximate, and when more is learned about the species they may be revised. For details of the distribution, see Table 81.

Abundance and Density

The range of *L. flavicauda* is small and shrinking and its populations are endangered (Table 81). Until 1974 only five specimens from two localities were known, and almost fifty years had passed since their collection. But between 1974 and 1978, captive individuals and natural populations were rediscovered and are now under study (Macedo-Ruíz et al. 1974; Mittermeier, pers. comm. 1978). No population data are available.

Habitat

L. flavicauda appears to be restricted to montane rain forest between 500 and 2,500–3,000 m elevation (Mittermeier, Macedo-Ruíz, Luscombe, and Cassidy 1977). No further details are known.

Factors Affecting Populations

Habitat Alteration

Considerable habitat destruction is occurring within the range of *L. flavicauda* (Macedo-Ruíz et al. 1974). This includes the clearing of forest to provide land for agriculture and cattle ranching, and road construction by the army in order to open the area for colonization and defense purposes (Mittermeier, Macedo-Ruíz, Luscombe, and Cassidy 1977). The rapid rate of deforestation is threatening this species (Freese 1975).

Human Predation

Hunting. L. flavicauda is heavily hunted for its flesh by local people and by hunters hired to supply meat for army construction crews, and for its skin (Mittermeier, Macedo-Ruíz, Luscombe, and Cassidy 1977).

Pest Control. No information.

Collection. Mother *L. flavicauda* are shot to obtain juveniles for sale as pets in Peru (Macedo-Ruíz et al. 1974). No *L. flavicauda* were imported into the United States between 1968 and 1973 (Appendix) or in 1977 (Bittner et al. 1978). One-hundred were listed as having been imported into the United Kingdom between 1965 and 1975 (Burton 1978), but considering the rarity of the species this is difficult to believe.

Conservation Action

L. flavicauda is not found within any protected areas in Peru, and although capture, hunting, and export of it are prohibited by law (Supreme Decree 934-73-AG) (Castro 1976), the provisions are very difficult to enforce in remote areas (Freese 1975).

Conservation measures that have been proposed for this species include special protection from hunting and inclusion in a series of wildlife stamps (Macedo-Ruíz et al. 1974), creation of a reserve within its range and a captive breeding program in

Peru (Mittermeier, Macedo-Ruíz, Luscombe, and Cassidy 1977), establishment of guard posts in a new reserve in the Río Utcubamba, Río Ciriaco, Río Alto Mayo area, and a conservation education program (Mittermeier 1977b).

L. flavicauda is listed as endangered under the U.S. Endangered Species Act (U.S. Department of Interior 1977b).

Table 81. Distribution of *Lagothrix flavicauda*

Country	Presence and Abundance	Reference
Peru	In Andes of extreme N	Cabrera 1958
	Localities listed and mapped at Puca Tambo (on upper R. Mayo, N San Martin, 80 km E of Chachapoyas) and La Lejia (on upper R. Huambo, S Amazonas, 25 km W of first locality)	Fooden 1963
	Threatened with extinction	Hershkovitz 1972
	Endangered; known only from Puca Tambo (6°09′S, 77°11′W) and La Lejia (6°07′S, 77°28′W); recently seen at Alva (5°56′S, 77°56′W); diminishing	I.U.C.N. 1974
	Rediscovered near Pedro Ruiz Gallo (near Alva), S Amazonas	Macedo-Ruíz et al. 1974, Perleche M. 1974
	Range mapped; rediscovered near Chachapoyas, Rodriguez, and Pedro Ruiz (Gallo)	Lüling 1975
	Three new localities within previously known range; one of smallest ranges of any New World monkey	Mittermeier, Macedo-Ruíz, and Luscombe 1975
	Status undetermined; not in any protected areas	Castro 1976
	Only in small pocket on E slope of central Cordillera; localities listed at and near Puca Tambo, San Martin	Napier 1976
	Endangered; localities listed and mapped include La Lejia, Puca Tambo, and Alva near Pedro Ruiz Gallo; restricted to broken intermountain plateau on E slope of Andes in S Amazonas and W San Martin; range shrinking	Mittermeier, Macedo-Ruíz, Luscombe, and Cassidy 1977
	Threatened	Castro 1978
	Endangered	I.U.C.N. 1978
	New localities have been discovered; field work is in progress	Mittermeier, pers. comm. 1978

Lagothrix lagothricha Humboldt's Wooly Monkey

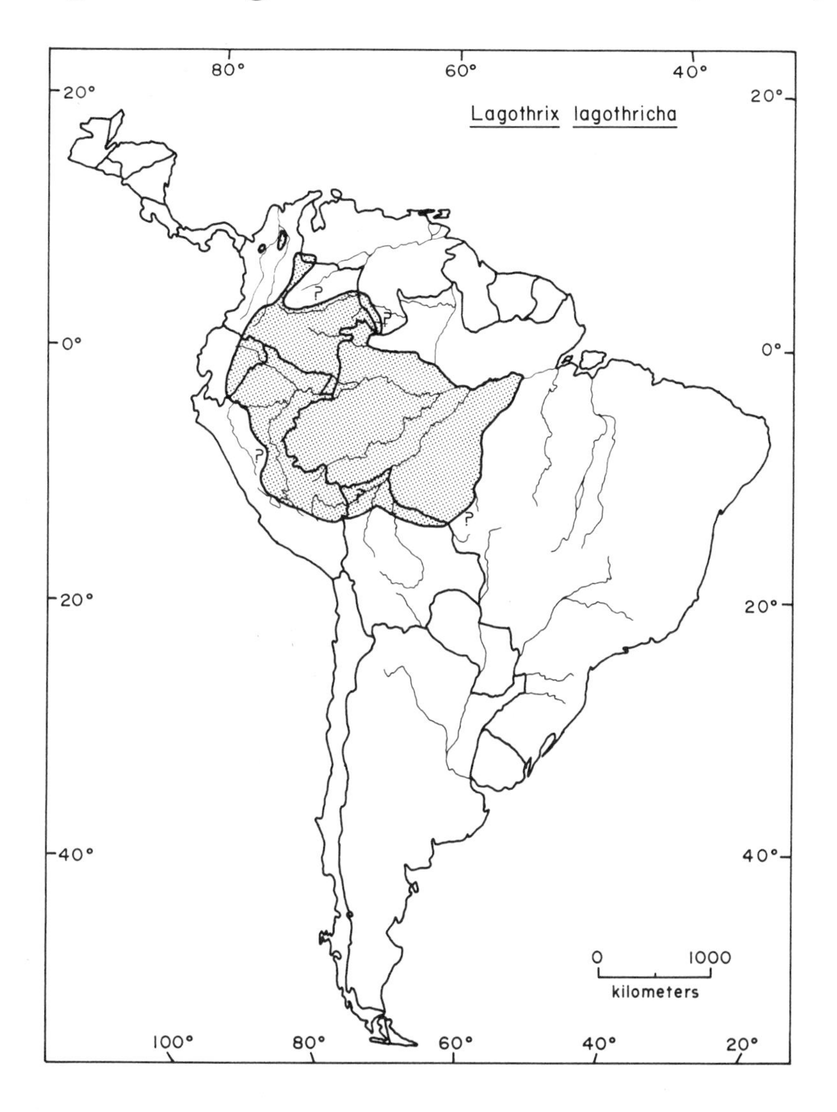

Taxonomic Note

Four subspecies are recognized (Fooden 1963, Napier and Napier 1967), including the southeastern form (*cana*), which is treated as a full species by some authors (Cabrera 1958, Grimwood 1969).

Geographic Range

Lagothrix lagothricha is found in northwestern and central South America from northern Colombia through Amazonian Brazil. It occurs as far north as about 9°N in Bolívar, Colombia (Hernández-Camacho and Cooper 1976), and as far south as about 14°S, 69°W, on the Río Tambopata, Puno, Peru (Fooden 1963). The range extends as far west as Macas, Santiago, Ecuador (2°22′S, 78°08′W), and as far east as the mouth of the Rio Tapajós at about 55°W, in western Pará, Brazil (Fooden 1963). For details of the distribution, see Table 82.

Abundance and Density

L. lagothricha is classified as vulnerable by the I.U.C.N. (1978). It has not recently been described as abundant in any country, has declined in parts of Colombia and throughout Peru, and is considered vulnerable in Peru and Brazil (Table 82). Its status in Ecuador, Bolivia, and Venezuela is not known.

Population density in this species varies from 6 to 45 individuals per km^2 (Table 83). Groups include 2 to 70 members, with averages of 5.5 to 43 per group, and home ranges of 400 to 1,100 ha (Table 83).

Habitat

All of the preferred habitats of *L. lagothricha* are types of humid primary forest. Colombian populations, for example, occupy primary humid gallery and palm associations, seasonally flooded or nonflooding rain forest and cloud forest, but not secondary growth or young secondary forest (Hernández-Camacho and Cooper 1976). They have also been recorded in lakeside and riverine forests, and in *Cecropia* stands (Klein and Klein 1976), but are generally absent from isolated forest patches (Bernstein et al. 1976a).

In southeastern Peru, *L. lagothricha* is found in lowland and highland semideciduous wet forest, including riparian growth in montane areas, and can occupy secondary forest if large trees and some nut and fruit bearing trees are present (Durham, pers. comm. 1975). In northeastern Peru this species has been observed in evergreen, mesophytic, broadleaf forest on high ground and on flood plains of blackwater and mixed black and white-water rivers (Freese et al., in prep.).

L. lagothricha occurs up to 1,800 m elevation in Peru (Grimwood 1969) and 3,000 m in Colombia (Hernández-Camacho and Cooper 1976). No information is available regarding its habitats in Ecuador, Bolivia, Brazil, or Venezuela.

Factors Affecting Populations

Habitat Alteration

Since *L. lagothricha* is restricted to primary forest in Colombia, its future there is dependent on the survival of undisturbed forest habitat (Hernández-Camacho and Cooper 1976). This species is expected to be one of those most adversely affected by reduction of forest for agricultural development (Bernstein et al. 1976a). Yet no reports were located regarding the extent of alteration of its habitat or changes in population due to habitat disturbance.

Human Predation

Hunting. Hunting is the major cause of the decimation of *L. lagothricha* populations in many areas. Its flesh is especially relished as food, because the all dark meat has a delicate flavor and the subcutaneous fat renders to a golden orange grease used for frying, seasoning, and medicinal purposes (Hershkovitz 1972).

As early as 1863, Bates calculated that 1,200 *L. lagothricha* were killed annually for food by 200 Indians living near Tabatinga, western Brazil; in 1958, Lehmann reported that this species was used for food by the natives almost everywhere, and for this reason was becoming scarce in central Colombia (Fooden 1963). More recently, *L. lagothricha* has been considered "probably the most persecuted primate species in Colombia" (Hernández-Camacho and Cooper 1976). Its meat is prized by people in the mountains and the Amazon region (Hernández-Camacho and Cooper 1976) and is a favorite food near La Macarena, where it "provides a major source of protein for settlers, especially during the first few years of colonization" (Klein and Klein 1976).

Humboldt's wooly monkeys are also hunted heavily in Peru, where their flesh is highly esteemed and local populations are easily exterminated near settlements and large rivers (Grimwood 1969). For example, in the Río Curanja region, *L. lagothricha* is usually the first species to be eliminated when hunters move into an area (Gardner, pers. comm. 1974). In several parts of northeastern Peru, populations have suffered greatly because of hunting, and even light hunting can eliminate them from an area (Freese 1975). In one seven-month period in 1973, the carcasses of 937 *L. lagothricha* were seen on sale in the markets of Iquitos (Neville 1976b). This consumption as food has probably more significantly affected populations than exportation has (Freese et al. 1976).

L. lagothricha is also heavily hunted for food and for bait for large cat traps in Brazilian Amazonia (Mittermeier and Coimbra-Filho 1977), northern Peru (Encarnación et al. 1978), and southern Colombia (Nishimura and Izawa 1975).

Pest Control. L. lagothricha feeds in banana plantations (Durham 1972), orange trees, and sugar cane fields, and is usually killed or chased away from fields rather than tolerated (Durham, pers. comm. 1974).

Collection. Young Humboldt's wooly monkeys are very popular as pets, and in Colombia are often collected by shooting the mother to obtain her clinging infant (Bernstein et al. 1976b, Hernández-Camacho and Cooper 1976). They are kept as pets in Peru (Soini 1972, Neville 1975, Castro 1976, Encarnación et al. 1978) and in Brazil (Mittermeier and Coimbra-Filho 1977).

This species has also been heavily collected for export, especially from Peru and Colombia. For example, between 1964 and 1974, a total of 17,829 were exported from Peru (Castro 1976). In 1970, 305 were exported legally from Colombia (Green 1976), yet 490 were imported into the United States from Colombia (Paradiso and Fisher 1972). Similarly, in 1971, 382 were exported legally from Colombia (Cooper, pers. comm. 1974), but 442 were imported into the United States from Colombia (Clapp and Paradiso 1973).

Between 1968 and 1973 a total of 14,083 *L. lagothricha* were imported into the United States (Appendix), primarily from Peru and Colombia. None entered the United States in 1977 (Bittner et al. 1978). In 1965–75, 1,609 were imported into the United Kingdom (Burton 1978).

Conservation Action

L. lagothricha occurs in three national parks in Colombia and one national park and two reserves in Peru (Table 82). But enforcement of hunting regulations is inadequate in the reserves (Castro 1976), and La Macarena National Park in Colombia is being steadily invaded by agriculturists and must have improved protection if it is to survive (Struhsaker 1976b).

Peruvian law (Supreme Decree 934-73-AG, 1973) prohibits the killing, capture, or commercialization of *L. lagothricha* (Castro 1976), but the antihunting provision is not enforced, and should be, to protect remaining populations (Freese 1975, Neville et al. 1976).

No special laws protect *L. lagothricha* in Colombia (Cooper and Hernández-Camacho 1975), and detailed ecological and distributional information on the species there is urgently needed (Struhsaker et al. 1975). Brazilian law prohibits commerce in *L. lagothricha* (Mittermeier and Coimbra-Filho 1977) and Venezuelan law prohibits export (Mondolfi 1976). The I.U.C.N. (1978) recommends adding *L. lagothricha* to Appendix I of the Convention on International Trade in Endangered Species of Wild Fauna and Flora, in order to regulate trade in this species.

Table 82. Distribution of *Lagothrix lagothricha* (countries are listed in counterclockwise order)

Country	Presence and Abundance	Reference
Colombia	In and E of Andes and S of R. Guaviare basin	Cabrera 1958
	Localities listed and mapped in central, S, and SE as far N as 7°30′N and as far W as 77°W; questionable in S Meta	Fooden 1963
	In Puracé N.P., Cauca-Huila; Cueva de los Guácharos N.P., Huila; and La Macarena N.P., Meta	Green 1972
	In Serranía de San Lucas, S Bolívar	Kavanagh and Dresdale 1975
	On R. Peneya, branch of R. Caquetá	Nishimura and Izawa 1975, Izawa 1976
	Near Ventura, Bolívar, between R. Cauca and R. Magdalena	Bernstein et al. 1976a
	S of Puerto Rico, near Tequisio, Bolívar	Green 1976
	Throughout SE and E except E Vichada and E Arauca; questionable in S Vichada, E Meta, E Antioquia, and S Tolima; two newly discovered isolated populations in upper San Jorge valley, SE Córdoba, and Serranía de San Lucas, SE Bolívar (map); has disappeared from southern trapezium region	Hernández-Camacho and Cooper 1976
	On S bank R. Guayabero, La Macarena N.P.	Klein and Klein 1976
	In Caquetá	Moynihan 1976a

Table 82 (continued)

Country	Presence and Abundance	Reference
Colombia (cont.)	Localities listed near Bogotá and in Tolima at 2°20′N (but this is S of Tolima border)	Napier 1976
	In good numbers in La Macarena N.P.	Struhsaker 1976b
	On right bank R. Duda, W boundary La Macarena N.P.	Izawa and Mizuno 1977
Ecuador	E of Andes	Cabrera 1958
	Localities listed and mapped throughout E	Fooden 1963
	Localities listed include Oriente, R. Pastaza, R. Copataza, Macas, Mindo, and Quito district	Napier 1976
Peru	In basins of middle R. Ucayali and lower R. Urubamba; between R. Napo and lower R. Ucayali and R. Javarí (Yavari)	Cabrera 1958
	Localities listed and mapped throughout NE and E	Fooden 1963
	In Loreto, San Martin, Huánuco, Pasco, Junín, Ayacucho, Cuzco, Puno, and Madre de Dios (including Manú N.P.); has disappeared near settlements and large rivers; probably declining everywhere except remote areas	Grimwood 1969
	Population decreased 1966–71 at R. Curanja, SE Loreto	Gardner, pers. comm. 1974
	At San Gaban–R. San Juan, Karos Pilcopata, and Manú N.P., Madre de Dios (all in SE)	Durham 1975
	Along R. Samiria (in NE); extinct over large areas along rivers in Loreto; probably rare throughout Amazonian region; decreasing	Freese 1975
	Rapidly disappearing in NE	Neville 1975
	In Pacaya and Samiria national reserves and Manú N.P.; vulnerable	Castro 1976
	Locally abundant in R. Samiria basin	Freese et al. 1976
	Localities listed in Cuzco, Huánuco, Junín, Loreto, Andoas, and R. Ucayali	Napier 1976
	Along middle R. Samiria and lower and middle R. Pacaya	Neville et al. 1976
	On R. Yavineto (R. Putumayo), N Loreto	Encarnación et al. 1978
	In Manú N.P.	Janson and Terborgh, in press
	Along Ríos Ampiyacu, Orosa, Nanay, and Samiria	Freese et al., in prep.
Bolivia	Probably in N (map)	Fooden 1963
Brazil	In Amazonian region from R. Negro and R. Madeira E, and from S bank of R.	Cabrera 1958

Table 82 (continued)

Country	Presence and Abundance	Reference
Brazil (cont.)	Solimões W and S to Serra dos Parecis, Mato Grosso (near 13°S, 60°W)	Cabrera 1958 (cont.)
	Localities listed and mapped throughout Amazonas; in Acre, W Pará W of R. Tapajós, SE Rondônia, NW Mato Grosso; questionable at 13°S in Mato Grosso	Fooden 1963
	Localities mapped in Amazonas and Rondônia	Deane et al. 1969
	Localities mapped in Amazonas, Pará, Rondônia, and Acre	Deane 1976
	Localities listed in Amazonas, R. Juruá, and R. Amazonas	Napier 1976
	Vulnerable	Mittermeier and Coimbra-Filho 1977
Venezuela	Recently discovered in extreme W Apure, S of R. Sarare (extreme S of W region)	Hernández-Camacho and Cooper 1976
	Locally distributed and rare in basin of upper R. Orinoco near Santa Bárbara and R. Guainía (along W border of S region, 2°–4°N)	Mondolfi 1976

Table 83. Population parameters of *Lagothrix lagothricha*

Population Density (individuals/km²)	Group Size $\bar{X}$	Group Size Range	Group Size No. of Groups	Home Range Area (ha)	Study Site	Reference
	10.6	2–18	12		N Colombia	Kavanagh and Dresdale 1975
	29–43	21–70	7	400–1,100	S Colombia	Nishimura and Izawa 1975
9					N Colombia	Bernstein et al. 1976a
		4–6			Colombia	Hernández-Camacho and Cooper 1976
	40	25–70	21	1,100	S Colombia	Izawa 1976
		6–20			SE Peru	Durham 1975
7	10	3–30			NE Peru	Freese 1975
6–45		4–30			NE Peru	Neville et al. 1976
		6	1		N Peru	Encarnación et al. 1978
		10–15			SE Peru	Janson and Terborgh, in press
7	5.5	5–6	2		NE Peru	Freese et al., in prep.
		≥50			General	Fooden 1963

Pithecia monachus Monk Saki

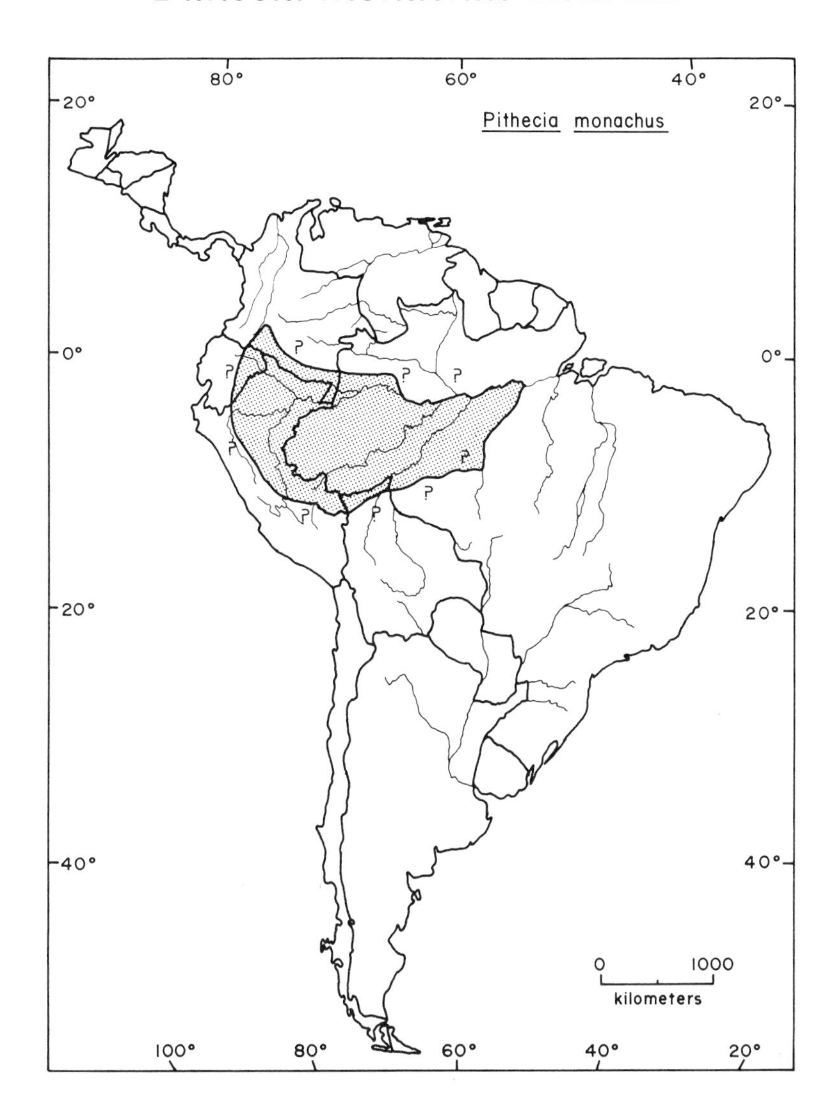

Taxonomic Note

Two subspecies are recognized (Cabrera 1958, Napier and Napier 1967).

Geographic Range

Pithecia monachus is found in western and central South America from southern Colombia to northern Bolivia and central Brazil. Its range extends north as far as about 2°20′N, 75°W in Colombia (Hernández-Camacho and Cooper 1976) and south at least as far as the Manú National Park (11°50′S, 71°23′W) in Peru (Grimwood 1969), and northern Mato Grosso, Brazil (Cabrera 1958). *P. monachus* occurs as far west as the Río Marañón (Grimwood 1969), but its western and southern limits are not known. The eastern limits in Brazil are thought to be the Rio Tapajós south of the Rio Amazonas and the Rio Japurá north of the Rio Amazonas (Napier 1976). Cabrera (1958) also included the Guianas and northern Brazil within the range of this species, and although recent field studies in the Guianas (Muckenhirn et al. 1975, Mittermeier 1977a) have not confirmed the presence of *P. monachus* in those countries, old records indicate that it may occur near the Rio Negro, in northern Brazil (Napier 1976). For details of the distribution, see Table 84.

Abundance and Density

P. monachus has been exterminated in many parts of Peru, and although it is locally abundant in some areas, it is rare in others, and in general its populations there are vulnerable (Table 84). In Brazil, *P. monachus* is not endangered (Table 84). Its status in Colombia, Ecuador, and Bolivia is not known.

Population densities for *P. monachus* vary from 0.62 to 30 individuals per km^2 (Table 85). Group sizes of 2 to 9 have been reported, with average sizes of 2.4 to 6 individuals (Table 85). No home range estimates are available.

Habitat

P. monachus is a forest species typical of humid, lowland primary forest. In Colombia it occupies only primary, well-developed rain forest (Hernández-Camacho and Cooper 1976). In Peru and Bolivia it has been observed in evergreen, mesophytic, broadleaf forests on high ground in white-water drainages and on flood plains of black-water rivers (Freese et al., in prep.). In Brazil it is found in primary *terra firme* (nonflooding) and primary *várzea* (flooded by white-water) forests (Mittermeier and Coimbra-Filho 1977). This species has been collected as high as 762 m elevation in Peru (Napier 1976).

Factors Affecting Populations

Habitat Alteration

No information.

Human Predation

Hunting. Monk sakis are often hunted for their meat in Peru (Soini 1972), and consumption as food has probably had a significant effect on their numbers (Freese et al. 1976, Neville 1975), but the populations seem to be able to survive hunting pressure fairly well (Freese 1975). *P. monachus* is also hunted for its flesh and for its tail to be used as a duster in Brazil (Mittermeier and Coimbra-Filho 1977). In Peru the tails are sold as dusters and decorations (Soini 1972, Castro 1976), the fur and teeth are used as ornaments (Gardner, pers. comm. 1974), and the body is used as bait for large cat traps (Encarnación et al. 1978).

Pest Control. No information.

Collection. P. monachus is kept as a pet in Peru (Castro 1976) and occasionally in Brazil (Mittermeier, Coimbra-Filho, and Van Roosmalen, in press). A total of 1,091 were exported from Peru between 1964 and 1974 (Castro 1976) and a total of 50 were exported from Colombia in 1970–71 (Cooper, pers. comm. 1974).

Between 1968 and 1973, a total of 382 *P. monachus* were imported into the United States (Appendix), primarily from Peru and Colombia. None entered the United States in 1977 (Bittner et al. 1978). Thirty-two were imported into the United Kingdom in 1965–75 (Burton 1978).

Conservation Action

P. monachus occurs in one national park and two national reserves in Peru (Table 84), but the latter areas are not effectively patrolled against hunting (Castro 1976). This species is not found within protected areas in other countries.

P. monachus is legally protected against hunting, capture, and export in Peru (Supreme Decree 934-73-AG), but enforcement of the hunting provision is inadequate and should be improved (Freese 1975, Neville et al. 1976). Monk sakis are also protected from commercial trade in Brazil (Mittermeier and Coimbra-Filho 1977).

Table 84. Distribution of *Pithecia monachus* (countries are listed in counterclockwise order)

Country	Presence and Abundance	Reference
Colombia	In SE to R. Caquetá	Cabrera 1958
	In S in Amazonas, Putumayo, and Caquetá, as far NE as R. Yarí and R. Caquetá, questionable farther NE; W to piedmont of E Andes (map)	Hernández-Camacho and Cooper 1976
	Along R. Peneya, tributary of R. Caquetá	Izawa 1976
	In Caquetá region	Moynihan 1976a
Ecuador	In E	Cabrera 1958
	Localities listed on R. Napo, R. Copataza, and in Napo-Pastaza	Napier 1976
Peru	In NE in basins of R. Napo and R. Ucayali	Cabrera 1958
	In Loreto (Ríos Yavarí, Maniti, Marañón, Ucayali, Curanja); Huánuco (R. Pachitea); and Madre de Dios (Manú N.P.); exterminated everywhere near settlement and heavily traveled rivers, in fair numbers elsewhere	Grimwood 1969
	Not rare in N Loreto near Iquitos (in N)	Soini 1972
	Population decreased 1966–71 along R. Curanja, SE Loreto	Gardner, pers. comm. 1974
	Along R. Nanay near Iquitos, R. Orosa, R. Samiria (all in N) and R. Pucallpa (central)	Freese 1975
	In Manú N.P. (in SE) and Pacaya and Samiria national reserves (in N): vulnerable	Castro 1976
	Locally abundant in R. Samiria basin; common in R. Pacaya basin (localities mapped)	Freese et al. 1976
	Localities listed in Loreto, including R. Ucayali, Pebas, Iquitos, and Cumeria	Napier 1976

Table 84 (continued)

Country	Presence and Abundance	Reference
Peru (cont.)	Relatively abundant along R. Samiria and R. Pacaya between lower R. Ucayali and R. Huallaga/Marañón	Neville et al. 1976
	In SE between R. Acre and R. Madre de Dios near E border	Encarnación and Castro 1978
	In N Loreto at R. Yavineto (R. Putumayo)	Encarnación et al. 1978
	Rare at Cocha Cashu, Manú N.P.	Janson and Terborgh, in press
Bolivia	Relatively uncommon at Cobija, R. Acre, on NW border (locality mapped)	Heltne et al. 1976
Brazil	Locality mapped at Benjamin Constant, W Amazonas	Kumm and Laemmert 1950
	In W Acre at Cruzeiro do Sul and Seringal Oriente (in W)	Carvalho 1957
	In N, N of lower R. Amazonas in Pará; in central, W of R. Tapajós as far S as N Mato Grosso	Cabrera 1958
	In Rondônia (in SW)	Deane et al. 1969
	As far E as R. Madeira (map)	Avila-Pires 1974
	Locality mapped in W Pará	Deane 1976
	Localities listed in Amazonas along Rios Purus, Içá, Juruá, Tefé, Tonantins, Madeira, Tapajós, and Negro	Napier 1976
	At R. Quixito, and between R. Japurá, R. Auati-Paraná, and R. Solimões; not endangered	Mittermeier and Coimbra-Filho 1977

Table 85. Population parameters of *Pithecia monachus*

Population Density (individuals/km^2)	Group Size $\bar{X}$	Group Size Range	Group Size No. of Groups	Home Range Area (ha)	Study Site	Reference
	2.4	2–5			S Colombia	Izawa 1976
	2.9	2–6	19		S Colombia	Moynihan 1976a
2.5–9	5	3–8			N Peru	Freese 1975
7.5–30	5				N Peru	Neville et al. 1976
0.62	5				SE Peru	Encarnación and Castro 1978
		3–4	2		N Peru	Encarnación et al. 1978
		4–6			SE Peru	Janson and Terborgh, in press
6.6	6	6–7	3		N Bolivia	Heltne et al. 1976
4.5–13.5	5	2–9	20		Peru and Bolivia	Freese et al., in prep.

Pithecia pithecia Pale-headed or White-faced Saki

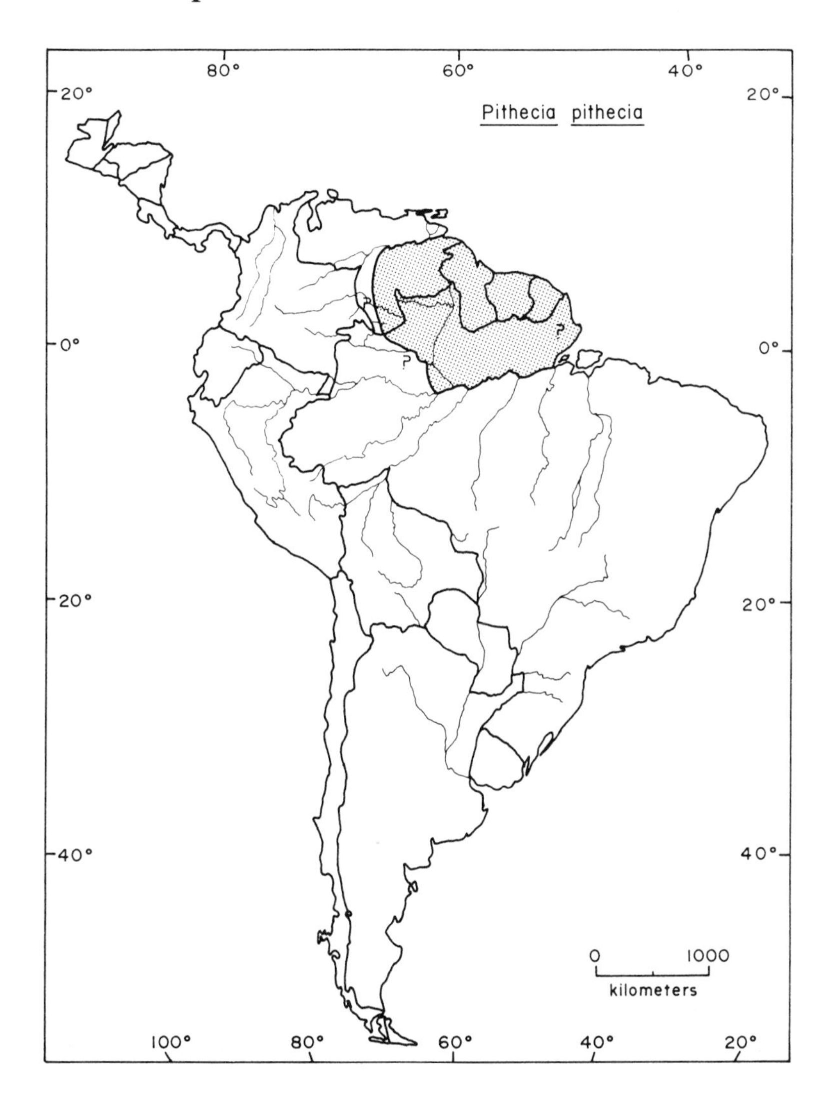

Taxonomic Note

This is a monotypic species (Cabrera 1958, Napier and Napier 1967).

Geographic Range

Pithecia pithecia is found in northern South America in Venezuela, the Guianas, and northern Brazil. It occurs as far north as the delta of the Río Orinoco (Mondolfi 1976), about 8°30′N, and as far south as the Rio Solimões at the mouth of the Rio Purus (Cabrera 1958, Napier 1976), about 3°50′S. The western and eastern limits are not known, but the range extends at least as far west as about 66°W along the Río Cunucunuma, Venezuela (Handley, pers. comm. 1974), and may extend as far east as 50°W at the eastern coast of Amapá, Brazil (Avila-Pires 1974). For details of the distribution, see Table 86.

Abundance and Density

P. pithecia is considered moderately vulnerable in Guyana and Surinam, and not endangered in Brazil (Table 86). Its status in Venezuela and French Guiana is not known.

Population densities of 2.5 to 12 individuals per km^2 (Muckenhirn et al. 1975) and 3.57 individuals per km^2 (Mittermeier 1977a) have been reported. Group size varies from 2 to 5 ($\bar{X} = 3.2$, $N = 10$) (Muckenhirn et al. 1975) and 2 to 4 ($\bar{X} = 2.67$, $N = 9$), with home ranges of 4 to 10 ha (Mittermeier 1977a).

Habitat

P. pithecia is a tropical humid forest monkey that occupies a variety of forest types. In Surinam, for example, it is found in inland high tropical rain forest, ridge forest, savanna forest, mountain savanna forest, liane forest, riverbank marsh levee forest, swamp wood, and secondary forest (Mittermeier 1977a), as well as gallery forest, open canopy *Mora* forest, palm forest, and coastal scrub (Durham and Durham, pers. comm. 1974).

In Venezuela, 93% of *P. pithecia* were found in moist evergreen forest and 7% in dry deciduous forest, all between 150 and 350 m elevation (Handley, pers. comm. 1974). Eighty-two percent of these were in tropical humid forest, 7% in tropical very humid forest, and 11% in premontane humid forest (life zones of Ewel and Madriz 1968). In Brazil, *P. pithecia* has been recorded in *terra firme* (nonflooding) secondary forest (Mittermeier and Coimbra-Filho 1977).

Factors Affecting Populations

Habitat Alteration

White-faced sakis are slightly adaptable to human encroachment (Muckenhirn and Eisenberg 1978) and can survive in areas of low-intensity selective logging in Guyana (Muckenhirn et al. 1975). In Surinam and French Guiana they are able to live near farms and even near an airport, and do not seem to be threatened by habitat destruction, which is still minimal in those two countries (Durham and Durham, pers. comm. 1974). No reports are available concerning the impact of habitat alteration on *P. pithecia* in Venezuela or Brazil.

Human Predation

Hunting. P. pithecia is hunted for its flesh in Venezuela (Mondolfi 1976) and for its meat and its tail, which is used as a duster, in Surinam (Husson 1957, Mittermeier 1977a) and Brazil (Mittermeier and Coimbra-Filho 1977).

Pest Control. No information.

Collection. White-faced sakis are sometimes kept as pets in Surinam (Husson 1957, Mittermeier 1977a).

Between 1968 and 1973 a total of 25 *P. pithecia* were imported into the United States (Appendix). None entered the United States in 1977 (Bittner et al. 1978). One-hundred ninety-four were imported into the United Kingdom between 1965 and 1975 (Burton 1978).

Conservation Action

P. pithecia occurs in two forest reserves in Guyana and one nature park and four nature reserves in Surinam (Table 86). No protected habitats are listed for it in Venezuela, French Guiana, or Brazil.

This species is legally protected against export from Venezuela (Mondolfi 1976), hunting in northern Surinam (Mittermeier 1977a), and commercial trade in Brazil (Mittermeier and Coimbra-Filho 1977).

Table 86. Distribution of *Pithecia pithecia* (countries are listed in clockwise order)

Country	Presence and Abundance	Reference
Venezuela	Localities listed and mapped in Amazonas NW of Esmeralda along R. Cunucunuma (in S); and in Bolívar on R. Supamo, near El Manteco, and near El Dorado (in E)	Handley, pers. comm. 1974
	Locally distributed in Amazonas near R. Negro and upper R. Orinoco to R. Cunucunuma; in Bolívar near El Manteco, El Palmar, Altiplanicie de Nuria, and SE of El Dorado (all in E), and in delta of R. Orinoco	Mondolfi 1976
Guyana	Present	Cabrera 1958
	Localities listed and mapped in N and central including 2 forest reserves (H.M.P.S. and Moraballi), Apoteri, and Matthews Ridge	Muckenhirn et al. 1975
	Localities listed include Berbice (in NE), Bonnesique Creek, R. Essequibo, R. Supinaam, R. Rupununi, and coast of Demerara	Napier 1976
	Moderately vulnerable; locally in high numbers	Muckenhirn and Eisenberg 1978
Surinam	Quite common; localities listed N of Moengo Tapoe (in NE) and in Nassau Mts. (E central)	Husson 1957
	Present	Cabrera 1958, Fooden 1964
	Very abundant; throughout including R. Maroni (in E), R. Saramacca (central), and R. Coppename (N central)	Durham and Durham, pers. comm. 1974
	At Zanderij (5°26′N, 55°14′W)	Napier 1976
	Rare, somewhat vulnerable; localities listed and mapped throughout; not common; at low densities; in Brownsberg Nature Park and 4 nature reserves of Raleighvallen-Voltzberg, Tafelberg, Eilerts de Haan Gebergte, and Sipaliwini Savanna	Mittermeier 1977a

Table 86 (continued)

Country	Presence and Abundance	Reference
French Guiana	Present	Cabrera 1958
	Throughout, including R. Maroni (W border) and R. Cayenne	Durham and Durham, pers. comm. 1974
	At R. Approuague (in E)	Napier 1976
Brazil	In N, N of R. Amazonas as far W as opposite mouth of R. Purus	Cabrera 1958
	In R. Negro region near Manaus	Avila-Pires 1964
	In Amazonas	Deane et al. 1969
	In Amazonas along Manaus-Itacoatiara Road	Deane et al. 1971
	Localities listed include Manacapuru, R. Solimões; R. Negro; Faro (2°10′S, 56°39′W) on R. Nhamundá; Forte de R. Branco	Napier 1976
	At Oriximiná (1°45′S, 55°50′W), lower R. Trombetas, NW Pará; not in danger	Mittermeier and Coimbra-Filho 1977

Saimiri sciureus Squirrel Monkey

Taxonomic Note

Some authors treat the Central American squirrel monkey as a distinct species, *Saimiri oerstedii* (e.g., Hall and Kelson 1959, Napier and Napier 1967). But several recent workers have concluded that all squirrel monkeys are conspecific (Hershkovitz 1969, 1972; Moynihan 1976a; Thorington 1975, 1976), and this arrangement is followed here. Eight subspecies are recognized (Napier and Napier 1967).

Geographic Range

Squirrel monkeys are found in southern Costa Rica and Panama, and from central Colombia to Bolivia and northeastern Brazil. The northwestern limit of their range is near 9°30′N, 84°30′W in western Costa Rica (Hall and Kelson 1959; Freese, pers. comm. 1974), and the easternmost locality is at about 47°W in eastern Pará, Brazil (Kumm and Laemmert 1950). The species has been recorded as far south as 17°28′S, 63°37′W in southern Bolivia (Soria 1959, Napier 1976). It is also reported from about 22°S, 48°W in São Paulo, Brazil (Soria 1959), but no other author shows it this far south. For details of the distribution, see Table 87.

Abundance and Density

S. sciureus has recently been described as locally abundant in parts of Colombia, Peru, Bolivia, Brazil, French Guiana, and Surinam, and is not threatened in Brazil, Surinam, or Guyana (Table 87). But this species is known to be declining and endangered in Costa Rica and Panama, is suspected to be decreasing in some parts of southern Colombia, and is considered by one author to be vulnerable in Peru. No information is available regarding its status in Ecuador or Venezuela.

Squirrel monkey populations have been studied at densities of 7.5 to 585 individuals per km^2 (Table 88). Groups vary in size from 2 to 200, with averages of 12.6 to 50 per group and home ranges of 17.5 to 300 ha (Table 88).

Habitat

Squirrel monkeys occupy a "wide range of forest types from gallery to low canopy sclerophyllous and hillside forests to palm forests and both seasonally flooded and nonflooded rainforests" (Hernández-Camacho and Cooper 1976). They are quite flexible in their choice of habitats, utilizing high and low rain forest, riverbank forest, savanna forest, liane forest, swamp, marsh, and mangrove forests (Mittermeier 1977a). In fact, Moynihan (1976a) states that they "seem to flourish in more kinds of habitats than any other New World primate with which I am familiar."

In many areas, *S. sciureus* seems to prefer riverine habitats. In Colombia, for example, it has been studied in creek levee bank heterogeneous forest, *Cecropia* stands, willow type trees near lakes and rivers, and swamp forest (Klein and Klein 1976). In Peru it has been recorded in gallery forest (Durham 1975) and in evergreen, mesophytic, broadleaf forests on high ground and on black and white-water flood plains (Freese et al., in prep.). And in Brazil it occupies swamp forest (Deane et al. 1971), *terra firme* (nonflooding) forest, *várzea* (flooded with a white-water river), and *igapó* (flooded with black or clear-water river) forests, and seems to prefer river margin habitats (Mittermeier and Coimbra-Filho 1977).

Elsewhere in South America, *S. sciureus* is also found most frequently near rivers. In one study in Venezuela, 89% were seen in trees near streams, and 11% were in other moist areas (Handley, pers. comm. 1974). All were in evergreen forest, 88% in tropical humid forest and 12% in tropical very humid forest (life zones of Ewel and Madriz 1968). Similarly, in Surinam, *S. sciureus* is especially abundant along rivers and in swamp forest in the coastal region, and seems to prefer seasonally and permanently flooded and marsh forests (Mittermeier 1977a). This is in contrast to the Panamanian population, which usually does not enter and even tends to avoid waterlogged swamp forest and thickets (Moynihan 1976a).

S. sciureus is not found in dry areas such as the dry forest of Costa Rica (Freese, pers. comm. 1974), and its range into northeastern Colombia is limited by decreasing gallery forest and decreasing rainfall (Hernández-Camacho and Cooper 1976). But it does occur in several nonriverine and nonflooded habitats, and the predominance of records along rivers may be a reflection of the greater accessibility of these forests. According to Thorington (1975), "we have little understanding of the habitat requirements of *Saimiri* and of the validity of the distribution map."

S. sciureus is found primarily at low elevations (Mendez 1970; Handley, pers. comm. 1974; Sousa et al. 1974; Durham 1975; Hernández-Camacho and Cooper 1976), but has been recorded as high as 914 m in Panama (Napier 1976) and 2,000 m in Peru (Grimwood 1969). It can utilize secondary forest as well as mature primary forest. For example, in Panama it has been observed in low bushy second growth (Carpenter 1935, Moynihan 1967) and other disturbed habitats (Bennett 1968). It has also been recorded in secondary forest in Costa Rica (Freese, pers. comm. 1974), Brazil (Mittermeier and Coimbra-Filho 1977, Ayres 1978), and Surinam (Mittermeier 1977a).

Factors Affecting Populations

Habitat Alteration

The main factor responsible for the decline of squirrel monkey populations in Costa Rica, Panama, and Colombia is deforestation. In Costa Rica the replacement of forest with banana plantations and the widespread slash and burn agriculture by the local people have eliminated most suitable squirrel monkey habitat and greatly reduced and fragmented the range of this species (Wilson, pers. comm. 1974).

In Panama, *S. sciureus* takes advantage of artificial habitats, including all kinds of edges and hedges near or within towns, crop fields, and plantations (Moynihan 1976a). Nevertheless, the "recent removal of all shrub cover for pasturage" and the "destruction of forests and woodlands" are responsible for the absence of *S. sciureus* in many areas (Bennett 1968). Populations of this species have diminished greatly during the last 20 years as "many acres of forest have been cleared for banana plantations, cattle ranches, sugar cane, and rice farms" (Baldwin and Baldwin 1972b). The Panamanian human population is growing rapidly (31% per decade), and the implementation of an old land reform law has led to increasing pressure to make profitable use of forested lands or forfeit them to the government (Baldwin and Baldwin 1976). Between 1950 and 1960, for example, forest cover in Chiriquí was reduced from 61% to 50% of the land surface, and *S. sciureus* can now be found only in remote, uncleared areas (Baldwin and Baldwin 1976).

The extensive spraying of insecticides to control mosquitoes has probably reduced *S. sciureus* populations in Panama (Baldwin and Baldwin 1972b). Pesticides have been used heavily on both cleared and forested land, and local people in Chiriquí report that squirrel monkeys were much more plentiful before spraying began (Baldwin and Baldwin 1976). Squirrel monkeys eat many insects (Thorington 1968a) and may be more vulnerable than primarily herbivorous primates to the adverse effects of insecticides.

In Colombia, habitat destruction is especially severe in the upper Magdalena valley, and *S. sciureus* populations are probably suffering from this disturbance

(Hernández-Camacho and Cooper 1976). Near San Martín, the extent of forested area has been reduced by burning to increase cattle pasture land (Thorington 1968a).

In Peru, *S. sciureus* has been able to tolerate limited disturbances such as selective logging, but in some areas of the northeast habitat alteration has had a striking effect on its populations, depressing them to levels one-third to one-half of those found in pristine areas (Freese 1975). At one study site in northern Bolivia, extensive land clearing has contributed to the sharp decline of monkey populations (Heltne et al. 1976).

In Surinam and Guyana, *S. sciureus* is adaptable to human encroachment and seems to benefit from limited habitat alteration (Mittermeier 1977a, Muckenhirn and Eisenberg 1978). But there, as in other countries, extensive deforestation would eliminate this species. No data are available concerning the effect of habitat alteration on *S. sciureus* in Ecuador, Brazil, French Guiana, or Venezuela.

Human Predation

Hunting. Hunting of *S. sciureus* for its meat is not a major factor affecting populations in Costa Rica (Freese, pers. comm. 1974), Panama (Baldwin and Baldwin 1976), or Colombia (Hernández-Camacho and Cooper 1976, Klein and Klein 1976). In Surinam, this species is seldom hunted for food (Mittermeier 1977a) because of its small size and peculiar taste (Husson 1957).

Squirrel monkeys are hunted in some parts of Peru (Gardner, pers. comm. 1974), and are an important source of meat, especially in populated areas where larger species have been hunted out (Soini 1972). Necklaces made from their skulls, paws, and teeth and stuffed squirrel monkey skins are sold in markets in Iquitos (Castro 1976). In Brazil, *S. sciureus* is occasionally shot for bait or for human consumption (Mittermeier and Coimbra-Filho 1977).

Pest Control. Squirrel monkeys feed on crops in Costa Rica (Freese, pers. comm. 1974) and Panama (Baldwin and Baldwin 1976). In Peru they occasionally damage cacao (Grimwood 1969) and also feed in cultivated banana and lime trees (Durham 1972). And in Surinam they are considered destructive to cacao and citrus plantations (Husson 1957). None of the authors consulted mentioned any attempts to control this species.

Collection. Squirrel monkeys are kept as pets by people in Brazil (Mittermeier and Coimbra-Filho 1977) and Surinam (Mittermeier 1977a).

This species has been captured in large numbers for export to laboratories, zoos, and pet dealers. One trader in Panama exported about 500 *S. sciureus* between 1952 and 1962, but the species then became too scarce for profitable exploitation (Baldwin and Baldwin 1976). Larger numbers have been exported from Peru, Colombia, and Guyana. For example, a total of 263,607 were exported from Peru between 1964 and 1974 (Castro 1976), 5,546 were exported from Colombia in 1970 (Green 1976), and about 2,700 were exported from Guyana during the first nine months of 1975 (Muckenhirn et al. 1975). In 1976–77, 370 *S. sciureus* were exported from Peru (Castro 1978).

Many of these squirrel monkeys were imported into the United States. Between 1968 and 1973, 187,119 *S. sciureus* were imported into the United States (Appendix), primarily from Peru, Colombia, Brazil, Guyana, and Bolivia. In 1977, 150 entered the United States from Bolivia and Peru (Bittner et al. 1978). In 1965–75, 15,265 squirrel monkeys were imported into the United Kingdom (Burton 1978). *S. sciureus* is frequently used in biomedical research in the United States (Thorington 1972, Comm. Cons. Nonhuman Prim. 1975a).

Some authors believe that collection, if continued, could lead to a severe depletion of *S. sciureus* (Cooper 1968), and that the annual average in 1962–71 of nearly 25,000 exported from Peru and thousands more lost in transit (Grimwood 1969, Soini 1972) represented a significant drain on the population. Others report that even the large-scale trapping done near Leticia, Colombia, had only a local effect (Mittermeier and Coimbra-Filho 1977), and that the Peruvian population can probably stand an annual exploitation of more than 1,000 individuals (Neville 1976b).

Conservation Action

S. sciureus is found in three national parks and eleven reserves (Table 87). But La Macarena National Park in Colombia is not adequately protected against illegal encroachment by agriculturists and will disappear within a few years unless effective protective measures are taken (Struhsaker 1976b). Also, the two reserves in Peru are not well patrolled against hunters (Castro 1976).

In Costa Rica, *S. sciureus* is legally (but not totally effectively) protected from hunting, trapping, and export, and a small forest south of Quepos where it occurs has been proposed as a new national park (Freese, pers. comm. 1974). The preservation of a large portion of the Osa Peninsula, the last sizable forested area within its range in Costa Rica, is necessary if the species is to survive in that country (Freese, pers. comm. 1974; Wilson, pers. comm. 1974).

In Panama, a national park or wildlife refuge should be established to preserve the coastal lowland fauna, including *S. sciureus* (Baldwin and Baldwin 1976). In Colombia, protective measures are needed to save some of the habitat in the upper Río Magdalena valley (Hernández-Camacho and Cooper 1976) and La Macarena National Park (Struhsaker 1976b).

In Peru, legal restrictions now prohibit hunting, capture, and commerce in squirrel monkeys, but the provision against hunting is not well enforced (Neville et al. 1976). Enforcement should be strengthened and export quotas imposed (Freese 1975). In Bolivia it may be possible to collect 500 to 1,000 *S. sciureus* per year from forest where clearcutting is more than 50% complete and continuing, but this trapping should be done by a professional team under strict regulations and should be preceded by a thorough field study (Heltne et al. 1976). Brazilian law prohibits commercial exploitation of *S. sciureus* (Mittermeier and Coimbra-Filho 1977), and Surinam laws protect this species from hunting in the north and transport from the interior to population centers (Mittermeier 1977a). In Venezuela, legal restrictions prevent the export of *S. sciureus* (Mondolfi 1976).

To avoid excessive depletion of *S. sciureus* populations by collection for research, several conservation measures have been recommended, including minimization of wastage through education and more exacting laws regarding treatment, shipment, and sale of captured animals (Cooper 1968). One dealer attempted to produce *S.*

sciureus for export on an island in the Río Amazonas near Leticia, Colombia. He released 5,690 individuals there between 1967 and 1970, but by 1972 the population had decreased to 850–966 and was still declining (Mittermeier, Bailey, Sponsel, and Wolf 1977).

The Costa Rican–Panamanian race of squirrel monkeys (*S. s. oerstedii*) is listed in Appendix I of the Convention on International Trade in Endangered Species of Wild Fauna and Flora, and as endangered under the U.S. Endangered Species Act (U.S. Department of Interior 1977a,b).

Table 87. Distribution of *Saimiri sciureus* (countries are listed in order from north to south and counterclockwise within South America)

Country	Presence and Abundance	Reference
Costa Rica	Along W coast S of 9°30′N (map)	Hall and Kelson 1959
	Restricted to narrow coastal zone in SW; may have been introduced by pre-Columbian man	Hershkovitz 1969
	Confined to coastal Pacific lowlands, including Osa Peninsula and as far N as 5 km S of Puerto Quepos (9°28′N, 84°10′W); scarce and declining	Freese, pers. comm. 1974
	Previously probably from R. Parris in N to S border; distribution now discontinuous; endangered	Wilson, pers. comm. 1974
	In SW corner (map)	Thorington 1975
	In Corcovado National Biological Reserve; endangered	I.U.C.N. 1978
Panama	Fairly plentiful in Coto region of W	Carpenter 1935
	Along S coast W of 81°W (map)	Hall and Kelson 1959
	Common on Pacific coast of W; localities listed include Boquerón, Bugaba, David, R. Coto region (Chiriquí), Almijas Island, Sevilla Island	Handley 1966
	Between 82°W and 83°W in SW and on islands S of David; previously common elsewhere, including lower slopes of El Baru; now absent in many areas, range greatly reduced; may have been introduced by Indians bringing pets	Bennett 1968
	Peculiar distribution pattern (absent in E) suggests introduction by pre-Columbian man	Hershkovitz 1969
	Along S coast of W in SW Veraguas and S Chiriquí (map); also islands of Almijas and Sevilla	Mendez 1970
	In SW Chiriquí (map); range and populations are reduced	Baldwin and Baldwin 1972b
	In W Chiriquí (map)	Sousa et al. 1974

Table 87 (continued)

Country	Presence and Abundance	Reference
Panama (cont.)	Localities mapped in SW Chiriquí; common in early 1950s, now scarce; only in scattered remote areas of SW	Baldwin and Baldwin 1976
	Near Puerto Armuelles and abundant on Burica Peninsula (both in SW); absent in E	Moynihan 1976a
	Localities listed include Cerro Campana (8°41′N, 79°55′W), Bocaron (Chiriquí), and Veragua	Napier 1976
	Endangered	I.U.C.N. 1978
Colombia	In E	Cabrera 1958
	E of San Martín (3°43′N, 73°42′W) (central)	Thorington 1967
	Absent in Chocó region and elsewhere W of Andes except in upper R. Magdalena valley, Huila	Hershkovitz 1969
	In La Macarena N.P., Meta	Green 1972, Hunsaker 1972
	In Putumayo Umbria region of S near R. Uchapayaco, R. Guineo, R. Putumayo	Whittemore 1972
	N bank R. Guayabero, La Macarena N.P.	Klein and Klein 1975, 1976
	In central and SE (map)	Thorington 1975
	2 localities mapped in N and central	Deane 1976
	In central and SE in Amazon region, piedmont of Andes, and S part of E plains (map); questionable in NE; N limit in R. Magdalena valley not well defined; probably decreasing in S near Leticia and perhaps elsewhere	Hernández-Camacho and Cooper 1976
	Along R. Peneya, tributary of R. Caquetá (in S)	Izawa 1976
	Abundant in parts of Caquetá; in Putumayo near Santa Rosa; in Meta, near San Martín	Moynihan 1976a
	Locality listed at Bogotá, Cundinamarca	Napier 1976
	In good numbers in La Macarena N.P.	Struhsaker 1976b
	Along right bank of R. Duda, W boundary of La Macarena N.P.	Izawa and Mizuno 1977
Ecuador	In E	Cabrera 1958
	E of about 77°W (map)	Thorington 1975
	Localities listed include R. Napo, Quito district, R. Pastaza (2°30′S, 76°W), and R. Copataza	Napier 1976
Peru	In N, E of Andes, and in E in basins of R. Ucayali and R. Urubamba	Cabrera 1958
	Common in Amazon region; not endangered	Grimwood 1969

Table 87 (continued)

Country	Presence and Abundance	Reference
Peru (cont.)	At R. Llullapichis, tributary of R. Pachitea (central)	Koepcke 1972
	Abundant in Iquitos area (in NE)	Soini 1972
	Common near R. Curanja, SE Loreto; stable there 1966–71	Gardner, pers. comm. 1974
	At Salvación, Madre de Dios	Durham 1975
	Along R. Nanay, R. Orosa, R. Samiria, R. Pucallpa (in NE); not common along R. Ampiyacu; in Manú N.P. (in SE); generally abundant	Freese 1975
	Throughout E (map)	Thorington 1975
	In Manú N.P., Madre de Dios and Cuzco; Tingo Maria N.P., Huánuco; and Pacaya and Samiria national reserves, Loreto; vulnerable	Castro 1976
	Common near Iquitos (along Nanay R.), R. Orosa basin, R. Samiria basin, and R. Pacaya basin	Freese et al. 1976
	Localities listed in Cuzco, Huánuco, Loreto, San Martin, R. Pachitea, R. Ucayali	Napier 1976
	Relatively abundant along R. Samiria and R. Pacaya between lower R. Ucayali and R. Huallaga/Marañón	Neville et al. 1976
	In E Madre de Dios along R. Tahuamanu, R. Nareuda, and R. Acre; high densities near R. Yaverija-Primavera and Iberia-Humaitá	Encarnación and Castro 1978
	Along R. Yavineto (R. Putumayo), N Loreto	Encarnación et al. 1978
	N of Isla de Iquitos, R. Amazonas (in NE)	Moro 1978
	Abundant at Cocha Cashu, Manú N.P.	Janson and Terborgh, in press
	Abundant in upper R. Nanay area, R. Orosa area, Panguana area, and Cocha Cashu (Manú N.P.)	Freese et al., in prep.
Bolivia	Throughout E	Cabrera 1958
	At R. Surutu and Buena Vista, Santa Cruz de la Sierra (locality mapped)	Soria 1959
	Common at Cobija, on R. Acre (N border in W); abundant at El Triunfo (central, at 15°15′S, 64°15′W)	Heltne et al. 1976
	Near Santa Cruz de la Sierra (at 17°27′S, 63°34′W)	Cambefort and Moro 1978
	Abundant in El Triunfo area (central); also seen in Cobija	Freese et al., in prep.

Table 87 (continued)

Country	Presence and Abundance	Reference
Brazil	Very abundant throughout R. Amazonas valley; localities mapped in W and central Amazonas and E Pará including Marajó Island	Kumm and Laemmert 1950
	In E basin of lower R. Juruá, Amazonas; in Amazon region S to Goiás state; in NW in R. Solimões basin	Cabrera 1958
	In SE at Juruá, São Paulo, about 22°S, 48°W (locality mapped)	Soria 1959
	At Utinga, near Belém (E Pará)	Carvalho and Toccheton 1969
	In Amazonas, Pará, and N Rondônia	Deane et al. 1969
	At Pôrto Mauá, Amazonas	Deane et al. 1971
	In urban parks in Belém	Pine 1973
	Throughout W and Amazon basin to mouth; absent in area E of R. Negro and W of R. Branco (map)	Thorington 1975
	Localities mapped in W Acre and central Amazonas	Deane 1976
	Localities listed include R. Juruá, R. Madeira, upper R. Solimões, R. Negro, in Amazonas; lower R. Amazonas, Marajó, R. Tocantins, Pará	Napier 1976
	Abundant; not threatened in Amazonian region	Mittermeier and Coimbra-Filho 1977
	In E Pará E of R. Xingu, including near Belém	Ayres 1978
French Guiana	Present	Cabrera 1958
	Low to moderately abundant along R. Cayenne and R. Maroni (W border)	Durham, pers. comm. 1974
	Throughout (map)	Thorington 1975
Surinam	Common in interior and coastal plain, including N of Moengo Tapoe in NE	Husson 1957
	Present	Cabrera 1958, Fooden 1964
	S of Affobakka Dam (in NE)	Walsh and Gannon 1967
	Low to moderately abundant along R. Maroni (E border)	Durham, pers. comm. 1974
	Localities listed include Paramaribo and R. Coppename (in N)	Napier 1976
	Localities mapped throughout; especially abundant in coastal region; in 6 nature reserves including Raleighvallen-Voltzberg, Tafelberg, Eilerts de Haan Gebergte, Sipaliwini Savanna, Wia-Wia, and Galibi; not in danger	Mittermeier 1977a

Table 87 (continued)

Country	Presence and Abundance	Reference
Guyana	Localities mapped in E including Berbice (in NE), H.M.P.S. Forest Reserve (N central), Moraballi Forest Reserve (N central), and Apoteri (central)	Muckenhirn et al. 1975
	Throughout except extreme E (map)	Thorington 1975
	Localities listed include R. Essequibo (central), Moon Mts. (S), Demerara (NE), and coastal	Napier 1976
	Widespread; not vulnerable	Muckenhirn and Eisenberg 1978
Venezuela	Localities mapped in S central in Amazonas SE, SW, and NE of Esmeralda, on R. Orinoco, R. Cunucunuma, R. Mavaca, R. Manapiare, and Brazo Casiquiare; as far N as about 5°20′N	Handley, pers. comm. 1974
	In extreme S tip only (map)	Thorington 1975
	Locally distributed S of R. Orinoco in Amazonas from R. Negro and Brazo Casiquiare to R. Manapiare; probably also in Bolívar, recorded at Bajo Esequibo	Mondolfi 1976

Table 88. Population parameters of *Saimiri sciureus*

Population Density (individuals/km²)	Group Size X̄	Group Size Range	Group Size No. of Groups	Home Range Area (ha)	Study Site	Reference
		≥15			Costa Rica	Freese, pers. comm. 1974
		10–35			Panama	Baldwin and Baldwin 1971
		23–27	2	17.5	Panama	Baldwin and Baldwin 1972b
130		10–30		17.5–40	Panama	Baldwin and Baldwin 1976
		18–40	2		Colombia	Thorington 1967
19–31		25–35	3–6		Colombia	Klein and Klein 1975
	40	42–200	3	300	Colombia	Izawa 1976
		30–40			Peru	Grimwood 1969
	13.2	8–23	5		Peru	Durham 1972
		25–30			Peru	Gardner, pers. comm. 1974
16–84	40				Peru	Freese 1975
151–528	20–50		5–7		Peru	Neville et al. 1976
7.5	30	10–30	6		Peru	Encarnación and Castro 1978
	12.6+	5–20+	5		Peru	Encarnación et al. 1978
52	20				Peru	Moro 1978
33–49	25	30–40	43	300	Peru	Janson and Terborgh, in press
8–100	40	2–50	61		Peru and Bolivia	Freese et al., in prep.
24–100	≥40	30–43	12		Bolivia	Heltne et al. 1976
		20–70	29		Brazil	Mittermeier and Coimbra-Filho 1977
		>100			Surinam	Husson 1957
27.7	27.7	17–42	3	76	Surinam	Mittermeier 1977a
12–585	17–31	8–60	15		Guyana	Muckenhirn et al. 1975

CERCOPITHECIDAE
Old World Monkeys

Allenopithecus nigroviridis Allen's Swamp Monkey

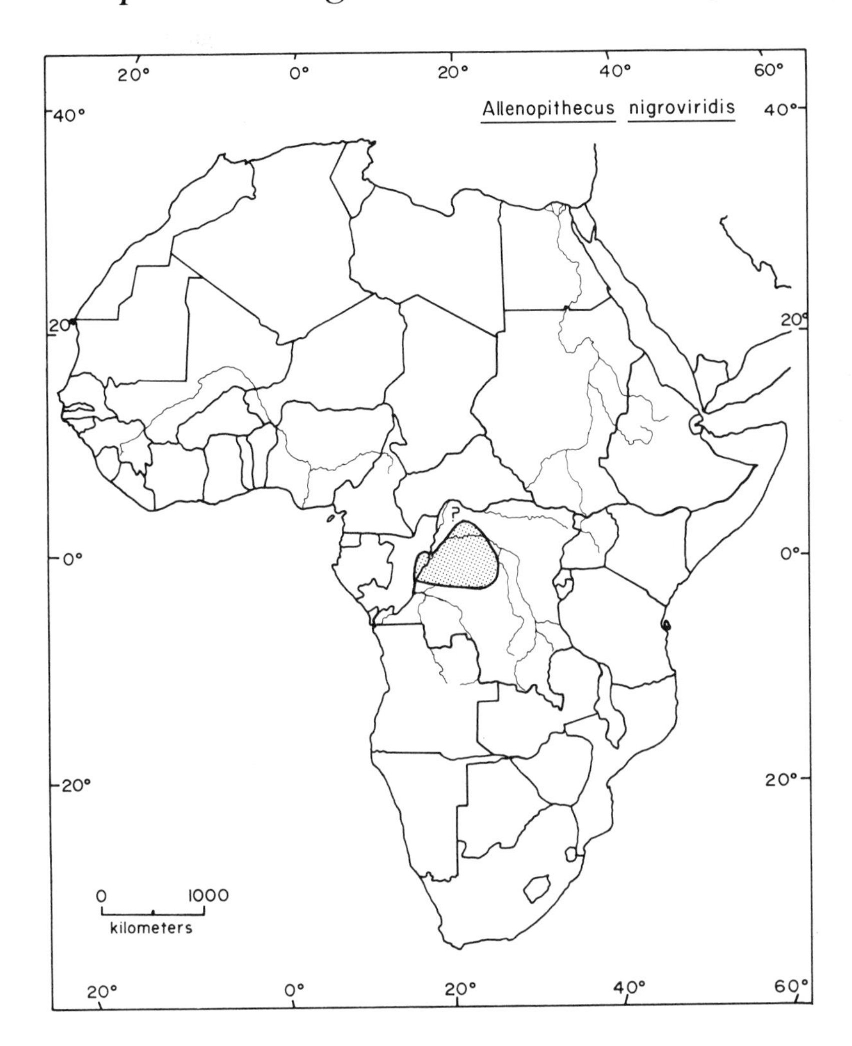

Taxonomic Note

Allenopithecus nigroviridis is a monotypic species (Napier and Napier 1967).

Geographic Range

A. nigroviridis is found in the central Congo basin in western Zaire and eastern Congo. It has been recorded as far north as 3°17′N and as far south as 2°17′S, east to 23°05′E, and west to 16°14′E on the west side of the Congo River (Verheyen 1963). For details of the distribution, see Table 89.

Abundance and Density

A. nigroviridis is reported to be "sparsely distributed" over its range (Hill 1966). Near Lake Tumba, a group of 3 to 6 individuals and a solitary male were seen (Horn, pers. comm. 1978). Otherwise no information was located regarding the abundance, density, group size, or home range of this species.

Habitat

A. nigroviridis lives in swampy areas (Malbrant and Maclatchy 1949; Horn, pers. comm. 1978). Verheyen (1963) stated that it was restricted to regularly inundated riverine forests.

Factors Affecting Populations

No information.

Conservation Action

No information.

Table 89. Distribution of *Allenopithecus nigroviridis*

Country	Presence and Abundance	Reference
Zaire	Along Zaire R. N of 3°S to 23°E near Tshuapa R., and N to Karawa, 3°–4°N	Schouteden 1947 cited in Tappen 1960
	Not rare near Bolobo (2°10′S, 16°17′E)	Malbrant and Maclatchy 1949
	In NW along Zaire R., Tshuapa R., and Momboyo R. (localities mapped)	Verheyen 1963
	W of Lake Tumba (in W near 3°S)	Horn, pers. comm. 1978
Congo	In E near lower Likouala-Mossaka R., Sangha R., and Likouala aux Herbes R.	Malbrant and Maclatchy 1949
	In E central at Pakama and Butando (localities mapped)	Verheyen 1963

Cercocebus albigena Gray-cheeked Mangabey

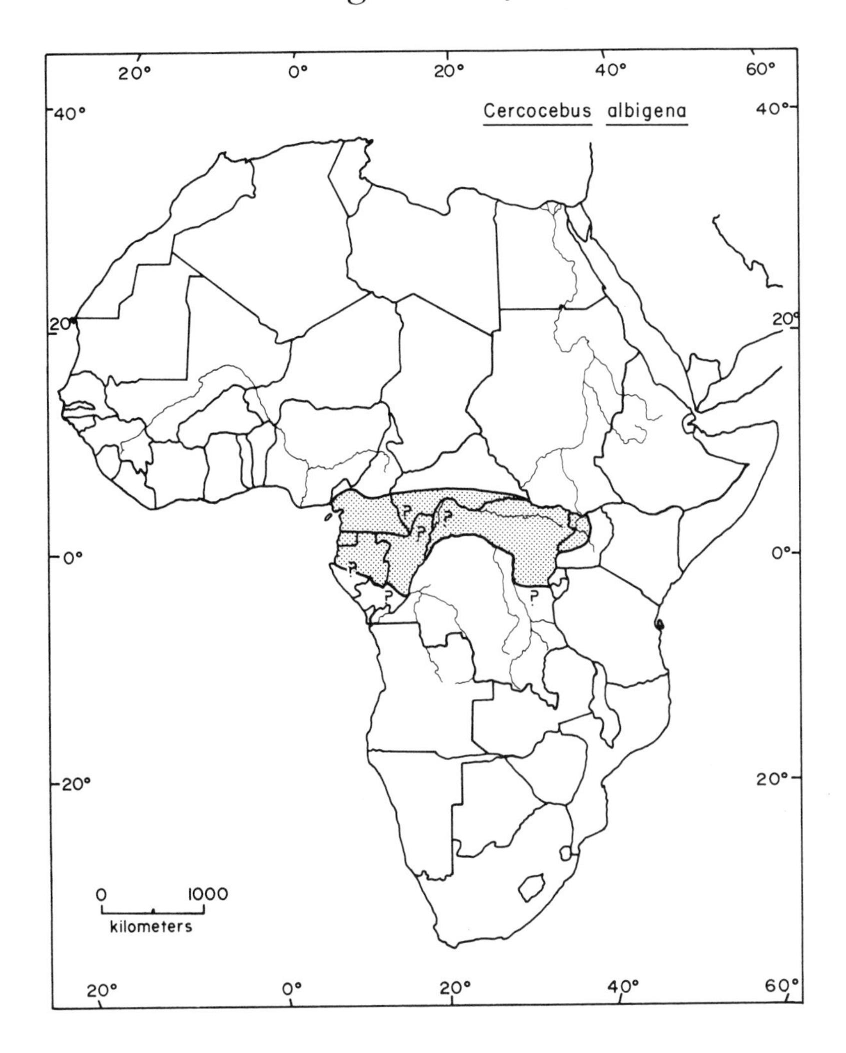

Taxonomic Note

Three races are recognized (Napier and Napier 1967).

Geographic Range

Gray-cheeked mangabeys are found in equatorial Africa, from the Atlantic coast of Cameroon to central Uganda. In the west they occur as far north as the Mamfe area (about 5°46′N, 9°18′E) (Gartlan and Struhsaker 1972), and as far south as at least Booué, Gabon (0°03′S, 11°58′E) (Quris 1976). Their range extends as far east as the Mabira Forest (about 0°30′N, 33°E) near the Victoria Nile (Tappen 1960; Kingdon 1971; Waser, pers. comm. 1974). The southern limits of the range are not known in either the west or the east. For details of the distribution, see Table 90.

Abundance and Density

Cercocebus albigena seems to be locally abundant in several parts of its range but is described as scarce or declining in parts of Uganda, Cameroon, and Equatorial Guinea (Table 90).

Population density estimates are generally less than 20 per km^2, although Chalmers's (1968) study in a small area ($0.6 km^2$) is an exception (Table 91). Typical group sizes in both East and West Africa are generally 14 to 17, and home ranges vary from 13 to 400 ha. Low population densities, large home ranges, and widely spaced groups are probably the rule in this species (Waser 1976).

Habitat

C. albigena is a forest species, typical of high, dense, primary evergreen forest (Sabater Pi 1966; Chalmers 1968; Gartlan, pers. comm. 1974; Waser, pers. comm. 1974). It also inhabits swamp forest (Lumsden 1951), coastal river-mouth forests (Jones and Sabater Pi 1968), and flooded forests (Gautier and Gautier-Hion 1969, Quris 1976), and is never found far from water (Kingdon 1971).

These mangabeys can utilize high secondary forest (Malbrant and Maclatchy 1949, Jones and Sabater Pi 1968, Gautier and Gautier-Hion 1969), but only where it is adjacent to relatively undisturbed forest (Jones, pers. comm. 1974). In East Africa they are also found in isolated or semi-isolated patches of forest (Chalmers 1968). Gartlan (pers. comm. 1974) found that they do not survive in disturbed habitats or near plantations, although Malbrant and Maclatchy (1949) reported them in brush and shrubs around old plantations. Tall trees seem to be a necessary component of their habitat (Jones, pers. comm. 1974; Rowell, pers. comm. 1978).

C. albigena has been studied in mountainous areas up to 900 m elevation (Hendrichs 1977) and at 1,600 m elevation (Mizuno et al. 1976). It probably does not occur above 1,700 m (Waser, pers. comm. 1974).

Factors Affecting Populations

Habitat Alteration

Since *C. albigena* is largely dependent on undisturbed primary forest, it is especially vulnerable to deforestation. Logging has been cited as the principal reason for its decline in Cameroon (Gartlan, pers. comm. 1974) and Equatorial Guinea (Jones, pers. comm. 1974). In the latter country, extensive clearcut timbering has reduced the range of this species and is further damaging the remaining forest in the northwest (Sabater Pi and Jones 1967, Jones and Sabater Pi 1968). Less than 33% of their habitat still remained in Equatorial Guinea in 1968 (Jones, pers. comm. 1974).

The selective logging being practiced in Uganda does not remove the trees of primary importance to *C. albigena*, but the application of arboricides to logged areas in order to eliminate commercially undesirable trees reduces mangabey food sources and has resulted in the disappearance of *C. albigena* from treated areas (Waser, pers. comm. 1974). Populations are likely to decrease further if the planned logging and treatment of all Ugandan forests on a 70 to 80 year cycle are carried out (Waser, pers. comm. 1974). At one study site, slightly fewer *C. albigena* (0.1 clusters per walk) were found in felled and herbicide treated areas than in undisturbed forests (0.3 clusters per walk) (Oates 1977a).

Smaller-scale clearing for agriculture is also reducing the habitat of this species in Cameroon (Gartlan, pers. comm. 1974), Equatorial Guinea (Jones, pers. comm. 1974), and Uganda (Waser, pers. comm. 1974). In general, gray-cheeked mangabeys seem to be less adaptable to habitat changes than are many forest monkeys.

Human Predation

Hunting. This species is hunted for food in Equatorial Guinea (Jones, pers. comm. 1974) and Cameroon (Gartlan, pers. comm. 1974), but not in Uganda (Waser, pers. comm. 1974).

Pest Control. Gray-cheeked mangabeys do raid crops (Chalmers 1968; Gartlan, pers. comm. 1974), and in East Africa steal maize, sweet potatoes, peanuts, and cassava (Kingdon 1971). In Uganda they are shot as pests to cultivation, but they only raid crops where forests have recently been felled and probably cannot maintain high population levels in long established agricultural areas (Waser, pers. comm. 1974).

Collection. C. albigena is not in demand for biomedical research and is difficult to trap (Rowell, pers. comm. 1974). Between 1958 and 1968 one was exported from Equatorial Guinea (Jones, pers. comm. 1974), and the number exported from Uganda is low (Waser, pers. comm. 1974). None were imported into the United States between 1968 and 1973 (Appendix) or in 1977 (Bittner et al. 1978). Nineteen entered the United Kingdom in 1965–75 (Burton 1978).

Conservation Action

C. albigena is vulnerable because of its dependence on primary forest and its naturally low population densities. Large habitat reserves are essential to insure its survival. In Equatorial Guinea this is especially urgent. If a preserve is not achieved, the species will probably soon be extinct in that country.

In Cameroon, *C. albigena* occurs in three forest reserves (Table 90). In Uganda it is not specifically protected by any game laws and does not occur in any national parks (Waser, pers. comm. 1974). It does occur in forest reserves, but these are being systematically logged and thus are not secure (see above). The inclusion of primary rain forest in Ugandan parks is imperative for the survival of gray-cheeked mangabeys as well as other forest primates there, and efforts are being made to obtain protection for several small forest preserves (Waser, pers. comm. 1974).

Table 90. Distribution of *Cercocebus albigena* (countries are listed in order from northwest to south and then east)

Country	Presence and Abundance	Reference
Cameroon	In Dja Reserve, Dipikar Island Reserve, Lac Tisongo (S of mouth of Sanaga R.), and Mamfe Overside near Maku R. and Makweli R.	Gartlan and Struhsaker 1972
	Common in Douala-Edéa Reserve	Struhsaker 1972
	Throughout S of 4°N (map); not abundant, declining; very rare N of Sanaga R.	Gartlan, pers. comm. 1974
	Common in Dja R. area (in S)	Rowell, pers. comm. 1978
Equatorial Guinea	Numerous in N near Campo R. and near coast	Sabater Pi 1966
	Patchy distribution in W and N (maps); range being reduced	Sabater Pi and Jones 1967, Jones and Sabater Pi 1968
	Locally abundant, declining in all areas	Jones, pers. comm. 1974
Gabon	Throughout	Malbrant and Maclatchy 1949
	In Ogooué-Ivindo region (NE)	Gautier and Gautier-Hion 1969
	Along Liboui R. (rarely seen) and at Booué (in N)	Quris 1976
Congo	Throughout	Malbrant and Maclatchy 1949
	In N half (map)	Dandelot 1965
Central African Republic	In SW (map)	Dandelot 1965
Zaire	On right banks of Zaire and Lualaba rivers	Tappen 1960
	In N, NE, and E	Rahm 1966
	Abundant throughout Haut-Zaire Province (N)	Heymans 1975
	W of Lake Kivu	Hendrichs 1977
Uganda	In Mongiro Forest, Bwamba (W)	Lumsden 1951
	In Bwamba and Mabira forests	Tappen 1960
	Near Bujoko and Nagojji (S central)	Chalmers 1968
	In S, central, and W (localities mapped)	Kingdon 1971
	Common in Kibale Forest (W)	Struhsaker 1972 [1975; Mizuno et al. 1976; Waser 1976, 1977; Waser and Floody 1974; Oates 1977a]
	In Mabira, Kifu, Lwamunda, Kibale, Bwamba, and Bugoma forests; scarce and probably decreasing	Waser, pers. comm. 1974

Table 91. Population parameters of *Cercocebus albigena*

Population Density (individuals/km^2)	Group Size $\bar{X}$	Group Size Range	Group Size No. of Groups	Home Range Area (ha)	Study Site	Reference
		13	1		Cameroon	Struhsaker 1969
22.1±3.5	17.3	10–24	3	200–300	Equatorial Guinea	Jones, pers. comm. 1974
		9	1		Gabon	Gautier and Gautier-Hion 1969
		17	1		Gabon	Quris 1976
77	16.8	7–25	5	13–26	Uganda	Chalmers 1968
		12–30			Uganda	Kingdon 1971
10	14.4	6–28	5	400	Uganda	Waser, pers. comm. 1974
9	15			400	Uganda	Struhsaker 1975
		11	1	67.4	Uganda	Mizuno et al. 1976

Cercocebus aterrimus Black Mangabey

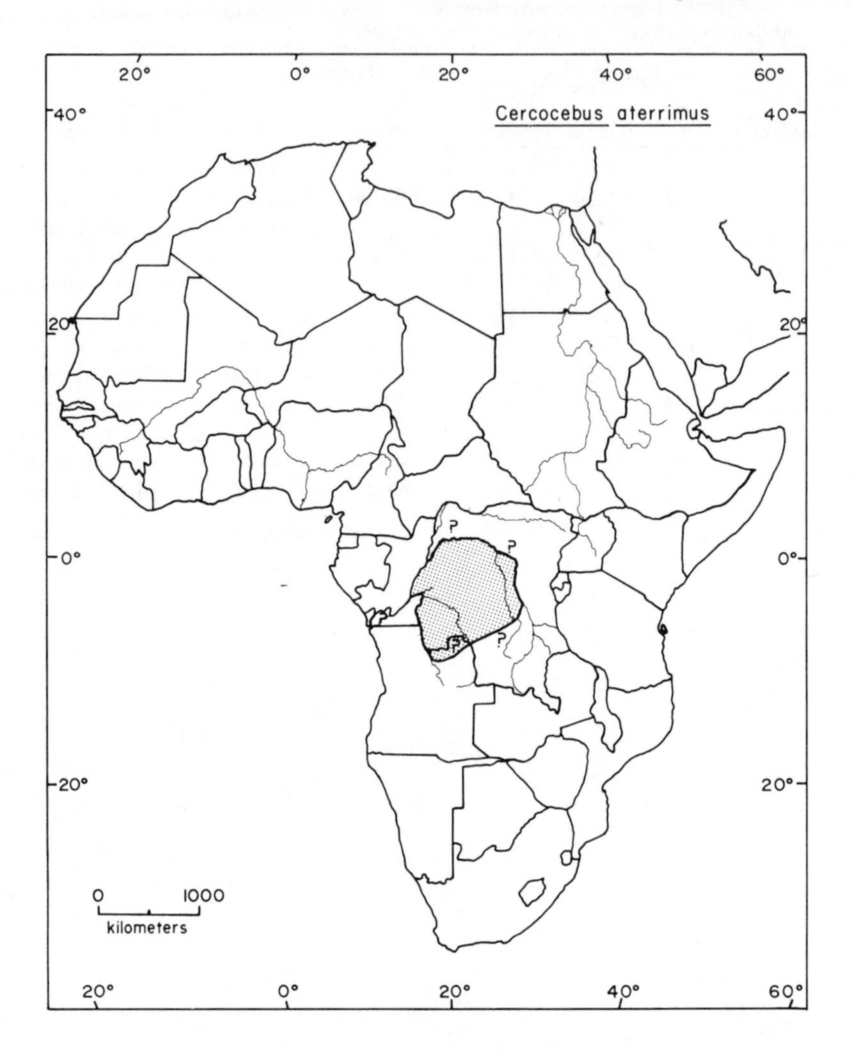

Taxonomic Note

Two subspecies are recognized (Dorst and Dandelot 1969).

Geographic Range

Cercocebus aterrimus is found in Zaire and possibly northern Angola. The limits of its range have not been precisely delineated. It occurs as far west as the Congo River near 2°S, at least as far east as the Lualaba River, and north to the Zaire River and perhaps to the Ubangi River (Dandelot 1965). The southernmost collecting locality is near 7°30′S, 17°30′E in southern Zaire, but local people have reported it as far south as 9°S in northern Angola (Barros Machado 1969). For details of the distribution, see Table 92.

Abundance and Density

Black mangabeys have recently been reported to be locally common in two areas, one in western and one in central Zaire (Table 92). No information is available regarding their overall abundance. No data pertaining to population density, group size, or home range area were located.

Habitat

C. aterrimus is a forest species (Heymans 1975), which has been observed in marsh forest (Dorst and Dandelot 1969) and in secondary lowland rain forest (Horn, pers. comm. 1978). No details of its ecology are known.

Factors Affecting Populations

Habitat Alteration

The forests of the western Congo basin have not yet been extensively logged, but large-scale timbering operations are being planned which would decrease the habitat available to *C. aterrimus* west of Lake Tumba (Horn, pers. comm. 1978).

Mangabeys are generally less able to adjust to habitat changes than are most guenons, and black mangabeys seem to be especially dependent on tall trees and hence vulnerable to habitat disturbance (Gartlan, pers. comm. 1974).

Human Predation

Hunting. C. aterrimus is hunted intensively for its flesh in the Basankusu Territory (Ladnyj et al. 1972) and near Lake Tumba (Horn, pers. comm. 1978).

Pest Control. Black mangabeys do enter plantations and are killed during crop raiding whenever possible (Horn, pers. comm. 1978).

Collection. C. aterrimus is not used in medical research. Two were imported into the United States between 1968 and 1973 (Appendix). None entered the United States in 1977 (Bittner et al. 1978). Three were imported into the United Kingdom in 1965–75 (Burton 1978).

Conservation Action

No information.

Table 92. Distribution of *Cercocebus aterrimus*

Country	Presence and Abundance	Reference
Zaire	In central and W (map)	Dandelot 1965
	One collecting locality (mapped) at about 7°30′S, 17°30′E (in SW)	Barros Machado 1969
	In Basankusu Territory, Equateur Province (in NW)	Ladnyj et al. 1972
	Fairly common on left bank of Zaire R., Haut-Zaire Province (central)	Heymans 1975

Table 92 (continued)

Country	Presence and Abundance	Reference
Zaire (cont.)	In Equateur Province S of Lisala (left bank of Zaire R. in NW), near Boende (0°15′S, 20°51′E), and common W of Lake Tumba (near 2°S, 18°E) (in W)	Horn, pers. comm. 1978
Angola	Reported in N, N of 9°S and between 17°30′ and 20°10′E (localities mapped)	Barros Machado 1969

Cercocebus galeritus Agile or Crested Mangabey

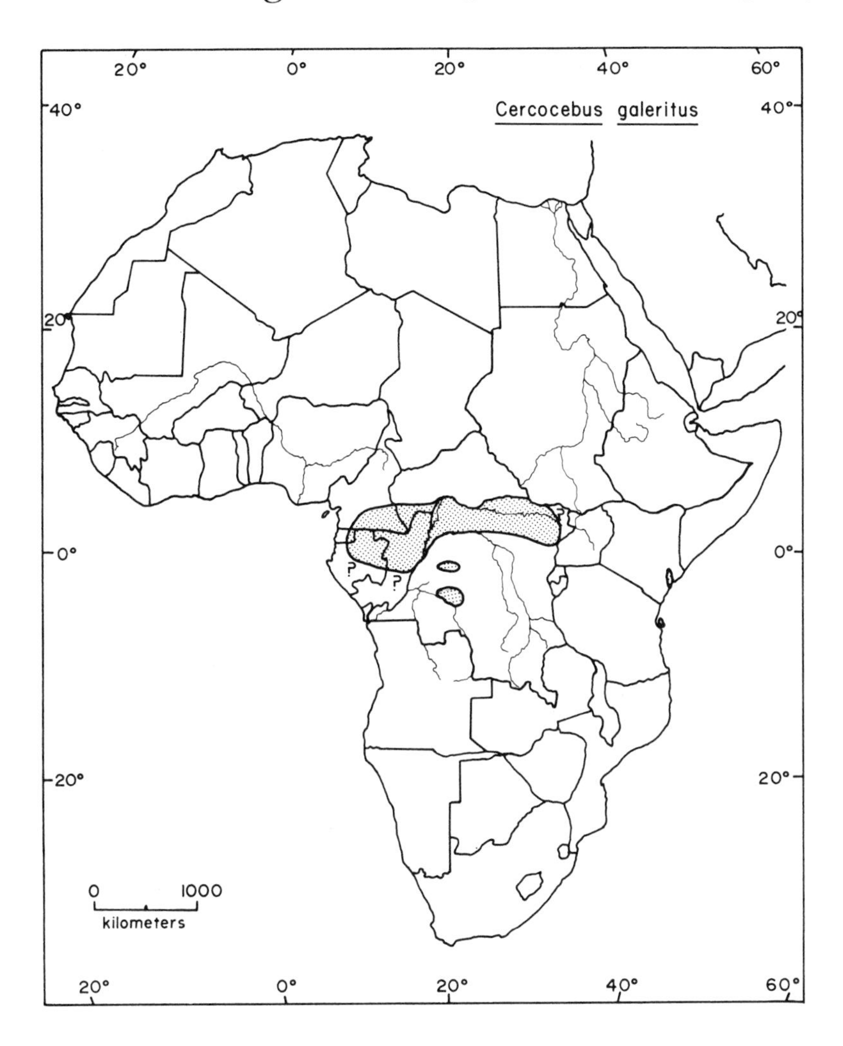

Taxonomic Note

Cercocebus galeritus consists of three races (Napier and Napier 1967).

Geographic Range

Agile mangabeys are found in equatorial Africa from Cameroon to Kenya. Their distribution is discontinuous, with a gap of 1,200 km between the Zairean and eastern Kenyan populations (Tappen 1960) and isolated populations reported in central Zaire (Dorst and Dandelot 1969). They occur as far northwest as Sangmelima, Cameroon (2°57′N, 11°56′E) (Rowell, pers. comm. 1978), and as far southeast as 2°15′S, 40°15′E near Garsen, Kenya (Groves et al. 1974). The other limits of the range have not been precisely determined, but the northeast limit has been estimated at near 29°E in northeast Zaire (Tappen 1960), and the southern limit is at about 4°S in Zaire

(Dorst and Dandelot 1969). This species occurs at least as far southwest as 13°E (near the equator) in northern Gabon (Quris 1975). For details of the distribution, see Table 93.

Abundance and Density

C. galeritus is clearly in a precarious position in the Tana River region of Kenya (Table 93). All authors consulted agree that its population there consists of 900–2,900 individuals in less than 2,000 ha of available habitat, and that both the population and the range are shrinking. This form is obviously endangered.

Although *C. galeritus* was judged to be common in part of northern Zaire, very little is known about its status in the western portion of its range (Table 93).

Typical group sizes for this species are between 10 and 20, with larger groups, smaller home ranges, and much higher population densities reported for the Kenyan than the Gabonaise form (Table 94), perhaps owing to the severe limitation of available habitat in Kenya.

Habitat

Agile mangabeys are forest monkeys, typically found near water in swampy or riparian forests (Malbrant and Maclatchy 1949; Gartlan, pers. comm. 1974). The western populations often utilize seasonally flooded forest (Gautier and Gautier-Hion 1969) and may spend up to 95% of their time in this habitat (Quris 1976). In Zaire this species is found in forests along rivers (Heymans 1975).

The eastern race occupies gallery forests with an open canopy, herbaceous undergrowth, periodic flooding which inhibits sapling growth, and a thin leaf litter (Groves, pers. comm. 1974). These forests are either mixed evergreen forest, *Garcinia* forest, *Pachystela* forest, or regenerating partially cleared cultivation forest (Groves et al. 1974). The Tana River agile mangabey also utilizes *Acacia* woodland and bushland (Groves et al. 1974, Andrews et al. 1975). Its highest population densities are found in the lowland evergreen and secondary regrowth forests, with lower densities and larger home ranges in the woodland habitats (Groves, pers. comm. 1974; Homewood, pers. comm. 1974).

Factors Affecting Populations

Habitat Alteration

In West Africa, forest clearance has reduced the habitat of *C. galeritus* (Sabater Pi and Jones 1967; Gartlan, pers. comm. 1974).

In Kenya, clearing for agriculture and logging have destroyed about 25% of the evergreen forest and 17% of the deciduous woodland, leaving about 75% of the range available to *C. galeritus* in 1960 (Homewood, pers. comm. 1974). The rapidly growing human population is practicing shifting agriculture, which is destroying forest faster than it can regenerate, and has already fragmented the agile mangabey range there into more than 70 discrete patches (Marsh and Homewood 1975). In addition, regular burning of grassland areas for pasture prevents the regeneration of forest and woodland after flooding (Andrews et al. 1975).

The second major threat to the Kenyan *C. galeritus* population is the planned hydroelectric and irrigation schemes that will eliminate flooding and the customary shifts in river course which have led to the continually changing, colonizing forests most suitable for *C. galeritus* (Homewood 1975). The planned dams will also reduce the river flow by one-half, lowering the ground water level and decreasing the total area of woodland and forest, and will eliminate much of the silt content of the lower Tana. The intensified agriculture upstream will also dump large quantities of chemical fertilizers and pesticides into the river (Homewood 1975). All these activities will reduce the carrying capacity of the habitat for *C. galeritus*.

Human Predation

Hunting. C. galeritus is killed for its meat in Gabon (Quris 1975) and Cameroon (Rowell, pers. comm. 1978). Hunting is not a serious factor affecting the Tana River population (Groves et al.1974; Homewood, pers. comm. 1974).

Pest Control. Agile mangabeys feed in rice farms (Tappen 1960) and plantations (Kingdon 1971; Groves, pers. comm. 1974; Homewood, pers. comm. 1974). In Cameroon they are killed and eaten by farmers when caught raiding crops (Rowell, pers. comm. 1978). In Kenya they eat crops only when the fields are inside the forest, and are sometimes killed as pests to cultivation (Groves, pers. comm. 1974). But many farmers chase them from fields rather than kill them (Homewood, pers. comm. 1974), and some people tolerate and even feed the Tana River mangabeys (Andrews, pers. comm. 1974).

Collection. C. galeritus is not captured in large numbers for research or other purposes. Three are known to have been captured in the Tana River area during the last 10 years (Groves, pers. comm. 1974). None were imported into the United States between 1968 and 1973 (Appendix) or in 1977 (Bittner et al. 1978), or into the United Kingdom in 1965–75 (Burton 1978).

Conservation Action

C. galeritus occurs in the Dja Reserve in southern Cameroon (Table 93). No further information about conservation of the western or central populations was located.

The Tana River mangabey (*C. galeritus galeritus*) is protected within the 175 km^2 Tana River Primate Game Reserve, established in 1976 in an effort to conserve the best remaining forest patches and to develop the area for tourism as a more profitable alternative than exploitation of the forest by subsistence agriculture (Marsh and Homewood 1975, Marsh 1977). Among other actions believed necessary in order to make the reserve viable are (1) studies and scientific management of other wildlife in the area such as elephants (Groves et al. 1974), and (2) the provision of economic assistance and alternatives to the resident Pokomo people, in compensation for resources unavailable to them because of establishment of the reserve (Homewood 1975).

The Tana River mangabey is "fully protected by the Kenyan Game Department" and no capture permits are given, and is included in Class A of the African Convention of 1969 (I.U.C.N. 1972, 1976). It is also listed in Appendix I of the Convention on International Trade in Endangered Species of Wild Fauna and Flora and as endangered under the U.S. Endangered Species Act (U.S. Department of Interior 1977a,b).

Table 93. Distribution of *Cercocebus galeritus* (countries are listed in order from west to east)

Country	Presence and Abundance	Reference
Cameroon	In E Dja Reserve (S central)	Gartlan and Struhsaker 1972
	As far W as Sangmelima	Rowell, pers. comm. 1978
Equatorial Guinea	1 record from N central (locality mapped)	Sabater Pi 1966, Sabater Pi and Jones 1967
	Specimen not personally collected; some doubt as to collecting site	Jones, pers. comm. 1974
Gabon	In extreme NE	Malbrant and Maclatchy 1949
	In Ogooué-Ivindo region (NE)	Gautier and Gautier-Hion 1969
	Along Mounianghi, Liboui, and Njadié rivers; reported along Liboumba, Djoua, and Ivindo rivers (map) (in NE)	Quris 1975
Congo	In N and E	Malbrant and Maclatchy 1949
	Throughout N, N of 2°S (map)	Dandelot 1965
Central African Republic	In SW corner (map)	Dandelot 1965
Zaire	Along Zaire R., Uele R., Itimbi R., Ituri R. (in N and NE) to 29°E, and Kasai R. (central)	Tappen 1960
	N of Zaire R. and along Kasai R. and Sankuru R. (central) (map)	Kingdon 1971
	Fairly common throughout length of Aruwimi and Itimbiri R., Haut-Zaire Province (in N, E of 24°E)	Heymans 1975
Kenya	Threatened; lower Tana R. region (SE)	Dorst and Dandelot 1969
	Endangered; lower Tana R. region	I.U.C.N. 1972
	Scarce, remaining stable	Andrews, pers. comm. 1974
	Rare; along Tana R. between Hewani (W of Garsen) and Wenje (1°45′S) and other nearby forest patches; total population about 1,500–2,500; remaining stable	Groves, pers. comm. 1974
	Total population about 2,245; total range about 733 ha and shrinking	Groves et al. 1974
	Scarce; range decreasing	Homewood, pers. comm. 1974
	One of most endangered African monkeys	Struhsaker, pers. comm. 1974

Table 93 (continued)

Country	Presence and Abundance	Reference
Kenya (cont.)	18 forest patches with resident populations listed; maximum population estimated at 1,466–2,932 in 733 ha range	Andrews et al. 1975
	Total population 1,000–1,500; habitat shrinking	Homewood 1975
	Total population 900–1,500; total range about 2,000 ha along 60 km stretch of Tana R.; decreasing	Marsh and Homewood 1975
	Endangered	I.U.C.N. 1976
	In Tana River Reserve (50 km N of Garsen)	Marsh 1977

Table 94. Population parameters of *Cercocebus galeritus*

Population Density (individuals/km²)	Group Size			Home Range Area (ha)	Study Site	Reference
	$\bar{X}$	Range	No. of Groups			
7.7–12.5	10.5	7–18	3	200	NE Gabon	Quris 1976
100–500					SE Kenya	Andrews, pers. comm. 1974
200–500					SE Kenya	Groves, pers. comm. 1974
100–300		≤12			SE Kenya	Groves et al. 1974
50–200	22–30			15–65	SE Kenya	Homewood, pers. comm. 1974
200–400					SE Kenya	Andrews et al. 1975
<50–300		13–36		15–100+	SE Kenya	Homewood 1975

Cercocebus torquatus

White-collared, Red-capped, and Sooty Mangabey

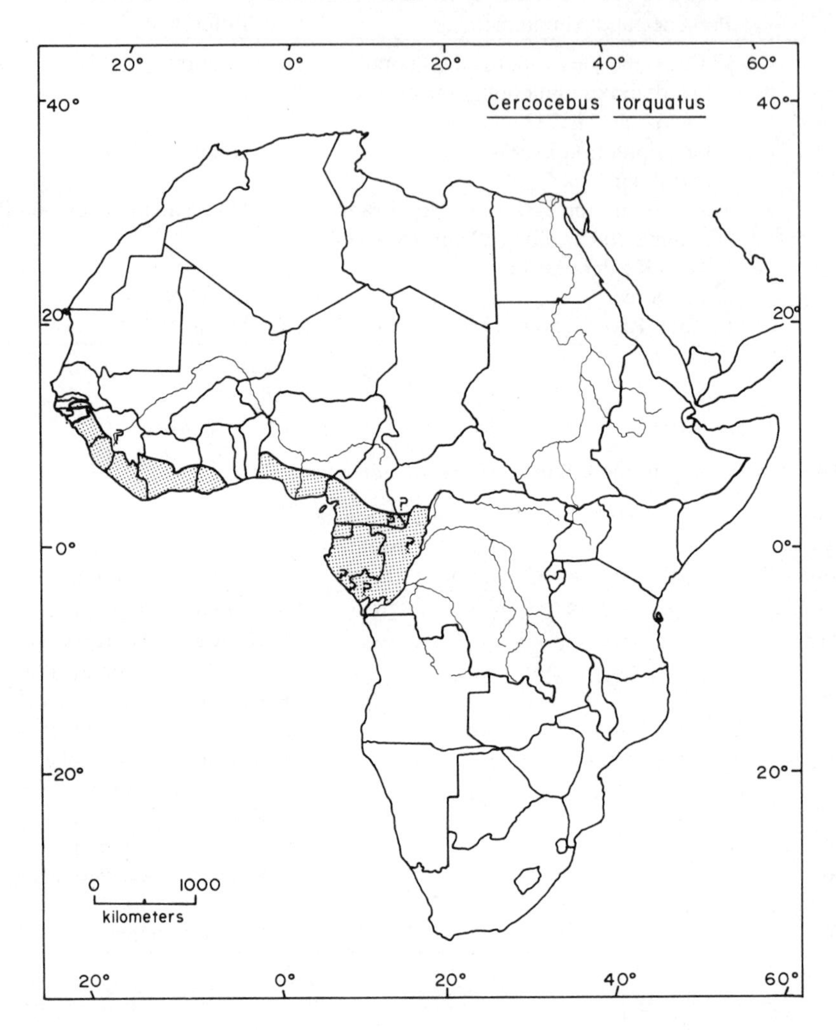

Taxonomic Note

Three races are recognized (Dorst and Dandelot 1969). Several authors have treated the westernmost populations as a separate species, *Cercocebus atys* or *C. fuligunosis* (Booth 1956a,b; Tappen 1960; Napier and Napier 1967; Dupuy 1971b; Struhsaker 1971). But others consider this form a subspecies of *Cercocebus torquatus* (Kuhn 1965; Jones 1966; Gartlan, pers. comm. 1974), and it is treated as such here.

Geographic Range

C. torquatus is found from Senegal to Ghana and from Nigeria to Gabon, with an 800 km gap separating the Ghanaian and Nigerian populations (Booth 1956b). In the west it extends as far north as 12°30′N, 16°05′W in Senegal (Struhsaker 1971). It has been reported as far east as northern Congo, extending southward along the right bank of the Congo River (Malbrant and Maclatchy 1949), but the eastern and southern limits of the range are not known. For details of the distribution, see Table 95.

Abundance and Density

C. torquatus has recently been described as common or abundant in parts of Liberia, Ivory Coast, Ghana, Cameroon, and Equatorial Guinea, but is known to be declining in the latter two countries (Table 95). Information regarding its status in the western and eastern portions of its range is lacking.

The only population density estimate available indicates local densities of 133 *C. torquatus* per km^2 in Equatorial Guinea (Jones, pers. comm. 1974). Recorded group sizes include 14 to 23 (Jones and Sabater Pi 1968), 3 and 18 (Struhsaker 1969), 12 (Dupuy 1971b), and 20 to 60 ($\bar{X} = 35$, N = 5 groups) (Jones, pers. comm. 1974). This species is said to have "fairly extensive" home ranges (Jones and Sabater Pi 1968), but no range size estimates were located.

Habitat

C. torquatus is a moist forest species, typical of mangrove, coastal, gallery, and inland swamp forest (Booth 1956a, Jones and Sabater Pi 1968, Gartlan and Struhsaker 1972). It also utilizes dry deciduous forest (Olson, pers. comm. 1978) and forest relicts in the derived savanna zone (Happold 1973).

This species has been found in secondary forest (Malbrant and Maclatchy 1949, Booth 1956a, Tahiri-Zagrët 1976) and active as well as regenerating plantations (Jones and Sabater Pi 1968, Schlitter et al. 1973). In Equatorial Guinea it usually occupies lowland habitats below 350 m elevation (Jones and Sabater Pi 1968) but has also been recorded in montane forest (Jones, pers. comm. 1974). It seems to be most frequent near water and to be dependent on the presence of water and undisturbed areas for its survival (Jones, pers. comm. 1974).

Factors Affecting Populations

Habitat Alteration

C. torquatus is not adaptable to changes in its environment and would be expected to suffer from any habitat disturbance (Gartlan, pers. comm. 1974). Deforestation is leading to a serious reduction of its habitat throughout much of its range and is the principal reason for the decline of this species in Cameroon (Gartlan, pers. comm. 1974) and Equatorial Guinea (Jones, pers. comm. 1974).

Large-scale logging is also occurring in Ivory Coast, where *C. torquatus* is rare in logged forests even where it is relatively common in adjacent undisturbed forest (Struhsaker 1972). In one area in Sierra Leone, *C. torquatus* occupied a forest which had been extensively logged 16 years earlier and subsequently treated with arboricides (Tappen 1964), indicating that this species can recolonize logged forests that are allowed to regenerate.

The replacement of forest with cacao, rubber, oil palm, and other plantations has greatly decreased *C. torquatus* habitat in Ghana (Jeffrey 1975b, Asibey 1978). Slash and burn clearing and other small-scale forest cutting have destroyed large areas of habitat in Ivory Coast (Lanly 1969), Cameroon (Gartlan, pers. comm. 1974), and Equatorial Guinea, where only about 33% of *C. torquatus* habitat was still intact in 1968 (Jones, pers. comm. 1974). And as early as 1963, clearing for agriculture had reduced the forested area of Sierra Leone to less than 4% of the land surface (Tappen 1964).

Human Predation

Hunting. C. torquatus is hunted for its flesh in Ivory Coast (Tahiri-Zagrët 1976), Ghana (Jeffrey 1974, 1975b; Asibey 1978), Nigeria (Happold 1973), Cameroon (Gartlan, pers. comm. 1974), and Equatorial Guinea (Jones, pers. comm. 1974).

Pest Control. C. torquatus does steal food, including maize and rice from farms in Ghana (Booth 1958, Jeffrey 1970, Asibey 1978), Cameroon (Gartlan, pers. comm. 1974), and Equatorial Guinea (Jones, pers. comm. 1974). It is regarded as an especially serious pest to the cacao crop in Sierra Leone, where it is considered one of the most destructive of all monkeys (Mackenzie 1952, Lowes 1970), and is heavily persecuted in monkey extermination campaigns (Robinson 1971).

Collection. Twenty-nine *C. torquatus* were exported from Equatorial Guinea between 1958 and 1968 (Jones, pers. comm. 1974), and a total of 45 were imported into the United States between 1968 and 1973 (Appendix). None entered the United States in 1977 (Bittner et al. 1978). In 1965–75, 166 were imported into the United Kingdom (Burton 1978).

Conservation Action

C. torquatus is found in three national parks and seven reserves (Table 95). It is legally protected from hunting in the forest reserves of Nigeria and Ivory Coast, although considerable poaching occurs (Happold 1973, Tahiri-Zagrët 1976).

This species is listed in Class B of the African Convention, and lactating females are protected in Liberia (Curry-Lindahl 1969). In Gongola State, Nigeria, it may be hunted only with a special license with a bag limit of four (Hall 1976). The Moslem prohibition against eating monkeys also protects it in some parts of its range (Lowes 1970).

C. torquatus is listed as endangered under the U.S. Endangered Species Act (U.S. Department of Interior 1977b).

Table 95. Distribution of *Cercocebus torquatus* (countries are listed in order from west to east)

Country	Presence and Abundance	Reference
Senegal	On S border in W, S of Casamance R. (map)	Dupuy 1971b, 1973
	S of Casamance R., 2 localities 12°30′N at 16°W and 16°5′W	Struhsaker 1971
Guinea-Bissau	No information	
Guinea	As far W as 15°W (W coast)	Tappen 1960
Sierra Leone	Throughout (map)	Booth 1958
	In Kasewe Forest, S central	Tappen 1964
	Present	Jones 1966
Liberia	Throughout (map)	Booth 1958
	Present	Kuhn 1965
	Abundant	Schlitter, pers. comm. 1974
Ivory Coast	In S (map)	Booth 1956b, 1958
	Common in Tai Reserve (now N.P.) (in W between Cavally and Sassandra rivers)	Struhsaker 1971 [Monfort and Monfort 1973]
	At Goudi, near Lamto (S central, along Bandama R.)	Bourlière et al. 1974
	In W and S (map); N limit is Biankouman	Tahiri-Zagrët 1976
Ghana	In SW (localities mapped)	Booth 1956a
	In Bia Tributaries Forest Reserve (SW)	Jeffrey 1970
	Also at Sefwi Wiawso and Sukusuku (in SW)	Jeffrey 1975b
	Common in Ankasa Game Production Reserve and Nini-Suhien N.P.; rare in Bia N.P.	Asibey 1978
Togo	Absent	
Benin	Absent	
Nigeria	In extreme SE, E of Cross R. (map); questionable in W	Booth 1958
	In S central and SW, E and W of Niger R. (localities mapped); including Ohosu, Omo, Inangan, and Oluwa forest reserves; probably rare in W	Happold 1973
	In SW (locality mapped)	Monath and Kemp 1973
	Throughout S (localities mapped)	Schlitter et al. 1973
Cameroon	Throughout S (map)	Booth 1958
	Extreme S and SW (map)	Sabater Pi and Jones 1967
	At Elongongo and Nsépé (along Sanaga R. in W), Campo Reserve and Dipikar (SW), and Dja Reserve (S central)	Gartlan and Struhsaker 1972
	Common in Douala-Edéa and Korup reserves (in W)	Struhsaker 1972
	Not abundant; declining	Gartlan, pers. comm. 1974

Table 95 (continued)

Country	Presence and Abundance	Reference
Equatorial Guinea	Along Benito, Utamboni, Utonde, and Mbia rivers near coast	Sabater Pi 1966
	In NW, central, and SW (map)	Sabater Pi and Jones 1967
	In NW (localities mapped); rare in interior	Jones and Sabater Pi 1968
	Locally abundant; declining in all areas	Jones, pers. comm. 1974
Gabon	Abundant near coast, less so inland	Malbrant and Maclatchy 1949
	To 5°S along coast	Tappen 1960
	Throughout (map)	Dorst and Dandelot 1969
Congo	Abundant near coast, less so inland; S border is Congo River	Malbrant and Maclatchy 1949
	Throughout (map)	Dorst and Dandelot 1969
Central African Republic	In SW corner (map)	Dorst and Dandelot 1969

Cercopithecus aethiops Vervet, Grivet, and Green Monkey

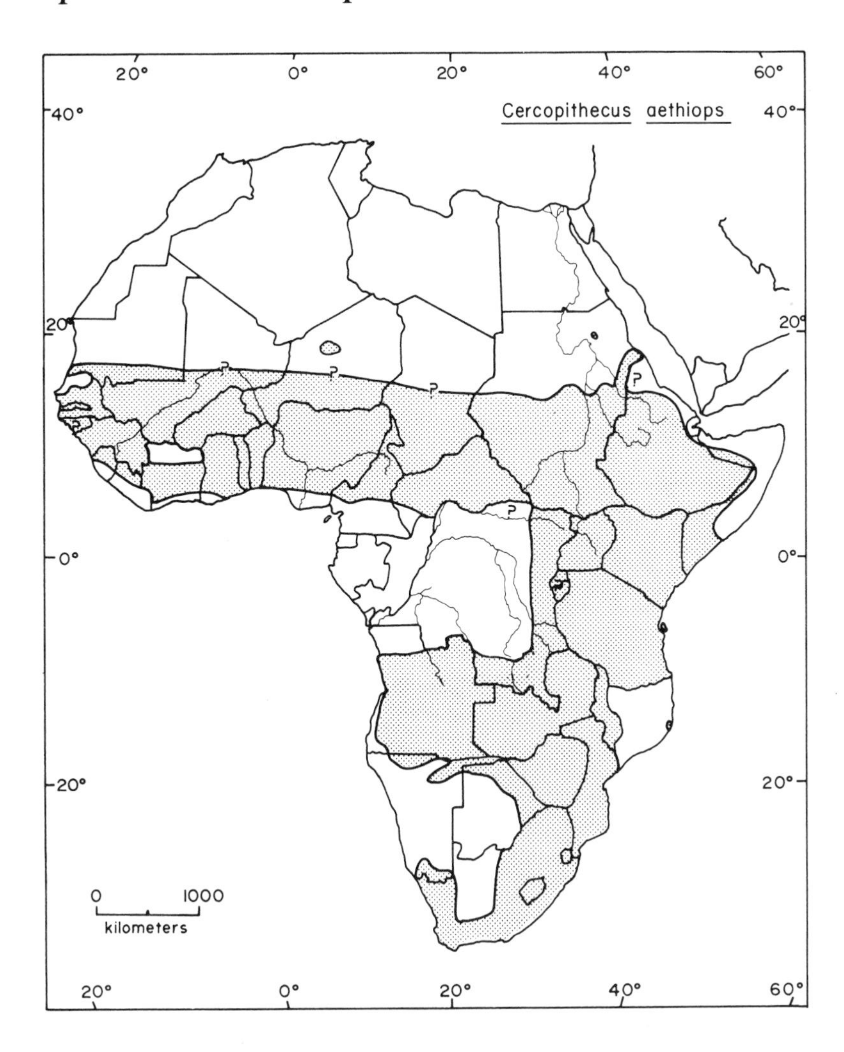

Taxonomic Note

Twenty-one subspecies have been described, and some authors recognize three separate species (Napier and Napier 1967) or five distinctive forms (Dorst and Dandelot 1969). All are considered together here, following Gartlan and Brain (1968) and Kingdon (1971).

Geographic Range

Cercopithecus aethiops is widely distributed across subsaharan Africa from Mauritania to Ethiopia and south to South Africa. It occurs from the western tip of Senegal at 17°27′W (Dupuy 1971b) to at least Mustayel (Mustahil), Ethiopia (5°15′N, 44°45′E) (Bolton 1973), or even farther east into Somalia (Gartlan and Brain 1968, Dorst and Dandelot 1969). The northernmost records of *C. aethiops* are near

19°30′N, 33°E in northern Sudan (Kock 1969). This species is apparently absent in the forests of the Congo basin (Tappen 1960), the deserts of southwest Africa, and several coastal areas, but otherwise occurs throughout the southern two-thirds of the continent, to the southern tip of South Africa (Gartlan and Brain 1968, Dorst and Dandelot 1969). For details of the distribution, see Table 96.

Abundance and Density

C. aethiops has been described as "perhaps the most abundant monkey in Africa" (Struhsaker 1967a). In many parts of its range it is plentiful, especially compared with other monkeys. It has recently been considered abundant or common throughout, or in parts of, 12 of the 36 countries within its geographic range (Table 96). But *C. aethiops* was also judged to be relatively scarce in five countries and declining in two. In five countries it is known to be widespread, but no estimates of abundance were located, and there is little or no information on its status in 12 of the countries within its range (Table 96).

Density estimates for *C. aethiops* vary from 0.87 to 153.7 individuals per km^2 (Table 97). Troops of up to 140 members have been reported, with average group sizes between 8 and 46. Home ranges may be as small as 9.4 ha and as large as 518 ha (Table 97).

Habitat

The characteristic habitats of *C. aethiops* are Guinea and Sudan savanna, woodland, forest edges near savanna, thickets, and riparian woodland (Booth 1958, Gartlan and Brain 1968). Many authors have mentioned its preference for riverine vegetation (e.g., Booth 1956a; Pienaar 1967; Howell 1968; DeMoor and Steffens 1972; Henshaw and Child 1972; Monath and Kemp 1973; Nagel 1973; Ansell, pers. comm. 1974; Dunbar 1974; Joubert and Mostert 1975). In some areas it is dependent on these gallery forests or woodlands and seldom travels far from rivers or streams (Nagel 1973, Poché 1976).

But vervet, grivet, and green monkeys also have a broad tolerance of environmental conditions and can occupy habitats ranging in humidity from semiarid to very wet (Brain 1965). They have been observed in Sahel savanna, small patches of dry deciduous forest (Booth 1958), swamp forest (Sikes 1964a), mangrove forests along saline rivers (Dupuy 1973, Galat and Galat-Luong 1976), *Acacia nilotica* thorn forest (Galat and Galat-Luong 1977), dry *Terminalia* scrub on Kalihari sand (Wilson 1975), and seven distinct, semiarid veld types (Lynch 1975).

This species also utilizes tropical rain forest, lowland evergreen forest, and mediterranean evergreen vegetation (Gartlan and Brain 1968). It is also found in montane forest and alpine moorland up to 4,500 m in elevation (Bernahu 1975). But it seldom invades the depths of forest, tending to remain on the edges (Tappen 1960, Gartlan and Brain 1968, Dunbar and Dunbar 1974a), and is more typical of wooded than forested areas (Booth 1958, Tahiri-Zagrët 1976).

C. aethiops is rarely found in completely open grassland where there are no trees nearby (Bourquin et al. 1971; Gartlan, pers. comm. 1974), but is typically restricted to the close proximity of trees or dense thickets (Struhsaker 1967a). It does range extensively into open savanna when moving between wooded areas (Tappen 1960, Hall 1965b) but generally feeds in, sleeps in, and retreats to woodland or forest (Booth 1956a, Butler 1966).

This species can also utilize secondary bush (Tappen 1964), secondary forest (Brown and Urban 1970, Gartlan 1973, Tahiri-Zagrët 1976), cultivated areas (Booth 1956a, Monath et al. 1974, Moreno-Black and Maples 1977), and urban habitats (Basckin and Krige 1973, Pringle 1974).

Factors Affecting Populations

Habitat Alteration

Despite its overall abundance, *C. aethiops* populations are declining in some parts of their range. The chief cause of this decline is the cutting of the trees on which they depend for sleeping, refuge, and food. In Sudan, for example, the northern boundary of their range has receded from 19°N to 15°N during the last hundred years, because the trees along the Nile in this area have been felled for firewood (Butler 1966).

Similarly, in Ghana (Asibey 1978), Cameroon (Gartlan, pers. comm. 1974), Kenya (Rowell, pers. comm. 1974), Niger (Poché 1973), Namibia (Coetzee and Joubert, pers. comm. 1974), and Ethiopia (Bernahu 1975) the cutting of trees for wood and to clear farmland has resulted in the decline of *C. aethiops* populations. In Ethiopia the burning of forest and grassland, the ploughing of steep slopes, and the lack of reforestation have led to further serious deterioration of *C. aethiops* habitat (Bernahu 1975). Fire, drought, and the destruction of savanna by cattle grazing are also problems in Niger (Poché 1973).

In the Masai-Amboseli Reserve in Kenya, *C. aethiops* populations declined 43% between 1964 and 1975 (Struhsaker 1976a). This decrease was correlated with a drastic reduction in thicket cover and fever trees, which provided the main food source for *C. aethiops* in the reserve. Struhsaker (1973) listed three possible reasons for this change in vegetation: (1) a rising water table and salt layer (Western and Sindiyo 1972), (2) overgrazing by the domestic livestock of the Masai people, and (3) tree damage by elephants.

Some kinds of habitat disturbance may be beneficial to populations of *C. aethiops*, since they thrive in secondary growth and agricultural areas as long as trees and water are available nearby. In Zambia (Ansell, pers. comm. 1974) and Zimbabwe (Attwell, pers. comm. 1974) vervet populations are probably increasing, because the spread of cultivation has raised the carrying capacity of the habitat by increasing the food supply. In Cameroon the savanna habitat is spreading (at the expense of forest) because of drought and the cutting of forest for timber and agriculture. Within this enlarged range the *C. aethiops* population density seems to be remaining stable, thus overall numbers are probably increasing (Gartlan, pers. comm. 1974).

Human Predation

Hunting. *C. aethiops* meat is considered to be tasty food and is eaten by non-Muslim human populations in many areas of Africa, including parts of Senegal (Galat and Galat-Luong 1976), Sierra Leone (Lowes 1970), Ivory Coast (Tahiri-Zagrët 1976), Ghana (Asibey 1974), Nigeria (Rice, pers. comm. 1978), Niger (Poché 1973),

Namibia (Coetzee and Joubert, pers. comm. 1974), and Cameroon (Gartlan, pers. comm. 1974).

In Zambia vervets are generally not hunted for either food or fur (Ansell, pers. comm. 1974), and in Zimbabwe their fur has little or no value (Ansell 1960a), while in Kenya it does have some value and is offered for sale (Rowell, pers. comm. 1978).

Pest Control. C. aethiops is a frequent agricultural pest, and crops constitute a large part of its diet in many regions (Tappen 1960). It does extensive damage to cacao plantations in Sierra Leone (Mackenzie 1952, Robinson 1971), feeds in maize patches in Ethiopia (Dandelot and Prévost 1972), and is a notorious crop raider in Senegal (Dunbar 1974), Sierra Leone (Lowes 1970), Ghana (Asibey 1978), Nigeria (Howell 1968), Uganda (Rowell 1968), Kenya (Maples 1972b), Zimbabwe (Smithers 1966), Malawi (Long 1973), Zambia (Ansell 1960a), and Cameroon (Gartlan, pers. comm. 1974).

This crop raiding has led to extermination programs in several countries. In Uganda about 6,000 vervets were killed as pests in 1964 (Rowell 1968). In Sierra Leone, massive eradication efforts have been directed at all destructive monkeys (Mackenzie 1952, Robinson 1971), including *C. aethiops*, although Rowell (1968) doubted the effectiveness of such programs in eliminating monkey populations. In other areas, *C. aethiops* is harassed, shot, or trapped by farmers protecting their crops (Howell 1968; Kingdon 1971; Ansell, pers. comm. 1974; Dunbar 1974; Dunbar and Dunbar 1974a; Rowell, pers. comm. 1974; Bernahu 1975).

In addition, vervet monkeys living in towns eat cultivated plants and are even fed by people (Basckin and Krige 1973, Moreno-Black and Maples 1977). *C. aethiops* sometimes steals food from gardens and houses, takes refuge in exotic tree plantations (Pringle 1974), and is considered a nuisance at picnic sites (Wilson 1975). But none of these authors mentioned human efforts to prevent *C. aethiops* from performing these activities.

Collection. C. aethiops is used as a subject for medical research and vaccine production. In the 1960s, 4,000 were exported per year from Uganda alone (Rowell 1968). But in 1969 the exportation of vervets from Uganda ceased because of fear of the Marburg virus which some populations carried (Rowell, pers. comm. 1974). Some authors have blamed *C. aethiops* carried viruses for the deaths of as many as 17 laboratory workers (Hayflick 1970).

In addition to Uganda, *C. aethiops* has been trapped for export from Cameroon (Gartlan, pers. comm. 1974), Nigeria (Howell 1968), Ethiopia (Bernahu 1975), and Kenya (Rowell, pers. comm. 1974). Hayflick (1970) believed that "this once plentiful species is now threatened with extinction . . . as a result of extensive trapping of hundreds of thousands" of them for polio vaccine production and other projects. But Gartlan (1975a) concluded that *C. aethiops* was not yet threatened by the export trade, provided that any increase in collection was carefully controlled. No information was located regarding collection of this species for the zoo and pet trade.

Between 1968 and 1973 a total of 21,779 *C. aethiops* were imported into the United States, originating primarily from Ethiopia, Somalia, and Tanzania (Appendix). In 1977, 40 individuals and 6 shipments of unspecified sizes were imported into the United States from Somalia (Bittner et al. 1978). In 1965–75, 11,323 *C. aethiops* were imported into the United Kingdom (Burton 1978).

Conservation Action

C. aethiops is present in 47 national parks, game reserves, or nature reserves throughout its range (Table 96). Additional reserves for its conservation have been proposed for the Ducuta area of northern Sierra Leone (Wilkinson 1974) and the Mahali Mountains of eastern Tanzania (Itani and Nishida, pers. comm. 1974).

But the enforcement of regulations in many of the protected areas is inadequate and an increase in funding, better management, stricter control of poaching and habitat destruction, popular education (Poché 1973, Tahiri-Zagrët 1976, Asibey 1978), scientific surveys, and trade restrictions in Ethiopia (Bernahu 1975) have all been advocated to insure protection of *C. aethiops* and other wildlife in these reserves.

Laws regarding the hunting of *C. aethiops* vary greatly from country to country — from total prohibition (Ivory Coast, Tahiri-Zagrët 1976) to permission with a special license (Northeastern State, Nigeria, Hall 1976), to no controls or restrictions (Ethiopia, Bolton 1972; Zambia, Ansell, pers. comm. 1974). In some areas the Moslem prohibition on eating monkey flesh has afforded some protection (Howell 1968; Robinson 1971; Gartlan, pers. comm. 1974; Galat and Galat-Luong 1976).

Several governments have instituted legal restrictions on the capture and export of *C. aethiops*. For example, Kenyan exports are fairly well controlled by the government (Rowell, pers. comm. 1974) and Namibia has totally banned the collection of this species for experimental purposes (Coetzee and Joubert, pers. comm. 1974).

Table 96. Distribution of *Cercopithecus aethiops* (countries are listed in order from west to east then north to south and north again, in a clockwise direction)

Country	Presence and Abundance	Reference
Senegal	Throughout except in N (map); especially common in Niokolo-Koba N.P. (in SE)	Dupuy 1971b [Dupuy 1970b, 1971c; Dunbar 1974, Dupuy and Verschuren 1977]
	S of Casamance R. (in SW)	Struhsaker 1971, Dupuy 1973
	Widespread and common	Boese, pers. comm. 1974
	In Basse Casamance N.P. (SW)	I.U.C.N. 1975
	Widespread on both banks in lower valley of Senegal R. (between Podor and St. Louis)	Bourlière et al. 1976
	At mouth of Saloum R.	Galat and Galat-Luong 1976
	On Morfil I (between Senegal R. and Doué R.) and in N'Dioum Forest (extreme N)	Galat and Galat-Luong 1977
	Common in Niokolo-Koba N.P. and Basse Casamance N.P.	Asibey 1978
Gambia	Present (including in Abuko Nature Reserve)	Gunderson 1977
Guinea-Bissau	No information	
Mauritania	In S (maps)	Gartlan and Brain 1968, Dorst and Dandelot 1969
	About 50 km NNE of Rosso (16°29′N, 15°53′W) (in SW)	Bourlière et al. 1976

Table 96 (continued)

Country	Presence and Abundance	Reference
Mali	In S (maps)	Gartlan and Brain 1968, Dorst and Dandelot 1969
	Widespread and common	Sayer 1977
Guinea	Throughout (map)	Booth 1958
Sierra Leone	In "substantial numbers" in W and N	Tappen 1964
	Very common	Jones 1966
	Abundant	Robinson 1971
	In Bombali District, Northern Province (extreme N)	Wilkinson 1974
Liberia	Present	Kuhn 1965
Ivory Coast	Throughout N	Booth 1956b
	Estimate several thousand in Comoe N.P. in NE	Geerling and Bokdam 1973
	Near Lamto Station (S central)	Bourlière et al. 1974
	Common in central (localities mapped) including Tai N.P.	Tahiri-Zagrët 1976
Ghana	Throughout except SW (map)	Booth 1956a
	Abundant in Mole N.P., Digya N.P., common in Bui N.P., Gbele, Kalakpa, and Shai Hills game production reserves, rare in Kogyae Strict Nature Reserve	Asibey 1978
	Common and stable in Shai Hills Game Production Reserve (NE of Accra)	Rice, pers. comm. 1978
Upper Volta	Rare in "W" N.P.	Asibey 1978
Togo	Throughout (map)	Booth 1958
	Rare in Keran and Malfacassa reserves	Asibey 1978
Benin	Throughout (map)	Booth 1958
	In Pendjari N.P. (NW)	Happold and Philp 1971
	Near Parákou and Save (E central)	Gorgas 1974
	Rare in "W" (in N) and Pendjari national parks	Asibey 1978
Niger	At Air Massif (18°N, 8°E)	Bigourdan and Prunier 1937 cited by Tappen 1960
	Absent at Air Massif	Dekeyser 1950 cited by Tappen 1960
	Infrequent in "W" N.P. (in SW)	Poché 1973
	Density low in SW (between Tapoa R. and 5 km S of Say), and in "W" N.P.; questionable in Air Mts.	Poché 1976
Nigeria	Increasing (population 150–500+) in Yankari Game Reserve (near Yuli in North Eastern State)	Sikes 1964 a,b [Henshaw and Ayeni 1971]
	Widely distributed in and near Borgu Game Reserve (to W of Kainji Reservoir);	Howell 1968, 1969

Table 96 (continued)

Country	Presence and Abundance	Reference
Nigeria (cont.)	abundant; total population in reserve 800–1,000 and increasing steadily	Howell 1968, 1969 (cont.)
	Few records but fairly common in Olokemeji Forest Reserve (about 50 km W of Ibadan)	Hopkins 1970
	In Upper Ogun/Old Oyo Game Reserve, Western State (about 75 km N of Ibadan)	Adekunle 1971, Geerling 1974
	Common in many areas N of high forest zone	Henshaw and Child 1972
	Recorded over wide range from Maiduguri in NE to Western State, large populations only along upper and middle Niger R. (map); common	Monath and Kemp 1973
	Large population in Nupeko Forest near Bida (9°06′N, 6°00E)	Monath et al. 1974
Chad	In S (map)	Gartlan and Brain 1968, Dorst and Dandelot 1969
Sudan	Common and widely distributed (map); previous N limit 19°N, now 15°N; has receded S during last 100 years	Butler 1966
	At Jebel Marra (W central)	Happold 1966
	In S half (localities mapped)	Kock 1969
	In Dinder N.P. (near E border between Rahad and Blue Nile R.)	Cloudsley-Thompson 1973
Ethiopia	As far N as 15°N	Starck and Frick 1958
	In E	Glass 1965
	Near Godare and Bagadur (in SW)	Brown and Urban 1970
	Plentiful near Webbe Shibeli R., Ogaden (in SE)	Bolton 1972
	Abundant around Agaro, Bedele, Gondar (all in W)	Dandelot and Prévost 1972
	In E sighted at Mustayel (Mustahil) on Webbe Shibeli R. and N of Negele on Ganale R.; none seen in N, E of Taccaze R.; in W reported W of Gondar along Dinder R., N of Goc in Gambella salient and S of Gilo R.	Bolton 1973
	Near Awash R. E of Addis Ababa	Nagel 1973
	On Mount Abaro, Shashemenē District (S central)	Bolton 1974
	Population 70–80 in Bole Valley (central); other mapped localities in N central, central, SW, and SE	Dunbar and Dunbar 1974a
	In Lake Stephanie (Chow Bahr or Chew Bahir) Wildlife Reserve (extreme S)	Tyler 1974

Table 96 (continued)

Country	Presence and Abundance	Reference
Ethiopia (cont.)	Considerable numbers throughout central highlands E of Rift Valley; abundant in Rift Valley (central and S) and in highlands of SW; common in W lowlands and N plateau	Bernahu 1975
Somalia	In W and S (maps)	Gartlan and Brain 1968, Dorst and Dandelot 1969
Uganda	N and S of Victoria Nile and between lakes Victoria and Albert	Hall 1965b
	Population of 2,514 on Lolui Island, Lake Victoria; at Chobi	Gartlan and Brain 1968
	Throughout (map)	Kingdon 1971
	On Entebbe Peninsula	McCrae et al. 1971
	In N outskirts Kibale Forest	Hayashi 1975
Rwanda	No information	
Burundi	No information	
Kenya	Abundant in Masai-Amboseli Reserve	Struhsaker 1967a, b
	Throughout except in N central (map)	Kingdon 1971
	Rare in S Turkana (E and W of Lake Rudolf, in NW)	Coe 1972
	In Olambwe Valley Game Reserve (near Kavirondo Gulf, Lake Victoria)	Sutherland 1972
	33% decline 1964–71, Masai-Amboseli Reserve	Struhsaker 1973
	In Lake Nakuru N.P.	Kutilek 1974
	Ubiquitous but declining	Rowell, pers. comm. 1974
	Along lower Tana R. (SE)	Andrews et al. 1975, Homewood 1975
	43% decline 1964–75, Masai-Amboseli Reserve	Struhsaker 1976a
Tanzania	In Serengeti N.P.	Swynnerton 1958, Hendrichs 1970
	In Tarangire Game Reserve (Masai Steppe) (in N)	Lamprey 1964
	Not common along NE shore Lake Tanganyika (between Kigoma and Mpanda)	Kano 1971a
	Throughout (map)	Kingdon 1971
	Present in Mahali Mts. proposed N.P., E shore Lake Tanganyika	Itani and Nishida, pers. comm. 1974
Mozambique	Uncommon near Zinave, Save R. (S)	Dalquest 1965, 1968
	Widely distributed throughout W from Tete District (central) to Gaza District; 1 record in NE (map)	Smithers, pers. comm. 1974
	In Maputo Game Reserve (extreme SE corner)	Tinley et al. 1976

Table 96 (continued)

Country	Presence and Abundance	Reference
Malawi	Widespread	Smithers 1966
	Very common in Nsanje (Port Herald) District (S)	Long 1973
Zambia	Common in most areas	Ansell 1960a
	Records from near Lusaka	Brain 1965
	Widespread	Smithers 1966
	Collected on Fort Jameson (Chipata) Plateau (SE)	Wilson 1968
	Near Zambezi R.	Lancaster 1971, Sheppe and Osborne 1971
	Common in most places, including Kafue N.P. and Lochinvar N.P.; populations probably stable	Ansell, pers. comm. 1974
Zimbabwe	Recorded throughout	Child and Savory 1964
	Records from Umtali, Salisbury, Zambezi Valley, Sinoia, SE Lowveld, Hartley, and Umvuma	Brain 1965
	Widespread	Smithers 1966
	In Victoria Falls N.P. (in NW corner)	I.U.C.N. 1971
	In Mana Pools Game Reserve, middle Zambezi Valley	Jarman 1972b
	Throughout except mountains; stable or increasing	Attwell, pers. comm. 1974
	Throughout (map)	Smithers, pers. comm. 1974
	In Wankie N.P. (W), abundant along Lukozi R.	Wilson 1975
Botswana	Restricted to N and E (map)	Smithers 1968
	In Chobe N.P. (N)	Sheppe and Haas 1976
South Africa	In Lake Simbu area, Tongaland	Davis 1957
	In Ndumu Game Reserve (NE Natal)	Paterson et al. 1964, DeMoor and Steffens 1972
	Common in Kruger N.P. (map)	Pienaar 1967; Pitchford et al. 1973, 1974
	Throughout NE, E, S, and SW (map)	Gartlan and Brain 1968
	Common throughout Hluhluwe and Umfolozi Game Reserve and corridor (E Natal)	Bourquin et al. 1971
	In Oribi Gorge Nature Reserve (in SE)	I.U.C.N. 1971
	In Mountain Zebra N.P. (in S near 32°S, 25°E)	Nel and Pretorius 1971
	In Durban town	Basckin and Krige 1973
	Widely distributed in Natal along coast and midlands to foothills of Drakensberg (map)	Pringle 1974
	In W Orange Free State mainly along Vaal R. and tributaries (map)	Lynch 1975

Table 96 (continued)

Country	Presence and Abundance	Reference
South Africa (cont.)	In Addo Elephant N.P. (55 km NNE of Port Elizabeth)	Swanepoel 1975
	In Tsitsikama N.P. (in S, E of Knysna)	Robinson 1976
Swaziland	Throughout (map)	Gartlan and Brain 1968
Lesotho	Throughout (map)	Gartlan and Brain 1968
Namibia	On inland plateau and along Kunene and Orange Rivers	Coetzee 1969
	Scarce in general; stable along Kunene R. (NW); declining slowly (5% since 1963) in Okavango R. area (N); also in Caprivi Strip (NE) and along Orange R. (in S) (map)	Coetzee and Joubert, pers. comm. 1974
	Recorded along Orange R., Kunene Valley, Omuhonga R., Hondoto R., and Namutoni area of E Etosha N.P.; total population probably not greater than 500	Joubert and Mostert 1975
Angola	Not frequent in Luando Reserve	Cabral 1967
	Throughout S except coastal strip (map)	Gartlan and Brain 1968
Zaire	W of Lake Edward (map)	Dandelot 1965
	In extreme E and S (maps)	Gartlan and Brain 1968, Dorst and Dandelot 1969
	Abundant E and NE, rare elsewhere in Haut-Zaire Province (in N)	Heymans 1975
	In Kundelungu N.P. (in SE)	I.U.C.N. 1975
Central African Republic	Throughout except SW (map)	Gartlan and Brain 1968
Congo	In S, recorded at Mindouli (W of Brazzaville)	Malbrant and Maclatchy 1949
Gabon	Recorded at Mayumba on SE coast	Malbrant and Maclatchy 1949
	Along coast (map)	Gartlan and Brain 1968
Cameroon	Present from Kak (4°52′N) in S to Waza (11°25′N) in N	Gartlan 1973
	Fairly abundant at center of range, rarer at N and S ends (map); range may be expanding; population stable or increasing; in Waza N.P., Bouba-Ndjida N.P., Bénoué N.P., and Faro Reserve	Gartlan, pers. comm. 1974
	Infrequent in Bouba-Ndjida N.P. (NE)	Lavieren and Bosch 1977
	Common in Kimbe R. and Mbi Crater game reserves	Asibey 1978

Table 97. Population parameters of *Cercopithecus aethiops*

Population Density (individuals /km^2)	Group Size X̄	Group Size Range	Group Size No. of Groups	Home Range Area (ha)	Study Site	Reference
	11.8	4–16	5	19–20	SE Senegal	Dunbar 1974
	45.8	10–140			S Senegal	Galat and Galat-Luong 1976
		33	1	138	N Senegal	Galat and Galat-Luong 1977
		≤15	8–10		SE Ghana	Rice, pers. comm. 1978
		1–14			SW Niger	Poché 1976
		≤35			N Nigeria	Howell 1968
73 (text) 135 (table)			4	19.5–29.4	Ethiopia	Dunbar and Dunbar 1974a
	20–30	6–60			Uganda, Kenya, Tanzania	Kingdon 1971
	16	2–25	9		W Uganda	Hall 1965b
88.8		6–21	46		S Uganda	Gartlan and Brain 1968
22–87	12–18		49	15.5	Uganda	Gartlan 1973
		43	1		W Uganda	Hayashi 1975
16.9–153.7	24	7–53	9	18.1–77.7	S Kenya	Struhsaker 1967a,b
19.6–108.5	17.9	13–26	9	18.1–77.7	S Kenya	Struhsaker 1973
86					NW Kenya	Coe 1972
0.87					Central Kenya	Kutilek 1974
		14	1		SE Kenya	Moreno-Black and Maples 1977
		9	1		N Tanzania	Lamprey 1964
		5–8			W Tanzania	Kano 1971a
		3–10			Central Mozambique	Dalquest 1965
		55	1	518	S Zambia	Lancaster 1971
				130	Zambia	Osborne cited by Ansell, pers. comm. 1974
5–40					Zambia	Swingland cited by Ansell, pers. comm. 1974
		20–30			Zimbabwe, Zambia, Malawi	Smithers 1966
		25–30			South Africa	Bourquin et al. 1971

Table 97 (continued)

Population Density (individuals /km²)	Group Size			Home Range Area (ha)	Study Site	Reference
	$\bar{X}$	Range	No. of Groups			
				9.4–57.8	South Africa	DeMoor and Steffens 1972
		19–38	2		South Africa	Basckin and Krige 1973
		15–25			South Africa	Robinson 1976
		5–30			Namibia	Coetzee and Joubert, pers. comm. 1974
	8–23		5		Cameroon	Gartlan 1973

Cercopithecus ascanius

Redtail, Coppertail, or Black-cheeked White Nosed Monkey

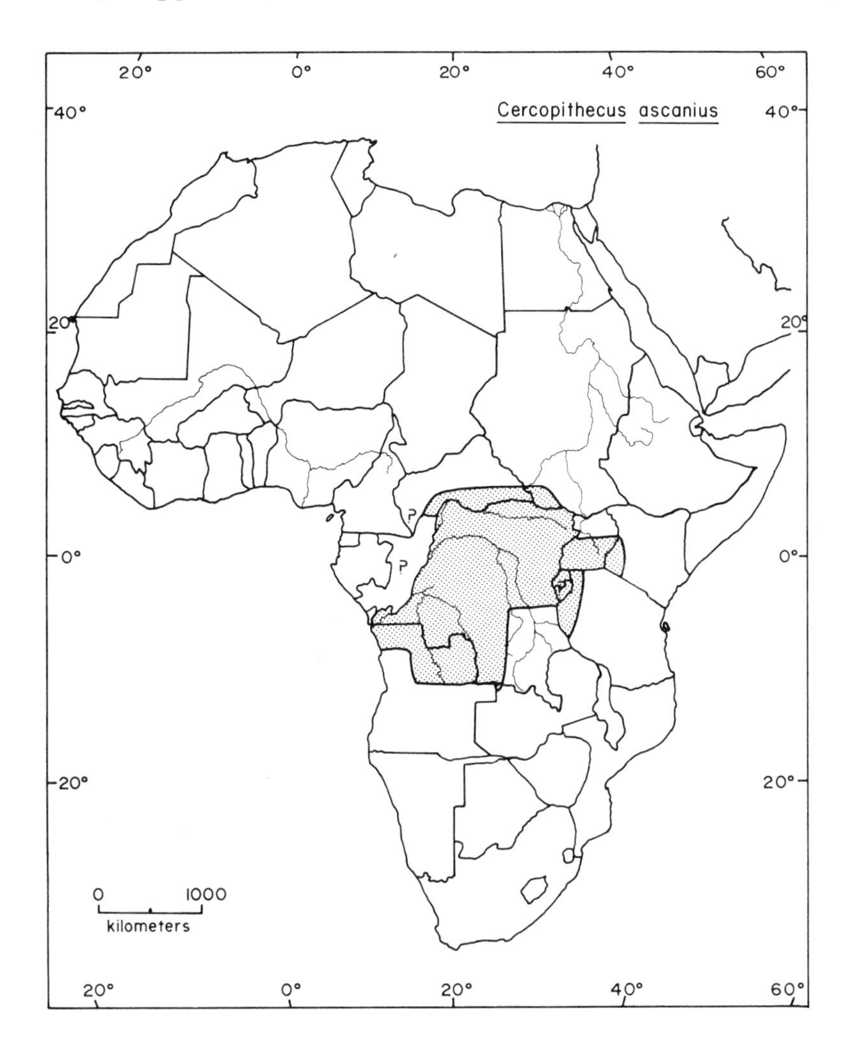

Taxonomic Note

Five races are recognized (Napier and Napier 1967).

Geographic Range

Cercopithecus ascanius is found in the Congo basin of central Africa from southern Sudan and Central African Republic to northern Zambia. Its northernmost record is near 6°15′N, 28°15′E (Kock 1969), and its southernmost populations are at about 11°30′S, 24°30′E (Ansell 1960a). *C. ascanius* occurs as far west as 13°30′E (at 8°S) in Angola (Barros Machado 1969) and as far east as 35°05′E (at 0°12′N) in Kenya (Pitman 1954). For details of the distribution, see Table 98.

Abundance and Density

Within the last 10 years, redtail monkeys have been described as abundant or common in parts of Uganda, Tanzania, and Zaire, and declining in Kenya (Table 98). Information is lacking regarding their abundance in the rest of their range.

Three estimates of population density are available, all for small localized populations (Table 99). Average group sizes in this species are between 7 and 35, and home ranges vary from 13 to 100 ha (Table 99).

Habitat

C. ascanius is a forest species, especially typical of lowland rain forest, and preferring dense tangled areas, secondary growth, riverine, lakeshore, and swamp forest (Haddow 1952). Several authors have observed it in wet forest (Curry-Lindahl 1956, Heymans 1975) and riparian forest (Rahm 1966, Smithers 1966, Barros Machado 1969, Kano 1971a). In Uganda the highest densities of this species are found in areas with a mixture of high forest and swamp forest (Struhsaker, pers. comm. 1978).

Redtail monkeys also occur in dry evergreen forest (Ansell 1960a), *Acacia* woodland and woodland-forest mosaic (Itani and Nishida, pers. comm. 1974), and disturbed deciduous forest near old plantations (Galat-Luong 1975). They may also be found in young forest, in plantations, and in montane areas up to about 2,000 m, as long as tall trees are present (Haddow 1952).

Factors Affecting Populations

Habitat Alteration

C. ascanius populations are declining in Kenya, because the forest areas available to them are being logged, cleared for farms, and replaced with exotic vegetation (Rowell, pers. comm. 1974). Near Lake Tumba in western Zaire their habitats are being decreased by slash and burn agriculture, but by 1974 no large-scale commercial logging had seriously affected them (Horn, pers. comm. 1978). In Uganda, redtail monkeys are much less frequently encountered in felled and herbicide treated forests than in undisturbed areas (Oates 1977a).

Haddow (1952) reported that one group of *C. ascanius* established itself in a eucalyptus plantation several miles away from any true forest.

Human Predation

Hunting. Redtail monkeys are hunted for their meat in parts of Uganda (Haddow 1952) and Zaire (Ladnyj et al. 1972, Heymans and Meurice 1973). The skins of their tails are used for covering bow staves by several tribes in Uganda (Haddow 1952), and they are hunted for their fur in Tanzania (Kano 1971a).

Pest Control. C. ascanius is notorious as a crop raider and can derive its entire subsistence from crops in places where there is little forest left (Haddow 1952). It has been reported to steal crops including bananas, millet, maize, beans, pumpkins, pineapples, passion fruits, and roots (Kingdon 1971) in Uganda (Haddow 1952; Pitman 1954; Struhsaker, pers. comm. 1974), Tanzania (Kano 1971a) (although seldom), Zambia (Smithers 1966), and Angola (Barros Machado 1969). It also feeds in old mango and guava plantations in the Central African Republic (Galat-Luong 1975).

Crop raiding has led to efforts at control, especially in East Africa where it is considered vermin (Kingdon 1971; Struhsaker, pers. comm. 1974).

Collection. This species has been used as a laboratory subject in studies of viral diseases at the East African Virus Research Institute (Kingdon 1971). It is not collected for any purpose in Zambia (Ansell, pers. comm. 1974).

Between 1968 and 1973, 10 *C. ascanius* were imported into the United States (Appendix). None entered the United States in 1977 (Bittner et al. 1978) or the United Kingdom in 1965–75 (Burton 1978).

Conservation Action

C. ascanius is protected within two areas in Uganda and one proposed national park in Tanzania (Table 98). In Kenya the government is attempting to protect the species at the eastern limit of its range by establishing forest reserves such as the Kakamega Nature Reserve and the Mount Elgon Game Reserve (Rowell, pers. comm. 1978). The redtail monkey's range occurs outside the national parks in Zambia, and no laws protect it there (Ansell, pers. comm. 1974). No information was located regarding its occurrence in habitat reserves in other countries, nor laws enacted concerning it.

Table 98. Distribution of *Cercopithecus ascanius* (countries are listed in clockwise order)

Country	Presence and Abundance	Reference
Sudan	In extreme S (localities mapped)	Kock 1969
Uganda	At Bwamba (in W)	Lumsden 1951
	Throughout S (localities mapped); abundant	Haddow 1952
	Abundant on Entebbe Peninsula, Lake Victoria	Buxton 1952 [McCrae et al. 1971]
	Numerous in S and W (localities mapped)	Kingdon 1971
	In Budongo Forest (in W) and Mabira Forest (in S)	Rahm 1972
	Fairly abundant in Kibale Forest (in W)	Struhsaker 1972, 1975 [Hayashi 1975, Mizuno et al. 1976]
	N limit is 2°N (map)	Struhsaker, pers. comm. 1974
	Fair numbers in all larger forests of W; status: good; protected in nature reserve in Kibale Forest and in Maramagambo Forest of Ruwenzori N.P.	Struhsaker, pers. comm. 1978
Kenya	In SW (locality mapped)	Haddow 1952
	E limit is Kapsabet (0°12′N, 35°05′E), E edge Kakamega Forest	Pitman 1954
	In Kakamega Forest (in W)	Rahm 1972
	Only in 1 or 2 forests; easternmost is Kakamega; declining	Rowell, pers. comm. 1974

Table 98 (continued)

Country	Presence and Abundance	Reference
Tanzania	Abundant in Kasakati Basin, Kasoge and Kobogo, uncommon elsewhere in Kigoma and Mpanda districts (in W) (localities mapped)	Kano 1971a
	Only in NW and W (localities mapped)	Kingdon 1971
	In Mahali Mts. proposed N.P. S of Kigoma	Itani and Nishida, pers. comm. 1974
Rwanda	No information	
Burundi	No information	
Zambia	Fairly common in NW Mwinilunga District (in N)	Ansell 1960a
	Confined to Mwinilunga District (map)	Smithers 1966
	Confined to small corner of Mwinilunga District; population probably stable	Ansell, pers. comm. 1974
Angola	In N half (N of 11°S) (localities mapped)	Barros Machado 1969
Congo	Questionable W of Congo R. (map)	Dorst and Dandelot 1969
Zaire	In N and E (localities mapped)	Haddow 1952
	On Mount Kahuzi, W of Lake Kivu (in E)	Curry-Lindahl 1956, Rahm 1965
	W of Lake Kivu	Rahm and Christiaensen 1963
	In N, NE, and E; fairly common in Irangi region (W of Lake Kivu)	Rahm 1966
	Throughout except in SE (map)	Rahm 1970
	In Basankusu Territory, Equateur Province (in NW)	Ladnyj et al. 1972
	Abundant throughout Haut-Zaire Province (in N); fairly common on left bank of Zaire R.	Heymans 1975
	In mountains W of Lake Kivu	Hendrichs 1977
	Common W of Lake Tumba (in W); less common E of lake	Horn, pers. comm. 1978
Central African Republic	Present	Kingdon 1971
	Near town of Bangui, on Ubangi R. (in S)	Galat-Luong 1975

Table 99. Population parameters of *Cercopithecus ascanius*

Population Density (individuals /km^2)	Group Size			Home Range Area (ha)	Study Site	Reference
	X̄	Range	No. of Groups			
190		35	1		S Uganda	Buxton 1952
		12–23	2	13–15	W Uganda	Hayashi 1975
65	15			25	W Uganda	Struhsaker 1975
	11.6–13.3	8–>20	6	13.5–21	W Uganda	Mizuno et al. 1976
125–140	30–35			25	W Uganda	Struhsaker, pers. comm. 1978
		18–22	2	100	W Kenya	Rowell, pers. comm. 1974
	7–8	4–15			W Tanzania	Kano 1971a
		>50			E Zaire	Rahm 1966
		15–20		15	C.A.R.	Galat-Luong 1975

Cercopithecus cephus Moustached Monkey

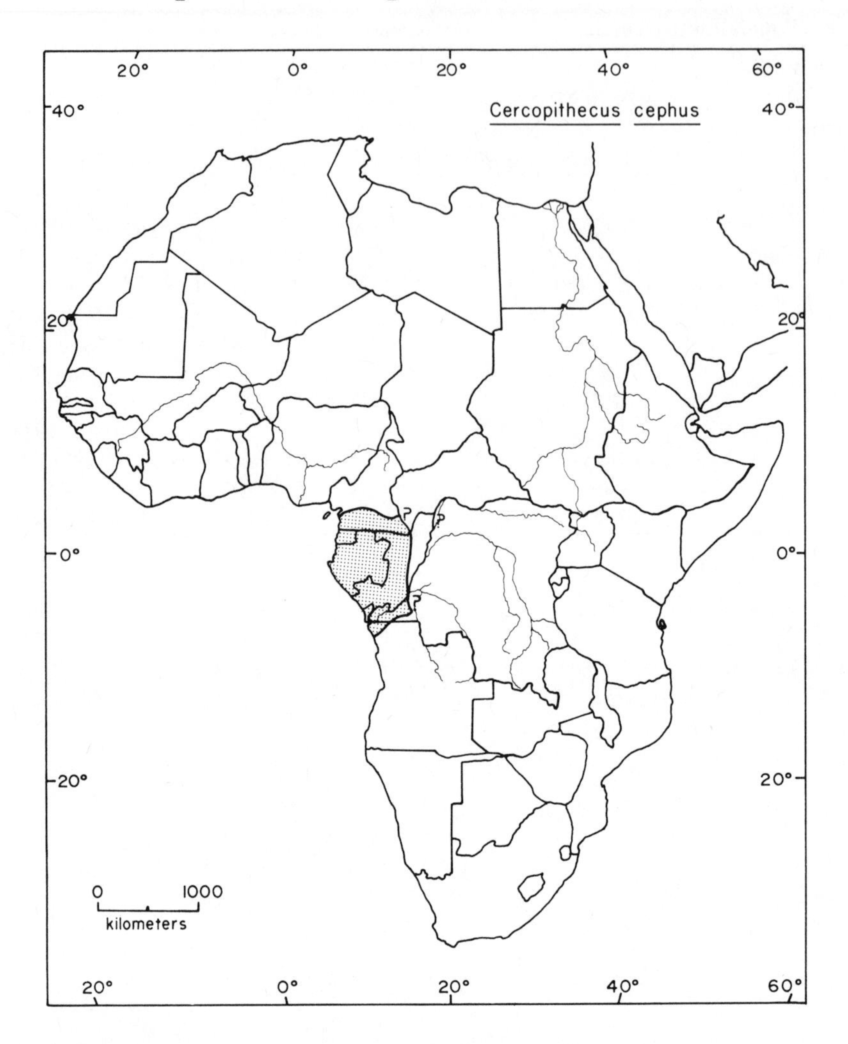

Taxonomic Note

Booth (1958) included *Cercopithecus erythrotis* within the *Cercopithecus cephus* group. The zone of hybridization between these two forms in Cameroon indicates that they are closely related and perhaps conspecific, but the hybridization may be recent and due to habitat disturbance (Gartlan, pers. comm. 1974). Thus *C. cephus* and *C. erythrotis* are treated separately in this report. Two subspecies are recognized (Napier and Napier 1967).

Geographic Range

C. cephus is found from southern Cameroon to northern Angola. Its northern limit is the Nyong River at about 3°30′N (Gartlan, pers. comm. 1974), and the southernmost record is near 7°S on the coast of Angola (Barros Machado 1969). In the north the eastern limit of the range is not known. It may be near the southward bend of the Ubangi River (Tappen 1960) or farther east in northern Zaire. Most authors have reported that *C. cephus* is restricted to the right bank of the Congo River, but Malbrant and Maclatchy (1949) showed the range extending into Zaire. For details of the distribution, see Table 100.

Abundance and Density

C. cephus has been described as the most common monkey in Gabon and Congo (Malbrant and Maclatchy 1949) and as generally "very common" (Tappen 1960).

More recently this species has been judged to be declining in numbers in Cameroon and parts of northeast Gabon, and its range has been reduced in Equatorial Guinea (Table 100). Its status in the other countries of its range is not known, and its very presence in the eastern areas has not been verified.

In Gabon, three groups of 5, 6, and 10 individuals each ($\bar{X} = 7$) (Gautier and Gautier-Hion 1969), and two groups of 3 to 5 and 8 monkeys, with home ranges of 18 and 45 ha and population densities of 39 and 22 individuals per km^2 (Gautier-Hion and Gautier 1974), have been reported. In Cameroon three groups with an average size of 3.7 members (range 2 to 5) were seen (Struhsaker 1969), but this species often lives in groups of 30 to 35 individuals (Gartlan, pers. comm. 1974).

Habitat

C. cephus is a rain forest species (Struhsaker 1969, Quris 1976). It is found in primary forest (Gautier and Gautier-Hion 1969) and also in gallery and secondary forest (Malbrant and Maclatchy 1949) and seems to prefer younger forest to more mature types (Tappen 1960, Struhsaker 1969). It also utilizes periodically flooded forest (Gautier and Gautier-Hion 1969, Quris 1976) and dense regenerating forest (Sabater Pi and Jones 1967). Malbrant and Maclatchy (1949) contended that it was dependent on oil palm (*Elias*) fruit and was rare or absent in areas lacking these trees.

Factors Affecting Populations

Habitat Alteration

The range of *C. cephus* in Equatorial Guinea has been reduced by the activities of man (Sabater Pi and Jones 1967). In the other countries of its range it is probably also being affected by deforestation, but perhaps not as severely as other species because of its ability to utilize secondary growth (Gartlan, pers. comm. 1974).

Human Predation

Hunting. Extensive hunting has been responsible for the decline of *C. cephus* in northeast Gabon (Gautier and Gautier-Hion 1969). This species is also shot for meat in Cameroon (Gartlan, pers. comm. 1974) and in Equatorial Guinea (Sabater Pi and Jones 1967, Sabater Pi and Groves 1972).

Pest Control. *C. cephus* is not reputed to be a crop raider (Tappen 1960), but it does steal food from farms in Cameroon (Gartlan, pers. comm. 1974). No information was located regarding efforts to control this activity.

Collection. *C. cephus* is not a common subject in medical research, but is occasionally exported from Cameroon for zoo exhibits (Gartlan, pers. comm. 1974). A total of 288 were exported from Equatorial Guinea between 1958 and 1968 by two animal dealers (Jones, pers. comm. 1974). Fifteen were imported into the United States between 1968 and 1973 (Appendix). None entered the United States in 1977 (Bittner et al. 1978). In 1965–75, 278 were imported into the United Kingdom (Burton 1978).

Conservation Action

C. cephus occurs in three forest reserves in Cameroon (Table 100).

Table 100. Distribution of *Cercopithecus cephus* (countries are listed in counterclockwise order)

Country	Presence and Abundance	Reference
Cameroon	In S and E (map)	Malbrant and Maclatchy 1949
	At Campo Reserve and Dipikar Island (in SW) and at Dja Reserve (S central)	Gartlan and Struhsaker 1972
	Throughout S, as far N as Nyong R.; between Nyong and Sanaga R. hybridizing with *C. erythrotis*; in Douala-Edéa, Dipikar, and Dja reserves; declining but not sharply	Gartlan, pers. comm. 1974
Equatorial Guinea	Throughout (map); common	Malbrant and Maclatchy 1949
	Throughout except coastal areas (map); range shrinking	Sabater Pi and Jones 1967
	In Monte Raices Territorial Park (in NW, E of Bata)	I.U.C.N. 1971
Gabon	Throughout (map); common	Malbrant and Maclatchy 1949
	Declining in Ntsi-Belong area (in NE)	Gautier and Gautier-Hion 1969
	In NE, 10 km SW of Makokou	Gautier-Hion and Gautier 1974
	Along Liboui River (in NE)	Quris 1976
Congo	Throughout (map); common	Malbrant and Maclatchy 1949
Angola	In extreme NW and Cabinda (localities mapped)	Barros Machado 1969
Zaire	Throughout W (map)	Malbrant and Maclatchy 1949
Central African Republic	In SW (map)	Malbrant and Maclatchy 1949

Cercopithecus diana Diana Monkey

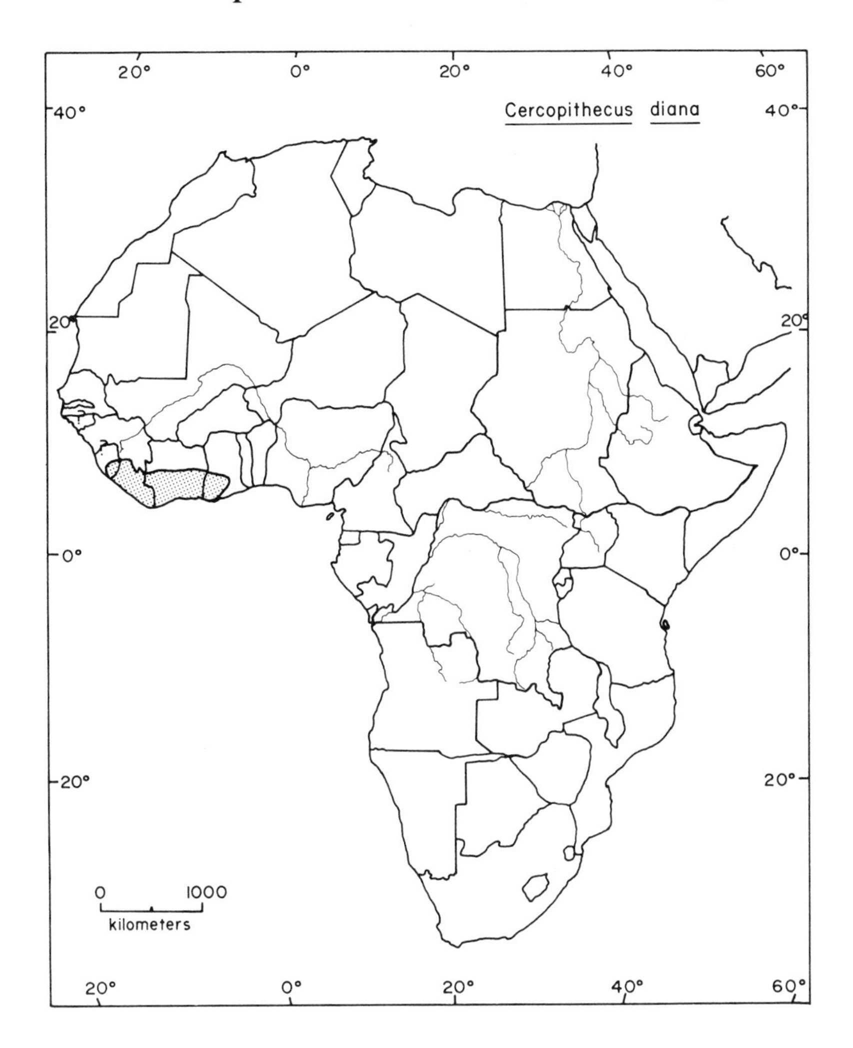

Taxonomic Note

Most authors recognize two races of diana monkeys (Booth 1956a,b; 1958).

Geographic Range

Cercopithecus diana is found in West Africa, between Gambia and Ghana. Tappen (1960) placed its northwestern limit at 13°N, 17°W in Gambia, but no other authors mention it farther north or west than 8°20′N, 12°W in south central Sierra Leone (Tappen 1964). The species occurs as far east as 96.5 km west of the Volta River (about 1°W) (Booth 1956b). Isolated old reports of diana monkeys in Cameroon and Zaire have never been confirmed and should probably be discounted. For details of the distribution, see Table 101.

Abundance and Density

Diana monkeys were already considered rare in Ghana, Ivory Coast, and Liberia more than 20 years ago. Recent information indicates that they have now become even more rare in these countries and are also rare in Sierra Leone (Table 101). Struhsaker (pers. comm. 1974) believed that this species was one of the most threatened in Africa, and the present evidence supports his conclusion.

C. diana is often found in fairly large groups of 10 to 50 (Monfort and Monfort 1973), about 30 (Tahiri-Zagrët 1976), and 14, 20, and 35, with the group of 14 occupying a home range of 250 ha (Curtin and Curtin, pers. comm. 1978). No other population data were located.

Habitat

The typical habitat of *C. diana* is high, dense, undisturbed forest. Booth (1956a) found that this species was restricted entirely to mature primary forest, while Bourlière et al. (1974) observed it in both primary and secondary forest. There are records of *C. diana* in newly cleared farmland (Jeffrey 1975b), gallery forest (Tahiri-Zagrët 1976), and undisturbed semideciduous forest and rain forest (Curtin and Curtin, pers. comm. 1978).

Factors Affecting Populations

Habitat Alteration

The decline in *C. diana* populations is due in large degree to the destruction of the forest on which the species depends. Booth (1956a) described eastern Ghana as already "densely populated and heavily exploited for timber and agriculture." Deforestation proceeded rapidly in Ghana, and by the 1950s only about one-third of the original forest cover remained (Asibey 1978). During the 1960s felling of forest for farms continued, and Ghanaian forest outside reserves was reduced from 80% to 20% between 1960 and 1973 (Jeffrey 1974). Even in the western part of the country, this process, combined with extensive burning, road building, and erosion, will soon destroy all primary forest outside reserves (Jeffrey 1975b). This deforestation is eliminating *C. diana* from large areas of Ghana (Jeffrey 1975b; Asibey 1978; Curtin and Curtin, pers. comm. 1978). Similarly, the high rate of destruction of forest in Ivory Coast (Lanly 1969) poses a major threat to the survival of the species there.

Human Predation

Hunting. *C. diana* is hunted for its flesh in Liberia (Coe 1975) and Ghana (Asibey 1974, Jeffrey 1975b, Rucks 1976), and for its flesh and fur in Ivory Coast (Tahiri-Zagrët 1976).

Pest Control. This species was never mentioned as a crop raider; in fact, it was cited as being one monkey that is *not* destructive to crops in Sierra Leone (Robinson 1971). Nevertheless, it was occasionally captured in the government-sponsored drives to eliminate monkeys from coffee and cacao growing regions in that country (Tappen 1964).

Collection. The striking coloration of diana monkeys makes them more likely to be sought for zoological display than dull colored species. A total of 56 were imported into the United States between 1968 and 1973 (Appendix). None entered the United States in 1977 (Bittner et al. 1978). In 1965–75, 443 were imported into the United Kingdom (Burton 1978).

Conservation Action

C. diana occurs in three national parks and three forest reserves (Table 101). The Tai National Park in Ivory Coast is the last remaining sizable area of high forest in that country, and is especially crucial for the survival of *C. diana* (Struhsaker 1972, Monfort and Monfort 1973). Similarly, the continued existence of dianas in Ghana is wholly dependent on the preservation of the protected areas where viable populations remain (Curtin and Curtin, pers. comm. 1978).

In Liberia it is illegal to shoot or trap a lactating female *C. diana* inside a forest reserve (Curry-Lindahl 1969). In 1973 Ivory Coast banned hunting, but badly needs better surveillance, popular education, and park management if its conservation measures are to succeed (Tahiri-Zagrët 1976).

The diana monkey is in Class A of the African Convention (Curry-Lindahl 1969) and is listed as endangered under the U.S. Endangered Species Act (U.S. Department of Interior 1977b).

Table 101. Distribution of *Cercopithecus diana* (countries are listed from west to east)

Country	Presence and Abundance	Reference
Gambia	To 13°N, 17°W	Tappen 1960
Senegal	No information	
Guinea-Bissau	No information	
Guinea	No information	
Sierre Leone	Present	Mackenzie 1952
	In Kasewe Forest, about 160 km E of Freetown (S central)	Tappen 1964
	Rare	Jones 1966
Liberia	To W border (map); rare	Booth 1956b, 1958
	Present	Kuhn 1965
	Previously reported common, now rare in Mount Nimba area (extreme E)	Coe 1975
Ivory Coast	Throughout S (map); rare	Booth 1956b, 1958
	Abundant in Tai Reserve (in W between Cavally R. and Sassandra R.)	Struhsaker 1972 [Monfort and Monfort 1973]
	Previously in Lamto region (S central), now absent	Bourlière et al. 1974
	Threatened	Struhsaker, pers. comm. 1974
	In S between Sassandra and Bandama rivers (map); becoming rare; going extinct	Tahiri-Zagrët 1976

Table 101 (continued)

Country	Presence and Abundance	Reference
Ghana	Only in SW; rarest monkey in Ghana (localities mapped)	Booth 1956a,b; 1958
	In Sukusuku Forest Reserve (in SW)	Jeffrey 1970
	Rare	Jeffrey 1974
	In Bia Tributaries North and Area 14 forest reserves (in SW)	Jeffrey 1975a,b
	Numerous in Bia N.P.	Rucks 1976
	Abundant in Bia N.P., Nini-Suhien N.P., and Ankasa Game Production Reserve	Asibey 1978
	Scarce and declining in Bia N.P.	Curtin and Curtin, pers. comm. 1978

Cercopithecus erythrogaster Red-bellied Monkey

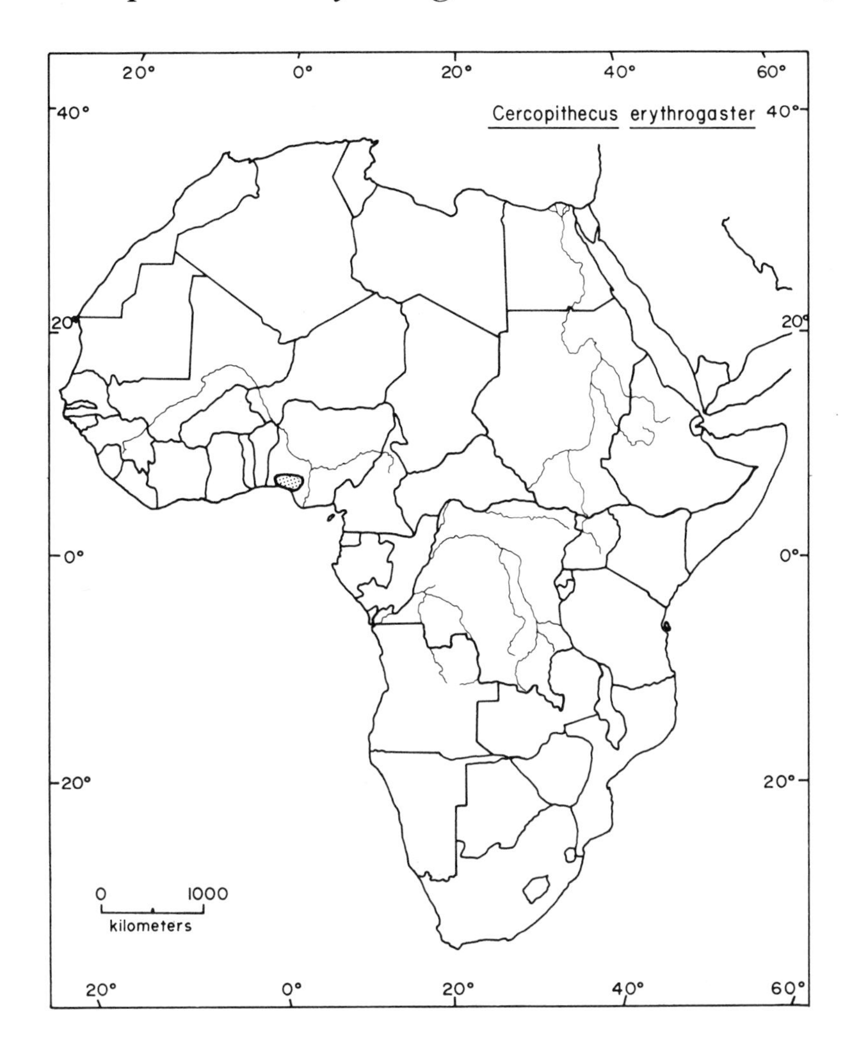

Taxonomic Note

This is a monotypic species (Napier and Napier 1967).

Geographic Range

Cercopithecus erythrogaster is confined to western Nigeria, west of the Niger River. The limits of the range have been given as 3°E to 7°E (Sanderson 1940) and 4°E to 6°E (Rosevear 1953; both cited in Tappen 1960). The northern limit is about 7°N (Booth 1958). No details of the distribution are available.

Abundance and Density

No information was located regarding either the past or present abundance of this species. It is "a rare and puzzling monkey . . . known only from a few skins"

(Dorst and Dandelot 1969). There seems to be no evidence regarding *C. erythrogaster*'s present existence, and this form may now be extinct (Gartlan, pers. comm. 1974).

Habitat

The habitats of *C. erythrogaster* have been described as moist evergreen and semideciduous forests (Booth 1958), coastal forest (Tappen 1960), and secondary high forest (Dorst and Dandelot 1969).

Factors Affecting Populations

No information.

Conservation Action

C. erythrogaster is listed as endangered under the U.S. Endangered Species Act (U.S. Department of Interior 1977b).

Cercopithecus erythrotis Red-eared Nose-spotted Monkey

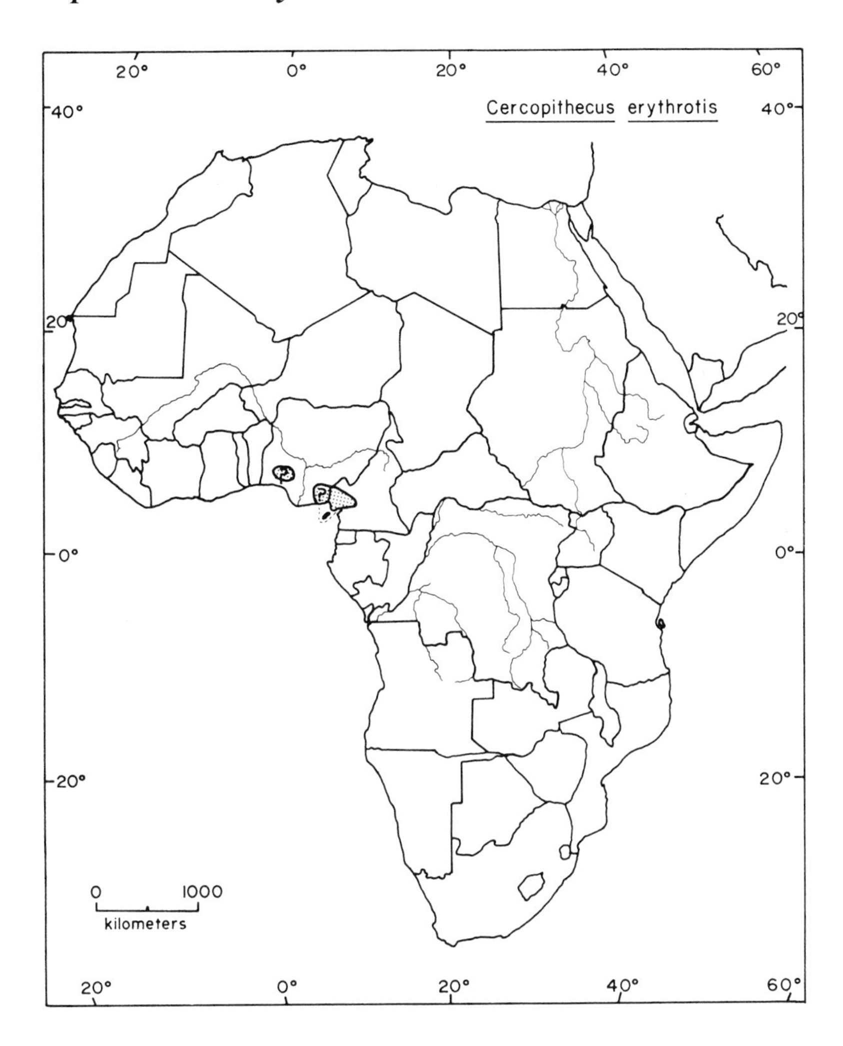

Taxonomic Note

Cercopithecus erythrotis consists of three subspecies (Napier and Napier 1967). The zone of hybridization between *C. erythrotis* and *C. cephus* may indicate that these two forms are conspecific or may be due to habitat disturbance (Gartlan, pers. comm. 1974).

Geographic Range

C. erythrotis is found in southern Nigeria and western Cameroon. It has also been recorded on Macias Nguema (formerly Fernando Po) (Sanderson 1940, cited in Tappen 1960). Booth (1958) showed the northern and western limits at about 8°N, 4°E, but expressed doubts about the Nigerian records. In Cameroon the southern limit is near 3°30′N near the mouth of the Sanaga River, and the easternmost record is near 4°N, 11°E (Gartlan, pers. comm. 1974). For details of the distribution, see Table 102.

Abundance and Density

In 1958, Booth suggested that since so little information was available regarding *C. erythrotis*, it must not be a very common form. It is reputed to be rare, and suspected to have declined recently in Nigeria (Monath and Kemp 1973; Gartlan, pers. comm. 1974), but no direct evidence was located regarding its present abundance there. It is declining, but not drastically, in Cameroon (Table 102).

Group sizes in *C. erythrotis* have been reported as 4 to more than 29 (N = 4) (Struhsaker 1969), and 30 to 35 (Gartlan, pers. comm. 1974). No estimates of population density or home range sizes are available.

Habitat

C. erythrotis is a rain forest species which utilizes both primary and secondary forest (Gartlan, pers. comm. 1974). It can be very abundant in mature forest and older secondary forest, and also occurs in seasonally flooded forest (Gartlan and Struhsaker 1972). It seems to prefer younger types of forest to more mature ones (Struhsaker 1969).

Factors Affecting Populations

Habitat Alteration

The habitat of *C. erythrotis* is being destroyed in Cameroon by "excessive and inefficient timber operations" (Howard, pers. comm. 1974) and rapid deforestation, especially in coastal areas (Gartlan, pers. comm. 1974). In eastern Nigeria, deforestation and the encroachment of human populations have resulted in the decline of all forest primate populations (Monath and Kemp 1973).

Human Predation

Hunting. This species is hunted for meat in Cameroon (Gartlan, pers. comm. 1974; Howard, pers. comm. 1974). Monkey populations in eastern Nigeria were greatly reduced by hunting during the civil war of 1967–70 (Monath and Kemp 1973).

Pest Control. *C. erythrotis* does raid crops, including cacao (Gartlan, pers. comm. 1974), but no information was located concerning efforts to control this species.

Collection. Three *C. erythrotis* were imported into the United States between 1968 and 1973 (Appendix). None entered the United States in 1977 (Bittner et al. 1978) or the United Kingdom in 1965–75 (Burton 1978).

Conservation Action

C. erythrotis occurs within at least four forest reserves in Cameroon (Table 102). It is also listed as endangered under the U.S. Endangered Species Act (U.S. Department of Interior 1977b).

Table 102. Distribution of *Cercopithecus erythrotis*

Country	Presence and Abundance	Reference
Nigeria	In W and SE (map)	Rosevear 1953 cited in Booth 1958
	In Ikom-Mamfe district, 5–7°N, 8–10°E (extreme SE)	Tappen 1960
	Considered rare	Monath and Kemp 1973
Cameroon	In W as far S as Sanaga R. (map)	Rosevear 1953 cited in Booth 1958
	At Idenau, Elongongo and Nsépé on Sanaga R., Lake Tisongo (mouth of Sanaga R.), and Bakundu Reserve (N of Mount Cameroon)	Gartlan and Struhsaker 1972
	Abundant in Douala-Edéa and Korup reserves (in W)	Struhsaker 1972
	In W as far S as Nyong R. (map); hybridizing with *C. cephus* between Nyong R. and Sanaga R.; declining but not sharply	Gartlan, pers. comm. 1974
	Gradually declining in Mungo Forest Reserve (in W)	Howard, pers. comm. 1974
Equatorial Guinea	Present on Macias Nguema	Sanderson 1940 cited in Tappen 1960; Gartlan, pers. comm. 1974

Cercopithecus hamlyni Owl-faced or Hamlyn's Monkey

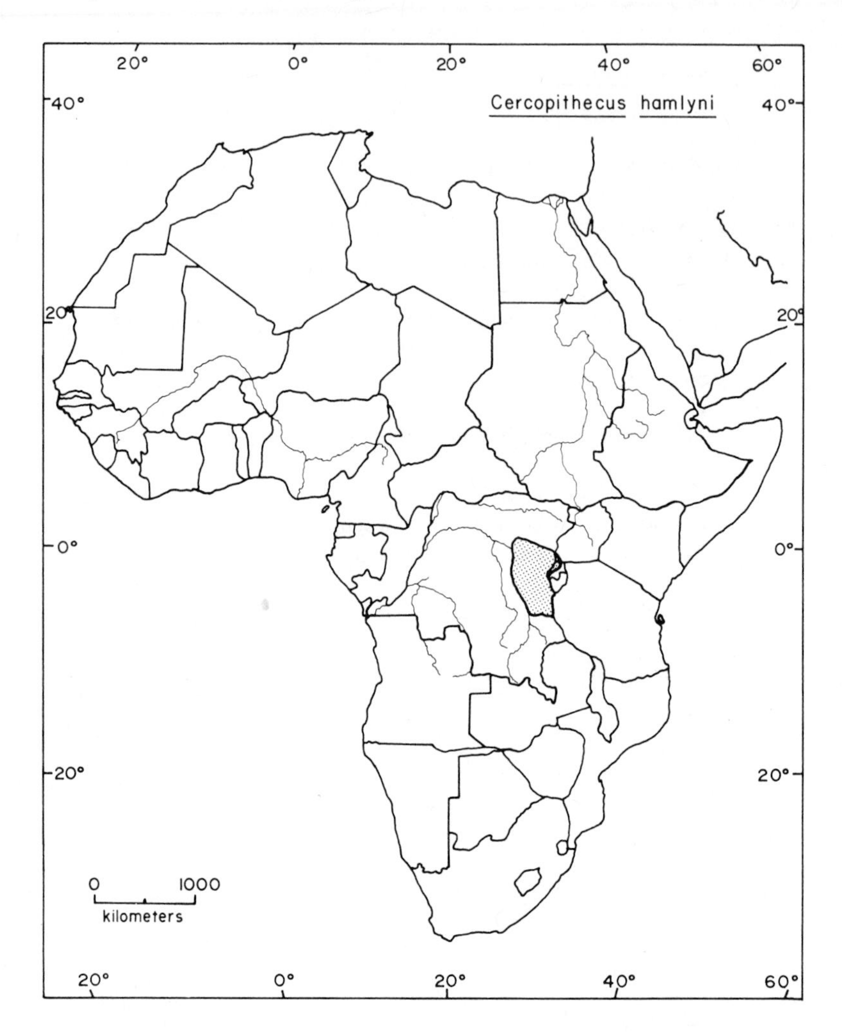

Taxonomic Note

This is a monotypic species (Napier and Napier 1967).

Geographic Range

Cercopithecus hamlyni is found in the eastern Congo basin, from the Zaire-Lualaba River to southwestern Uganda. The northern limit of its range is at the Epulu River (1°30′N), the southern limit is the Lukuga River, near 6°S, the western boundary is the Zaire River (near 26°E), and the eastern limit is the Kabale Forest, Uganda (1°15′S, 29°58′E) (Rahm 1970). For details of the distribution, see Table 103.

Abundance and Density

C. hamlyni seems to be fairly rare in parts of Zaire (Table 103), but information is lacking regarding its abundance in most of its range. This species is said to travel in small groups of 10 members or less (Rahm 1970). No estimates of population density or home range size are available.

Habitat

Owl-faced monkeys are forest dwellers. They have been observed in bamboo and montane forest (Rahm and Christiaensen 1963), lowland forest (Rahm 1970), and "very dense forest" (Heymans 1975). This species seems characteristic of higher rather than lower elevations. It has been reported present only above 900 m (Hendrichs 1977), between 1,000 and 3,000 m (Rahm 1965), and up to 4,600 m (Dorst and Dandelot 1969).

Factors Affecting Populations

Habitat Alteration

No information.

Human Predation

Hunting. No information.

Pest Control. No information.

Collection. Two *C. hamlyni* were imported into the United States between 1968 and 1973 (Appendix). None entered the United States in 1977 (Bittner et al. 1978) or the United Kingdom in 1965–75 (Burton 1978).

Conservation Action

No information.

Table 103. Distribution of *Cercopithecus hamlyni*

Country	Presence and Abundance	Reference
Zaire	Between 26°–29°E and 2°N–4°S	Schouteden 1947 cited in Tappen 1960
	Not common in Kahuzi region, W of Lake Kivu	Rahm and Christiaensen 1963
	Much less rare than was thought, in Kahuzi region (localities mapped)	Rahm 1965
	Present between Lowa-Osa R. (1°S) and Elila R. (3°30′S), Lualaba-Zaire and Rift	Rahm 1966
	Between Zaire-Lualaba and Rift, Epulu and Lukuga rivers (localities mapped)	Rahm 1970
	Fairly rare in central Haut-Zaire Province (in N)	Heymans 1975
	Present in mountains W of Lake Kivu	Hendrichs 1977
Rwanda	In extreme NW (localities mapped)	Rahm 1965, 1970
Uganda	Present in Kabale Forest, extreme SW (locality mapped)	Rahm 1970

Cercopithecus l'hoesti L'Hoest's or Mountain Monkey

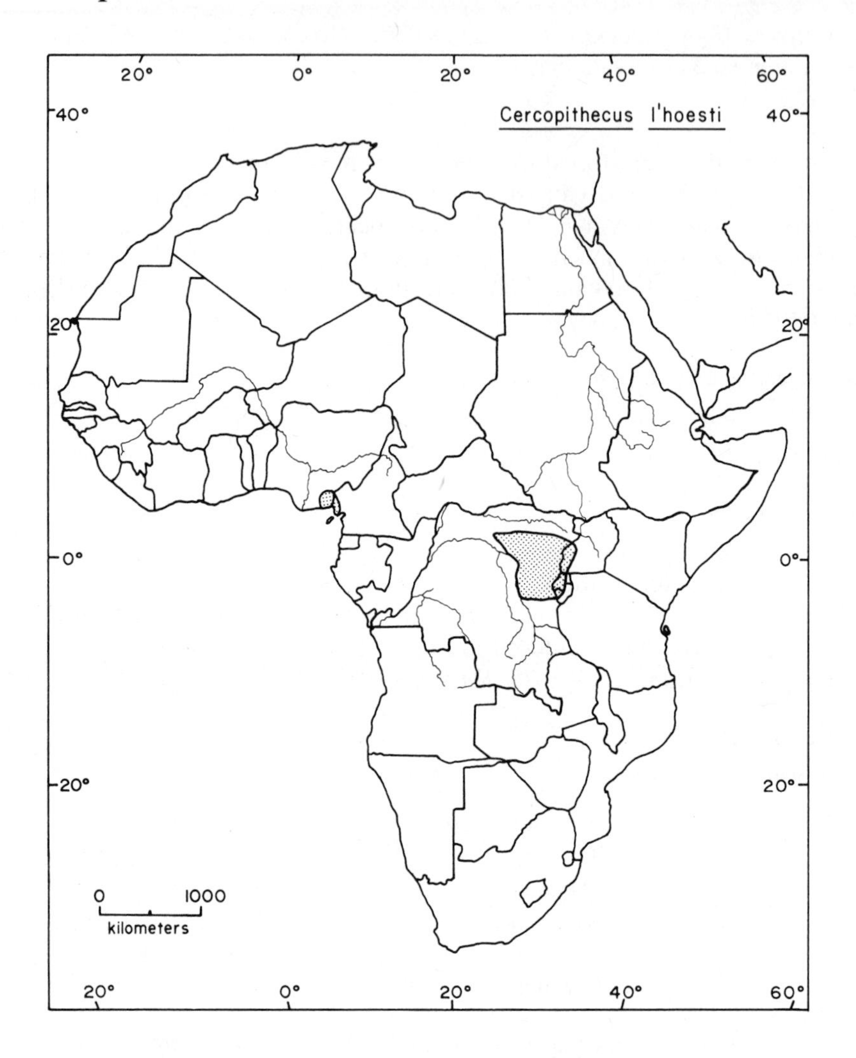

Taxonomic Note

Some authors consider the western race of this species (*Cercopithecus l'hoesti preussi*) and the eastern race (*C. l'hoesti l'hoesti*) to be separate species (Tappen 1960, Napier and Napier 1967). But most workers (e.g., Rahm 1970, Kingdon 1971) list them as conspecific, and they are treated as such here.

Geographic Range

C. l'hoesti has a disjunct distribution, with one population in the upper eastern Congo basin and southwest Uganda, and the other in western Cameroon and eastern Nigeria. Its eastern limit is about 31°E (Rahm 1970), and its western limit is near the Cross River at about 8°E (Booth 1958). In the east it occurs as far north as 2°N, just north of the Ituri River, and as far south as 2°30′S, south of the Ulindi River (Rahm 1970). The eastern and western populations are separated by a gap more than 1,609 km wide, in which no *C. l'hoesti* are known to occur. For details of the distribution, see Table 104.

Abundance and Density

The status of *C. l'hoesti* is unknown over most of its range (Table 104). In Cameroon it is locally common but in a precarious position because of its small and shrinking range. In the east it has been described as common in one area and uncommon in another, but its overall status has not been determined (Table 104).

The only population density estimate available indicates 5 to 8 *C. l'hoesti* per km^2 in parts of the Kibale Forest, Uganda (Struhsaker 1975). Group sizes for this species have been reported as 5 to 8 and rarely larger (Rahm and Christiaensen 1963), 2 to 7 ($\bar{X} = 3.7$, modes 2 and 3, $N = 15$) (Struhsaker 1969), average of 10 (Struhsaker 1975), and average of 17 (Møller cited by Kingdon 1971). This last study also revealed a home range of 700 to 1,000 ha for each *C. l'hoesti* group.

Habitat

C. l'hoesti is a forest monkey, typical of moist high primary forest (Booth 1958). The western population is essentially a montane form, found in high altitude rain forest above 1,000 m elevation, and older secondary as well as mature forest (Gartlan and Struhsaker 1972, Gartlan 1973). It has also been recorded in isolated forest patches in mountain grassland (Sanderson 1940 cited in Tappen 1960).

The eastern race occupies a wider range of habitat types. It is found in mature lowland rain forest (Struhsaker 1969), gallery forest (Rahm 1970), wooded savanna on mountain slopes, forest borders, and cultivated lands (Rahm and Christiaensen 1963). In the lowland forest, eastern *C. l'hoesti* seems to prefer areas of regenerating forest, and in montane areas it is also frequently found in tangled undergrowth beneath broken canopy (Kingdon 1971). It has been seen at 1,500 to 2,500 m elevation (Rahm 1965), and Hendrichs (1977) found it only above 900 m.

Factors Affecting Populations

Habitat Alteration

The forests near Mount Cameroon are presently being logged, drastically reducing the habitat available to *C. l'hoesti*, further compressing its range, and driving it farther up the mountain (Gartlan, pers. comm. 1974). No reports are available regarding effects of habitat alteration within the range of the Congo basin population.

Human Predation

Hunting. In Cameroon, where compression of its range has made *C. l'hoesti* more vulnerable to human predation, hunting and trapping for food are reducing its remaining populations (Gartlan, pers. comm. 1974).

In East Africa this species is also hunted for its skins, which are used as shoulder bags, but is not threatened by this hunting, because the rugged terrain makes access difficult (Kingdon 1971).

Pest Control. C. l'hoesti does eat cultivated foods (Pitman 1954, Tappen 1960), and the eastern race frequently steals bananas, peas, cassava, and black wattle (Kingdon 1971). In West Africa this species is not known as a crop pest (Gartlan, pers. comm. 1974).

Collection. L'Hoest's monkey is not in great demand as a zoo exhibit or subject for medical research. None were imported into the United States between 1968 and 1973 (Appendix) or in 1977 (Bittner et al. 1978). Six were imported into the United Kingdom in 1965–75 (Burton 1978).

Conservation Action

The western race of *C. l'hoesti* is in a precarious position and faces extinction unless some of its habitat on Mount Cameroon is protected. It is forbidden to hunt the species in the Lake Barombi Mbo Forest, but this law is not being enforced (Gartlan, pers. comm. 1974). *C. l'hoesti* may occur within the boundaries of the Korup Reserve in Cameroon (Gartlan, pers. comm. 1974). No information is available regarding conservation of the eastern form.

This species is listed as endangered under the U.S. Endangered Species Act (U.S. Department of Interior 1977b).

Table 104. Distribution of *Cercopithecus l'hoesti* (countries are listed in order from west to east)

Country	Presence and Abundance	Reference
Nigeria	In SE, E of Cross R. (map)	Booth 1958
	In E, S of about 7°N	Tappen 1960, Happold 1971
Cameroon	In W as far S as Sanaga R. (map)	Booth 1958
	On Mount Cameroon and Mount Victoria	Tappen 1960
	At Idenau and on Mount Cameroon	Gartlan and Struhsaker 1972
	Area of distribution less than 120 by 120 km	Gartlan 1973
	Fairly common around Mount Cameroon only, otherwise absent; may be in Rumpi Hills (near Kumba) (map); range being reduced; population probably decreasing; very vulnerable	Gartlan, pers. comm. 1974
Equatorial Guinea	On Macias Nguema (formerly Fernando Po)	Malbrant and Maclatchy 1949
Zaire	Between 3°N–4°S, 26–29°E	Tappen 1960
	W of Lake Kivu (in E)	Rahm and Christiaensen 1963
	In E Kahuzi (in E)	Rahm 1965
	E of Zaire-Lualaba R., from N of Ituri R. to Elila R. (localities mapped)	Rahm 1970
	Common throughout Haut-Zaire Province (in N)	Heymans 1975
	W of Lake Kivu	Hendrichs 1977
Burundi	Central and W (localities mapped)	Rahm 1970

Table 104 (continued)

Country	Presence and Abundance	Reference
Rwanda	In Rugege Forest, E of Lake Kivu	Rahm 1965
	In NW	Elbl et al. 1966
	In W (localities mapped)	Rahm 1970
Uganda	In isolated Ruwenzori and Kayonza forests in W; E limit about 31°E	Haddow et al. 1951 cited in Tappen 1960
	Common in Impenetrable (Kayonza) Forest and on N spurs of Ruwenzori	Pitman 1954
	In SW (localities mapped)	Rahm 1970, Kingdon 1971
	Uncommon in Kibale Forest (in W)	Struhsaker 1972, 1975
	In Kibale, Maramagambo, and Impenetrable forests and Ruwenzori Mts.	Struhsaker, pers. comm. 1974
	Infrequent in Kibale Forest	Oates 1977a

Cercopithecus mitis Mitis, Blue, Sykes', and Samango Monkey

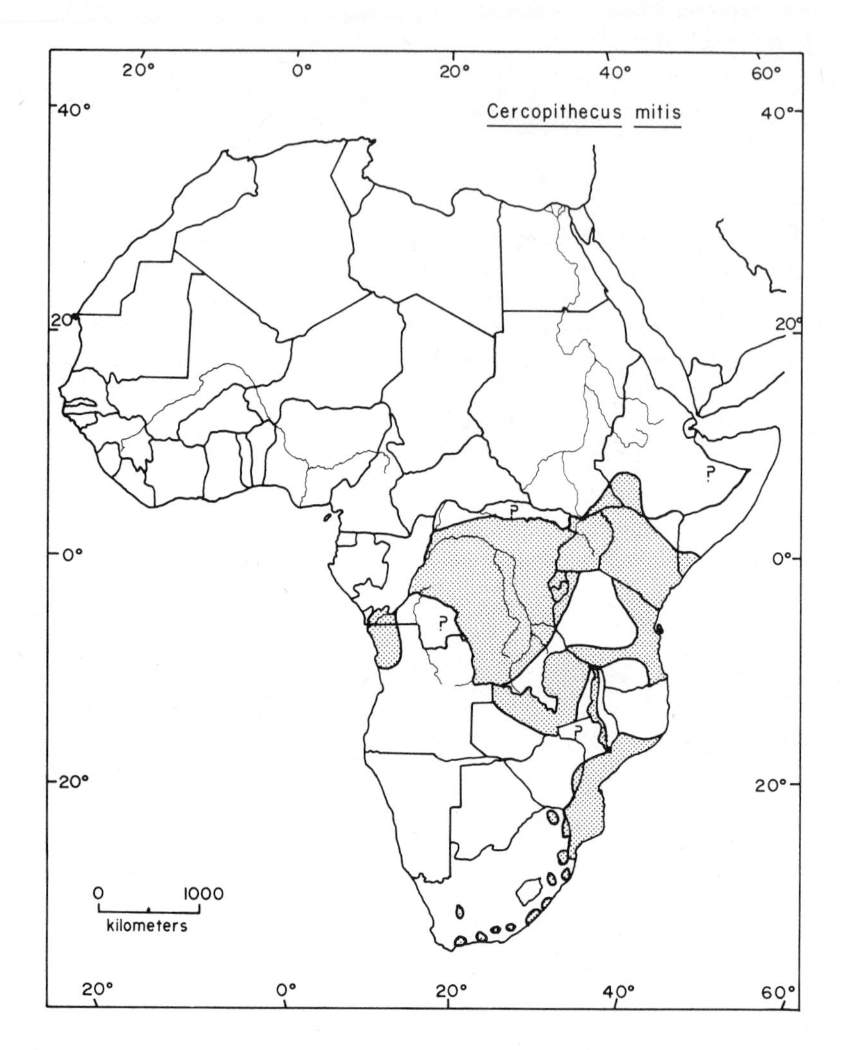

Taxonomic Note

This is a large and complex group, consisting of 20 to 25 races, many of which occur in single isolated populations (Rahm 1970, 1972). Some authors regard the 12 races located in southeast Africa as members of a separate species, *Cercopithecus albogularis* (Hill 1966).

Geographic Range

Cercopithecus mitis is found in central, eastern, and southern Africa. It occurs in the Congo basin and in forest patches to the east and south in every country except Namibia and Botswana. Mitis monkeys have been reported as far north as 7°50′N, 36°38′E in Ethiopia (Dandelot and Prévost 1972), and as far south as about 34°30′S, 21°E along the southern coast of South Africa (Gartlan and Brain 1968). The westernmost record is at about 9°S, 13°30′E on the Angolan coast, and the easternmost population is found near 0°N, 43°E in Somalia (Rahm 1970, 1972). Wilkinson (1974) reported this species in Sierra Leone, but this is almost certainly a case of misidentification. For details of the distribution, see Table 105.

Abundance and Density

Within the last 10 years, *C. mitis* was judged to be abundant or common in parts of six countries within its range and possibly increasing in numbers in one area of South Africa. But it was also described as scarce or rare in parts of four countries and declining in some regions of South Africa and Uganda (Table 105).

Mitis monkeys have been studied at population densities of 42 to 183.4 monkeys per km^2 (Table 106). Their group sizes are usually between 12 and 20, although groups as small as 2 and as large as 70 have been reported. Home range estimates vary from 5.2 to 100 ha (Table 106).

Habitat

C. mitis is a rain forest species, typically found in evergreen or semideciduous forest (Gartlan and Brain 1968, Aldrich-Blake 1970a). It utilizes riparian forest (Swynnerton 1958; Ansell, pers. comm. 1974), lowland forest (Dutton 1974), montane cloud and bamboo forest (Curry-Lindahl 1956, Omar and de Vos 1970), dune and coastal forests (Pringle 1974), and papyrus swamps (Pitman 1954, Kano 1971a).

Mitis monkeys are characteristic of humid forests (Bourquin et al. 1971, Bernahu 1975), but also occur in dry scrub forest (Booth 1968), dry upland evergreen forest (de Vos and Omar 1971), and dry montane forest (*Podocarpus* and *Juniperus*) (Aldrich-Blake 1970a). They have been reported as high as 3,000 m elevation in eastern Zaire (Rahm 1965) and 3,300 m in southwestern Uganda (Haltenorth 1975).

Several factors have been proposed as crucial components of mitis monkey habitat. Water seems to be very important, especially in the southern portion of the range. *C. mitis* is never found far from water in Zambia (Ansell 1960a), and in western Tanzania cannot survive even in riverine forests without swampy areas (Kano 1971a). *C. mitis* is dependent on shade and is thus excluded from dry open habitats (Kingdon 1971). Tall trees are important for it for both feeding and shelter (Ansell, pers. comm. 1974).

Forest seems to be the most important prerequisite for *C. mitis*, and most authors agree that this species is closely confined to forested areas (Aldrich-Blake 1970a; Attwell, pers. comm. 1974; Scorer, pers. comm. 1974; Smithers, pers. comm. 1974). It can utilize secondary forest (Moreno-Black and Maples 1977), although it prefers virgin forest when available (Brown and Urban 1970). It can also move through and feed in exotic tree plantations (Kingdon 1971), but prefers native trees to introduced vegetation (Omar and de Vos 1970, de Vos and Omar 1971).

Factors Affecting Populations

Habitat Alteration

The destruction of forests is causing the decline of *C. mitis* habitat and populations in several countries. In South Africa, for example, habitat destruction has led to substantial decreases in *C. mitis*, especially in northern Zululand (Richter 1974) and elsewhere in Natal (Pringle 1974). In Transvaal, logging for hardwoods has almost ceased and *C. mitis* populations seem to be recovering (Scorer, pers. comm. 1974). In Zambia some erosion of gallery forest is occurring, but is not yet a threat to *C. mitis* populations (Ansell, pers. comm. 1974).

In Ethiopia large areas of forest have been cleared and burned for the cultivation of crops, especially coffee, thus destroying much *C. mitis* habitat (Bernahu 1975). The same is true in Kenya, where the Kakamega Forest is shrinking because of agricultural clearing and charcoal burning (Zimmerman 1972). *C. mitis* on Mount Kenya is being driven to progressively higher elevations by timbering on lower slopes (Rowell, pers. comm. 1974).

In Uganda, high forest outside forest reserves is disappearing rapidly (Aldrich-Blake, pers. comm. 1974). Inside reserves, clearcutting of all usable trees is carried out on a rotation plan to provide maximum sustained timber yield (Karani, senior conservator, Forest Department, pers. comm. 1974). After logging, agricultural encroachment in the reserves is prevented, but remaining vegetation is treated with arboricides to allow regeneration of commercially valuable timber trees (Aldrich-Blake, pers. comm. 1974). The long-term effects of this treatment on *C. mitis* are unknown but probably detrimental. In one study slightly fewer *C. mitis* were found in felled and herbicide-treated forests (1.0 sightings per walk) than in undisturbed forests (1.7 sightings per walk) (Oates 1977a). Struhsaker (pers. comm. 1974) is studying this problem with relation to other primates, and trying to persuade the Forest Department to cease poisoning in certain areas on an experimental basis, to test the effects of arboricides on the regenerating tree species and on the wildlife.

Human Predation

Hunting. Mitis monkeys are hunted for their meat in Uganda (Aldrich-Blake, pers. comm. 1974), Mozambique (Dalquest 1968), Zaire (Heymans and Meurice 1973), South Africa (Richter 1974), and Kenya (Rowell, pers. comm. 1974). They were previously hunted for their skins in Zambia (Ansell 1960a) and in East Africa, where before 1925 the large slaughter of blue monkeys to obtain skins for export led to the extermination of the species in several areas (Kingdon 1971).

Pest Control. C. mitis seems to feed in cultivated areas relatively rarely (Aldrich-Blake, pers. comm. 1974). It is not a pest to agriculture in Zambia (Ansell 1960a) or in Zimbabwe, although it is somewhat destructive to gardens there (Smithers, pers. comm. 1974). This species is an occasional crop raider in Kenya (Moreno-Black and Maples 1977), where it eats peas, beans, potatoes (Omar and de Vos 1970), maize, and bananas (Kingdon 1971). It may also steal eggs from chicken houses (Hill 1966), oranges from orchards (Scorer, pers. comm. 1974), and other cultivated foods in South Africa (Pringle 1974). But overall mitis monkeys are usually not considered serious food crop pests, and are not harassed for this reason.

But in some areas they are destructive to commercially grown exotic trees, and are thus subject to control efforts. In parts of Uganda, Kenya, and South Africa they enter plantations of exotic softwoods (e.g., *Pinus* and *Cupressus*) and eat the growing shoots or the cambium layer under the bark (Kingdon 1971, de Vos and Omar

1971). The damage can be quite serious, killing the tree directly by ring-barking or allowing the entry of destructive bark beetles, and in some cases *C. mitis* has damaged 30 to 60% of the trees in one plantation (Omar and de Vos 1970). This behavior has led to attempts to exterminate mitis monkeys, resulting in about 1,500 per year being trapped and killed in southwest Uganda (Struhsaker, pers. comm. 1974), and over 1,000 per year being destroyed during several years in Kenya (Omar and de Vos 1970). These last authors have pointed out that these efforts have not been successful, since tree damage has actually increased recently.

In some parts of Uganda (Struhsaker, pers. comm. 1974) and Kenya (de Vos and Omar 1971) *C. mitis* is not destructive to softwood plantations, and de Vos (pers. comm. 1974) believes that these monkeys only damage exotic trees in areas where their natural food is scarce and they are forced to feed on the introduced trees.

Collection. C. mitis is not often used in medical research. Between 1968 and 1973 a total of 33 were imported into the United States (Appendix); none entered the United States in 1977 (Bittner et al. 1978). In 1965–75, 109 were imported into the United Kingdom (Burton 1978).

Conservation Action

C. mitis occurs in 15 national parks and several protected forests and game reserves (Table 105), and in the proposed Mahali Mountains National Park in western Tanzania (Itani and Nishida, pers. comm. 1974). In Uganda, additional and larger reserves for this species and further research on the effects of arboricides on it are needed (Aldrich-Blake, pers. comm. 1974). In South Africa, continued habitat protection is necessary for *C. mitis* survival (Richter 1974), and in Ethiopia, habitat reserves, status surveys and conservation education have been recommended (Bernahu 1975).

Mitis monkeys are legally protected from hunting in Zambia (Ansell 1957) and South Africa (Scorer, pers. comm. 1974). The trade in their skins has also been outlawed in East Africa (Kingdon 1971). In Zimbabwe their classification as Royal Game has been proposed (Attwell, pers. comm. 1974). In Ethiopia they may only be hunted under special license for scientific or educational purposes (Bernahu 1975), and in Zambia scientific collection is also permitted only under sparingly issued licenses (Ansell, pers. comm. 1974).

In order to avoid heavy damage to exotic tree species and consequent efforts to control *C. mitis* populations, Omar and de Vos (1970) recommend (1) leaving indigenous trees within the plantations to provide food for the monkeys, (2) consolidation of plantations (since small patches surrounded by forest are the most vulnerable to damage); (3) maintenance of strips of open grassland around plantations (since mitis monkeys are reluctant to cross these); (4) vigilant patrols morning and evenings, especially during the peak damage months of July to October, and (5) the planting of less palatable tree species (*Pinus* rather than *Cupressus*) in areas where protection is difficult. In South Africa, Forestry Department officials claiming damage by mitis monkeys, rather than being given permission to shoot the monkeys, are advised by provincial administrators to replant with trees the monkeys do not like; animals causing a nuisance are caught and relocated (Scorer, pers. comm. 1974).

Table 105. Distribution of *Cercopithecus mitis* (countries are listed in clockwise order)

Country	Presence and Abundance	Reference
Zaire	Common in E	Curry-Lindahl 1956
	W of Lake Kivu (in E) (localities mapped)	Rahm and Christiaensen 1963 [Hendrichs 1977]
	Throughout N and E (maps)	Dandelot 1965
	In Kahuzi Province (in E)	Rahm 1965
	Between Zaire-Lualaba R. and Rift, and Lowa-Osa R. and Elila R.	Elbl et al. 1966, Rahm 1966
	On Idjwi Island, Lake Kivu	Rahm and Christiaensen 1966
	Throughout S of Ubangi R. (map)	Gartlan and Brain 1968
	Throughout NW, NE, and SE (map)	Rahm 1970
	Common between Kindu and Kisangani along Zaire R.	Mallinson 1974
	Common throughout Haut-Zaire Province (in N); rather rare near Lomami, Kasai-Oriental Province (central)	Heymans 1975
	In Kundelungu N.P. (in SE, 10°S, 27°E)	I.U.C.N. 1975
	In Kahuzi-Biega N.P. (W of Lake Kivu)	Schlichte 1975
Sudan	Quite common S of Torit, near S border (map)	Butler 1966
	In extreme S (localities mapped)	Kock 1969
Ethiopia	Isolated populations primarily S of 10°N	Starck and Frick 1958
	In E	Glass 1965
	Common at Godare, Bagadur, and upper Godare R. (in SW)	Brown and Urban 1970
	In Ilubabor and Kaffa provinces (in SW)	Dandelot and Prévost 1972 [Bolton 1973]
	Locally in small numbers in highlands of SW	Bernahu 1975
Somalia	In extreme SW corner (maps)	Gartlan and Brain 1968; Rahm 1970, 1972
Kenya	In W and S (localities mapped)	Booth 1962
	Fairly common in Aberdare and Mount Kenya forest reserves	Omar and de Vos 1970
	In Nyere Hill Forest and 35 km NW of Nairobi	de Vos and Omar 1971
	In SW, central, and SE (localities mapped)	Kingdon 1971
	In Kakamega Forest (in W)	Zimmerman 1972
	Range shrinking in Kakamega Forest	Rowell, pers. comm. 1974
	At lower Tana River (in SE)	Andrews et al. 1975
	In Aberdare N.P.	Dupuy 1975
	In Tutla area (n.l.)	Haltenorth 1975
	At Diani Beach	Moreno-Black and Maples 1977
Uganda	In Bwamba Forest (in W, NE corner Semliki Valley)	Lumsden 1951, Kingston 1971

Table 105 (continued)

Country	Presence and Abundance	Reference
Uganda (cont.)	In Ruwenzori, Mount Elgon, Virunga Volcanoes, and Malabigamo Forest	Pitman 1954
	In Kayonsa Forest	Stott 1960
	Very common	Butler 1966
	In Impenetrable (Kayonza), Maramagambo, Mafuga, and Budongo forests; abundant but possibly declining slowly	Aldrich-Blake 1970a,b; pers. comm. 1974
	In W, SW, and NE (localities mapped)	Kingdon 1971
	Fairly abundant in Kibale, Ruwenzori, and Budongo forests; overall declining but not drastically	Struhsaker, pers. comm. 1974, 1975
	In Ruwenzori and Virunga Volcanoes	Haltenorth 1975
	In N edge Kibale Forest	Hayashi 1975, Mizuno et al. 1976
	In Kibale Forest	Rudran 1978
Rwanda	Common	Curry-Lindahl 1956
	E of Lake Kivu (Rugege Forest)	Rahm 1965, Elbl et al. 1966
	E of Lake Kivu (map)	Rahm 1970
	In Volcanoes N.P. (N central border)	I.U.C.N. 1971
Burundi	Common	Curry-Lindahl 1956
	Present (map)	Rahm 1970
Tanzania	Abundant in E Serengeti N.P.	Swynnerton 1958
	Rather rare in area between 5°–6°30′S and 30°–31°E; abundant near Kasoge (E central shore Lake Tanganyika) (localities mapped)	Kano 1971a
	Primarily in N, E, and S and on Zanzibar (localities mapped)	Kingdon 1971
	In Amani District (E Usambara Mts. in NE)	Groves 1973
	In Mahali Mts. proposed N.P. S of Kigoma (in W)	Itani and Nishida, pers. comm. 1974
Malawi	From Mount Waller to Mount Nchisi (n.1.)	Ansell et al. 1962
	Throughout central (map)	Smithers 1966
	In Nsanje (Port Herald) District (in S) including Mwabvi Game Reserve	Long 1973
Zambia	In Kasanka Game Reserve	Ansell 1957
	Not uncommon in N and NE (localities mapped)	Ansell 1960a
	In N and E, S to Kabompo and Kasempa districts in W and Zambezi Valley in SE (map)	Smithers 1966
	In Lunga Game Reserve (in NW)	I.U.C.N. 1971
	In West Lunga, Kasanka, Lusenga Plain,	Ansell, pers. comm. 1974

Table 105 (continued)

Country	Presence and Abundance	Reference
Zambia (cont.)	South Luangwa, and Nyika national parks; common, populations probably stable	Ansell, pers. comm. 1974 (cont.)
Zimbabwe	Limited to E border mountains	Child and Savory 1964
	In E districts (map)	Smithers 1966
	Near Binga, Chimanimani Mts. (E border)	Jackson 1973a
	Confined to E, 18°–20°S; population stable; in Inyanga, Mtarasi, Vumba, and Chimanimani national parks	Attwell, pers. comm. 1974
	Confined to E districts between 18°30′S and 20°31′S, E of 32°30′E (localities mapped)	Smithers, pers. comm. 1974
Mozambique	Rare at Zinave R. (central)	Dalquest 1968
	In Chimanimani Mts. (in W)	Dutton 1974
	Widespread and common in Manica and Sofala districts (central); present in S and N Inhambane District (in SE); probably in E Tete District (in NW) and W Zambezia District (E central) (localities mapped)	Smithers, pers. comm. 1974
	In Bazaruto N.P. and Pomene and Maputo Reserves (all in SE)	Tinley et al. 1976
South Africa	Near Lake Simbu, Tongaland	Davis 1957
	In Ndumu Game Reserve (NE Natal border)	Paterson et al. 1964
	Along E coast to Drakensberg	Brain 1965
	Enter Kruger N.P. from E (map)	Pienaar 1967
	In Pilgrim's Rest District, Woodbush, Pondoland, and Kaffraria E to Natal and Zululand (map)	Rahm 1970
	Quite common in NW Hluhluwe Game Reserve, E Natal (NE Kwazulu)	Bourquin et al. 1971
	In Oribi Gorge Nature Reserve (in SE, SW of Durban)	I.U.C.N. 1971
	Throughout N and E Natal (localities mapped); range shrinking and numbers decreasing	Pringle 1974
	Reasonable numbers in E Cape, Natal, Zululand, and E and NE Transvaal; has decreased substantially outside conserved areas; not common but secure if habitat protected	Richter 1974
	In E Cape and Natal midlands, E Transvaal escarpment and Zululand, and N to NE borders; scarce overall, probably increasing	Scorer, pers. comm. 1974

Table 105 (continued)

Country	Presence and Abundance	Reference
Swaziland	Absent (map)	Rahm 1970, 1972
	Present	Scorer, pers. comm. 1974
Lesotho	Absent (maps)	Gartlan and Brain 1968; Rahm 1970, 1972
Angola	In NW between Congo R. and Cuanza R. (map)	Gartlan and Brain 1968
	Only along W Cuanza R. (maps)	Rahm 1970, 1972

Table 106. Population parameters of *Cercopithecus mitis*

Population Density (individuals /km²)	Group Size			Home Range Area (ha)	Study Site	Reference
	$\bar{X}$	Range	No. of Groups			
		2–10			E Zaire	Rahm and Christiaensen 1966
	13	11–16	3		E Zaire	Schlichte 1975
		5–8			SW Ethiopia	Brown and Urban 1970
183.4	14	10–17	6	5.2–11.7	W Uganda	Aldrich-Blake 1970b
		13–20	2	6–8	W Uganda	Hayashi 1975
42	15			100	W Uganda	Struhsaker 1975
63	10–13	3–12	5	19–20.5	W Uganda	Mizuno et al. 1976
35	20.8	13–27	5	72.5	W Uganda	Rudran 1978
				13.2–14.2	S Kenya	Omar and de Vos 1970
118	16	15–18	3	13.2–16	S Kenya	de Vos and Omar 1971
		6–12			S Kenya	Andrews et al. 1975
		18–22			S Kenya	Moreno-Black and Maples 1977
		≤70			W Kenya	Rowell, pers. comm. 1978
		1–15			W Tanzania	Kano 1971a
		19	1		NE Tanzania	Groves 1973
		≤20			Zimbabwe	Smithers, pers. comm. 1974
	20–25				E South Africa	Bourquin et al. 1971
		22	1	67	NE South Africa	Scorer, pers. comm. 1974

Cercopithecus mona Mona Monkey

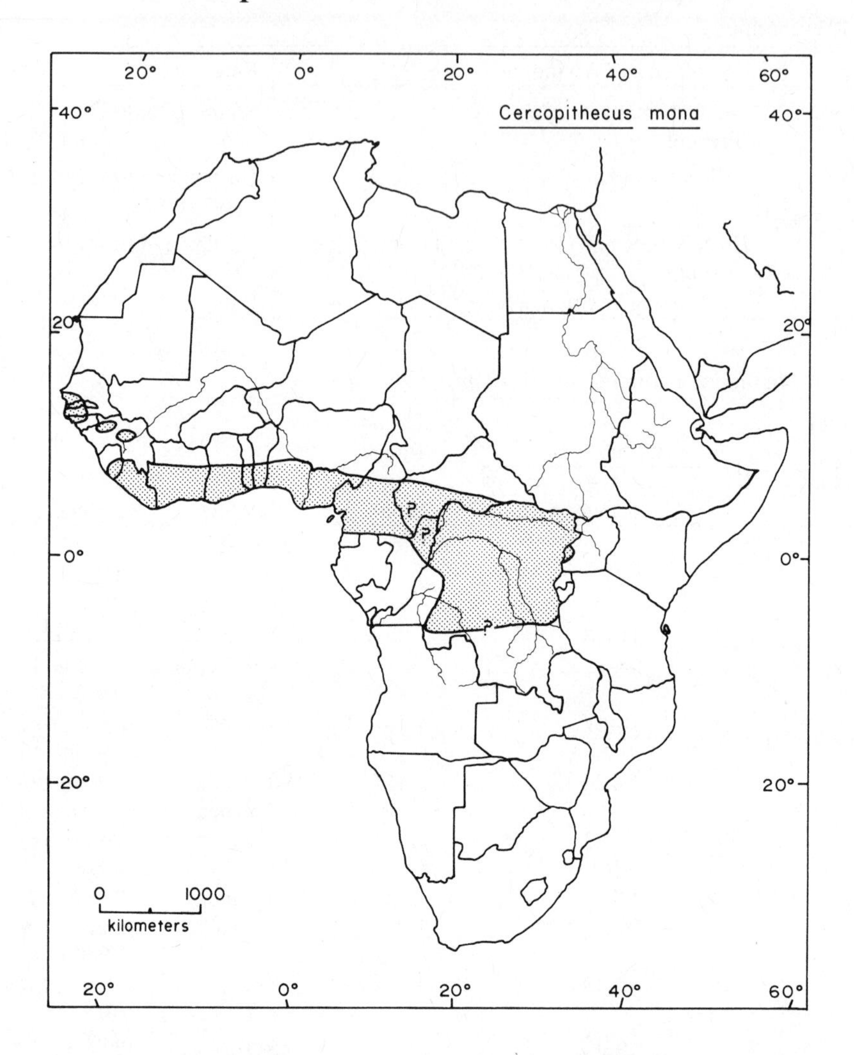

Taxonomic Note

This is a large and widely distributed group consisting of seven to eleven distinct forms (Booth 1955). There is no agreement on how many species constitute the group. Since very little is known about the forms from the Congo basin (*wolfi*, *denti*, *elegans*, *pyrogaster*), they are considered here together with the western representatives (*mona*, *lowei*, and *campbelli*), even though they may constitute one or more separate species. The south central forms (*Cercopithecus pogonias*) are also considered by some authors to be members of the mona group. But because they occupy a different niche than do others of this group, and because recent data are available regarding their status in Cameroon (Gartlan, pers. comm. 1974), *C. pogonias* will be treated in a separate section.

Geographic Range

Mona monkeys occur from Senegal through coastal west and central Africa to western Uganda. They are found as far northwest as 13°50′N, 16°30′W in Senegal (Dupuy 1971b) and as far east as about 30°30′E in Uganda (Kingdon 1971). No recent data are available regarding their southern limit in Zaire, but Booth (1955) showed the southernmost locality near 7°S,16°30′E. For details of the distribution, see Table 107.

Abundance and Density

Within the last 10 years, *Cercopithecus mona* was described as abundant or common in parts of seven countries and as rare in two countries (Table 107). There is no information available regarding its abundance in the other seven countries within its range.

Estimates of group size for *C. mona* include 8 to 20 (Monath and Kemp 1973) and up to 38 (Gartlan 1973). Struhsaker (1969) counted 6 groups and found a mean size of 8.8 (range 3 to 13, mode 11). Bourlière et al. (1969) studied a small group with a home range of about 3 ha. No measures of population density were located.

Habitat

C. mona is a rain forest species typical of primary and secondary high forest and riparian forest (Booth 1956a, 1958; Dupuy 1973; Bourlière et al. 1974). It also utilizes islands of dense forest in savanna (Bourlière et al. 1969, Monath and Kemp 1973) and mangrove swamps (Struhsaker 1969; Dupuy 1971b, 1973). In Cameroon this species is most abundant in mangrove swamp, second most numerous in secondary forest, and least abundant in mature lowland forest (Gartlan and Struhsaker 1972).

Mona monkeys are commonly seen at forest fringes (Booth 1956a, Jeffrey 1974) and in dense, tangled, colonizing secondary growth at the edges of old forest (Haddow 1952; Howard, pers. comm. 1974). They can occupy farmland, including shaded cacao plantations and gardens (Monath et al. 1974, Jeffrey 1975b), and have been described as ecologically tolerant (Booth 1958). Rahm (1965) reported *C. mona* up to 2,000 m elevation, while Hendrichs (1977) found this species only below 900 m.

Factors Affecting Populations

Habitat Alteration

In much of the western portion of its range *C. mona* populations are decreasing, owing to deforestation. For example, in Sierra Leone, extensive logging and poisoning of noneconomic tree species have greatly reduced forested areas (Tappen 1964), and in Ivory Coast the Tai National Park is the only large tract of undisturbed rain forest remaining (Struhsaker 1972). Similarly, in Ghana, large-scale timbering, burning, planting with cacao and other cash crops, road building, and soil erosion have resulted in a drastic decrease in *C. mona* habitats (Jeffrey 1975b, Asibey 1978). And in Nigeria, deforestation long ago resulted in the depletion of mona monkey populations in East Central State, and continuing encroachment by the expanding human population has led to further decreases and the isolation of forest reserves and remnants in southern Nigeria (Monath and Kemp 1973).

Several authors have reported that *C. mona* is adaptable to limited changes in its habitat (Jeffrey 1974, Rucks 1976). In Cameroon this species seems to be increasing its range, because it thrives in secondary forest and logging is increasing the extent of this vegetation type (Gartlan, pers. comm. 1974; Howard, pers. comm. 1974).

About 10 years ago, more than half of *C. mona*'s range in Uganda (the section west of the Kirimia River) was scheduled for felling and replanting with timber trees (Kingdon 1971). The effect of this action on the *C. mona* population is not known. Near Lake Tumba, western Zaire, slash and burn agriculture has not greatly degraded *C. mona* habitats, but large-scale commercial logging is being planned for this area (Horn, pers. comm. 1978).

Human Predation

Hunting. Mona monkeys are hunted for their meat in Ivory Coast (Bourlière et al. 1969, Tahiri-Zagrët 1976), Ghana (Asibey 1974, Jeffrey 1975b), Cameroon (Gartlan, pers. comm. 1974), and Zaire (Ladnyj et al. 1972; Heymans and Meurice 1973; Horn, pers. comm. 1978).

Pest Control. Mona monkeys do raid crops and are considered pests to cash crops, such as cacao (Mackenzie 1952; Gartlan, pers. comm. 1974). They are destructive to maize fields in Ghana (Jeffrey 1974) and newly cleared farms in Uganda (Kingdon 1971), and may represent a real economic drain on subsistence agriculture in western Cameroon (Howard, pers. comm. 1974). They are frequently shot as pests in Cameroon (Gartlan, pers. comm. 1974) and Zaire (Horn, pers. comm. 1978), and killed along with other monkeys in the extermination drives of Sierra Leone (Mackenzie 1952, Tappen 1964).

Collection. *C. mona* is often kept as a pet in Ghana (Jeffrey 1974). It has been collected for a few medical studies (Kingdon 1971, Monath et al. 1974) but is not a common subject for research. It is exported from Cameroon occasionally for zoos (Gartlan, pers. comm. 1974).

Between 1968 and 1973 a total of 87 *C. mona* were imported into the United States, primarily from Liberia, Benin, Ghana, and Sierra Leone (Appendix). None entered the United States in 1977 (Bittner et al. 1978). In 1965–75, 602 were imported into the United Kingdom (Burton 1978).

Conservation Action

Mona monkeys are found in at least five national parks and fifteen forest, nature, or game reserves (Table 107).

They are also in Class B of the African Convention (Curry-Lindahl 1969). In Ivory Coast they are legally protected from hunting, but better law enforcement, public education, and national park management are needed (Tahiri-Zagrët 1976). In Ghana improved regulation of hunting inside forest reserves and scientific management of park and reserve ecosystems are necessary if the forest primates are to survive (Asibey 1978).

Table 107. Distribution of *Cercopithecus mona* (countries are listed in order from west to east)

Country	Presence and Abundance	Reference
Senegal	Common near mouth of Casamance R. (in SW) (map); abundant in Casamance N.P.	Dupuy 1971b, 1973 [I.U.C.N. 1975, Dupuy and Verschuren 1977]
	N and S of Casamance R. (localities mapped)	Gatinot 1974
Gambia	N and S of Gambia R. (in W) (map)	Dupuy 1971b
	In Abuko Nature Reserve (2 individuals)	Gunderson 1977
Guinea-Bissau	In N (map)	Booth 1958
Guinea	In NW and central (map)	Booth 1958
	To 10°N and 15°W	Tappen 1960
Sierra Leone	Present	Mackenzie 1952, Jones 1966
	In Kasewe Forest, 160 km E of Freetown (S central)	Tappen 1964
Liberia	Present	Kuhn 1965
Ivory Coast	Throughout S (localities mapped)	Booth 1955, 1958
	Widely distributed and very common	Bourlière et al. 1969, 1970
	Common in Tai Reserve (now N.P.) (in W between Cavally and Sassandra rivers)	Struhsaker 1972 [Monfort and Monfort 1973]
	Near Lamto (in S, near Bandama R.)	Bourlière et al. 1974
	Throughout S from Grand Béréby to Ayamé; seen frequently at San Pedro, Grand Lahou, and Adiopodoumé (map)	Tahiri-Zagrët 1976
Ghana	Throughout S (localities mapped); N limit is Bamboi, scarce there	Booth 1955; 1956a,b; 1958
	Common in Bia Tributaries North and Sefwi Wiawso Forest Reserve (in SW)	Jeffrey 1975b
	Abundant in Boabeng-Fiena Monkey Sanctuary; common in Bia N.P., Nini-Suhien N.P., Bomfobiri Wildlife Sanctuary, Willi Falls Reserve, Ankasa Game Production Reserve; rare in Kalakpa Game Production Reserve and Digya N.P.	Asibey 1978
Togo	In N and central (maps)	Booth 1955, 1958
	Rare in Malfacassa Reserve	Asibey 1978
Benin	In S (map)	Booth 1958
Nigeria	In S (localities mapped)	Booth 1955, 1958
	Few records but common in Olokemeji Forest Reserve (in SW)	Hopkins 1970
	Commonly encountered	Henshaw and Child 1972
	Recorded in SE, W, near Nupeko, and Jos Plateau (localities mapped); abundant	Monath and Kemp 1973
	Very large population in Nupeko Forest near Bida (W central)	Monath et al. 1974

Table 107 (continued)

Country	Presence and Abundance	Reference
Cameroon	In W and S (localities mapped)	Booth 1955, 1958
	At Idenau and Bakundu Reserve (in W); Elongongo and Nsépé (on Sanaga R.); Lake Tisongo (mouth of Sanaga R.)	Gartlan and Struhsaker 1972
	Abundant in Douala-Edéa and Korup reserves (in W)	Struhsaker 1972
	In Korup, Douala-Edéa, Dipikar, and Dja reserves; locally abundant and increasing	Gartlan, pers. comm. 1974
	In Mungo Forest Reserve near Kumba (in W); range increasing	Howard, pers. comm. 1974
	Common in Kimbe River Game Reserve	Asibey 1978
Central African Republic	To 7°N, 20°E (central)	Tappen 1960
Congo	No information	
Zaire	Throughout except SE (localities mapped)	Booth 1955
	Near Epulu, Ituri Forest (in NE)	Curry-Lindahl 1956
	W of Lake Kivu	Rahm and Christiaensen 1963, Rahm 1965, Hendrichs 1977
	In N, NE, and E, abundant near Irangi (W of Lake Kivu)	Rahm 1966
	In Basankusu Territory, N Equateur Province (in NW)	Ladnyj et al. 1972
	denti: abundant throughout Haut-Zaire Province; *elegans:* rather rare between Zaire R. and Lomami R.; *woffi:* rather rare W of Lomami R.	Heymans 1975
	Common W and E of Lake Tumba in W; also S of Zaire R. across from Lisala; near Boende	Horn, pers. comm. 1978
Uganda	In Bwamba Forest, E Semliki extremity of Ituri Forest	Lumsden 1951
	"Sparingly" in Bwamba Forest	Pitman 1954
	Fairly common near Bwamba (maps) including nature sanctuary E of Kirimia R.	Kingdon 1971
	Restricted to Bwamba Forest	Rowell, pers. comm. 1978

Cercopithecus neglectus DeBrazza's Monkey

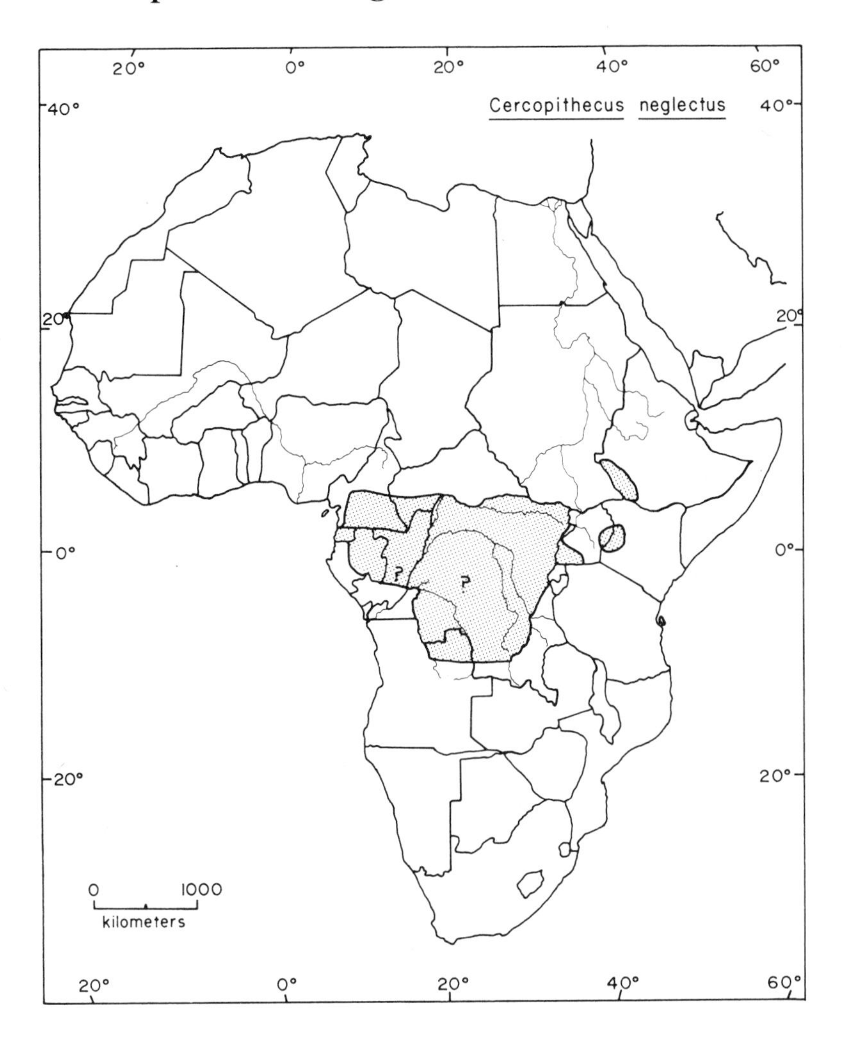

Taxonomic Note

This is a monotypic species (Napier and Napier 1967).

Geographic Range

Cercopithecus neglectus is found in central Africa from Cameroon to southern Ethiopia and southward through Zaire and northern Angola. In the west it occurs as far north as 4°40′N along the Sanaga River (Gartlan, pers. comm. 1974), as far south as 10°S in Angola (Barros Machado 1969), and as far west as the Atlantic coast in Equatorial Guinea (Sabater Pi and Jones 1967). In the east it has been reported as far north as 7°26′N, 35°02′E (Brown and Urban 1969) and as far east as 5°11′N, 36°12′E (Bolton 1973) in Ethiopia. The southeastern limit of its range is at about 10°S, 25°E along the Lualaba River in southeast Zaire (Dandelot 1965, Kingdon 1971). For details of the distribution, see Table 108.

Abundance and Density

C. neglectus has recently been described as common in parts of Angola and Zaire and uncommon in Ethiopia, Kenya, and northeastern Gabon (Table 108). Its status in Cameroon, Central African Republic, Uganda, Congo, and Equatorial Guinea is not known. Leakey (1969) believed that this species was declining and in danger of extinction in East Africa.

The only estimate of population density available is that of 28 *C. neglectus* per km^2 in northeastern Gabon (Quris 1976). The same author estimated the average home range area for each group to be 13 ha and counted eight groups with an average of 3.75 members (range 2 to 6).

Other workers have described *C. neglectus* groups as very small (Malbrant and Maclatchy 1949; Gartlan, pers. comm. 1974), fairly large (Barros Machado 1969), consisting of 6 to 10 members (Brown and Urban 1970), and 15 to 35 individuals (Kingdon 1971).

Habitat

C. neglectus is a humid forest species typical of swamp forest (Malbrant and Maclatchy 1949; Booth 1962; Gartlan, pers. comm. 1974) and seasonally flooded forest (Gautier and Gautier-Hion 1969, Quris 1976). It is most often found near rivers (Sabater Pi and Jones 1967, Barros Machado 1969, Heymans 1975), although Brown and Urban (1969) found it as far as 1 km from rivers in Ethiopia.

This species has been recorded along streams in dry montane forest up to 2,100 m elevation (Kingdon 1971). It can also utilize secondary forest (Brown and Urban 1970; Horn, pers. comm. 1978) and bamboo forest and palm trees (Malbrant and Maclatchy 1949). Kingdon (1971) believed that it probably originally occupied all types of forest.

Factors Affecting Populations

Habitat Alteration

Deforestation has led to a reduction in *C. neglectus* habitat in the eastern portion of its range. The species has declined in East Africa because of extensive timbering along streams to expand agriculture and settlement (Leakey 1969). The widespread destruction of forests has fragmented the range of *C. neglectus* and restricted it to isolated swampy areas in Kenya and Uganda (Kingdon 1971). And in Ethiopia, clearing of forest for coffee plantations has destroyed large areas of *C. neglectus* habitat (Bernahu 1975).

No specific information is available on the effect of habitat alteration on *C. neglectus* populations in the central and western parts of the range.

Human Predation

Hunting. *C. neglectus* is hunted for its flesh in Zaire (Horn, pers. comm. 1978) and in Cameroon.

Pest Control. C. neglectus does feed in millet plantations in Angola (Barros Machado 1969) and occasionally raids maize and cotton fields in East Africa (Kingdon 1971). But it was not described as a frequent crop raider and does not seem to be subject to control efforts.

Collection. This species is popular in zoo exhibits because of its striking facial pattern, and is exported from Cameroon (Gartlan, pers. comm. 1974) and Equatorial Guinea (Jones, pers. comm. 1974). It is also captured frequently as a pet in East Africa (Kingdon 1971). Its populations in Kenya have been greatly reduced by trapping (Booth 1962).

Between 1968 and 1973 a total of 152 *C. neglectus* were imported into the United States, primarily from Uganda and Kenya (Appendix). None entered the United States in 1977 (Bittner et al. 1978). In 1965–75 a total of 373 were imported into the United Kingdom (Burton 1978).

Conservation Action

C. neglectus occurs within the Dja Reserve in Cameroon, but was not reported in any other reserves or national parks (Table 108). It is legally protected from hunting or trapping in Ethiopia (Bernahu 1975).

Table 108. Distribution of *Cercopithecus neglectus* (countries are listed in clockwise order)

Country	Presence and Abundance	Reference
Cameroon	In SE, N at least to Nanga-Emboko (4°38′N, 12°21′E) on Sanaga R. (map); may occur along rivers to coast	Gartlan, pers. comm. 1974
	In Dja Reserve (in S)	Rowell, pers. comm. 1978
Central African Republic	In SW (maps)	Dandelot 1965, Hill 1966, Kingdon 1971
Sudan	Absent	Kock 1969
Ethiopia	Common near Godare Mission (7°26′N, 35°02′E); probably widespread in Kaffa and Ilubabor provinces (all in SW)	Brown and Urban 1969
	One locality in extreme SW corner (map)	Kock 1969
	In Gemu Gofa (Gemu Gwefa) Province (in SW) near Omo R. and Lake Dipa (5°11′N, 36°12′E)	Bolton 1973
	Locally in small numbers in SW	Bernahu 1975
Kenya	Fairly plentiful on Mount Kenya	Pitman 1954
	Only on lower slopes Cherangani Hills (W central); rare; may soon be extinct	Booth 1962
	On Mount Elgon and Cherangani Hills (localities mapped)	Kingdon 1971
	In W in Cherangani Hills and near Maralal (1°05′N, 36°42′E); not common	Rowell, pers. comm. 1978

Table 108 (continued)

Country	Presence and Abundance	Reference
Uganda	In Bwamba Forest (in W)	Lumsden 1951
	In Bwamba, Tero, Sango Bay (SE Masaka), and Mount Kadam (E Karamoja)	Pitman 1954
	In W, S, and E (localities mapped)	Kingdon 1971
	On lower slopes Mount Kadam and in Semliki Forest near N end Ruwenzori Mts.	Struhsaker, pers. comm. 1974
Zaire	In N and S, questionable in center (map)	Dandelot 1965
	One locality in S near 7°S, 21°E (map)	Barros Machado 1969
	Two localities in extreme NE near 4°N, 30°E (map)	Kock 1969
	Fairly common throughout Haut-Zaire Province to N of Aruwimi R. (1°N, 24°E)	Heymans 1975
	Near Lake Tumba (in W)	Horn, pers. comm. 1978
Angola	In NE near Dundo and Sombo (localities listed)	Hayman 1963
	In Lunda District (NE), especially abundant in NE portion (localities mapped); S limit 8°10′S in W, 10°S in E; presence in NW unknown	Barros Machado 1969
Congo	In N, NE, and E	Malbrant and Maclatchy 1949
	Present	Bassus 1975
Gabon	In E only	Malbrant and Maclatchy 1949
	Not common in NE (Ivindo basin)	Gautier and Gautier-Hion 1969
	10 km SW Makokou (in NE)	Gautier-Hion and Gautier 1974
	Along Liboui R. (in NE)	Quris 1976
Equatorial Guinea	In NW, central, and S (along rivers) (map); more common in coastal regions than interior	Sabater Pi and Jones 1967
	In Monte Raices Territorial Park (in NW, E of Bata)	I.U.C.N. 1971

Cercopithecus nictitans

Greater White-nosed or Putty-nosed Monkey

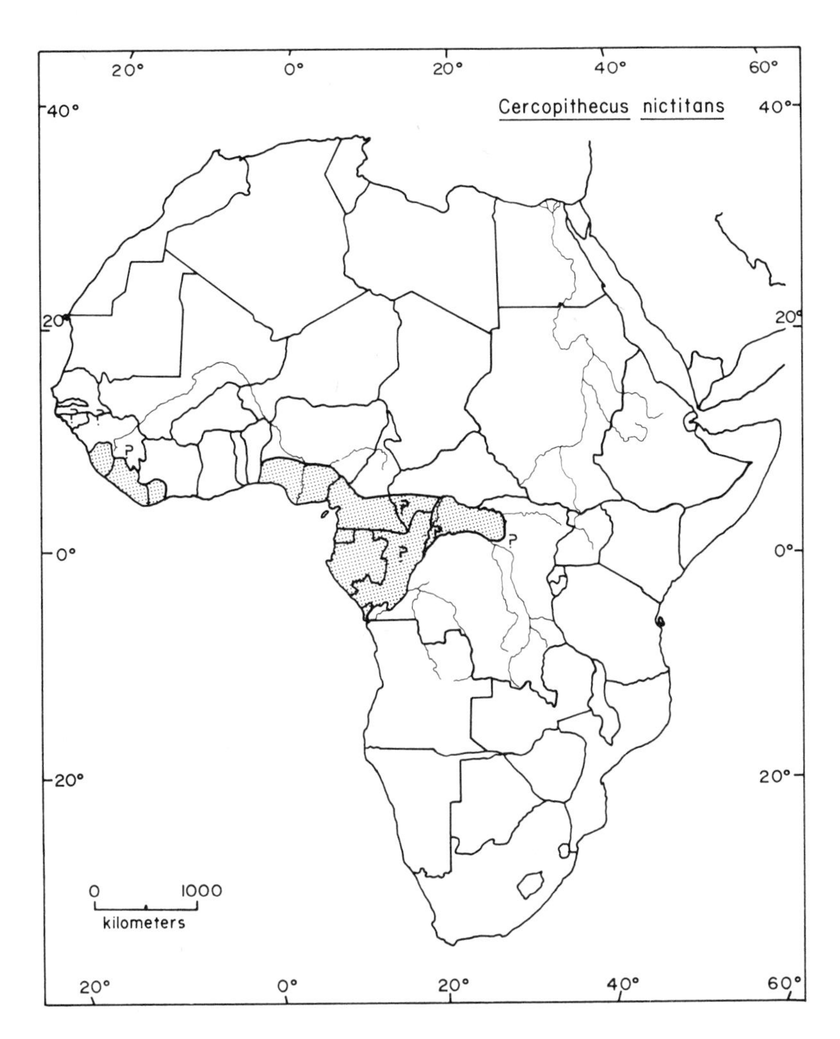

Taxonomic Note

This species consists of three subspecies (Booth 1958).

Geographic Range

Cercopithecus nictitans is found along the coast of West Africa from Sierra Leone to northern Angola. The northwesternmost record of it is near Selety, Senegal (Dupuy 1972a) (probably Seleti, about 13°N, 16°30′W), but this may represent a misidentification of *C. petaurista*. The southernmost locality is near 5°S, 12°E in Cabinda, Angola (Barros Machado 1969). The range extends east to at least Buta, Zaire (2°49′N, 24°50′E) (Heymans 1975). For details of the distribution, see Table 109.

Abundance and Density

C. nictitans has been described as locally abundant or common in four countries and declining in parts of two of these and two other countries (Table 109). No recent information is available regarding its abundance in much of Zaire, Gabon, eastern Cameroon, Congo, or Central African Republic.

No estimates of population density are available. In one study area, home range areas for eight groups were 37 to 75 ha (Quris 1976).

Groups of up to 60 individuals have been reported (Malbrant and Maclatchy 1949), but most recent authors have found smaller group sizes, such as 18 (Gautier and Gautier-Hion 1969) and averaging about 15 (Gartlan, pers. comm. 1974). In western Cameroon, 13 groups contained 7 to 17 members ($\bar{X}$ = 10.5, mode 7), while in southern Cameroon 8 groups consisted of 5 to 13 individuals ($\bar{X}$ = 9.3, modes 8 and 10) (Struhsaker 1969).

Habitat

C. nictitans is a rain forest species which is largely restricted to primary forest (Booth 1956b; Gartlan, pers. comm. 1974; Howard, pers. comm. 1974). It often utilizes riverine forest (Monath et al. 1974) and seasonally flooded forest (Gautier and Gautier-Hion 1969, Gartlan and Struhsaker 1972, Quris 1976).

In Cameroon, Equatorial Guinea, Gabon, and Congo, *C. nictitans* can also occupy old secondary forest (Malbrant and Maclatchy 1949, Sabater Pi and Jones 1967, Gautier and Gautier-Hion 1969, Gartlan and Struhsaker 1972). The Nigerian form is said to have a wide ecological tolerance and can utilize all types of closed forest as well as secondary and farm brush (Booth 1956b, 1958). But in Ivory Coast and Cameroon, *C. nictitans* is found only rarely in secondary growth (Booth 1956b; Gartlan, pers. comm. 1974), and, in general, this species is more successful in mature forest than in younger successional stages (Struhsaker 1969).

Factors Affecting Populations

Habitat Alteration

Deforestation by logging, clearing for plantations, and shifting agriculture has greatly reduced *C. nictitans* habitat in Cameroon (Gartlan, pers. comm. 1974; Howard, pers. comm. 1974), Equatorial Guinea (Sabater Pi and Jones 1967), Nigeria (Monath and Kemp 1973), and Ivory Coast (Lanly 1969).

Human Predation

Hunting. C. nictitans is commonly hunted for its flesh and is a popular item in the human diet in Equatorial Guinea (Sabater Pi and Groves 1972). It was greatly depleted by hunting during the Nigerian civil war of 1967–70 (Monath and Kemp 1973), and is also hunted in Ivory Coast (Tahiri-Zagrët 1976) and Cameroon (Gartlan, pers. comm. 1974; Howard, pers. comm. 1974).

Pest Control. This species is not a frequent crop raider in Cameroon (Gartlan, pers. comm. 1974). But in Sierra Leone it is considered destructive to crops and is shot as a pest (Mackenzie 1952, Robinson 1971). Malbrant and Maclatchy (1949) reported that it steals chickens from farmyards.

Collection. At least 60 *C. nictitans* were exported from Equatorial Guinea between 1958 and 1968 (Jones, pers. comm. 1974). This species is not often captured for export from Cameroon (Gartlan, pers. comm. 1974).

Between 1968 and 1973 a total of 113 *C. nictitans* were imported into the United States (Appendix). In the last four years, these were recorded as originating primarily from Ghana, where they are not supposed to occur, suggesting that they were misidentified *C. petaurista*. None entered the United States in 1977 (Bittner et al. 1978). In 1965–75, 269 *C. nictitans* were imported into the United Kingdom (Burton 1978).

Conservation Action

C. nictitans populations are present in at least one national park, one territorial park, and six forest reserves (Table 109). The western Cameroon race (*C. nictitans martini*) is especially in need of protection because its already small range is being further reduced by deforestation. Since it lives in relatively small groups with large home ranges and is completely dependent on mature rain forest, its survival can only be insured by protecting large reserves of this habitat (Gartlan, pers. comm. 1974). The available evidence indicates that this species is also in need of protection in Liberia, Ivory Coast, and Equatorial Guinea. Its status in the western, southern, and eastern extremities of its range is unknown.

Table 109. Distribution of *Cercopithecus nictitans* (countries are listed in order from west to east, then southward and northward in a counterclockwise direction)

Country	Presence and Abundance	Reference
Senegal	In Enarang Forest near Selety (later called *C. petaurista* and Seleti)	Dupuy 1972a, 1973
Guinea-Bissau	Present (later called *C. petaurista*)	Dupuy 1972a
Guinea	Present (later called *C. petaurista*)	Dupuy 1972a
Sierra Leone	Present	Mackenzie 1952, Jones 1966
Liberia	Throughout (map)	Booth 1958
	Throughout N (localities mapped); previously common in Gio Forest (NE), now extinct there	Kuhn 1965
	Locally abundant	Schlitter, pers. comm. 1974
Ivory Coast	In W, perhaps to Bandama R.	Booth 1956b, 1958
	In W (2 localities mapped)	Kuhn 1965
	In Tai N.P. (extreme W)	Monfort and Monfort 1973
	Reported along Sassandra R. and Nzi R. (localities mapped)	Tahiri-Zagrët 1976
Ghana, Togo, Benin	Absent (map)	Booth 1958

Table 109 (continued)

Country	Presence and Abundance	Reference
Nigeria	Throughout S (map)	Booth 1958
	In large numbers at Nupeko Forest near Bida (9°06′N, 6°00′E)	Monath and Kemp 1973, Monath et al. 1974
Cameroon	Throughout S (map)	Booth 1958
	At Idenau, Elongongo, and Nsépé (on Sanaga River), Lake Tisongo (near mouth of Sanaga), and Bakundu Reserve (all in W), Campo Reserve and Dipikar (in SW), and Dja Reserve (S central); more abundant in W than E	Gartlan and Struhsaker 1972
	Abundant in Douala-Edéa and Korup reserves	Struhsaker 1972
	Fairly common in SW between border S of 6°N and Sanaga R. W of 11°E; throughout S of Sanaga R. as far E as Dja R., possibly farther; overall declining but not drastically	Gartlan, pers. comm. 1974
	Gradually declining in Mungo Forest Reserve (in SW)	Howard, pers. comm. 1974
	Common in Dja R. area (S central)	Rowell, pers. comm. 1978
Equatorial Guinea	On Fernando Po (now Macias Nguema)	Cabrera 1929 cited in Tappen 1960
	Throughout (map); range is shrinking	Sabater Pi and Jones 1967
	In Monte Raices Territorial Park (in NW, E of Bata)	I.U.C.N. 1971
Gabon	In NE (Ogooué-Ivindo basin); declining in Ntsi-Belong region	Gautier and Gautier-Hion 1969, Gautier-Hion and Gautier 1974
	On Liboui R., Ivindo watershed (NE)	Quris 1976
Congo	Common from coast to Oubangi R.	Malbrant and Maclatchy 1949
Angola	In Cabinda (localities mapped)	Barros Machado 1969
Zaire	E limit is 26°E, S limit 2°N	Schouteden 1947 cited in Tappen 1960
	Common between Aketi and Buta (2°49′N, 24°50′E) in Haut-Zaire Province	Heymans 1975
Central African Republic	Common in Ubangi Territory	Malbrant 1952 cited in Hill 1966

Cercopithecus petaurista
Lesser White-nosed or Spot-nosed Monkey

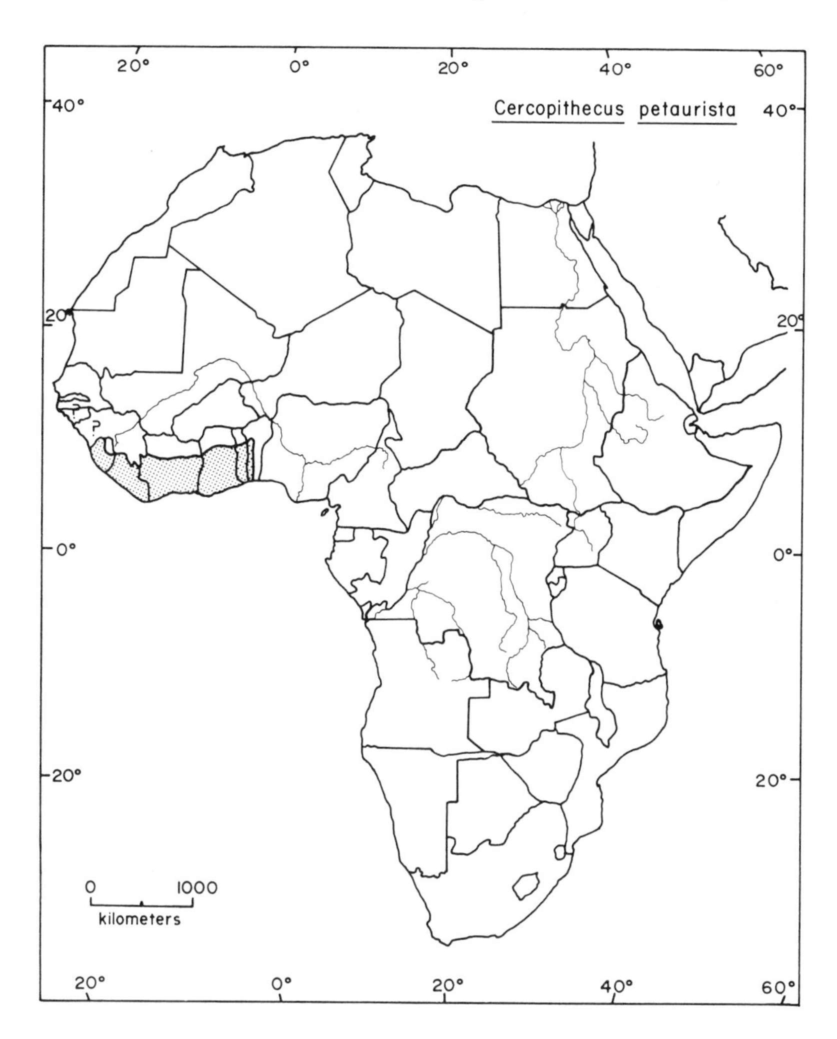

Taxonomic Note

Two races are recognized (Booth 1958).

Geographic Range

Cercopithecus petaurista is found along the southern coast of West Africa, from at least Sierra Leone to Benin. It has been recorded as far north and west as southern Senegal, but the records north and west of Sierra Leone may result from confusion with the similar *C. nictitans* (Dupuy 1971b, 1972a, 1973). The eastern limit of *C. petaurista* is near 2°E in Benin (Booth 1958). For details of the distribution, see Table 110.

Abundance and Density

C. petaurista was reported to be abundant or common in parts of Ivory Coast and Ghana, but also rare in regions of these countries, and declining in northeast Liberia (Table 110). Its status in the western and eastern portions of its range is unknown.

Bourlière et al. (1974) saw groups of about 20 individuals, and Rice (pers. comm. 1978) reported a maximum group size of 15. No further information regarding group size, population density, or home range could be located.

Habitat

C. petaurista is a forest species found in high forest, fringing forest, and coastal scrub (Booth 1956a). It is especially typical of thickets (Booth 1956a; Jeffrey 1974; Rice, pers. comm. 1978), swampy areas, and riparian forest (Booth 1956a, Bourlière et al. 1974, Tahiri-Zagrët 1976).

This species can also utilize secondary forest (Booth 1956a; Struhsaker, pers. comm. 1974; Asibey 1978) and mature cacao plantations and gardens (Jeffrey 1975b). It is an "ecologically plastic" species, able to utilize many forest habitats, including those marginal for other forest monkeys (Booth 1956a, 1958).

Factors Affecting Populations

Habitat Alteration

Large-scale deforestation and conversion of forest lands to plantations or farms has greatly decreased the area of *C. petaurista* habitat in Ghana (Jeffrey 1975b, Asibey 1978) and Ivory Coast (Lanly 1969). In Sierra Leone, logging has been followed by extensive poisoning of noneconomic tree species, further reducing forest habitats (Tappen 1964).

Human Predation

Hunting. *C. petaurista* is killed for its meat in Ivory Coast (Tahiri-Zagrët 1976). On Mount Nimba, northeast Liberia, mining has brought 20,000 additional people to the area, and the consequent intensive hunting has resulted in the decline of *C. petaurista* and all other monkeys there (Coe 1975). Monkeys are not considered acceptable food and are not hunted in the Bombali District, northern Sierra Leone (Wilkinson 1974).

In Ghana, *C. petaurista* is hunted heavily and forms a regular item of human diet wherever it is available (Asibey 1974, Jeffrey 1974).

Pest Control. *C. petaurista* was a common victim of the monkey extermination drives carried out in the coffee and cacao growing regions of Sierra Leone (Tappen 1964). It is also considered a pest to maize crops in Ghana (Jeffrey 1975b).

Collection. Between 1968 and 1973 a total of two *C. petaurista* were imported into the United States (Appendix), both from Ghana. None entered the United States in 1977 (Bittner et al. 1978). Five were imported into the United Kingdom in 1965–75 (Burton 1978).

Conservation Action

C. petaurista has been reported in three national parks in Ivory Coast and three national parks and six reserves and sanctuaries in Ghana (Table 110). It is also listed in Class B of the African Convention, and it is illegal to shoot or trap a lactating female inside a forest reserve in Liberia (Curry-Lindahl 1969).

Table 110. Distribution of *Cercopithecus petaurista* (countries are listed in order from west to east)

Country	Presence and Abundance	Reference
Senegal	In Casamance, S of Gambia R. (map)	Booth 1958
	Might be in Casamance at Seleti (about 13°N, 16°30′W) (previously reported as *C. nictitans* and Selety)	Dupuy 1973, 1972a
Guinea-Bissau	Present	Dupuy 1971b
Guinea	In N (map)	Booth 1958
	Present	Dupuy 1971b
Sierra Leone	In NW and throughout S (map)	Booth 1958
	In Kasewe Forest, 160 km E Freetown (S central)	Tappen 1964
	At Ducuta, Bombali District (N)	Wilkinson 1974
Liberia	Throughout (map)	Booth 1958
	Present	Kuhn 1965
	Previously common, now rare in Mount Nimba area (NE)	Coe 1975
Ivory Coast	In S (map)	Booth 1958
	Abundant in Tai Reserve (now N.P.) (in W between Cavally R. and Sassandra R.)	Struhsaker 1972
	In S part of Comoe N.P. (in NE)	Geerling and Bokdam 1973
	Near Lamto, on Bandama R. (S central)	Bourlière et al. 1974
	Throughout S (map)	Tahiri-Zagrët 1976
	Abundant in Tai N.P., rare in Banco N.P. (n.l.)	Asibey 1978
Ghana	Throughout S (localities mapped); commonest monkey; abundant at Mount Krobo (n.l.) and Shai Hills (SE)	Booth 1956a, 1958
	Common	Jeffrey 1970, 1974
	Common in Bia Tributaries Forest Reserve and at Sefwi Wiawso (all in W)	Jeffrey 1975b
	Common in Digya N.P., Bia N.P., Shai Hills Game Production Reserve, Bomfobiri Wildlife Sanctuary, Willi Falls, Nini-Suhien N.P., Ankasa Game Production Reserve; rare in Kogyae Strict Nature Reserve	Asibey 1978
	Common and stable in Shai Hills Game Production Reserve (in SE)	Rice, pers. comm. 1978
Togo	Present (map)	Booth 1958
Benin	Present (map)	Booth 1958

Cercopithecus pogonias Crowned Guenon

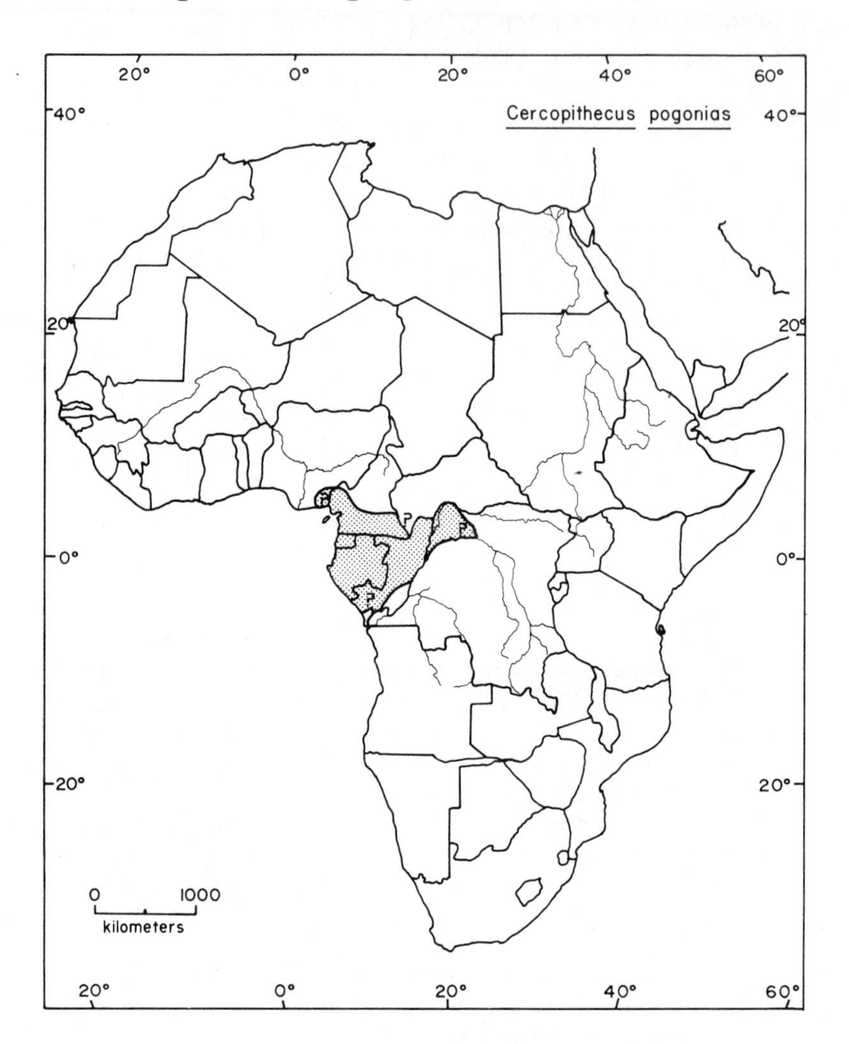

Taxonomic Note

Some authors consider *Cercopithecus pogonias* to be a subspecies of *C. mona* (Sabater Pi and Jones 1967, Dorst and Dandelot 1969, Gautier and Gautier-Hion 1969). But others consider it a separate species consisting of three or four races (Booth 1955, Napier and Napier 1967, Struhsaker 1969). Since it is sympatric with *C. mona* in Cameroon, and occupies a different ecological niche, thus facing different problems than *C. mona* (Gartlan, pers. comm. 1974), it is treated separately here.

Geographic Range

C. pogonias is found in West Africa from western Cameroon to southern Congo. It originally ranged as far west as the Cross River (about 8°E) in eastern Nigeria (Booth

1958), but today its western limit is probably near the western border of Cameroon (about 8°40′E) (Gartlan, pers. comm. 1974). This species has been reported as far south as almost 5°S along the west coast of southern Congo and as far east as 22°E along the right bank of the Zaire River in Zaire (Booth 1955), but no recent confirmation of its presence there was located. For details of the distribution, see Table 111.

Abundance and Density

C. pogonias populations are declining in western Cameroon, Equatorial Guinea, and parts of Gabon (Table 111). Their status in the eastern part of the range is not known.

No estimates of population density are available. Seven groups of the northern subspecies included an average of 13.3 individuals (range 9 to 19, modes 10 and 12) while three groups of the southern Cameroon form had an average of at least 15 members each (range 11 to 19, incomplete counts) (Struhsaker 1969). In Gabon two groups contained 5 and 11 *C. pogonias* (Gautier and Gautier-Hion 1969).

The northern race of this species makes seasonal movements over long distances, bringing them into areas with seasonally high food concentrations (Gartlan, pers. comm. 1974). These movements make it especially difficult to determine population densities and home range areas and suggest that a large reserve area would be necessary to protect these monkeys.

Habitat

C. pogonias is generally restricted to moist high primary forest. This is particularly true of the northernmost race, which is unable to utilize or move through patches of low forest (Gartlan, pers. comm. 1974). The forms to the south seem to be more ecologically tolerant (Booth 1958) and have been reported in secondary forest (Malbrant and Maclatchy 1949, Sabater Pi and Jones 1967, Gautier and Gautier-Hion 1969, Gartlan and Struhsaker 1972) and flooded forest (Gartlan and Struhsaker 1972, Quris 1976). In general, this species is more successful in mature forest than in younger regenerating vegetation (Struhsaker 1969, Bourlière et al. 1970).

Factors Affecting Populations

Habitat Alteration

In Cameroon and Equatorial Guinea logging has already drastically reduced the habitat available to *C. pogonias* (Gartlan, pers. comm. 1974; Sabater Pi and Jones 1967). The northern race is especially hampered by timbering, because it cannot move through cut areas to complete is seasonal movements (see above). Thus, it will certainly be adversely affected by planned logging operations in western Cameroon (Gartlan, pers. comm. 1974).

Human Predation

Hunting. Hunting is responsible for the decline of this species in northeastern Gabon (Gautier and Gautier-Hion 1969) and, to some extent, in Cameroon (Gartlan, pers. comm. 1974).

Pest Control. C. pogonias does not raid crops in Cameroon (Gartlan, pers. comm. 1974). But Hill (1966) described it as being "addicted to raiding plantations."

Collection. C. pogonias is not often captured for research or zoos in Cameroon (Gartlan, pers. comm. 1974). At least 23 were exported from Equatorial Guinea between 1958 and 1968 (Jones, pers. comm. 1974). None were recorded as entering the United States between 1968 and 1973 (Appendix) or in 1977 (Bittner et al. 1978), or the United Kingdom between 1965 and 1975 (Burton 1978).

Conservation Action

C. pogonias occurs in five forest reserves in Cameroon (Table 111).

Table 111. Distribution of *Cercopithecus pogonias* (countries are listed in order from north to south then west to east)

Country	Presence and Abundance	Reference
Nigeria	In extreme SE, E, and N of Cross R. (maps)	Booth 1955, 1958
Cameroon	In W and S (maps)	Booth 1955, 1958
	At Idenau, Elongongo, and Nsépé (on Sanaga River), Lake Tisongo (at mouth of Sanaga R.), and Bukundu Reserve (all in W); Campo Reserve and Dipikar (in SW), and Dja Reserve (S central); more numerous in E than W	Gartlan and Struhsaker 1972
	Abundant in Korup and Douala-Edéa reserves (in W)	Struhsaker 1972
	Throughout SW; declining in S; stable in W but vulnerable (map)	Gartlan, pers. comm. 1974
	In Dja Reserve (S central)	Rowell, pers. comm. 1978
Equatorial Guinea	Throughout, including Fernando Po (map) (now Macias Nguema)	Booth 1955
	In NW and central (localities mapped); range shrinking	Sabater Pi and Jones 1967
Gabon	Throughout	Malbrant and Maclatchy 1949
	In W (map)	Booth 1955
	In Ogooué-Ivindo district (in NE); declining near Ntsi-Belong	Gautier and Gautier-Hion 1969
	Along Ivindo R., 10 km SW Makokou	Gautier-Hion and Gautier 1974
	Along Liboui R., 36 km from Makokou (in NE)	Quris 1976
Congo	In NE and E near Sangha R., Likouala aux Herbes R., Mossaka; not rare in Etoumbi-Ewo region (in W near 0°, 14°49′E)	Malbrant and Maclatchy 1949
	In E and SW (map)	Booth 1955

Table 111 (continued)

Country	Presence and Abundance	Reference
Central African Republic	No information	
Zaire	Between Ubangi R. and Zaire R., E to about 22°E in Stanleyville District (map)	Booth 1955

Colobus angolensis Angolan Black-and-White Colobus Monkey

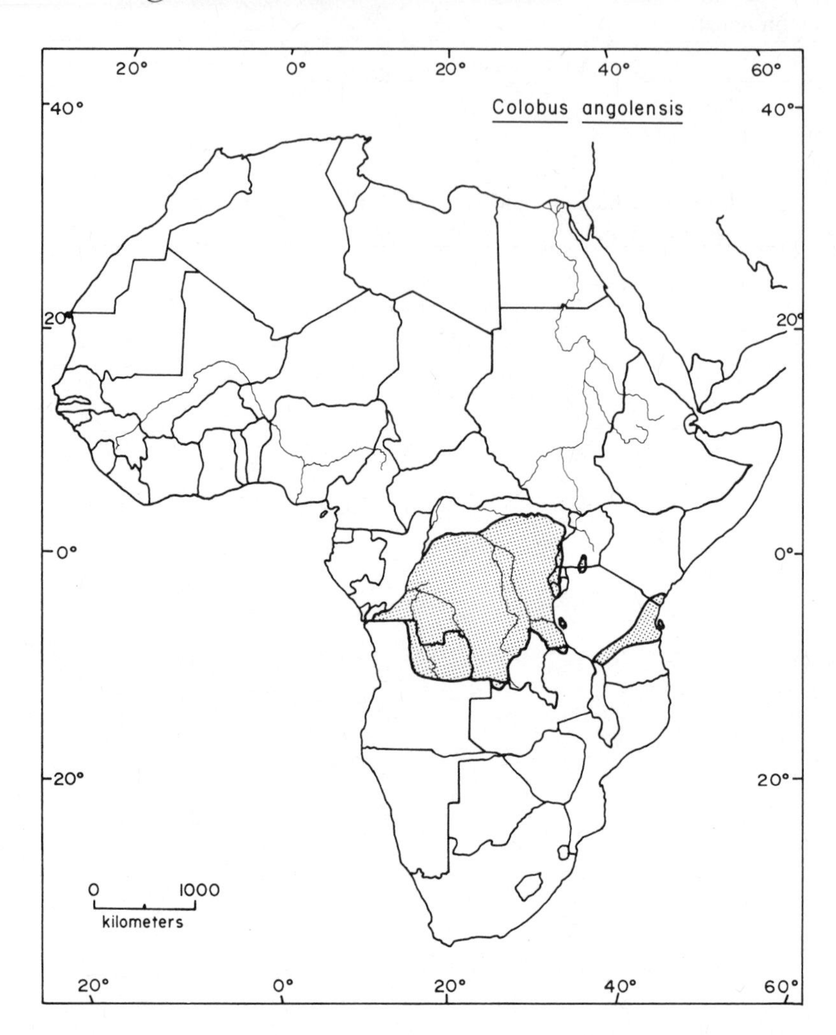

Taxonomic Note

Many earlier workers and some recent ones have included the south central African black-and-white colobus monkey within *Colobus polykomos* (Curry-Lindahl 1956; Rahm and Christiaensen 1960, 1963; Tappen 1960; Napier and Napier 1967; Kano 1971a; Kingdon 1971). But others consider *Colobus angolensis* a separate species (Dandelot 1965; Rahm 1970, 1972; Oates 1977a), and it is treated as such here. Eight races are recognized (Rahm 1970).

Geographic Range

C. angolensis is found in equatorial Africa from the Congo River to the coast of the Indian Ocean in Tanzania. It occurs as far north as the northern limit of the Congo forest at about 3°N (Tappen 1960) and south to about 11°15′S in northwestern Zambia (Ansell 1959). Its distribution is discontinuous east of about 30°E, with patches in northern, southern, and eastern Tanzania, southern Uganda, and southern Kenya (Kingdon 1971). For details of the distribution, see Table 112.

Abundance and Density

C. angolensis has recently been described as common only in some regions of Zaire (Table 112). It is thought to be threatened with extinction in Angola and on Mount Ruwenzori, Uganda, and only a very small population resides in Zambia. Its status throughout most of Zaire, Rwanda, Burundi, Kenya, and Tanzania is not known (Table 112).

The only estimate of population density available is that of 180 to 200 *C. angolensis* per km^2 in the southern highlands of Tanzania (Kingdon 1971). Group sizes vary from 2 to 16 ($\bar{X}$ = 4.3 to 5.5, N = 37 groups) (Groves 1973), up to 12 (Rahm and Christiaensen 1963), and 8 to 10 (N =2) (Moreno-Black and Maples 1977). Some authors have concluded that the maximum group size is about 20 (Rahm and Christiaensen 1960, Rahm 1966), but others have reported group sizes of 2 to 50 (Kingdon 1971).

Habitat

C. angolensis is a forest species, found primarily in lowland and montane rain forest. It can utilize forest patches in savanna (Tappen 1960), lakeside forest borders (Rahm and Christiaensen 1963), riparian forest (Rahm 1966), and coastal forest (Rahm 1970).

Kingdon (1971) characterized this species as being able to occupy a wide range of vegetation, including dry thicket forests, moist savanna woodland, lowland swamp, and coastal, bamboo, and montane forests. In the Tanzania highlands, *C. angolensis* has been observed in montane forest and grassland, alpine bamboo forest, and solid-stemmed bamboo brush (Itani and Nishida, pers. comm. 1974). It has often been reported at high elevations, up to 3,000 m in Rwanda (Rahm 1965) and on Mount Ruwenzori, Uganda (Haltenorth 1975).

C. angolensis does occupy mature secondary lowland rain forest (Horn, pers. comm. 1978). But it does not frequent agricultural areas (Rahm and Christiaensen 1960) or disturbed forest or young secondary vegetation (Moreno-Black and Maples 1977), and seems dependent on the presence of mature forest for its survival. Some populations, such as those in eastern Zaire, are never found far from rivers or streams (Rahm and Christiaensen 1960, Rahm 1966). Those that venture into Zambia are restricted to well-developed high rainfall and montane forests (Smithers 1966).

Factors Affecting Populations

Habitat Alteration

The future outlook is poor for *C. angolensis* populations in the Jadani forest of southern Kenya, owing to extensive forest clearance to make way for hotel development (Oates 1977a). Because of the dependence of this species on mature forest, it is very likely that deforestation is adversely affecting it in other areas of its range, but no direct evidence is available.

Human Predation

Hunting. C. angolensis is hunted intensively in some parts of Zaire (Rahm and Christiaensen 1960; Ladnyj et al. 1972; Horn, pers. comm. 1978). In Uganda it was formerly severely persecuted for its handsome pelt, and poaching is still continuing (Pitman 1954, Kingdon 1971).

In Tanzania, *C. angolensis* is not greatly affected by the fur trade (Mittermeier 1973a). In the past a few skins were imported into Zambia from Zaire and Tanzania, but there is no regular trade in these skins in Zambia, although local villagers sometimes kill *C. angolensis* (Ansell, pers. comm. 1974).

Pest Control. C. angolensis does not enter agricultural fields (Rahm and Christiaensen 1960), and even where it occupies areas adjoining human gardens, it does not eat cultivated foods (Moreno-Black and Maples 1977).

Collection. No *Colobus angolensis* were recorded as entering the United States between 1968 and 1973 (Appendix) or in 1977 (Bittner et al. 1978). None were imported into the United Kingdom in 1965–75 (Burton 1978).

Conservation Action

C. angolensis occurs in one national park in Zaire and one reserve in Kenya (Table 112). It sometimes travels through one national park in Zambia, although it is not resident there, and occurs within one proposed park in Tanzania (Table 112).

This species is legally protected from hunting in Uganda (Pitman 1954) and Zambia (Ansell, pers. comm. 1974), but enforcement is very difficult in all areas.

Table 112. Distribution of *Colobus angolensis* (countries are listed from south to north to south in a clockwise direction)

Country	Presence and Abundance	Reference
Angola	In NE (localities mapped)	Barros Machado 1969
	Threatened	Bothma 1975
Congo	Present (probably confused with *C. guereza*)	Malbrant and Maclatchy 1949
Zaire	W of Lake Tanganyika and on Mount Kahuzi (in E)	Curry-Lindahl 1956
	Recorded from Kanaima, Kapanga, Kinda, Masimba, and Lusiji R. (all in S, 22°–25°E)	Verheyen cited by Ansell 1959
	Throughout S of Uele R. except NW and SE (localities mapped)	Rahm and Christiaensen 1960; Rahm 1970, 1972
	W of Lake Kivu, near Musisi and Lemera (in E)	Rahm and Christiaensen 1963
	Between Lualaba R., Lowa-Osa R., Elila R., and Luama R. (in E)	Elbl et al. 1966, Rahm 1966, Hendrichs 1977
	In Basankusu Territory, Equateur Province (S of Zaire R. near 20°E)	Ladnyj et al. 1972
	Common in center of Haut-Zaire Province (N)	Heymans 1975
	In Salonga N.P. (central)	I.U.C.N. 1975

Table 112 (continued)

Country	Presence and Abundance	Reference
Zaire (cont.)	Fairly common W of Lake Tumba (in W, near 2°S)	Horn, pers. comm. 1978
Burundi	NE of Lake Tanganyika (in W) (locality mapped)	Rahm and Christiaensen 1960, 1963
Rwanda	In Rugege Forest, E of Lake Kivu (in W)	Curry-Lindahl 1956
	In W and S (localities mapped)	Rahm and Christiaensen 1960, 1963
	In Rugege Forest; locally common	Elbl et al. 1966
Uganda	In SE Masaka District, N of Kagera R. (in S) and on Mount Ruwenzori (in W)	Pitman 1954
	In S and W (localities mapped)	Kingdon 1971
	Threatened on Mount Ruwenzori	Struhsaker, pers. comm. 1974
	On Mount Ruwenzori	Haltenorth 1975
	Small numbers on Mount Ruwenzori, may be extinct there; fairly large numbers at Sango Bay	Oates 1977a
Kenya	In extreme SE (localities mapped)	Kingdon 1971
	As far N as Kilifi Creek only	Andrews et al. 1975
	In Shimba Hills National Reserve, Jadani and Shimoni forests	Oates 1977a
Tanzania	In NE, central, and S (N of Lake Malawi) (localities mapped)	Kingdon 1971
	Along lower Pangani R., in Msumbugwe, Matatoro, and Zanzibar forests; E Usambara Mts., Amani District; W Usumbara Mts., Balangai, Muzumbai, and Shume Escarpment (all in NE between 4°45′S and 5°30′S, and 38°13′E and 38°50′E)	Groves 1973
	In Mahali Mts. proposed N.P., E shore Lake Tanganyika, S of Kigoma (in W)	Itani and Nishida, pers. comm. 1974
	Not in any national parks or game reserves	Oates 1977a
	Fairly common in Uzungwa Mts. (S central)	Struhsaker, pers. comm. 1978
Zambia	In NW Mwinilunga District (in NW); probably never occurred S of Songwe R. (in NE)	Ansell 1959
	Occasional vagrant in NW, probably not resident	Ansell 1960a,b; Smithers 1966
	Occasional in NW and in Mweru Marsh Game Reserve	Ansell 1969
	Resident only near Kasombu stream, NW Mwinilunga District; population small but stable; possibly also present at NE end of Lake Mweru, and occasional in Mweru Wantipa N.P.	Ansell, pers. comm. 1974

Colobus badius Red Colobus Monkey

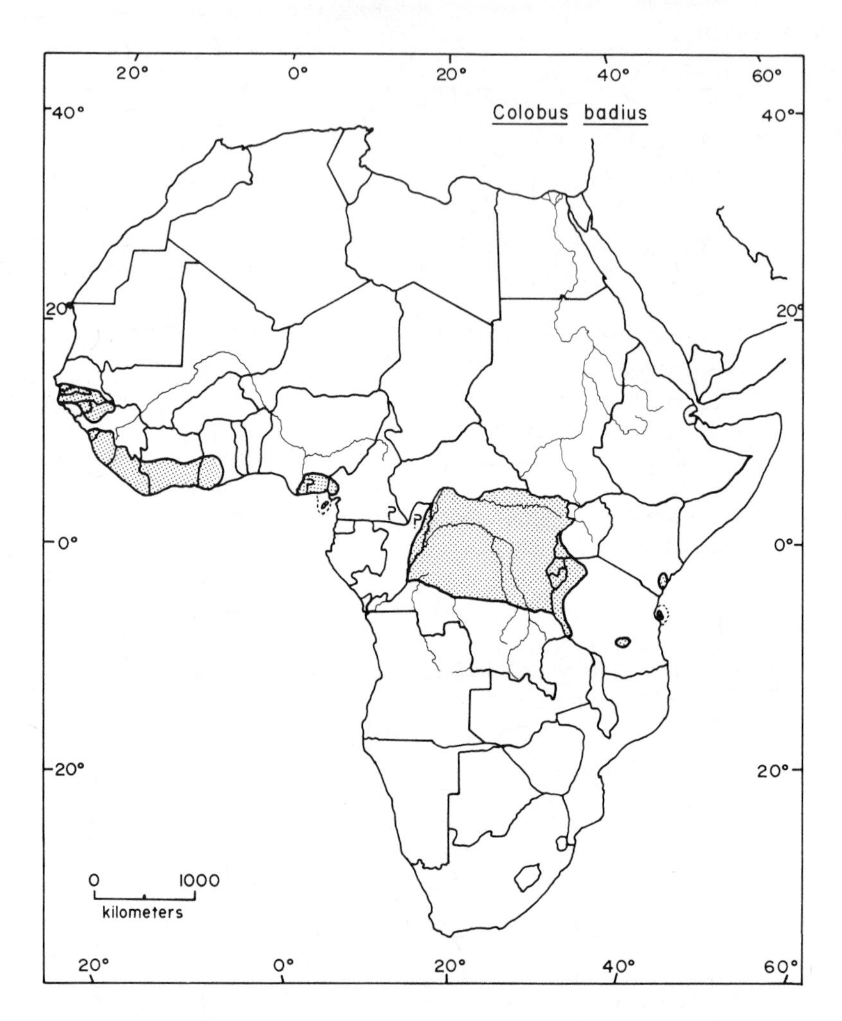

Taxonomic Note

Red colobus monkeys form a large and complex group; as many as 20 subspecies are recognized (Napier and Napier 1967). Dorst and Dandelot (1969) classify the extreme West African forms as one species (*Colobus badius*) and the central and eastern races as another (*C. pennanti*), while Groves (pers. comm. 1974) finds the Tana River form (*C. rufomitratus*) distinctive enough to deserve specific status. Because the systematic questions have not been resolved, all red colobus monkeys are treated together as *C. badius* in this report, and distinctions between populations are made only when they differ in their status.

Geographic Range

C. badius has a discontinuous distribution in western, central, and eastern Africa. It is found from Senegal to Ghana, in Nigeria and Cameroon, from Congo to Uganda, and in parts of Kenya and Tanzania. It occurs as far north as 13°45′N, 13°16′W, and as far west as 12°23′N, 16°34′W in western Senegal (Gatinot 1974), as far east as Zanzibar Island, and as far south as about 8°S, 36°E in Tanzania (Tappen 1960). For details of the distribution, see Table 113.

Abundance and Density

In almost every country for which information is available, red colobus monkeys are locally abundant in a few protected areas but otherwise rare or threatened with extinction (Table 113). This is especially true in Ivory Coast, Ghana, Cameroon, Uganda, Kenya, and Tanzania. The Tana River (Kenya) race, the Zanzibar Island race, and the southern Tanzania race are all quite localized and threatened. Only in Zaire is the red colobus probably still widespread and plentiful, but detailed information is not available. Data are also lacking from Congo, Rwanda, Burundi, and the countries between Senegal and Ivory Coast.

Estimates of population density range from 80 to 880 per km^2 (Table 114), but these are local densities and not indicative of a high density for *C. badius* throughout its range.

Group sizes of 5 to 100 have been reported, with averages of 8 to 40 members per group. Group sizes are generally larger at lower elevations (Nishida 1972), in larger forest patches (Groves et al. 1974), and in ecologically richer habitats (Struhsaker 1975). Home ranges are often less than 20 ha in area, although a range as large as 259 ha has been reported (Table 114).

Habitat

C. badius is a forest monkey, typical of moist lowland or gallery forest. If the whole group is considered together, it appears to be a versatile species, able to utilize a wide variety of habitats. But some of the forms of *C. badius* inhabit rather specialized forest types not utilized by other red colobus races. For example, the Senegalese race (*C. b. temmencki*) is adapted to dry deciduous forest (Booth 1958, Dupuy and Verschuren 1977), gallery forest (Dupuy 1971b, Gatinot 1976), and dry savanna woodland (Struhsaker, pers. comm. 1974; Gatinot 1976). But the populations in the rest of West Africa are restricted to undisturbed, mature moist primary forest (Booth 1956a; Jeffrey 1974; Gartlan, pers. comm. 1974; Struhsaker 1975).

Similarly, in East Africa, the Kibale Forest population occurs primarily in undisturbed, mature, moist, evergreen primary forest (Struhsaker 1975). In contrast, the Tana River race (*C. b. rufomitratus*) inhabits high gallery forest but is most abundant in regenerating forests of abandoned cultivation, because these have many young trees with young foliage, a large variety of food plant species, and tall emergent trees (Groves et al. 1974, Andrews et al. 1975). This race also utilizes evergreen forests, *Acacia* woodland, and woodland-evergreen mixtures (Groves et al. 1974), and occasionally ventures into low bush habitat with scattered trees (Marsh, pers. comm. 1974). On Zanzibar, *C. badius* is found in gallery forest (Kingston, pers. comm. 1974), ground-water forest, and low-stature evergreen scrub forest (Struhsaker 1977b).

Red colobus monkeys in Zaire inhabit swamp, gallery, and rain forests (Rahm

1970) and montane forests up to 2,000 m elevation (Rahm and Christiaensen 1963, Rahm 1965). In Tanzania they occur in semideciduous and evergreen forest (Clutton-Brock 1974), montane forest, riparian and lakeside forest, woodland, and alpine bamboo (Kano 1971a, Nishida 1972a).

Although *C. badius* in Zaire does utilize mature secondary lowland rain forest (Horn, pers. comm. 1978), the populations in Ghana and Cameroon do not occupy secondary growth (Booth 1958; Gartlan, pers. comm. 1974). Several sources cited the presence of tall trees as a prerequisite for *C. badius* survival (Booth 1956a; Gartlan, pers. comm. 1974; Groves, pers. comm. 1974; Kingston, pers. comm. 1974; Marsh, pers. comm. 1974).

Most races of red colobus monkeys are typically found in high forest in the vicinity of water or with high humidity. In East Africa the species is never far from lakes, rivers, or swamps, and the presence of water seems to be important for it (Kingdon 1971). The Senegalese form may be able to survive in savanna woodland, but all other populations seem dependent on dense moist forest with tall emergent trees.

Factors Affecting Populations

Habitat Alteration

Red colobus monkeys are adversely affected by timbering and are not common in disturbed forest (Struksaker 1972, 1975). Logging and the destruction of forest are rapidly reducing their habitats throughout most of their range. For example, as early as 1956, *C. badius* was absent in eastern Ghana, owing to heavy exploitation for timber and clearing for agriculture, and Booth (1956a) predicted that the development of the rest of Ghana would result in its complete elimination. More recently several workers in Ghana have noted the high rate of deforestation, burning, conversion to large-scale agriculture, and road building occurring there (Jeffrey 1975b, Asibey 1978). Forest land outside reserves was reduced by cultivation from 80% to 20% of its original extent between 1960 and 1973 (Jeffrey 1974). *C. badius* is strongly affected by logging because of its dietary dependence on timber trees (Rucks 1976, Asibey 1978). At present, roughly 10% or less of the original *C. badius* habitat in Ghana remains intact (Rucks, pers. comm. 1978).

In other countries of West Africa, deforestation is also rapidly destroying *C. badius* habitat. In Ivory Coast, for example, the Tai National Park is the last remaining sizable area of natural forest remaining (Monfort and Monfort 1973), and in Cameroon, logging has greatly reduced the range of *C. badius* (Gartlan, pers. comm. 1974).

The Miranga and Kasenda forests of Uganda are being rapidly cleared and replaced by cultivation, and the Kibale Forest is being progressively felled and replanted with pine, eucalyptus, and cypress (Struhsaker 1975). After logging, Ugandan forests are treated with herbicides to eliminate undesirable tree species. *C. badius* is much less frequently encountered in these felled and treated areas than in unfelled forest (Oates 1977a). The conservation of red colobus monkeys depends to a large degree on protection of large areas of mature rain forest from timber exploitation (Struhsaker 1975).

The Tana River race of red colobus can occupy the secondary forest which regenerates after cultivation has been abandoned (Groves et al. 1974). But the expanding human population in the area is engaging in extensive permanent clearing for agriculture, cutting for firewood, and burning for charcoal and pasture land (Groves et al. 1974). Between 1961 and 1969 the forests within the range of *C. badius* along the Tana River were reduced by 13.6%, and continued clearing and burning pose a severe threat to the existence of the Tana River red colobus over the next 10 to 20 years (Marsh and Homewood 1975). In addition, planned irrigation projects upriver will reduce river flow and flooding, and may further decrease the extent of the flood plain forest habitat of *C. badius* (Andrews, pers. comm. 1974).

On Zanzibar Island most of the forest has been cleared for agriculture, and in 1972 logging was still continuing at a low rate (Kingston, pers. comm. 1974). More recently the greatest threat to the Zanzibar red colobus is still habitat destruction: timber felling, charcoal burning, cultivation, and bush burning by hunters (Struhsaker 1977b).

In southern Tanzania, illegal agricultural encroachment and clearing of forest for railway access have destroyed large areas of *C. badius* habitat (Struhsaker 1977a), as has clearance for settlements and tea estates (I.U.C.N. 1972).

At Nishida's (1972) study area in western Tanzania, all forest areas had been intensively cultivated and were in various stages of regeneration, but still supported *C. badius* populations. In western Zaire, rotational agriculture was widespread and *C. badius* successfully occupied mature secondary forest (Horn, pers. comm. 1978). Thus some populations of *C. badius* can tolerate limited habitat disturbance, and adults may recolonize secondary growth if intact forest remains nearby. But reproduction is often unsuccessful in these altered habitats, and the populations frequently fail because of low recruitment (Struhsaker, pers. comm. 1978).

Human Predation

Hunting. C. badius is hunted for its flesh in many parts of its range. It is killed for food in Liberia (Leutenegger 1976), Ivory Coast (Struhsaker 1975), Ghana (Asibey 1974, Jeffrey 1975b), Cameroon (Gartlan and Struhsaker 1972, Gartlan 1975a), and Zaire (Ladnyj et al. 1972; Horn, pers. comm. 1978). It is also considered a delicacy by those who eat monkey in East Africa (Kingdon 1971), and is shot for its meat and skins for ceremonial costumes and bellows in western Tanzania (Kano 1971a, Nishida 1972a). It is not hunted for food by people in the Casamance region of Senegal, the Kibale Forest of Uganda (Struhsaker 1975), or the Tana River, Kenya (Groves et al. 1974). The Zanzibar Island race is likely to be shot, because its range is in a military training zone (Kingdon 1971).

In areas where *C. badius* is heavily hunted, this predation can have a serious effect on its populations. In Cameroon, hunting is the second most serious threat to its survival (after deforestation) (Gartlan, pers. comm. 1974) and has probably been responsible for its extinction in the Kumba area (Gartlan and Struhsaker 1972). In Ghana, *C. badius* is the first species to become extinct in areas where hunting is common (Asibey 1978). *C. badius* is especially easy to hunt, because it ascends to tree tops and remains immobile when frightened (Gartlan 1975a). The pressure of hunting is intensified by timbering, which provides roads and hence access by hunters to previously remote areas (Struhsaker 1975).

Pest Control. This species is not regarded as a crop raider, and was described as "not damaging to agriculture" by Kano (1971a), Robinson (1971), Gartlan (pers. comm. 1974), and Groves et al. (1974). On Zanzibar Island, farmers shoot *C. badius* for crop damage that was probably committed by vervet monkeys (Leakey 1969).

C. badius does enter plantations of exotic pine trees (*Pinus carrabae*) contiguous to natural forest in Uganda (Struhsaker, pers. comm. 1974). Since the monkeys break the branches of these trees while eating the needles, foresters are beginning to worry about damage to the trees, and this could lead to control programs in the future (Struhsaker, pers. comm. 1974).

Collection. Red colobus monkeys do not survive well in captivity and thus are not used in research or zoo exhibits. None were imported into the United States between 1968 and 1973 (Appendix) or in 1977 (Bittner et al. 1978). None entered the United Kingdom in 1965–75 (Burton 1978).

Conservation Action

C. badius is found in seven national parks and six forest reserves (Table 113). Nevertheless, populations in many parts of its range are not adequately protected and are in need of conservation action. For example, this species is fully protected by law in Ghana but is still heavily poached outside national parks and game reserves (Asibey 1978). Many of the forest reserves are still being depleted by settling and farming (Jeffrey 1975b). The Department of Game and Wildlife is making a strong effort to establish and maintain parks and enforce hunting laws, but local political pressure and popular resistance have made conservation difficult (Rucks, pers. comm. 1978).

The Cameroon race (*C. b. preussi*) is legally protected under the African Convention of 1969 and Cameroon law (I.U.C.N. 1978). Its last remaining habitat, the Korup Reserve, has been proposed for upgrading to national park status — this would protect it from further reduction by logging — and the Cameroon government is trying to improve the enforcement of hunting regulations within the reserve (Gartlan, pers. comm. 1974).

In Uganda the only viable population of *C. badius* is in the Kibale Forest Reserve, of which 8 km^2 have been declared a research plot and 20 km^2 a nature reserve — both to be completely protected from all forms of exploitation or disturbance (Struhsaker, pers. comm. 1974, 1978). Also, all forms of colobus are protected from hunting by Ugandan law (Struhsaker, pers. comm. 1978). The long-term conservation of most subspecies of *C. badius* can only be achieved by the protection of large areas of mature rain forest, and these could be promoted as tourist attractions and biological research sites to enable economic self-sufficiency (Struhsaker 1975).

The Tana River race (*C. b. rufomitratus*) has recently been given protection in the 175 km^2 Tana River Primate Game Reserve (Marsh 1977). This reserve encompasses about 60% of the remaining habitat, and if it can be protected from agricultural encroachment and water control projects, should be an effective conservation area (Marsh and Homewood 1975). It is being developed for tourism and game viewing (Marsh and Homewood 1975), and efforts are being made to provide alternatives for the local people to compensate for the changes in their economy because of the reserve (Marsh 1977).

The Jozani Forest Reserve, Zanzibar Island, is being protected by the Tanzanian government (Kingston, pers. comm. 1974). Struhsaker (1977a) recommended that this reserve be expanded northward, eastward, and southward and be upgraded to national park status in which all forms of human activity would be prohibited. He also suggested that another national park be established on southwest Zanzibar Island and nearby Usi Island, to increase the area of protected habitat.

On Tanzania's mainland, *C. b. tephrosceles* is protected in the Mahali Mountains (proposed) and Gombe Stream National Parks (Table 113). But the southeastern form (*C. b. gordonorum*) is in need of immediate conservation action (Struhsaker, pers. comm. 1978). The forest patches of the Uzungwa Mountains are listed as forest reserves but are difficult to protect (I.U.C.N. 1978). Struhsaker (1977a) has recommended that the Magombero Forest be incorporated into the nearby Selous Game Reserve, that laws against agricultural encroachment be enforced in order to protect this and other forest patches in the area, and that an extensive survey by conducted to provide data for a management plan for this form of red colobus and its habitats.

The Tana River and Zanzibar Island races of red colobus (*C. b. rufomitratus* and *C. b. kirkii*) are listed in Appendix I of the Convention on International Trade in Endangered Species of Wild Fauna and Flora and as endangered under the U.S. Endangered Species Act (U.S. Department of Interior 1977a,b). They and the southern Tanzanian race (*C. b. gordonorum*) are included in Class A of the African Convention and are prohibited from import into Great Britain (I.U.C.N. 1972).

Table 113. Distribution of *Colobus badius* (countries are listed in order from west to east and north to south)

Country	Presence and Abundance	Reference
Senegal	Along Casamance R. (map)	Booth 1958
	In Niokolo-Koba N.P. (in SE)	Dupuy 1969
	Rare in Niokolo-Koba N.P.	Dupuy 1970b, 1971c
	Throughout S, as far N as Sine-Saloum (map); fairly common near mouth of Casamance R.	Dupuy 1971b
	In Casamance Province, S of Casamance R. (in SW)	Struhsaker 1971
	Common in Basse Casamance N.P. (in SW)	Dupuy 1973
	Confined to S of Gambia R. except for small population N of river near mouth (localities mapped)	Gatinot 1974
	At Niadio and Narangs Forest (in SW)	Struhsaker 1975
	Population more than 600 in Saloum Delta N.P. (in W near 14°N); also in Niokolo-Koba N.P.	Dupuy and Verschuren 1977
Gambia	Along Gambia R. (localities mapped)	Gatinot 1974
	At Dembakunda (extreme E)	Struhsaker 1975
	In Abuko Nature Reserve (in W)	Gunderson 1977
Guinea-Bissau	In E (maps)	Rahm 1970, Kingdon 1971

Table 113 (continued)

Country	Presence and Abundance	Reference
Guinea	In N and S (maps)	Booth 1958, Rahm 1970
Sierra Leone	In Kasewe Forest (S central)	Tappen 1964
	Present	Jones 1966
	Throughout (map)	Rahm 1970
Liberia	Throughout (map)	Booth 1958, Rahm 1970
	Present	Kuhn 1965
	Abundant	Schlitter, pers. comm. 1974
	Along Cess R. between Potogle and Kpeaple	Leutenegger 1976
Ivory Coast	In S near Sassandra R., Bandama R., and Nzi R. (localities mapped)	Booth 1954, 1958
	In Banco N.P. (in SE)	Happold 1971
	Abundant in Tai Reserve (now N.P.)	Struhsaker 1972 [Monfort and Monfort 1973]
	In Goudi Forest; disappearing near Lamto (S central near Bandama R.)	Bourlière et al. 1974
	Near Troya and Sakré (in SW) and Béréby-Tabou Road and Itragi (near Atlantic Coast)	Struhsaker 1975
	Abundant in Tai N.P., rare in Asagny Fauna Reserve	Asibey 1978
Ghana	In SW (localities mapped)	Booth 1954
	In SW (localities mapped); locally abundant; declining, range shrinking, probably will face extinction	Booth 1956a
	Declining; in need of protection	Jeffrey 1974
	Rare in Bia Forest Reserves, Krokosua Hills, Kwambi Krom (all in SW); susceptible to extinction	Jeffrey 1975b
	Common in Bia N.P.; rare in Nini-Suhien N.P. and Ankasa Game Production Reserve	Asibey 1978
	Scarce and declining	Rucks, pers. comm. 1978
Nigeria	E of Niger R.; in need of protection	Happold 1971
Cameroon	In extreme W (map)	Booth 1958
	In W and in SE (map)	Kingdon 1971
	Abundant in Korup Reserve (extreme W); the last population of this race (*C. b. preussi*)	Struhsaker 1972
	Restricted to 1 or 2 small areas in W, perhaps in Rumpi Hills (map); locally numerous; previously throughout coastal areas; declining, threatened	Gartlan, pers. comm. 1974
	Seriously threatened	Struhsaker, pers. comm. 1974

Table 113 (continued)

Country	Presence and Abundance	Reference
Cameroon (cont.)	Last surviving population in Korup Reserve	Gartlan 1975a
	On the verge of extinction	Struhsaker 1975
	Endangered; total population believed to be less than 8,000; total range about 7,200 km^2	I.U.C.N. 1978
Equatorial Guinea	On Macias Nguema (formerly Fernando Po) (maps)	Rahm 1970, Kingdon 1971
	Threatened on Macias Nguema	Gartlan, pers. comm. 1974
Congo	Localized populations in N and NE	Malbrant and Maclatchy 1949
	In E (map)	Rahm 1970
	Throughout N (map)	Kingdon 1971
Zaire	In N between Zaire R. and Ubangi R.	Malbrant and Maclatchy 1949
	W of Lake Tanganyika (in E)	Curry-Lindahl 1956
	Very abundant in Ituri Forest (in NE)	Hayman cited in Tappen 1960
	W of Lake Kivu, S of Mount Kahuzi and near Irangi (in E)	Rahm and Christiaensen 1963
	Between Ulindi, Lugulu, Luama, Aruwimi, and Ituri rivers (E and NE)	Rahm 1966
	Throughout N as far S as Kasai R. in W and almost 6°S in E (map)	Rahm 1970
	In Basankusu Territory, Equateur Province (in W, S of Zaire R.)	Ladnyj et al. 1972
	Common; in Haut-Zaire Province in N: E and N of Aruwimi R., S of Aruwimi to Ituri R.; along Lomami R., S of Zaire R. in Equateur Province	Heymans 1975
	In Salonga N.P. (central)	I.U.C.N. 1975
	In Mts. W of Lake Kivu	Hendrichs 1977
	Common W and S of Lake Tumba (in W), and S of Zaire R. across from Lisala	Horn, pers. comm. 1978
Uganda	Only in small area in W Toro	Pitman 1954
	In SW (localities mapped)	Kingdon 1971
	One viable population: Kibale Forest; potentially endangered	Struhsaker, pers. comm. 1974
	Common near Bigodi and Kanyawara Station, Kibale Forest	Clutton-Brock 1975a
	In Kibale Forest	Hayashi 1975, Struhsaker and Oates 1975, Mizuno et al. 1976
	Abundant in Kibale Forest; in small patches of Miranga Forest, Kasenda Forest, Mpanga R. forest	Struhsaker 1975
Rwanda	Present (maps)	Rahm 1970, Kingdon 1971
Burundi	Present (maps)	Rahm 1970, Kingdon 1971

Table 113 (continued)

Country	Presence and Abundance	Reference
Kenya	At mouth of Tana R. and Galana R. (in SE) (localities mapped) (*C. b. rufomitratus*)	Kingdon 1971
	Rare; confined to lower Tana R.	I.U.C.N. 1972
	Scarce	Andrews, pers. comm. 1974
	Locally abundant in patches between Garsen-Lamu road and Kipende, S of Wenje, all near Tana R.; total population about 1,860; stable but likely to decline	Groves, pers. comm. 1974
	Locally abundant but declining	Marsh, pers. comm. 1974
	Total range about 6.48 km^2; total population about 1,860	Groves et al. 1974
	Recorded in 14 forest patches, total area about 3.13 km^2, may be in 7 more patches, additional area of 1.85 km^2; in relatively large numbers in Home Forest and Mnazini Forest I	Andrews et al. 1975
	In Lake Nakuru N.P. (central)	Dupuy 1975
	Total population about 1,500–2,000; total range about 20 km^2 along 60 km stretch of Tana R.; habitat shrinking; part now in Tana River Primate Game Reserve	Marsh and Homewood 1975
	Endangered	I.U.C.N. 1976
Tanzania	Zanzibar Island race (*C. b. kirkii*) declining, endangered	Leakey 1969
	Locally plentiful in isolated populations in mountains E of Lake Tanganyika, E of Kasoge (in W, near 6°S) (localities mapped) (*C. b. tephrosceles*)	Kano 1971a
	In NW, W, central, and Zanzibar Island (localities mapped)	Kingdon 1971
	In Uzungwa Mts. in SE; this race (*C. b. gordonorum*) rare; habitat shrinking	I.U.C.N. 1972
	In Mahali Mts. (proposed N.P.) E of Lake Tanganyika (in W, at about 6°S)	Nishida 1972; Itani and Nishida, pers. comm. 1974
	Zanzibar Island race: scarce but remaining stable; limited to Jozani Forest, area about 2 km^2; total population about 200	Kingston, pers. comm. 1974
	In Gombe Stream Reserve (now N.P.) E of Lake Tanganyika (about 4°40′S); total population at least 1,000	Clutton-Brock 1975a
	On Zanzibar Island: in Jozani Forest Reserve, Uzi Island, Muyuni, Muungwi, and Pete Forest	Struhsaker 1977a
	On S mainland: viable population (at least 150–200) in Magombero Forest Reserve; present near Mangula in Mwanihana Forest Reserve	Struhsaker 1977b

Table 113 (continued)

Country	Presence and Abundance	Reference
Tanzania (cont.)	Rare on Zanzibar Island: total population 150–200 and stable; rare in S and SE mainland: 200–500 in Magombero Reserve; scattered populations from Dabaga E to Ifakara and N along Great Ruaha and Msolwa rivers	I.U.C.N. 1978

Table 114. Population parameters of *Colobus badius*

Population Density (individuals /km²)	Group Size $\bar{X}$	Group Size Range	Group Size No. of Groups	Home Range Area (ha)	Study Site	Reference
	12				SE Senegal	Dupuy 1971b
45–480	29	14–62	22	9–19.7	W Senegal	Gatinot 1975
310–880		24–40	3	4.3–12.8	W Gambia	Gunderson 1977
		≤50			Sierra Leone	Jones 1966
	40	24–46		259	W Ghana	Rucks, pers. comm. 1978
	30				W Cameroon	I.U.C.N. 1978
		50–100			Uganda	Kingdon 1971
		58–64	2	49–96	W Uganda	Clutton-Brock 1975a
	37	25–49	2	7.6–8	W Uganda	Hayashi 1975
125–450	50	12–80		8.3–35.3	W Uganda	Struhsaker 1975
		60–70	1	31	W Uganda	Mizuno et al. 1976
260–575	8–15	5–24			SE Kenya	Groves et al. 1974
		50+			SE Kenya	Marsh and Homewood 1975
	20–25			4.86	SE Kenya	Marsh 1977
	10–15	5–30			W Tanzania	Kano 1971a
80–100	40	10–50	8	40–60	W Tanzania	Nishida 1972a
		40–80		114	W Tanzania	Clutton-Brock 1975b
100					Zanzibar Island, Tanzania	Kingston, pers. comm. 1974

Colobus guereza

Guereza; Abyssinian Black-and-White Colobus Monkey

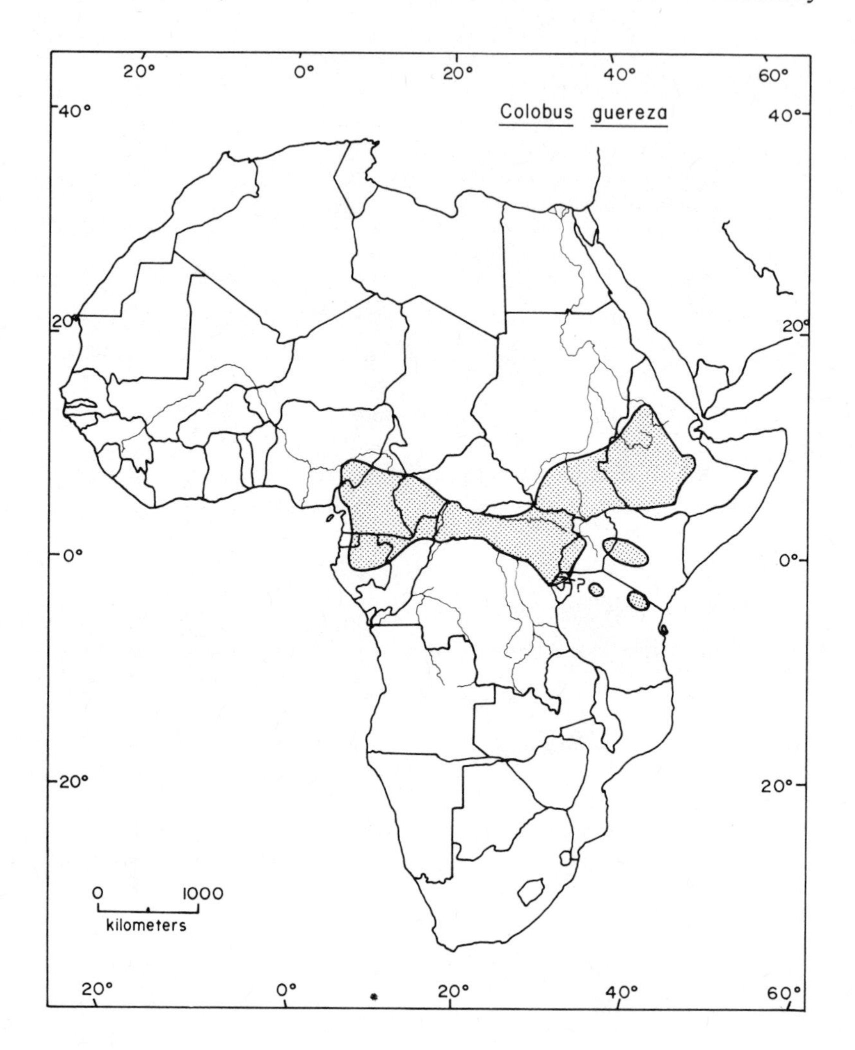

Taxonomic Note

Some authors (e.g., Butler 1966, Gautier and Gautier-Hion 1969) include these monkeys within *Colobus polykomos*, but most recent workers agree that the guereza is specifically distinct from both *C. polykomos* and *C. angolensis*. Rahm (1970) lists nine subspecies.

Geographic Range

Colobus guereza is found in central Africa from eastern Nigeria to eastern Ethiopia. It occurs as far west as Fotabong, Cameroon (5°35′N, 9°55′E), as far east as Harrar, Ethiopia (9°18′N, 42°10′E), as far north as Adai Arkai, Ethiopia (13°50′N,

38°00′E), and as far south as Lake Jipe, Tanzania (3°35′S, 37°45′E) (records and their sources in Oates 1977a). In the west its range is narrower, extending from 9°N in eastern Nigeria to about 1°S in Gabon (Oates 1977a). The maximum size of the range is about 1.28 million km^2 (Oates 1977a). For details of the distribution, see Table 115.

Abundance and Density

C. guereza has recently been described as locally abundant or locally common in parts of Congo, Zaire, Sudan, Uganda, and Ethiopia (Table 115). But it is also known to be declining in the last three countries as well as in Cameroon. Its status in the southwestern and southeastern portions of its range is not known.

Population density estimates for this species range from 0.10 to 300 per km^2, depending on the habitat. Group sizes vary from 2 to 18, with average sizes of 4 to 10. Home ranges of 1.5 to 39 ha have been reported (Table 116).

Habitat

C. guereza is a forest species which can utilize many different habitats. It is found in all types of closed forest formations, including primary rain forest (Haddow 1952, Schenkel and Schenkel-Hulliger 1967, Brown and Urban 1970, Dandelot and Prévost 1972, Bernahu 1975, Clutton-Brock 1975a, Dunbar and Dunbar 1975a, Mizuno et al. 1976, Quris 1976), swamp forest and flooded forest (Malbrant and Maclatchy 1949, Gautier and Gautier-Hion 1969), and gallery forest (Dunbar and Dunbar 1974b, Struhsaker and Oates 1975, Lavieren and Bosch 1977).

The majority of *C. guereza* habitats are moist colonizing, riparian, or upland forests, although lowland forests are also frequently utilized (Oates 1977a). Guerezas often occupy secondary growth (Brown and Urban 1970), and seem to be most successful in, and best adapted to, relatively young secondary forest (Struhsaker and Oates 1975).

Guerezas can also utilize several other habitat types, including *Acacia* and *Combretum* brushland (Swynnerton 1958), *Acacia* woodland (Bernahu 1975), bamboo forest (Omar and de Vos 1970), and thickets (Kingdon 1971, Dunbar and Dunbar 1974b). They have often been found in montane forests (Ullrich 1961, Rahm 1970, Dunbar and Dunbar 1974b) at elevations up to 4,500 m (Bernahu 1975).

In Ethiopia, population density of *C. guereza* seems to be closely related to both the amount of tree cover and the presence of preferred food tree genera (*Ficus* and *Celtis*) (Dunbar and Dunbar 1974b). All authors agree that guerezas are dependent on trees, and although they do come to the ground (Dunbar and Dunbar 1974b), and can travel up to 2 km over open country (Rowell, pers. comm. 1978), they cannot survive in the absence of trees.

Factors Affecting Populations

Habitat Alteration

Intensive logging is destroying *C. guereza* habitat in Cameroon (Gartlan, pers. comm. 1974), Ethiopia (Dunbar and Dunbar 1975a), and Kenya (Zimmerman 1972; Rowell, pers. comm. 1978). In Uganda, selective felling is practiced in the forest reserves. If regeneration is permitted, this logging may actually benefit *C. guereza* by promoting secondary growth and hence increasing its food supply (Struhsaker 1972). But in many felled areas, arboricides are applied to kill undesirable trees. This results in "the complete removal of large trees over wide areas and . . . is almost certainly deleterious to the guereza" (Oates 1977a).

A comparison of guereza populations in forests in Kenya showed significantly lower densities in disturbed areas in which forests had been cleared for plantations than in undisturbed forests (Kingston 1971). Similarly, lower population densities were found in selectively felled than in unfelled forest areas in Uganda (Oates 1977a). Although selective logging may be less deleterious to *C. guereza* than to other species more dependent on large trees, the alteration of forest structure, the killing of individuals during tree felling, and the disruptions to ranging patterns may cause long-term declines in *C. guereza* populations (Oates 1977a).

Large portions of *C. guereza* habitat have been replaced with coffee plantations (Bernahu 1975) and other agricultural infringement (Dandelot and Prévost 1972) in Ethiopia, and tea estates and food crop fields in Uganda (Oates 1977a). In addition, natural forest vegetation is being replaced by exotic conifers and *Eucalyptus* forests, uninhabitable by guereza, in many parts of East Africa, including the Mount Elgon and Mafuga forest reserves in Uganda and many areas of the Kenya Highlands (Oates 1977a).

In Ethiopia the destruction of vast tracts of forests in Kefa, Ilubabor, and Sidamo provinces has already eliminated much of the best *C. guereza* habitats and is threatening the continued existence of the species (Dunbar and Dunbar 1975a). The demand by local people for wood for fuel and land to grow food, and the demand in the cities for charcoal, have also led to widespread cutting of vegetation in the Amhara Highlands, riverine forests, and *Acacia* savanna, and the further decline of guereza habitats (Dunbar and Dunbar 1975a).

Repeated burning of cleared areas and grasslands to maintain pastures has prevented the regeneration of forest vegetation in Ethiopia (Bernahu 1975) and Uganda (Oates 1977a), and is restricting guerezas to continually decreasing forest patches. As Oates (1977a) has pointed out, "small and isolated patches are unlikely to favor the long-term survival of their [guereza] inhabitants."

In Kenya the Kakamega Forest is steadily retreating, owing to the activities of charcoal burners and agriculturists (Zimmerman 1972), and guereza habitat on Mount Kenya is shrinking rapidly as a result of timbering (Rowell, pers. comm. 1978). Deforestation is slowly decreasing guereza habitat in southern Central African Republic (Oates 1977a). Overall, habitat alteration is having a detrimental effect on guereza populations in every country for which information is available.

Human Predation

Hunting. C. guereza is threatened by hunting for its beautiful fur, especially in East Africa, where the pelts are in demand as ornaments for warriors and weapons (Tappen 1960), for dance costumes, rugs, furnishings, coats, hats, bicycle saddles, and trophies (Kingdon 1971). *C. guereza* skins were seen by Marco Polo on capes of the mongol khans of central Asia, and the current absence of guerezas in several East African forests is attributed to the demand for their fur which resulted in more than two million skins reaching Europe in the late 1800s (Kingdon 1971).

Today many guereza skins are made into rugs and wall hangings and sold to tourists. In 1972 a survey of 60 tourist shops in Kenya revealed rugs on display representing 5,002 *C. guereza*, and undisplayed stock in these stores probably brought the total to at least 27,500 guereza skins (Mittermeier 1973a). In 1974 considerably fewer skins were being offered for sale in some of the same shops, and the annual turnover of *C. guereza* skins in Nairobi was estimated at 20,000 or fewer (Oates 1977a). Both of these investigators found evidence that many of the skins sold in Kenya had been smuggled into that country from Ethiopia, and that a large number of skins were being exported from Kenya to Europe and the United States for sale as rugs and clothing.

In Ethiopia about 20,000 guerezas were killed annually in the past for the sale of their skins to tourists, and this hunting is responsible for the local extermination of the species in some areas such as the one near Shashemenē in the Rift Valley (Bernahu 1975). During 1972 about 200,000 guereza skins were being offered for sale in Ethiopia, representing the deaths of 40,000 *C. guereza* per year over the preceding five years, primarily in Kefa Province (Dunbar and Dunbar 1975a). These authors calculate that at this rate of hunting the guereza would become extinct in Ethiopia in about 25 years, but that with the losses of skins due to wasteful hunting and tanning methods, and the additional heavy pressure of deforestation, extinction may take as few as 10 to 15 years.

No guereza skins were discovered being offered for sale in Tanzania in 1972 (Mittermeier 1973a), but local people do kill this species for its fur, used for ornaments and clothing (Ullrich 1961). Poaching for skins is still occurring in Tanzania and Uganda (Oates 1977a).

Guerezas are hunted for their flesh in some parts of Uganda (e.g., Semliki and Ruwenzori) (Kingdon 1971; Oates 1977a; Rowell, pers. comm. 1978). In Sudan they are in danger of extinction from hunting, for their meat and skins (Butler 1966). They are not hunted extensively in Cameroon (Gartlan, pers. comm. 1974) or Central African Republic (Oates 1977a). They were not heavily hunted in the Congo-Gabon area 30 years ago (Malbrant and Maclatchy 1949), but no recent information is available from this part of their range or from Zaire or Nigeria.

Pest Control. Local residents often report that guerezas damage crops (Tappen 1960, Mittermeier 1973a). But in Ethiopia this species does not raid fields (Dunbar and Dunbar 1975a), and in Uganda it may do so occasionally but not frequently (Oates 1977a). Although *C. guereza* is sometimes shot in conifer plantations in Kenya and Uganda, no evidence of their actually damaging conifers could be found (Oates 1977a). It is likely that guerezas have been falsely incriminated in conifer damage because of their association with destructive species such as *Cercopithecus mitis*, and also that crop and tree damage is used as an excuse for shooting guerezas to obtain their skins for sale (Mittermeier 1973a, Oates 1977a).

Collection. C. guereza is not often used in medical research. It makes a striking zoo exhibit and is represented in many zoos. Although captive breeding may provide some of the zoo stock, a large percentage of zoo animals are probably from the wild (Oates 1977a).

Two black-and-white colobus were exported from Equatorial Guinea between 1958 and 1968 (Jones, pers. comm. 1974). Between 1968 and 1973 a total of 65 *C. guereza* were imported into the United States (Appendix). In addition, the *C. polykomos* originating from Kenya were probably *C. guereza*. In 1977, one shipment (of unspecified size) of *C. guereza* entered the United States from Kenya (Bittner et al. 1978). Three individuals entered the United Kingdom in 1965–75 (Burton 1978).

Conservation Action

Colobus guereza is found in 18 national parks and 16 reserves (Table 115). But in several of these it is rare or it or its habitats are not completely protected. Ethiopia is planning to declare 5 new national parks in which guerezas occur, but has no plans to establish wildlife reserves in the best guereza habitats in Kefa and Ilubabor provinces (Dunbar and Dunbar 1975a). No information was located about habitat reserves for guerezas in Nigeria, Gabon, Congo, Sudan, or Rwanda.

This species is legally protected from hunting in many countries, including Cameroon (Gartlan, pers. comm. 1974), Tanzania, Kenya, and Uganda (Oates 1977a). But these protective laws are difficult to enforce, and hunting, smuggling, and sale of skins are all occurring. For example, the curtailment of the influx of Ethiopian guereza skins into Kenya is necessary for effective enforcement of Kenya's protective laws (Mittermeier 1973a).

The Ethiopian skin trade is now about 80% under government control, with only three dealers legally authorized to trade in guereza pelts (Bernahu 1975). The success of antihunting conservation laws depends largely on cessation of public demand for guereza skin products, and unless these laws are effective, the guereza will survive in Ethiopia only in isolated pockets in national parks (Dunbar and Dunbar 1975a). The end of commercial exploitation would greatly improve the outlook for *C. guereza*, and could perhaps be aided by a ban on importation of all black-and-white colobus skins by western countries and Japan (Oates 1977a).

Table 115. Distribution of *Colobus guereza* (countries are listed in order from west to east)

Country	Presence and Abundance	Reference
Nigeria	Only in upper Benue River valley and nearby drainage (in E, near 9°N, 12°E); status unknown	Oates 1977a
Cameroon	In W and E Dja Reserve (S central)	Gartlan and Struhsaker 1972
	In S two-thirds, S of about 7°N and E of about 11°E (absent in coastal area); declining	Gartlan, pers. comm. 1974
	Infrequent in Bouba-Ndjida N.P. (in NE 8°25′–9° on Chad border)	Lavieren and Bosch 1977
	In Faro N.P. and Benue (Bénoué) N.P.	Oates 1977a
	Rare in Dja Reserve	Rowell, pers. comm. 1978
Equatorial Guinea	Reported in Mikomeseng area; not verified	Oates 1977a

Table 115 (continued)

Country	Presence and Abundance	Reference
Gabon	In Ivindo basin (NE) as far W as Voung R. (= Mvoung?) and S to confluence of Ivindo and Ogooué R. (near equator)	Malbrant and Maclatchy 1949
	In NE (Ogooué-Ivindo basin)	Gautier and Gautier-Hion 1969
	In NE on Liboui R.	Quris 1976
Congo	Locally abundant in N and E	Malbrant and Maclatchy 1949
	Abundant in N and E in basins of Alima, Sangha, and Likouala rivers	Vattier-Bernard cited by Oates 1977a
Central African Republic	Widespread in W and S; in St. Floris N.P. in N	Bahuchet cited by Oates 1977a
Zaire	In N and NE (localities mapped)	Rahm and Christiaensen 1960
	In Garamba N.P. (NE border with Sudan)	I.U.C.N. 1971
	Common throughout E and NE Haut-Zaire Province (N)	Heymans 1975
	Probably in Virunga N.P. and Ituri Forest	Oates 1977a
Sudan	Locally common in Imatong and Dindinga Mts. to E of White Nile and Lui (in SE) (map); declining; in danger of extermination	Butler 1966
	In S and SW (localities mapped)	Kock 1969
Uganda	In Bwamba Forest (in W)	Lumsden 1951
	In W; on Mount Elgon and Mount Kadam	Pitman 1954
	Locally common	Tappen 1960
	In Budongo Forest (in W)	Marler 1969
	In W and E; extinct in central and S (localities mapped)	Kingdon 1971
	In Murchison Falls N.P. (NW)	Leskes and Acheson 1971
	Common in Kibale Forest (W)	Struhsaker 1972 [1975, Struhsaker and Oates 1975, Clutton-Brock 1975a, Hayashi 1975, Mizuno et al. 1976]
	Locally abundant; declining in Semliki Forest and probably elsewhere	Oates, pers. comm. 1974
	In 2 national parks (Ruwenzori and Kabalega); 9 forest reserves (Kibale, Itwara, Budongo, Bugoma, Kalinzu, and Kasyoha-Kitomi all in W; Bwindi in SW; Mount Elgon, Mount Kadam in E); Toro Game Reserve, Debasien Animal Sanctuary; and Kigezi Gorilla Sanctuary (localities mapped); declining in some areas	Oates 1977a
Rwanda	Previously in N, now uncertain	Oates 1977a

Table 115 (continued)

Country	Presence and Abundance	Reference
Tanzania	Hundreds in W Serengeti N.P. (in N)	Swynnerton 1958
	On Mount Meru (in N)	Ullrich 1961
	In Serengeti N.P. near Orangi R.	Hendrichs 1970, 1972
	In Arusha N.P.	I. U.C.N. 1971
	In N (localities mapped)	Kingdon 1971
	In Arusha N.P., Mount Meru, and Ngurdoto Crater	Groves 1973
	In Arusha N.P., Serengeti N.P., and Kilamanjaro Game Reserve (all in N) (localities mapped)	Oates 1977a
Kenya	Near Limuru (central)	Booth 1962, Schenkel and Schenkel-Hulliger 1967
	In W and central (localities mapped)	Kingdon 1971
	Rather numerous in Kakamega Forest (W)	Zimmerman 1972
	In Lake Nakuru N.P. (central)	Kutilek 1974
	In Lake Nakuru N.P. and Aberdare N.P.	Dupuy 1975
	At Limuru, Mount Warges, Masai-Mara Game Reserve and 4 national parks (Aberdare, Mount Elgon, Mount Kenya, and Lake Nakuru)	Oates 1977a
Ethiopia	Near Gudela and Djam Djam, Schoa (Shewa Province?, central)	Starck and Frick 1958
	In E	Glass 1965
	In proposed Rift Valley Lakes N.P. (150 km S of Addis Ababa)	Bolton 1970
	Abundant in Kefa and Ilubabor provinces (in SW); very common in Bale and Sidamo provinces (in S); total population estimated at 200,000–500,000	Brown and Urban 1970
	In Menagasha N.P. (W of Addis Ababa)	I.U.C.N. 1971
	Locally abundant in SW and S (localities mapped)	Dandelot and Prévost 1972
	Widespread, especially abundant in SW; W at least to Goc (34°22′E); S to Lake Rudolf and Lake Stephanie; in N Sidamo Province; along Awash R. at least to Awash; near Rift lakes; in W plateau as far N as Simien Mts. (13°13′N); on E plateau in Arussi and Bale Mts. and Chercher range	Bolton 1973
	Abundant on Mount Abaro, E of Lake Awassa (central); population in Abaro Range may exceed 2,000	Bolton 1974
	Increasing at 7.6% per year at Bole Valley; also at Lake Shala, Simien Mts., Menegash Forest Reserve, Gondar, and Bale N.P. (localities mapped)	Dunbar and Dunbar 1974b

Table 115 (continued)

Country	Presence and Abundance	Reference
Ethiopia (cont.)	Throughout central highlands E of Rift in considerable numbers; abundant in Rift central and S and in SW; common in W lowlands and N plateau; declining near Shashemenē	Bernahu 1975
	Overall declining, may be extinct in 10–25 years; at Bole Valley increasing at 3–5% per year; small populations in Awash N.P. and Simien N.P.	Dunbar and Dunbar 1975a

Table 116. Population parameters of *Colobus guereza*

Population Density (individuals /km²)	Group Size			Home Range Area (ha)	Study Site	Reference
	X̄	Range	No. of Groups			
	5	2–8	2		Cameroon	Struhsaker 1969
	8	2–13	14	16.1	W Uganda	Marler 1969
		12	1	1.5	W Uganda	Leskes and Acheson 1971
	9.3	6–12	4	24–39	W Uganda	Clutton-Brock 1975a
	6.9	4–10	8	4–6	W Uganda	Hayashi 1975
11.9					W Uganda	Struhsaker 1975
	10.5	8–15		15–16	W Uganda	Struhsaker and Oates 1975
	8.3–8.9	5–13	9	12–17	W Uganda	Mizuno et al. 1976
50–100	8.6–10.7	3–15			W Uganda	Oates 1977a
	9.9	6–13	9	15	W Uganda	Oates 1977b
	8	5–15	6		S Kenya	Schenkel and Schenkel-Hulliger 1967
180	4.5–5.5		110		Kenya	Kingston 1971
0.10					Central Kenya	Kutilek 1974
		13	1	15	NE Tanzania	Ullrich 1961
	8.3	5–18	9		NE Tanzania	Groves 1973
7.7	4.25	3–6	4		SW Ethiopia	Brown and Urban 1970
		8–12			Ethiopia	Dandelot and Prévost 1972
50–140	6–8	3–11	40	2.01	Ethiopia	Dunbar and Dunbar 1974b
20–300				2–15	Ethiopia	Dunbar and Dunbar 1975a

Colobus polykomos Western Black-and-White Colobus Monkey

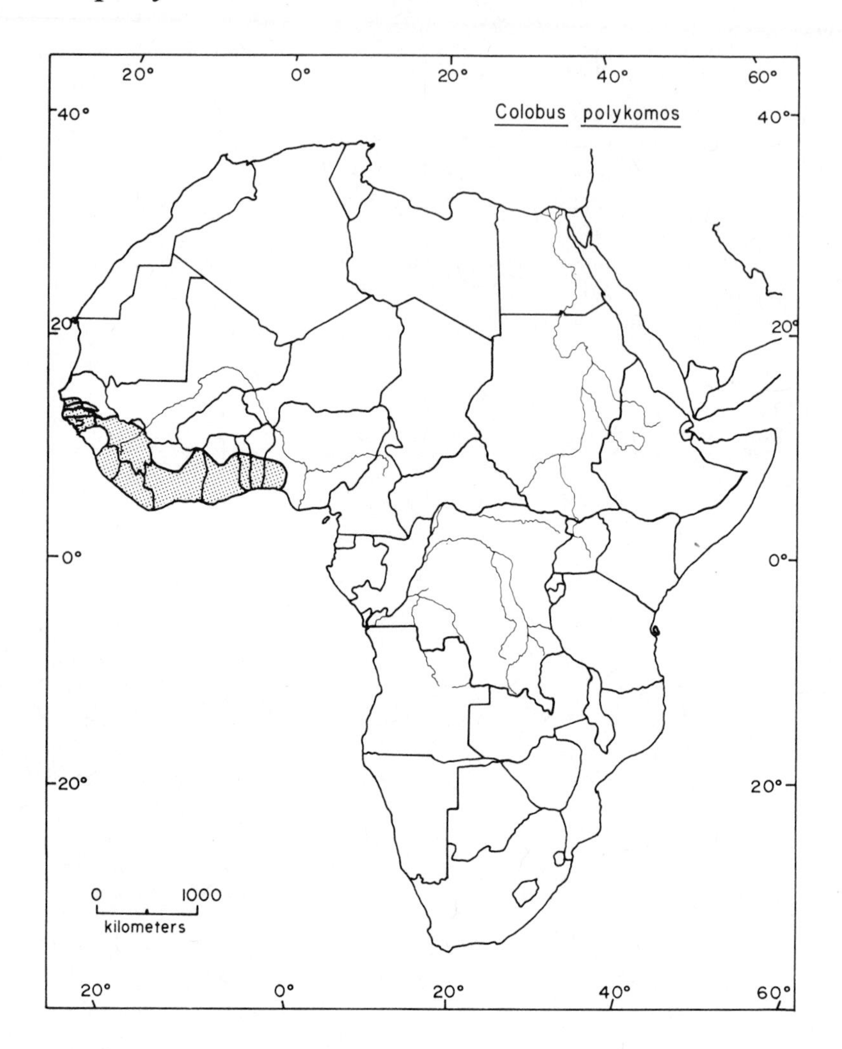

Taxonomic Note

Some authors include the central African populations of black-and-white colobus monkeys in *Colobus polykomos* (Napier and Napier 1967, Kingdon 1971). Here Rahm (1970) is followed, and *C. polykomos* includes only the three subspecies found in West Africa. The central African forms are treated under *C. angolensis*.

Geographic Range

C. polykomos is found along the southern coast of West Africa from Senegal to Nigeria. It occurs as far north as about 14°N along the coast of Senegal (Booth 1958). The easternmost record is in the Olokemeji Forest Reserve near 7°20′N, 3°30′E in western Nigeria (Hopkins 1970). For details of the distribution, see Table 117.

Abundance and Density

C. polykomos has recently been described as common in parts of Ivory Coast and Ghana, but is also known to be declining, rare, or extinct in some areas of these countries and in Togo and Nigeria (Table 117). No information is available about its present status in the western portion of its range.

The only *C. polykomos* demographic data available are for two groups in the Bia National Park, Ghana (Olson, pers. comm. 1978). These groups underwent fission and fusion and hence fluctuated in size from 5 to 21 animals per group. They utilized home ranges of 18 to 22 ha and their population density was about 40 individuals per km^2.

Habitat

C. polykomos is a forest species, which prefers primary wet evergreen, moist evergreen, moist semideciduous, and dry semideciduous forest (Olson, pers. comm. 1978). It can occupy all types of closed forest formations except thicket (Booth 1958), and is found in patches of riverine forest in the savanna zone (Asibey 1978).

This species also utilizes mature secondary forest (Booth 1956a) and has been seen foraging in secondary roadside bush (Jeffrey 1974, 1975b). It has been described as "ecologically tolerant" because it can utilize many types of forest (Booth 1958).

Factors Affecting Populations

Habitat Alteration

Deforestation has led to a drastic decline in *C. polykomos* habitat in several countries. By 1964 the increasing demand of the human population for agricultural land in Sierra Leone had already reduced forested areas to less than 4% of that country's surface (Tappen 1964). In Ivory Coast, logging and slash and burn agriculture have deforested most of the country (Lanly 1969); the Tai National Park is the last remaining area of forest of any significant size (Monfort and Monfort 1973).

Ghanaian forests have been cleared at a rapid rate for their timber and to provide land for agriculture and cacao farming (Asibey 1978). Aside from the 25% of forest in forest reserves, the forest zone was reduced from 80% forest cover to 20% between 1960 and 1973 (Jeffrey 1974). Extensive logging, burning, road building, and cultivation are continuing to decimate the remaining forested areas (Jeffrey 1975b).

Limited selective logging can be tolerated by *C. polykomos*, because much of its food supply is obtained from the least commercially valuable tree species and climber plants (Olson, pers. comm. 1978). Thus, regulated selective cutting would have a much less severe effect than clearcutting on *C. polykomos* populations.

No information was located regarding the extent of habitat alteration east of Ghana or west of Sierra Leone.

Human Predation

Hunting. *C. polykomos* has been heavily hunted for its beautiful pelt in Sierra Leone (Jones 1966, Wilkinson 1974), Ghana (Booth 1956a, Asibey 1972), and Nigeria (Happold 1972). This species has been the major target of international trade in monkey fur because of the length and glossiness of its hair (Oates 1977a). In addition, *C. polykomos* is eaten in Sierra Leone (Lowes 1970), Liberia (Leutenegger 1976), Ivory Coast, and Ghana (Booth 1954).

In some parts of Ghana, *C. polykomos* had already been exterminated by hunting by 1957 (Booth 1957). More recently, about one *C. polykomos* was shot per day in each of Ghana's western forest reserves (Jeffrey 1970). This hunting has had a significant effect on colobus populations (Jeffrey 1975b).

Pest Control. C. polykomos is not considered damaging to agriculture (Mackenzie 1952; Robinson 1971; Olson, pers. comm. 1978).

Collection. Fifteen *C. polykomos* were imported into the United States between 1968 and 1973 (Appendix), but since their origin was given as Kenya, these animals were probably *C. guereza.* None entered the United States in 1977 (Bittner et al. 1978). In 1965–75, 150 *C. polykomos* were imported into the United Kingdom (Burton 1978).

Conservation Action

C. polykomos is found in eight national parks and twelve reserves (Table 117). It is completely protected by law in Ghana (Booth 1954, Asibey 1972), but the law is widely disregarded and *C. polykomos* is heavily poached (Jeffrey 1970, Asibey 1978).

This species is also protected under the African Convention (Class B) (Curry-Lindahl 1969) and has been proposed for protected status in Sierra Leone (Lowes 1970). Religious traditions have protected it from hunting in Muslim areas of Sierra Leone (Lowes 1970) and the villages of Boabeng and Fiema in Ghana (Olson, pers. comm. 1978). Nevertheless, widespread deforestation and hunting are threatening its survival throughout much of its range. Oates (1977a) proposed a ban on all trade in black-and-white colobus fur, and a system of adequate reserves to protect the habitat of this species.

Table 117. Distribution of *Colobus polykomos* (countries are listed in order from west to east)

Country	Presence and Abundance	Reference
Senegal	In central, N and S of Gambia R. (map)	Booth 1958
	Possible in Casamance N.P. but not verified	Dupuy 1971b, 1973
Gambia	Localities mapped N and S of mouth of Gambia R.	Dupuy 1971b
Guinea-Bissau	Reported present	Dupuy 1971b
Guinea	In N and E (maps)	Booth 1958, Rahm 1970
	Reported present	Dupuy 1971b
Sierra Leone	Throughout (maps)	Booth 1958, Rahm 1970
	In Kasewe Forest (S central)	Tappen 1964
	Present	Jones 1966, Robinson 1971
	At Ducuta, Bombali District (in N)	Wilkinson 1974
Liberia	Throughout (map)	Booth 1958
	Present	Kuhn 1965

Table 117 (continued)

Country	Presence and Abundance	Reference
Liberia (cont.)	In S (map)	Rahm 1970
	Abundant	Schlitter, pers. comm. 1974
	Along Cess R. between Potogle and Kpeaple	Leutenegger 1976
Ivory Coast	In S (localities mapped)	Booth 1954, 1958; Rahm 1970
	In Banco N.P. (in S, near Abidjan)	Happold 1971
	Common in Tai Reserve (now N.P.) (in W between Cavally R. and Sassandra R.)	Struhsaker 1972 [Monfort and Monfort 1973]
	Common in S part of Comoe N.P. (in NE)	Geerling and Bokdam 1973
	Disappearing along Bandama R. near Lamto (S central)	Bourlière et al. 1974
	Abundant in Tai N.P., common in Comoe N.P., rare in Banco N.P. and Asagny Reserve	Asibey 1978
Ghana	Throughout S half except central area (localities mapped); locally common but rare in inhabited areas; extinct over large areas	Booth 1954, 1956a
	Extinct in Kwahu-Mankrong area	Booth 1957
	In SW and NE (map)	Booth 1958
	Locally common in Bia Tributaries and Sukusuku forest reserves (in SW); rare in some areas	Jeffrey 1970, 1975b
	In SW (map)	Rahm 1970
	Abundant in Boabeng-Fiena Monkey Sanctuary; common in Bia N.P., Nini-Suhien N.P., Ankasa Game Production Reserve; rare in Mole N.P., Digya N.P., Bui N.P., Kogyae Strict Nature Reserve, Kalakpa Game Production Reserve, Bomfobiri Wildlife Sanctuary and Willi Falls Reserve	Asibey 1978
	N to about 8°N, along rivers to 9°N; locally abundant but probably overall declining	Olson, pers. comm. 1978
Togo	In S (maps)	Booth 1958, Rahm 1970
	Rare in Keran and Malfacassa Reserves	Asibey 1978
Benin	In S (maps)	Booth 1958, Rahm 1970
Nigeria	In SW corner (maps)	Booth 1958, Rahm 1970
	Seen infrequently in Olokemeji Forest Reserve (in SW)	Hopkins 1970
	Previously frequent but no definite records for several years	Happold 1972
	In S but nowhere abundant	Henshaw and Child 1972
	In Upper Ogun Game Reserve (in SW)	Geerling 1974

Colobus satanas Black Colobus Monkey

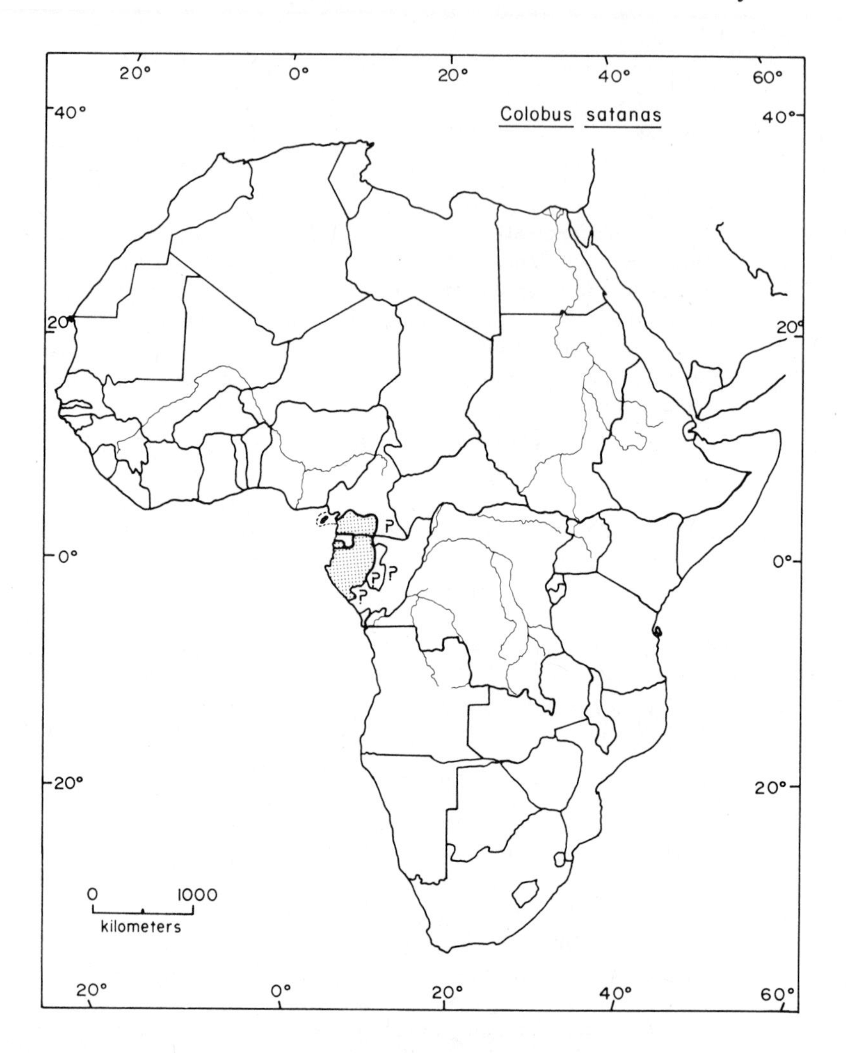

Taxonomic Note

Colobus satanas is a monotypic species which some authors include within *C. polykomos* (Napier and Napier 1967, Sabater Pi 1973), but most accord full specific status.

Geographic Range

C. satanas is found in western equatorial Africa from southern Cameroon through Gabon (Rahm 1970). It occurs as far north as about 4°N along the Sanaga River in western Cameroon (Gartlan, pers. comm. 1974), and as far west as Macias Nguema (formerly Fernando Po) (Sabater Pi 1973). Old records also place it in the Congo (Malbrant and Maclatchy 1949), but the eastern and southern limits of its range are not known. For details of the distribution, see Table 118.

Abundance and Density

C. satanas is known to be declining in parts of Cameroon and Equatorial Guinea (Table 118). Its status in Gabon and Congo is not known. This species is considered to be one of the most threatened of African monkeys (Struhsaker, pers. comm. 1974). It is classified as "vulnerable," with a total range of 240,000 km^2 (1,000 km by 240 km) (I.U.C.N. 1978).

No estimates of population density or home range area are available. Early workers reported groups of 60 to 80 *C. satanas*, but later authors have found smaller group sizes of 2 to 8 ($\bar{X}$ = 5.8, mode 8, N = 5 groups) (Struhsaker 1969), 5 to 30 ($\bar{X}$ = 12.8, N = 5 groups) (Sabater Pi 1973), and an average of 7 to 9 per group (N = 6 groups) (McKey, pers. comm. 1974).

Habitat

C. satanas is a rain forest species most typical of high, dense primary forests (Malbrant and Maclatchy 1949, Sabater Pi and Jones 1967). It is found in lowland rain forest (Gartlan, pers. comm. 1974) and swamp forest (McKey, pers. comm. 1974), as well as montane forests (Malbrant and Maclatchy 1949, Sabater Pi 1973). On Macias Nguema it occupies bushes in high meadows between 2,500 and 3,000 m elevation (Sabater Pi 1973).

Black colobus monkeys have been observed in mature secondary forest (Sabater Pi 1973), but are generally not found near human habitation (Sabater Pi and Jones 1967). They evidently do not utilize gallery forest extending into savanna (Malbrant and Maclatchy 1949). They are essentially arboreal, and tall trees seem to be a vital component of their niche (Sabater Pi 1973; Gartlan, pers. comm. 1974).

Factors Affecting Populations

Habitat Alteration

Logging has already reduced the area of habitat available to *C. satanas* in Cameroon (Gartlan, pers. comm. 1974), and will become a threat to its survival there "if effective protection of the remaining reserves is not secured very soon" (McKey, pers. comm. 1974).

Timbering and clearing for agriculture have also destroyed large amounts of *C. satanas* habitat in Equatorial Guinea (Sabater Pi and Jones 1967; Sabater Pi, pers. comm. 1974). The montane forests are still relatively intact only because their rough terrain makes logging there difficult (Sabater Pi 1973).

Human Predation

Hunting. Black colobus monkeys are shot frequently for their meat, and are being seriously affected by this hunting in Cameroon (Gartlan, pers. comm. 1974). They have been hunted nearly to extinction in accessible parts of the Douala-Edéa Reserve, and will continue to decline there unless hunting is prevented (McKey, pers. comm. 1974).

C. satanas is also hunted in Equatorial Guinea, where an estimated 1,000 to 1,500 are killed annually (Sabater Pi, pers. comm. 1974). The flesh is eaten, but informants interviewed in that country stated that the meat was dry and bitter and was not a preferred food (Sabater Pi and Groves 1972). Intensive hunting has also eliminated black colobus from some parts of Macias Nguema (Oates 1977a).

Pest Control. C. satanas generally avoids areas of human settlement (Tappen 1960, Sabater Pi and Jones 1967), and does not utilize food from plantations (Malbrant and Maclatchy 1949; McKey, pers. comm. 1974). Thus it is not involved in crop raiding and should not be subject to control efforts.

Collection. This species is not in demand for research and does not survive well in captivity (Gartlan, pers. comm. 1974). None were imported into the United States between 1968 and 1973 (Appendix) or in 1977 (Bittner et al. 1978), or into the United Kingdom between 1965 and 1975 (Burton 1978).

Conservation Action

C. satanas occurs in at least two forest reserves in Cameroon (Table 118). In the past it has not been protected from hunting in these reserves, but recently extra game guards have been assigned to the Douala-Edéa Reserve, and the Cameroon Department des Eaux et Forêts is trying to help protect the monkeys there (McKey, pers. comm. 1974). No information was located regarding efforts to conserve this species in Equatorial Guinea, Gabon, or Congo.

C. satanas is listed as endangered under the U.S. Endangered Species Act (U.S. Department of Interior 1977b).

Table 118. Distribution of *Colobus satanas* (countries are listed in order from north to south)

Country	Presence and Abundance	Reference
Cameroon	At Lake Tisongo (near mouth of Sanaga R.)	Gartlan and Struhsaker 1972
	Common in Douala-Edéa Reserve	Struhsaker 1972
	Throughout S of Sanaga R. in W, not confirmed in SE; in Douala-Edéa, Dipikar, and perhaps Dja Reserve; locally abundant but overall threatened	Gartlan, pers. comm. 1974
	Locally common in some parts of Douala-Edéa Reserve, declining and almost extinct in other parts	McKey, pers. comm. 1974
	Inland record from 4°15′N, 14°15′E; rare and seriously threatened	Oates 1977a
Equatorial Guinea	Confined to 100 km long strip parallel to coast in mountains; previously had larger range	Sabater Pi 1966
	In SW (map); range shrinking	Sabater Pi and Jones 1967
	In W and central (localities mapped) and on Mount St. Isabel (Pico de Santa Isabel) on Fernando Po (now Macias Nguema)	Sabater Pi 1973
	Declining	Sabater Pi, pers. comm. 1974
Gabon	Abundant in mountains near Mimongo (in W, about 1°S, 11°30′E)	Malbrant and Maclatchy 1949
	Throughout except SE (map)	Rahm 1970

Table 118 (continued)

Country	Presence and Abundance	Reference
Congo	Present	Malbrant and Maclatchy 1949
	Probably in N and SW (map)	Rahm 1970
	Present	Vattier-Bernard cited by Oates 1977a

Colobus verus Olive Colobus Monkey

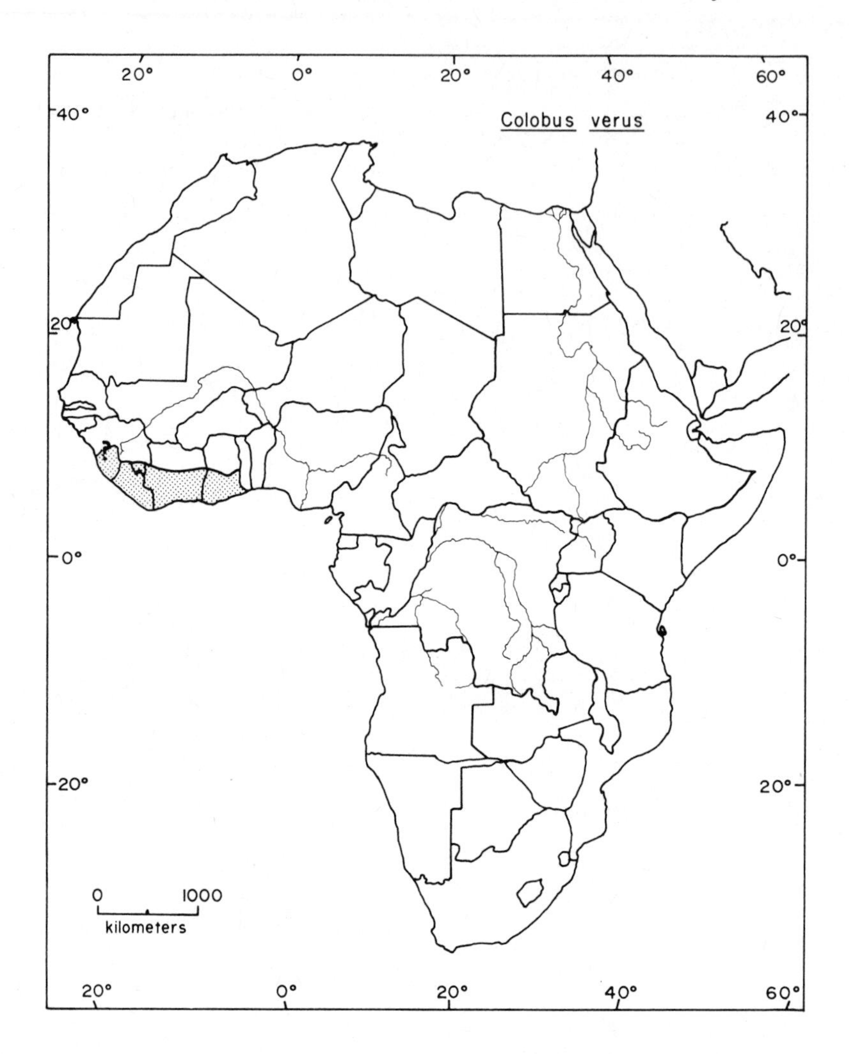

Taxonomic Note

Colobus verus is a monotypic species (Napier and Napier 1967). Many older authors have placed it in the genus *Procolobus* (e.g., Booth 1956a, 1957, 1958; Tappen 1960), but most modern workers consider it *Colobus* (e.g., Struhsaker 1972, Bourlière et al. 1974).

Geographic Range

C. verus is found in West Africa, from Sierra Leone to eastern Ghana. The northern and western limits of its range are not known but appear from Booth's (1958) map to lie near the northwest border of Sierra Leone, near 10°N, 13°W. In Ghana this species has been recorded as far east as the eastern border with Togo at about 7°N, 1°E (Booth 1956a). For details of the distribution, see Table 119.

Abundance and Density

C. verus has recently been described as common in parts of Ghana and uncommon in one area of Ivory Coast (Table 119). Its status west of Ivory Coast is not known. The I.U.C.N. (1972) classified *C. verus* as rare, with "extremely rare and localized populations," and in 1976 described it as "rare but locally numerous."

No estimates of population density or home range size are available. Group size for this species varies from 6 to 20, with an average of 10 to 15 members per group (Booth 1957).

Habitat

C. verus is a forest species which can inhabit many kinds of forest vegetation, including primary moist semideciduous forest and primary evergreen rain forest (Olson, pers. comm. 1978). It is most often found in thickets or dense tangled undergrowth near rivers or swamps (Booth 1956a, 1957; Monfort and Monfort 1973). It tends to favor areas of broken canopy or sparse trees, where light penetration permits the growth of lush underbrush near the ground (Booth 1957), and has frequently been recorded in "secondary bush" (Booth 1956a, Jones 1966) or "low dense bush" (Jeffrey 1975a).

C. verus can also utilize regenerating forest, dry semideciduous forest, and riparian forest in the savanna zone (Booth 1957). It has only been found at low and medium elevations (Booth 1958). Water and thick cover seem to be crucial components of the habitat of this species (Booth 1956a, 1957).

Factors Affecting Populations

Habitat Alteration

Extensive deforestation has occurred in all of the countries within the range of *C. verus*. For example, in Ivory Coast the Tai National Park is the last remaining sizable area of forest (Monfort and Monfort 1973). And in Ghana large-scale logging, conversion of forest to cacao, oil palm, rubber, and food plantations have greatly decreased forest habitats (Jeffrey 1975b, Asibey 1978). But direct evidence regarding the effect of these activities on *C. verus* populations is not available.

Human Predation

Hunting. Olive colobus are killed for food in Liberia (Leutenegger 1976) and Ghana (Asibey 1974). They are not eaten in some parts of northern Sierra Leone (Wilkinson 1974).

Pest Control. *C. verus* was not described as a crop raider by any of the authors consulted. Robinson (1971) specifically listed it as "not damaging to agriculture" in Sierra Leone. Nevertheless, this species has sometimes been captured in the government sponsored monkey drives to reduce damage to coffee and cacao plantations in Sierra Leone (Tappen 1964).

Collection. *C. verus* is not used in biomedical research. None were imported into the United States between 1968 and 1973 (Appendix) or in 1977 (Bittner et al. 1978) or into the United Kingdom between 1965 and 1975 (Burton 1978).

Conservation Action

C. verus occurs within three national parks and three reserves (Table 119). It is fully protected in Ghana, but is heavily poached outside the national parks and game reserves (Asibey 1978). It is included in Class A of the African Convention (Organization of African Unity 1968), and lactating females may not be trapped or shot inside forest reserves in Liberia (Curry-Lindahl 1969).

Table 119. Distribution of *Colobus verus* (countries are listed in order from west to east)

Country	Presence and Abundance	Reference
Guinea	In SE	Booth 1958
Sierra Leone	Throughout (map)	Booth 1958
	Present	Jones 1966
	At Ducuta, Bombali District (in N)	Wilkinson 1974
Liberia	Throughout (map)	Booth 1958
	Present	Kuhn 1965
	Abundant	Schlitter, pers. comm. 1974
	Along Cess R. between Potogle and Kpeaple	Leutenegger 1976
Ivory Coast	Throughout S (map)	Booth 1958
	Uncommon in Tai Reserve (now N.P.) (between Cavally and Sassandra rivers)	Struhsaker 1972
	In Tai N.P.	Monfort and Monfort 1973
	In Goudi Forest near Lamto (S central, near Bandama R.)	Bourlière et al. 1974
Ghana	Throughout S (localities mapped); not rare	Booth 1956a
	Fairly common; N limit 8°30′N	Booth 1957 [1958]
	Probably quite common	Jeffrey 1970
	Endangered	Asibey 1974
	Not rare	Jeffrey 1974
	At Bia and Sukusuku reserves, Prampramase and Fantima (all in SW)	Jeffrey 1975b
	Common in Bia N.P., Nini-Suhien N.P., and Ankasa Game Production Reserve; rare in Bomfobiri Wildlife Sanctuary	Asibey 1978

Erythrocebus patas Patas Monkey

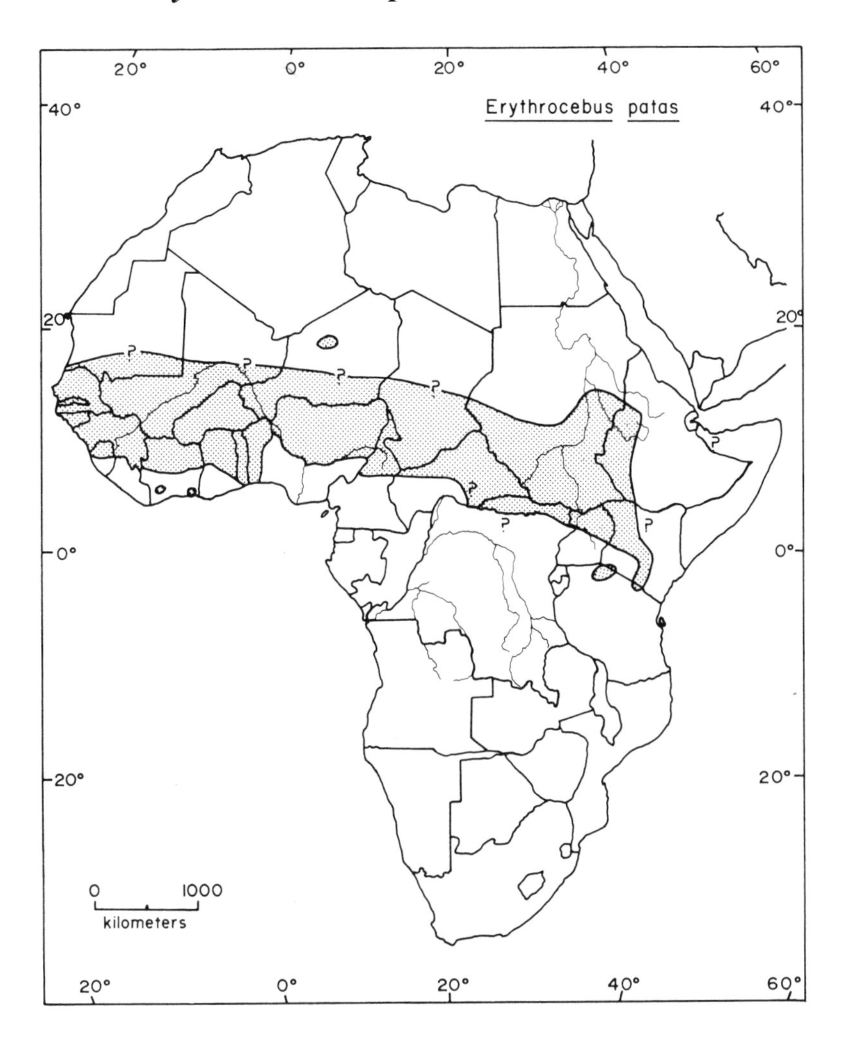

Taxonomic Note

Erythrocebus is a monotypic genus closely related to *Cercopithecus*. Four subspecies are recognized (Hill 1966, Napier and Napier 1967).

Geographic Range

Erythrocebus patas has a wide distribution across subsaharan Africa from the western tip of Senegal to East Africa. Most authors have reported the northeastern limit of its range as being in southwest Ethiopia (e.g., Dorst and Dandelot 1969), but others show it extending into northwestern Somalia (Kingdon 1971). The species has been recorded as far north as Air Massif (18°N, 8°E) (Bigourdan and Prunier 1937 cited in Tappen 1960). Elsewhere the northern limit is about 15°–17°N (Hall 1965a). Patas monkeys occur as far south as about 7°N in Cameroon (Gartlan, pers. comm.

1974) and 3°S in northern Tanzania (Hall et al. 1965). For details of the distribution, see Table 120.

Abundance and Density

Patas monkeys have been described as "fairly common" across Africa (Hall et al. 1965), but their population density is probably low over much of this range (Hall 1965a). Within the last 10 years, *E. patas* has been reported as common or abundant in parts of 11 countries, and increasing in population in 2 of these. But it was also described as rare or in low numbers in parts of 6 countries and declining in at least one (Table 120). Information regarding its present status is lacking for many countries.

Patas monkeys live in relatively small groups compared with other terrestrial species. Most groups number between 8 and 38, with average sizes of 15 to 26, although larger aggregations do occur (Table 121). Home ranges may be quite large (up to 8,000 ha), and population densities low (Table 121).

Habitat

E. patas is open country species which is restricted to savanna and woodland habitats. It is typically found in grass savanna, woodland savanna, grass steppe with thicket clumps (Hall et al. 1965), and both dry and dense woodlands with tall dense growths of grass (Hall 1965a). It has been most frequently reported in Guinea savanna (Booth 1956b, Happold and Philp 1971, Poché 1973), Sudan savanna (Service 1974, Sayer 1977), and Sahelian savanna (Gartlan 1975b, Bourlière et al. 1976).

This species has been observed in moist woodland (Bernahu 1975) but is more often seen in drier woodlands and savanna (Howell 1968, Henshaw and Child 1972, Tahiri-Zagrët 1976). It also utilizes man-made clearings within forest (Dupuy 1973) and the zone of flooding in the delta of the Senegal River (Bourlière et al. 1976).

Patas monkeys can tolerate dry conditions well and may go for long periods without water (Hall 1965b). But in dry areas and during dry seasons, water is a limiting factor for them, and suitable habitat must contain water sources (Gartlan 1975a, b).

E. patas avoids dense cover (Booth 1956a; Gartlan, pers. comm. 1974), riverine vegetation (Hall 1965b), and rocky areas (Kingdon 1971). It is best adapted to open habitats with good visibility and usually enters only the fringes of forests and thickets (Gartlan, pers. comm. 1974).

Factors Affecting Populations

Habitat Alteration

E. patas habitat is being reduced by human activities in some parts of the range. For example, in Niger, patas populations have decreased along the Niger River because of heavy cattle grazing and the conversion of savanna to farmland (Poché 1976). And in Ethiopia, burning of grassland for pasture reduces available habitat (Bernahu 1975). These factors are probably operating in many other countries, but no documentation is available. In Ghana the loss of habitat to agriculture is not yet a serious threat to *E. patas* (Asibey 1978).

Some kinds of habitat alteration can actually increase the extent of habitat suitable for *E. patas*. Deforestation has led to an expansion of patas habitat in Senegal (Dupuy 1973) and East Africa (Kingdon 1971). And drought, which has resulted in the southward incursion of arid savanna zones into previously more humid areas, has increased patas monkey habitat in Niger (Poché 1976) and Cameroon (Gartlan, pers. comm. 1974). But in the northern parts of the range the drought has also transformed savanna areas into desert, resulting in loss of habitat and deaths of patas monkeys (Gartlan 1975b, Poché 1976).

Human Predation

Hunting. Patas monkeys are hunted for their flesh in Ivory Coast (Bourlière et al. 1974), Ghana (Asibey 1974), and, infrequently, in Cameroon (Gartlan, pers. comm. 1974). They are hunted heavily by the mountain Nuba people of Sudan (Kock 1969) and by tribal groups hunting with dogs in East Africa (Kingdon 1971). But this species is a fast runner, attaining sprint speeds up to 55 km per hour (Kingdon 1971), and is skilled at hiding in grassland and low bushes (Hall 1965b). Thus it is "justly reported to be one of the most difficult animals in the world to hunt" (Tappen 1960). Its silence, wariness, and elusiveness have enabled it to survive in areas in which the more obvious baboons have been shot or driven away (Hall 1965b). Also, much of its range is within Muslim regions where monkeys are not eaten (Blancou 1961, Howell 1968, Wilkinson 1974), thus it is protected from hunting to some extent.

Pest Control. Patas monkeys frequently raid crops and consequently are shot as pests in many parts of their range (Hall 1965a,b; Howell 1968; Robinson 1971; Bourlière et al. 1974; Asibey 1978). They steal millet, bananas, peanuts, cassava, wheat (Kingdon 1971), maize, and dates (Hill 1966). In Sudan they feed in pineapple plantations and damage growing cotton plants by eating the flower buds and young capsules (Kock 1969).

In Cameroon, *E. patas* eats crops from small farms more often than from large cash crop plantations of peanuts and cotton (Gartlan, pers. comm. 1974). And in East Africa it is quite shy and is easily driven from the fields (Kingdon 1971). In both areas it is considered a less serious crop pest than the vervet or baboon, and can survive even in fairly densely populated farmland in Uganda (Hall 1965b).

Collection. *E. patas* is used in medical research, and as early as 1965 was in demand for this purpose (Hall 1965a). For example, it has been exported from Cameroon for vaccine culture and research (Gartlan, pers. comm. 1974) and captured in Nigeria for yellow fever research (Monath and Kemp 1973). Recently, however, the once extensive trapping of live patas monkeys in northern Nigeria for sale as pets or for research has been stopped (Howell 1968). And in East Africa a few have been captured for export to zoos, but populations have not been heavily exploited, because of their limited numbers (Kingdon 1971).

Between 1968 and 1973 a total of 964 *E. patas* were imported into the United States, primarily from Nigeria (Appendix). In 1977 one shipment of unspecified size entered the United States from Nigeria (Bittner et al. 1978). In 1965–75, 14,262 *E. patas* were imported into the United Kingdom (Burton 1978).

Conservation Action

E. patas is found in 18 national parks and 11 reserves (Table 120). A permit is required for the exportation of *E. patas* from Cameroon, but no quotas are imposed (Gartlan, pers. comm. 1974). Since its populations are so vulnerable to fluctuations in rainfall, any cropping of this species should be carefully controlled and monitored to detect changes in the status of local populations (Gartlan 1975a).

Table 120. Distribution of *Erythrocebus patas* (countries are listed in order from northwest to southeast)

Country	Presence and Abundance	Reference
Mauritania	In S and SE (maps)	Monod 1961 cited by Huard 1962, Hall 1965a
Senegal	Throughout (maps)	Booth 1958, Hall 1965a
	Throughout; common in Niokolo-Koba N.P. (in SE)	Dupuy 1971b, c [Dupuy and Verschuren 1977]
	Near Bissine, S of Casamance R.	Struhsaker 1971
	In Oiseaux du Djoudj N.P., Senegal R. delta (in NW)	Dupuy 1972b
	In N Ferlo (N central)	Poulet 1972
	Not frequent in Basse Casamance N.P. (in SW)	Dupuy 1973
	Widespread and common	Boese, pers. comm. 1974
	At Badi (in SE)	Dunbar 1974
	In Oiseaux du Djoudj N.P. (in NW) and Basse Casamance N.P. (in SW)	I.U.C.N. 1975
	Widespread betwen Morfil Island, Podor and St. Louis, lower valley of Senegal R. (NW)	Bourlière et al. 1976
Gambia	40–50 in Abuko Nature Reserve seasonally	Gunderson 1977
Mali	Throughout S (map)	Booth 1958
	Throughout S to Adrar des Iforas in NE (maps)	Monod 1961 cited by Huard 1962, Hall 1965a
	Widespread and common; as far N as Nioro du Sahel in W (15°12′N) and Niger R. in E (17°N)	Sayer 1977
Guinea-Bissau	Throughout (maps)	Booth 1958, Hall 1965a
Guinea	Throughout (maps)	Booth 1958, Hall 1965a
Sierra Leone	In N (map)	Booth 1958
	At Ducuta, Bombali District (extreme N)	Wilkinson 1974
Liberia	Absent (map)	Booth 1958
Ivory Coast	In N (maps)	Booth 1956b, 1958; Hall 1965a
	Limited numbers in Comoe N.P. (in NE) (S limit in E part of country)	Geerling and Bokdam 1973

Table 120 (continued)

Country	Presence and Abundance	Reference
Ivory Coast (cont.)	Formerly near Ayéremou; now S only to Toumodi (S central, near Bandama R.)	Bourlière et al. 1974
	Throughout N half and around Soubré (SW near Sassandra R.) and Ayamé (SE corner) (map)	Tahiri-Zagrët 1976
Upper Volta	Throughout (maps)	Booth 1958, Hall 1965a
	Common in "W" N.P. (in SE)	Asibey 1978
Ghana	In N and E (localities mapped)	Booth 1956a, 1958
	Abundant in Mole, Digya, and Bui national parks; common in Gbele and Kalakpa game production reserves; rare in Kogyae Strict Nature Reserve	Asibey 1978
Togo	Throughout (map)	Booth 1958
	In Koué Reserve (central, W of Sokodé)	I.U.C.N. 1971
	Common in Keran and Malfacassa reserves	Asibey 1978
Benin	Throughout (map)	Booth 1958
	In Pendjari N.P. (NW)	Happold and Philp 1971, Gorgas 1974
	Common in "W" and Pendjari national parks	Asibey 1978
Niger	To Air Massif 18°N, 8°E (isolated population)	Bigourdan and Prunier 1937 cited in Tappen 1960, Dekeyser 1950 cited in Hall 1965a, Monod 1961 cited in Huard 1962
	Throughout S (maps)	Booth 1958, Hall 1965a
	Populations depleted but viable in "W" N.P. (in SW corner)	Jones 1973
	Common in "W" N.P.	Poché 1973
	In S and SW and Air Mts., lower density along Niger R. (in W)	Poché 1976
Nigeria	In N two-thirds (maps)	Booth 1958, Hall 1965a
	Occasional in Yankari Game Reserve (NE)	Sikes 1964a
	Abundant and increasing (total 250–400) in Borgu Game Reserve (W central)	Howell 1968
	In Upper Ogun and Old Oyo reserves (SW)	Adekunle 1971
	Fair numbers in Yankari Game Reserve	Henshaw and Ayeni 1971
	Common	Henshaw and Child 1972
	Near Jos, Jos Plateau (central)	Lee 1972
	Central and NE (localities mapped)	Monath and Kemp 1973
	In Nupeko Forest near Bida (W central)	Monath et al. 1974
	Common in NW near Sokoto, Bernin Kebbi, and other localities	Service 1974

Table 120 (continued)

Country	Presence and Abundance	Reference
Cameroon	Throughout N of 6°N (map)	Booth 1958
	Throughout N of 9°N (map)	Hall 1965a
	In Waza Reserve (now N.P.) in N	Struhsaker 1969, Struhsaker and Gartlan 1970, Gartlan 1975b
	Abundant N of 7°N; range may be increasing	Gartlan, pers. comm. 1974
	Rare in Bouba-Ndjida N.P. (8°30′ on E border)	Lavieren and Bosch 1977
	Abundant in Benove (Bénoué) N.P. (8°N, central) and Bouba-Ndjida N.P.; common in Waza N.P.	Asibey 1978
Chad	In S half to S edge of Ennedi (16°N) (map)	Hall 1965a
	In Zakouma N.P. (in SE, W of Am Timan)	I.U.C.N. 1971
Central African Republic	In N (maps)	Booth 1958, Hall 1965a
Zaire	In extreme NE (localities mapped)	Kingdon 1971
	Common in N and NE Haut-Zaire Province	Heymans 1975
Sudan	At Torit and Imatong Mts. (extreme S) and S Jebel Marra (in W, 13°N, 24°E)	Setzer 1959
	Central and S (map)	Butler 1966
	Central and S to 14°N (localities mapped); abundant in Kordofan	Kock 1969
	In Dinder N.P. (SE of Khartoum on E border between Rahad and Blue Nile R.)	Cloudsley-Thompson 1973
Ethiopia	In W Begemder (NW), W Gojjam (Gwejam) (W), and Ilubabor (SW) provinces; probably sporadically along W border	Bolton 1973
	In small numbers in W lowlands	Bernahu 1975
Somalia	In extreme NW corner (map)	Kingdon 1971
Kenya	In W, doubtful E of 36°E (map)	Hall 1965a
	In W and S central (localities mapped)	Kingdon 1971
Uganda	To right bank of Victoria Nile and lake complex (2°N, 33°E)	Tappen 1960
	Common in Teso, Acholi, and W Nile districts; in N and NE	Hall 1965b
	N of Lake Kyoga (localities mapped)	Kingdon 1971
	In Murchison Falls N.P. and Kidepo N.P.	Rowell, pers. comm. 1978
Tanzania	In Serengeti N.P.	Swynnerton 1958
	S limit is about 3°S; E and W of Lake Victoria	Hall et al. 1965
	In Serengeti and Mount Kilamanjaro area in N (localities mapped)	Kingdon 1971

Table 121. Population parameters of *Erythrocebus patas*

Population Density (individuals/km²)	Group Size $\bar{X}$	Range	No. of Groups	Home Range Area (ha)	Study Site	Reference
		8–15	3		SW Senegal	Dupuy 1973
		10–30			Ivory Coast	Tahiri-Zagrët 1976
		3–38			SW Niger	Poché 1976
0.08					N Nigeria	Sikes 1964b
		≤12			N Nigeria	Howell 1968
	18–26	7–34	8		N Cameroon	Struhsaker 1969
	21		20		N Cameroon	Gartlan, pers. comm. 1974
	15	9–31		5,200	W Uganda	Hall 1965a
	17	2–25	7	777–2,072	W Uganda	Hall 1965b
		4–31		5,180	W Uganda	Hall et al. 1965
		5–50		≤8,000	Uganda, Kenya, Tanzania	Kingdon 1971

Macaca arctoides Stumptailed Macaque

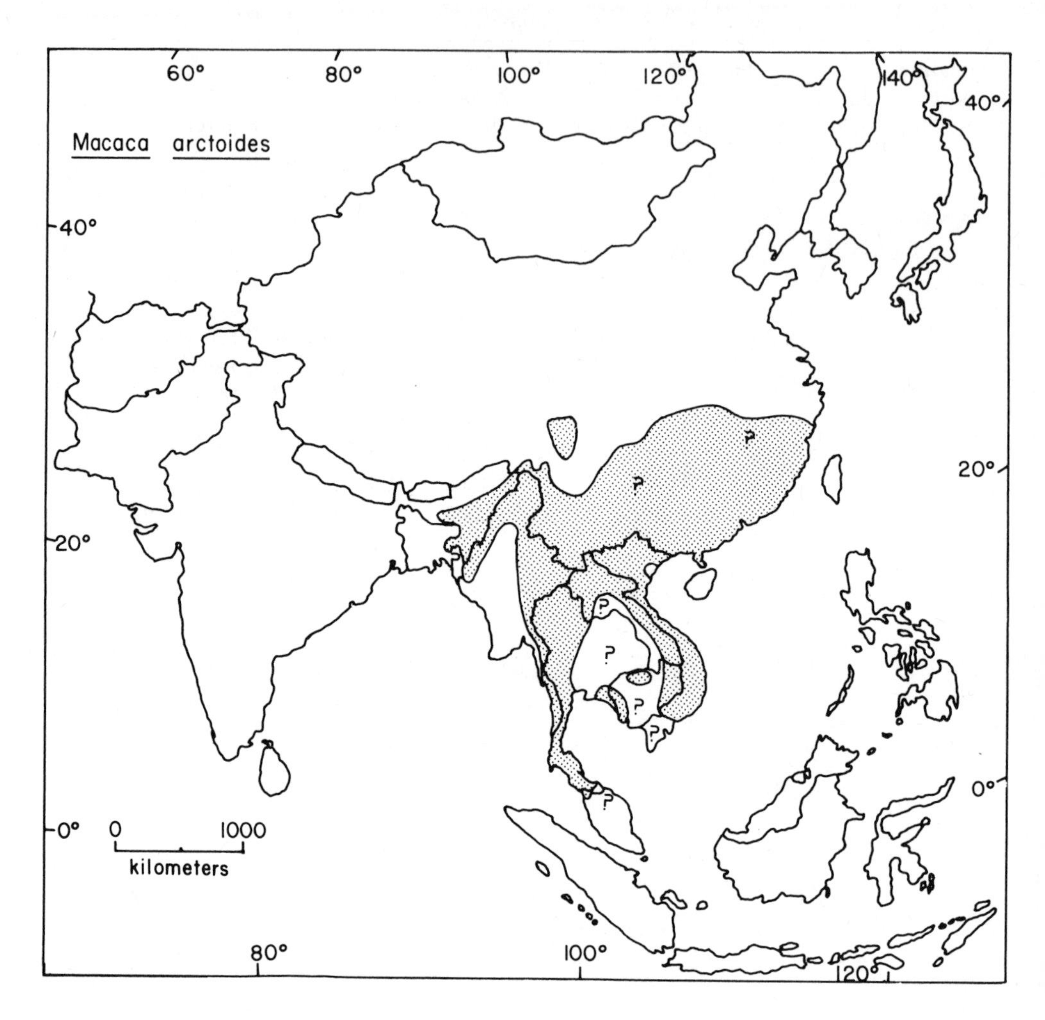

Taxonomic Note

The stumptailed macaque was previously called *Macaca speciosa*, but most recent workers use the name *Macaca arctoides* (Fooden 1967, Napier and Napier 1967, Medway 1970, Thorington and Groves 1970). The northernmost race has been considered a separate species, *M. thibetana* (Fooden 1967, 1971a; Thorington and Groves 1970), but so little is known of the ecology of this form that it is treated here with *M. arctoides*. In addition to *M. a. thibetana*, three races are recognized (Dobroruka 1967).

Geographic Range

M. arctoides is found from Assam, India, eastward through southern China and south on the Malay and Indochinese peninsulas. Its western limit is thought to be near 92°E in Assam and southeastern Bangladesh (Bertrand 1969, Green 1978a). The range extends as far north as about 31°N, 100°E in Szechuan Province, China, and east to the coast at about 28°N in Chekiang Province (Fiedler 1956, Bertrand 1969). It extends as far south as about 10°N in southern Vietnam and about 6°N in peninsular Malaysia (Bertrand 1969, Lim 1969). For details of the distribution, see Table 122.

Abundance and Density

As early as 1937, *M. arctoides* was in danger of extinction in China (Sowerby 1937). Bertrand (1969) predicted that it would be threatened with extinction in India and Thailand in the near future. Within the last 10 years *M. arctoides* has been described as declining or rare in India, Thailand, and Malaysia (Table 122). Its status in Bangladesh, Burma, China, Cambodia, Laos, and Vietnam is not known.

No estimates of population density or home range size of *M. arctoides* could be located. Group sizes of 25 to 30 (McCann 1933b), up to 20 (9 troops; Bertrand 1969), up to 50 (Medway 1969a), 5 and 50 (Fooden 1971a), and 14 to 15 (Fooden 1976) have been reported.

Habitat

M. arctoides occupies monsoon and dry forests from sea level to 2,400 m (Bertrand 1969, Mohnot 1978). It has been recorded in evergreen forests (Fooden 1971a). No further details are available.

Factors Affecting Populations

Habitat Alteration

M. arctoides is "feeling the pressure of a deteriorating environment" (Southwick et al. 1970). Deforestation (by logging and agriculturalization) and military activity (bombing and defoliants) (Orians and Pfeiffer 1970, S.I.P.R.I. 1976) have altered many natural habitats within *M. arctoides*'s range, but no information is available about the effect of these habitat changes on this species.

Human Predation

Hunting. *M. arctoides* is killed for its flesh in the Naga Hills of Assam, India (McCann 1933b), and in southern Thailand (Lekagul 1968).

Pest Control. Stumptailed macaques feed on crops (Medway 1969a), and this habit has aroused public opinion against them in Thailand (Lekagul 1968). In Assam they do considerable damage to crops such as potatoes (McCann 1933b). Their crop raiding increases their vulnerability to trapping, and they are often trapped during forays into rice fields (Bertrand 1969).

Collection. Bertrand (1969) blamed the scarcity of *M. arctoides* in Assam and southern Thailand partially on trapping and export for research. She reported that "uncontrolled demand for subjects to be used in . . . research, as well as changes in local traditions, are beginning to endanger the existence of the species." In addition to a large demand for *M. arctoides* for research and the preparation of vaccines, villagers shoot adult stumptails which they cannot sell, rather than releasing them.

Bertrand could find no groups in southern Thailand that had not been trapped or shot, except in remote, protected, dense forests.

Export for production of polio vaccines in the United States is cited as a factor causing the decline of *M. arctoides* in Thailand (Lekagul 1968). And "wasteful trapping and shooting" are blamed for decimating several populations (Southwick et al. 1970). Stumptailed macaques have been recommended as a laboratory animal because of their "remarkable docility" (Kling and Orbach 1963), but this seems to be a false reputation based on experience with young animals only (Bernstein and Guilloud 1965).

Between 1968 and 1973, 7,595 *M. arctoides* were imported into the United States (Appendix). In 1970–73 most of these originated in Thailand. None entered the United States in 1977 (Bittner et al. 1978). A total of 1,415 *M. arctoides* were imported into the United Kingdom in 1965–75 (Burton 1978). This species is also exported from Vietnam to the Soviet Union for research at the Sukhumi Primate Center (Lapin et al. 1965).

Conservation Action

In Thailand, *M. arctoides* is found in one sanctuary (Table 122), may be hunted, captured, or kept in captivity only under license, and its export is regulated by a quota system (Lekagul and McNeely 1977). Prompt and effective conservation action is necessary if this species is to survive (Bertrand 1969, Southwick et al. 1970).

M. arctoides is listed as threatened under the U.S. Endangered Species Act (U.S. Department of Interior 1977b).

Table 122. Distribution of *Macaca arctoides* (countries are listed in order from west to east)

Country	Presence and Abundance	Reference
Bangladesh	In E to 90°E (map)	Fiedler 1956
	Likely to occur in Chittagong Hill Tracts (extreme SE)	Green 1978a
India	In upper and lower Assam (in E); fairly common in Naga Hills	McCann 1933b
	In E Assam (maps)	Fiedler 1956, Dobroruka 1967, Roonwal and Mohnot 1977
	Reported in E Assam but could not find any in 1965; Yoshiba could not find any in 1959–62	Bertrand 1969
	Almost none left	Kurt 1970
	Formerly abundant; has diminished sharply	Southwick and Siddiqi 1970
	Rare if still present	Southwick, pers. comm. 1974
	Threatened in Meghalaya State (S of Brahmaputra R., 90–92°E) and in Arunachal Pradesh Territory (extreme NE)	Mohnot 1978
Burma	In N, NE, and SE (peninsular) (maps)	Dobroruka 1967, Bertrand 1969

Table 122 (continued)

Country	Presence and Abundance	Reference
China	In SE and in Tibet; in need of protection if to avoid extermination	Sowerby 1937
	Throughout SE to 31°N, 100°E (map)	Fiedler 1956
	In Szechuan and Yunnan provinces; E to Fukien and adjacent provinces	Ellerman and Morrison-Scott 1966
	In SE and S, questionable between 100°E and 103°E (map)	Dobroruka 1967
	As far N as Yangtze R.; questionable in Kweichow and Kiangsi provinces (map)	Bertrand 1969
	In SE to Fukien and Kwangtung provinces	Fooden 1971a
Thailand	In extreme N, S, and peninsular; questionable in E (map)	Dobroruka 1967
	Declining in S	Lekagul 1968
	In W and peninsular (map); becoming scarce in S; questionable in E	Bertrand 1969
	Very common	Lim 1969
	Difficult to find	Southwick and Siddiqi 1970
	Rare in W	Fooden 1971a
	Not abundant and under heavy pressure in S	Bernstein, pers. comm. 1974
	Very scarce	Southwick, pers. comm. 1974
	At Ban Palian (S peninsula) and in W central; rarely encountered	Fooden 1976
	Throughout (map)	Lekagul and McNeely 1977
	In Huay Kha Khaeng Game Sanctuary (in W near 15°29′N)	Mittermeier 1977b
Malaysia	S to about 3°N in peninsular (map)	Dobroruka 1967
	Recently migrated into Perlis and Kedah states (in extreme NW peninsular) (map)	Bertrand 1969
	In extreme NW peninsular (map)	Lim 1969
	Reported in NW Perlis (extreme N peninsular)	Medway 1969a
	Reported from Perlis, N Perak, and Kelantan (all in N peninsular)	Medway 1970
	No reports since 1950; probably extinct there	Southwick, pers. comm. 1974
Cambodia	In N	Pfeffer 1969
	In NE, N central, and SW; questionable in center (map)	Bertrand 1969
Laos	Throughout (map)	Dobroruka 1967
	Throughout except in SW; questionable in S central (map)	Bertrand 1969
Vietnam	In Tonkin region (in N), at Col des Nuages near Hue and Lao Bao (central)	Delacour 1940

Table 122 (continued)

Country	Presence and Abundance	Reference
Vietnam (cont.)	Captured in "various regions" of N including Tiuen Kuang (probably Tuyen Quang) Province in Tiem Khoa district	Lapin et al. 1965
	In Annam and Cochin China	Ellerman and Morrison-Scott 1966
	Throughout except in N central and extreme S (S of about 10°N) (map)	Dobroruka 1967
	Throughout except near Hanoi (in NE); questionable in extreme S (map)	Bertrand 1969

Macaca assamensis Assamese Macaque

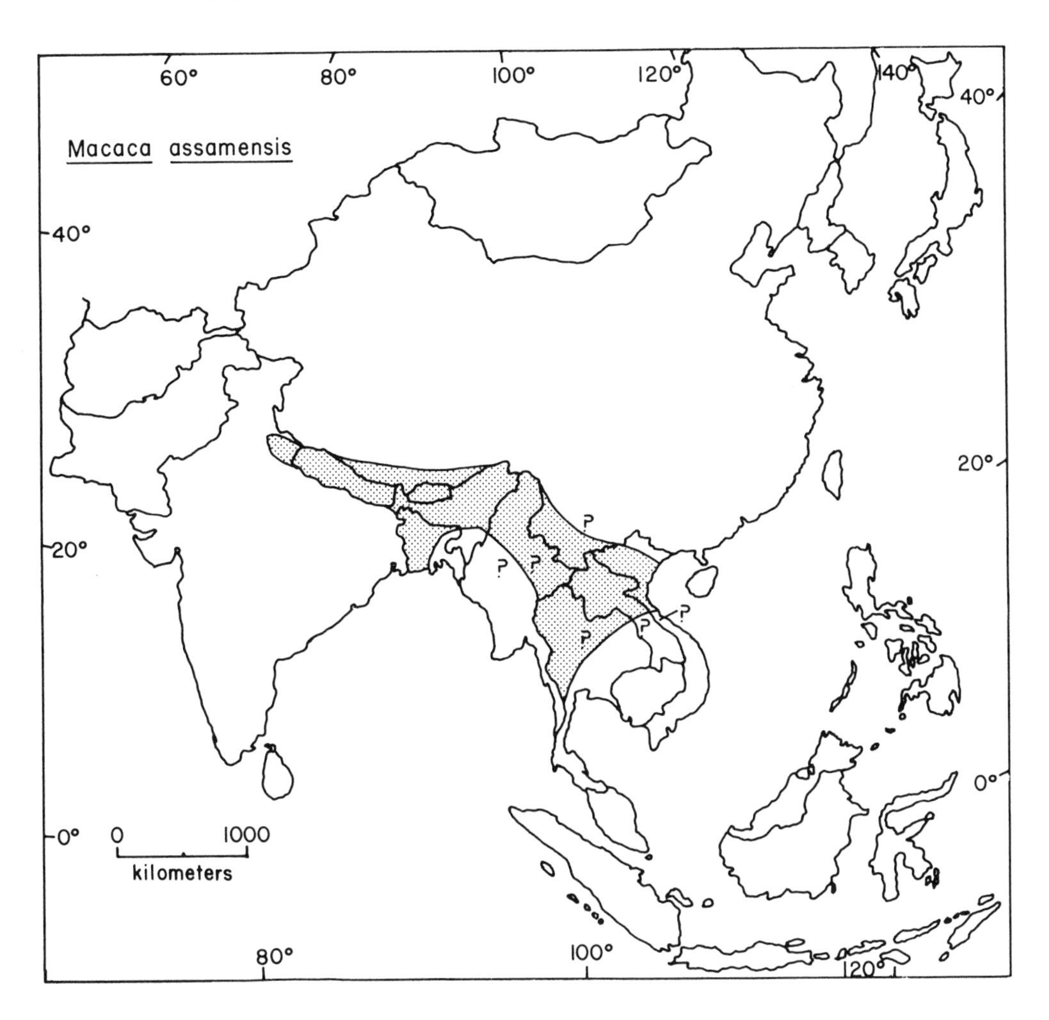

Taxonomic Note

Two subspecies are recognized (Napier and Napier 1967).

Geographic Range

Macaca assamensis is found from northern India and Nepal to Vietnam. It occurs as far north and west as Mussoorie, Uttar Pradesh (30°26′N, 78°04′E) (Prater 1971), and at least as far east as Hoi Xuan, Vietnam (20°22′N, 105°09′E) (Delacour 1940). In the west it is found as far south as the shore of the Sundarbans at the mouth of the Ganges River near 22°N in Bangladesh (Ellerman and Morrison-Scott 1966). In the eastern part of its range, the northern and southern limits are not known. For details of the distribution, see Table 123.

Abundance and Density

Recently *M. assamensis* has been considered abundant in India (Mohnot 1978), but no information is available regarding its status elsewhere (Table 123). No estimates of population density or home range area were located.

Counts of individuals per group in this species revealed groups of 12 and 26 (Carpenter 1942), 10 to 25 (N = 2) (Southwick et al. 1964), 10 to 50 ($\bar{X}$ = 22.8, N = 11) (Fooden 1971a), and about 20 (Chakraborty 1975).

Habitat

M. assamensis is a forest monkey which utilizes many types of forest vegetation. It has been recorded in sal forest, young teak forest, and semideciduous mixed wet oak, chestnut, maple, walnut, and conifer forests in India (Khajuria and Ghose 1970). In Nepal it inhabits tropical dry deciduous forest, most sal forest, monsoon forest, and tropical evergreen montane forest (Frick 1969), and in Thailand it has been observed in evergreen and bamboo forest (Fooden 1971a). It is found over a wide altitudinal range, from 300 m (Dào Van Tien 1967, Khajuria and Ghose 1970) to 3,500 m above sea level (Lekagul and McNeely 1977).

Factors Affecting Populations

Habitat Alteration

No specific information could be located. The military activity (bombing, defoliant use, etc.) in Laos and Vietnam has undoubtedly destroyed some of its habitat (Orians and Pfeiffer 1970, S.I.P.R.I. 1976), but the extent of this damage is not known.

Human Predation

Hunting. M. assamensis is hunted in the Darjeeling area for its flesh, which is eaten and is reputed to have medicinal value (Prater 1971).

Pest Control. This species does feed in cultivated fields (Prater 1971) and visits army camp kitchens (Roonwal and Mohnot 1977), but none of the authors consulted mentioned any measures taken to control these activities.

Collection. M. assamensis has been captured in northern Vietnam and shipped to the Soviet Union for research at the Sukhumi Primate Center (Lapin et al. 1965). No information regarding the magnitude of this collection is available.

No *M. assamensis* were imported into the United States between 1968 and 1973 (Appendix) or in 1977 (Bittner et al. 1978), nor into the United Kingdom between 1965 and 1975 (Burton 1978).

Conservation Action

In Thailand, hunting, capture, or keeping *M. assamensis* in captivity is allowed only under license, and export is regulated by a quota system (Lekagul and McNeely 1977).

Table 123. Distribution of *Macaca assamensis* (countries are listed in order from northwest to southeast)

Country	Presence and Abundance	Reference
Nepal	In E (map)	Fiedler 1956
	Present	Ellerman and Morrison-Scott 1966, Frick 1969, Mitchell 1975
Bhutan	Present	Ellerman and Morrison-Scott 1966
	Near Shamgong (27°13′N, 90°40′E), S central	Chakraborty 1975
India	In N West Bengal near Darjeeling	Southwick et al. 1964
	In Sikkim and Assam	Ellerman and Morrison-Scott 1966
	In Uttar Pradesh, Assam, Naga Hills, Mishmi Hills	Napier and Napier 1967
	At Tarkhola and Takdah, Darjeeling District, West Bengal	Khajuria and Ghose 1970
	From Mussoorie to Assam and Sundarbans	Prater 1971
	Considered abundant; from parts of Uttar Pradesh to Assam and S to Sundarbans in Bengal	Mohnot 1978
Bangladesh	To Sundarban swamps	Ellerman and Morrison-Scott 1966
Burma	In N (map)	Fiedler 1956, Ellerman and Morrison-Scott 1966
China	In SE Tibet and SW Yunnan (map)	Fiedler 1956
	In Yunnan Province	Ellerman and Morrison-Scott 1966
Thailand	In NW	Carpenter 1942
	In W between 14°41′N and 16°45′N (localities mapped)	Fooden 1971a
	In Huay Kha Khaeng Game Sanctuary (in W near 15°29′N)	Eudey, pers. comm. 1977
	In N and W as far S as Tenasserim (map)	Lekagul and McNeely 1977
Laos	In N	Delacour 1940
Vietnam	At Hoi Xuan and Chapa	Delacour 1940
	In N to 17°N (map)	Fiedler 1956
	Present	Lapin et al. 1965
	In Yen Bai Province (21°40′N, 104°48′E) in N	Dào Van Tien 1967

Macaca cyclopis Formosan Rock Macaque

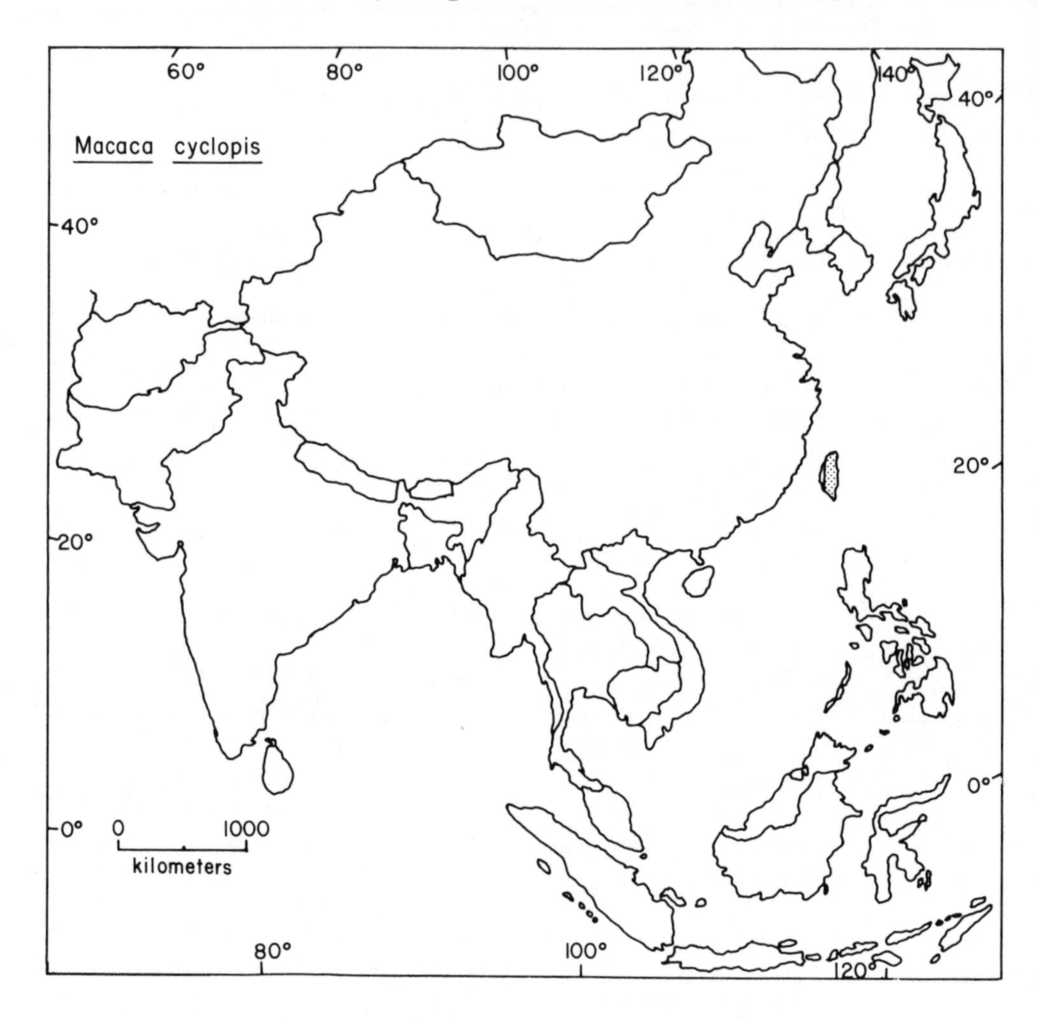

Taxonomic Note

Macaca cyclopis is a monotypic species (Ellerman and Morrison-Scott 1966).

Geographic Range

M. cyclopis is found on the island of Taiwan (Formosa), Nationalist China (Fiedler 1956, Ellerman and Morrison-Scott 1966). In 1962–64 it occurred in every part of Taiwan except along the west coast, and was most frequent in the northeastern and southwestern regions of the island (localities mapped by Bergner and Jachowski 1968).

Recently, *M. cyclopis* has been described as "generally distributed in the mountain ranges" (Wayre 1969) and present in the highlands which span the length of the island (Kuntz and Myers 1969). Natural populations apparently exist in the east

(Southwick and Siddiqi 1970), but no information is available from the interior or west, or regarding the details of the present distribution.

Abundance and Density

Ruhle (1966) reported that *M. cyclopis* was "seriously threatened," while Wayre (1969) contended that it was "not seriously threatened but in need of some protection." But no population survey of *M. cyclopis* has ever been done (Peng et al. 1973). And no information concerning population density, group size, or home range area was located.

Habitat

The preferred habitat of Formosan rock macaques has been described as "rocks and grassy hills near the sea, generally at low altitudes" (Swinhoe 1862, cited by Kuntz and Myers 1969), broadleaf forest less than 1,000 m above sea level (Bergner and Jachowski 1968), and caves and beaches (Kuntz and Myers 1969). The species is most common between 609 m and 1,830 m elevation, although it occurs up to 3,048 m (Kano 1940, cited by Kuntz and Myers 1969).

Factors Affecting Populations

Habitat Alteration

M. cyclopis has been pushed into highland regions by human populations and is now absent from many low elevation areas which it previously occupied. Extensive agricultural and other development in the lowlands, as the result of a rapidly expanding human population, is partly responsible for *M. cyclopis*'s present occupation of mountainous regions (Kuntz and Myers 1969).

Deforestation has also stopped *M. cyclopis* from traveling from west to east at the northern and southern ends of the island (Bergner and Jachowski 1968). These authors reported that Formosan rock macaque populations from opposite sides of the central north-south mountain range could previously achieve contact and gene exchange through the lowland areas at the tips of the island. But the removal of the forest from these areas now prevents this movement and has isolated certain segments of the population from each other.

Human Predation

Hunting. *M. cyclopis* is shot and trapped as a food item (Wayre 1969), although some local people report that it is forbidden to kill this animal because of its resemblance to man (Kuntz and Myers 1969). These authors add that the "skeletal system is used to prepare Chinese medicinal mixtures with aphrodisiacal attributes." About 200 *M. cyclopis* are killed annually to prepare this "skeletal jelly" (Peng et al. 1973).

Pest Control. No information.

Collection. Live Formosan rock macaques are kept as pets and are considered good luck (Kuntz and Myers 1969). They are used extensively for research in Taiwan (Southwick and Siddiqi 1970), but few references are available in non-Oriental languages (Kuntz and Myers 1969). About 760 *M. cyclopis* are collected yearly for research and education, including 300 for medical research, 400 for the preparation of skeletons to be sent to Japan as teaching materials, and 60 "babies" to be exported to Japan for study (Peng et al. 1973). Wayre (1969) stated that *M. cyclopis* was in need of protection from unrestricted shooting and trapping.

Seventy *M. cyclopis* were imported into the United States between 1968 and 1973 (Appendix). Thirty-one were imported into the United Kingdom in 1965–75 (Burton 1978). None entered the United States in 1977 (Bittner et al. 1978).

Conservation Action

M. cyclopis is listed as threatened under the U.S Endangered Species Act (U.S. Department of Interior 1977b).

Macaca fascicularis Long-tailed or Crab-eating Macaque

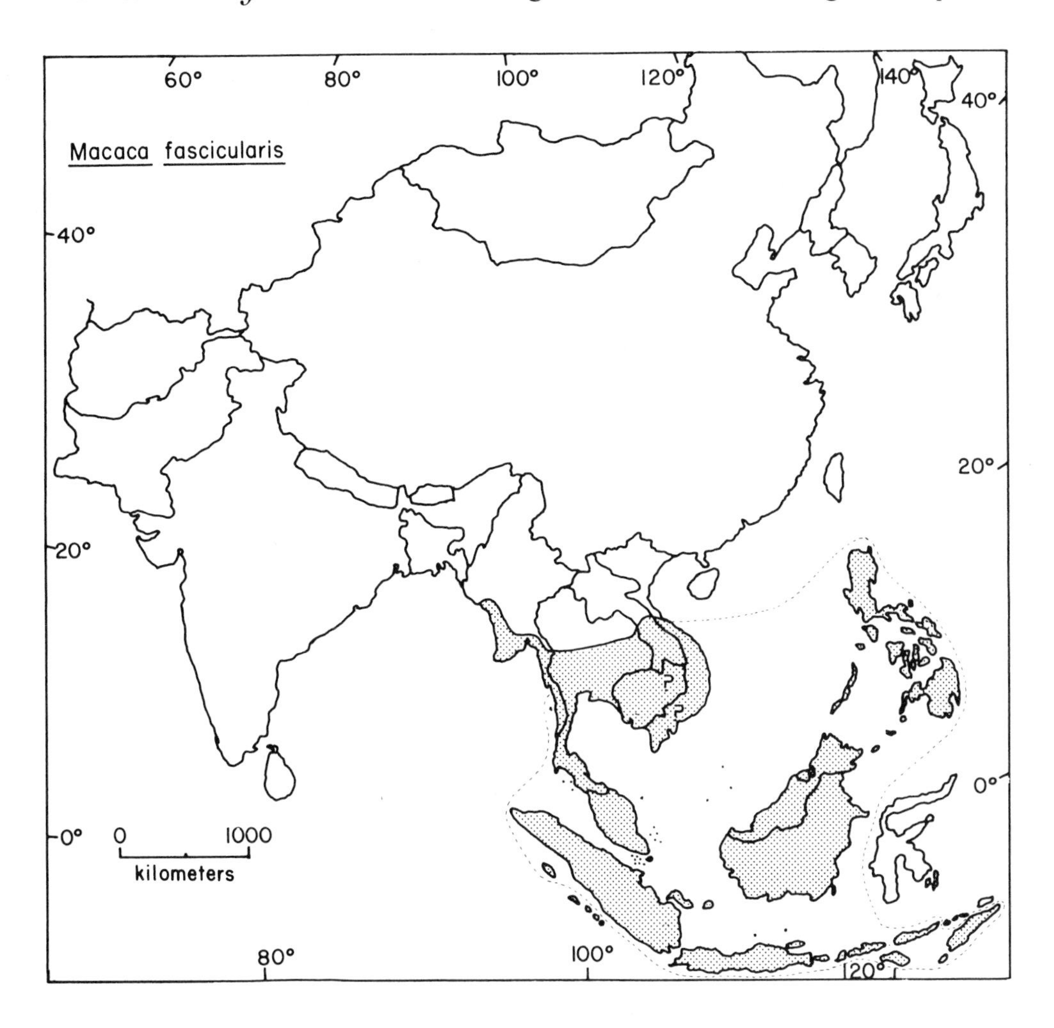

Taxonomic Note

This form has previously been called *Macaca irus* and *M. cynomologous*, but most modern workers now use the name *Macaca fascicularis*. Twenty-one subspecies are recognized (Napier and Napier 1967). Fooden (1964) proposed that *M. fascicularis* be considered conspecific with *M. mulatta*, based on a population with intermediate characteristics in Indochina. But he later concluded that this population probably represents secondary contact and hybridization between two previously geographically isolated forms which should be retained as separate species (Fooden 1971a).

Geographic Range

M. fascicularis is found in southeast Asia from Burma to the Philippines and southward through Indochina, Malaysia, and Indonesia. Its northern limit is at about

20°N,94°E in southwest Burma (Fooden 1971a). *M. fascicularis* is found as far south as about 11°S and as far east as about 125°E on Timor Island (Fooden 1969a), but is absent from Sulawesi (Celebes Island). In Vietnam, *M. fascicularis* occurs as far north as about 17°N, and in the Philippines its range extends north to northern Luzon (Rabor 1955), at about 18°N. For details of the distribution, see Table 124.

Abundance and Density

M. fascicularis has recently been described as abundant in parts of Malaysia, Thailand, and Indonesia, but is also thought to be declining in all three of these countries as well as in Vietnam and the Philippines (Table 124). No information is available regarding its present status in Burma, China, Laos, Cambodia, or the Nicobar Islands.

M. fascicularis population densities vary from 10 to more than 400 per km^2 (Table 125). Average group sizes are between 10 and 48, although groups as large as 100 have been reported. The few estimates available indicate home ranges of 25 to 200 ha (Table 125).

Based on their estimate of population density in the Krau Game Reserve, Malaysia, Medway and Wells (1971) concluded that 500 km^2 of habitat are necessary to support a viable population of 5,000 *M. fascicularis*. They postulated that this is the minimum size for a reserve for this species.

Habitat

M. fascicularis is an "ecologically opportunistic" species which can "exploit a great variety of habitats" including primary forest, disturbed and secondary forest, and riverine and coastal forests of nipa palm and mangrove (Medway 1970). It is primarily an "edge" species which is most successful in disturbed habitats and on the periphery rather than the interior of forest areas (Medway 1970, Kurland 1973, Lekagul and McNeely 1977).

In Sumatra, *M. fascicularis* achieves its highest population densities in rhizophora and mixed mangrove swamps, secondary hill forest, and lowland riverine forest (Wilson and Wilson 1977). These authors also observed it in many other habitats, including freshwater swamp, scrub grassland, lowland primary forest, rubber groves, and villages. Also in Sumatra, *M. fascicularis* utilizes pockets of jungle within savanna, and the edges of fields and plantations (Kurt 1973).

In western Thailand, *M. fascicularis* occurs in evergreen forest, bamboo forest, and, less frequently, deciduous forest (Fooden 1971a). In peninsular Malaysia it is very abundant in coastal lowland forest and less so in inland primary forest (Lim 1969) and disturbed primary forest (Harrison 1969). It is frequently found in urban forests and parks, and much more often in secondary (36% of groups) than in primary forest (3% of groups) (Southwick and Cadigan 1972). In Vietnam it is also seen most frequently in secondary forest and apparently does not utilize primary rain forest (Van Peenen et al. 1971).

Many workers have stressed this species' dependence on water and its occurrence in wet habitats such as swamps or forests near rivers on the sea coast (Yoshiba 1964; Harrison 1966; Lekagul 1968; Chivers 1971; Kawabe and Mano 1972; Kurland 1973;

Lindburg and Fittinghoff, pers. comm. 1974; Lekagul and McNeely 1977). For example, in one study in North Borneo it was found only in nipa-mangrove associations in tidal zones and near rivers (Davis 1962). It has been described as most common in estuarine swamps (Burgess 1961), and as well adapted to damp forest, inhabiting mainly mangrove sea swamp and tropical rain forests (Furuya 1965).

M. fascicularis is found from sea level to 1,524 m (Medway 1970) and 2,000 m elevation (Lekagul and McNeely 1977). It is more common at lower than higher elevations in northern Sumatra (Kurt 1973), but can achieve high densities in secondary hill forest between 458 and 915 m (Wilson and Wilson 1977). It occurs up to 1,219 m in North Borneo (Burgess 1961) and in "highland" areas in northern Philippines (Rabor 1955).

In summary, the evidence indicates that *M. fascicularis* mainly occupies secondary, swampy, and riparian forests, prefers wet areas and edge habitats, and can utilize many different types of vegetation.

Factors Affecting Populations

Habitat Alteration

Because *M. fascicularis* can utilize disturbed and secondary forest, it benefits from some kinds of human alteration of its habitats. For example, in peninsular Malaysia it readily colonizes disturbed or partly cleared land, including plantations, parks, and gardens (Medway 1970), and troop and population sizes are larger in disturbed habitats than in pristine forest (Kawabe and Mano 1972). Southwick and Cadigan (1972) observed 61% of the *M. fascicularis* they saw in close association with man in urban forests and parks, and other authors have also reported this species in urban areas such as Kuala Lumpur and Singapore (Medway 1969a), and near villages (Chivers 1971).

In Sumatra, rubber groves often support a larger population of *M. fascicularis* than an equivalent area of undisturbed forest nearby (Wilson and Wilson 1975a), and groups may range farther from rivers where the primary forest has been partly cleared than where the forest is uncut (Wilson and Wilson, pers. comm. 1974). Similarly, in eastern Kalimantan, *M. fascicularis* flourishes in both naturally and artificially disturbed habitats (Kurland 1973; Lindburg and Fittinghoff, pers. comm. 1974). And in eastern Sabah these macaques occupy logged areas without showing ill effects (Yoshiba 1964).

But some changes do destroy *M. fascicularis* habitat, and these monkeys are able to occupy certain "artificial but not totally deforested habitats" (Medway 1970). Extensive timbering operations are occurring throughout much of *M. fascicularis*'s range. Clearcutting is proceeding in peninsular Malaysia at the rate of 22,000 ha per year (Southwick and Cadigan 1972), and forestry officials there are now considering the feasibility of log importation (Mei 1977). *M. fascicularis* will not be able to occupy these logged areas until or unless they are allowed to regenerate into secondary forest.

In Indonesia, selective logging is practiced, and although Wilson and Wilson (1975a) believed that *M. fascicularis* was "relatively unaffected by selective logging," MacKinnon (1973) found lower population densities in selectively logged forest (about 50% tree removal) than in undamaged forest in northern Sumatra. Elsewhere in Sumatra the range of *M. fascicularis* has been reduced by deforestation and the takeover by the local alang-alang grass (Wilson and Wilson 1972–73).

Throughout most of southeast Asia the degree of habitat alteration and its effect on *M. fascicularis* populations are not known. But five brief reports provide clues to the situation. In 1968 Marshall and Nongngork (1970) tried to find *M. fascicularis* on Samui Island, Thailand, and found none. In 1913 the species had been seen on the island, but already there was also "relentless deforestation and persecution of native mammals such as monkeys by the human population of 8,000" (Robinson and Kloss 1915, cited by Marshall and Nongngork 1970). By 1968 this island had a human population of 30,000 and no original forests (Marshall and Nongngork 1970), so the absence of *M. fascicularis* is not surprising.

On mainland Thailand, long-tailed macaques are absent from large parts of their former range because of "progressively intensifying human activity" (Fooden 1971a), and have been "forced into a marginal mountainous area due to man's activities" (Prychodko, cited by Weiss and Goodman 1972).

In peninsular Malaysia, the 23.4% decrease in the estimated population of *M. fascicularis* from 1958 to 1975 is blamed in part on the clearing of forest for agriculture and rural development (Khan 1978).

Near Mount Sontra, Vietnam, *M. fascicularis* was "common" in 1965–66 but had become "unusual" by 1967–69 (Van Peenen et al. 1971). These authors noted that the whole area had been heavily disturbed by military activity, road building, clearing and burning of slopes, and woodcutting for charcoal. Since much of Vietnam, Laos, and Cambodia have undergone these pressures as well as bombing, herbicide treatment, and other forms of destruction (Orians and Pfeiffer 1970, S.I.P.R.I. 1976), it is very probable that monkey populations have declined drastically in many areas.

Human Predation

Hunting. M. fascicularis is hunted for food in Thailand (Lekagul 1968), Malaysia (Southwick and Cadigan 1972), and on Borneo (Kurland 1973). Hunting is not a serious factor in most areas of Indonesia, because most of the people are Moslems and do not eat monkeys (Lindburg and Fittinghoff, pers. comm. 1974; Wilson and Wilson, pers. comm. 1974). Long-tailed macaques are considered sacred on Bali and are not hunted there at all (Brotoisworo 1978).

Pest Control. M. fascicularis often feeds in cultivated fields and is known to eat young dry rice, cassava leaves, rubber fruit (Wilson and Wilson, pers. comm. 1974), taro plants, and many other crops (Poirier, pers. comm. 1974). It is considered a serious pest to agriculture and is often killed as such in Malaysia (Lim 1969, Medway 1969a), Sumatra, Java (Brotoisworo 1978), Kalimantan (Wilson and Wilson, pers. comm. 1974), and Thailand (Lekagul 1968). In Thailand, although *M. fascicularis* populations have greatly declined, the species has not been specially protected because of public sentiment against it for its crop-raiding habits (Lekagul 1968).

In addition to feeding in farms, long-tailed macaques also take food from graveyards (Wilson and Wilson, pers. comm. 1974), garbage cans (Medway 1969a), and garbage pits (Kurland 1973). They do considerable damage to the Botanical

Gardens in Singapore (Harrison 1966), and have become involved in aggressive interactions with people in the Botanical Gardens at Penang (Spencer 1975). No information regarding measures taken to control these activities was located.

Collection. M. fascicularis is one of the five species of primates most extensively used in laboratory research (Southwick and Cadigan 1972, Comm. Cons. Nonhuman Prim. 1975a). In the late 1950s and early 1960s, 70,000 per year were exported from peninsular Malaysia; by 1970 this number had decreased to 10,000 per year (Chin, cited by Southwick and Cadigan 1972).

Between 1968 and 1973 a total of 11,042 *M. fascicularis* were imported into the United States (Appendix). In 1970–73 the majority of these came from Malaysia, with large numbers also coming from the Philippines and Thailand. In 1977 a total of 1,900 individuals plus six shipments (of unspecified sizes) of *M. fascicularis* were imported into the United States (Bittner et al. 1978). In 1965–75, 53,432 were imported into the United Kingdom (Burton 1978).

Scientists in all three of these source countries blame trapping for export for the decline of *M. fascicularis*. Rabor (1968) reported that the species had been greatly depleted in the Philippines owing to the demand in the United States. He found that "the ruthlessness with which they were hunted (for export) . . . has been responsible for their decrease. Trade has practically ceased because of the radical depletion of the animals." In Thailand "tens of thousands" of *M. fascicularis* are said to have been exported yearly to the United States for the production of polio vaccines (Lekagul 1968). And in Malaysia, heavy trapping for export has contributed to the decline of the species (Khan 1978).

Recently the export volume from Indonesia was described as small but increasing (Lindburg and Fittinghoff, pers. comm. 1974). One Japanese company has been exporting large numbers from Java (Wilson and Wilson, pers. comm. 1974), and great numbers are reported to have been exported from Lombok in the past few years (Angst, pers. comm. 1974). The *M. fascicularis* populations on Sumatra are thought to be "large enough to maintain themselves in the face of considerable trapping pressure" (Wilson and Wilson 1972–73).

Long-tailed macaques are occasionally captured for the pet trade in Sumatra (Wilson and Wilson, pers. comm. 1974).

Conservation Action

M. fascicularis occurs in nine national parks, nine reserves, and two sanctuaries (Table 124). Weber (1972) proposed the protection of Kuala Selangor as a national monument which would create another reserve for *M. fascicularis* in Malaysia. This species also receives some protection in temple ruins in Thailand (Fooden 1971a) and protection and food in temples in Bali (Angst 1975).

In Malaysia, *M. fascicularis* is legally protected (Yong 1973), and is fed and protected in urban forests and parks (Southwick and Cadigan 1972). In the Philippines the government has been apathetic in enforcing proper conservation practices to protect this species (Rabor 1968). In Indonesia, *M. fascicularis* seems to be relatively well protected (Wilson and Wilson 1972–73; Angst, pers. comm. 1974), but some of the reserves containing it are being considered for oil drilling and timber harvesting, and may not remain protected in the future (Lindburg and Fittinghoff, pers. comm. 1974). In Thailand, *M. fascicularis* may be hunted, captured, or kept in captivity only under license, and its export is regulated by a quota system (Lekagul and McNeely 1977).

Table 124. Distribution of *Macaca fascicularis* (countries are listed in order from west to east)

Country	Presence and Abundance	Reference
India	On Nicobar Islands (7–9°N, 93–94°E)	Ellerman and Morrison-Scott 1966
Burma	In S and peninsular (localities mapped)	Fooden 1971a
	In Moscos Island Game Sanctuary (Mergui Archipelago off west coast of Tenasserim)	I.U.C.N. 1975
China	Present	Medway 1969a
Thailand	In SW, SE, peninsular, Terutao Island, Ko Phangan Island, Ko Samui Island	Napier and Napier 1967
	Declining	Lekagul 1968
	Extinct on Samui Island	Marshall and Nongngork 1970
	Throughout S as far N as 16°25′N (localities mapped); absent from large areas of former range	Fooden 1971a
	S and W of Bangkok; range shrinking	Weiss and Goodman 1972
	In W central; fairly abundant in S peninsular near Ban Thap Plik and Ban Palian	Fooden 1976
	In Huay Kha Khaeng Game Sanctuary (in W, near 15°29′N)	Eudey, pers. comm. 1977
	In central and S (map)	Lekagul and McNeely 1977
Laos	S of about 18°N; 1 locality mapped at Plateau des Bolovens (in S)	Fooden 1971a
Cambodia	In S half (map)	Lekagul and McNeely 1977
Vietnam	Common around Hue (central) and Saigon (in S)	Delacour 1940
	In Quang Nam Province (central), Darlac Province, and near Saigon (in S) (localities listed)	Van Peenen et al. 1969
	On Con Son Island (off SE coast)	Van Peenen et al. 1970
	On Mount Sontra (near Da Nang); common 1965–66; unusual 1967–69	Van Peenen et al. 1971 [Lippold 1977]
	In S, S of about 17°N and on Con Son Island (localities mapped)	Fooden 1971a
Malaysia	Throughout N Borneo; not in danger of becoming rare	Burgess 1961
	Common in N Borneo	Davis 1962
	In Ulu Bernam area, Selangor (W peninsular)	Milton 1963

Table 124 (continued)

Country	Presence and Abundance	Reference
Malaysia (cont.)	Common along Kinabatangan R., E Sabah (N Borneo)	Yoshiba 1964
	At Ratau Panjang, near Kuala Lumpur (W peninsular)	Furuya 1965
	In peninsular	Bernstein 1967b, Harrison 1969
	In Sabah (N Borneo)	de Silva 1968b
	Along W and E coast peninsular (map)	Lim 1969
	Throughout peninsular, often common; on islands of Langkawi, Penang, Tioman, Pemanggil, Aur, Tinggi, and Redang	Medway 1969a
	Throughout peninsular, adjacent islands and N Borneo (localities mapped)	Medway 1969b
	Widespread	Medway 1970
	In central and E peninsular (localities mapped)	Chivers 1971
	Throughout peninsular (localities mapped)	Fooden 1971a
	In Ulu Segama Reserve, near junction of Bole R. and Segama R., Sabah (N Borneo)	MacKinnon 1971
	Numerous at Kuala Lompat, Krau Game Reserve (central peninsular)	Medway and Wells 1971 [Chivers 1973, Curtin 1977]
	Very common in Bako N.P., Sarawak (N Borneo)	Rothschild 1971
	Common in Sabah (N Borneo), including Padas Bay (in W), Kinabatangan R., and Sepilok Forest Reserve (in E)	Kawabe and Mano 1972
	At Gunong Benom (central W peninsular)	Medway 1972
	Not as abundant as expected in peninsular	Southwick and Cadigan 1972
	At Kuala Selangor (W coast peninsular)	Weber 1972
	In Negri Sembilan, SE of Kuala Lumpur (W peninsular)	Weiss and Goodman 1972
	Abundant in peninsular 1965–66	Bernstein, pers. comm. 1974
	At Penang (W coast peninsular)	Spencer 1975
	Peninsular population has decreased 23.4% (from estimated 415,000 to 318,000) between 1958 and 1975	Khan 1978
Singapore	Present	Harrison 1966, Medway 1969a, Fooden 1971a
Indonesia	In Udjung Kulon Nature Park, extreme W Java	Satmoko 1961
	On Sumatra, Java, Kalimantan (Borneo), Bali, Lombok, Sumbawa, Sumba, Bangka, Belitung, Littung, and Riau archipelagos; localities mapped	Medway 1969a,b

Table 124 (continued)

Country	Presence and Abundance	Reference
Indonesia (cont.)	On Simalur and Nias islands, W of Sumatra	Medway 1970
	As far E as Timor Island	Fooden 1971a
	In Takokak Reserve, W Java, and Dolok Tinggi Radja Reserve, W Sumatra	I.U.C.N. 1971
	Abundant throughout Sumatra (localities mapped); not endangered	Wilson and Wilson 1972–73
	In Udjung Kulon Nature Reserve, W Java and on Peutjang Island off W tip of Java	Angst 1973
	In Kutai Reserve, E Kalimantan	Kurland 1973, Rodman 1973
	In Gunung Leuser Reserve, N Sumatra	Kurt 1973
	At Ranun R., Sikundur, and in W Langkat Reserve (N Sumatra)	MacKinnon 1973
	Abundant but slowly declining	Angst, pers. comm. 1974
	Abundant, perhaps increasing in Kutai Reserve, E Kalimantan	Lindburg and Fittinghoff, pers. comm. 1974
	In E Kalimantan	Van Peenen et al. 1974
	At Udjung Kulon, W Java; Baluran, E Java; Peutjang Island, W of Java; and Pulaki and Sangeh on Bali	Angst 1975
	At Sepaku, Renggang Road, and Kutai Reserve, E Kalimantan	Wilson and Wilson 1975a
	Abundant on Java, locally abundant on Sumatra, numerous in Sangeh and Pulaki temples on Bali	Brotoisworo 1978
Philippines	Rather common in Abra Highlands, central N Luzon	Rabor 1955
	On Luzon, Mindanao, Basilan, and Mindoro islands	Napier and Napier 1967
	Widely distributed, greatly depleted	Rabor 1968
	On Negros Island	Rabor et al. 1970
	In 8 national parks: Bataan, Mount Isarog, Aurora Memorial, and Baik-Na-Bato, on Luzon; Canlaon, Negros; Sohoton Natural Bridge, Samar; Naujan Lake, Mindoro; and Mainit Hot Springs, Mindanao	I.U.C.N. 1971

Table 125. Population parameters of *Macaca fascicularis*

Population Density (individuals/km²)	Group Size			Home Range Area (ha)	Study Site	Reference
	X̄	Range	No. of Groups			
	31	7–100	11		Thailand	Fooden 1971a
	13	10–20	5		Thailand	Fooden 1976
		2–3			Vietnam	Lippold 1977
					Malaysia	
		10–12			N Borneo	Davis 1962
	47	30–73	5		Peninsular	Furuya 1961–62a
		20–30			N Borneo	Yoshiba 1964
	41	13–72	6	200	Peninsular	Furuya 1965
	30	14–70	4		Peninsular	Bernstein 1967b
		20–50			Peninsular	Lim 1969
		8–>40			Peninsular	Medway 1969a
11		3–6	2		Peninsular	Medway and Wells 1971
	24	7–44			Peninsular	Southwick and Cadigan 1972
					Indonesia	
		10–50			Sumatra	Wilson and Wilson 1972–73
13–90	18	15–30	12	80	E Kalimantan	Kurland 1973
10–90	10.7	2–30	14		N Sumatra	Kurt 1973
5–50					N Sumatra	MacKinnon 1973
>400	48	17–>85	11		Java	Angst 1975
19–27	10	10–40			E Kalimantan	Wilson and Wilson 1975a
11.2–143.4	18.6		33	25–100	Sumatra	Wilson and Wilson 1977
		12–24			N Philippines	Rabor 1955

Macaca fuscata Japanese Macaque

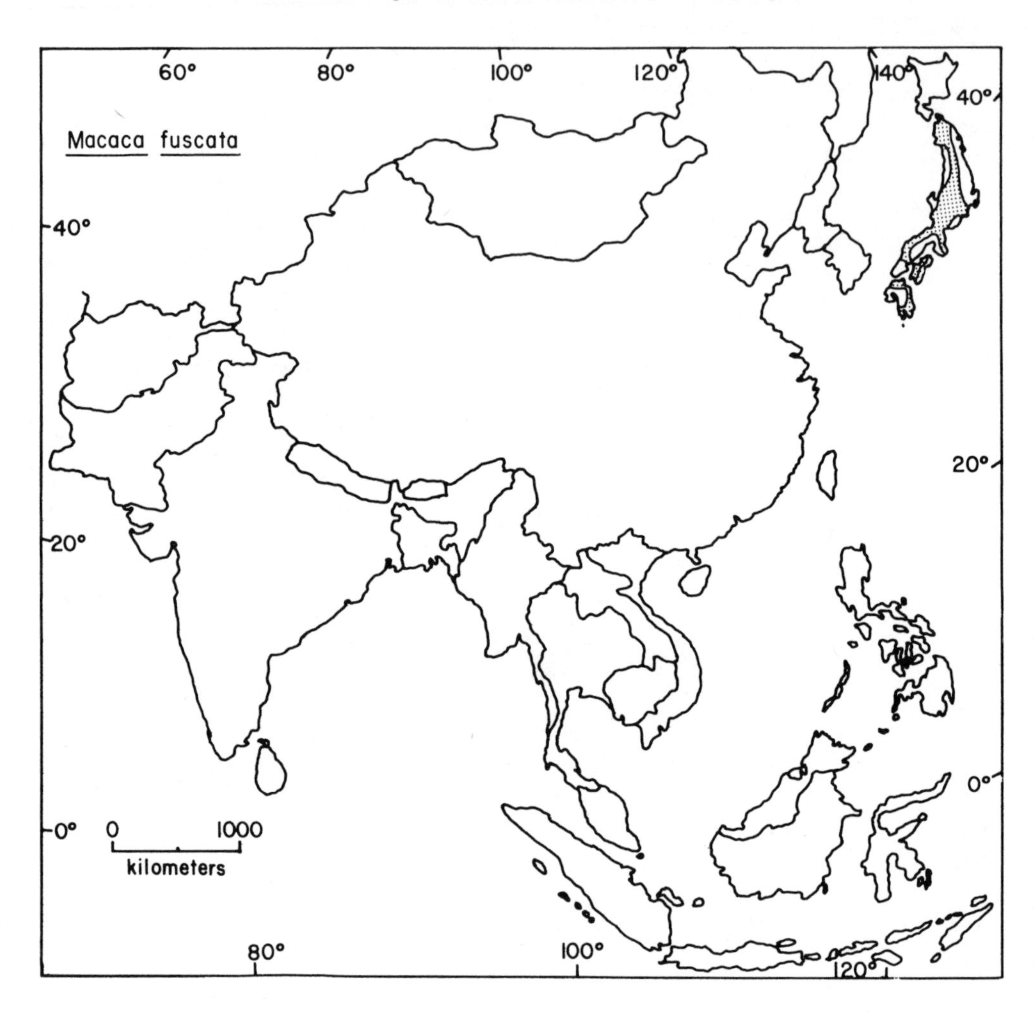

Taxonomic Note

Two subspecies are recognized (Ellerman and Morrison-Scott 1966, Napier and Napier 1967).

Geographic Range

Macaca fuscata is found on three of the four major islands of Japan: Shikoku, Kyushu, and Honshu, but not on the northernmost island of Hokkaido (Iwano 1975). It also inhabits some of the smaller islands including Yakushima (south of Kyushu), Awajishima and Shodoshima (both northeast of Shikoku), and Koshima (south of Honshu) (Kawanaka 1973). It occurs farther north than any other primate except man, at 41°30′N, 141°E on the Shimokita Peninsula (Izawa and Nishida 1963) and as far south as 30°30′N on Yakushima (Koford 1969b, Fukuda 1975, Uehara 1975). For details of the distribution, see Table 126.

Abundance and Density

The present total population of *M. fuscata* is estimated to be between 50,000 and 100,000 individuals (Kawamura 1975). But, most recent authors (except Kawanaka 1973) believe that the population and its range are both decreasing rapidly. *M. fuscata* has been eliminated in many areas where it previously occurred, and could face extinction in the next few decades (Iwano 1975) (Table 126).

Group sizes in this species range from 10 to 890, with average sizes of 40 to 194 (Table 127). Population densities of 2 to 104 per km^2 and home ranges of 220 to 2,700 ha have been reported (Table 127). Generally, smaller groups and larger home ranges are found in the northern part of the range, where climatic conditions are severe (Koford 1969b, Kawamura 1975).

In many areas of Japan the monkeys are fed ("provisioned") by people, leading to increased group sizes, decreased home range areas, and increased population densities (Yamada 1966; Izawa, Sato, Masui, and Mizuno, pers. comm. 1974; Itani 1975; Koyama et al. 1975; Masui et al. 1975). But the data presented by Kawamura (1975) for groups under natural conditions show that this species can achieve large group sizes and population densities under nonprovisioned circumstances.

Habitat

M. fuscata occupies a wide range of habitats from subtropical lowland forest to subalpine vegetation in snowy regions (Koford 1969b, Uehara 1975). It utilizes evergreen broadleaf forest, secondary evergreen and deciduous broadleaf forest, and mixed coniferous and deciduous forest (Kawamura 1975).

This species appears to prefer primary or mixed primary and secondary forests containing a variety of broadleaf and conifer evergreens as well as deciduous trees, in submontane and montane areas, but this apparent altitudinal preference may be due to human activities at lower elevations (Kurland, pers. comm. 1974). *M. fuscata* has been recorded from sea level to 1,500 m elevation, and most of its present habitat is in mountainous, forested areas (Kawanaka 1973).

Factors Affecting Populations

Habitat Alteration

Japan is an industrialized, rapidly developing country in which the natural vegetation is being destroyed by many different human activities. Deforestation is the most prominent factor decreasing the habitat available to *M. fuscata* (Suzuki 1972, Kawamura 1975). Most of the primary forests in Japan have already been logged, and although *M. fuscata* can occupy secondary forest, large-scale mechanized clear cutting has denuded large areas since 1950 and has eliminated monkeys in many parts of their former range (Izawa and Nishida 1963, Suzuki 1972, Kawamura 1975).

Many of the lowland areas are heavily settled and cultivated, and this may be partly responsible for *M. fuscata*'s restriction to highland regions (Kurland, pers. comm. 1974). In addition, widespread timbering in the mountains tends to force the monkeys to lower elevations in search of food, leading to conflicts with farmers (Itani 1975). Shifting cultivation is also destroying *M. fuscata* habitat in mountainous areas (Izawa, Sato, Masui, and Mizuno, pers. comm. 1974).

In some regions, natural forest vegetation is replaced with plantations of economically valuable Japanese cedar, which provide only a small quantity of food for *M. fuscata* and hence are not occupied by monkeys (Kurland, pers. comm. 1974; Kawamura 1975). In other regions the herbicide "Brushkiller" (2,4,D and 2,4,5T) has been sprayed to kill undesirable plants near commercial tree plantations (Azuma 1970, Izawa 1971). When 12.6 tons of herbicide were sprayed over a 300 ha area on the Shimokita Peninsula, many Japanese scientists became concerned for the monkey populations at that northern limit of their distribution, and voiced a strong protest to the authorities (Azuma 1970, Tamura 1970). Follow-up studies of the *M. fuscata* there revealed that at first they deserted the area but then returned the following year and fed on bark, twigs, and buds of defoliated trees (Izawa 1972). No evidence of mortality, birth defects, or other ill effects from the herbicide has yet been discovered. The spraying was eventually prohibited, but lumbering continued in the area (Editors of *Primates* 1971).

Another kind of habitat alteration that is destroying *M. fuscata* habitat is the building of highways. A study of the effect of the proposed Boso Skyline toll highway on the Japanese macaque population of Mount Takago showed that the highway would result in significant destruction of the monkeys' habitat, partitioning of their range area (prevention of movements), killing of monkeys by automobiles and malicious people, and overfeeding by tourists, as well as direct adverse effects of the construction activity (Suzuki 1972).

Most workers agree with Itani (1975) that "environmental disruption [in Japan] is reaching a critical level." Suzuki (1972) lamented that the mountains have been destroyed by the development of cities, the forests have been desolated, and the habitat of *M. fuscata* has been excessively "aggravated." He added that these processes are proceeding all over Japan and that *M. fuscata* habitat is rapidly disappearing. Even many of the monkey parks are seriously degraded, with leafless and dying trees, eroded or compacted and laterized soil, or natural vegetation replaced with cedar monoculture (Kurland, pers. comm. 1974).

Human Predation

Hunting. In the past, heavy hunting has been responsible for reductions in *M. fuscata* populations (Kawanaka 1973, Itani 1975). Some of this hunting was for meat for food, and some was for medicinal uses — for example, the head was charred and made into a headache medicine, the intestines were dried and used to prepare a potion for delivery pain (Izawa, Sato, Masui, and Mizuno, pers. comm. 1974). Since 1947 it has been forbidden to hunt *M. fuscata* except in special cases, but some poaching has continued (Izawa and Nishida 1963). Hunting may be partly responsible for *M. fuscata*'s present preference for inaccessible highland regions.

Pest Control. *M. fuscata* enters fields and damages crops such as potatoes, peas, and melons, especially in areas where deforestation has destroyed much of its natural food supply (Izawa and Nishida 1963). In some areas it is considered a pest to crops and may be legally shot to prevent damage (Suzuki 1972, Kawamura 1975).

The provisioning of *M. fuscata* at more than 30 artificial feeding grounds throughout Japan has allowed populations to grow beyond the normal carrying capacity of their habitats and has resulted in increased crop damage to farms adjacent to the feeding grounds (Itani 1975). But provisioning also shrinks the home range of troops, thus decreasing the overall amount of farm acreage damaged (Itani 1975). Nevertheless, complaints of crop raiding by *M. fuscata* have greatly increased since 1960, and many monkeys are killed annually to prevent damage (Itani 1975).

Provisioning has also resulted in a loss of wariness of man by *M. fuscata*, and in several instances tourists have been bitten by monkeys (Itani 1975). Many problems relating to the management of the artificially fed population of about 7,000 monkeys are still to be solved (Kawamura 1975).

Collection. The Japan Monkey Center at Inuyama is a large and active research institution where studies of many primates, including *M. fuscata*, are carried out. The number of wild monkeys used in this research was not learned. Kawamura (1975) reported that the collection of *M. fuscata* for science has not increased significantly since 1955.

Thirty-five Japanese macaques were imported into the United States in 1968–69, but none in 1970–73 (Appendix) or in 1977 (Bittner et al. 1978), nor were any imported into the United Kingdom between 1965 and 1975 (Burton 1978).

Conservation Action

Since 1947 the official policy of the Japanese government has been to protect and conserve *M. fuscata* (Suzuki 1972), and hunting has been permitted only to control crop raiding, for scientific purposes, or for special public needs such as zoos (Kawamura 1975). Efforts to conserve specific populations, such as those dosed with herbicides and those threatened with freeway construction, have been organized by concerned Japanese workers (Azuma 1970, Suzuki 1972).

M. fuscata is found in 16 national parks throughout the country (Table 126) and would appear to be well protected. But many of these areas are operated to produce maximum economic gain from tourism and other uses to the detriment of the monkey populations (Kurland, pers. comm. 1974), and Japanese scientists have called for a sweeping reappraisal of the basic policies for the protection of *M. fuscata* (Itani 1975).

Kawamura (1975) outlined five specific recommendations: (1) establishment of at least 20 national reservations with minimum areas of 200 km^2 each and several prefectural reservations of 50 km^2 each where all forest utilization is prohibited, (2) use of a shifting mosaic pattern of cutting and planting in the forests, so that natural succession will occur, (3) use of protective fences (of thorned metal plates) to prevent crop raiding, (4) introduction of ecological supervision of provisioning; and (5) creation of breeding colonies to produce laboratory monkeys, and culling of provisioned troops. Some scientists also believe that no new provisioning places should be established (Itani 1975).

M. fuscata is listed as threatened under the U.S. Endangered Species Act (U.S. Department of Interior 1977b).

Table 126. Distribution of *Macaca fuscata*

Country	Presence and Abundance	Reference
Japan	Population of 550–600 on Shodoshima Island, Seto Inland Sea N.P. (35°N, 134°E)	Yamada 1966
	Population of 100 or less and decreasing near Okoppe R., Shimokita Peninsula, N Honshu	Azuma 1970
	In 13 national parks: Daisetsuzan, on Hokkaido (outside range); Bandai-Asahi, Jo-shin-Etsu Kogen, Chubu-Sangaku, Nikko, Fuji-Hakone-Izu, Towada-Hachimantai, Hakusan, Minami Alps, and San-In Kaigan, all on Honshu; Aso, Karishima-Yaku, on Kyushu; Seto Naikai, on Inland Sea	I.U.C.N. 1971
	Population of 120 in Okoppe basin	Izawa 1971
	On Mount Takago, Boso Peninsula, SE Honshu	Suzuki 1972
	Population in 1953 was 15,600, in 1964 was 28,100 (localities mapped); range shrank 1940–60, but during last 10 years has remained stable and population has increased	Kawanaka 1973
	Population of 276–351 on N part of Mount Hakusan, W Honshu; stable	Izawa, Sato, Masui, and Mizuno, pers. comm. 1974
	Overall abundant; population 50,000–100,000 but declining	Kurland, pers. comm. 1974
	Only in large numbers in inner Kii Peninsula, Yaku Island, Boso Mts.; decreasing and range shrinking in Boso Mts.	Fukuda 1975
	Population of about 1,400 and increasing in Mount Takasaki N.P., NE Kyushu; throughout Japan population in 1953 was 15,614; in 1964 was 22,000–34,000; range shrinking	Itani 1975
	In 3 more national parks: Chubu, on Honshu; Ashizuri-Uwakai, Shikoku; and Nichinan Kaigan Quasi, Kyushu	I.U.C.N. 1975
	Population has decreased greatly since 1923: rapid decrease at every locality and extinct at some (previous and present ranges mapped); could become extinct in next few decades	Iwano 1975
	Total population 50,000–100,000; numbers decreasing, range shrinking; localities mapped in central Honshu	Kawamura 1975

Table 126 (continued)

Country	Presence and Abundance	Reference
Japan (cont.)	Study sites: Koshima Islet, Mount Takasakiyama, Mount Arashiyama, Boso Mts., Shiga Heights, Shimokita Peninsula	Uehara 1975

Table 127. Population parameters of *Macaca fuscata*

Population Density (individuals /km²)	Group Size			Home Range Area (ha)	Study Site	Reference
	X̄	Range	No. of Groups			
	194	26–570	10		General	Itani et al. 1963, cited by Napier and Napier 1967
		10–70			Honshu	Izawa and Nishida 1963
	104	52–160	5	450–1,550	Shodoshima (provisioned)	Yamada 1966
		20–100			N and S limits	Koford 1969b
	40		3	1,500–1,800	N Honshu	Izawa 1971
			1	250	N Honshu	Izawa 1972
2–16	44.7	20–80	7	300–2,700	W Honshu (provisioned)	Izawa, Sato, Masui, Mizuno, pers. comm. 1974
		250–890			NE Kyushu (provisioned)	Itani 1975
5.6–104.5	40–110	≤230	21	220–802	Various (natural)	Kawamura 1975

Macaca mulatta Rhesus Monkey

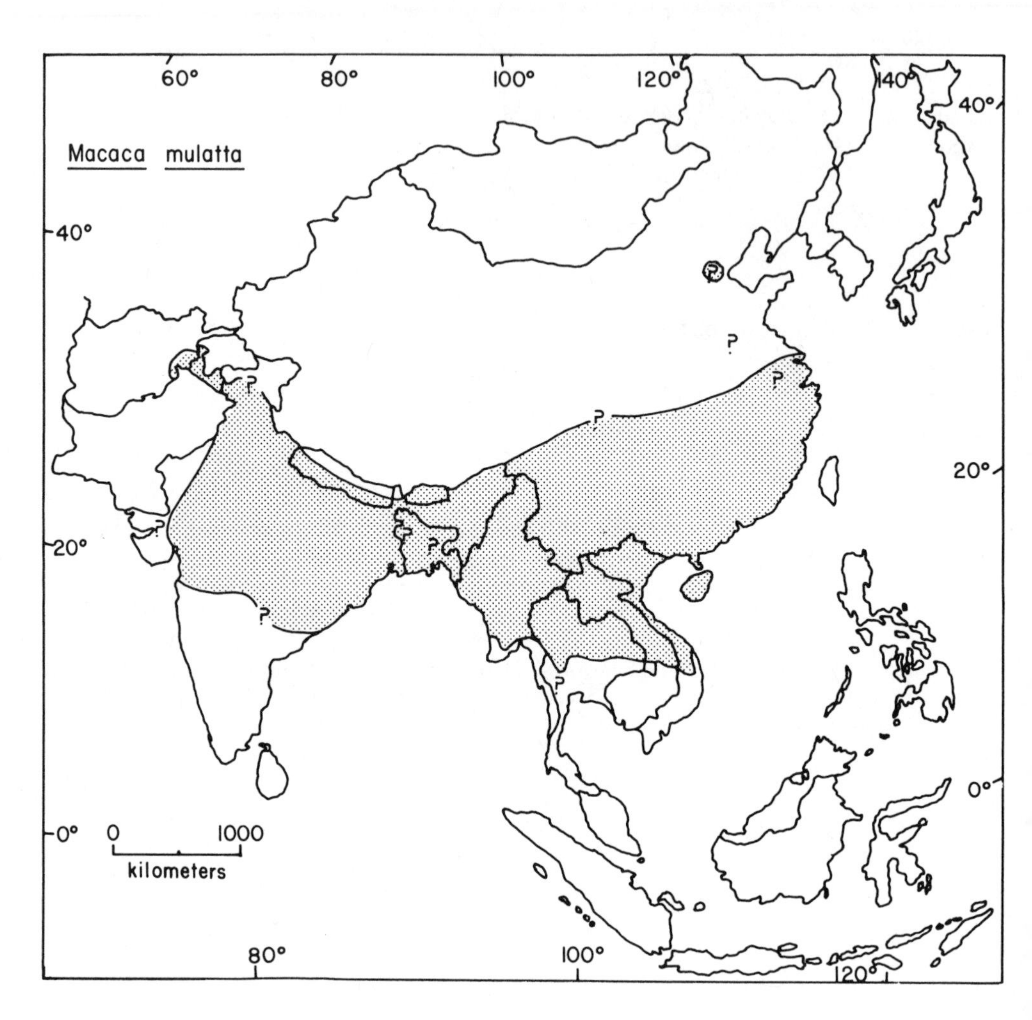

Taxonomic Note

Four subspecies are recognized (Napier and Napier 1967).

Geographic Range

Macaca mulatta is found in southern Asia from eastern Afghanistan to southern Vietnam. The westernmost record for the species is in Paktia, Afghanistan (Puget 1971, Hassinger 1973), at about 33°55′N, 69°46′E, and the northernmost locality in the west is in Chitral, Pakistan (Roberts 1977), near 35°35′N, 71°41′E. In the east the northern limit of distribution is thought to be near 32°N in southern China, but Fiedler (1956) also reported an isolated population much farther north near 40°N, and Annenkov et al. (1972 citing Weber) showed this northern range continuous with that in southern China. The northern Chinese population may be an introduced one (Ellerman and Morrison-Scott 1966).

The range of *M. mulatta* extends at least as far south as Dak Sut in Kontum Province, Vietnam (14°56′N, 107°44′E) (Van Peenen et al. 1969), and to 16°N in Thailand (Fooden 1964). In India the southern limit is usually given as the Godavari River (Fiedler 1956, Prater 1971, Roonwal and Mohnot 1977), but the species is also said to be common near Hyderabad (17°22′N, 78°26′E), 160 km south of the Godavari (Southwick et al. 1961a). For details of the distribution, see Table 128.

Abundance and Density

Despite the large amount of data available on many aspects of the biology of *M. mulatta*, relatively little is known of its distribution or abundance over most of its range (Table 128). It is known to be abundant in eastern Afghanistan, in some parts of northern India, southern Nepal, and southern Bangladesh. It has been eliminated from much of its former range in Thailand and probably Indochina, and is now rare in parts of northern Pakistan. But its status in central India, most of Nepal, Bhutan, Burma, Laos, Vietnam, and China is not known (Table 128).

Southwick and his coworkers have studied the rhesus populations in Uttar Pradesh, India, since 1954. In 1961 they reported that there were at most one million *M. mulatta* in this area (instead of the 10 to 20 million previously reported), and that this population was declining. Since then they have documented the pattern of this decline in detail, showing an average decrease of 5% per year in number of groups and 3% per year in individuals (Southwick and Siddiqi 1966). In 1970 they estimated the total *M. mulatta* population in Uttar Pradesh to be about 500,000 (Southwick et al. 1970).

These same workers have demonstrated that while the rural populations have decreased, the urban ones have generally increased (Southwick and Siddiqi 1968). This overall trend toward urbanization of the Indian *M. mulatta* population has had significant effects on the behavior and health of the monkeys and will be discussed below.

Population density estimates for *M. mulatta* vary from 0.008 per km^2 to 1,000 per km^2 (Table 129) depending on the habitat. Similarly, home range areas as small as 0.05 ha and as large as 149,433 ha have been described (Table 129). Group size varies from very small to as large as 236 members. Average group sizes are between 9 and 58 individuals (Table 129).

Habitat

M. mulatta occupies many types of habitats, including tropical moist deciduous forest, chir pine, oak and scrub formations, and sheesham forest (Neville 1968). It has been observed in upper monsoon forest, tropical deciduous dry forest, and northern thorn forest dominated by *Acacia* and *Euphorbia* (Vogel 1976). Although Krishnan (1971) reported that it was not present where sal (*Shorea robusta*) dominated the flora, other authors have found it most frequently in sal forest (Lindburg 1977, Green 1978a). In some areas at least, it seems to prefer open country and forest edges to the interior of dense forest (McCann 1933b, Prater 1971).

In Afghanistan, *M. mulatta* is found mainly in the cedar and oak forests of mountainous areas (Kullman 1970; Puget, pers. comm. 1974). In Pakistan it also is confined to forested montane areas, is typically associated with Himalayan moist temperate forests, and also utilizes dry forests dominated by *Cedrus deodara*, *Pinus gerardiana*, and *Quercus ilex* (Roberts 1977). In Nepal it occupies deciduous forest, sal forest, subtropical monsoon forest, tropical evergreen montane forest (Frick 1969), and pine-oak-spruce forest (Richie et al. 1978). And in Thailand it was encountered equally often in bamboo and evergreen forests (Fooden 1971a). In the Sundarbans of West Bengal and Bangladesh it lives in mangrove forests on swampy deltas where no fresh water is available except rainwater, and the submerged, muddy, slippery soil and regular cyclones create unpredictable conditions (Mukherjee and Gupta 1965).

Rhesus monkeys have a wide temperature tolerance (from below freezing to above 50°C), can occupy habitats ranging in humidity from dry, near-desert to more than 100 cm of rainfall annually (Lindburg 1971), and have been reported from sea level to 3,050 m (Richie et al. 1978) and 4,300 m elevation (Haltenorth 1975).

In northern India, much of the natural vegetation has been removed (see below) and more one-half of all *M. mulatta* are now found in villages and towns (Southwick et al. 1965). These authors reported that rhesus monkeys have lived in close contact with man for centuries and that the relationship between man and *M. mulatta* represents a natural one in both frequency and persistence. Urban habitats provide troops of *M. mulatta* with large trees for cover and food, buildings and walls for shelter, abundant food in fields and gardens, and water in irrigation ditches and ponds (Southwick et al. 1961b). The wide ecological tolerance shown by *M. mulatta* in its ability to utilize many niches, from that of remote forest to commensalism, makes it difficult to assess what the present "natural" habitat is for this species. In mixed forest and agricultural areas, *M. mulatta* groups appear to be living in forests secondarily, because they have been driven from human habitation, and in some cases they may actually prefer to live in domestic habitats (Southwick et al. 1965). Among the man-made habitats utilized by *M. mulatta* are markets, bazaars, temples, railroad stations, roadsides, and canal banks (Southwick et al. 1961b, Southwick and Siddiqi 1968, Krishnan 1971).

In Uttar Pradesh, northern India, rhesus monkeys are most abundant in villages and towns and second most plentiful in tropical mixed deciduous forests which are dense secondary forest with abundant cover and food (Southwick et al. 1961b). Water is a critical factor in their distribution, and they are often concentrated around both natural and artificial water sources during dry periods (Southwick et al. 1961b, 1965). Cover, providing protection from dust storms and monsoon rains, is also a vital component of rhesus monkey habitat (Southwick et al. 1961b). Cover is obtained from forest trees, roadside, temple, or town trees, or urban buildings and walls (Southwick et al. 1961b, Southwick and Siddiqi 1966).

Factors Affecting Populations

Habitat Alteration

In northern India three kinds of habitat alteration have reduced the available habitat for *M. mulatta*: cultivation, woodcutting, and commercial forestry (Southwick et al. 1961a,b). In the early 1960s these authors found that 65% of the land in

Uttar Pradesh was already under cultivation, 21% was fallow, barren, or uncultivatable land, and 14% was forested area. In addition to clearing for agriculture, roadside trees and groves were being cut extensively for goat fodder, lumber, and firewood. And the remaining forest was being cleared and replaced with pure stands of commercially valuable sal and sheesham trees which are not favorable habitats for *M. mulatta* (Southwick et al. 1961a,b, 1970).

In 1973 these activities were continuing and the biggest threats to rhesus monkey populations were "deforestation for the extension of terraced agriculture and specialized reforestation with stands of eucalyptus, conifers or sal" (Southwick and Siddiqi 1973). Forest destruction is now proceeding at the rate of 2% per year in some parts of northern India (Southwick, pers. comm. 1974).

Heavy grazing by cattle and the cutting of branches for cattle fodder have also reduced the habitat and food supply for *M. mulatta* in northern India (Lindburg 1971). Outside of northern India, less detailed information is available, but similar kinds of habitat changes are occurring. In some areas of Afghanistan, forest clearance has greatly decreased *M. mulatta* habitat (Puget, pers. comm. 1974). In northern Pakistan, the opening of new regions for timber extraction is expected to cause the rhesus population there to decline further (Roberts 1977). In at least one part of Nepal, *M. mulatta* has moved to high elevations "in response to deforestation and loss of habitat at lower altitudes" (Richie et al. 1978).

The rapidly expanding human population in Bangladesh is generating a great need for firewood and building materials, resulting in widespread destruction of forest (Green 1978a). And in western Thailand, "progressively intensifying human activity" has eliminated much former *M. mulatta* habitat (Fooden 1971a).

Although no data are available regarding the Vietnamese populations, the widespread bombing, defoliation, and other destruction in this area (Orians and Pfeiffer 1970, S.I.P.R.I. 1976) have almost certainly reduced monkey populations. No information is available from Laos, China, or Burma.

In summary, although this species has a "wide ecological tolerance" which enables it to "thrive in areas of major human modification of the habitat" (Lindburg 1971), it is being adversely affected by deforestation and other destruction of natural vegetation in many parts of its range.

Human Predation

Hunting. No information.

Pest Control. The second major reason for changes in *M. mulatta* populations in northern India is their habit of obtaining food from farms and gardens. In some cases this concentrated food source has increased *M. mulatta* populations (Neville 1968). But often crop raiding brings *M. mulatta* into competition with people for food, and has resulted in attempts to eliminate the monkeys as pests. *M. mulatta* is known to damage such crops as sugar cane, wheat, grain, pulses, millet, and vegetables, but the extent of this damage is not known (Southwick et al. 1961a). The damage is probably not serious on a national scale, and is not as great as that caused by insects, rodents, and plant diseases, but because damage done by monkeys is more obvious than that caused by other pests, it arouses more concern, and most villagers are anxious to drive *M. mulatta* away from their fields or have them trapped and removed (Southwick and Siddiqi 1970).

The decline in rural rhesus populations in Uttar Pradesh is due in part to the change in the attitudes of the local people from tolerance of monkeys to increasing intolerance of crop depredations (Southwick et al. 1961a,b, 1965). Tradition and a religious association with the monkey god Hanuman previously protected *M. mulatta* from extermination efforts by local farmers. But now many villagers have decided that they cannot afford the losses caused by *M. mulatta* and are trapping, shooting, or driving away the monkeys. In certain areas where devout local people still protect monkeys, the populations were denser than in other areas (Southwick et al. 1961a), and had increased 87% in individuals during the same period that unprotected populations decreased 70% (Southwick and Siddiqi 1973). Over a longer census period, protected populations increased 95% while unprotected populations decreased by 76.2% (Siddiqi and Southwick 1977).

Thus human practices and attitudes are important in determining the patterns of abundance, distribution, and population trends of *M. mulatta* (Southwick et al. 1961a). The future agricultural development of India will probably tend toward larger and larger "agribusiness" which will be less tolerant of monkeys than are small local farmers, and will tend to eiliminate monkeys as pests to crops (Southwick and Siddiqi 1973).

Already the intolerance of villagers has resulted in the elimination of *M. mulatta* from certain areas. In some places rhesus monkeys obtain almost all their food from agricultural crops (Southwick and Siddiqi 1973) and are forced to leave an area if they are prevented from crop raiding. *M. mulatta* was formerly present throughout Punjab but has now been chased out of this prime agricultural area (Lindburg, pers. comm. 1974). Elsewhere in India, *M. mulatta* has not been completely eliminated, but the rural populations have declined markedly, owing in part to the extermination efforts of farmers (Southwick et al. 1961a).

In Bangladesh, *M. mulatta* is also considered a serious pest to crops such as rice (Green 1978a). In Afghanistan, rhesus monkeys are killed because of the damage they cause to commercially valuable cedar trees, and the number of monkeys killed each year is increasing because of the use of modern weapons and the rising value of the trees (Puget, pers. comm. 1974). *M. mulatta* also feeds on crops such as maize, raisins, mulberries, and pomegranates in Afghanistan (Puget 1971, Naumann and Nogge 1973), but this behavior is generally tolerated by the local people (Puget, pers. comm. 1974).

In northern Pakistan, rhesus monkeys are especially fond of ripening maize cobs, hence are destroyed by the local people (via shooting, trapping, amd stone throwing) whenever possible (Roberts 1977). At one high altitude site in Nepal, rhesus monkeys probably obtain a significant portion of their food from cultivated fields (Richie et al. 1978), but no information about control efforts there, or in the eastern part of the range, was located.

Collection. The third major factor affecting populations of rhesus monkeys is the large-scale trapping and export for biomedical research in other countries. For example, in 1938, 250,000 *M. mulatta* were exported from India (Mohnot 1978), and between 1950 and 1960, 100,000 to 200,000 were exported annually from India (Southwick et al. 1970). In 1959 commercial exporters reported that they were having increasing difficulty obtaining sufficient monkeys to meet the demand (Southwick et al. 1961a). In 1962 one exporter in Calcutta believed that the Bengal population had been reduced below the point where it was commercially feasible to trap *M. mulatta* (Southwick et al. 1964).

In 1964–66 the number being exported had decreased to less than 50,000 per year (Southwick et al. 1970). By 1971, capture for export had depopulated many areas of India (Prater 1971). In 1973 India exported about 60,000 *M. mulatta*, but announced that they would reduce exports to 30,000 per year in 1974 (Southwick, pers. comm. 1974). In 1975 the export rate was cut to 20,000 per year and further restrictions were predicted (Mohnot 1978).

Thus more than two million *M. mulatta* had been exported from India since 1938, and the population seemed to be decreasing, possibly jeopardizing the stocks necessary to supply the 5,000 rhesus monkeys per year that India needs for its own uses (Mohnot 1978). The Indian government also concluded that the original agreement regarding the uses to which imported *M. mulatta* would be put (medical research with humanitarian benefits rather than military or production related projects) had been violated by the United States, and as of April 1978, India prohibited export of *M. mulatta* to the United States (Wade 1978a).

Between 1968 and 1973 a total of 147,742 *M. mulatta* were imported into the United States (Appendix). These were primarily from India, with 100 to 500 per year also supplied by Thailand and Pakistan and a few from other countries. In 1977, 1,795 individual *M. mulatta* and 24 shipments of unspecified sizes were imported into the United States (Bittner et al. 1978). From 1965 to 1975 a total of 39,263 *M. mulatta* were imported into the United Kingdom (Burton 1978).

Because older infants and juveniles are most often captured for the export trade, the population structure of *M. mulatta* can be changed greatly by trapping (Southwick and Siddiqi 1966). These workers found that the percentage of infants and juveniles in groups dropped to as low as 31% in heavy trapping years, and showed that this "juvenile gap" would make future decline of the populations inevitable if trapping continued unabated. In 1973 trapping had declined and the proportion of immatures had risen to an average of 44.6% for the groups in the Aligarh district, but 50% immatures is necessary if a population is to remain stable (Southwick and Siddiqi 1973). Thus although the rate of trapping had decreased, the long-term effects on population structure were still being felt.

Southwick et al. (1970) calculated that of the 176,000 *M. mulatta* born yearly in Uttar Pradesh, 60,000 (35%) could be harvested yearly without severely affecting the populations if other factors were not operating to change and decrease *M. mulatta*'s numbers and habitats.

M. mulatta is not exported from Afghanistan (Puget, pers. comm. 1974) but is exported from Vietnam to the Soviet Union for work at the Sukhumi Primate Center (Lapin et al. 1965). Ruhle (1964) concluded that *M. mulatta* was threatened in Thailand because of the "lucrative return from sales for the manufacture of Salk polio vaccine." In Pakistan, infants secured after their mothers are shot as crop pests are sold as pets (Roberts 1977).

One of the most prominent characteristics of *M. mulatta* is its ability to live as a commensal of man: an estimated 60% of the *M. mulatta* in India are in villages and towns (Southwick and Siddiqi 1968). These monkeys occupy the niche of commensals in the slums, bazaars, and commercial areas, feeding on scraps of food from shops, storage areas, and houses. In this environment they are exposed to diseases, including tuberculosis, shigella, and many other respiratory and enteric pathogens (Southwick and Siddiqi 1968). *M. mulatta* has been suspected of transmitting some of these diseases to man, including dengue fever via arthropod vectors, and bacillary dysentery, leptospirosis, and dracontiasis via contaminated water or food (Southwick et al. 1961a).

Southwick et al. (1965) postulate that the commensal habit of *M. mulatta* may be (1) a species characteristic (as is the case with *Mus musculus*), (2) an ecological necessity dictated by the availability of food and water, or (3) a product of human social tradition. Whatever the cause of this phenomenon, the result is that in combination with the alteration of the rural habitat (agriculturalization, woodcutting, and commercial forestry), extermination of monkeys from farm areas because of crop raiding, and trapping of rural monkeys for export, *M. mulatta* is becoming concentrated and localized in urban areas (Southwick and Siddiqi 1968).

Rhesus monkeys are not trapped for export in towns, because their high incidence of disease makes them undesirable for research (Southwick and Siddiqi 1968). For example, in 1961, 60 to 70% of the rhesus monkeys exported from India were obtained in Uttar Pradesh (Southwick et al. 1961a). But by 1973 the majority and highest quality *M. mulatta* were being trapped in the more remote and less settled Kashmir and Jammu areas (Southwick, pers. comm. 1974). In addition, urban monkey populations tend to increase owing to their abundant food supply, and urban people do not cooperate to drive monkeys away from gardens as village people do (Southwick and Siddiqi 1968).

The result of this combination of factors is that urban populations of *M. mulatta* are increasing while rural populations decrease, leading to an unhealthy, urbanized *M. mulatta* population in India (Southwick and Siddiqi 1968). Urban rhesus monkeys also differ in behavior from forest populations, showing more intertroop aggressive behavior, smaller home ranges, less fear of humans, and other adaptations to city life (Southwick et al. 1965, Singh 1969).

In rural areas rhesus monkeys are also affected by human activities. In addition to obtaining some or all of their food from crops, they are fed by many people at temples, roadsides, and railway stations (Southwick et al. 1961b). Through this human contact and because they often drink water contaminated with human wastes, rhesus monkeys contract and spread human diseases, and disease is a significant factor in infant and juvenile monkey mortality (Southwick and Siddiqi 1973).

Groups that live in or near villages persist in frequenting human habitations despite harassment from people, and do not desert villages unless they are hunted or trapped heavily (Southwick et al. 1965). Thus, although some forest groups are well adapted to forest life and occur naturally in wild areas, some populations are dependent on man and are reluctant to utilize forest habitats (Southwick et al. 1965).

Conservation Action

M. mulatta occurs within at least one national park, one reserve, and five sanctuaries (Table 128), including the Kaziranga Wildlife Sanctuary in Assam, which has been proposed for national park status (Lahan and Sonowal 1973). This is a small number of protected areas for such a wide-ranging species.

The future of *M. mulatta* in India is entirely dependent on human tolerance of them: if all Indians turned against the monkeys, they would survive only in the few forest reserves that remain (Southwick, pers. comm. 1974).

To achieve conservation of *M. mulatta* populations in India, Southwick and Siddiqi (1973) recommended (1) educational programs aimed at demonstrating to Indians that rhesus monkeys are an economic and scientific asset, (2) the development of primate reserves and managed areas in India, (3) management of food and water resources to reduce mortality due to disease and produce a harvestable supply of juvenile rhesus, and (4) long-term development of breeding colonies in the United States to produce a defined stock of high quality monkeys.

More recently these same authors have observed that local management of rhesus populations by villagers can be an effective way of conserving the monkeys while permitting a sustainable yield to prevent excessive population growth (Siddiqi and Southwick 1977). They concluded that the best policy would be controlled trapping of limited numbers of monkeys at planned intervals. This would prevent excessive population growth and hence serious damage to crops, while leaving at least 50% immatures in the population to insure its continuation (Southwick and Siddiqi 1977a,b).

In Thailand, hunting, capture, and keeping of *M. mulatta* in captivity are allowed only under license, and the number exported is regulated under a quota system (Lekagul and McNeely 1977). No reports of legal protection or recommendations for conservation measures were located for the other countries in which *M. mulatta* occurs.

Table 128. Distribution of *Macaca mulatta* (countries are listed in order from west to east)

Country	Presence and Abundance	Reference
Afghanistan	In Landai-Sin Valley	Kullman 1970
	In NE in provinces of Paktia, Nangarhar, Kounara, and Laghman (localities mapped); population in Kamdech Valley 4,000	Puget 1971
	Along E border only (map); NW as far as Nuristan	Hassinger 1973
	Abundant in Paktia and Nuristan	Naumann and Nogge 1973
	Population stable	Puget, pers. comm. 1974

Table 128 (continued)

Country	Presence and Abundance	Reference
Pakistan	In N from Kafir valleys of S Chitral, S through Dir and E through Swat Kohistan and Hazara District, lower Kaghan Valley, Neelum Valley of Azad Kashmir and throughout Murree Hills (map); tiny population in outer foothills; widespread but has become rare in Chitral and Swat Kohistan; status fairly secure	Roberts 1977
India	*Southwick and coworkers*	
	From W Kashmir through Assam, S to Hyderabad and to Tapti R. in W; less abundant than commonly believed; declining in Uttar Pradesh	Southwick et al. 1961a
	Throughout N (map); total population in Uttar Pradesh estimated at less than 1 million; declining; concentrated near Lucknow, Faizabad, and Gorakhpur	Southwick et al. 1961b
	Have declined in West Bengal; rare in S and central West Bengal; abundant in extreme N West Bengal	Southwick et al. 1964
	Abundant in central, N, and E Uttar Pradesh and N West Bengal (map); decline of 50% 1954–59 in Uttar Pradesh village populations	Southwick et al. 1965
	In Uttar Pradesh 1950–60: substantial decline; 1960–65: 5% decline per year in number of groups, 3% decline per year in number of individuals	Southwick and Siddiqi 1966
	Decline in Uttar Pradesh continuing especially in rural areas	Southwick and Siddiqi 1968
	Declining in rural areas and increasing in urban areas; overall in India not in danger of extinction or further serious depletion, but distribution, habitat, and population quality changing; probably stable in Corbett N.P.	Southwick and Siddiqi 1970
	Decline in rural Uttar Pradesh continued 1964–66; stable in forest; may be increasing in cities; total population in Uttar Pradesh about 500,000; no adequate study of distribution has ever been made	Southwick et al. 1970
	1962–72: total population near Aligarh decreased 48% in individuals, 43% in number of groups	Southwick and Siddiqi 1973
	Common and abundant populations in Kashmir and Jammu (in N); still present	Southwick, pers. comm. 1974

Table 128 (continued)

Country	Presence and Abundance	Reference
India (cont.)	in Assam; abundant over much of India but declining and becoming rare in some areas	Southwick, pers. comm. 1974 (cont.)
	Population near Aligarh has declined 50% in individuals and 59% in number of groups between 1961 and 1975; protected populations have increased	Southwick and Siddiqi 1977a
	Most populations in N India have been declining for more than 20 years; the few protected populations are increasing	Siddiqi and Southwick 1977, Southwick and Siddiqi 1977b
	Other authors	
	In Assam and "United Provinces"	McCann 1933b
	In Rajasthan Desert	Prakash 1960
	Locally abundant in N; highest densities in Uttar Pradesh; declining on Gangetic Plain	Jay and Lindburg 1965
	In Basirhat Forest Reserve and several other forest blocks along Bangladesh border, E West Bengal	Mukherjee and Gupta 1965
	In Jaldapara Wildlife Sanctuary, West Bengal	Spillett 1966a
	Near Haldwani, N central Uttar Pradesh	Neville 1968
	N of 20°N in W and 17°N in E, from Deccan Plateau to Himalaya Mts. (map)	Singh 1969
	Throughout N of Godavari R. and slightly S of it in Andhra Pradesh; in Jaldapara Sanctuary, West Bengal; Orissa around Balimela; Bilhar: Palamau, Karkatnagar; Madhya Pradesh, Uttar Pradesh, and Assam; distribution discontinuous	Krishnan 1971
	Abundant at Asarori Forest NE of New Delhi; range shrinking	Lindburg 1971 [1977]
	Declining in some parts of N	Mukherjee and Mukherjee 1972
	In Kaziranga Wildlife Sanctuary, Brahmaputra R., Assam	Lahan and Sonowal 1973
	Formerly throughout Punjab (in N); now range greatly restricted there	Lindburg, pers. comm. 1974
	Between Hooghly R. and Meghna R. (E West Bengal)	Mukherjee 1975
	In Simla Region, Himachal Pradesh	Sugiyama 1976
	At Bhimtal, Uttar Pradesh, and in Sariska Wildlife Sanctuary, Rajasthan State (both in N)	Vogel 1976
	Abundant (population about 500) at Asarori Forest (30°15′N, 77°68′E) (in NE)	Makwana 1978
	Abundant but vulnerable	Mohnot 1978

Table 128 (continued)

Country	Presence and Abundance	Reference
Nepal	Throughout S (maps)	Southwick et al. 1961a, 1965; Singh 1969
	Present	Frick 1969, Mitchell 1975
	Common throughout Terai, Himalayas area	Blower 1973
	Locally common in Terai area of S; distribution disjunct	Southwick, pers. comm. 1974
	At Rara Lake (29°35′N, 82°05′E), NW Karnali Zone	Richie et al. 1978
Bhutan	In S (map)	Southwick et al. 1965
	Throughout (map)	Singh 1969
Bangladesh	Throughout (maps)	Southwick et al. 1965, Singh 1969
	Population of about 40,000 in Sundarbans region (in S); frequent to abundant there	Hendrichs 1975
	In Dacca and some villages	Oppenheimer 1977
	In central (map); questionable in NE, NW, and S; seen infrequently	Green 1978a
Burma	In Pidaung Wildlife Sanctuary, Kachin State (in N, on Irrawaddy R.)	Milton and Estes 1963
	Throughout except in S and Tenasserim (localities mapped)	Fooden 1971a
Thailand	Threatened	Ruhle 1964
	N of about 16°N (localities mapped); absent from large portion of former range	Fooden 1971a
	In Huay Kha Khaeng Game Sanctuary (in W, near 15°29′N)	Eudey, pers. comm. 1977
	In N (map)	Lekagul and McNeely 1977
Laos	Present (maps)	Fiedler 1956, Annenkov et al. 1972 (citing Weber)
Vietnam	Throughout N as far S as Hue (16°28′N, 107°35′E)	Delacour 1940
	Near Ninh Binh (20°14′N, 106°E)	Dào Van Tien 1963
	In Tuyen Quang Province (in N)	Lapin et al. 1965
	At Dak Sut, Kontum Province (14°56′N, 107°44′E) (central)	Van Peenen et al. 1969
	In N and central (localities mapped)	Fooden 1971a
China	Once plentiful in Tung Ling area in NE but now almost extinct there; in need of protection or will face extermination	Sowerby 1937
	As far N as about 32°N and isolated population near 40°N, 115°E (map); also on Hainan Island	Fiedler 1956
	In S Yunnan Province (extreme S)	Kao et al. 1962
	In Chinghai (Tsinghai) Province (central) near Yellow R. (Huang Ho) (about 34–36°N, 99–103°E)	Chang and Wang 1963

Table 128 (continued)

Country	Presence and Abundance	Reference
China (cont.)	In Tibet, Szechuan, central and S Yunnan, E to Fukien and adjacent states in S; on Hainan Island; perhaps introduced near Peking	Ellerman and Morrison-Scott 1966
	In SE (localities mapped)	Fooden 1971a
	In Dupleix Mts. (n.l.) in Szechuan	Haltenorth 1975
Hong Kong Province	On Kowloon Peninsula 16 to 24 km from Hong Kong; population of several hundred	Southwick, pers. comm. 1974

Table 129. Population parameters of *Macaca mulatta*

Population Density (individuals /km²)	Group Size			Home Range Area (ha)	Study Site	Reference
	X̄	Range	No. of Groups			
		80–180			Afghanistan	Puget 1971
		30–250			Afghanistan	Naumann and Nogge 1973
		12–80			N Pakistan	Roberts 1977
		50–60			E India	McCann 1933b
	17.6		399		N India	Southwick et al. 1961b, 1965
		20–120	5		E India	Southwick et al. 1964
	24.3	3–98	131		India	Jay and Lindburg 1965
		20–30			E India	Mukherjee and Gupta 1965
	12.7				N India	Southwick and Siddiqi 1966
5–753	27	10–77	12	0.05–310	N India	Neville 1968
	21	2–70			India	Singh 1969
0.008–0.05	9.3–13.3	21–236	83	17,862–149,433	N India	Mukherjee and Mukherjee 1972
50–1,000					N India	Southwick, pers. comm. 1974
		35–85	2	1,500	N India	Lindburg 1977
	9–57.5		22		N India	Siddiqi and Southwick 1977
	17	6–36	22		N India	Southwick and Siddiqi 1977a
	32.8	6–90	6	130–1,340	NE India	Makwana 1978
5–25					S Bangladesh	Hendrichs 1975
6.8					S Bangladesh	Green 1978a
		20–50			Thailand	Fooden 1971a

Macaca nemestrina Pigtailed Macaque

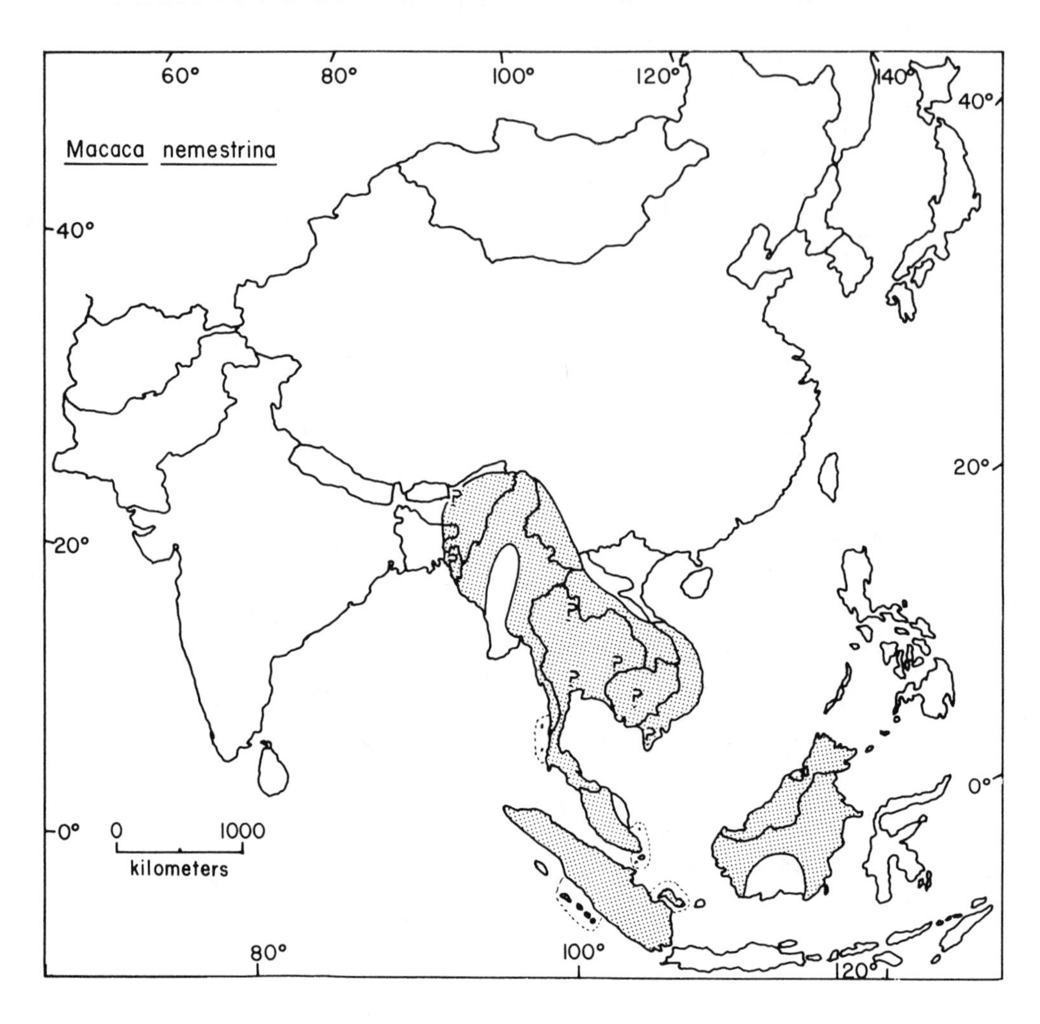

Taxonomic Note

Four subspecies are recognized (Napier and Napier 1967). Some recent workers consider the Mentawi Island form to represent a distinct species, *Macaca pagensis* (Tenaza, pers. comm. 1974; Wilson and Wilson, pers. comm. 1974; Tilson, pers. comm. 1975). But most other authors still view it as a race of *Macaca nemestrina*, and it is treated as such here.

Geographic Range

M. nemestrina is found from Assam in northeastern India, through the Malaysian and Indochinese peninsulas, and on Borneo, Sumatra, and the Mentawi Islands. Fiedler (1956) showed the northwestern limit at about 28°N, 91°E in western Assam, and the western limit at about 91°E in eastern Bangladesh. The northernmost record in the east is in southern Yunnan Province, China (Kao et al. 1962). The southern and eastern limits are formed by the southern coast of Sumatra and the eastern coast of Borneo. For details of the distribution, see Table 130.

Abundance and Density

M. nemestrina is known to be declining in India, Thailand, peninsular Malaysia, and the Mentawi Islands and is locally abundant but may also be declining in Sumatra (Table 130). Virtually nothing is known of its present status throughout the rest of its range.

Densities of 1.5 to 51 per km^2 and home ranges of 100 to 300 ha have been reported, and several authors have concluded that *M. nemestrina* groups are widely dispersed and wide ranging (Medway 1970; Wilson and Wilson, pers. comm. 1974; Tilson, pers. comm. 1975). Groups are often small (3 to 47 members), with average sizes of 5 to 22 (Table 131).

Medway and Wells (1971) estimated that the minimum size of a reserve capable of supporting a population of 5,000 *M. nemestrina* in Malaysia would be 3,333 km^2, based on 3 individuals which inhabited a 2 km^2 area.

Habitat

M. nemestrina is a forest-dwelling monkey which utilizes both primary and secondary, inland and coastal, lowland and highland forests. Its specific habitat requirements are not well understood.

In Assam and Thailand, *M. nemestrina* is found more frequently in evergreen than deciduous forest (McCann 1933b; Fooden 1971a, 1975). On the Malaysian peninsula it occupies woodland and inland forests from the coastal swamps to the foothills (Medway 1969a, 1970) and has been described as abundant in inland primary forest and rare in coastal lowland forest (Lim 1969). Southwick and Cadigan (1972) encountered it only rarely in submontane dipterocarp forest.

On Sumatra, pigtailed macaques have been reported in rain forest and pockets of jungle within savanna, and are most frequent near the edges of fields (Kurt 1973). They reach their highest population densities in lowland and hill primary forest, but also utilize secondary hill forest, scrub grassland, and lowland riverine vegetation, and only rarely enter swampy areas (Wilson and Wilson 1977).

Mentawi Island pigtailed macaques have been seen in primary forest, mangrove swamps (Tenaza, pers. comm. 1974), hilly and lowland primary forest, and lowland secondary forest (Tilson, pers. comm. 1974). And on Borneo their habitat has been described as primary and secondary forest (Davis 1962) and inland, lowland rain forest (Kawabe and Mano 1972).

M. nemestrina seems to show a preference for lowland rather than highland forest (Burgess 1961, Harrison 1969, Kurt 1973) but has been recorded as high as 1,700 m elevation in Borneo (Fooden 1975).

Factors Affecting Populations

Habitat Alteration

M. nemestrina is "feeling the pressure . . . of a deteriorating environment" (Southwick et al. 1970). In Assam, India, its populations are being reduced by removal of forests (Mohnot 1978). In peninsular Malaysia, deforestation is proceeding so rapidly (Aiken and Moss 1975) that officials are considering the importation of logs (Mei 1977). In 1966 only about 75,000 km^2 of rain forest remained in peninsular Malaysia (Bernstein 1967a), and in 1970 this was being cut at the rate of 222 km^2 per year (Southwick and Cadigan 1972). Since *M. nemestrina* is "not normally resident in deforested land" (Medway 1970), logging is undoubtedly contributing to its decline in Malaysia. Clearing for agriculture and rural development are also partly to blame for the decrease in *M. nemestrina* populations in Malaysia (Khan 1978).

On Sumatra, government regulations forbid the cutting of trees less than 50 cm in diameter, and although this law is not completely enforceable, it does prevent large-scale clearcut logging (Wilson and Wilson, pers. comm. 1974). Nevertheless, "persistent expansion of farmlands and logging is slowly taking its toll" of the forest (Wilson and Wilson 1972–73). In selectively logged areas in northern Sumatra, in which about one-half of the trees have been removed, *M. nemestrina* is found in lower densities than in undamaged forest (MacKinnon 1973). And, in general, "the most persistent threat to primates on Sumatra is deforestation for lumber or new farms" (Wilson and Wilson 1972–73).

On Siberut Island, logging commenced in 1970 and proceeded at the initial rate of 10% of the forest area per year (Tenaza, pers. comm. 1974). The long-term effect on primate populations has not been determined, but will probably be detrimental (Tilson, pers. comm. 1974).

Although no direct information is available, the military activity and the ecological destruction by bombing and defoliants (Orians and Pfeiffer 1970, S.I.P.R.I. 1976) have undoubtedly had an adverse effect on the *M. nemestrina* populations in Vietnam.

Human Predation

Hunting. Hunting and trapping of *M. nemestrina* for meat occur frequently in Thailand (Lekagul 1968), peninsular Malaysia (Southwick and Cadigan 1972), and the Mentawi Islands (Tenaza, pers. comm. 1974; Tilson, pers. comm. 1974), but only rarely in predominantly Moslem Sumatra (Wilson and Wilson, pers. comm. 1974).

In peninsular Malaysia, extensive human predation is responsible for *M. nemestrina*'s absence in many areas (Bernstein 1967a). The meat is highly valued and certain parts of the body are sought as delicacies or for their reputed body-building powers. Although the aborigines have hunted pigtails with blowguns for many years, the introduction of shotguns as part of a government program has greatly increased hunting pressure, which is reflected in the rising cost of *M. nemestrina* meat in local markets and the absence of this species near centers of human population (Bernstein 1967a). Malaysian hunters also kill the feared and respected large adult male *M. nemestrina* whenever possible (Bernstein 1967a).

On Sipora and the Pagai Islands the human population has increased fivefold since 1820 and is posing the threat of excessive hunting pressure for *M. nemestrina* (Tenaza, pers. comm. 1974).

Pest Control. M. nemestrina is considered the most destructive primate species in Sumatra, and since these monkeys are large and clever, they are capable of doing a great deal of damage to crops such as corn, papaya, and oil palm (Wilson and Wilson

1972–73). In hilly areas they sleep in the forest and make periodic raids into the nearby farms and retreat to forests when chased. The damage is often large enough to prompt farmers to shoot or trap the monkeys or to hire people with guns to kill them (Wilson and Wilson, pers. comm. 1974).

M. nemestrina is also a pest to farms in Borneo (Burgess 1961, Wilson and Wilson 1975a). In Thailand its crop-raiding habits have aroused public sentiment against it, so that although it is declining, it has not been included on the list of protected species (Lekagul 1968). In peninsular Malaysia, pigtailed macaques feed on paddy (Medway 1969a) and grain (Medway 1970), and permission to shoot those accused of raiding crops, coming too close to houses, or otherwise creating a nuisance is easy to obtain even in protected forests (Bernstein 1967a).

Collection. M. nemestrina is one of the primates which have been used extensively in research (Southwick and Cadigan 1972). The species has felt heavy pressure from "wasteful trapping or shooting" (Southwick et al. 1970), and exportation for "making polio vaccines in the U.S." has been cited as one reason for its decline in Thailand (Lekagul 1968). *M. nemestrina* is collected in Vietnam for export to the Soviet Union (Lapin et al. 1965), and is also trapped for export from peninsular Malaysia (Khan 1978) and Indonesia (Brotoisworo 1978).

Between 1968 and 1973 a total of 3,099 *M. nemestrina* were imported into the United States (Appendix). In 1970–73 most of these came from Malaysia, Thailand, and Indonesia. In 1977, 23 individual *M. nemestrina* and six shipments of unspecified sizes were imported into the United States (Bittner et al. 1978). A total of 1,163 *M. nemestrina* were imported into the United Kingdom between 1965 and 1975 (Burton 1978).

Young *M. nemestrina* are also captured for use and sale as coconut pickers in Malaysia (Bernstein 1967a, Khan 1978), Sumatra (Wilson and Wilson, pers. comm. 1974), and Borneo (Burgess 1961). This species is preferred as a picker over other macaques because of its greater size and strength and its reputation as a long and consistent worker (Bernstein 1967a). In the course of this trade it is often transported across zoogeographic barriers (Medway 1970).

Capture and trade in young *M. nemestrina* as pets also occurs on a small scale on Sumatra (Wilson and Wilson, pers. comm. 1974) and on the Mentawi Islands (Tilson, pers. comm. 1974).

Conservation Action

M. nemestrina is found in four reserves and two sanctuaries (Table 130), but no evidence of its occurrence in any fully protected national parks was located.

Southwick and Cadigan (1972) advocated "specific conservation measures" for *M. nemestrina* in Malaysia. The species is classified as protected in Malaysia (Yong 1973). In Thailand, *M. nemestrina* may be hunted, captured, and kept in captivity only under license, and its export is regulated by a quota system (Lekagul and McNeely 1977).

The Sumatran population can withstand limited cropping but not large-scale exploitation (Wilson and Wilson 1972–73). These authors recommended (1) that *M. nemestrina* be used in research only when particularly suitable for a certain experiment, (2) that it be bred in captivity wherever possible, and (3) that crop-raiding monkeys be live-trapped for export instead of being killed as pests.

The Mentawi Island pigtailed macaque has been the subject of a status and conservation study which resulted in the establishment of a 6,600 ha reserve in central Siberut and plans for several smaller reserves on the other Mentawi Islands (Tilson, pers. comm. 1975).

Table 130. Distribution of *Macaca nemestrina* (countries are listed in order from northwest to southeast)

Country	Presence and Abundance	Reference
India	In Naga Hills, E Assam	McCann 1933b
	In Assam	Ellerman and Morrison-Scott 1966
	In Hollangapar Reserve Forest (26°30′N, 94°20′E), E Assam	Tilson, pers. comm. 1975
	In parts of Meghalaya (S of Brahmaputra R., 90–92°E) and Nagaland (E Assam, N of 26°N); populations declining, status uncertain	Mohnot 1978
Bangladesh	In E (map)	Fiedler 1956
	Likely to be in Chittagong Hill Tracts (in E)	Green 1978a
Burma	Throughout, as far N as 28°N (map)	Fiedler 1956
	In Pidaung Wildlife Sanctuary (in N)	Milton and Estes 1963
	In N, SW, SE, and Tenasserim (localities mapped); on islands off W coast of Tenasserim	Fooden 1975
China	Only in Hsi-Shuan-Pan-Na area, S Yunnan Province (in S)	Kao et al. 1962
Thailand	Declining	Lekagul 1968, Medway 1969a
	Declining in W	Medway 1970
	In W between 11°N and 13°N (localities mapped)	Fooden 1971a
	Throughout (localities mapped); on islands off W coast	Fooden 1975
	Relatively uncommon near Ban Thap Plik and Ban Kachong (S peninsular)	Fooden 1976
	In Huay Kha Khaeng Game Sanctuary (in W near 15°29′N)	Eudey, pers. comm. 1977
	In W, N, and central (map)	Lekagul and McNeely 1977
Laos	In S and W (localities mapped)	Fooden 1975
Cambodia	No information	
Vietnam	In "various regions"	Lapin et al. 1965
	Records from Hue (central), Tay Ninh (SE), and Bien Hoa Province (SE)	Van Peenen et al. 1969
	Central and S (localities mapped)	Fooden 1975

Table 130 (continued)

Country	Presence and Abundance	Reference
Malaysia	Not rare or in danger in N Borneo	Burgess 1961
	In N Borneo	Davis 1962
	At Ulu Bernam, Selangor (W peninsular)	Milton 1963
	Occasional E of Kuala Lumpur (W peninsular)	McClure 1964
	W of Sandakan, E Sabah (N Borneo)	Yoshiba 1964
	Badly disturbed in peninsular: distribution uneven, absent in many areas	Bernstein 1967a
	In Sabah (N Borneo)	de Silva 1968b
	In peninsular	Harrison 1969
	On N Borneo	Kuntz 1969
	In inland E areas of peninsular (map)	Lim 1969
	Widespread; in S peninsular and N Borneo (localities mapped)	Medway 1969a,b
	In N, central, and S interior of peninsula (localities mapped)	Chivers 1971
	In Ulu Segama Reserve (Sabah)	MacKinnon 1971
	Not common at Kuala Lompat (central peninsular)	Medway and Wells 1971 [Chivers 1973]
	On inland plain of Sabah, near Kinabatangan R.	Kawabe and Mano 1972
	At Gunong Benom (central W peninsular, just S of 4°N)	Medway 1972
	Becoming rare in peninsular	Southwick and Cadigan 1972
	Scarce and declining in peninsular	Bernstein, pers. comm. 1974
	Throughout (except E coast of peninsular) including small islands off E and S coasts of peninsular (localities mapped)	Fooden 1975
	1958–75: population decreased 43.8% (from estimated 80,000 to 45,000) in peninsular	Khan 1978
Singapore	Unlikely that any remain	Burkill 1961
	No longer occurs wild there	Harrison 1966
	Not native but introduced	Medway 1969a
Indonesia	As far E as Makassar Strait (map)	Fooden 1969a
	On Sumatra, Borneo, Bangka, and Pagai islands (localities mapped)	Medway 1969b
	Throughout Sumatra (localities mapped) but rarely seen; not abundant; may be declining in S	Wilson and Wilson 1972–73
	In Gunung Leuser Reserve, N Sumatra	Kurt 1973
	In Langkat Reserve and at Ranun R. and Sikundur (all in N Sumatra)	MacKinnon 1973
	Declining on Mentawi Islands (Siberut, Sipora, N and S Pagai)	Tenaza, pers. comm. 1974; Tilson, pers. comm. 1974

Table 130 (continued)

Country	Presence and Abundance	Reference
Indonesia (cont.)	In E Kalimantan	Van Peenen et al. 1974
	Throughout Sumatra and Borneo (except S central) on Sipora, N and S Pagai and Bangka Island (localities mapped)	Fooden 1975
	In Kutai Reserve, E Kalimantan	Wilson and Wilson 1975a
	Locally abundant in Sumatra	Brotoisworo 1978
	Status of Mentawi form: indeterminate; in Teiteibatti Wildlife Reserve	I.U.C.N. 1978

Table 131. Population parameters of *Macaca nemestrina*

Population Density (individuals/km²)	Group Size			Home Range Area (ha)	Study Site	Reference
	X̄	Range	No. of Groups			
	22	12–40	8		W Thailand	Fooden 1971a
	6	3–15	3		S Thailand	Fooden 1976
					Malaysia	
		30–47	2		Peninsular	Bernstein 1967b
		5–20			Peninsular	Lim 1969
		3–15	4		Peninsular	Medway 1970
1.5		3	1		Peninsular	Medway and Wells 1971
					Indonesia	
		20–30			N Sumatra	Kurt 1973
27–51					N Sumatra	MacKinnon 1973
10–15		7	1		Mentawi Islands	Tenaza, pers. comm. 1974
	5	5	2		Siberut Island	Tilson, pers. comm. 1974
4.3–28.3	18.3			100–300	Sumatra	Wilson and Wilson 1977

Macaca nigra Celebes Macaque

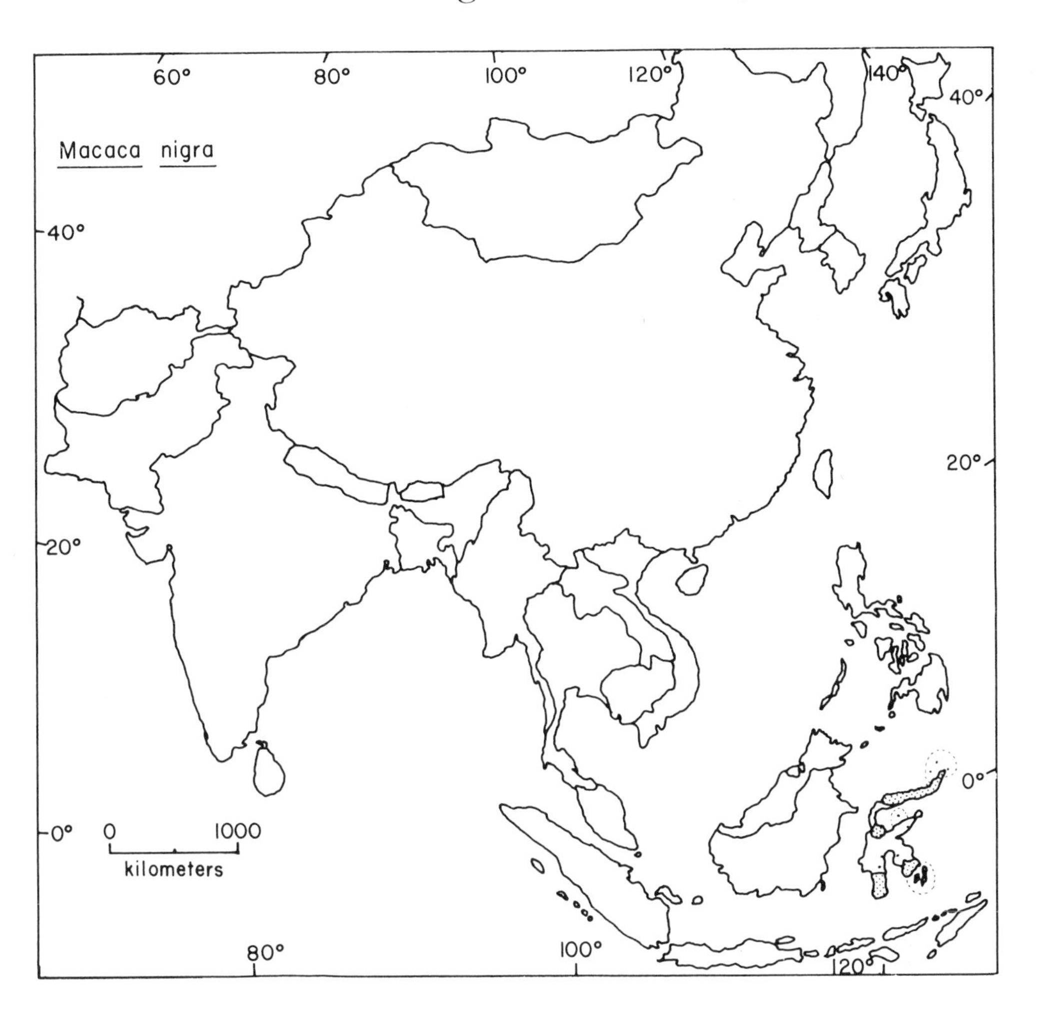

Taxonomic Note

All of the macaques occurring on Sulawesi (Celebes) are considered here as *Macaca nigra* following Thorington and Groves (1970). Fooden (1969a) split this group into seven distinct species: *M. nigra* (previously *Cynopithecus niger*), *M. nigrescens*, *M. hecki*, *M. tonkeana*, *M. ochreata*, *M. brunnescens*, and *M. maura*.

Geographic Range

M. nigra is found on Sulawesi, Indonesia, and a few adjacent smaller islands. It occurs on the northern long arm of Sulawesi from about 121° E to the northeastern tip, and on the small islands of Manadotua and Talise (Talisei) to the north and Lembeh to the east of the tip. Other populations occur on the large island in the north central area between the equator and 2°S, on the distal portion of the southwest peninsula south of 4°S, on the southeast peninsula primarily south of 4°S., on Muna and Butung islands off the southeast shore, and at scattered other localities (all from Fooden 1969a, localities listed and mapped). No further details of the distribution are available.

Abundance and Density

No data regarding the abundance of *M. nigra* in any part of its range could be located. Population density is suspected to be fairly high "judging from the comments of field observers and the large scientific collections which have been assembled in a short time in relatively small areas" (Fooden 1969a).

Typical group size in Celebes macaques is 5 to 25, with field observers reporting large bands of several hundred, and groups of 6 to 10 (Fooden 1969a). No estimates of home range size are available.

Habitat

The northernmost form of *M. nigra* is said to be confined to rain forest, while the southern and central populations utilize deciduous forest and arid vegetation (Jolly 1966). The explorers who sighted *M. nigra* between 1869 and 1930 saw it in humid lowland forest, mountain forest, low forest of screw pines, near the seashore, in mangrove swamps, spiny bamboo forest, and riverine forest (Fooden 1969a).

No studies of the preferred habitats or requirements of this species could be located. The present vegetation of most of Sulawesi is thick evergreen forest from sea level to 1,000 m, cloud forest from 1,000 to 2,500 m, and "nocturnal forest" and subalpine vegetation above 2,500 m (Fooden 1969a). *M. nigra* has been recorded in grass and scrubland but is absent above 2,500 m, probably because of cool nighttime temperatures; it is also absent from the high and cool portions of cloud forest between 2,000 and 2,500 m (Fooden 1969a).

Factors Affecting Populations

Habitat Alteration

Celebes macaques do inhabit deforested areas on the southwestern peninsula of the island, which has been converted into scrub and grassland as the result of continual cultivation (Fooden 1969a). But earlier workers reported that this species was absent in old cultivated land which had been denuded of primitive forest (Fooden 1969a). No other information was located regarding the effect of habitat alteration on *M. nigra* populations.

Human Predation

Hunting. M. nigra was reported at the turn of the century to be "hunted for food . . . only by non-Moslems who are in the minority in most parts of the island" (Fooden 1969a). No more recent data are available.

Pest Control. Celebes macaques feed in orchards, gardens, and cornfields and are caught in traps baited with corn (Fooden 1969a). "They may obtain a substantial part of their food by raiding cultivated gardens" (Fooden 1969a). Other authors have also reported crop raiding (Jolly 1966, Napier and Napier 1967) but have not mentioned efforts to control monkey populations.

Collection. Between 1968 and 1973 a total of 221 *M. nigra* were imported into the United States (Appendix). From 1965 to 1975, 215 *M. nigra* were imported into the United Kingdom (Burton 1978). None entered the United States in 1977 (Bittner et al. 1978).

Conservation Action

M. nigra is found in the Bantimurung Reserve, a small reserve less than 500 ha in size near the southern end of the southwest peninsula of Sulawesi (I.U.C.N. 1971). It is also protected by law in Indonesia (Brotoisworo 1978).

Macaca radiata Bonnet Macaque

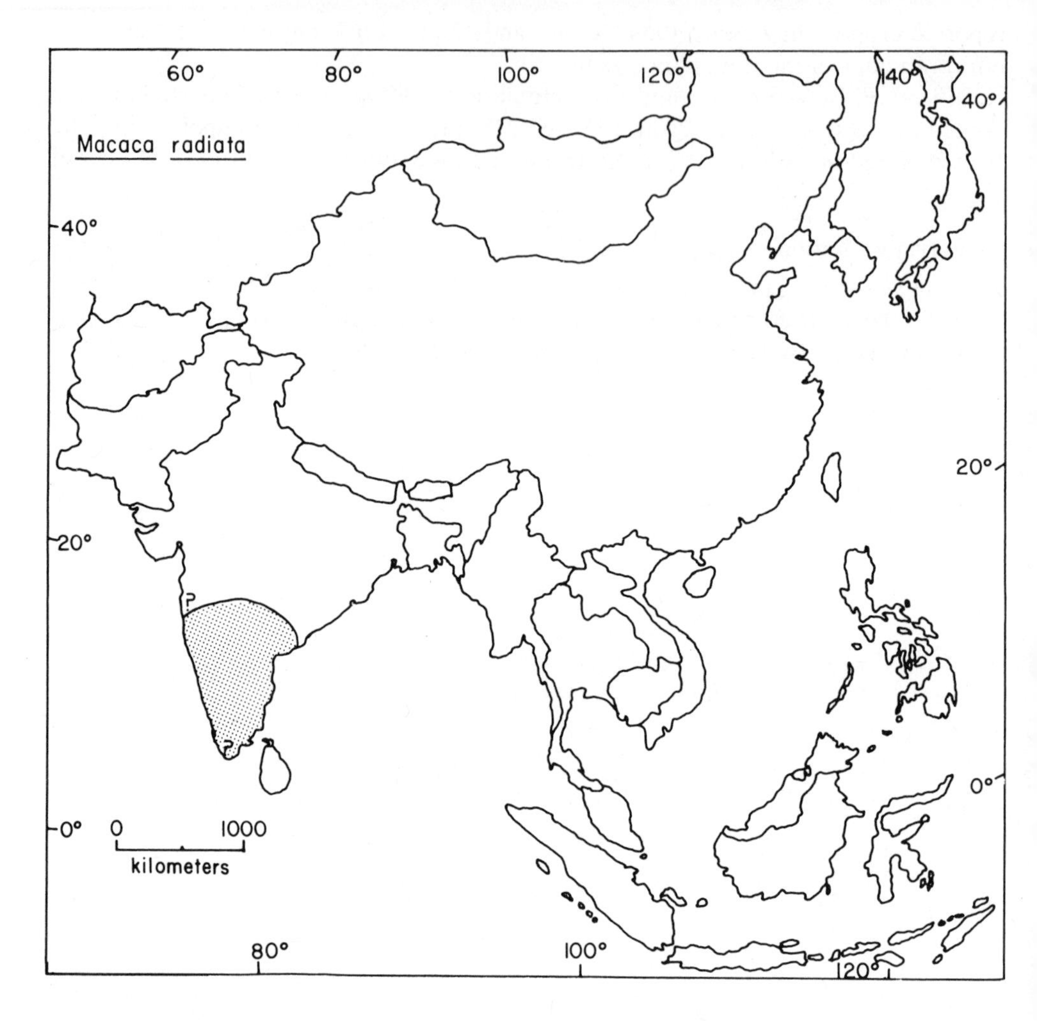

Taxonomic Note

Two subspecies are recognized (Ellerman and Morrison-Scott 1966, Napier and Napier 1967).

Geographic Range

Macaca radiata is found in southern India, as far north as the Godavari River, which extends to about 19°N. In the west its northern limit is given as Satara (17°43′N) (Simonds 1965, Ellerman and Morrison-Scott 1966) but other authors have recorded this species as far north as Bombay (18°56′N) (Prater 1971). Similarly, it has been reported to be absent in the southern portion of the Indian peninsula (Roonwal and Mohnot 1977), and as present as far south as Cape Comorin on the southern tip of the peninsula (Simonds 1965). For details of the distribution, see Table 132.

Abundance and Density

Neither the distribution nor the status of *M. radiata* has been well delineated (Table 132). The species has been described as common in several areas and as declining in at least one, but evidence about it is scanty (Table 132).

No estimates of population density are available. Average group size is between 15 and 40, with groups as small as 3 and as large as 76 members (Table 133). The two home range area estimates which have been published (Table 133) differ greatly, perhaps because of habitat differences or varying methods of measurement.

Habitat

M. radiata is found in evergreen forests and along rivers in wet deciduous forest (Webb-Peploe 1947; Simonds, pers. comm. 1974; Parthasarathy 1977). It also occupies banyan trees along roadsides in cultivated areas and the semidesert of the central Deccan Plateau (Simonds 1965), dry deciduous forest (Sugiyama 1968a), and scrub jungles around villages and towns (Krishnan 1971). This last author has reported that *M. radiata* is not a typical forest animal and is commoner in comparatively open areas, forest edges, and around human settlements than in the interior of forests.

M. radiata occurs in lowland areas and highlands up to 2,134 m elevation (Simonds 1965). It has been seen on rocky upper mountain slopes (Webb-Peploe 1947) but is said to be rare in montane forests (Krishnan 1971).

Bonnet macaques inhabit urban areas, including botanical gardens, temples, markets, residential neighborhoods (Rahaman and Parthasarathy 1967), and railroad stations and shrines (Krishnan 1971). Either tall trees (at least 15 to 30 m tall) or buildings that provide escape routes are a necessary component of their habitat and they are not found in areas lacking these (Simonds, pers. comm. 1974).

Factors Affecting Populations

Habitat Alteration

M. radiata successfully exploits many habitats that have been altered by man's activities, including agricultural areas (Simonds 1965) and towns (Rahaman and Parthasarathy 1967). Where bonnet macaque groups live in close association with man, their groups are larger than in undisturbed (and less food-rich) areas (Simonds, pers. comm. 1974). This species is less likely to be adversely affected by habitat alteration than are other animals of the plains forest, because it adapts so readily to life around human settlements (Krishnan 1971). It has "lost ground" in some areas, but has also acquired "new territory" (by introduction) elsewhere (Krishnan 1971).

In the Palni Hills of Tamil Nadu State a 50% human population increase between 1949 and 1971 has led to extensive deforestation for agriculture, introduction of many exotic tree species, and fires, which have contributed to the decline of *M. radiata* there (Matthew et al. 1975).

Human Predation

Hunting. Hunting has been blamed for the decline of *M. radiata* (Kurt 1970), but another worker has concluded that this species has not been heavily affected by either shooting or trapping (Simonds, pers. comm. 1974).

Pest Control. M. radiata commonly enters cultivated fields and eats peanuts, beans, squash, rice, and grain (Simonds 1965), and coffee and coconuts (Roonwal and Mohnot 1977). In some areas it obtains most of its food this way (Krishnan 1971). But it is not generally considered a pest to agriculture; instead it is tolerated, revered, and sometimes fed by the local people, although its crop depredations are extensive (Simonds, pers. comm. 1974).

In the town of Bangalore, *M. radiata* forages in gardens and garbage heaps, and steals food from market stalls and people (Rahaman and Parthasarathy 1967). These authors found that "the number of monkeys here is not so large as to cause any serious damage, although their nuisance value and depredations in fruit and vegetable gardens is not inconsiderable. Attempts are being made to collect statistics regarding the extent of damage done." This suggests that citizens are becoming concerned about monkey depredations.

M. radiata is a host for the ticks that carry Kyasanur Forest Disease, a viral infection that reached epidemic proportions and caused the deaths of many monkeys and men in the Shimoga District, Mysore (Karnataka) State (Rajagopalan and Anderson 1971). No mention was made of efforts to control these monkey vectors.

Collection. This species is reported to be difficult to ship successfully, hence has not been collected extensively for export (Simonds, pers. comm. 1974). But its use in research laboratories is now increasing (Mohnot 1978).

Between 1968 and 1973 a total of 93 *M. radiata* were imported into the United States (Appendix). From 1965 to 1975, 13 *M. radiata* were imported into the United Kingdom (Burton 1978). None entered the United States in 1977 (Bittner et al. 1978).

Conservation Action

M. radiata occurs in five reserves, wildlife sanctuaries, or parks (Table 132). As far as could be learned, no other protective measures have been taken.

Table 132. Distribution of *Macaca radiata*

Country	Presence and Abundance	Reference
India	Common in S and in Western Ghats; also at Khandala, Trombay, Salsette Island	McCann 1933b
	In S Tinnevelly, Madras (S Tamil Nadu) (extreme S)	Webb-Peploe 1947
	Very common in High Wavy Mts., Madura District (in S)	Hutton 1949
	In Somanathapur Sandal Reserve, S Mysore (Karnataka) State (in SW)	Simonds 1965
	At Bangalore (S central)	Rahaman and Parthasarathy 1967, 1969
	Frequent in Mudumalai Wildlife Sanctuary, extreme W Tamil Nadu; in Guindy Deer Park, NE Tamil Nadu	Spillett 1968
	At Panniar, Kerala (extreme SW)	Sugiyama 1968a
	Common in Nilgiri District, Madras (in SW)	Poirier 1970

Table 132 (continued)

Country	Presence and Abundance	Reference
India (cont.)	Common; in Mudumalai Sanctuary, NE Nilgiris; Ranganathittoo Sanctuary (n.l.); Bandipur Sanctuary (n.l.); and other localities listed; distribution discontinuous; absent in many areas	Krishnan 1971
	Common	Prater 1971
	In Shimoga District, Mysore (Karnataka) State	Rajagopalan and Anderson 1971
	Status not known; no evidence of marked change in population during past 10 years but some areas have not been sampled	Simonds, pers. comm. 1974
	Becoming scarce in Palni Hills, Madurai District, Tamil Nadu	Matthew et al. 1975
	Common in Segur Range, Mudumalai, Bandipur, and Nagarhole wildlife sanctuaries; occasional in higher areas of Nilgiris	Reza Khan 1976
	In and near Bangalore	Parathasarathy 1977
	Considered abundant	Mohnot 1978

Table 133. Population parameters of *Macaca radiata*

Population Density (individuals/km²)	Group Size X̄	Group Size Range	Group Size No. of Groups	Home Range Area (ha)	Study Site	Reference
		>50			S India	Hutton 1949
	34.5	6–58	4	518	S India	Simonds 1965
	21.1	<50	14		S India	Rahaman and Parthasarathy 1967
		3–12			S India	Krishnan 1971
	30		12	40	S India	Sugiyama 1971
	15–40	6–76			S India	Simonds, pers. comm. 1974

Macaca silenus Lion-tailed Macaque

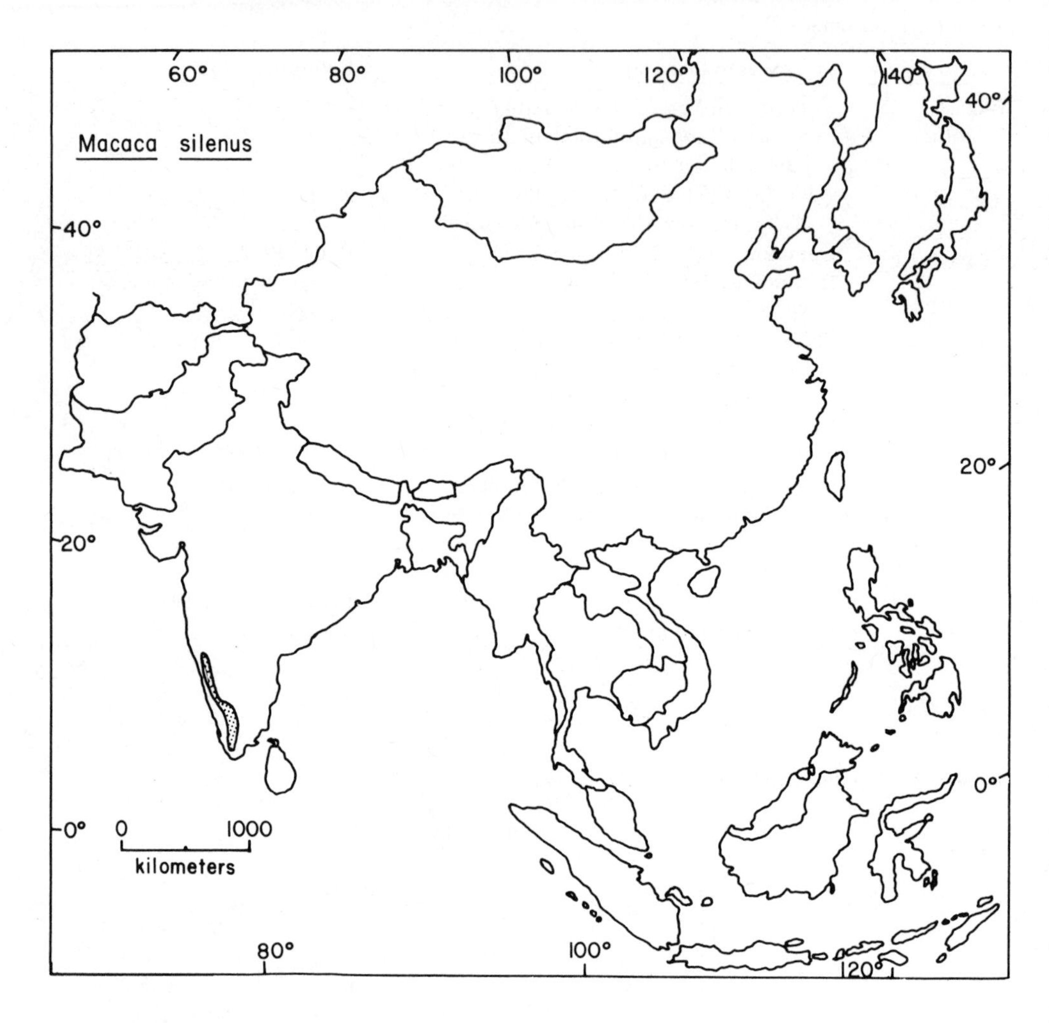

Taxonomic Note

This is a monotypic species (Napier and Napier 1967).

Geographic Range

Macaca silenus is found in the Western Ghats Mountains of southern India. It has been recorded as far north and west as Anshi Ghat (14°55′N, 74°22′E) near Goa, and as far south and east as the Kalakkadu Hills (8°25′N, 77°25′E) near Cape Comorin (Fooden 1975, Green and Minkowski 1977, Kurup 1977). The total maximum area of its available range has been estimated at 2,810 km^2 (Kurup 1977). For details of the distribution, see Table 134.

Abundance and Density

M. silenus is a seriously endangered species with a restricted and shrinking range and a small and decreasing population. Its total population was estimated at less than 1,000 in 1968, 800 in 1974, and 405 in 1975 (Table 134).

Groups of *M. silenus* usually contain 15 to 20 members, and home range areas are 200 to 500 ha (Table 135). In its optimal habitat this species can achieve a population density of 3.75 individuals per ha (1 group per 400 ha) (Green and Minkowski 1977). These authors calculate that a minimum viable deme for the species would consist of about 500 to 2,000 individuals, and that the minimum area needed to support 500 individuals under the best of conditions would be 132 km^2. This is the minimum size of a reserve capable of protecting a remaining population of *M. silenus*.

Habitat

The typical habitats of *M. silenus* are tropical wet evergreen and semievergreen broadleaf forests (Sugiyama 1968a, Kurup 1977). The evergreen forests, called "sholas," are generally found between 500 and 1,500 m in elevation, but may extend as low as 100 m along rivers (Green and Minkowski 1977).

The forests most favorable for *M. silenus* are those in which mature *Cullenia exarillata* trees dominate the vegetation, in association with *Palaquium*, *Calophyllum*, and other species (Green and Minkowski 1977). These *Cullenia* forests are found in the southern part of *M. silenus*'s range, while in the north the forests are dominated by *Dipterocarpus* and do not support as high densities of *M. silenus* (Green and Minkowski 1977).

Factors Affecting Populations

Habitat Alteration

M. silenus's habitat has been greatly reduced by expanding human settlements and conversion of forested areas to monotypic plantations principally of teak, eucalyptus, cardamom, coffee, and tea (Kurup 1977). Lion-tailed macaques do not utilize or travel through nonforested artificial vegetation, and any gaps in the sholas due to plantations, shifting cultivation, or grassland can effectively block their movements (Green and Minkowski 1977). Extensive clearcutting has already isolated populations of *M. silenus* occupying hill ranges separated by inhabited plains or plantations, since the monkeys will not cross these gaps. Selective logging is also seriously detrimental, since it destroys the character of the dense, shade tolerant climber-covered evergreen forests (Green and Minkowski 1977).

All types of deforestation are continuing at a rapid rate and are gravely threatening the small remaining area of *M. silenus* habitat. Clearcutting has been proposed for the heart of the remaining range and would seriously affect the status of the lion-tail (Kurup 1977). In addition, the construction of dams in the Papanasam area, road building, timber extraction, cultivation, grazing, and firewood cutting associated with the opening of the area near the Kodayar hydroelectric project are increasingly threatening the major remaining habitat in the Ashambu Hills (Green and Minkowski 1977).

Human Predation

Hunting. M. silenus has been exterminated over much of its original range by hunting for its meat and its fur (Sugiyama 1968a, Kurt 1970, Krishnan 1971). Although some of the local people will not eat the flesh of this species, but consider it sacred (Hutton 1949), some tribes in the Nilgiri Hills do eat it (Roonwal and Mohnot 1977), and *M. silenus* is also heavily poached for its meat in the northern part of the Ashambu Hills (Green and Minkowski 1977).

M. silenus is also frequently killed by people mistaking it for *Presbytis johnii*, another black monkey which inhabits this area and whose flesh is valued for its purported medicinal or aphrodisiac properties (Poirier 1971; Kurup, pers. comm. 1974; Green and Minkowski 1977).

Pest Control. None of the sources consulted mentioned crop raiding by this species.

Collection. M. silenus is said to make an excellent pet, and in the past many were kept by shopkeepers, presumably to promote business (Hutton 1949). More recently, infants were still being captured by killing their mothers (I.U.C.N. 1972), and being displayed for sale as pets in the markets of Bombay and Calcutta (Roonwal and Mohnot 1977). This trade has been blamed in part for the decline in *M. silenus* populations (Karr 1973). Southwick et al. (1970) mention the "waste in non-scientific commercial trade," by which they probably mean the pet trade and perhaps also the zoo trade, as a factor affecting this species.

Between 1968 and 1973 a total of 20 *M. silenus* were imported into the United States (Appendix). From 1965 to 1975, 65 *M. silenus* were imported into the United Kingdom (Burton 1978). None entered the United States in 1977 (Bittner et al. 1978).

Conservation Action

M. silenus is found in four sanctuaries (Table 134), but these do not offer sufficient protection to insure its survival. "Only immediate and strenuous measures to preserve its disappearing habitat and to protect the monkeys that remain can save the lion-tailed macaque" (Green and Minkowski 1977). These authors outlined six steps necessary to conserve *M. silenus* and its Ashambu Hills habitat: (1) stricter regulation of felling in government leased lands, (2) no further use of plantation or felling permits, (3) enforcement of antihunting laws, (4) restricting access to the Ashambu Hills, (5) stricter regulation of use of privately owned lands, and (6) a force of patrolling forest and wildlife guards.

For the long-term protection of *M. silenus*, Green and Minkowski (1977) recommended the protection of a large reserve in the Ashambu Hills, consisting of a 100 km^2 core area in the Kalakkadu, Singampatti, Papanasam, and adjacent forests, surrounded by a 60 km^2 buffer zone of already partly disturbed forest. They also suggested the establishment of a second refuge in the Silent Valley–Bhavani River valley region. They expressed the hope that the conservation effort started by the establishment of the Kalakkadu Forest Sanctuary in 1976 would be the beginning of an effective program to conserve all the remaining habitat in these hills — a program that would continue to be promoted by the Forest Department and approved by the government.

Other areas that have been recommended as reserves for *M. silenus* include the Nilambur forests, the Kanjirapuzha River valley, the Kallar Valley, and the Papanasam-Singampatty Kalakkad areas in the Agasthya-Ashambu hills (Kurup 1977). The enforcement of regulations within existing sanctuaries is badly in need of improvement, since habitat alteration by squatters and poaching are problems in several reserves (Hill 1971). Mohnot (1978) reported, "Fortunately its habitat has now been protected by law and all forest-felling operations in its range . . . have been stopped."

Dr. John Oates has submitted a proposal to the I.U.C.N. for an ecological study of the Ashambu Hills area, with the goals of determining the population density of *M. silenus* there and producing proposals for conserving portions of this area (Mittermeier 1977b).

Within India the Wildlife Protection Act of 1972 prohibits the hunting of *M. silenus* except by special permit for educational or research purposes, but this law is not well enforced (Green and Minkowski 1977). Export for zoos or as pets is allowed "on merits" by the Export Trade Control Handbook of Policy and Procedures, 1969. *M. silenus* is listed in Appendix I of the Convention on International Trade in Endangered Species of Wild Fauna and Flora and as endangered under the U.S. Endangered Species Act (U.S. Department of Interior 1977a,b).

Table 134. Distribution of *Macaca silenus*

Country	Presence and Abundance	Reference
India	Not uncommon near Dohnavur, S Tinnevelly, Madras (Tamil Nadu)	Webb-Peploe 1947
	Rare in High Wavy Mts., Madura	Hutton 1949
	In Muthukuzhivayal and Mundandurai wildlife sanctuaries, S Tamil Nadu	Spillett 1968
	In Nilgiri Hills, Anaimalai Hills, Cardamom Hills, and near Periyar Lake (map); range shrinking; less than 1,000 individuals remain, facing extinction	Sugiyama 1968a
	Declining	Kurt 1970
	Common in Nilgiri District, Madras (Tamil Nadu)	Poirier 1970
	In serious danger of extinction	Southwick et al. 1970
	Restricted to S part of range in Tamil Nadu and Kerala; absent from Thirunelveli District and other parts of N; greatly depleted, situation "precarious"	Krishnan 1971
	Endangered	Prater 1971
	Endangered; distribution localized and shrinking	I.U.C.N. 1972, 1976
	In Thirunelveli District, Tamil Nadu	Karr 1973
	Range narrowly restricted and still being reduced (localities mapped)	Fooden 1975
	In 1975 surveyed all localities (listed); estimate populations at Ashambu Hills,	Green and Minkowski 1977

Table 134 (continued)

Country	Presence and Abundance	Reference
India (cont.)	195; Silent Valley–Bhavani River valley, 60; all others, 150; total about 405 individuals (195–570); protected in Kalakkadu Forest Sanctuary; never have been common; are still declining; extinction is imminent	Green and Minkowski 1977 (cont.)
	Populations in Nilgiri, Amarambalam, Silent Valley, Attapadi, Nilambur Hills, Palaghat Hills, Nelliampathi, Parambikulam-Varagaliyar, Panniar, Periyar Sanctuary–Kallar Valley, and Agasthya–Ashambu Hills; in 1974: total of about 750 individuals in 2,510 km^2; assume another 50 individuals in 300 km^2 in Karnataka region (map)	Kurup 1977
	Threatened with extinction	Mohnot 1978

Table 135. Population parameters of *Macaca silenus*

Population Density (individuals/km^2)	Group Size			Home Range Area (ha)	Study Site	Reference
	$\bar{X}$	Range	No. of Groups			
	20				W India	Webb-Peploe 1947
		16–22	2	200	W India	Sugiyama 1968a
		6–>12			W India	Krishnan 1971
		12–>20			W India	Prater 1971
3.75	15	6–34	13–38	500	W India	Green and Minkowski 1977
	15				W India	Kurup 1977

Macaca sinica Toque Monkey

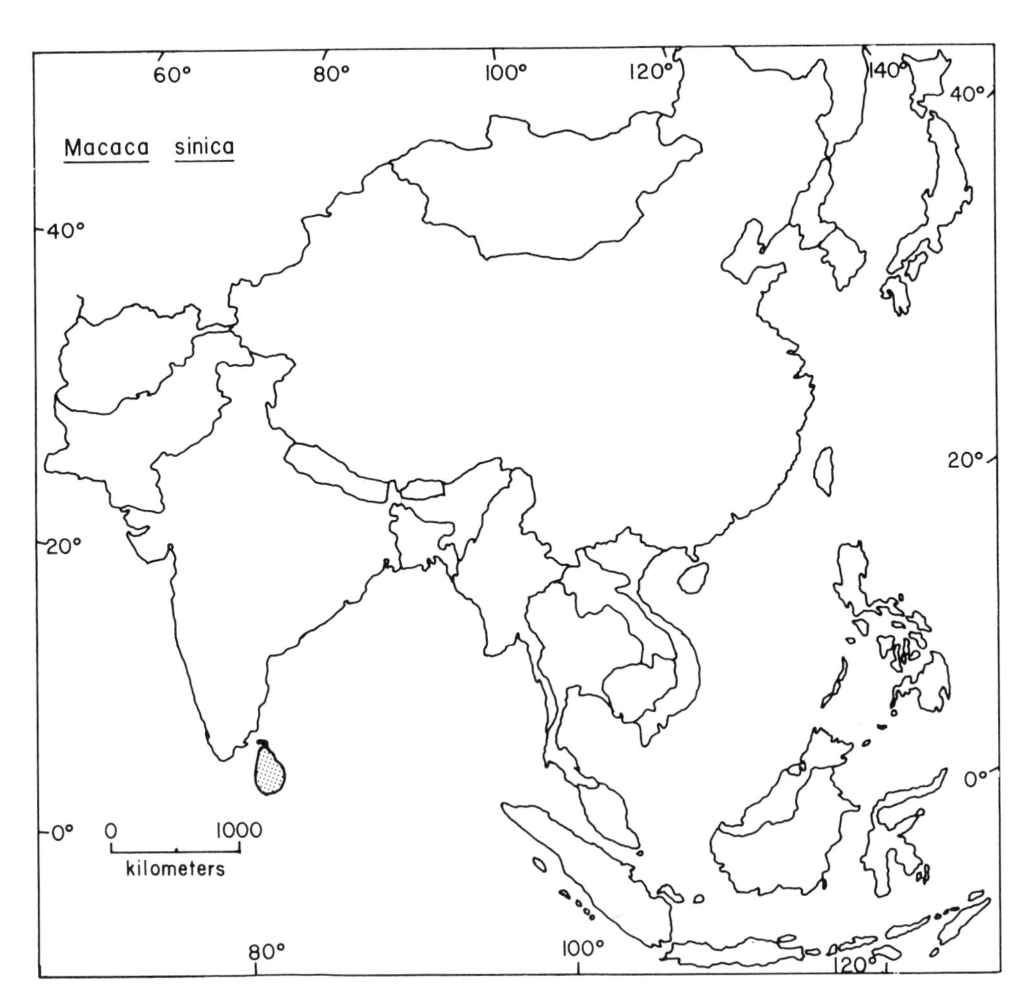

Taxonomic Note

Three subspecies are recognized (Ellerman and Morrison-Scott 1966, Napier and Napier 1967).

Geographic Range

Macaca sinica is found on Sri Lanka (Ceylon). It occurs throughout the island, but is very rare in the arid areas of the northern, northwestern, and southeastern coasts (Dittus, pers. comm. 1974). In 1976 the total area of habitat available to it was estimated to be 22,546 km^2 (Dittus 1977a). For details of the distribution, see Table 136.

Abundance and Density

The total population of *M. sinica* in 1976 was estimated at 590,678 individuals (Dittus 1977a). Some local populations are abundant, but at least one has recently declined significantly. The area of habitat available has decreased by about 5,000 km^2 since 1959, and the population of the montane race is small and threatened (Table 136).

Hladik and Hladik (1972) observed two groups of 12 and 13 *M. sinica*, with home ranges of 30 and 45 ha. The toque monkey population at Polonnaruwa has been studied extensively and found to have a density of 100 individuals per km^2, consisting of 18 troops of 8 to 43 members with an average size of 20.6 (including first-year infants = 24.7), occupying home ranges of 17 to 115 ha (Dittus 1977a). This high density in optimal conditions in Polonnaruwa and other prime habitats may represent the saturation level of *M. sinica* habitat (Dittus 1975). In other, less rich habitats, population densities were 0.3 to 30 *M. sinica* per km^2 (Dittus 1977a).

Habitat

M. sinica is a forest monkey found in semideciduous forest (intermediate between tropical evergreen rain forest and monsoon, seasonally defoliated forest), in riverine forest in arid areas, in low-stature scrub in seasonally dry areas, and in montane rain forest (Dittus 1977a). It has also been reported in monsoon scrub jungle, intermonsoon forest, monsoon forest, and grassland, from sea level to 1,524 m elevation (Eisenberg and McKay 1970). This species is able to occupy a wide variety of forest types, and is found "wherever a semblance of natural forest with a permanent source of water is available" (Dittus 1975).

Factors Affecting Populations

Habitat Alteration

The population decrease at Polonnaruwa between 1972 and 1975 was due to (1) a severe drought in 1974 which probably led to a general scarcity of vegetable food, and (2) the closure of a rice mill which had been a major food source for the resident *M. sinica* (Dittus 1977b). At other study sites, cattle grazing, cultivation, and the clearing of underbrush have reduced the natural vegetation available to toque monkeys (Dittus 1975).

In Sri Lanka as a whole, "the greatest threat to natural habitats is the rapidly expanding human population and its concomitant demands for land for agriculture, plantation, and forest products" (Dittus 1977a). The annual rate of deforestation since 1956 has been 1.06%, and is expected to "increase under present pressures for greater timber exploitation in the last remnants of rain forest and for agricultural expansion in the dry zone. At present rates the estimated 22,546 km^2 of remaining forests will vanish within one human life-span" (Dittus 1977a).

Much of the habitat of *M. sinica* has already been converted into tea plantations or potato farms, or destroyed by slash and burn agriculture and timber exploitation (Dittus 1977a). Very little lowland rain forest remains intact, having been replaced with rubber, coconut, and rice plantations (Dittus, pers. comm. 1974). The semideciduous forests have been heavily logged, often by clearcutting, and in some areas replaced with commercial monoculture of species, such as teak, which provide no food for *M. sinica* (Dittus, pers. comm. 1974).

In areas where cultivation has replaced forest, the food supply for toque monkeys may be increased, but if all trees are removed, *M. sinica* will not utilize the open habitats devoid of cover (Dittus, pers. comm. 1974). In many areas, *M. sinica* is already confined to remaining forest patches, and its available habitat, which has been reduced by 5,000 km^2 since 1959, continues to decrease (Dittus 1977a).

Human Predation

Hunting. M. sinica exudes a strong and unpleasant odor, and its flesh is reputed to be unpalatable, thus it is not hunted for its meat (Dittus, pers. comm. 1974, 1977a). The main predators of toque macaques are domestic dogs (Dittus 1977b).

Pest Control. M. sinica enters fields and eats cultivated foods, as well as feeding from rice mills and garbage dumps (Dittus 1977a,b). But its crop raiding occurs primarily where its natural food supplies have been destroyed by deforestation or consumed by livestock, and during seasons of scarcity of forest foods (Dittus 1977a). In some areas, toque monkeys surviving in remnant patches of forest rely heavily on crops for their food, but seem to be in poor health (Dittus 1977a).

Efforts to control *M. sinica*'s crop raiding include poisoning, trapping, shooting, mutilation (by the amputation of a hand as punishment for thievery) and other forms of harassment (Dittus 1975, 1977a). *M. sinica* receives some degree of religious protection, and occurs in some Buddhist and Hindu shrines where it is not molested (Dittus, pers. comm. 1974), but this protection does not extend to cultivated areas (Dittus 1977a). Toque monkeys do not occupy towns or live as commensals with man as do some other macaques (Dittus 1975).

Collection. M. sinica is not collected for export for research and is captured only infrequently for exhibit in zoos (Dittus, pers. comm. 1974). None were imported into the United States in 1968–73 (Appendix) or 1977 (Bittner et al. 1978). Five were imported into the United Kingdom between 1965 and 1975 (Burton 1978).

Conservation Action

M. sinica occurs in three national parks, one reserve, and one sanctuary (Table 136). But more of its natural habitat must be protected if this species is to survive. The montane form (*M.s. opisthomelas*) is especially in need of legal protection and habitat preservation (Dittus 1977a). The existing parks and forest reserves are also inadequate for conservation of significant populations of the other two races (Dittus 1977a).

Dittus (1977a) recommended the establishment of several large national wildlife reserves, enclosing areas of optimal remaining habitat and bordered by 200 to 300 m wide zones of treeless grassland, which *M. sinica* is reluctant to cross. He also advocated the protection of small town sanctuaries, especially as aids in the education of local students and citizens.

M. sinica is listed as threatened under the U.S. Endangered Species Act (U.S. Department of Interior 1977b).

Table 136. Distribution of *Macaca sinica*

Country	Presence and Abundance	Reference
Sri Lanka	Locally abundant in Wilpattu N.P. (in NW about 8°30′N, 80°E)	Eisenberg and Lockhart 1972
	In Gal Oya N.P. (in E about 7°15′N, 81°30′E)	Atapattu and Wickremasinghe 1974
	Population dense and stable at Polonnaruwa (in E, about 8°N, 81°E); at Anuradhapura (N central); Wilpattu N.P.; Ruhunu N.P. (in SE); Sinharaje Forest Reserve (SW); Udawattekelle Sanctuary (near Kandy, central); and Ohiya and Horton Plains region (central)	Dittus 1975
	Population still stable at Polonnaruwa; total population of species in 1976 estimated at 590,678; area of available habitat in 1959, 27,495 km^2; in 1976, 22,546 km^2; montane race threatened, population 1,500	Dittus 1977a
	Population at Polonnaruwa decreased significantly (by 15.3%) 1972–75	Dittus 1977b

Macaca sylvanus Barbary Macaque

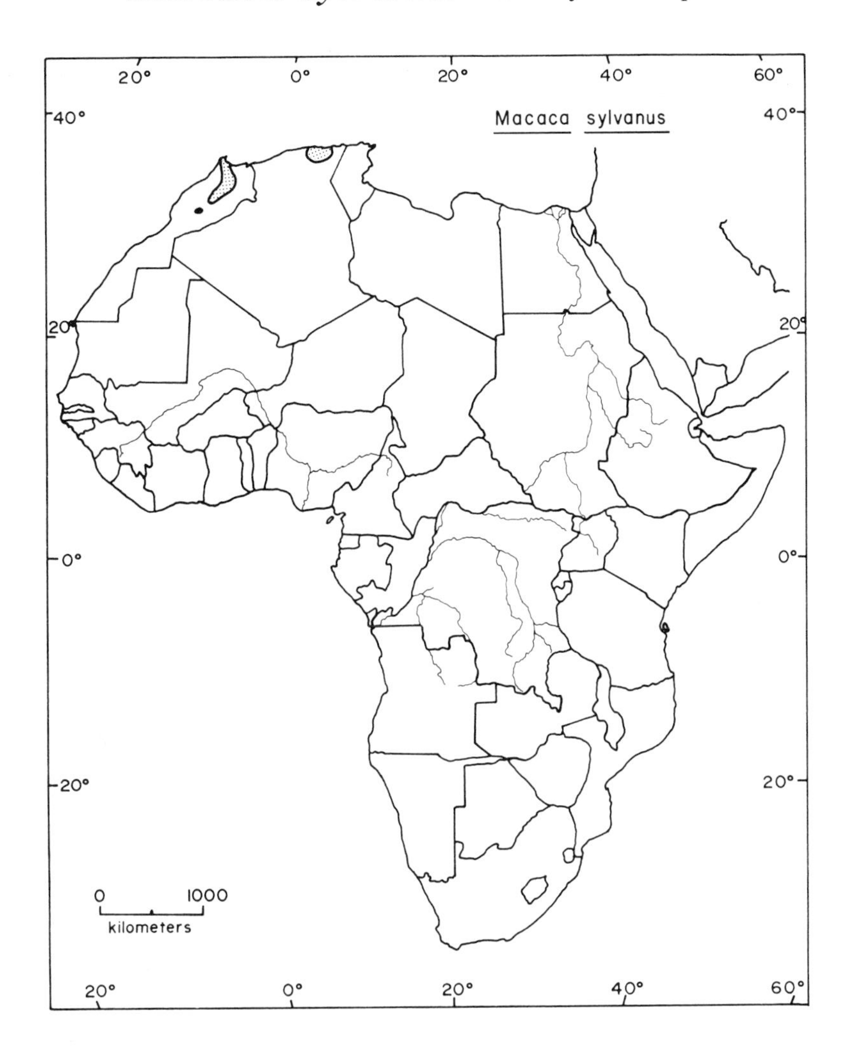

Taxonomic Note

Macaca sylvanus is a monotypic species (Napier and Napier 1967).

Geographic Range

M. sylvanus is found in northern Morocco and Algeria. An isolated population at 31°18′N, 7°47′W in the Ourika Valley, 50 km southeast of Marrakech, probably is at the southwestern limit of the range (Deag and Crook 1971). Barbary macaques formerly occurred as far east as Tunisia, and within historical times were found near Skikda, Algeria (36°53′N, 6°54′E), but today their easternmost population is at 5°41′E (at 36°44′N), southwest of Djijel (Taub 1977). A small population has been introduced on Gibraltar (Ellerman and Morrison-Scott 1966), and escapees are reproducing ferally in Spain (Deag 1977). For details of the distribution, see Table 137.

Abundance and Density

The total population of *M. sylvanus* in Morocco and Algeria is estimated at 12,840 to 21,340 individuals (Table 137). This species exists today only in localized, mainly restricted areas, in small and shrinking populations. Only one continuous expanse of high quality habitat with an abundant population of *M. sylvanus* remains, and 65% of the worldwide total is found there, in the central Middle Atlas Mountains of Morocco (Taub 1977). Populations and ranges in both Morocco and Algeria are decreasing and are potentially threatened. The I.U.C.N. (1976) lists *M. sylvanus* as "vulnerable."

The introduced population on Gibraltar is kept to a maximum of 30 monkeys by periodic removal to zoos (MacRoberts and MacRoberts 1971).

Population densities in this species vary from 12 to 70 individuals per km^2 (Deag 1977, Taub 1977). Group sizes of 12 to 25 ($\bar{X}$ = 18.3, N = 6 groups) (Deag and Crook 1971), 25 and 30 (N = 2 groups) (Whiten and Rumsey 1973), and 20 to 35 (Taub 1977) have been reported. No estimates of home range area were located.

Habitat

M. sylvanus is found in mixed cedar and holm oak forests, humid Portuguese and cork oak mixtures, gorges with scrub vegetation, high cedar forests, and mixed oak forest without cedars (Taub 1977). In Morocco its preferred habitat seems to be the Portuguese and cork oak mixtures, but this may be because of the destruction of cedar forests; in Algeria it achieves its greatest population densities in high mixed cedar forest (Taub 1977). Cedar seems to be a critical resource for *M. sylvanus*, since the trees provide a food source during severe snowy winter months when no other food is available to barbary macaques (Taub 1977).

At one study site at 2,010 m elevation, *M. sylvanus* utilized cedar forest, oak woodland, and open grassland "bleds" devoid of trees, while at lower elevations it occupied almost pure oak stands, cedar oak and juniper forests, and hot semiarid regions with oaks, junipers, and dwarf palm trees (Deag and Crook 1971). The most dense populations of *M. sylvanus* have been reported in mixed cedar and oak forests about 1,600 m in elevation (Deag 1977). The study site of Whiten and Rumsey (1973) was at 1,600 to 2,160 m and consisted of steep cliffs and gorges with oak, fir, and juniper forests dissected by sparse grassy vegetation.

Factors Affecting Populations

Habitat Alteration

The greatest threat to the continued survival of *M. sylvanus* is severe habitat disturbance due to intensive logging in both Algeria and Morocco (Taub 1975). In Algeria the deforesting of large areas has led to the present disjunct distribution of the remaining forest and macaque populations, and Portuguese, evergreen, and cork oak are being commercially logged at a rapid rate (Taub 1977). In Morocco, evergreen oak is being clearcut, and the prime *M. sylvanus* cedar forests in the Middle Atlas are being intensively timbered (Taub 1977). This ecosystem is quite fragile, and since the rate of exploitation is too rapid, and the efforts at reforestation are insufficient to prevent deterioration of the habitat, this deforestation could lead to the extinction of *M. sylvanus* in the near future (Taub 1975).

The grazing of bleds and forest areas by goats, sheep, and cattle, the cutting of forests for timber, firewood, and charcoal, and the increase in cultivation at low elevations are also reducing the extent of *M. sylvanus* habitat (Deag and Crook 1971, Deag 1977).

Human Predation

Hunting. M. sylvanus has become rare or extinct in some areas, such as eastern Kabylie, where it was used as food (Deag 1977), but in general hunting is not having a major impact on its populations (Taub 1975).

Pest Control. M. sylvanus is reputed to raid crops and garbage dumps, and is shot as a crop pest in the High Atlas and Rif areas (Deag 1977). It is also shot in Morocco because of the damage it does to cedar trees by stripping the bark and eating the cambium layer (Deag 1977). In Algeria neither crop raiding nor tree damage is considered a serious problem, and even in Morocco the damage is probably small but may be considered important because of the scarcity of forests (Deag 1977), and could result in the death of many Barbary macaques if controls were increased (Taub 1975).

On Gibraltar this species feeds from gardens and garbage cans and has often been regarded as a pest by the local people (MacRoberts and MacRoberts 1971).

Collection. Between 1969 and 1974 at least 425 *M. sylvanus* were captured for sale in Morocco, and several dozen were captured yearly in Algeria (Deag 1977). Many of these were destined for European wildlife parks, some may have been used in research, and a small number were sold as pets or entertainers (Deag 1977). Those captured for export were not damaging cedar trees, but were from oak forest areas, since *M. sylvanus* is much easier to capture in this habitat than in tall cedar forests (Deag 1977).

Deag (1977) concluded that "this species has never been used in quantity for laboratory research but future trapping of large numbers for this purpose always remains a possibility." He added that in the past, international trade in *M. sylvanus* has involved animals culled from Gibraltar.

Between 1968 and 1973 a total of 23 *M. sylvanus* were imported into the United States (Appendix). None entered the United States in 1977 (Bittner et al. 1978). Eight were imported into the United Kingdom in 1965–75 (Burton 1978).

Conservation Action

In Algeria several small national parks contain *M. sylvanus* populations, and the monkeys are legally protected in all localities (Taub 1977). In Morocco there are no national parks, but a U.N. Development Project is under way to assist in establishing national parks (Taub 1977). *M. sylvanus* is not protected in Morocco, and reclassification to protected status, and conservation of its last major refuge in the central Middle Atlas Zone, are crucial to its long-term survival there (Taub 1975, 1977; Deag 1977).

This species is included in Class A of the African Convention, to which both Algeria and Morocco are parties (Organization of African Unity 1968).

Table 137. Distribution of *Macaca sylvanus*

Country	Presence and Abundance	Reference
Morocco	In N and central (map)	Huard 1962
	At Ain Kahla (33°15′N, 5°13′W) and Ourika Valley	Deag and Crook 1971
	At Ain Kahla	Deag 1973
	At Jbel Lakra and other parts of Rif region (localities mapped)	Whiten and Rumsey 1973
	In Rif, High Atlas; most abundant in Middle Atlas; range being fragmented and reduced; locally extinct or decreasing	Deag 1977
	Small relict population in Ourika Valley; 5–8 small disjunct populations in Rif; abundant in central Middle Atlas (population 7,000–14,000); rare in E Middle Atlas; about 2,000 in S Middle Atlas; localities mapped and listed; total population 9,490–16,490; decreasing; range shrinking; potentially threatened	Taub 1977
Algeria	In extreme N only (map)	Huard 1962
	7 disjunct populations in Grande and Petite Kabylie (localities mapped and listed); only 2 (these near Guerrouch and Agfadou) are sizable (1,000–2,000); also near Chiffa (S of Blida); in several national parks, including Pic des Singes and Djebel Babor; total populations 3,350–4,850; decreasing; range shrinking; potentially threatened	Taub 1977

Miopithecus talapoin Talapoin Monkey

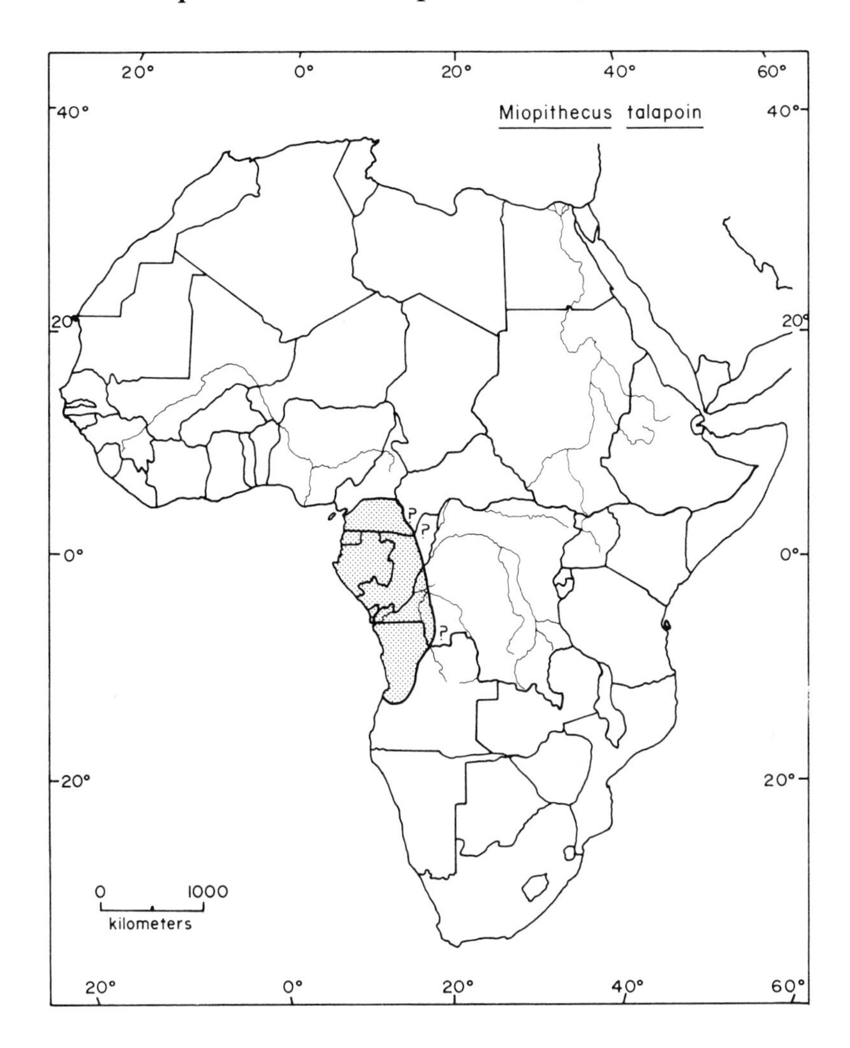

Taxonomic Note

Miopithecus talapoin is closely related to the *Cercopithecus* monkeys and has been placed in that genus by some taxonomists (Napier and Napier 1967). Three or four subspecies are recognized (Napier and Napier 1967).

Geographic Range

M. talapoin is found in equatorial West Africa from southern Cameroon to Angola. Its northern limit is at about 4°30′N, 11°30′E along the Sanaga River (Gartlan, pers. comm. 1974) and its southernmost record near 12°30′S, 14°E in Angola (Barros Machado 1969). The species has been collected as far east as about 17°45′E in southwestern Zaire (Barros Machado 1969), but the eastward extent of the range is not known. One specimen has been reported from Mount Ruwenzori (Lönnberg 1919 cited in Tappen 1960), but no confirming evidence of this species in East Africa has been found. For details of the distribution, see Table 138.

Abundance and Density

M. talapoin populations have recently been described as locally abundant and remaining stable in Cameroon and northeastern Gabon, but rare in Angola. In 1966 their range seemed to be increasing in Equatorial Guinea (Table 138). No information is available regarding their status in Central African Republic, Congo, or Zaire.

Typical group sizes for talapoins are larger than those reported for other African forest monkeys (Table 139). Groups as large as 125 members, densities up to 92 per km^2, and home ranges of 100 to 500 ha have been reported (Table 139). Populations living near human settlements have densities three times as great as those found distant from villages (Gautier-Hion 1973) and group sizes double those living far from man's influence (Gautier-Hion 1971).

Habitat

M. talapoin is a forest species found primarily in swamp, mangrove, coastal, and riverine forest. It is common in flooded primary forest but not in nonflooded mature forest without thick underbrush (Gautier-Hion 1971, Gartlan and Struhsaker 1972).

Talapoins frequently inhabit secondary forest (Gautier-Hion 1966) and low brush around old overgrown plantations (Malbrant and Maclatchy 1949). They can utilize agricultural lands and have been reported in plantations, coffee forests, and cultivated fruit trees (Barros Machado 1969). They are absent above 700 m elevation (Gautier-Hion 1971).

This species is never found in dry areas but is always close to marshes, swamps, rivers, or other bodies of water. It sleeps at river edges or in seasonally flooded forest (Quris 1976), and the presence and distribution of thick riverine vegetation for sleeping sites is an important determinant of suitable habitat (Gautier-Hion 1973). It seems quite dependent on water and thick underbrush for its survival (Gautier-Hion 1971).

Factors Affecting Populations

Habitat Alteration

Because *M. talapoin* flourishes in secondary growth, it can reinvade selectively logged forest and other disturbed habitats. In Equatorial Guinea it was common in regenerating forests and near plantations, and its range has probably been increased by human habitat alteration (Sabater Pi and Jones 1967). In the northeast section of this country where agriculture and other habitat disturbance have eliminated most natural vegetation, talapoins and man were the only primates remaining (Sabater Pi and Jones 1967).

This species also utilizes disturbed habitats in Gabon (Gautier-Hion 1966), Angola (Barros Machado 1969), and Cameroon, but no direct evidence of the effects of habitat alteration on its populations in these countries was located.

Human Predation

Hunting. Talapoin monkeys are not hunted extensively for food, because their small size makes them not worth the price of a bullet (Gautier-Hion 1970). But they are killed, probably with slingshots, and sold for meat in Cameroon, and are snared in Equatorial Guinea, Gabon, and Cameroon (Gartlan, pers. comm. 1974).

Pest Control. M. talapoin is considered to be a frequent crop raider (Tappen 1960; Gautier-Hion 1966; Gartlan, pers. comm. 1974). In fact, the larger sizes of groups living near human settlements could be due to the extra food the talapoins obtain from plantations and cassava soaking sites (Gautier-Hion 1971). This increased food supply could increase survival or reproductive success or both. Talapoins also may be dependent on this food during seasons when less fruit is available in the forest (Gautier-Hion 1973).

Collection. Although they do not adjust well to laboratory conditions and are difficult to maintain in captivity, talapoins are used in medical research to some extent. Because of their small size they are easier to handle and less expensive to house than larger monkeys. A recent increase in the number of requests for exportation of talapoins from Cameroon may be due to an increasing demand for them as laboratory animals (Gartlan, pers. comm. 1974). At least 2,225 *M. talapoin* were exported from Equatorial Guinea between 1958 and 1968 (Jones, pers. comm. 1974).

Talapoins do not make good zoo exhibits or pets because of their shyness and nervousness.

Between 1968 and 1973 a total of 113 *M. talapoin* were imported into the United States (Appendix). None entered the United States in 1977 (Bittner et al. 1978). In 1965–75, 600 *M. talapoin* were imported into the United Kingdom (Burton 1978).

Conservation Action

M. talapoin is protected within two reserves in Cameroon (Table 138).

Table 138. Distribution of *Miopithecus talapoin* (countries are listed from north to south)

Country	Presence and Abundance	Reference
Cameroon	Common in Douala-Edéa Reserve (in W)	Struhsaker 1969
	At Elongongo and Nsépé (on Sanaga R.) and in Dja Reserve (S central)	Gartlan and Struhsaker 1972
	Near M'Balmayo (3°30′N, 11°30′E), Nyong R.	Rowell 1972, 1973; Wolfheim 1975
	Throughout S, S of 4°30′N; N and S of Sanaga R. (map); absent N of Wouri R. (NW of Douala); in Douala-Edéa and Dja Reserves; locally abundant; not declining	Gartlan, pers. comm. 1974
Central African Republic	Present	Blancou 1958 cited in Barros Machado 1969

Table 138 (continued)

Country	Presence and Abundance	Reference
Equatorial Guinea	Throughout including Corisco Island (map); locally abundant, range increasing	Sabater Pi and Jones 1967
Gabon	Throughout	Malbrant and Maclatchy 1949
	In Ogooué-Ivindo region, W of Makokou (in NE)	Gautier-Hion 1966, 1970, 1973; Gautier-Hion and Gautier 1974
	Stable in Ntsi-Belong region (NE)	Gautier and Gautier-Hion 1969
	Abundant in NE	Gautier-Hion 1971
	On Liboui R. (in NE)	Quris 1976
Congo	In W, N, and E	Malbrant and Maclatchy 1949
Zaire	Just N of mouth of Congo R.	Schouteden 1947 cited in Tappen 1960
	In SW (localities mapped)	Barros Machado 1969
Angola	In W to at least 12°30′S (localities mapped)	Barros Machado 1969
	Rare	Bothma 1975

Table 139. Population parameters of *Miopithecus talapoin*

Population Density (individuals /km^2)	Group Size			Home Range Area (ha)	Study Site	Reference
	$\bar{X}$	Range	No. of Groups			
		≤80		400–500	NE Gabon	Gautier-Hion 1966
		40–100			NE Gabon	Gautier and Gautier-Hion 1969
		82–115			NE Gabon	Gautier-Hion 1970
26–92	63–112	59–125		100–140	NE Gabon	Gautier-Hion 1971
30–92	67–112		10	121	NE Gabon	Gautier-Hion 1973
		≤100			NE Gabon	Quris 1976
		60–80			Gabon and Congo	Malbrant and Maclatchy 1949
30		70	1	220	S Cameroon	Rowell 1972
		85	1		S Cameroon	Wolfheim 1975

Nasalis larvatus Proboscis Monkey

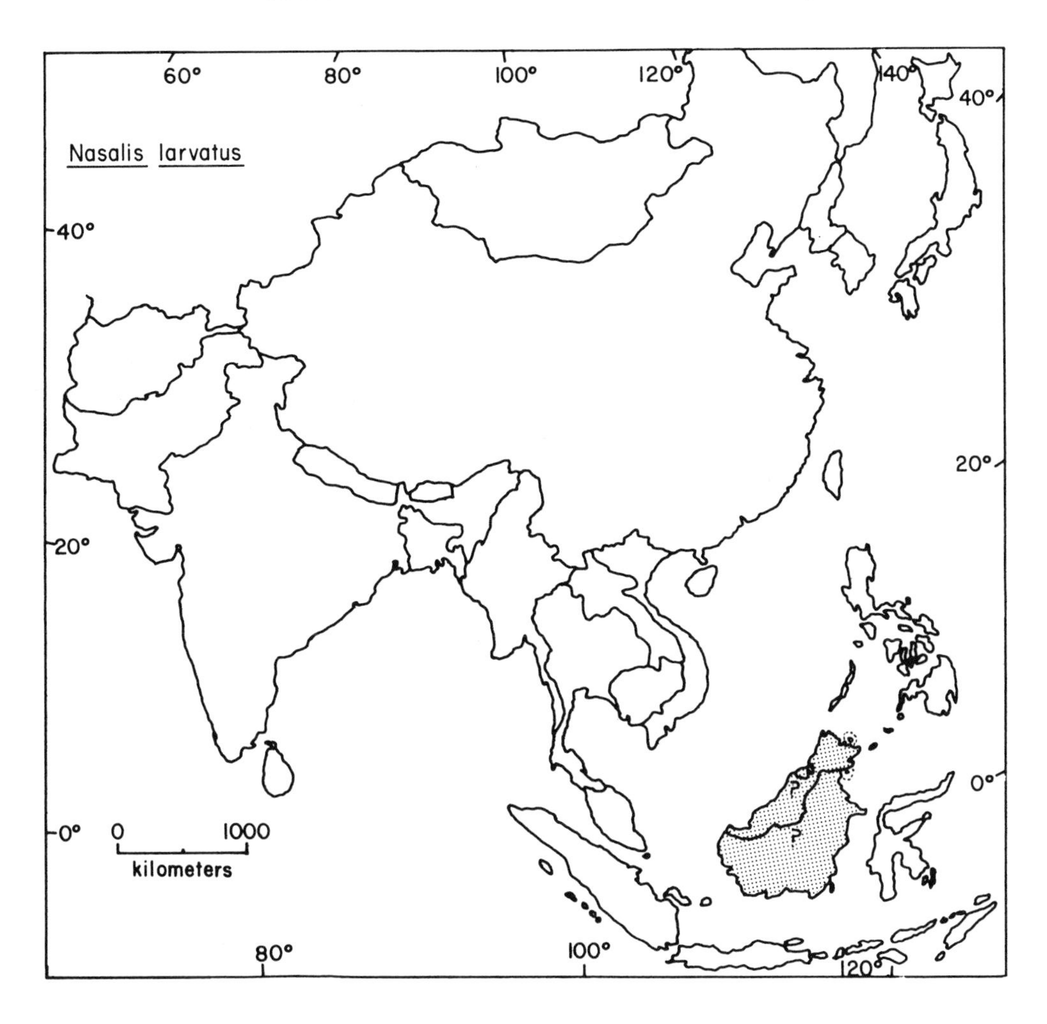

Taxonomic Note

Two races of *Nasalis larvatus* have been described (Kern 1964), but these may not be valid subspecies (Groves 1970d).

Geographic Range

Proboscis monkeys are found on the island of Borneo and on two small islands close to its northeastern coast: Berhala in Sandakan Bay, northeast Sabah, and Sebatik Island, about 4°N, southeast Sabah (Medway 1970). On Borneo, *N. larvatus* has been reported in Sabah and Sarawak, Malaysia (Kawabe and Mano 1972), in Brunei (Kern 1964), and in Kalimantan, Indonesia (Groves 1970d). Its presence along the northwest coast and in the interior of Borneo has not been documented. For details of the distribution, see Table 140.

Abundance and Density

Recently, *N. larvatus* has been described as common in Sabah, with an estimated total population of 9,400 in Sabah and Sarawak (Table 140). No information is available regarding its status in Brunei or Kalimantan. Overall, this species is thought to be declining and vulnerable (I.U.C.N. 1978).

The scant population data for *N. larvatus* indicate that it lives in groups of 10 to 32 with mean sizes of 10 to 20 individuals, home ranges of 100 to 150 ha, and, in one area, a population density of 8.9 individuals per km^2 (Table 141).

Habitat

N. larvatus is a forest monkey, most often found in riverine or coastal forest. It occupies nipa palm–mangrove (*Rhizophora* sp.) forest along rivers and estuaries (Stott and Selsor 1961a, Davis 1962), nipa-pedada (*Sonneratia alba*) swamps (Kern 1964, Wilson and Wilson 1975a), and riparian habitats in lowland tropical rain forest on the inland plain up to 245 m elevation (Kawabe and Mano 1972).

Proboscis monkeys are frequently found in coastal mangrove swamps (Burgess 1961, Yoshiba 1964, Okano 1965), and tidal creeks at river mouths, but not in pure nipa forest (Kawabe and Mano 1972). One record was obtained from a site 16 km inland and near no large rivers, but most sightings have been close to major rivers or salt water (Medway 1970).

Factors Affecting Populations

Habitat Alteration

In the past, *N. larvatus* habitat was considered inaccessible owing to the "impenetrable mass of mangrove roots and soft slippery mud," and of little economic value to man (Kern 1964). Now, however, the pulp of the common mangrove is known to be useful in the manufacture of rayon (Wilson and Wilson 1975a), paper pulp, chipboard, charcoal, carbon, house frames, and scaffolding poles, and new techniques have been developed to enable clearcutting of mangrove forests (I.U.C.N. 1978).

Large areas of Sabah, Sarawak, Brunei, and Kalimantan are either already being logged or have been allocated to timber companies (many of them Japanese) for future exploitation (I.U.C.N. 1978). This deforestation is rapidly reducing the habitat of *N. larvatus* (I.U.C.N. 1978), and any areas which are clearcut would not be expected to support this species (Wilson and Wilson 1975a).

Human Predation

Hunting. Proboscis monkeys are considered a delicacy by Chinese people in Brunei, but no evidence of heavy hunting pressure was found (Kern 1964).

Pest Control. No information.

Collection. *N. larvatus* is not used in medical research, but its appearance makes it a novel zoo exhibit. Between 1968 and 1973 a total of 20 individuals were imported into the United States (Appendix). None entered the United States in 1977 (Bittner et al. 1978). From 1965 to 1975, 2 were imported into the United Kingdom (Burton 1975).

Conservation Action

N. larvatus occurs within one extant and one proposed national park, six established reserves, and one proposed reserve (Table 140). It is protected from hunting and capture in Sabah (except in defense of life or crops) (de Silva 1971), in Sarawak (except by license) (Chin 1971), and in Kalimantan (Kern 1964, Brotoisworo 1978), and protection has been proposed for it in Brunei (I.U.C.N. 1978).

This species is listed in Appendix I of the Convention on International Trade in Endangered Species of Wild Fauna and Flora, and as endangered under the U.S. Endangered Species Act (U.S. Department of Interior 1977a,b).

Table 140. Distribution of *Nasalis larvatus*

Country	Presence and Abundance	Reference
Malaysia: North Borneo	Fairly common along coast of Sarawak; in Bako N.P. at mouth of Sarawak R. near Kuching, W Sarawak	Anderson 1961
	Not rare along E and W coasts; not in danger of extinction	Burgess 1961
	Plentiful along Segaliud R., SW of Sandakan, NE Sabah	Stott and Selsor 1961a
	Common along coast and rivers	Davis 1962
	Near mouth and along Kinabatangan R., E Sabah	Yoshiba 1964
	Near Weston (S of Jesselton), W Sabah	Okano 1965
	In Sabah	de Silva 1968b
	In Bako N.P., W Sarawak	I.U.C.N. 1971, Rothschild 1971
	In Ulu Segama Reserve, E Sabah	MacKinnon 1971
	Common and widely distributed in Sabah including Padas Bay in W and Kinabatangan R. in E; also in Sarawak	Kawabe and Mano 1972
	In Sungei Samunsam Wildlife Sanctuary, SW Sarawak; other localities in Sarawak mapped	Mittermeier 1977b
	Population estimates: 3,000 in Sabah, 6,400 in Sarawak in less than 1,500 km^2 range; in Kabili-Sepilok Forest Reserve and proposed Klias N.P. (Sabah); in Gunung Pueh Forest Reserve, on Samunsam R., Sarawak	I.U.C.N. 1978
Brunei	Common near Brunei Bay and at mouths of Brunei, Limbang, Rangau, and Labu rivers	Kern 1964
Indonesia: Kalimantan	At Pontianak and along Kapuas R. (in W), Karangmumus R. (E central), Balikpapan Bay, Pamukan Bay, Buntok, and Klumpang Bay (all in SE), Berau, Birang,	Groves 1970d

Table 140 (continued)

Country	Presence and Abundance	Reference
Indonesia (cont.)	Domaring, and Segah rivers (all in NE); also at Sampang, Mankol, Baai, and Djambajan rivers, and Telok Asam	Groves 1970d (cont.)
	In Mandor Reserve (in SW)	I.U.C.N. 1971
	In Kutai Reserve (in E along Kutai R.)	Rodman 1973
	In E	Van Peenen et al. 1974
	In Kutai Reserve and near Mahakam R. and E coast between equator and 1°S	Wilson and Wilson 1975a
	In proposed Kayan Delta Nature Reserve (in E)	I.U.C.N. 1978

Table 141. Population parameters of *Nasalis larvatus*

Population Density (individuals/km^2)	Group Size			Home Range Area (ha)	Study Site	Reference
	$\bar{X}$	Range	No. of Groups			
8.9	20	12–27	8	129.5	Brunei	Kern 1964
		15–20	2		Sabah	Yoshiba 1964
	10		3		Sabah	Okano 1965
	19	11–32	8	100–150	Sabah	Kawabe and Mano 1972

Papio anubis Anubis or Olive Baboon

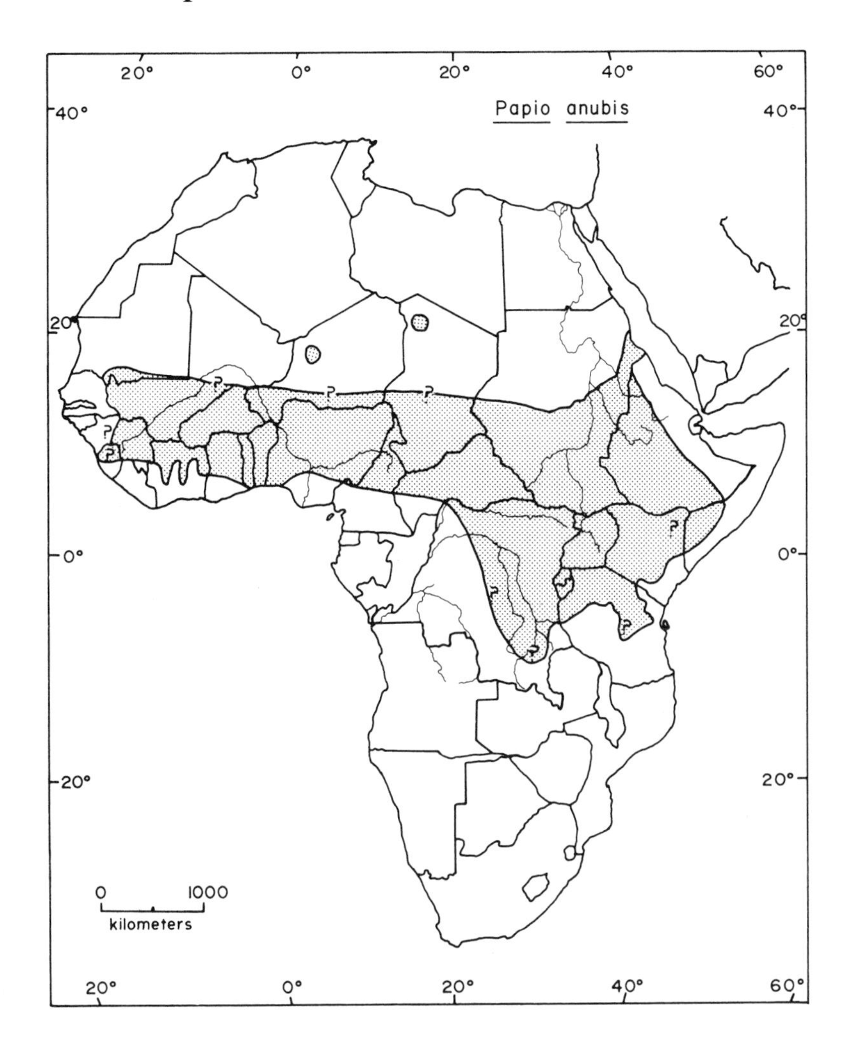

Taxonomic Note

Some workers regard *Papio anubis* as a subspecies of *P. cynocephalus* (DeVore 1965, Kock 1969, Kingdon 1971), but most authors consider *P. anubis* to be a distinct species, and it is treated as such here. It has also frequently been called *P. doguera* (Curry-Lindahl 1956, Butler 1966, Rahm 1966). Four to seven races are recognized (Napier and Napier 1967; Hill 1967a, 1970).

Geographic Range

Papio anubis occurs within a broad belt extending across subsaharan Africa from Mauritania to Ethiopia, and in isolated areas in the Sahara Desert. Populations are found as far north as 20°N, 16°E at Tibesti, Chad (Bigourdan and Prunier 1937, cited in Tappen 1960), and about 19°N, 36°E in northeastern Sudan (Butler 1966). Otherwise the northern limit of the range is at about 15°N (Huard 1962). Olive baboons occur as far south as the Gulf of Guinea at the Dahomey Gap (Booth 1958), and as far south as about 5°15′S in northern Tanzania (Kano 1971a). They have been recorded as far east as 45°01′E (at 4°58′N) along the Webbe Shibeli in southeastern Ethiopia (Bolton 1973). The western limit of the range is not known. For details of the distribution, see Table 142.

Abundance and Density

Baboons in general have been considered "the most widespread and abundant non-human primates on the African continent" (Altmann and Altmann 1970). *P. anubis*, specifically, has been described as abundant or common in parts of eight countries within its range (Table 142). But as early as 1963 this species was found to be declining in some areas in eastern Zaire, and during the last ten year its numbers have also decreased in parts of Niger, Sudan, Kenya, and Uganda (Table 142). In several areas this species is only locally abundant and is generally declining outside of reserves and national parks.

Population density estimates for *P. anubis* range from 1.12 to 63 per km^2 (Table 143). Most authors have found an average group size of between 20 and 60 individuals, although groups as large as 140 have been reported. Home ranges vary from 74.5 to 4,014 ha (Table 143).

In general most workers agree that high baboon densities (due to large group sizes and small home ranges) are correlated with habitat quality, specifically water and food availability (DeVore and Hall 1965, Rowell 1966). But this relationship is not always clear-cut (Altmann and Altmann 1970). The high density reported by Coe (1972) (Table 143) in an arid habitat where all large mammals were rare contradicts this generalization so strikingly that it suggests a typographical error.

Habitat

P. anubis is found in a wide variety of habitat types. Among its relatively open habitats are semidesert steppe (Altmann and Altmann 1970), arid thorn scrub (Aldrich-Blake et al. 1971; Paterson, pers. comm. 1974), open grassland with patches of dense scrub (Harding 1976), plains, rocky hills, and guinea savanna (Booth 1956a, Happold and Philp 1971, Poché 1973), Sudan and Sahel savanna (Sayer 1977), and wooded savanna (Rahm and Christiaensen 1963, Hall 1965b, Geerling and Bokdam 1973). It also occupies woodlands and forests, including *Brachystegia* woodland (Kano 1971a), *Acacia* forest (Hill 1970, Coe 1972), *Olea* forest (Crook and Aldrich-Blake 1968), gallery forest and woodland (Hall 1965c, Rowell 1966, Dunbar and Dunbar 1974d, Poché 1976), and tropical rain forest (Rahm 1966, Mallinson 1974).

Many authors consider the primary habitat of *P. anubis* to be savanna (Booth 1956a, Hall 1965b, Altmann and Altmann 1970, Monath and Kemp 1973). But some populations in Uganda live in high forest or dense bush, and, at Ishasha, spent up to 60% of their time in forest (Rowell 1966). At Debra Libanos, Ethiopia, *P. anubis* also preferred wooded areas to open country (Crook and Aldrich-Blake 1968) and spent about 50% of its time in forest (Crook 1970).

In some areas, forest edge with neighboring open grassland or cultivated clearings is the preferred habitat of olive baboons (Paterson 1973, Dunbar and Dunbar 1974a). This species can utilize secondary growth and quickly colonizes abandoned clearings (Tappen 1965; Brown and Urban 1970; Gartlan, pers. comm. 1974).

Water is a limiting factor for *P. anubis*, and several authors note that it is restricted to areas where water is available, or concentrated near water in dry areas (Sikes 1964a; Hausfater, pers. comm. 1974; Poché 1976). It cannot survive without regular access to water, and migrates over long distances to reach it (Hall 1966), so that the home range of an olive baboon group always includes a stream, river, water hole, or other water supply (Hall 1965a).

Another requirement for *P. anubis* is the availability of sleeping sites. If trees are present, they are used for sleeping (Hall 1966, Rowell 1966, Crook and Aldrich-Blake 1968). The number of suitable sleeping trees can limit the size of *P. anubis* populations (Paterson 1973). In some regions, such as the dry, open country of Sudan and Ethiopia, olive baboons also sleep on rocky outcrops (Butler 1966, Dunbar and Dunbar 1974a) or on cliff faces (Hall 1966, Crook and Aldrich-Blake 1968, Harding 1976).

P. anubis is found at low, medium, and high elevations (Starck and Frick 1958) and has been studied in lowland forest (Tappen 1965, Bernahu 1975) and montane habitats (Omar and de Vos 1970, Dandelot and Prévost 1972, Lavieren and Bosch 1977), as high as 4,500 m (Bernahu 1975). At one site in Ethiopia it was found more frequently in habitats at low elevations than in higher zones (Dunbar and Dunbar 1974a). But in East Africa, Maples (1972a) considered it to be primarily a highland form, since it occupied higher, cooler, and wetter areas than the closely related yellow baboon, which occurred in the hot arid lowlands. This may be a case of character displacement in areas of sympatry of these two species.

Factors Affecting Populations

Habitat Alteration

Because it is a very adaptable animal which adjusts well to new conditions and can occupy many habitat types (Hall 1966), some kinds of limited habitat alteration do not adversely affect *P. anubis*. It can utilize clearings and cultivated fields, and can survive in areas in which up to 50% of the land is occupied by humans (Paterson, pers. comm. 1974). Since it can inhabit secondary vegetation, the amount of suitable habitat available to it may be increased by limited timbering, slash and burn cultivation, and burning (Tappen 1965; Gartlan, pers. comm. 1974).

Nevertheless, the expansion of agriculture and other man-caused habitat disturbances are some of the main causes of the decline of *P. anubis* populations. In Uganda, for example, the spread of farming has driven baboons from all areas of the Teso district except isolated thickets and rocky hills (Hall 1965b) and is continuing to encroach upon baboon habitats (Paterson, pers. comm. 1974). In Ethiopia, deforestation, a dense human population, and clearing and burning for cultivation and pasture have eliminated baboon habitat in many regions (Dandelot and Prévost 1972, Bernahu 1975). Nomadic herds also denude areas of vegetation and compete with *P. anubis* for food (Aldrich-Blake et al. 1971).

In Niger, rapidly expanding agriculture, overgrazing by livestock, and fire have combined to greatly reduce the range of *P. anubis* (Poché 1973, 1976). The "W" National Park is being threatened by illegal grazing, heavy pressure for land use by migrants from drought stricken areas to the north, and exploitation of iron ore deposits by mining firms (Jones 1973).

In Ghana, *P. anubis* habitat has been decreased by agricultural expansion (Asibey 1978). For example, in the Accra Plains area, about 30% of its habitat has been destroyed (Rice, pers. comm. 1978). In Kenya the clearing of bush cover to improve range conditions for cattle is continuing (Harding 1976). These processes are probably occurring in every country that *P. anubis* inhabits.

Maples (1972a) suspected that the spread of agriculture was responsible for the breakdown in isolation of yellow from olive baboons. He theorized that the habitat destruction caused by the extensive human occupation of the Kenya highlands in the early 1900s forced olive baboons to occupy lowland habitats, establishing contact and interbreeding with yellow baboons.

Drought would be expected to have an adverse effect on *P. anubis* populations, and is known to be responsible for the deaths of many baboons and the reduction of the carrying capacity of their habitats in Niger (Poché 1976). No direct information is available regarding the effects of the recent subsaharan drought on *P. anubis* populations in other countries.

In 1960, *P. anubis* populations in Kenya actually increased during a severe drought, because of their ability to utilize roots and rhizomes as food sources (DeVore 1965). Other authors suggested that baboons survived by eating carcasses of dead ungulates, but DeVore contended that they generally avoid carrion. In some areas of Kenya, *P. anubis* drink from cattle water troughs, increasing their ability to withstand dry periods (Harding 1976).

Human Predation

Hunting. Anubis baboons are shot for their flesh in Ivory Coast (Tahiri-Zagrët 1976) and in Ghana (Asibey 1974). In Ghana, game production reserves such as the one in the Shai Hills were established to raise animals for human consumption, but so far no cropping schemes have been implemented (Rice, pers. comm. 1978).

P. anubis are also hunted for sport in Cameroon (Gartlan, pers. comm. 1974) and in the Narok district of Kenya, where they "are shot indiscriminately by hunting parties, as target practice, or for leopard bait" (Hausfater, pers. comm. 1974). In Ethiopia there is a local trade in the fatty matter from baboon callosities, which is used as a remedy for rheumatism and stiff muscles (Bernahu 1975).

Pest Control. Olive baboons frequently raid crops and are pests to agriculture throughout their range (Booth 1956a, Hall 1966, Rahm 1966, Nagel 1973, Bourlière et al. 1974, Poché 1976). In some areas, such as southwest Ethiopia, they are probably dependent on man's crops for their food (Brown and Urban 1970), and prefer raiding plantations to foraging (Paterson, pers. comm. 1974).

Crops that *P. anubis* eats include potatoes, yams, bananas, peanuts, beans, sugar cane, and maize (DeVore and Washburn 1963). In Ivory Coast it does considerable damage to the yam crop, and also eats palm hearts, thus affecting production of the local palm wine (Tahiri-Zagrët 1976).

This damage to crops has resulted in efforts to control olive baboon populations in many areas (Rahm and Christiaensen 1963; Hall 1966; Rowell 1968; Dunbar and Dunbar 1974a; Gartlan, pers. comm. 1974). As early as 1936–37 massive poisoning campaigns in Sudan destroyed at least 7,000 baboons in one year (Beaton 1942). In western Tanzania the Kigoma Agricultural Office employs hunters to shoot baboons because of their damage to agriculture (Kano 1971a), and in Kenya and Ethiopia, *P. anubis* is classified as vermin and may be shot without license (Maples 1972b, Bernahu 1975).

In Ghana, *P. anubis* is frequently killed as an agricultural pest, and the ready market for its meat provides an additional incentive for control measures (Asibey 1978). In this case, poison is not used to kill *P. anubis*, since the meat is eaten by humans, but in other countries, such as Nigeria, anubis baboons are poisoned in large numbers to protect crops (Howell 1968).

The severity of persecution of crop raiders is often inversely correlated with the supply of human food. For example, olive baboons were previously tolerated on the large ranches of Kenya, where their impact was not considered significant. But in areas where the human population has increased rapidly and the ranches have been divided into smaller maize farms, the local farmers are now at the subsistence level, see baboons as competitors for their food, and shoot them whenever possible (Rowell, pers. comm. 1974).

P. anubis can also be destructive to other human property and resources in addition to crops. For example, in Niger, it creates deep ruts in dirt roads by digging holes to uncover edible roots and tubers (Poché 1976). In 1965–67 several dozen *P. anubis* were shot in Ethiopia because they raided tents and huts in the Awash National Park (Nagel 1973). And in 1962, 700 *P. anubis* were killed in Lake Manyara National Park, Tanzania, with the justification that they were detrimental to resident bird populations; no evidence was found to substantiate this claim (Altmann and Altmann 1970).

There is some disagreement about the long-term effect of extermination programs on baboon populations. Although 20,000 have been killed in Uganda during the last 30 years, baboons are still numerous in many areas (Kingdon 1971). But they have been eliminated over large areas of southern Mengo and other heavily populated regions (Rowell, pers. comm. 1978). Near Lake Manyara, Tanzania, baboons are repeatedly "controlled" by shooting, but appear to recover when control operations cease (Hausfater, pers. comm. 1974). But present levels of shooting, poisoning, and trapping are probably greater than replacement rates in East African populations (Altmann, pers. comm. 1973), and could eventually lead to their decline.

Collection. *P. anubis* is used frequently in biomedical research, and many are trapped and exported yearly for this purpose. In Kenya, commercial trapping has increased despite the scarcity of baboons (Altmann, pers. comm. 1973), and has put heavy pressure on *P. anubis* populations (Harding 1976). Especially intensive trapping has occurred in the Kajiado District, and, if continued, could exterminate the populations in this area (Hausfater, pers. comm. 1974).

Relatively large numbers of *P. anubis* are exported from Ethiopia to Europe and America for research and other uses (Bernahu 1975). Few baboons are now being exported from Uganda (Paterson, pers. comm. 1974) or Cameroon (Gartlan, pers. comm. 1974).

Howard and Gresham (1966) reported that baboons were readily available and relatively inexpensive, and that although breeding colonies had been established, medical researchers believed that "there are still at present sufficient trapped animals from Africa available for experimental purposes, and for many years in the future it will be possible to employ these very useful animals in many different areas of research."

P. anubis are imported into the United States more frequently than any other species of baboon. Between 1968 and 1973 a total of 4,158 were imported (Appendix), the majority from Kenya and Tanzania. In 1977, 50 individuals plus 10 shipments of unspecified sizes entered the United States from Kenya (Bittner et al. 1978). Between 1965 and 1975, 3,604 were imported into the United Kingdom (Burton 1978).

Conservation Action

P. anubis has been recorded in 27 national parks and 21 reserves (Table 142). But poaching and habitat deterioration are contributing to the steady decline of populations in several of the areas where baboons and their habitats are nominally protected, such as the "W" National Park in Niger, Benin, and Upper Volta (Poché 1973).

Altmann (pers. comm. 1973) concluded, "It seems unlikely that baboons will survive [in East Africa] except in protected areas or in very barren areas seldom occupied by people." He recommended support and strengthening of wildlife protection areas, and promotion of the concept of harvesting *P. anubis* on a sustained yield basis as an alternative to uncontrolled trapping for research.

Trapping of *P. anubis* should be especially strongly discouraged in the parks of northern Tanzania and in the Kajiado District of Kenya (Hausfater, pers. comm. 1974). In the Kajiado District, natural habitat changes have already reduced the baboon populations to the point where "human interference through trapping could easily tip the scales toward extinction in this area" (Hausfater, pers. comm. 1974).

Crop raiding *P. anubis* could be captured and exported to fill research needs, rather than being shot (Rowell 1968). Since the baboons may prove to be worth more than the crops they damage, growing crops to attract baboons for capture and sale could prove to be economically profitable (Rowell 1968).

Table 142. Distribution of *Papio anubis* (countries are listed in order from west to east and from north to south)

Country	Presence and Abundance	Reference
Mauritania	In S (maps)	Huard 1962, Baldwin and Teleki 1972
Mali	Widespread and common throughout S as far N as Hombori Mts. (about 15°N, 2°W)	Sayer 1977
Guinea	In NE (map)	Booth 1958
Sierra Leone	In N (other authors mention only *P. papio* there)	Jones 1966
Ivory Coast	In N (map)	Booth 1958
	Abundant in Comoe N.P. (in NE)	Geerling and Bokdam 1973
	Near Lamto Station (S central, near Bandama R.)	Bourlière et al. 1974
	Throughout N half and in S near Soubré (in W) and Ayamé (in E) (map)	Tahiri-Zagrët 1976
	Abundant in Comoe N.P. and Mount Peko N.P.	Asibey 1978
Upper Volta	Throughout (map)	Booth 1958
	Abundant in "W" N.P. (in SE)	Asibey 1978
Ghana	Throughout except in SW (localities mapped)	Booth 1956
	Abundant in Mole N.P., Digya N.P., Bui N.P., Kalakpa and Shai Hills game production reserves, and Willi Falls Reserve; common in Bomfobiri Wildlife Sanctuary, Gbele Game Production Reserve, and Kogyae Strict Nature Preserve	Asibey 1978
	Abundant and stable in Shai Hills Game Production Reserve (in SE); total population 190–240	Rice, pers. comm. 1978
Togo	Throughout (map)	Booth 1958
	In Koué Reserve (central, W of Sokodé)	I.U.C.N. 1971
	Common in Keran and Malfacassa reserves	Asibey 1978
Benin	Throughout (map)	Booth 1958
	In Pendjari N.P. (in NW) and "W" N.P. (in N)	Happold and Philp 1972, Gorgas 1974
	Abundant in Pendjari N.P. and "W" N.P.	Asibey 1978
Niger	Isolated populations at Air Massif (18°N, 8°E)	Bigourdan and Prunier 1937 cited in Tappen 1960, Huard 1962
	Populations viable but depleted in "W" N.P.	Jones 1973
	Abundant in "W" N.P. (in SW)	Poché 1973
	In much of S; range being reduced	Poché 1976

Table 142 (continued)

Country	Presence and Abundance	Reference
Nigeria	Throughout, N of about 7°N (map)	Booth 1958
	Numerous and increasing in Yankari Game Reserve (near 10°N, 10°E); total population about 1,500–2,500	Sikes 1964b
	Very numerous throughout NW including 1,500–2,000 in Borgu Game Reserve; increasing steadily	Howell 1968, 1969
	Infrequent in Olokemeji Forest Reserve (in SW, W of Ibadan)	Hopkins 1970
	In Upper Ogun and Oyo reserves (in SW, N of Ibadan)	Adekunle 1971
	Common in Yankari Game Reserve	Henshaw and Ayeni 1971
	Common in N; widely but sparsely distributed; abundant in Yankari Game Reserve; very common Borgu Game Reserve	Henshaw and Child 1972
	On Jos Plateau (central)	Lee 1972
	Widely distributed and abundant, especially in W and E central (localities mapped)	Monath and Kemp 1973
	In Yankari Game Reserve	Ajayi 1974
	In Upper Ogun Game Reserve	Geerling 1974
	In Nupeko Forest (N of Bida)	Monath et al. 1974
	In Kontagora area (NW)	Service 1974
Cameroon	Throughout N (map)	Booth 1958
	At Waza N.P.	Struhsaker 1969
	Throughout N as far W as Wum (about 10°E), as far S as Bamenda (about 6°S), including Faro and Bénoué Reserves; abundant and stable; range may be increasing	Gartlan, pers. comm. 1974
	Common in Bouba-Ndjida N.P. (near 9°N on E border)	Lavieren and Bosch 1977
	Abundant in Benove (Bénoué) N.P., Bouba-Ndjida N.P., and Mbi Crater Game Reserve; common in Waza N.P. and Kimbe River Game Reserve	Asibey 1978
Chad	At Tibesti (20°N, 16°E)	Bigourdan and Prunier 1937 cited in Tappen 1960
	As far S as Lake Chad	Huard 1962
	In Zakouma N.P. (in SE, W of Am Timan)	I.U.C.N. 1971
Libya	Rare visitors in extreme S (Tibesti Mts.)	Hufnagl 1972
Central African Republic	In N (map)	Booth 1958
	In N and central (map)	Baldwin and Teleki 1972

Table 142 (continued)

Country	Presence and Abundance	Reference
Zaire	Very common N of Lake Kivu; W slope of Mount Kahuzi, near Lubero, and in Albert N.P. (all in E)	Curry-Lindahl 1956
	Abundant near Lemera; declining in several other areas W of Lake Kivu (in E)	Rahm and Christiaensen 1963
	Throughout E between Lualaba-Zaire R. and Rift, and Lowa-Osa R. and Elila R.; rather common around Irangi	Rahm 1966
	In E along Zaire-Lualaba R. between Kalombo and Kindu	Mallinson 1974
	Abundant throughout Haute-Zaire Province (in N)	Heymans 1975
	In Kundelungu N.P. (in SE)	I.U.C.N. 1975
	In mountains W of Lake Kivu	Hendrichs 1977
Sudan	In SE (map)	Starck and Frick 1958
	In W, S, and E (map)	Butler 1966
	Throughout S, central, and E (localities mapped); nearly exterminated in Kordofan (central)	Kock 1969
	In Dinder N.P. (near 12°N on E border) and Nimule N.P. (near 32°E on S border)	Cloudsley-Thompson 1973
Ethiopia	Throughout SW (map)	Starck and Frick 1958
	In E	Glass 1965
	At Debra Libanos (N of Addis Ababa); population about 80	Crook and Aldrich-Blake 1968
	In proposed Rift Valley Lakes N.P. (S of Addis Ababa)	Bolton 1970
	In Kefa and Ilubabor provinces (in SW)	Brown and Urban 1970
	In Awash Valley (E of Addis Ababa); population about 287	Aldrich-Blake et al. 1971
	Near Webbe Shibeli, Ogaden (in SE)	Bolton 1972
	In SW and central (localities mapped); locally common	Dandelot and Prévost 1972
	Common throughout except in E; localities listed; abundant in Rift Valley	Bolton 1973
	In Awash N.P.	Nagel 1973
	Common at Mount Abaro, Shashemenē District (S central)	Bolton 1974
	Central and SW (localities mapped)	Dunbar and Dunbar 1974a
	As far N as Wolkefit Pass (about 13°N); (localities mapped)	Dunbar and Dunbar 1974d
	In Lake Stephanie Wildlife Reserve (S)	Tyler 1974
	Considerable numbers in central highlands E of Rift; abundant in Rift (central and S) and highlands of SW; common in W lowlands and N plateau	Bernahu 1975
	In proposed Bale Mts. N.P. (S central)	Waltermire 1975

Table 142 (continued)

Country	Presence and Abundance	Reference
Somalia	In SW (map)	Baldwin and Teleki 1972
Kenya	In Nairobi N.P.; net population increase of 40% 1959–63	DeVore 1965 [Hall and DeVore 1965, Altmann and Altmann 1970]
	S of Lake Naivasha to S of Sultan Hamud (map)	Maples and McKern 1967, Maples 1972a
	Throughout W (map)	Kingdon 1971
	Near Rumuruti and Nanyuki in Laikipia District (0°15′N, 36°30′E)	Berger 1972
	Rare in S Turkana area (E and W of Lake Rudolf in N)	Coe 1972
	Very scarce outside reserves and parks	Altmann, pers. comm. 1973
	Local populations probably declining in Kajiado District, Masai-Mara Game Reserve, and Narok District	Hausfater, pers. comm. 1974
	In Lake Nakuru N.P.	Kutilek 1974
	In Lake Nakuru N.P.; about 500 in Aberdare N.P.	Dupuy 1975
	Near Gilgil (SE of Nakuru)	Harding 1976
Uganda	At Bwamba (in W)	Lumsden 1951
	In Murchison Falls N.P.; rare in Teso District and range shrinking	Hall 1965b
	In Queen Elizabeth N.P.; 1 population increased 15% per year for 2 years, then stabilized	Rowell 1966, 1969
	Throughout (map); numerous	Kingdon 1971
	In several parts of Queen Elizabeth N.P.	Lock 1972
	Uncommon in Kibale Forest Reserve (in W)	Struhsaker 1972, 1975
	In Queen Elizabeth N.P. and Budongo Forest Reserve; relatively abundant but declining in most areas	Paterson 1973, pers. comm. 1974
Rwanda	In Kagera N.P. (near E border)	I.U.C.N. 1971
	Throughout (map)	Baldwin and Teleki 1972
Burundi	Throughout (map)	Baldwin and Teleki 1972
Tanzania	In Lake Manyara N.P., Ngorongoro Crater, and Serengeti N.P. (all in N)	Altmann and Altmann 1970
	In Serengeti N.P.	Hendrichs 1970
	On E shore Lake Tanzania as far S as Malagarasi R. (about 5°15′S)	Kano 1971a
	Throughout N except NE corner (map)	Kingdon 1971
	Abundant in Lake Manyara N.P., Serengeti N.P., and Tarangire N.P. (all in N)	Hausfater, pers. comm. 1974
	Abundant, perhaps declining at Gombe Stream N.P. (N of Kigoma, E shore Lake Tanganyika)	Nash, pers. comm. 1974

Table 143. Population parameters of *Papio anubis*

Population Density (individuals /km²)	Group Size X̄	Range	No. of Groups	Home Range Area (ha)	Study Site	Reference
		20–35	7–10		SE Ghana	Rice, pers. comm. 1978
	32	10–125	32		SW Niger	Poché 1976
1.12					N Nigeria	Sikes 1964b
	6.5				N Nigeria	Henshaw and Ayeni 1971
	7	2–30	48		N Nigeria	Ayeni 1972
		Up to 50+			N Nigeria	Henshaw and Child 1972
		45	1		N Cameroon	Struhsaker 1969
		25–30			N Cameroon	Gartlan, pers. comm. 1974
		20–30			N Cameroon	Lavieren and Bosch 1977
		Up to 100			Sudan	Butler 1966
	17.3	2–49	30		Central Ethiopia	Crook and Aldrich-Blake 1968
11.58				777	Central Ethiopia	Crook 1970
	44	14–87	8	430	Central Ethiopia	Aldrich-Blake et al. 1971
5.6					Central Ethiopia	Nagel 1973
26	20	15–24	7	74.5–112.0	Central Ethiopia	Dunbar and Dunbar 1974a
3.86	41	12–87	9	518–4,014	S Kenya	DeVore and Hall 1965
	34.5	30–39	2		S Kenya	Altmann and Altmann 1970
	96.5	60–140	4		Central Kenya	Berger 1972
63					N Kenya	Coe 1972
1.81					Central Kenya	Kutilek 1974
10.3	70.86	35–121	7	1,968.4	Central Kenya	Harding 1976
		16–40			W Uganda	Lumsden 1951
	27	14–48	8	777–3,885	W Uganda	Hall 1965b
10.8		32–58	2	388–518	W Uganda	Rowell 1966
53		38–80	3	200–600	W Uganda	Paterson, pers. comm. 1974
	37	8–81	5		N Tanzania	Altmann and Altmann 1970
		10–30			W Tanzania	Kano 1971a
38–58					W Tanzania	Nash, pers. comm. 1974

Papio cynocephalus Yellow Baboon

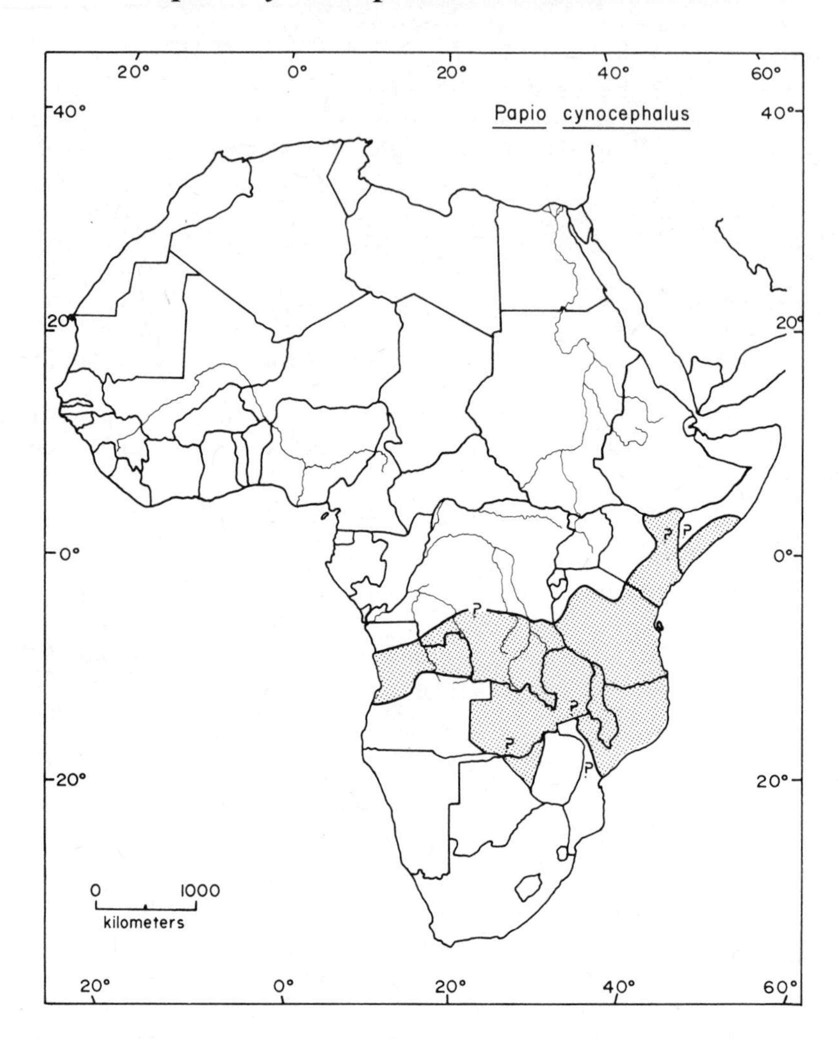

Taxonomic Note

Some authors consider *Papio cynocephalus* to be conspecific with *P. anubis* (Hall 1965b, Kock 1969, Maples 1972a), *P. ursinus* (Hall 1965c), or both (Buettner-Janusch 1966, Kingdon 1971). In the present report, *P. cynocephalus* refers only to the yellow baboon and does not include either *P. anubis* or *P. ursinus*, which are discussed in separate sections. Three or four races of *P. cynocephalus* are recognized (Napier and Napier 1967; Hill 1967a, 1970).

Geographic Range

P. cynocephalus is found in southern Africa from Angola to the Indian Ocean. It occurs as far west as 14°E (at about 13°30′S) in Angola (Barros Machado 1969) and as far east as the coasts of Mozambique, Tanzania, Kenya, and Somalia. It is found as far south as about 21°S in Zimbabwe, and at least to the Zambezi River in Mozambique (Smithers, pers. comm. 1974). It has been reported as far north as 4°S in Zaire (Schouteden 1947, cited in Tappen 1960), and about 4°N in Somalia (Hill 1967a), but recent records from these areas are lacking. For details of the distribution, see Table 144.

Abundance and Density

P. cynocephalus has recently been described as locally abundant in parts of Kenya and Malawi, and as having stable populations in Zambia (Table 144). But three workers agree that its populations have declined markedly in southern Kenya. Its status throughout most of its range is unknown.

This species has been studied at population densities of 9.65 to 60 individuals per km^2 (Table 145). Group sizes range from 7 to 198, with average sizes between 28.3 and 80 and home ranges of 518 to 2,409 ha (Table 145).

Habitat

P. cynocephalus is primarily found in savanna and woodland habitats. It is most typical of perennial grass-covered savanna with variable tree densities, but also utilizes many other types of vegetation (Altmann and Altmann 1970). In Kenya this species has also been studied in scrub bush and riparian woodland (Maxim and Buettner-Janusch 1963), forest and thick bush (Andrews et al. 1975), semidesert bushy grassland (Kuntz and Moore 1973), and in habitats varying from open grassland to thick low vegetation and true forest (Maples 1972a).

In Tanzania, *P. cynocephalus* occupies gallery forest, *Brachystegia*, *Uapaca*, *Combretum*, and *Diplorhynchus* woodland, and bamboo brush (Itani and Nishida, pers. comm. 1974). The Zambian populations are found mainly in woodlands and *Cryptosepalum* forest (Ansell 1960a), and venture into open country to feed (Smithers 1966). In Zimbabwe they inhabit savanna and thorn-scrub savanna (Hall 1965b), mopane deciduous woodland, and low scrub (Hall 1966).

P. cynocephalus has been characterized as typical of the hot, arid lowlands of Kenya in contrast to *P. anubis*, which was found in higher, cooler, more humid areas (Maples 1972a). Other authors have found *P. cynocephalus* as high as 1,000 m elevation in Kenya (Kuntz and Moore 1973), and common on the escarpment and in the foothills but rare on the lowland plains of Malawi (Long 1973). In Zambia this species is not found in montane areas (Ansell 1960a).

Yellow baboons are unable to survive without access to water and may migrate long distances to drink and congregate at water holes (Hall 1966). Their home range always includes a water source (Hall 1966), and they occur at reduced densities in arid regions (Maples, pers. comm. 1974).

Sleeping sites are another crucial component of *P. cynocephalus* habitat (DeVore and Washburn 1963). In some areas, tall trees are utilized (Hall 1966; Kano 1971a; Maples, pers. comm. 1974), and in Zimbabwe, cliffs also serve as sleeping sites (Hall 1965b).

P. cynocephalus can utilize secondary growth and can survive in cultivated regions as long as a large uncleared area remains nearby for refuge (Maples, pers.

comm. 1974). At one site in Kenya it often fed in disturbed growth and secondary vegetation (Moreno-Black and Maples 1977).

Factors Affecting Populations

Habitat Alteration

One of the most notable characteristics of yellow baboons is their "ability to adapt to man-created changes in their environment" (Hall 1965d). Nevertheless, the encroachment of agriculture is making many areas uninhabitable for this species. In Kenya, for example, 5 to 10% of suitable yellow baboon habitat is being converted to cultivation yearly, with higher annual rates of 10 to 20% in coastal forests (Maples, pers. comm. 1974). Although *P. cynocephalus* can tolerate moderate habitat disturbance, it cannot survive in areas where all large patches of natural vegetation are removed, and hence is disappearing from many parts of Kenya (Maples, pers. comm. 1974). No information was located regarding the effect of agricultural clearance on yellow baboons elsewhere in their range.

A combination of natural and human-induced ecological changes has led to the decline of *P. cynocephalus* populations in the Masai-Amboseli Reserve (Altmann, pers. comm. 1973). Heavy rains during the 1960s raised the underground water table, bringing a layer of salts to the surface and narrowing the habitable soil zone for *Acacia* trees and other vegetation (Western and Sindiyo 1972). At the same time, overgrazing by herds of Masai livestock greatly depleted the food resources for *P. cynocephalus* and probably increased the rate of soil surface evaporation. The high salt concentration, overgrazing, and severe flooding all contributed to the death of many *Acacia* trees necessary as a source of food, shade, and shelter for the yellow baboons of this area (Altmann, pers. comm. 1973). *P. cynocephalus* populations in the Amboseli had declined to about 10% of their 1963 level by 1969, and were continuing to decline in 1973 (Altmann, pers. comm. 1973). By 1978 the rate of decline had decreased but was still detectable (J. Altmann, pers. comm. 1978).

Human Predation

Hunting. No information.

Pest Control. Yellow baboons, well known as crop raiders, eat potatoes, yams, bananas, beans, maize, peanuts, and sugar cane (DeVore and Washburn 1963). They have also been observed feeding in coconut farms and cotton farms, in garbage dumps, and stealing chickens, and in some areas rely heavily on cultivated plants for their food supply (Maples et al. 1976, Moreno-Black and Maples 1977).

They are considered a pest to agriculture and are killed by farmers in Zambia (Ansell 1960a, pers. comm. 1974), Malawi (Hill 1970), and Kenya (Kuntz and Moore 1973; Maples, pers. comm. 1974). But even rather elaborate methods of preventing their access to crops are not very effective, and *P. cynocephalus* troops use several strategies for obtaining food from guarded fields (Maples et al. 1976).

When subjected to heavy trapping, yellow baboons learn to release each other from traps and chase juveniles away from bait (Maxim and Buettner-Janusch 1963). On the other hand, their large size and noisiness make them easier to exterminate in agricultural areas than smaller, less obtrusive monkeys (Hall 1966). Ansell (pers.

comm. 1974) has found them to be "extremely wary and well able to hold their own" against crop defenders in Zambia.

Yellow baboons also eat locusts, and, by removing destructive insects, may be more valuable to the farmer than the crops they consume (Altmann and Altmann 1970).

Collection. P. cynocephalus is used for biomedical research, and commercial trapping is increasing even in areas where baboons are already scarce (Altmann, pers. comm. 1973). In Kenya, yellow baboon populations in general are probably suffering significantly from trapping (Maples, pers. comm. 1974). In the Kajiado District, for example, populations have been reduced by a combination of natural ecological changes and heavy trapping, and could be eliminated by a year or two of sustained collection (Hausfater, pers. comm. 1974).

Between 1968 and 1973 a total of 356 *P. cynocephalus* were imported into the United States, primarily from Kenya (Appendix). In 1977 two shipments of unspecified size entered the United States from Kenya (Bittner et al. 1978). Between 1965 and 1975, 4,352 *P. cynocephalus* were imported into the United Kingdom (Burton 1978).

Conservation Action

P. cynocephalus occurs in five parks and reserves and one proposed national park (Table 144). Yellow baboons probably will not survive outside of parks and reserves in southern Kenya and northern Tanzania, and these protected areas should be supported and strengthened (Altmann, pers. comm. 1973). In addition, the "farming" of baboons on a sustained yield basis should be substituted for uncontrolled trapping in this area (Altmann, pers. comm. 1973). All trapping in the Kajiado District of Kenya south of the town of Kajiado should be discouraged, since it could cause the extinction of baboons in this area (Hausfater, pers. comm. 1974).

Maples (1972b) noted that the Kenyan government had shown relatively little interest in the protection of primates compared with their efforts on behalf of other wildlife, primarily large mammals. He proposed to do a two to three year survey of Kenya's primate populations to locate study and conservation areas and to enable the "development and proper utilization of the primate populations in Kenya."

Table 144. Distribution of *Papio cynocephalus* (countries are listed in order from west to east and then south to north)

Country	Presence and Abundance	Reference
Angola	In W and NE (localities mapped)	Barros Machado 1969
Zaire	Throughout S (maps)	Hill 1967a, Baldwin and Teleki 1972
	In Kundelungu N.P. (in SE)	I.U.C.N. 1975
Zambia	Numerous throughout most of N, NW, W, and N central; S and E limits unknown	Ansell 1960a
	Throughout N two-thirds (map)	Smithers 1966
	Populations stable; abundance variable	Ansell, pers. comm. 1974

Table 144 (continued)

Country	Presence and Abundance	Reference
Zimbabwe	In Wankie Game Reserve (in W)	DeVore and Washburn 1963
	At Kariba (in NW)	Hall 1965d
	In Victoria Falls N.P. (NW) and Rhodes-Matopos N.P.	I.U.C.N. 1971
	In W (map)	Baldwin and Teleki 1972
Mozambique	In N along coast and as far S as Zambezi R. (localities mapped)	Smithers, pers. comm. 1974
Malawi	Present	Ansell et al. 1962
	Throughout (map)	Smithers 1966
	Very common on escarpment of Nsanje District (in S) and Tangadzi	Long 1973
Tanzania	At Matete, Masai Steppe	Lamprey 1964
	Throughout except NW Province; extensive along E shore Lake Tanganyika, S of Malagarasi R. (localities mapped)	Kano 1971a
	Throughout except NW and N (map)	Kingdon 1971
	At Mahali Mts. proposed N.P., S of Kigoma, E shore Lake Tanganyika	Itani and Nishida, pers. comm. 1974
Kenya	Along Athi R. near Darajani (in S)	Maxim and Buettner-Janusch 1963
	In S as far W as Simba and Lake Amboseli (map)	Maples and McKern 1967
	Locally abundant in Masai-Amboseli Game Reserve; in Olnaigon Swamp	Altmann and Altmann 1970
	Throughout E (map)	Kingdon 1971
	Declining sharply in Masai-Amboseli Reserve (90% decline 1963–69); very scarce and decreasing outside of parks and reserves	Altmann, pers. comm. 1973
	Abundant near Marigat, Rift Valley Province (central)	Kuntz and Moore 1973
	Declining rapidly (5–10% annually since 1963) in Masai-Amboseli Reserve; present only locally and declining throughout Kajiado District; about 200–300 in Olnaigon area and 200–300 near Ol Tukai Village	Hausfater, pers. comm. 1974
	Locally dense and perhaps increasing in some areas, elsewhere declining and range shrinking; overall decreasing	Maples, pers. comm. 1974
	Abundant along lower Tana R. (SE)	Andrews et al. 1975
	At Roka (between Kilifi and Malindi); Shimoni and Diani Beach (all in SE)	Maples et al. 1976
Somalia	Along S coast to about 4°–5°N (maps)	Hill 1967a, 1970; Baldwin and Teleki 1972

Table 145. Population parameters of *Papio cynocephalus*

Population Density (individuals/km²)	Group Size X̄	Group Size Range	Group Size No. of Groups	Home Range Area (ha)	Study Site	Reference
40–60					Zambia	Swingland (cited by Ansell, pers. comm. 1974)
	46	12–109	18		NW Zimbabwe	Hall 1965d
		45	1		N Tanzania	Lamprey 1964
		10–30			W Tanzania	Kano 1971a
	80	13–185	15	777–1,554	S Kenya	Washburn and DeVore 1961a
	28.3	8–60	4	518–777	S Kenya	Maxim and Buettner-Janusch 1963
9.65	80	12–185	15		S Kenya	DeVore and Hall 1965
	51.4	16–198	51	2,409	S Kenya	Altmann and Altmann 1970
		30–100			S Kenya	Andrews et al. 1975
		7–37	4		S Kenya	Maples et al. 1976

Papio hamadryas Hamadryas or Sacred Baboon

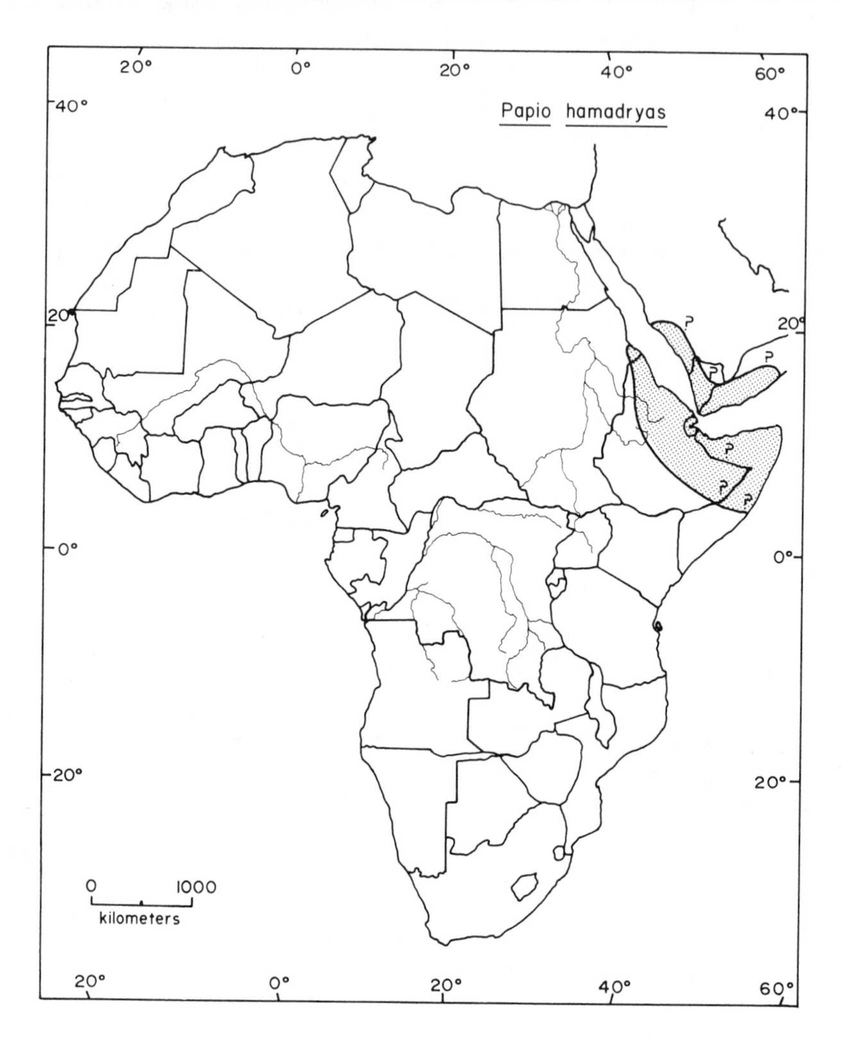

Taxonomic Note

Papio hamadryas is a monotypic species (Napier and Napier 1967). It hybridizes in nature with *Papio anubis* where these two forms are sympatric (Kummer 1968a, Nagel 1973), but is considered a distinct species.

Geographic Range

P. hamadryas is found in northeast Africa and the southwestern part of the Arabian peninsula. It has been reported from as far north as 18°27′N in eastern Sudan (Hill 1970) and 21°40′N in Saudi Arabia (Harrison 1964). The majority of its range is in Ethiopia, where it occurs as far west as about 37°E in the north, and throughout the eastern and central regions (Starck and Frick 1958). Its southern and eastern limits in Somalia are not known. It is found as far east as about 51°E in South Yemen (Harrison 1968). For details of the distribution, see Table 146.

Abundance and Density

P. hamadryas has recently been described as abundant in parts of Ethiopia and the Arabian peninsula, but its status over most of its range is not known (Table 146). Normal population densities are relatively low (Kummer, pers. comm. 1974). Thus overall numbers of this species are thought to be low (Bernahu 1975).

At one study site near Erer-Gota, population density of *P. hamadryas* was 1.8 per km^2 (Kummer, pers. comm. 1974). At the Awash River one group was observed at a density of 3.4 individuals per km^2 (Nagel 1973).

Daytime parties of *P. hamadryas* consist of 6 to 494 members, with average foraging groups of 25 to 38 in poorer habitats and 90 to 227 in food-rich habitats (Kummer 1968a). At night these groups congregate at sleeping cliffs, forming large groups of 12 to 750 animals ($\bar{X}$ = 136, N = 19 groups) (Kummer 1968b). In Saudi Arabia three daytime herds contained 20 to 30, 50, and 70 members (Ellis, pers. comm. 1978).

The home range size for this species is not known, but is thought to be large (Crook and Aldrich-Blake 1968). Day ranges for two groups varied from 9.98 to 19.6 km ($\bar{X}$ = 13.35) in length (Kummer 1968b). Another group had a mean day range of 6.5 km in length (Nagel 1973).

Habitat

The preferred habitat of *P. hamadryas* is arid subdesert steppe with *Acacia* trees and shrubs and *Dobera glabra* (Kummer, pers. comm. 1974). This species also utilizes dry short-grass plains (Bernahu 1975), and alpine grass meadows (Kummer, pers. comm. 1974). In Saudi Arabia it occupies rugged scrubby *Acacia* grass desert and montane grassland (Ellis, pers.comm. 1978).

P. hamadryas has often been observed in lowland areas (Kummer 1968a, Dandelot and Prévost 1972, Bernahu 1975). It also occurs in highland areas in Saudi Arabia (Harrison 1964, 1968; Ellis, pers. comm. 1978), and up to at least 2,300 m elevation in Ethiopia (Kummer, pers. comm. 1974).

Hamadryas baboons sleep on vertical rock faces 15 to 25 m in height (Kummer 1968a). These cliffs are a crucial component of their niche and a limiting factor in their distribution (Kummer, pers. comm. 1974).

Factors Affecting Populations

Habitat Alteration

In Ethiopia the arid habitat occupied by *P. hamadryas* is not yet heavily exploited by human populations. But agricultural development has been promoted by the Ethiopian government, is beginning to encroach on *P. hamadryas* habitat, and is expected to reduce *P. hamadryas* populations sharply (Kummer, pers. comm. 1974).

The cutting of small *Acacia* trees and local overgrazing by cattle are altering *P. hamadryas* habitat to some extent, but have not yet significantly reduced its populations (Kummer, pers. comm. 1974). Kummer estimated that more than 80% of the original range in Ethiopia was still suitable habitat, that the rate of habitat destruction was quite slow, and that since large areas of its range in eastern Ethiopia were inhabited solely by nomadic herdsmen, its habitat there was relatively secure.

In the Asir Mountains of Saudi Arabia much of the land occupied by *P. hamadryas* is severely overgrazed (Ellis, pers. comm. 1978). No information is available regarding habitat alteration in the Sudan, Somalia, Yemen, or South Yemen portions of the range.

Human Predation

Hunting. No author cited hunting for food as a factor affecting *P. hamadryas* populations. In Ethiopia this species is tolerated by the nomadic herdsmen with whom it shares its habitat (Kummer, pers. comm. 1974), and in Saudi Arabia hamadryas baboons are not eaten (Ellis, pers. comm. 1974). Large numbers of *P. hamadryas* are shot by Galla tribesmen, who use the skins for ceremonial cloaks and headdresses and also sell these to tourists (Bernahu 1975). The effect of this hunting on populations is not known.

Pest Control. *P. hamadryas* does raid plantations and eats chaff from grain mills (Kummer 1968a). It is considered a pest to agriculture and is frequently killed by farmers in Ethiopia (Kummer and Kurt 1963, Bernahu 1975). In Saudi Arabia it has been persecuted because of its crop-raiding habits, but has remained plentiful because it retreats to inaccessible mountain refuges (Harrison 1968). It is also reported to raid the city garbage dump at At Ta'if and to steal from gardens (Ellis, pers. comm. 1978).

Collection. Kummer (pers. comm. 1974) reported that trapping of *P. hamadryas* was negligible in Ethiopia, but Bernahu (1975) found that "relatively large numbers" were being exported to Europe and America. Hamadryas baboons are used in several types of medical research in the Soviet Union (Lapin 1965).

Between 1968 and 1973 a total of 394 *P. hamadryas* were imported into the United States (Appendix), primarily from Ethiopia and Tanzania. None entered the United States in 1977 (Bittner et al. 1978). In 1965–75, 1,043 were imported into the United Kingdom (Burton 1978).

Conservation Action

P. hamadryas occurs within two or three national parks in Ethiopia (Table 146). No other conservation actions have been taken for this species.

Table 146. Distribution of *Papio hamadryas* (countries are listed in order from west to east)

Country	Presence and Abundance	Reference
Sudan	Along E coast (map)	Starck and Frick 1958
	At Takar (18°27′N) (locality mapped)	Hill 1970
Ethiopia	Throughout N and E (map)	Starck and Frick 1958
	In E, E of Addis Ababa, from Sendafa to Harer	Kummer and Kurt 1963; Kummer 1968a,b
	Extensive populations in E central and SE (Danakil, Harer, Ogaden) and Simien Mts. in N	Crook and Aldrich-Blake 1968
	In Menagasha N.P. (W of Addis Ababa) (seems to be outside range)	I.U.C.N. 1971
	At Awash (E of Addis Ababa) (map)	Dandelot and Prévost 1972
	In provinces of Tigre, Begemder, Shoa, Harrar, and Bale (all in N or E); localities listed	Bolton 1973
	In Awash N.P. (E of Addis Ababa)	Nagel 1973
	In central, N, and NE (localities mapped)	Dunbar and Dunbar 1974a
	In and near Simien N.P.	Dunbar and Dunbar 1974c
	Abundant but low density between Miesso and Diredawa (E of Addis Ababa); population stable since 1961	Kummer, pers. comm. 1974
	In N and NE; low density	Bernahu 1975
Afars-Issas	Throughout (maps)	Starck and Frick 1958, Dorst and Dandelot 1969
Somalia	Throughout N and central (map)	Starck and Frick 1958
	Throughout; recorded in N at Berbera and Asseh Hills (localities mapped)	Hill 1970
Saudi Arabia	Localities listed and mapped in Red Sea hills (in SW)	Harrison 1964
	In SW from S of Jedda through Asir Mts.; still plentiful in some areas	Harrison 1968
	On E slope of Asir Mts. near Al Fooga, Al Tawah, and Bishah (about 20°N); reported throughout Asir Mts. from At Ta'if (about 21°N) to Khamis Mushayt (18°19′N); widely distributed and not rare	Ellis, pers. comm. 1978
Yemen	Still plentiful in some areas	Harrison 1968
South Yemen	Near Aden (in SW)	Tappen 1960, Harrison 1964
	E of Hadramaut as far E as Wadi Masilah R. (near 51°E); still plentiful in some areas	Harrison 1968

Papio leucophaeus Drill

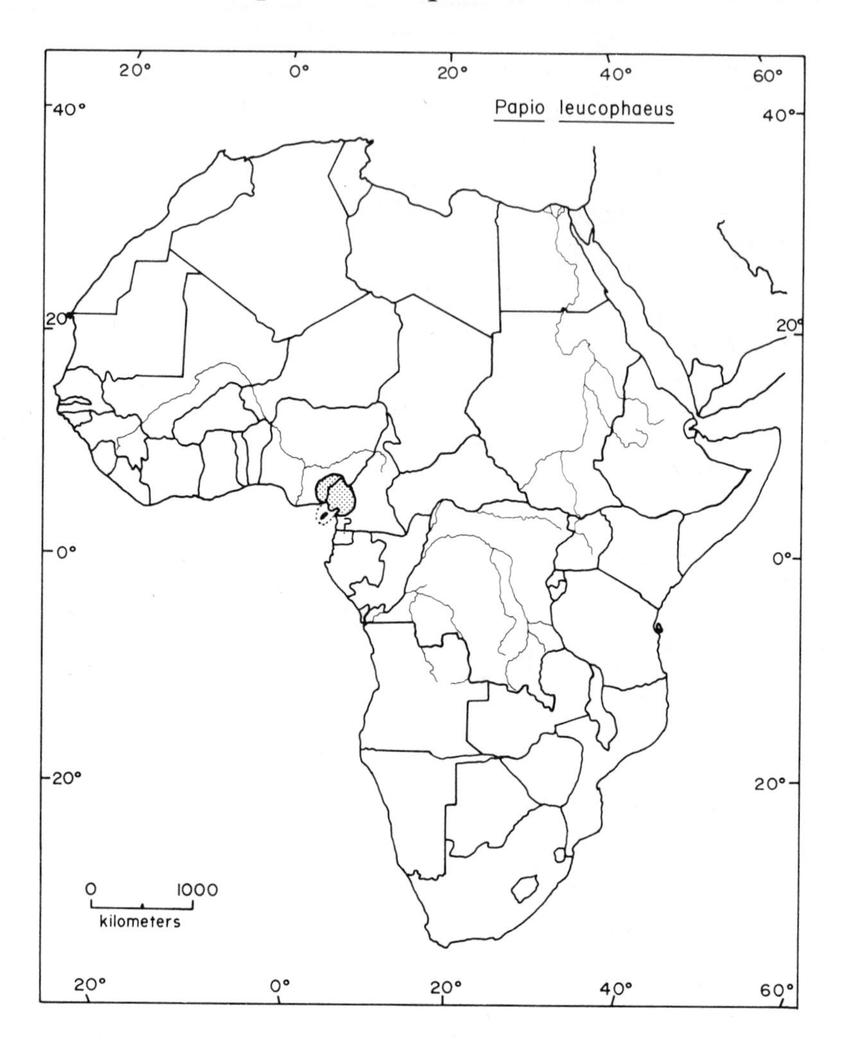

Taxonomic Note

The drill is often placed in the genus *Mandrillus*, but several older authors (Malbrant and Maclatchy 1949, Booth 1958) and recent ones (Thorington and Groves 1970, Gartlan 1975a) consider it a member of the genus *Papio*. Two subspecies have been described (Napier and Napier 1967).

Geographic Range

Papio leucophaeus is found in western Cameroon and eastern Nigeria, and on Macias Nguema (formerly Fernando Po), Equatorial Guinea. Previously its range was described as much more extensive, including central Cameroon, mainland Equatorial Guinea, Gabon, and Congo. But recent evidence indicates that drills do not now inhabit these southern areas and that earlier reports of them there may have been erroneous (Grubb 1973). *P. leucophaeus* has been recorded as far north as 6°17′N, 9°32′E and as far west as 8°25′E (at 5°45′N) (Grubb 1973). Its southern limit is probably the Sanaga River, along which it occurs as far east as about 10°30′E (Gartlan, pers. comm. 1974). For details of the distribution, see Table 147.

Abundance and Density

P. leucophaeus is an endangered species, with restricted and decreasing distribution and numbers. It is considered to be one of the most threatened monkey species in Africa (Struhsaker, pers. comm. 1974). Recent estimates of its total remaining range vary from 38,400 km^2 to 77,600 km^2 (Table 147). No calculations of population density or home range area are available.

Older workers have reported drills in groups of 15 to 20 (Jeannin 1936), up to 40 (Sanderson 1940), and 10 to 30 (Bourgoin 1955) (all cited in Gartlan 1970). More recent studies have revealed group sizes of 9 to 55 ($\bar{X}$ = 23.3, mode 12, N = 12 groups) (Struhsaker 1969), 14 to 179 ($\bar{X}$ = 63, N = 11 groups) (Gartlan 1970), and 20 to 200, some of which may have been temporary aggregations of smaller units (Gartlan, pers. comm. 1974).

Habitat

P. leucophaeus is a forest species found in lowland rain forest, coastal forest, and riverine forest (Gartlan, pers. comm. 1974). It utilizes mature secondary forest but is only infrequently found in young secondary forest (Gartlan and Struhsaker 1972). I.U.C.N. (1978) describes its typical habitat as "hilly, rock-strewn country in coastal, lowland evergreen rainforest."

This species may also utilize the forest-savanna mosaic and montane communities (Hall 1966). It does venture into clearings but never utilizes completely open country away from forest (Gartlan, pers. comm. 1974).

Factors Affecting Populations

Habitat Alteration

The coastal forest habitats of *P. leucophaeus* are especially accessible to loggers and are being clearcut for timber at a fast rate. Because their rich volcanic soil is also desirable for agriculture, the forested areas are being cleared for plantations and used for shifting agriculture. Deforestation is proceeding at a fast rate in Cameroon and is one of the principal causes for the decline of the species there (Gartlan, pers. comm. 1974). The remaining range of drills in Cameroon is "commercially and agriculturally vulnerable coastal forest" (Gartlan 1975c). No direct information is available regarding habitat alteration in eastern Nigeria.

Human Predation

Hunting. Hunting for meat is also threatening *P. leucophaeus* populations (Gartlan, pers. comm. 1974). Its flesh is preferred for its sweet flavor, and its large size (up to 36 kg) makes it a worthwhile target. Drills are often hunted with dogs, treed, and shot in groups of 20 or more; they are also caught in traps on the ground (Gartlan, pers. comm. 1974). Their long prereproductive period (five to six years) makes them especially vulnerable to decimation by hunting, since their population recruitment rate is quite slow (Gartlan, pers. comm. 1974). The present level of hunting is "almost certainly too high to be sustained by such a geographically restricted species without seriously affecting its chances of survival" (Gartlan 1975a).

Pest Control. Drills do enter farms to eat crops such as banana, plantain, and cacao, and are sometimes shot as pests (Gartlan, pers. comm. 1974).

Collection. P. leucophaeus populations are not being reduced by collection for research, zoos, or pets (Gartlan, pers. comm. 1974). Jolly (1966) stated, "there is no reason on present evidence to prefer *Mandrillus* as an experimental animal, while its restricted range and threatened status . . . are arguments encouraging further trade."

Ten *P. leucophaeus* were imported into the United States between 1968 and 1973 (Appendix). None entered the United States in 1977 (Bittner et al. 1978). In 1965–75, 144 were imported into the United Kingdom (Burton 1978).

Conservation Action

P. leucophaeus occurs within one national park and two reserves (Table 147).

The Cameroon government is now aware of the seriously threatened status of *P. leucophaeus* and has encouraged Gartlan's efforts to protect the 125,000 ha Korup Reserve as a preserve for this species. The protection of this area and its proposed upgrading to national park status will greatly aid in drill conservation, but control of deforestation and hunting outside the Korup are also necessary if *P. leucophaeus* is to survive. Improvement of the system for issuing hunting licenses and employment of more game wardens to enforce the hunting laws are essential (Gartlan, pers. comm. 1974).

Petrides (1965) recommended the etablishment of a national park near the Cross River for the protection of drills and their habitat there. He believed that the future of drills in Nigeria was uncertain unless active steps toward preservation were taken immediately.

P. leucophaeus is listed as endangered under the U.S. Endangered Species Act (U.S. Department of Interior 1977b).

Table 147. Distribution of *Papio leucophaeus* (countries are listed in order from north to south)

Country	Presence and Abundance	Reference
Nigeria	In small area E of Calabar (map) (in SE)	Rosevear 1953 cited in Gartlan 1970
	E of Cross R. (map)	Booth 1958
	Populations much reduced; scarce	Petrides 1965
	In SE (localities listed and mapped)	Grubb 1973
	Recorded E of Cross R. between 1966 and 1971	Monath and Kemp 1973
Cameroon	In W corner as far N as Mamfe and as far E as E of Mungo R. (map)	Rosevear 1953 cited in Gartlan 1970
	Throughout S (map)	Booth 1958
	Most abundant in W	Struhsaker 1969
	In S Bakundu Forest Reserve (in W)	Gartlan 1970
	At Idenau and in Bakundu Reserve	Gartlan and Struhsaker 1972
	Abundant in Korup Reserve (in W)	Struhsaker 1972
	Only in W; mainly N of Sanaga R. (localities mapped)	Grubb 1973
	Only in W corner, N of Sanaga R. (map); total range 77,600 km^2; previously locally common, now decreasing rapidly	Gartlan, pers. comm. 1974
	Total range 38,400 km^2; habitat vulnerable	Gartlan 1975c
	Endangered; total range 75,000 km^2; numbers unknown; populations and range being reduced	I.U.C.N. 1978
Equatorial Guinea	On mainland	Malbrant and Maclatchy 1949
	On mainland (map)	Booth 1958
	Not seen on mainland; absent according to local people	Sabater Pi 1972a
	Two records listed from Macias Nguema; none from mainland	Grubb 1973
	No recent evidence from Macias Nguema or mainland	I.U.C.N. 1978
Gabon	Relatively abundant locally	Malbrant and Maclatchy 1949
	In Wonga-Wongue N.P. (in W, W of estuary of Gabon and Ogooué rivers)	Harroy 1972
	Fair-sized populations in Wonga-Wongue N.P.	Curry-Lindahl 1974a
Congo	To Congo R.	Malbrant and Maclatchy 1949

Papio papio Guinea Baboon

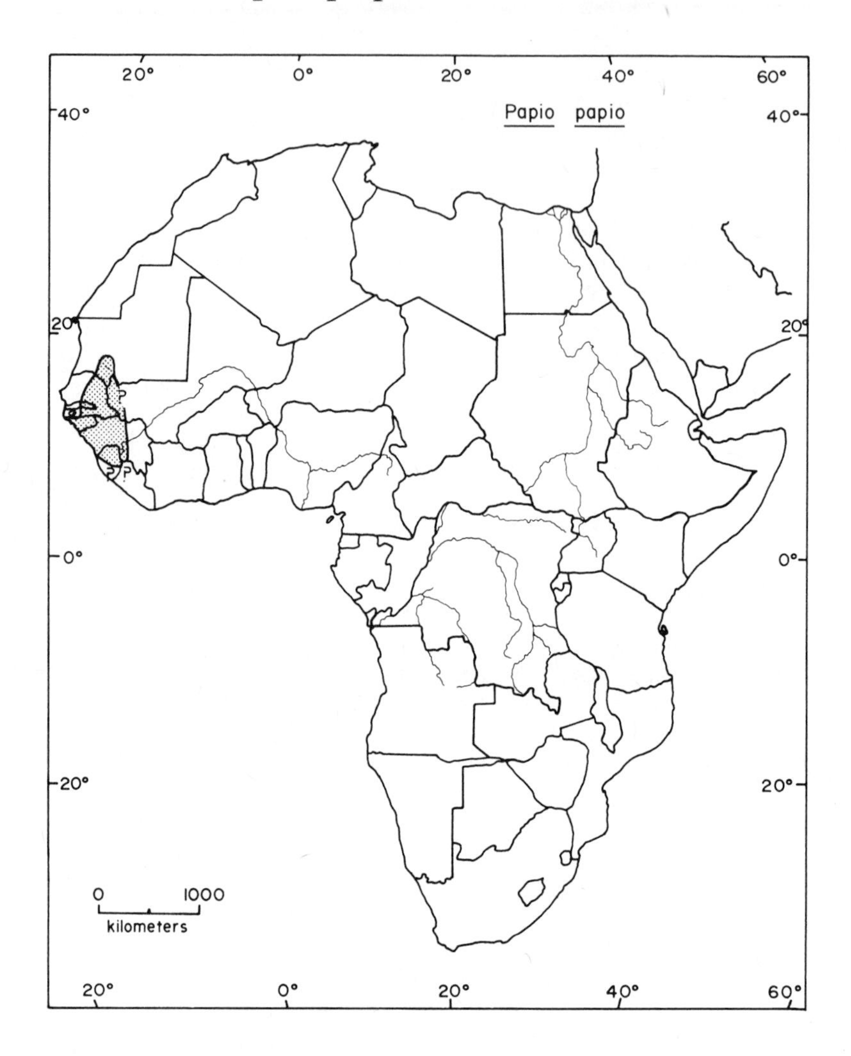

Taxonomic Note

Papio papio is a monotypic species (Napier and Napier 1967).

Geographic Range

P. papio is found in extreme West Africa, from Senegal to Sierra Leone. It has been reported as far north as 18°N in Mauritania (Dekeyser 1955, cited in Hill 1970), and as far west as Seleti, Senegal (about 13°N, 16°30′W), 20 km from the Atlantic Ocean (Boese, pers. comm. 1974). The range may extend east into Mali and south as far as Liberia, but the eastern and southern limits have not been determined. For details of the distribution, see Table 148.

Abundance and Density

P. papio has recently been described as locally abundant in parts of Senegal and Sierra Leone, increasing in the Niokolo-Koba National Park, and decreasing in the Casamance region (Table 148). Its status throughout the rest of its range is not known.

In an area larger than 8,000 km^2 in the Niokolo-Koba National Park, the population density was 12 to 15 *P. papio* per km^2, and troops of more than 500 individuals were not unusual (Dupuy and Verschuren 1977). Other authors have reported groups of 40 to 83 (N = 5 groups) (Dunbar and Nathan 1972) and 10 to 200 individuals ($\bar{X}$ = 69, N = 9 groups), of which two groups had home ranges of 1,036 and 1,554 ha (Boese, pers. comm. 1974).

Habitat

P. papio occupies wooded steppe, woodland, and savanna (Hall 1966, Dupuy 1971b), dry forest, and gallery forest (Bert et al. 1967b). It has also been observed in the daytime in dense bushes and trees, and, at night, sleeping in tall gallery forest (Dunbar and Nathan 1972).

In western Senegal, guinea baboons utilize primarily disturbed evergreen forest, whereas in the east their habitat consists of wooded savanna adjoining agricultural lands (Boese, pers. comm. 1974). In Niokolo-Koba National Park they favor wooded savanna and gallery forests and avoid tall (1 to 2 m) grass, although they utilize short (0.03 to 0.30 m) grass areas (Boese, pers. comm. 1974).

Factors Affecting Populations

Habitat Alteration

Senegal and Gambia are undergoing extensive habitat destruction owing to agricultural expansion, overgrazing by cattle and goats, woodcutting, and burning of woodland and savanna to produce pasture land (Boese, pers. comm. 1974). Although *P. papio* can utilize disturbed forest, secondary growth, clearings, and the short grass which regrows after burning, it does not occupy open grasslands, tall grass, or areas devoid of trees (Boese, pers. comm. 1974).

The conversion of *P. papio* habitat into rice plantations is partially responsible for the disappearance of this species from the Casamance region of Senegal (Dupuy 1971b).

The recent severe subsaharan drought probably reduced *P. papio*'s food and water supplies and increased its competition with man for these limited resources. No information was located regarding the effect of this habitat alteration on *P. papio* populations, nor regarding habitat changes in the other countries within *P. papio*'s range.

Human Predation

Hunting. P. papio is hunted with dogs by the local people in Senegal (Dupuy 1971b), and parts of eastern Gambia, and for sport by tourists in western Senegal

(Boese, pers. comm. 1974). In southern Senegal it is only found in parks or faunal reserves where hunting is prohibited (Boese, pers. comm. 1974).

In the Niokolo-Koba National Park the increase in *P. papio* populations is attributed to the reduction of large predators by hunters (Dupuy and Verschuren 1977).

Pest Control. Kingdon (1971) reported that "it was estimated that one-quarter of the total food production of Gambia was destroyed by baboons in 1948." *P. papio* is also considered a pest to peanut farms in northwestern Sierra Leone (Robinson 1971). In Senegal and Gambia it raids crops but is generally tolerated except during harvest seasons when farmers guard their crops and stores with clubs (Boese, pers. comm. 1974).

Collection. P. papio is used as a laboratory animal in Senegal (Bert, cited in Jolly 1966). Boese (pers. comm. 1974) observed collection for research and predicted that this activity could affect certain populations.

Between 1968 and 1973 a total of 602 *P. papio* were imported into the United States, primarily from Senegal (Appendix). None entered the United States in 1977 (Bittner et al. 1978). In 1965–75, 11,940 were imported into the United Kingdom (Burton 1978).

Conservation Action

In Senegal, *P. papio* is virtually restricted to parks and reserves, and the continued protection of these areas is vital to the survival of the species there (Boese, pers. comm. 1974).

In the Niokolo-Koba National Park, *P. papio* populations are dense, but Dupuy and Verschuren (1977) contend that the density is not excessive and that no attempt should be made to reduce these baboon populations unless a thorough scientific study has demonstrated a need for this. They add that this instance of local overpopulation provides an opportunity for studies of behavioral changes of an increasing population in a limited area, and that the culling of baboons within the national park could have a negative psychological effect on the local people who are now prohibited from killing animals there.

Table 148. Distribution of *Papio papio* (countries are listed in order from north to south)

Country	Presence and Abundance	Reference
Mauritania	At 18°N, 12°E (in S)	Dekeyser 1955 cited in Hill 1970
	In S (map)	Booth 1958
Senegal	Throughout (map)	Booth 1958
	In SE only (map)	Huard 1962
	In E near Dialakoto	Bert et al. 1967a
	Thousands in Niokolo-Koba N.P. (in SE)	Dupuy 1970b, 1971c
	In N, E, and S (map); declining in Casamance region (in S); common in Niokolo-Koba N.P.	Dupuy 1971b
	Near Badi, N border of Niokolo-Koba N.P.	Dunbar and Nathan 1972, Dunbar 1974
	At Seleti (in W); Kolda (central); Dialakoto, Simenti, and Niokolo-Koba (in E) (map); locally abundant but confined to parks and reserves	Boese, pers. comm. 1974
	Upstream from Matam in Senegal Valley	Bourlière et al. 1974
	Throughout Niokolo-Koba N.P.; at least 100,000 there; increasing	Dupuy and Verschuren 1977
Gambia	Throughout (map)	Booth 1958
	In E (maps)	Huard 1962, Dupuy 1971b
	At Diabugu (in E) (map)	Boese, pers. comm. 1974
Mali	In extreme W (map)	Baldwin and Teleki 1972
Guinea-Bissau	Throughout (map)	Booth 1958
Guinea	In W (map)	Booth 1958
Sierra Leone	In N (map)	Booth 1958
	In NW	Robinson 1971
	Ubiquitous and locally extremely common in extreme N (Ducuta, Bombali District, and Sainya Scarp)	Wilkinson 1974
Liberia	In W (map)	Baldwin and Teleki 1972

Papio sphinx Mandrill

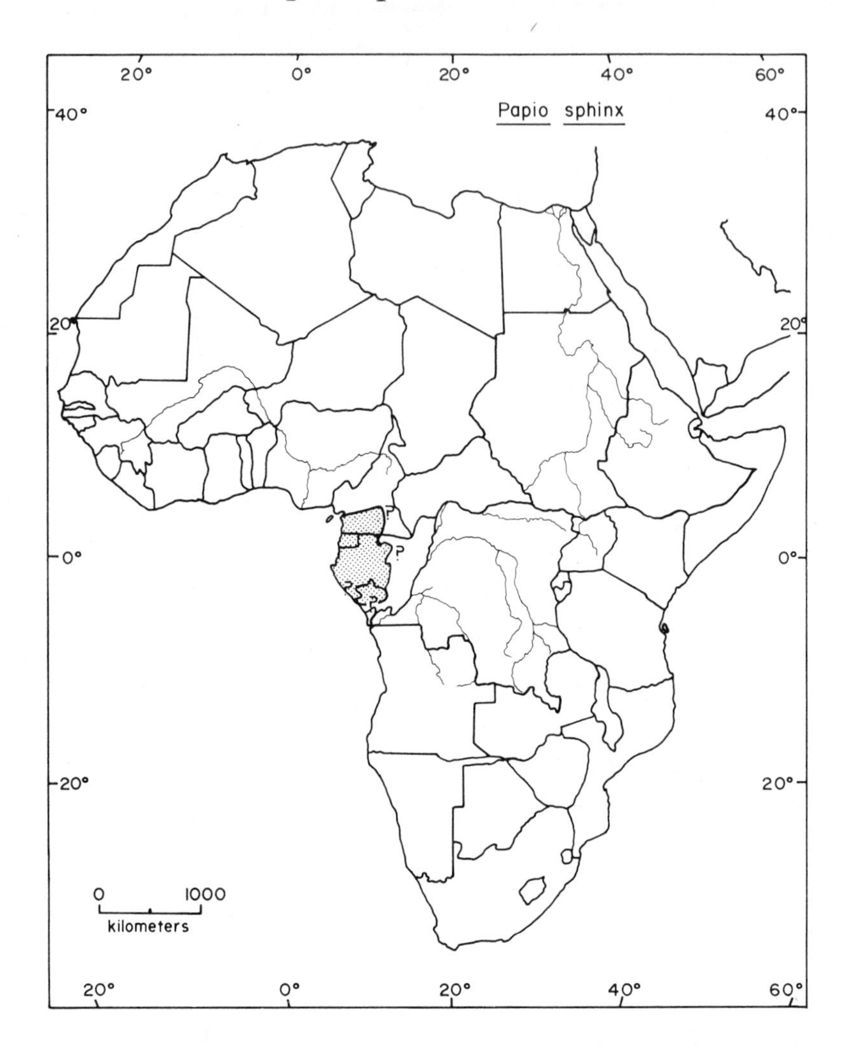

Taxonomic Note

Many authors place the mandrill in the genus *Mandrillus* (e.g., Tappen 1960, Hall 1966, Grubb 1973), but most workers regard it as a member of *Papio* (e.g., Booth 1958, Thorington and Groves 1970, Gartlan 1975c). *Papio sphinx* is a monotypic species (Napier and Napier 1967, Grubb 1973).

Geographic Range

P. sphinx is found in equatorial West Africa, from southern Cameroon to Congo. Its northern limit is the Sanaga River at about 4°N (Gartlan, pers. comm. 1974). It occurs as far west as the Atlantic coast in Cameroon (Gartlan, pers. comm. 1974), Equatorial Guinea (Sabater Pi 1972a), and Gabon (Grubb 1973). *P. sphinx* has been reported as far south and east as 3°12′S, 13°20′E in Congo (Grubb 1973), but the southern and eastward extents of its range are not known. For details of the distribution, see Table 149.

Abundance and Density

P. sphinx is declining in Cameroon and Equatorial Guinea; it is neither abundant nor rare in Gabon, and its status in Congo is not known (Table 149). *P. sphinx* is considered to be among the most threatened monkeys in Africa (Struhsaker, pers. comm. 1974).

No measures of population density are available for mandrills. Group sizes of 15 to 30 members (Jeannin 1936, cited in Hall 1966), up to 50 (Malbrant and Maclatchy 1949), and 20 to 120 (Sabater Pi 1972a) have been reported. In Gabon 29 groups consisted of 2 to 250 individuals ($\bar{X} = 95$); two of the groups had home ranges of about 4,000 and 5,000 ha (Jouventin 1975). In Equatorial Guinea two groups occupied home ranges of 1,000 ha each (Sabater Pi 1972a).

Habitat

Mandrills are forest monkeys, found in rain forest and coastal forest (Gartlan, pers. comm. 1974). In Equatorial Guinea they are especially common in regenerating forest (Sabater Pi and Jones 1967), thick brush, and secondary forest (Sabater Pi 1972a). In Gabon, mandrills have been recorded in primary rain forest, secondary forest, gallery forest, flooded primary forest, plantations, primary forest on the border of savanna, and very rarely in the savanna itself (Jouventin 1975).

P. sphinx has been seen in low elevation river basins and flat, mountainous, and plateau terrains (Jouventin 1975). It also occurs in fairly open areas on domed peaks and in meadows (Sabater Pi and Jones 1967).

Factors Affecting Populations

Habitat Alteration

The forests in which mandrills live are being logged for timber, cleared for plantations and shifting agriculture, and cut for logging access roads. In Cameroon the coastal forests are being destroyed especially rapidly because of their accessibility for logging and their fertile soil (Gartlan, pers. comm. 1974). In Equatorial Guinea the range of *P. sphinx* had already been significantly reduced by deforestation by 1966 (Sabater Pi and Jones 1967), and has now undoubtedly been further reduced.

Although *P. sphinx* can utilize secondary and brush vegetation and can probably tolerate limited clearing of forest, it does not thrive in areas of intensive cultivation (Sabater Pi and Jones 1967) nor occupy large deforested areas. If the human population in Gabon continues to expand at its present rate, so much of the forest will be destroyed for agriculture that soon no forested areas will remain for *P. sphinx* in Gabon (Jouventin 1975).

Human Predation

Hunting. Local people hunt mandrills for food in Equatorial Guinea, Cameroon, and Gabon (Sabater Pi 1972a; Gartlan, pers. comm. 1974; Jouventin 1975). In Equatorial Guinea they are captured in traps, shot with poisoned arrows, and hunted with dogs (Sabater Pi 1972a). Mandrill flesh is highly regarded, and 20% of the Fang people interviewed considered it superior to all other meat, although only 4% ate it more frequently than any other meat, probably because it has become so rare, owing to heavy hunting (Sabater Pi and Groves 1972).

Pest Control. P. sphinx does eat crops, and has been very destructive to plantations in Gabon (Jouventin 1975). Sabater Pi (1972a) reported that 20 out of 101 of his captures were made in cassava or banana fields. In Cameroon, *P. sphinx* is shot while feeding in farms (Gartlan, pers. comm. 1974).

Collection. P. sphinx is sometimes exported from Cameroon, and is in demand for zoo exhibits because of its striking coloration (Gartlan, pers. comm. 1974). A total of 237 were exported from Equatorial Guinea by two dealers between 1958 and 1968 (Jones, pers. comm. 1974). Jolly (1966) argued against the use of *P. sphinx* in laboratory research because of its restricted range and threatened status.

Between 1968 and 1973, 69 *P. sphinx* were imported into the United States (Appendix). Many of these came from Tanzania, where *P. sphinx* does not occur. None entered the United States in 1977 (Bittner et al. 1978). In 1965–75, 719 were imported into the United Kingdom (Burton 1978).

Conservation Action

Mandrills are found in at least one national park and two reserves (Table 149), and may also occur in the Dja Forest Reserve (Gartlan, pers. comm. 1974).

P. sphinx is listed as endangered under the U.S. Endangered Species Act (U.S. Department of Interior 1977b).

Table 149. Distribution of *Papio sphinx* (countries are listed in order from north to south)

Country	Presence and Abundance	Reference
Cameroon	S of Sanaga R. (map)	Booth 1958
	In SW, S of Sanaga R. as far E as 12°20′E (localities listed and mapped)	Grubb 1973
	In S, S of Sanaga R. near mouth and S of 4°N farther E (map); in Dipikar Reserve (in SW); may extend to E border; not known if in SE corner; declining	Gartlan, pers. comm. 1974
Equatorial Guinea	Throughout (map)	Booth 1958
	Throughout (map); locally abundant; range shrinking	Sabater Pi and Jones 1967
	Throughout (localities mapped); relatively scarce	Sabater Pi 1972a
	Rare	Sabater Pi and Groves 1972
	Throughout (localities mapped)	Grubb 1973
Gabon	Throughout	Malbrant and Maclatchy 1949
	In Wonga-Wongue N.P. (in W, S of estuary of Gabon R. and Ogooué R., near equator)	Harroy 1972
	In N and central (localities mapped)	Grubb 1973
	"Fair-sized population" in Wonga-Wongue N.P.	Curry-Lindahl 1974a
	On Mpassa Plateau, 10 km SW of Makokou (in NE)	Gautier-Hion and Gautier 1974
	In N and central (localities mapped); nowhere abundant nor rare	Jouventin 1975
Congo	Throughout	Malbrant and Maclatchy 1949
	At Mbila 3°12′S, 13°20′E (in SW); probably absent in N	Grubb 1973
	In Niari Valley Faunal Reserve (in SW)	Bassus 1975

Papio ursinus Chacma Baboon

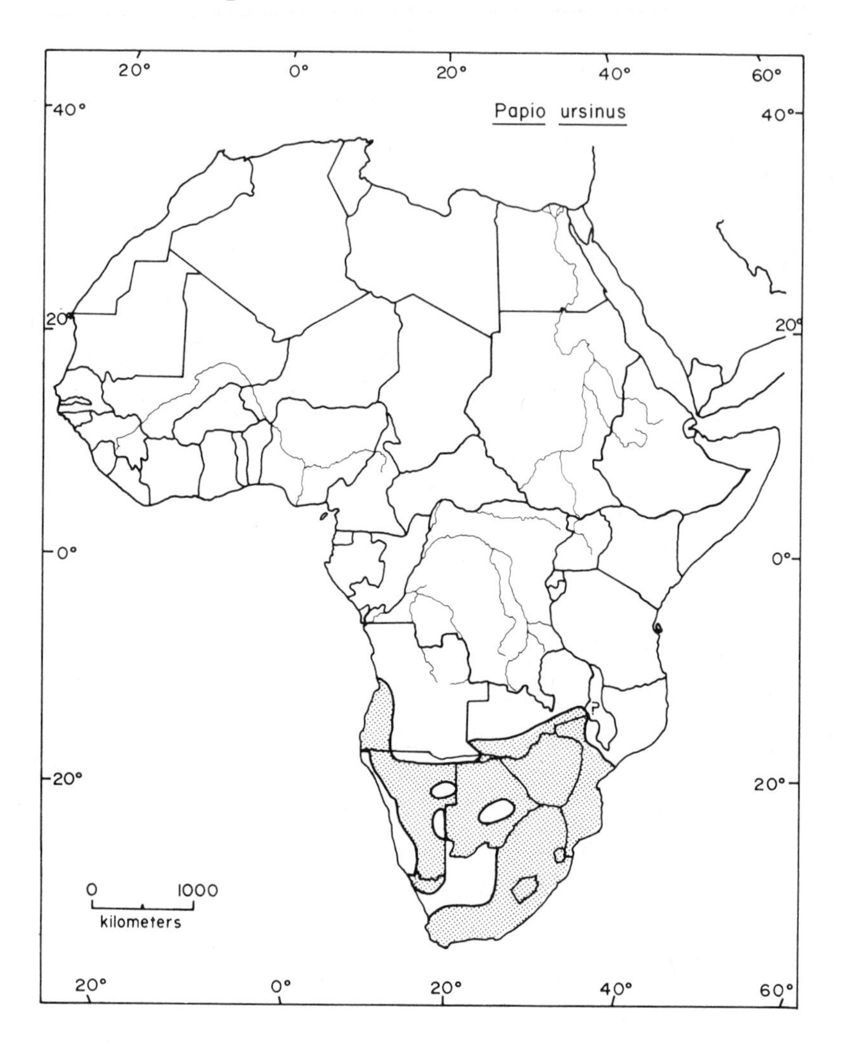

Taxonomic Note

Some authors classify the chacma baboon as a race of *Papio cynocephalus* (Hall 1965c, Kingdon 1971). But most workers consider *Papio ursinus* to be a separate species, and it is treated as such here. Napier and Napier (1967) recognize four subspecies, while Hill (1970) lists eight.

Geographic Range

P. ursinus occurs across southern Africa from the Atlantic coast to the Indian Ocean. It is found as far north as 11°30′S in western Angola (Barros Machado 1969) and about 13°S in eastern Zambia (Smithers 1966), and as far south as the Cape of Good Hope (Hall 1965c). For details of the distribution, see Table 150.

Abundance and Density

P. ursinus has recently been described as locally abundant in Namibia, Mozambique, Zimbabwe, and South Africa, but is also reported to be declining in parts of the last two countries and Botswana (Table 150). No information was located regarding its status in Zambia, Malawi, or Angola.

Estimates of population density vary from 2.31 to 43.2 individuals per km^2 (Table 151). Group sizes range from 7 to 128, with most studies revealing a mean group size of 20 to 50 members. Groups occupy home ranges of 210 to 3,367 ha (Table 151).

Habitat

P. ursinus is most typical of woodland habitat (Ansell 1960a; Dalquest 1965; Hall 1966; Smithers 1966; Pienaar 1967; Attwell, pers. comm. 1974). It is frequently found in mopane woodland, and has also been recorded in other woodland types dominated, for example, by *Combretum* or *Tamarindus* (Jarman 1972, Stoltz and Keith 1973, Buskirk et al. 1974, Wilson 1975). It also occupies riparian forests (Pienaar 1967, Coetzee 1969, McConnell et al. 1974, Sheppe and Haas 1976).

Chacma baboons also utilize more open habitats, including subdesert steppe, highland grassland (Hall 1966), and savanna, but less frequently than woodlands (Smithers 1966, Pienaar 1967, Nel and Pretorius 1971, Sheppe and Osborne 1971). Among the open habitats in which they are found are various types of veld and karoo (Stoltz and Keith 1973, Lynch 1975).

Chacma baboons seem to have a wide habitat tolerance. They are found at temperatures and humidities ranging from that of near desert conditions (Hamilton et al. 1976) and arid thornveld to Mediterranean climates (Hall 1962). They are found from sea level to mountains up to 3,000 m in elevation, where snow and nighttime temperatures of 0°C often occur (Hall 1966).

P. ursinus cannot survive without regular access to water and may migrate or congregate to obtain needed moisture (Hall 1966). In many areas, water sources are important determinants of *P. ursinus* distribution and movements (Smithers 1968, Stoltz and Saayman 1970, McConnell et al. 1974, Sheppe and Haas 1976).

This species often sleeps on rocky ledges, boulders, or cliffs (Hall 1963, Pienaar 1967, Stoltz and Saayman 1970, Jackson 1973a, McConnell et al. 1974, Hamilton et al. 1976). Thus, rocky outcrops, hills, canyons, or mountains are often listed as typical components of *P. ursinus* habitat (Ansell 1960a, Smithers 1966, Coetzee 1969). In other areas chacma baboons sleep in trees (Hall 1963, Pitchford et al. 1973, Hamilton et al. 1976). The availability of sleeping sites represents a limiting factor in their distribution (Stoltz and Saayman 1970).

Several authors have attempted to find correlations between various measures of habitat quality and population density in *P. ursinus*. They have predicted that groups should be smaller in food-scarce habitats at higher altitudes than in richer, lower areas (Hall 1963), and smaller in drier habitats than in lush ones (DeVore and Hall 1965). Recent studies have revealed that smaller groups and larger home ranges do occur in areas with low rainfall and low food plant diversities (Stoltz and Keith 1973; Joubert and Coetzee, pers. comm. 1974; Wilson 1975). The location of water supplies is also an important determinant of home range size in *P. ursinus* (Stoltz and Saayman 1970).

Factors Affecting Populations

Habitat Alteration

P. ursinus is very adaptable, and habitat alteration is not having a seriously detrimental effect on *P. ursinus* populations in most regions of the range. Although farming activities are reducing chacma baboon habitats, by 1968 they had not yet exerted heavy pressure on populations in the Transvaal (Stoltz and Saayman 1970). In Zimbabwe, *P. ursinus* can survive in "areas of intensive agricultural development, provided there are adjacent hills or undeveloped woodland to afford cover" (Smithers, pers. comm. 1974). In Namibia, 100% of the chacma baboon's original range is still suitable habitat (Joubert and Coetzee, pers. comm. 1974). Hall (1962) reported that *P. ursinus* was surviving "without notable diminution in numbers, in spite of human encroachment."

In some areas human activities have improved the quality of the habitat for *P. ursinus*. For example, artificial water sources such as cattle drinking troughs, irrigation canals, and reservoirs are utilized by chacma baboons in Namibia (Hall 1965c; Joubert and Coetzee, pers. comm. 1974) and South Africa (Stoltz and Saayman 1970). And in northern Transvaal the availability of vegetables and water on farms results in *P. ursinus* troops remaining permanently in otherwise marginal habitats (Stoltz and Keith 1973).

In Cape Province, South Africa, *P. ursinus* has adapted well to introduced conifer forests and feeds extensively on seeds from pine cones (Hall 1965c). In the Cape of Good Hope Nature Reserve these baboons have become so dependent on exotic plants that a planned program to eradicate the plants could remove a significant source of *P. ursinus* food (Davidge 1976).

Human Predation

Hunting. No information.

Pest Control. *P. ursinus* is considered a major pest to cultivation in Zambia (Ansell 1960a) and Zimbabwe (Attwell, pers. comm. 1974), and its distribution is becoming localized as it is eliminated in many parts of these two countries (Smithers 1966). In South Africa it is especially destructive to citrus and tomato crops (Stoltz and Saayman 1970), and farmers retaliate by building fires under the baboons' sleeping trees and shooting large numbers of them as they flee (Stoltz and Keith 1973). Their wasteful method of feeding often leads to much more damage in a field or orchard than what they actually eat (Bolwig 1959).

Chacma baboons also prey on livestock, evoking control measures by ranchers. Dart (1963) collected evidence of the magnitude of this damage in South Africa and received reports of 522 lambs lost by four farmers in a one-year period. One of these farmers had trapped 213 baboons in four months in efforts to protect his livestock. In 1967, baboon predation on lambs in the Transvaal was especially frequent during a severe drought, and included the killing of adult as well as young goats and sheep (Stoltz and Saayman 1970).

In Namibia, *P. ursinus* is also regarded as a predator of lambs (Joubert and Mostert 1975), but baboon populations seem to be remaining stable even in regions where they are killed to protect livestock (Joubert and Coetzee, pers. comm. 1974). Hall (1962) believed that pest control programs of poisoning, trapping, and shooting had not reduced *P. ursinus* populations in southern Africa, and that this species became sufficiently wary to make control measures only moderately effective (Hall 1966). More recently, *P. ursinus* populations in Natal have been "greatly reduced due to their depredation on crops and small stock" (Pringle 1974) and have been "exterminated as vermin in many parts of South Africa" (Doyle and Bearder 1977).

Chacma baboons also kill chickens and other small animals in Mozambique (Dalquest 1965). Dalquest believed that the abundance of baboons there was greater than normal, probably because of the reduction of leopards and lions by poachers. He suggests that control of poaching would allow these predators to increase, thus limiting baboon populations and their damage to livestock. One rancher in South Africa refused to permit baboons to be shot on his land, because he believed that leopards preferred baboons to calves (Stoltz and Keith 1973).

P. ursinus is also trapped by forestry officials because of the damage it does to pine plantations (Robinson 1976). It is suspected of destructiveness to nesting bird populations, but no solid evidence supports this suspicion (Bourquin et al. 1971). In populated areas it is often a nuisance, raiding garbage dumps and campsites and begging food from tourists (De Vos et al. 1973; Joubert and Coetzee, pers. comm. 1974).

In some of the parks of South Africa, chacma baboons become dependent on food obtained from humans, and may become bold and aggressive (McConnell et al. 1974). Over a five-year period, 54 dangerous baboons were destroyed near one camp area in Kruger National Park (McConnell et al. 1974). In one case *P. ursinus* populations declined because of a "policy of indiscriminate destruction carried out by the warden of the reserve . . . in an effort to reduce the damage caused by baboons to the cars of visitors" (Hall 1963). A concern for public safety could lead to the extermination of *P. ursinus* in the Cape of Good Hope Nature Reserve (Davidge 1976).

In addition, close contact between *P. ursinus* and humans leads to the transmission of diseases between the two species (McConnell et al. 1974).

Collection. *P. ursinus* is used in laboratory research in South Africa (Gilbert and Gillman 1951, Doyle and Bearder 1977). Only one individual was imported into the United States between 1968 and 1973 (Appendix), and none entered the United States in 1977 (Bittner et al. 1978). In 1965–75, 1,069 were imported into the United Kingdom (Burton 1978).

Conservation Action

P. ursinus occurs in at least nine national parks and twelve reserves (Table 150), although in some of these its populations are being reduced (see above).

Its export from Namibia is totally prohibited (Joubert and Coetzee, pers. comm. 1974). In 1967 one export permit was granted in Zimbabwe to a game rancher (Attwell, pers. comm. 1974), but no data regarding the number exported were located.

Table 150. Distribution of *Papio ursinus* (countries are listed in clockwise order)

Country	Presence and Abundance	Reference
Zambia	Throughout S; in E as far N as Fort Jameson District Plateau; common, locally very common	Ansell 1960a
	Locally throughout S, N to about 13°N in E (map)	Smithers 1966
	At Kafu Flats, Zambezi R. (in S)	Sheppe and Osborne 1971
Malawi	Present (but not shown on map)	Smithers 1966
Mozambique	Very abundant along Save R., midway between coast and W border	Dalquest 1965
	Extremely common in Chimanimani Mts. (in W), and W of Mucrera R.	Jackson 1973a,b
	Throughout S and central, N to Zambezi R. (localities mapped)	Smithers, pers. comm. 1974
	In Pomene Game Reserve, SE coast	Tinley et al. 1976
Zimbabwe	Near Lake Kariba and on S bank of Zambezi R. (in N)	Hall 1963
	Throughout (map)	Child and Savory 1964
	Locally throughout (map)	Smithers 1966
	In Mana Pools Game Reserve, middle Zambezi Valley (in N)	Jarman 1972
	Extremely common in Chimanimani Mts. N.P. (in E)	Jackson 1973a
	Widespread, locally exterminated	Attwell, pers. comm. 1974
	Widespread, locally extirpated (localities mapped)	Smithers, pers. comm. 1974
	Throughout Wankie N.P. (in W); locally common	Wilson 1975
Botswana	Widely distributed throughout except in center (localities mapped)	Smithers 1968
	In Okavango Swamp, NW of Maun (in N)	Buskirk et al. 1974, Hamilton et al. 1976
	Abundant along lower Chobe R., Chobe N.P. (in N); seems to have decreased since 1966	Sheppe and Haas 1976
South Africa	In Kruger N.P. (in NE)	Bolwig 1959
	In Cape of Good Hope Nature Reserve; very common	Hall 1962
	Also in E Cape Province and Drakensberg Mountains (in SE)	Hall 1963
	In Ndumu Game Reserve (NE border of Natal)	Paterson et al. 1964
	From coast of Cape Peninsula intermittently E to Natal	Hall 1965c
	Common in Kruger N.P.	Pienaar 1967
	High density in Honnet Nature Reserve, N Transvaal	Stoltz and Saayman 1970

Table 150 (continued)

Country	Presence and Abundance	Reference
South Africa (cont.)	Very common throughout Hluhluwe and Umfolozi game reserves and corridor (E Natal)	Bourquin et al. 1971
	Widespread and abundant in Mountain Zebra N.P. (in S near 32°S, 25°E)	Nel and Pretorius 1971
	In Kruger N.P.	Saayman 1971
	In Aughrabies Falls N.P. (along Orange R. at 20°E)	Harroy 1972
	Locally excessively common in Kruger N.P.	De Vos et al. 1973
	Several thousand in Kruger N.P.; throughout park	Pitchford et al. 1973
	Numerous N and E of Soutpansberg Mts., N Transvaal; population total: 27,348; of local farmers questioned, 50% reported population increasing, 37% remaining stable, 13% decreasing	Stoltz and Keith 1973
	Total population 5,100 in Kruger N.P.	McConnell et al. 1974
	Throughout Natal (localities mapped); previously widely distributed and numerous; now greatly reduced outside protected areas, increasing inside, especially in game reserves; S into E Cape Province	Pringle 1974
	In Orange Free State, primarily in mountains of E (localities mapped)	Lynch 1975
	Isolated population of 150 in Cape of Good Hope Nature Reserve	Davidge 1976
	In Golden Gate Highlands N.P., NE Orange Free State	Rautenbach 1976
	Widespread and abundant in Tsitsikama N.P. (in S)	Robinson 1976
Swaziland	Present	Pringle 1974
Lesotho	Present	Pringle 1974
Namibia	Within 129 km radius of Windhoek (central)	Hall 1962, 1963
	On plateau E of Namib Desert	Coetzee 1969
	Throughout central plateau and Caprivi Strip (NE) (map); total population 9,800; abundant and stable; in 5 game reserves and 1 N.P.	Joubert and Coetzee, pers. comm. 1974
	Throughout central and E, escarpment and plateau and isolated areas of Kavongo (map); total population about 30,000	Joubert and Mostert 1975
	Small isolated population along Kuiseb R., SE of Walvis Bay (in W)	Hamilton et al. 1976
Angola	W of Cunene R. (in SW)	Hall 1966
	In SW only, N to about 11°30′S (localities mapped)	Barros Machado 1969

Table 151. Population parameters of *Papio ursinus*

Population Density (individuals /km²)	Group Size X̄	Range	No. of Groups	Home Range Area (ha)	Study Site	Reference
		40–60	3		S Zambia	Sheppe and Osborne 1971
		10–100			Mozambique	Dalquest 1965
	46	12–109	18		N Zimbabwe	Hall 1963
		≤100+			Zimbabwe	Smithers 1966
	<25				Zimbabwe	Jackson 1973a
		15	1		Zimbabwe	Jackson 1975
	23–49	15–85	97		W Zimbabwe	Wilson 1975
		10–70			Botswana	Smithers 1968
	81.3	67–109	4		N Botswana	Buskirk et al. 1974
16.8–43.2	79.3	7–128	7	210–650	N Botswana	Hamilton et al. 1976
		≤15			N Botswana	Sheppe and Haas 1976
		8–68	31		South Africa and Namibia	Hall 1962
	33.5	15–80	15	1,424–3,367	South Africa	Hall 1963
	34.9	8–109	53		South Africa, Namibia, and Zimbabwe	Hall 1963
		≤100+			E South Africa	Pienaar 1967
	49	24–77	6	1,295–2,331	NE South Africa	Stoltz and Saayman 1970
	25–30				E South Africa	Bourquin et al. 1971
	47.8	33–60	4		NE South Africa	Saayman 1971
2.31	40	8–118	161	1,813	NE South Africa	Stoltz and Keith 1973
	19	<10≥100	273		NE South Africa	McConnell et al. 1974
		150	4	500	S South Africa	Davidge 1976
	27	8–65	20		Central Namibia	Hall 1963
		15–70		1,500	Namibia	Joubert and Coetzee, pers. comm. 1974
		12–35	3		W Namibia	Buskirk et al. 1974
	≤40				W Namibia	Joubert and Mostert 1975
3.5–7.0		12–35	3	402–941	W Namibia	Hamilton et al. 1976

Presbytis aygula-melalophus Group Banded Leaf-Monkey

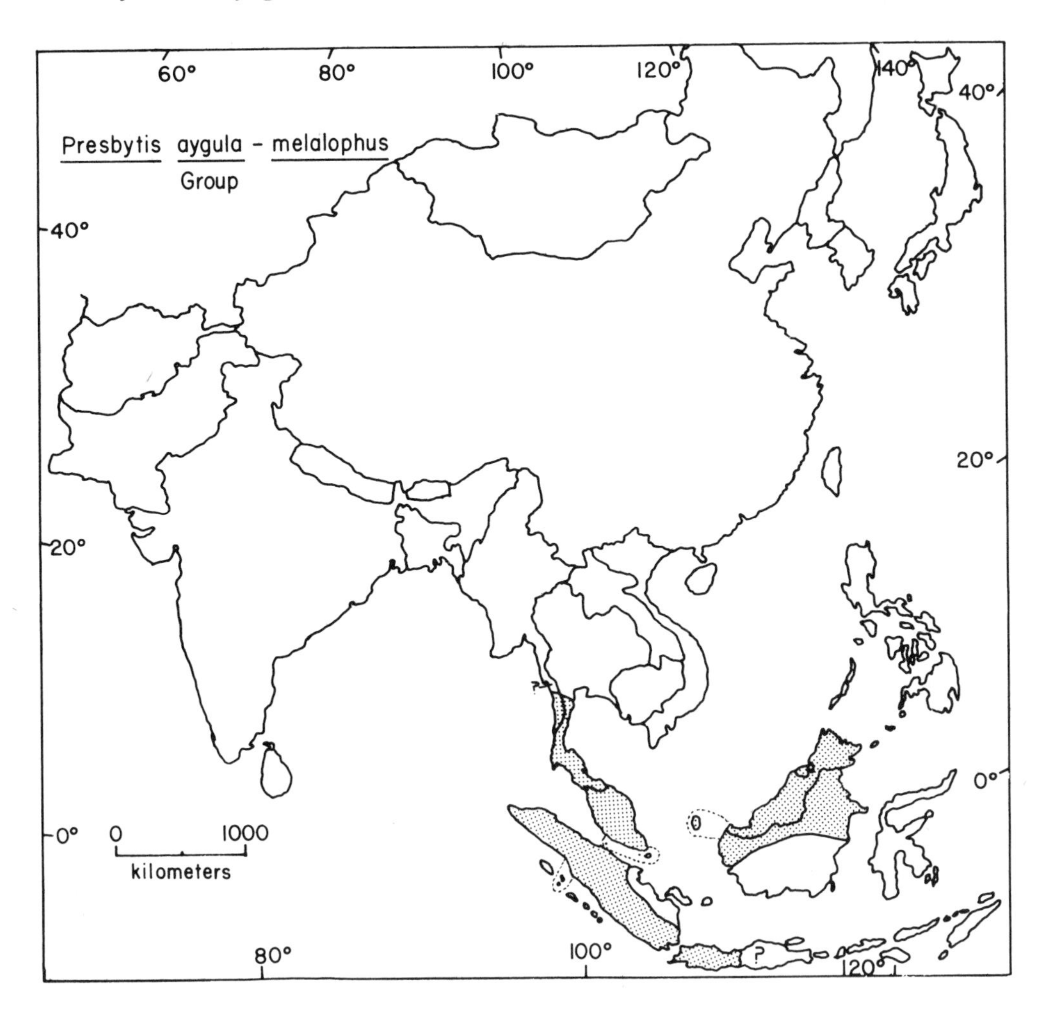

Taxonomic Note

This is a large and varied group and taxonomists do not agree on how many species constitute it. For example, some include *P. rubicundus* and *P. frontata* in the group (Davis 1962, Napier and Napier 1967), and list *P. hosei* as a species (Davis 1962). Here, Medway (1970) is followed: the 24 forms classified by other authors into the species *P. melalophus*, *P. aygula*, *P. femoralis*, *P. thomasi*, and *P. hosei* are considered together as members of the *Presbytis aygula-melalophus* group and *P. rubicundus* and *P. frontata* are treated as separate species.

Geographic Range

Banded leaf-monkeys are found in peninsular Burma and Thailand, the Malaysian peninsula, Sumatra, Java, northern Borneo, and several of the small Indonesian islands. They occur as far north as about 13°N in Tenasserim, Burma, and southern Thailand, and as far west as 97°10′E (at 4°05′N) on Sumatra (Medway 1970, Fooden 1976). Their range extends east to northeastern Borneo (about 5°N, 119°E) and south to about 7°S, 110° E on Java (Medway 1970). For details of the distribution, see Table 152.

Abundance and Density

Banded leaf-monkeys have recently been described as locally common in peninsular Malaysia and Sumatra, but they have declined 42.4% in Malaysia since 1958, are also declining in western Java, and are suspected to be declining in Sumatra (Table 152). They are relatively uncommon in southern Thailand, and their status in Burma and Borneo is not known (Table 152).

Population density estimates for *P. aygula-melalophus* range from 3.8 to 64 individuals per km^2 (Table 153). Medway and Wells (1971) calculated the density in a small (200 ha) area near the Krau Game Reserve, and concluded that the area needed to support 5,000 banded leaf-monkeys would be 250 km^2. They proposed this as the minimum size for a reserve for this species.

Average group sizes for *P. aygula-melalophus* are usually between 8 and 13 individuals, although groups of 2 to 20 have been reported (Table 153). Home range estimates vary from 9 to 30 ha (Table 153).

Habitat

P. aygula-melalophus is a rain forest species, able to utilize both primary and secondary forest. In Thailand it has been observed in evergreen forest including dipterocarp (Fooden 1976) and in inland and coastal forest and coastal swamps (Lekagul and McNeely 1977). On the Malaysian peninsula it occurs in all types of forest and woodland (Medway 1969a) and was recorded most often, 79% of all groups, in secondary forest and 21% in primary forest (Southwick and Cadigan 1972), but probably occurs at higher population densities in undisturbed rain forest than in secondary forest (Curtin 1976). It has been described as confined to inland forest (Medway 1970), but has also been reported as common in coastal lowland forest and coastal swamps as well as inland primary forest (Lim 1969).

In peninsular Malaysia, *P. aygula-melalophus* is said to be common at all elevations (Medway 1969a). Some workers have found it more common in lowland areas (below 610 m elevation) than in montane habitats (McClure 1964, Harrison 1969). But it has been observed in submontane dipterocarp forest (McClure 1964), and as high as 2,042 m (Medway 1970).

In Sumatra, banded leaf-monkeys are in rain forest and pockets of jungle within savanna, near fields, and along rivers, with the highest population densities between 250 and 500 m above sea level, and none above 1,200 m (Kurt 1973). In another study the highest densities were found in lowland (0 to 458 m) primary forest, selectively logged forest, and rubber groves, with large numbers also in primary hill forest and scrub grassland (both between 458 and 915 m) (Wilson and Wilson 1977). Other habitats in which *P. aygula-melalophus* often occurred included villages, secondary lowland forests, and riverine swamps (Wilson and Wilson 1977). These monkeys did not occupy coastal saline swamps but did utilize coastal hillside forest (Wilson and Wilson, pers. comm. 1974).

Bornean *P. aygula-melalophus* occurs in inland lowland rain forest (Kawabe and Mano 1972) and undisturbed primary forest (Wilson and Wilson 1975a). Banded leaf-monkeys have been reported by older authors as high as 1,220 m in Borneo, 1,000 m in Java, and 2,225 m in Sumatra (cited by Medway 1970).

In summary, banded leaf-monkeys utilize both primary and secondary inland forests and in some areas also occupy coastal forest. In peninsular Malaysia and Sumatra they are seen more often in lowland than highland areas, although they have been recorded as high as 2,225 m elevation.

Factors Affecting Populations

Habitat Alteration

Malaysian banded leaf-monkeys utilize rubber estates, plantations (Medway 1969a), and secondary forest (Southwick and Cadigan 1972, Curtin 1976). Nevertheless, their populations are declining because of widespread clearcut logging and clearing of forest for agriculture (Curtin, pers. comm. 1978; Khan 1978). In 1970 clearcutting was proceeding at the rate of 20,000 ha per year on the Malaysian peninsula, and was a potential threat to the habitat of this species (Southwick and Cadigan 1972).

Similarly, in Sumatra *P. aygula-melalophus* can occupy selectively logged areas, secondary forest, and rubber plantations (Wilson and Wilson 1977). But it cannot survive in totally deforested areas, and since large sections of western Sumatra have already been deforested, and slash and burn clearing for agriculture is proceeding rapidly in the southern part of the island, *P. aygula-melalophus* will probably decline drastically there in the near future (Wilson and Wilson, pers. comm. 1974).

Even selective logging seems to have an adverse effect on banded leaf-monkeys, and forests in which about half the trees have been felled support lower population densities than undisturbed forests do (MacKinnon 1973).

No information was located regarding the effects of habitat alteration on *P. aygula-melalophus* in Burma, Thailand, or Borneo.

Human Predation

Hunting. P. aygula-melalophus was formerly heavily hunted by local people in Sarawak (Malaysian North Borneo) for certain concretions from it alimentary tract used in medicinal preparations, but by the early 1960s this hunting was no longer occurring (Burgess 1961). It is still hunted in peninsular Malaysia by aborigines using blowpipes and poison arrows, but this hunting probably has no significant impact on populations (Curtin, pers. comm. 1978). No reports of hunting pressure in Burma, Thailand, or Indonesia were located.

Pest Control. Banded leaf-monkeys have been described as "culture followers" and are responsible for a large amount of damage to cultivated fields in Sumatra (Kurt 1973). They feed in mature rubber trees, rubber nurseries, rice fields, and other croplands (Wilson and Wilson 1975b). They are often chased away from fields, but because guns are rare in Sumatra, the monkeys are not often killed as pests (Wilson and Wilson, pers. comm. 1974). This species also raids crops in peninsular Malaysia and is shot whenever farmers can obtain guns (Curtin, pers comm. 1978). Its damage to rubber trees there is not considered serious (Khan 1978).

Collection. Since leaf-monkeys do not survive well in captivity (Jolly 1966; Curtin, pers. comm. 1978), they are not in demand for the export trade or the pet markets (Khan 1978). Three banded leaf-monkeys were imported into the United States between 1968 and 1973 (Appendix). Ten were imported into the United Kingdom between 1965 and 1975 (Burton 1978). None entered the United States in 1977 (Bittner et al. 1978).

Conservation Action

P. aygula-melalophus is found in five reserves (Table 152). In peninsular Malaysia the Game Department does an excellent job of managing reserves and protecting them from poachers, but these areas are under continual economic pressure from logging interests (Curtin, pers. comm. 1978).

Banded leaf-monkeys are classified as protected in Malaysia (Yong 1973) and Singapore (Burkill 1961). In Thailand they may be hunted, captured, or kept in captivity only with a license and may be exported only with a special permit from the Wildlife Advisory Board (Lekagul and McNeely 1977).

Table 152. Distribution of *Presbytis aygula-melalophus* group (countries are listed in order from west to east)

Country	Presence and Abundance	Reference
Burma	In Tenasserim to about 14°N; locality at 12°N mapped	Medway 1970
	At Bankachon (10°09′N, 98°36′E) (locality mapped)	Fooden 1976
Thailand	In peninsular S of 12°N (localities mapped)	Medway 1970
	In peninsular from 7°20′N to 13°N (localities mapped); relatively uncommon	Fooden 1976
	In peninsular only (map); especially common in Trang, Krabi, Phangnga, and Surat	Lekagul and McNeely 1977
Malaysia	In Sarawak (N Borneo)	Burgess 1961
	In peninsular and in N Borneo from Mount Kinabalu in Sabah S into Sarawak	Davis 1962
	E of Kuala Lumpur (peninsular)	McClure 1964
	W of Sandakan, Sabah (N Borneo)	Yoshiba 1964
	In peninsular	Bernstein 1967b
	In Sabah (N Borneo)	de Silva 1968b
	In peninsular	Harrison 1969
	In peninsular: on E coast, near W coast, isolated populations in interior (distribution mapped)	Lim 1969
	In peninsular: widespread and common	Medway 1969a

Table 152 (continued)

Country	Presence and Abundance	Reference
Malaysia (cont.)	Throughout peninsular and N Borneo (localities mapped)	Medway 1970
	In central peninsular (localities mapped)	Chivers 1971
	In Ulu Segama Reserve, Segama R., E Sabah (N Borneo)	MacKinnon 1971
	Numerous at Kuala Lompat, Pahang, W coast of peninsular	Medway and Wells 1971
	Near Kinabatangan R., NE Sabah (N Borneo)	Kawabe and Mano 1972
	At Gunong Benom, S of 4°N, central peninsular	Medway 1972
	Abundant in peninsular	Southwick and Cadigan 1972
	Throughout peninsular but population not dense	Bernstein, pers. comm. 1973
	Near Kuala Lompat, W coast of peninsular	Chivers 1973
	Throughout peninsular (localities mapped)	Fooden 1976
	Locally abundant but declining slowly near Kuala Lompat (peninsular)	Curtin, pers. comm. 1978
	Population in peninsular decreased 42.4% from estimated 962,000 in 1958 to estimated 554,000 in 1975	Khan 1978
Singapore	In Bukit Timah and Water Catchment Area	Burkill 1961
	In Bukit Timah Nature Reserve	Harrison 1966
	Native on Singapore Island	Medway 1969a
Indonesia	Throughout Sumatra; on Batu Island (but not on other islands W of Sumatra); on W half of Java; Bunguran (Natuna Island); and Borneo N of Kapuas R. and Mahakam R. (N of equator) (localities mapped)	Medway 1970
	In Gunung Leuser Reserve, N Sumatra	Kurt 1973
	At Ranun R., West Langkat Reserve, Sikundur, N Sumatra	MacKinnon 1973
	In Kutai Reserve, E Kalimantan	Rodman 1973, Wilson and Wilson 1975a
	Locally common in Sumatra but total population not large; may be declining but not drastically; not now endangered but likely to decrease in future	Wilson and Wilson, pers. comm. 1974
	Common inland from S Sumatra to N of Lake Toba and mainland and islands of Riau Province	Wilson and Wilson 1975b
	Throughout Sumatra except SE coast (localities mapped)	Wilson and Wilson 1977
	Rapidly becoming seriously endangered in W Java	Brotoisworo 1978

Table 153. Population parameters of *Presbytis aygula-melalophus*

Population Density (individuals /km^2)	Group Size $\bar{X}$	Group Size Range	Group Size No. of Groups	Home Range Area (ha)	Study Site	Reference
		5–8	3		Thailand	Fooden 1976
					Malaysia	
		3–12			N Borneo	Davis 1962
	10.7	10–20	5		Peninsular	Bernstein 1967b
		6–20			Peninsular	Medway 1970
20	10		4–5		Peninsular	Medway and Wells 1971
	20		3	13–22	Peninsular	Chivers 1973
	12.6	4–18		9–21	Peninsular	Curtin 1977
					Indonesia	
10–25	4.2	2–10	26		Sumatra	Kurt 1973
20–64					Sumatra	MacKinnon 1973
24.3		6–12			Kalimantan	Wilson and Wilson 1975a
		5–13			Sumatra	Wilson and Wilson 1975b
3.8–48.7	8–9.2		48	10–30	Sumatra	Wilson and Wilson 1977

Presbytis cristata Silvered Leaf-Monkey

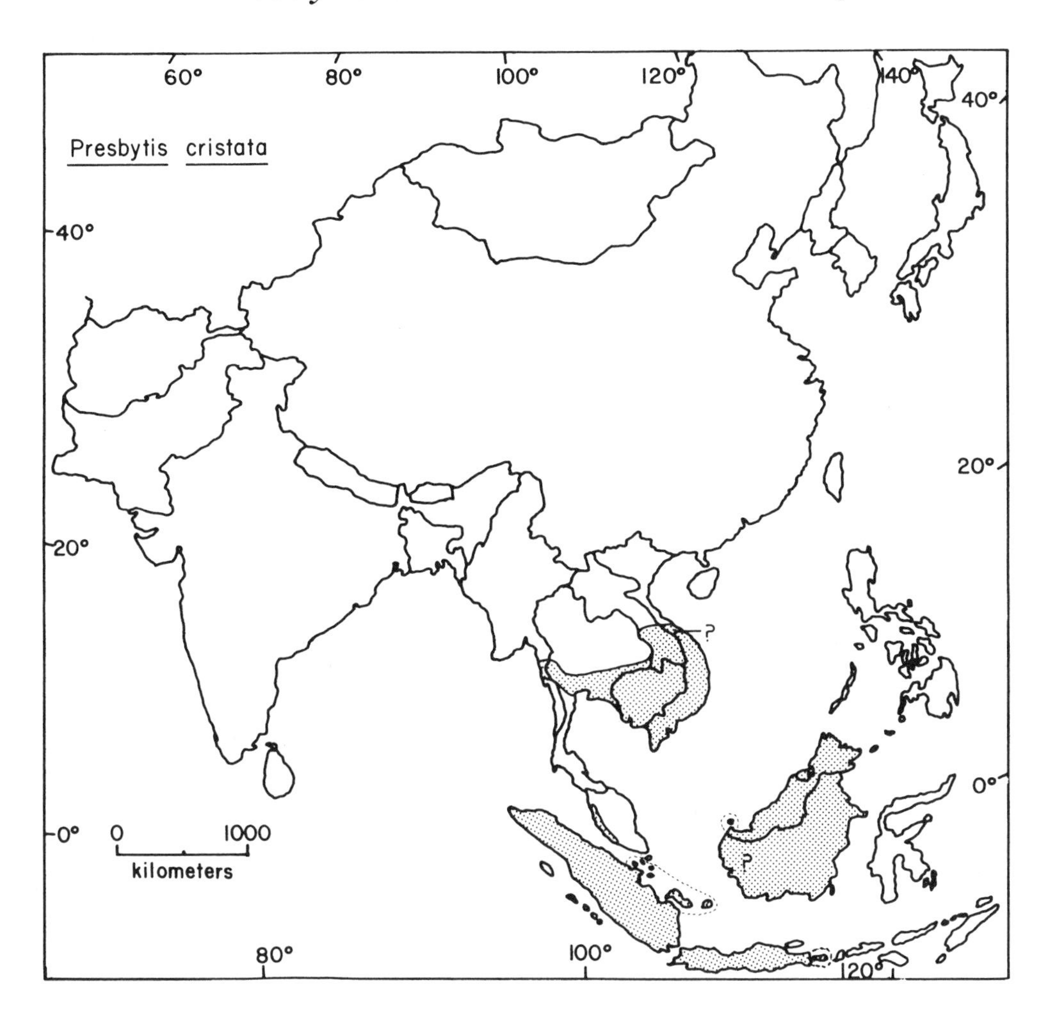

Taxonomic Note

Eight subspecies are recognized (Napier and Napier 1967).

Geographic Range

Presbytis cristata is found from southern Burma through Indochina, in Malaysia, and in Indonesia as far east as eastern Borneo and Bali. Its distribution is disjunct, with a gap of 800 km between populations in southern Thailand and those on the Malaysian peninsula (Fooden 1976). It occurs as far northwest as 15°15′N, 98°0′E in Burma and perhaps as far northeast as 16°37′N, 106°36′E in Vietnam (Fooden 1976) or Laos. Its westernmost locality is at the western tip of Sumatra (Fooden 1976), and its easternmost population is near 118°E in northeastern Borneo (Medway 1969b, 1970). For details of the distribution, see Table 154.

Abundance and Density

The peninsular Malaysian population of *P. cristata* declined an estimated 33.3% between 1958 and 1975, and has recently been described as relatively rare and not abundant (Table 154). The species was recently found to be abundant in southern Sumatra and Java; its status throughout the rest of its range has not been determined (Table 154).

P. cristata groups vary from 9 to 51 members, with average sizes of 10.6 to 32 (Table 155). The data are scant but indicate that perhaps groups are larger in peninsular Malaysia than in Thailand or Sumatra. Home range areas are between 3.5 and 20 ha (Table 155). This species has been observed at high and low population densities in several habitats on Sumatra (Table 155).

Habitat

P. cristata is a forest species typical of coastal, mangrove, or riverine forests. In peninsular Malaysia its most common habitat is coastal lowland and mangrove swamp forest (Furuya 1961–62b; Harrison 1966, 1969; Lim 1969; Medway 1969a, 1970; Weber 1972), and it also utilizes secondary forest (Southwick and Cadigan 1972), plantations (Medway 1969a), and urban areas (Bernstein 1968, Southwick and Cadigan 1972).

In Thailand, *P. cristata* is found in evergreen forest (Fooden 1971a) and in the junction between mangrove and terrestrial forest (Lekagul and McNeely 1977). In the Indochinese peninsula it occupies inland areas hundreds of kilometers from mangrove forests (Fooden 1976), but in Borneo this species is found in lowland swamp forest, mangroves, secondary forest (Medway 1970 citing older authors) and nipa-mangrove mixed forest (Kawabe and Mano 1972). It is common in mangrove swamps and riverine habitats periodically disturbed by tidal and seasonal flooding, and seems to prefer secondary to primary vegetation (Wilson and Wilson 1975a).

In Sumatra, silvered leaf-monkeys show a preference for disturbed lowland and riparian habitats (Wilson and Wilson 1977). These authors found the highest population densities of *P. cristata* in mixed mangrove swamps, riverine secondary forest, and rubber groves. They concluded that Sumatran *P. cristata* was rare in deep primary forest and at elevations above 458 m. On Java, *P. cristata* has been reported from as high as 1,737 m (Medway 1970).

Factors Affecting Populations

Habitat Alteration

P. cristata is able to utilize several types of moderately disturbed and secondary forests, and can live in close association with man in towns (see above). It seems to be relatively unaffected by selective logging, and in Sumatra, rubber groves often support larger populations of *P. cristata* than an equivalent area of primary forest nearby (Wilson and Wilson 1975a, 1977).

Nevertheless, like all forest monkeys, *P. cristata* is dependent on forest for survival; and deforestation for lumber and farms, and the spread of the local alang-alang grass, have probably reduced its range in Sumatra (Wilson and Wilson 1972–73). In peninsular Malaysia, where clearcutting is proceeding at the rate of 20,000 ha per year (Southwick and Cadigan 1972), logging is seriously diminishing *P. cristata*'s already limited range, and clearing of forest for agriculture and rural development is responsible for the sharp decline of the species between 1958 and 1975 (Khan 1978).

No information is available regarding the effect of habitat alteration on *P. cristata* populations in Thailand or Indochina, but the recent heavy military activity, including the use of defoliants, has adversely affected many forests (Orians and Pfeiffer 1970, S.I.P.R.I. 1976).

Human Predation

Hunting. Hunting pressure on *P. cristata* is rather high in western Bali, where the local people use some parts of this species for the manufacture of traditional medicine (Brotoisworo 1978). No reports were located regarding hunting of silvered leaf-monkeys in other areas.

Pest Control. *P. cristata* feeds in gardens in Kuala Selangor and is chased away and harassed with stone throwing and firecrackers (Bernstein 1968). In Sumatra it eats young leaves of rubber trees and is consequently chased from nurseries, but is seldom killed as a pest owing to the scarcity of guns (Wilson and Wilson, pers. comm. 1974). It is considered a serious pest to rubber in the state of Perak, peninsular Malaysia (Khan 1978), but no efforts to control it were mentioned.

Collection. *P. cristata* does not survive well in captivity (Jolly 1966) and is not in demand for research, zoo exhibits, or the pet trade (Khan 1978). It is captured and sold as a pet by the local people in Java (Brotoisworo 1978). Between 1968 and 1973, 74 *P. cristata* were imported into the United States (Appendix), primarily from Thailand. From 1965 to 1975 a total of 47 *P. cristata* were imported into the United Kingdom (Burton 1978). None entered the United States in 1977 (Bittner et al. 1978).

Conservation Action

P. cristata is found in one game sanctuary in Thailand and one reserve in Indonesia (Table 154), and is classified as protected in Malaysia (Yong 1973). In Thailand it may not be hunted, captured, or kept in captivity without a license and may only be exported with special permission from the Wildlife Advisory Committee (Lekagul and McNeely 1977).

Table 154. Distribution of *Presbytis cristata* (countries are listed in order from west to east, then north to south)

Country	Presence and Abundance	Reference
Burma	In S	Medway 1969a
	In SE as far NE as edge of Dawna Range (about 16°N, 98°E)	Fooden 1971a
	In Tenasserim between 15°–15°15′N and 98°–98°25′E (2 localities listed and mapped)	Fooden 1976
Thailand	In S	Medway 1969a
	In SW as far N as Ban Kerng Chada (15°08′N, 98°31′E)	Fooden 1971a
	Throughout S (localities listed and mapped)	Fooden 1976
	In Huay Kha Khaeng Game Sanctuary (in W near 15°29′N)	Eudey, pers. comm. 1977
	In SW and SE (map)	Lekagul and McNeely 1977
Laos	In S at Ban Phon, Saravan, and Plateau des Bolovens, Xedon (localities listed and mapped)	Fooden 1976
Vietnam	In provinces of Darlac, Tuyen Duc (both central), and Tay Ninh (SW)	Van Peenen et al. 1969
	In Darlac, Tuyen Duc, Tay Ninh; in central at Quang Tri (locality uncertain); (localities listed and mapped)	Fooden 1976
Cambodia	At Sambor, Kracheh, and Siemreab (Siem Reap?) (both central) (localities listed and mapped)	Fooden 1976
Malaysia	In N Borneo	Davis 1962
	In peninsular on W coast from Province Wellesley to Malacca State	Furuya 1961–62b
	In peninsular on W coast between Penang (5°N) and Malacca (2°N)	Harrison 1966
	At Kuala Selangor near mouth of Selangor R. (peninsular)	Bernstein 1968
	Abundant along W coast of peninsular	Lim 1969
	Confined to W coast of peninsular; also in N Borneo (map)	Medway 1969b
	At Padas Bay, NW Sabah (N Borneo)	Kawabe and Mano 1972
	Limited range along W coast of peninsular	Southwick and Cadigan 1972
	At Kuala Selangor (peninsular); relatively rare in general	Weber 1972
	Seen only on W coast of peninsula; not abundant	Bernstein, pers. comm. 1974
	Along W coast of peninsular from Pinang (5°25′N) to Kuala Kelang (3°N); (localities listed and mapped)	Fooden 1976

Table 154 (continued)

Country	Presence and Abundance	Reference
Malaysia (cont.)	Population in peninsular decreased 33.3% from estimated 6,000 in 1958 to estimated 4,000 in 1975	Khan 1978
Indonesia	On Sumatra, Borneo (Kalimantan), Java, Bali, Sirhasan Island (in S Natuna group), Bangka, Belitung, and most of islands in Riau-Lingga Archipelago (map)	Medway 1969b, 1970
	Abundant in S Sumatra especially near coast and rivers, scarcer in NW (localities mapped); not endangered	Wilson and Wilson 1972–73
	In E Kalimantan	Van Peenen et al. 1974
	Near Mahakam R. and E coast of E Kalimantan; on Sumatra	Wilson and Wilson 1975a, 1977
	In NW Sumatra and Kepulauan Riau (Archipelago) (localities listed and mapped)	Fooden 1976
	In West Bali Reserve; abundant in Java	Brotoisworo 1978

Table 155. Population parameters of *Presbytis cristata*

Population Density (individuals/km²)	Group Size $\bar{X}$	Group Size Range	Group Size No. of Groups	Home Range Area (ha)	Study Site	Reference
	17	9–30	5		Thailand	Fooden 1971a
		10–40			Thailand	Lekagul and McNeely 1977
					Malaysia	
		22–48	2	3.5	Peninsular	Furuya 1961–62a
	32	23–51	5	20	Peninsular	Bernstein 1968
		12–20			Peninsular	Medway 1969a
					Indonesia	
6.4–119.7	10.6		105	10–20	Sumatra	Wilson and Wilson 1977

Presbytis entellus Common, Hanuman, or Gray Langur

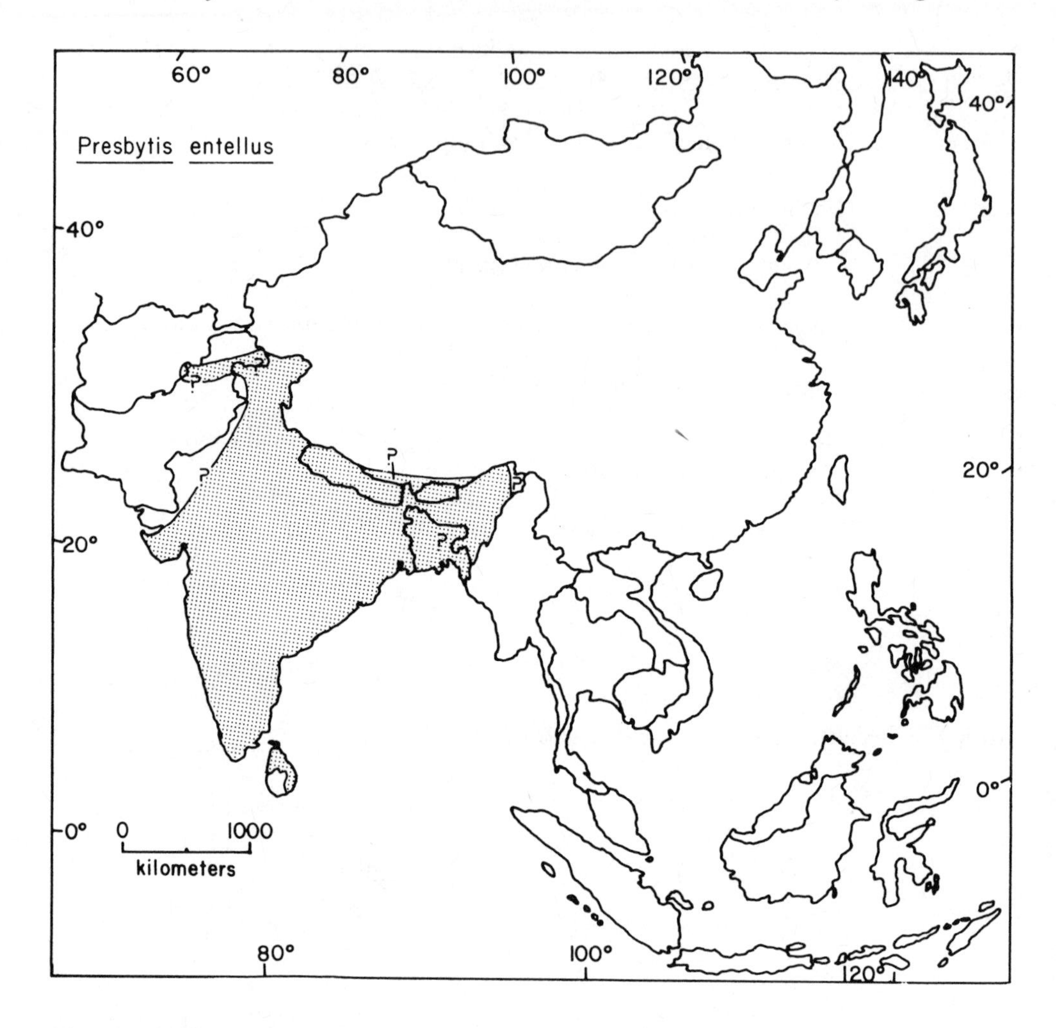

Taxonomic Note

Fifteen (Napier and Napier 1967) or sixteen (Roonwal and Mohnot 1977) subspecies are recognized.

Geographic Range

Presbytis entellus is found from Pakistan through India, on Sri Lanka, and north through Nepal possibly into southern Tibet. The westernmost reports of it are from the Paktia region of Afghanistan (about 33°N, 70°E) (Vogel and Weber, pers. comm. 1974) and around Shogran, Pakistan (34°37′N, 73°28′E) (Mirza 1965, Roberts 1977). It has been recorded as far east as about 93°E (Spillett 1966a) and 95°E in Assam (Oppenheimer 1977). It is found as far south as the southern tip of India, but its southern limit in Pakistan and its northern limits in Kashmir and Tibet are not known. For details of the distribution, see Table 156.

Abundance and Density

P. entellus has been described as common or abundant in India and in parts of Sri Lanka and Nepal, and as rare in Pakistan (Table 156). In 1970 the Indian population was judged to be reduced and declining by one author but suspected to be stable or increasing by others; more recently it has been estimated to number one million (Table 156). No evidence is available regarding populations in the other countries of the range.

Summarization of population parameters for *P. entellus* is complicated by the occurrence of single male, multimale, and all-male groups with varying home range areas and population densities (for details see Sugiyama 1976, Oppenheimer 1977, Vogel 1977). Heterosexual groups in general vary from 4 to 125 members, with average group sizes of 12.8 to 64 (Table 157). These occupy home ranges of 8.9 to 1,300 ha and constitute populations with densities of 2.7 to 134.7 individuals per km^2 (Table 157).

Several theories have been offered to explain the large variation in group size, home range area, and population density in *P. entellus*. Habitat quality, especially supplies of food, water, and cover, have generally been considered responsible for these differences (Jay 1965, Blaffer Hrdy 1977, Oppenheimer 1977). For example, large home range and large group size are thought to be necessary in order for *P. entellus* groups to utilize and store information about the location of their seasonally shifting, widely dispersed food sources in the dry zone of Sri Lanka (Ripley 1967). In another study, calculation of the number of food plants and hence the area needed to support each individual *P. entellus* led to the conclusion that the size, density, and productivity of the plants in an area are the main determinants of the population density of common langurs there (Yoshiba 1967).

But Vogel (1977) has examined data from many study sites and finds that (1) "there is no conclusive evidence whatsoever of a correlation between population density and the type of vegetation," (2) "troop size alone is generally independent of vegetation types," and (3) "from the reported data it is impossible to demonstrate any generally applicable relationship between vegetation and home range size."

Vogel (1977) has also calculated the area necessary for a reserve for *P. entellus*. He assumed an average troop size of 23, an average home range of 48 ha, and an average population density of 50 individuals per km^2. He concludes that a minimum of 120 km^2 would be required to protect a viable population of 5,000 *P. entellus*, and that at least four or five such sanctuaries should be established in various habitats to assure a heterogeneous gene pool for this species.

Habitat

P. entellus is a "very adaptable" species which is able to utilize a wide variety of habitats, including dry scrub, deciduous oak forest, thick wet forest (Jay 1965), semidesert, open cultivated areas, open park woods, dry and moist deciduous forest, and homogeneous oak-coniferous forest (Vogel 1977).

In India this species has been recorded in upper monsoon forest, tropical deciduous sal (*Shorea robusta*) forest, northern thorn forest of *Acacia* and *Euphorbia* (Vogel 1976), and mixed forest and bamboo jungle (Nagel and Lohri 1973). At one study site it spent most of its time in riverine vegetation consisting primarily of *Ficus*, *Holoptelia*, *Tamarindus*, and *Terminalia*, and a small proportion in tropical dry deciduous forest dominated by teak (*Tectona grandis*) (Starin 1978). In Pakistan it is found in coniferous forest (Roberts 1977).

P. entellus also occupies groves near water tanks, temples, towns, and villages (Prater 1971), and managed teak and *Acacia* forests (Rahaman 1973). It has been reported in planted roadside evergreen trees (Sugiyama 1964) and dense evergreen forest (Vogel 1977), although some authors have concluded that it does not enter evergreen forest (Krishnan 1971, Roonwal and Mohnot 1977).

In Sri Lanka, *P. entellus* occupies lowland mixed deciduous and evergreen forest and favors "a mixture of physiognomic types such as a combination of open and closed forest, herbaceous cover and open surfaces " (Ripley 1970). It is found only in the dry zones of the island, in arid forest and semideciduous forest, and is absent in high rainfall areas (Muckenhirn, pers. comm. 1973). Eisenberg and McKay (1970) reported it from monsoon scrub jungle, monsoon forest-grassland, and intermonsoon forest vegetation zones.

In Nepal, *P. entellus* utilizes tropical dry deciduous forest, sal forest, monsoon forest, tropical evergreen montane forest, and tropical evergreen high and cloud forest (Frick 1969). At one study site it occupied *Quercus semicarpifolia* forest and upper temperate mixed broadleaf forest (Bishop, pers. comm. 1974). No information is available regarding the habitats of this species in Afghanistan, Bangladesh, Bhutan, or Tibet.

P. entellus is never found far from trees, which it requires for sleeping sites and escape routes (Jay 1965), and "taller woody vegetation seems to be a critical factor in [its] distribution" (Ripley 1970). Some studies have indicated that population densities are higher in deciduous forests than in open or scrubby areas (Sugiyama 1964, Yoshiba 1968), and that this species prefers riverine deciduous forest to open habitats (Rahaman 1973).

Several fieldworkers have observed an association of *P. entellus* with water, and concluded that groups tend to live near permanent water sources (Mohnot 1971, Vogel 1971, Eisenberg and Lockhart 1972). In Sri Lanka the highest population densities of this species are found near rivers (Muckenhirn, pers. comm. 1973). Nevertheless, *P. entellus* can live for months without drinking and can survive a dry summer in extremely inhospitable areas (Jay 1965). It does not prefer humid habitats, and seems to be especially adapted to habitats that "experience a period of distinct dryness and an annual round of phenological changes" (Ripley 1970).

P. entellus is found at a wide variety of elevations, from 7 to 4,267 m above sea level (Oppenheimer 1977). Its high altitude habitats include monsoon forest at 1,524 m (Vogel 1971), cloud forest at 3,100 m (Frick 1969), and treeless alpine slopes at 4,084 m (Bishop, pers. comm. 1974).

Some *P. entellus* groups in Pakistan spend summers at about 2,400–3,050 m and winters at about 1,800 to 2,150 m elevation (Mirza 1965, Roberts 1977). In India, seasonal altitudinal movements in highland areas have also been observed (Jay 1965, Vogel 1971). But at one study site in India *P. entellus* groups remained at 3,200 m even during heavy winter snows (Sugiyama 1976), and in central Nepal, *P. entellus* living at 2,590 m also did not migrate to lower elevations (Bishop, pers. comm. 1974).

Factors Affecting Populations

Habitat Alteration

Destruction of the natural habitat is the primary cause for the decline of *P. entellus* in India. Natural vegetation has been reduced to less than 20% of its original area, and the rapidly increasing human population (which added 150 million members between 1935 and 1970) demands so much land for cultivation and grazing that even land in game reserves is released in order to gain votes (Kurt 1970).

This widespread deforestation is responsible for *P. entellus*'s present discontinuous distribution in India, where it is absent from "the desert regions and peripheral scrub which are characteristic of the human occupation of the plains forests for the past 30 years" (Krishnan 1971). In West Bengal, remnant *P. entellus* populations have been isolated by the cutting of trees, and are found primarily in villages because the only trees remain there (Oppenheimer, pers. comm. 1974). And at Dharwar the high density of *P. entellus* populations may be a concentration due to destruction of the surrounding forests (Yoshiba 1967).

India's remaining natural vegetation continues to be destroyed by clearing for agriculture, the introduction of exotic plant species, fires (Matthew et al. 1975), and the cutting of branches for firewood and fodder (Blaffer Hrdy 1977). Much of the vegetation now present consists of monocultural stands or scrubby second growth (Oppenheimer 1977). Although *P. entellus* can live in managed forests (Rahaman 1973) and in mature secondary forests (Vogel 1976), habitats suitable for it are being continually reduced and restricted.

In Nepal, deforestation is also accelerating, and most arable land up to 2,440 m elevation is being or has been cleared for agriculture (Bishop, pers. comm. 1974). In Sri Lanka much of the semideciduous forest is being clearcut and replaced with teak, which does not provide food for *P. entellus* (Dittus, pers. comm. 1974). This species can utilize partly cleared forest or forest disturbed by slash and burn agriculture (Ripley 1970). But in many areas natural vegetation is not allowed to regenerate after disruption, and at the present rate of deforestation the forests of Sri Lanka will be gone by the year 2050 (Dittus 1977a).

Human Predation

Hunting. P. entellus is not hunted for meat by many people of India because of its sacred status in the Hindu religion (Prater 1971), but at least four non-Hindu tribes do hunt and eat the common langur (Oppenheimer 1977). In some areas whole populations have been eliminated by hunting for the flesh's "therapeutic potency" (Krishnan 1971), while in other areas predation by humans is considered rare (Sugiyama 1976). In Bangladesh, hunting has been blamed for the absence of *P. entellus* near Moslem villages (Oppenheimer 1977).

The Nepalese people do not eat langurs (Bishop, pers. comm. 1974). Aborigines in Sri Lanka say that they do not eat monkeys, but monkey bones found in their villages indicate that they do so when other food is scarce (Muckenhirn, pers. comm. 1973), and the Veddas of Sri Lanka are reported to eat *P. entellus* (Roonwal and Mohnot 1977).

Pest Control. P. entellus is "a serious pest to agriculture and horticulture and a great nuisance in damaging home gardens and domestic installations'' (Sharma 1976). It eats potato leaves and tubers, maize kernels, pumpkin leaves, and wheat and buckwheat kernels in northern India (Vogel 1976), and causes significant damage to fruit crops (tamarind, blackberry, mango, banana, and hog plum) in West Bengal (Oppenheimer 1973).

In some areas *P. entellus* obtains as much as 90% of its food from cultivated plants (Yoshiba 1968), and even where the percentage is lower, this species is often dependent on crops for much of its diet (Blaffer Hrdy 1977). In addition to the food it takes, *P. entellus* causes damage to garden plants by trampling them, breaks tiles on roofs, steals objects from window ledges (Oppenheimer 1977), breaks branches on ornamental trees, breaks food containers, and causes a lot of resentment in local people (Blaffer Hrdy 1977).

The religious protection afforded by devout Hindus has allowed *P. entellus* to survive unmolested in many parts of India (Prater 1971, Sharma 1976). But farmers post watchmen to threaten, whistle, throw pebbles, and sometimes fire shots to scare it from the fields (Mohnot 1971). Dogs and children also try to chase common langurs from fields (Sugiyama 1976), but it is difficult to keep them away (Vogel 1976).

In Singur, West Bengal, each troop was harassed an average of once every two hours and police were summoned to shoot common langurs that stole food or damaged dwellings (Oppenheimer 1977). Elsewhere in India, harassment of *P. entellus* by stone throwing and slingshots is intense, and articles in local newspapers have suggested that ``something should be done about the monkey problem,'' although one case of langur poisoning was met with public disapproval (Blaffer, pers. comm. 1974). The increasing secularization of Indian society, and the severity of crop damage in the face of food shortages, could undermine the protection of *P. entellus* in India and lead to extermination efforts in populated areas (Blaffer Hrdy 1977).

The cultural traditions of Indian society still protect *P. entellus*, but these traditions may not be able to withstand the pressures of logarithmic population growth (Oppenheimer 1977). In addition to the possibility of pest control efforts, crop raiding can be detrimental to *P. entellus* because it increases the dependence of the species on introduced rather than natural foods and also promotes the transmission of diseases between man and langurs (Oppenheimer 1977). Further contact between man and *P. entellus* is encouraged by the common practice of feeding peanuts, chickpeas, and other foods to common langurs near temples and homes (Blaffer Hrdy 1977).

In Nepal *P. entellus* feeds on wheat and potato crops and is chased from fields and occasionally shot (Bishop, pers. comm. 1974). In Sri Lanka this species eats young crop leaves, shoots, buds, mangoes, and other fruit and is sometimes killed as a crop raider (Muckenhirn, pers. comm. 1973). In Pakistan, *P. entellus* does not raid crops (Roberts 1977).

Collection. Langurs do not survive well in captivity (Jolly 1966) and thus are not in demand for research, zoos, or the pet trade. But *P. entellus* is becoming a popular research subject among some biomedical workers (Mohnot 1978). Between 1968 and 1973 a total of 49 were imported into the United States (Appendix), primarily from Thailand, which is outside the range of *P. entellus*. Between 1965 and 1975, 55 *P.*

entellus were imported into the United Kingdom (Burton 1978). None entered the United States in 1977 (Bittner et al. 1978).

P. entellus carries ticks which are vectors for the viral Kyasanur Forest Disease, which sometimes reaches epidemic proportions and causes many deaths of both men and monkeys in the Shimoga District, Mysore (now Karnataka) State, India (Rajagopalan and Anderson 1971). In addition, *P. entellus* may act as a carrier for other viral diseases in West Bengal, where common langurs drink from village ponds in which human wastes accumulate during the hot, dry season (Oppenheimer 1973). A few *P. entellus* have been collected for the study of these diseases. A few were also collected in Nepal in the late 1960s for ectoparasite studies (Bishop, pers. comm. 1974).

Conservation Action

P. entellus is found in five national parks and nine wildlife sanctuaries in India, and two national parks in Sri Lanka (Table 156). But in India "most national parks within the range of *P. entellus* have resident populations of man and cattle," and destruction of vegetation for fodder and fuel is occurring there and in some sanctuaries (Oppenheimer 1977).

Based on his estimates of average population parameters (see above), Vogel (1977) recommended the establishment of four or five reserves, each 120 km^2 in area, to protect about 5,000 *P. entellus* each, in various habitats and regions of India. He added that further empirical data are necessary to enable planning of such reserves, and Oppenheimer (1977) agreed that "at minimum, studies are needed to establish the extent of genetic variability present in demes of *P. entellus* and to work out how large a deme must be so that this variability can be maintained over the long term."

The Nepal Conservation Society is attempting to increase awareness of conservation principles and to promote interest in local flora and fauna, including *P. entellus*, and the Nepalese government is making strong efforts to establish two new national parks and control hunting, scientific collection, and export from Nepal (Bishop, pers. comm. 1974).

P. entellus is listed in Appendix I of the Convention on International Trade in Endangered Species of Wild Fauna and Flora and as endangered under the U.S. Endangered Species Act (U.S. Department of Interior 1977a,b).

Table 156. Distribution of *Presbytis entellus* (countries are listed in order from west to east)

Country	Presence and Abundance	Reference
Afghanistan	Reported in Paktia region (in E)	Vogel and Weber, pers. comm. 1974
Pakistan	E of Shogran, Kaghan Valley, Hazara District (in NE)	Mirza 1965
	Extremely rare; restricted range and numbers; in NE in Hazara District near Shogran, Sharan, and Dhanial forests (map); total population probably much	Roberts 1977

Table 156 (continued)

Country	Presence and Abundance	Reference
Pakistan (cont.)	less than 200; additional population in Azad Kashmir, Jhelum Valley; unconfirmed reports in Amb State	Roberts 1977 (cont.)
India	In Rajasthan Desert	Prakash 1960
	Not abundant past or present in central and SE West Bengal (localities listed)	Southwick et al. 1964
	Near Dharwar (in SW)	Sugiyama 1964, 1967; Sugiyama et al. 1965
	At Orcha, Madhya Pradesh (E central), and Kaukori, Uttar Pradesh (in N)	Jay 1965
	Frequent in Kaziranga Wildlife Sanctuary, Assam (in E)	Spillett 1966a
	Common in Pakhal Wildlife Sanctuary, Andhra Pradesh (in SE); in Gir Wildlife Sanctuary, Gujarat (W central)	Spillett 1966b
	Common near Dharwar	Yoshiba 1967
	Common in Mudumalai Wildlife Sanctuary, extreme W Madras (now Tamil Nadu) State	Spillett 1968
	As far S as Cape Comorin, absent in N (map)	Yoshiba 1968
	Considerably reduced in numbers, declining, could recover	Kurt 1970
	Stable, perhaps increasing; no specific data	Southwick and Siddiqi 1970
	Distribution discontinuous; in Mudumalai Sanctuary, NE Nilgiris; Bandipur Sanctuary, in Venugopal N.P.; Raigoda Sanctuary; Palamau N.P.; Kanha N.P.; Taroba N.P.; Churna; Uttar Pradesh	Krishnan 1971
	At Jodhpur (in W)	Mohnot 1971
	From Himalayas to Cape Comorin except W deserts	Prater 1971
	At Shimoga District, Mysore (in SW)	Rajagopalan and Anderson 1971
	At Kumaon Hills near Bhimtal (in N) and Sariska Wildlife Sanctuary in Rajasthan (in NW)	Vogel 1971, 1975, 1976
	Common in Borivli N.P. (in W near Bombay)	Harroy 1972
	Fairly large population in Kanha N.P., Madhya Pradesh	Nagel and Lohri 1973
	Near Jalaghata and Apurbapur, near Singur, West Bengal	Oppenheimer 1973
	In Gir Wildlife Sanctuary, Gujarat	Rahaman 1973

Table 156 (continued)

Country	Presence and Abundance	Reference
India (cont.)	Abundant at Mount Abu, SW Rajasthan (E central)	Blaffer, pers. comm. 1974; 1977
	Small remnant discontinuous populations in West Bengal: in 24 Parganas, New Alipore, Hooghly, Howrah, and Purulia districts	Oppenheimer, pers. comm. 1974
	Throughout including parts of Kashmir; locally rather common	Vogel and Weber, pers. comm. 1974
	In Manas Sanctuary, N border with Bhutan	Ghosh and Biswas 1975
	Becoming scarce in Palni Hills, Madurai District, Tamil Nadu (SE)	Matthew et al. 1975
	Common in Segur Range and in Madumalai, Bandipur, and Nagarhole wildlife sanctuaries (all in SW)	Reza Khan 1976
	Near Simla (31°06′N, 77°10′E)	Sugiyama 1976
	Throughout except Sind (=SE Pakistan?) and Punjab regions of NW	Oppenheimer 1977
	Population dense at Dharwar	Parthasarathy 1977
	Throughout from Kashmir and Kathiawar to N Shan States (map)	Roonwal and Mohnot 1977
	Small population at Asarori Forest (in NE, 30°15′N, 77°68′E)	Makwana 1978
	Abundant; estimated population of 1 million	Mohnot 1978
	In Gir Forest, Gujarat (21°20′N, 70°00′E)	Starin 1978
Sri Lanka	At Polonnaruwa (in E, about 8°N, 81°E)	Ripley 1967, 1970
	In N and E	Eisenberg and McKay 1970
	In Wilpattu N.P. (in NW, about 8°30′N, 80°E)	Eisenberg and Lockhart 1972
	Abundant throughout dry zone (in N and E)	Muckenhirn, pers. comm. 1973
	In Gal Oya N.P. (in E, about 7°15′N, 81°30′E)	Atapattu and Wickremasinghe 1974
China	N of Nepal border, near 29°N, 88°E (map)	Fiedler 1956
	In extreme S Tibet	Ellerman and Morrison-Scott 1966
Nepal	Present	Fiedler 1956, Frick 1969, Mitchell 1975
	Fairly common throughout Terai, Himalayas area	Blower 1973
	At Melamchigaon, Sindu District (E central, 28°03′N, 85°33′E)	Bishop, pers. comm. 1974
Bhutan	Present	Fiedler 1956
Bangladesh	Present	Oppenheimer 1977
	Present (map)	Roonwal and Mohnot 1977
	Not listed	Green 1978a

Table 157. Population parameters of *Presbytis entellus* (heterosexual groups)

Population Density (individuals /km^2)	Group Size			Home Range Area (ha)	Study Site	Reference
	X̄	Range	No. of Groups			
		40–60			NE Pakistan	Roberts 1977
		12–18	4		E India	Southwick et al. 1964
11.4–134.6	15.1		38	10–30	SW India	Sugiyama 1964
85	15	10–24		10.3–31.5	SW India	Sugiyama et al. 1965
	18–30	5–120	39	129–1,295	India	Jay 1965
81	15–16	5–32		18–26	SW India	Yoshiba 1967
2.7–134.7	15–22	5–120	63	10.4–32.6	India	Yoshiba 1968
		4–18			India	Krishnan 1971
		29–68	20		W India	Mohnot 1971
	15–25			130–1,300	India	Prater 1971
97–104	23–64	15–125		10–80	N India	Vogel 1971, 1976
20–25		17–24			Central India	Nagel and Lohri 1973
	12.8	10–15	2	38.5–47.6	E India	Oppenheimer 1973
	30.4	16–48	9		W India	Rahaman 1973
24.6	46.8	19–98	8	66–491	N India	Sugiyama 1976
50	21	15–30		30	W India	Blaffer Hrdy 1977
4.6	12.8	10–14		43	E India	Oppenheimer 1977
	15.1		38	8.9	W India	Parthasarathy 1977
50				48	N India	Vogel 1977
115–128		25–31	2	17–23	W India	Starin 1978
	25			32–130	NE Sri Lanka	Ripley 1967
2.8					NW Sri Lanka	Eisenberg and Lockhart 1972
	21	12–25	4	9–15	Sri Lanka	Hladik and Hladik 1972
		33	1	259	E Nepal	Bishop, pers. comm. 1974

Presbytis francoisi François' Leaf-Monkey

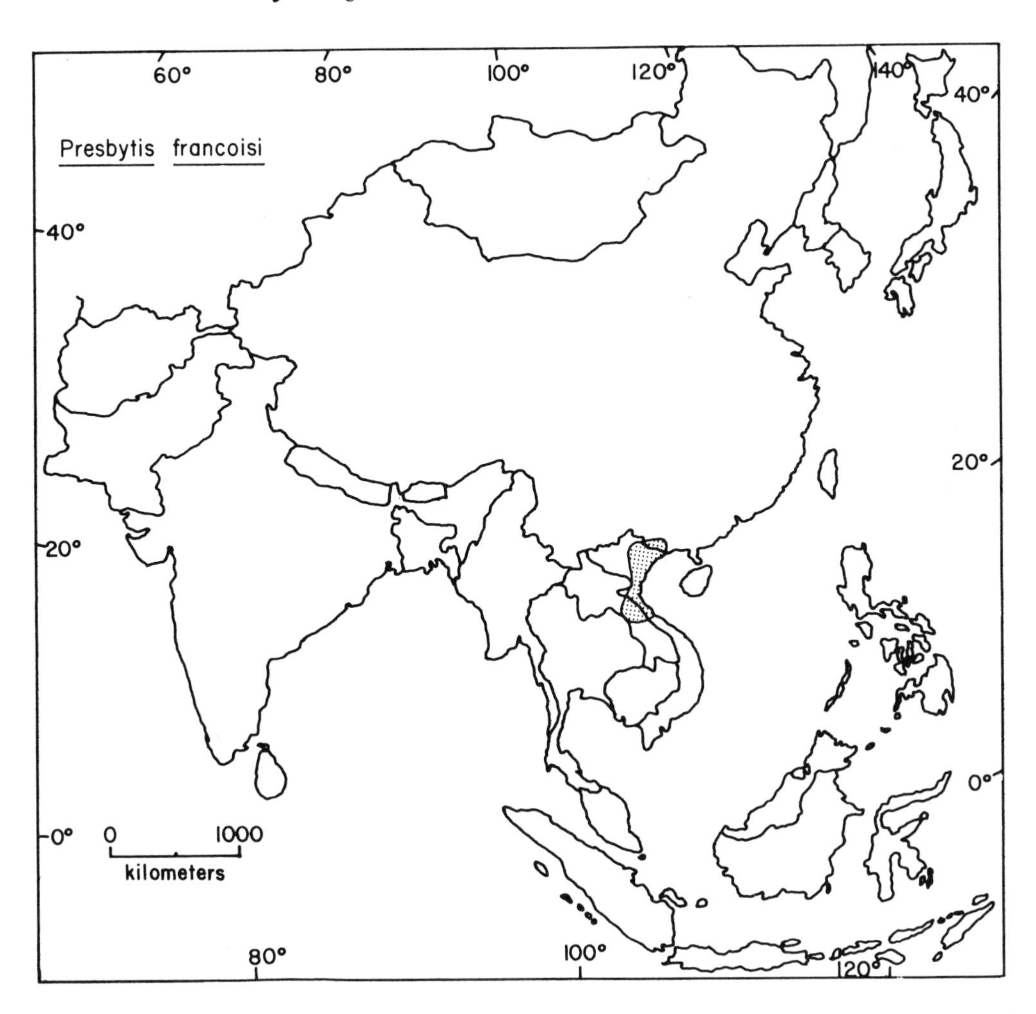

Taxonomic Note

Four subspecies are recognized (Ellerman and Morrison-Scott 1966, Napier and Napier 1967).

Geographic Range

Presbytis francoisi is found in southeast Asia from southeastern China to central Laos and Vietnam. Its northernmost record is in Kwangsi, China at 22°35′N, 107°57′E, and its southernmost locality is Quang Binh (17°50′N, 106°00′E) in central Vietnam (Fooden 1976). In Laos this species occurs as far west as 104°19′E (at 18°N) in Khammouan (Fooden 1976). For details of the distribution, see Table 158.

Abundance and Density

No information.

Habitat

P. francoisi seems to inhabit rocky areas (Delacour 1940, Fooden 1976). No further descriptions of its habitat could be located.

Factors Affecting Populations

Habitat Alteration

Virtually nothing is known about this species except that part of its range is in the area that suffered heavy bombing during the Vietnam war (Van Dyke 1972, Pfeiffer 1973). Almost certainly both leaf-monkeys and their habitats were affected.

Human Predation

Hunting. No information.

Pest Control. No information.

Collection. No *P. francoisi* were imported into the United States between 1968 and 1973 (Appendix) or in 1977 (Bittner et al. 1978). A total of 25 were imported into the United Kingdom between 1965 and 1975 (Burton 1978).

Conservation Action

P. francoisi is classified as endangered under the U.S. Endangered Species Act (U.S. Department of Interior 1977b).

Table 158. Distribution of *Presbytis francoisi*

Country	Presence and Abundance	Reference
China	In S	Ellerman and Morrison-Scott 1966
	In S Kwangsi (in SE) at Fu-sui and Lung-chou (localities listed and mapped)	Fooden 1976
Vietnam	At Bac Kan, Lang Son, and Hoi Xuan (in N)	Delacour 1940
	In Bac Can, Hong Quang (n.l.), Lang Son, Thanh Hoa, and Yinh Quang Ninh (all in N) and Ha Tinh and Quang Binh (central) (localities listed and mapped)	Fooden 1976
Laos	At Bannassao (n.l.)	Delacour 1940
	In Khammouan at Ban Naxao, Ban Phontiou, and Pha Som (all central) (localities listed and mapped)	Fooden 1976

Presbytis frontata White Fronted Leaf-Monkey

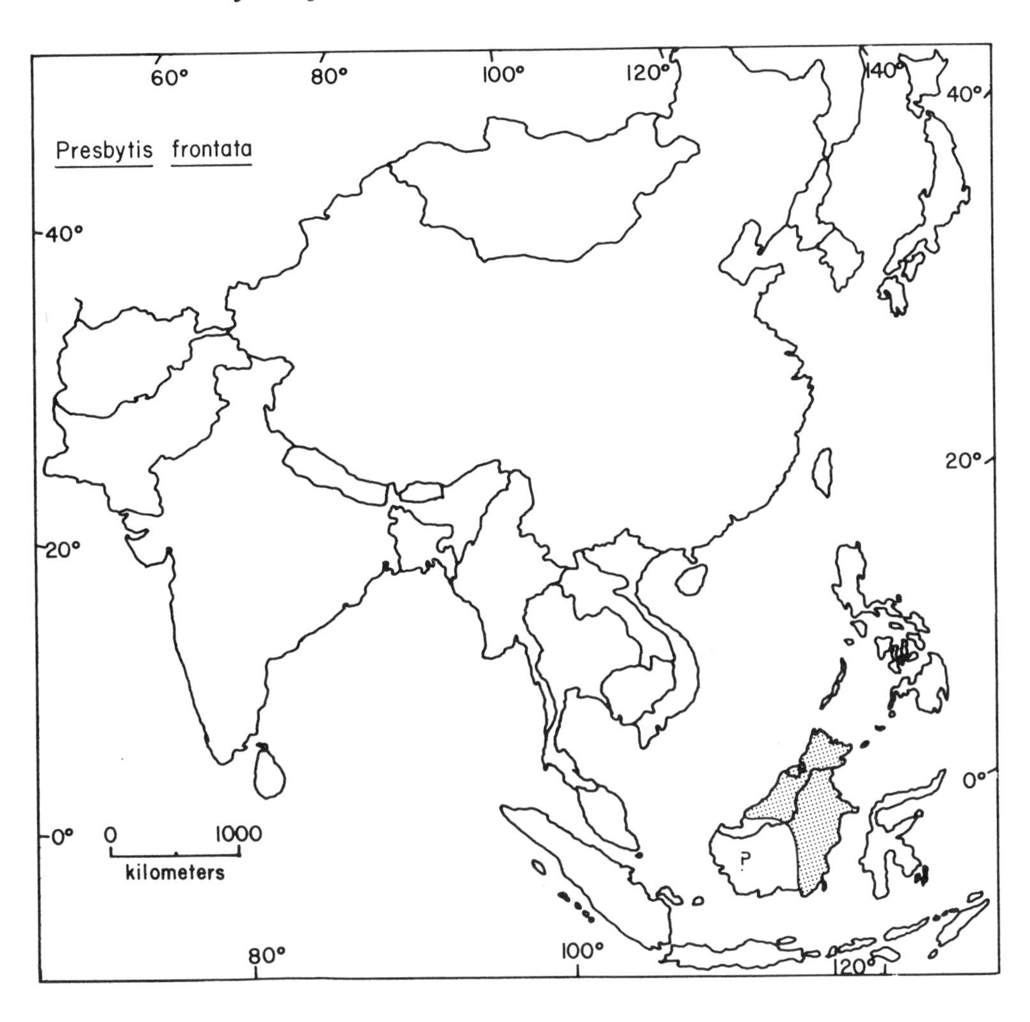

Taxonomic Note

Two subspecies are recognized (Napier and Napier 1967).

Geographic Range

Presbytis frontata is found on the island of Borneo: in eastern Kalimantan (Indonesia) and in Sarawak and Sabah (Malaysia). In Kalimantan it has been recorded only east of the Barito River (Medway 1970, localities mapped), including in the Kutai Reserve (near 0°, 117°E) (Rodman 1973) and near Renggang (near 1°S, 117°E) (Wilson and Wilson 1975a). In Sarawak this species occurs north of the Rejang (Rajang) River (about 2°N) (Medway 1970, localities mapped), and in Sabah it has been reported only as present (de Silva 1968b). No further details of the distribution are known.

Abundance and Density

In eastern Kalimantan, *P. frontata* was seen rarely, perhaps because it occurs at low population densities (Wilson and Wilson 1975a) or is highly nomadic (Rodman 1973). In one area 2 individuals were seen together, and population density was estimated at 5.7 individuals per km^2 (Wilson and Wilson 1975a). No other data are available.

Habitat

P. frontata is confined to lowland forest up to 300 m elevation (Medway 1970).

Factors Affecting Populations

Habitat Alteration

Some of the forests of Kalimantan are being selectively logged, and *P. frontata* has been sighted in these forests (Wilson and Wilson 1975a). But no information was located regarding the effect of this or other habitat disturbance on *P. frontata* populations.

Human Predation

Hunting. P. frontata was formerly hunted intensively in Sarawak for bezoar (a concretion from the digestive tract sought for magical and medical purposes and as a pigment), but in recent years this hunting has not been recorded (Burgess 1961).

Pest Control. No information.

Collection. No *P. frontata* were imported into the United States between 1968 and 1973 (Appendix) or in 1977 (Bittner et al. 1978), nor into the United Kingdom in 1965–75 (Burton 1978).

Conservation Action

No information.

Presbytis geei Golden Langur

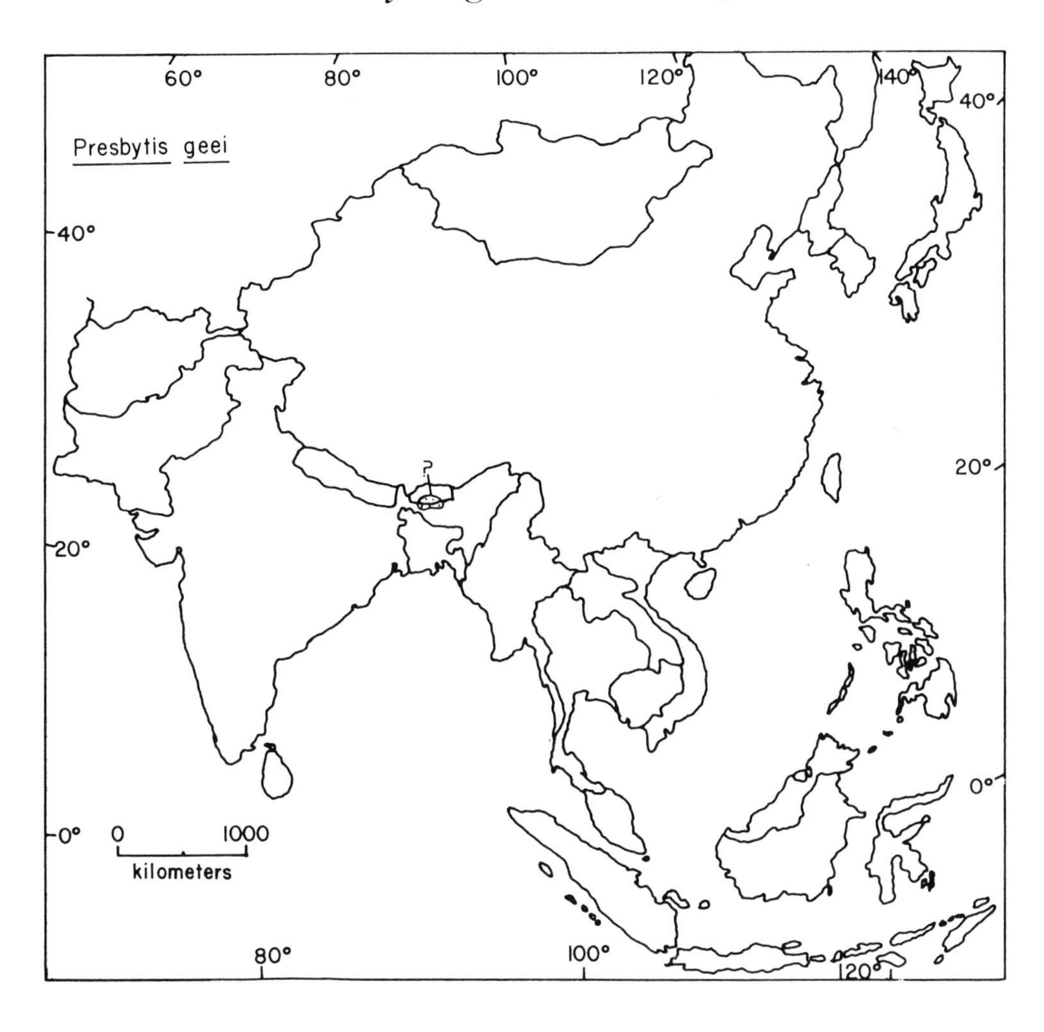

Taxonomic Note

This is a monotypic species (Napier and Napier 1967).

Geographic Range

Presbytis geei is found in northwestern Assam, India, and south central Bhutan, west of the Manas River and east of the Sankosh River (Gee 1961). Along the Bhutan-India border it occurs as far east as about 91°E and as far west as Jamduar at 89°50′E (Mukherjee and Saha 1974). Golden langurs are found as far south as 26°20′N at Raimona, India (Mukherjee and Saha 1974). Recent surveys have failed to reveal their presence any farther south or east (Ghosh and Biswas 1975, Khajuria 1977). The species is known to occur at several localities in southern Bhutan, but the northern boundary of its range has not been determined. For details of the distribution, see Table 159.

Abundance and Density

P. geei is found within a very small total range and seems to be uncommon in the Indian section of this area, with total population estimates varying from 540 to 6,000 (Table 159). A recent unpublished study (cited by I.U.C.N. 1978) has revealed that it is locally relatively abundant on the Bhutan side of the border, but in general this species is considered rare (Table 159).

No estimates of population density are available. Groups vary in size from 4 to 40, with average groups containing 6 to 16.7 individuals, and home ranges of 150 to 600 ha (Table 160).

Habitat

Golden langurs are tropical forest monkeys, found in deciduous (Khajuria 1956, Wayre 1968a) and evergreen forests (Oboussier and Maydell 1959, Ghosh and Biswas 1975). Several of the sites at which they have been studied were in forests dominated by sal (*Shorea robusta*) and described as mixed, moist, deciduous, or semideciduous (Mukherjee and Saha 1974, Khajuria 1977).

In the Manas Sanctuary, *P. geei* habitat consists of tropical moist deciduous forest with trees up to 45 m tall, a nearly closed canopy, many climbers, and a patchy, dense, shrubby understory (Wayre 1968a). In other areas, savanna forest and riverine forest are also components of this species' habitat (Mukherjee and Saha 1974).

P. geei has been observed in hilly habitats up to 1,600 m (Ghosh and Biswas 1975), 1,630 m (Mukherjee and Saha 1974), and 2,400 m (Roonwal and Mohnot 1977).

Factors Affecting Populations

Habitat Alteration

The range of *P. geei* is almost entirely within reserved forests (Gee 1961). Timber is extracted from at least some of these forests (Mukherjee and Saha 1974), but the effect of timbering on *P. geei* populations has not been described. There are no known settlements in these forests, and "forest exploitation does not seem to be on a detrimental level" (I.U.C.N. 1972, 1976). No specific information is available.

Human Predation

Hunting. Golden langurs are hunted for meat by some local tribes (Khajuria 1977). The impact of this hunting has not been determined.

Pest Control. Most authors state that this species does not raid crops (Gee 1961, Mukherjee and Saha 1974, Khajuria 1977), but local people in Bhutan reported that it caused considerable damage in a cardamom plantation (Chakraborty 1975). No evidence of control efforts was located.

Collection. *P. geei* is reported to make a good pet (Khajuria 1977), indicating that it is sometimes captured. None were imported into the United States between 1968 and 1973 (Appendix) or in 1977 (Bittner et al. 1978), nor into the United Kingdom in 1965–75 (Burton 1978).

Conservation Action

P. geei is found in one forest reserve in India and one wildlife sanctuary in Bhutan (Table 159), where it is said to be completely protected (Ghosh and Biswas 1975). The species is fully protected in India but not in Bhutan (I.U.C.N. 1978). It is listed in Appendix I of the Convention on International Trade in Endangered Species of Wild Fauna and Flora and as endangered under the U.S. Endangered Species Act (U.S. Department of Interior 1977a,b).

Recommendations for further conservation measures include declaration of the entire range as a sanctuary protected from habitat disturbance (I.U.C.N. 1972), protection from human predation, rearing in captivity for distribution to zoos (Khajuria 1977), and full protection in Bhutan (I.U.C.N. 1978). Khajuria (1977) also suggested the introduction of *P. geei* into adjoining areas and the provision of food and water, but these actions could upset balanced ecosystems and lead to unexpected problems.

Table 159. Distribution of *Presbytis geei*

Country	Presence and Abundance	Reference
India	Quite common near Jamduar, E bank of Sankosh R.	Khajuria 1956
	Range of about 160 km^2 between Sankosh R. and Ranga R. (localities mapped)	Oboussier and Maydell 1959
	In Goalpara and Kamrup districts of Assam between Sankosh R. and Manas R. (map); probably more common at W end than in E; total population estimated at 540; unconfirmed reports from Garo Hills and Kasai Hills	Gee 1961
	Rare	I.U.C.N. 1972, 1976, 1978
	In Ripu Reserve Forest near Jamduar, and Raimona (localities mapped)	Mukherjee and Saha 1974
	Very rare; total range 600 km^2; maximum population is 6,000 but probably fewer	Khajuria 1977
Bhutan	Two localities near S border (mapped)	Oboussier and Maydell 1959
	N limit probably near Maure	Gee 1961
	20 to 35 in Manas Wildlife Sanctuary	Wayre 1968a
	35 to 40 in Manas Wildlife Sanctuary	Waller 1972
	At Panjurmane, Tama, and Gaylegphung	Mukherjee and Saha 1974
	N of Gaylegphug on road to Tongsa (27°33′N, 90°30′E); near Panjurmane; "in plenty" on slope of Tama Khola	Chakraborty 1975
	In Manas Wildlife Sanctuary	Ghosh and Biswas 1975
	Relatively numerous; population at least 1,200; classification, rare	I.U.C.N. 1978

Table 160. Population parameters of *Presbytis geei*

Population Density (individuals/km^2)	Group Size $\bar{X}$	Range	No. of Groups	Home Range Area (ha)	Study Site	Reference
	6		4		India	Khajuria 1956
		4–40	8		India	Oboussier and Maydell 1959
	15	10–20	6		India	Gee 1961
	12.5	5–18	10		India	Mukherjee and Saha 1974
	9.7	7–15	6	150	India	Khajuria 1977
		7–11	2		Bhutan	Wayre 1968a,b
	16.7	10–20	3		Bhutan	Mukherjee and Saha 1974
		15–20	3		Bhutan	Chakraborty 1975
		6–9	2	600	Bhutan	Ghosh and Biswas 1975

Presbytis johnii Nilgiri Langur

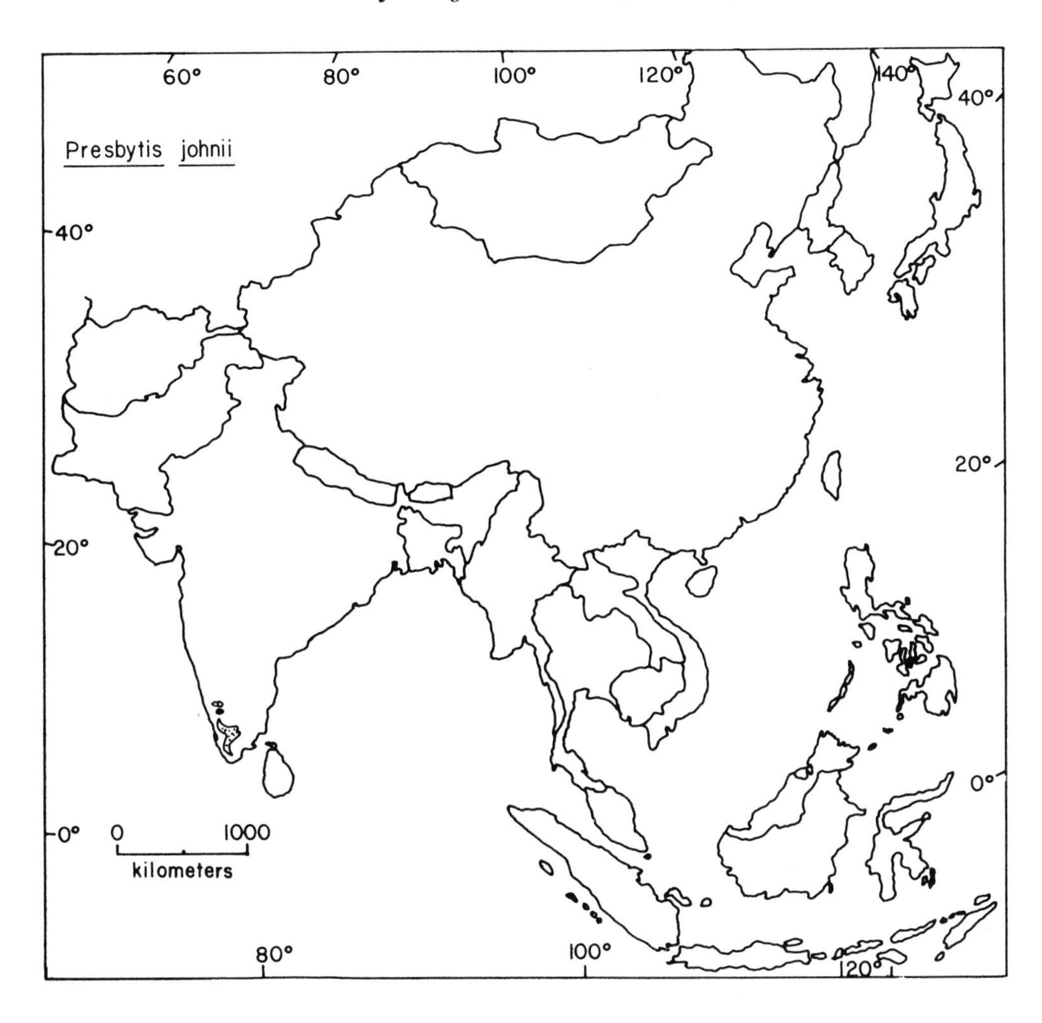

Taxonomic Note

This is a monotypic species (Napier and Napier 1967).

Geographic Range

Presbytis johnii is found in southwestern peninsular India from about 8°24′N to 12°10′N and 75°45′E to 77°45′E (Kurup 1977). Within this area it occurs in the Western Ghats region of southwestern Karnataka State, eastern Kerala State, and western Tamil Nadu State, in a series of hill ranges including the Nilgiri, Anaimalai, Palni, Pandalam, and Ashambu (or Agastyamalai) hills (Oates 1979). It had also been reported in the Shevaroy Hills (Poirier 1970), but no recent evidence is available regarding its presence there or its status in the other hill ranges where it has been recorded in the past (Oates 1979). For details of the distribution, see Table 161.

Abundance and Density

P. johnii populations have recovered recently in some areas and are considered abundant, with an estimated total of 5,000 to 15,150 individuals in the Ashambu Hills (Table 161). But the overall range of the species is very small, the species is declining and possibly extinct in several parts of the range, and it is generally considered to be in a vulnerable position.

This species lives in groups of 3 to 25, with average group sizes of 8.9 to 18 and home ranges of 2.3 to 259 ha (Table 162). In two study areas in the Ashambu Hills it was observed at population densities of 21 to 60 individuals per km^2 (Table 162).

Habitat

P. johnii is a forest monkey, most often found in evergreen, mixed evergreen and deciduous, and deciduous forest (Krishnan 1971). Its typical habitat is narrow tracts of evergreen forest (sholas) surrounded by grassland and drained by narrow, slow-moving streams (Poirier 1968a, 1970). It seems to be a rather specialized species confined to moist or wet forests with streams and open patches as crucial components of its habitat, and is absent in many seemingly suitable areas (Kurup 1975). But it has also been described as "highly adaptable" (Kurup 1977), and has been sighted in "malarious jungle," "woods of plateau areas" (Poirier 1968b), gardens and belts of cultivated woodland (Prater 1971), and open deciduous woodland far from riverine forest (Oates 1978).

The Nilgiri langur inhabits primarily hilly areas above 900 m elevation (Krishnan 1971, Prater 1971, Kurup 1977). It has been reported as low as 150 m (Oates, pers. comm. 1978) and as high as 2,400 m (Oates 1978), but is most common in wind sheltered valleys 1,524 to 2,134 m above sea level (Poirier 1969). Its hilly evergreen forest habitats are naturally patchily distributed in some areas (Oates 1978), and may be utilized mainly because many lowland forests have disappeared (Oates 1979).

Factors Affecting Populations

Habitat Alteration

The forest habitat on which *P. johnii* depends is rapidly being destroyed. The expanding human population is clearing large areas of shola for cultivation of food crops, coffee, tea, and cardamom (Poirier 1969, 1971; Matthew et al. 1975; Kurup 1977). Already much of *P. johnii*'s range has been deforested. For example, in Kerala, its habitats around 500 to 700 m elevation have been completely denuded, and most forests in the Nilgiri Hills and near Valparai have been cleared for plantations, resulting in a marked decrease in langurs (Kurup 1977).

Widespread forest destruction has also greatly reduced the range of this species in the Nelliampathi and Peermade hills and in the High Wavy Mountains (Oates 1979). Former Nilgiri langur habitats on the Wynaad Plateau are almost completely deforested, and burning, cattle grazing, tree felling, and road building are threatening large areas of the Ashambu Hills (Oates, pers. comm. 1978). Since *P. johnii* rarely crosses treeless areas, deforestation also isolates populations, preventing genetic exchange (Poirier 1969).

The establishment of extensive forestry plantations, especially of *Eucalyptus* and teak, is further reducing remaining natural vegetation (Poirier 1969, Kurup 1977). *P. johnii* occasionally eats *Eucalyptus* petioles (Poirier 1971, 1977) but gradually moves out of an area planted with *Eucalyptus* (Poirier 1968a). The replacement of natural forests with *Eucalyptus* forests is "rapidly causing the deterioration of the fauna indigenous to southern India," including *P. johnii* (Horwich 1972).

P. johnii can survive in teak plantations where other food-bearing tree species remain, and the protection of these plantations by foresters may account for the recent recovery of *P. johnii* in some regions. But further expansion of teak forests would reduce food variety and *P. johnii* populations (Kurup 1977). In general, Nilgiri langurs do not utilize exotic tree species and cannot occupy deforested areas, and habitat alteration of this sort has made many areas unsuitable for them (Krishnan 1971).

In remaining forests, harvesting of forest produce and flooding by hydroelectric projects are having a detrimental impact (Kurup 1977; Oates, pers. comm. 1978). *P. johnii* populations are being progressively restricted to small inaccessible forest patches, and their future will depend on the survival of these refuges (I.U.C.N. 1972).

Human Predation

Hunting. P. johnii is hunted for its flesh, glands, and blood, which are reputed to have rejuvenatory powers, and are used to prepare health tonics called "black monkey medicine," especially prescribed for liver and lung ailments (Hutton 1949, Poirier 1971) and as an aphrodisiac (Kurup 1975). It is also hunted for meat, for its beautiful fur, for its skin, which is used to make drumheads, and for sport (Poirier 1970, 1971).

This species has been heavily hunted for many years (McCann 1933a), and is hunted more than any other species of Indian monkey (Prater 1971). Since 1959, hunting seems to have decreased because *P. johnii* is becoming more difficult to find and protection from poachers has become better organized (Krishnan 1971). But recent sources still list hunting as a significant pressure affecting *P. johnii* populations (Kurup 1977; Mohnot 1978; Oates 1978, 1979).

Pest Control. Nilgiri langurs feed on crops including potatoes, cauliflower, ornamental flowers (Poirier 1971), and cardamom (Prater 1971). Local people attempt to scare them away from fields by chasing, throwing rocks, and by using scarecrows, fireworks, and dogs; and estate owners are "allowed to shoot any monkeys which raid their crops" (Poirier 1971). The effect of these control efforts on *P. johnii* populations is not known.

Collection. No *P. johnii* were imported into the United States between 1968 and 1973 (Appendix) or in 1977 (Bittner et al. 1978), nor into the United Kingdom in 1965–75 (Burton 1978).

Conservation Action

P. johnii is found in five wildlife sanctuaries, but neither habitat nor monkeys are adequately protected there (Oates, pers. comm. 1978). Recommendations to improve the situation include increased supervision of the reserves and protection from poaching, public education with regard to the myth of medicinal value of *P. johnii*, and the cessation of expansion of wattle and *Eucalyptus* plantations in the Naduvattom area (Kurup 1977).

Two recent status surveys in the Palni Hills and the Ashambu Hills reveal that stronger measures are needed in these areas to prevent hunting and protect the habitat from further destruction (Oates 1978; pers. comm. 1978; 1979). In addition, a detailed study of the Ashambu Hills is now being planned in order to arrive at recommendations for effective conservation of this area, which is crucial for the survival of *P. johnii* (Oates, pers. comm. 1978), and "if well protected these hills would support a *P. johnii* population of long-term viability" (Oates 1979).

P. johnii may not be exported from India without a permit, and Kerala and Tamil Nadu state laws prohibit its killing and capture, but they are poorly enforced (I.U.C.N. 1978).

Table 161. Distribution of *Presbytis johnii*

Country	Presence and Abundance	Reference
India	Near Periyar, SE Kerala	Tanaka 1965
	In Ootacamund area of Nilgiri Hills, Madras (now Tamil Nadu)	Poirier 1968a,b, 1969
	In Kodaikanal Hills, Kakkal Wildlife Sanctuary (W central Tamil Nadu) and Mundandurai Wildlife Sanctuary (S Tamil Nadu)	Spillett 1968
	At Panniar, Kerala	Sugiyama 1968a
	In Anaimalai, Brahmagiri, Cardamom, Nelliampathi, Nilgiri, Palni, Shevaroy, and Tinnevelly hills, 8°–13°N	Poirier 1970
	In Periyar Wildlife Sanctuary	I.U.C.N. 1971, Horwich 1972, Waller 1972
	Discontinuous distribution in hills of Western Ghats; previously locally threatened with extinction; now less rare in some areas; status has improved since 1959	Krishnan 1971
	Reduced in numbers; endangered	Prater 1971
	Vulnerable; steadily declining since 1875; populations viable in Anaimalai Hills and Cardamom Hills but not in Nilgiri Hills; about 80 troops in S Coimbatore Division and 100 troops in Periyar Sanctuary; capable of recovery if protected	I.U.C.N. 1972
	In Tirunelveli District, Tamil Nadu at 8°38′N, 77°25′E	Karr 1973
	In Palni Hills	Haltenorth 1975

Table 161 (continued)

Country	Presence and Abundance	Reference
India (cont.)	Alarming decline in Nilgiri Hills: nowhere common, last stronghold is in Paikara area; declining in Cardamom Hills: less than 100 troops in Periyar Sanctuary; recovering in Anaimalai Hills: about 80 troops in Thunacadavu and Punachi ranges in about 220 km^2 of preferred habitat; distribution patchy	Kurup 1975
	Becoming scarce near Kodaikanal, Palni Hills, Tamil Nadu	Matthew et al. 1975
	Distribution patchy but more common in Nilgiri Hills than elsewhere in Western Ghats; seen near Ootacamund, Mukurti, Sispora Pass, Bangitappal, Upper Bhavani, and other localities (listed)	Reza Khan 1976
	Decreased in past but trend now reversed and populations becoming reestablished; still declining in Nilgiri Hills (confined to Nudavattom, Paikara, Mukirthy, Glenmorgan, and Nilgiri peaks) and Palni Hills; in Anaimalai area status stable near Topslip but not in Nelliampathy-Parambikulam area or in Punachi range around Valparai (localities mapped)	Kurup 1977
	Vulnerable, numbers and range greatly reduced	I.U.C.N. 1978
	Population in Ashambu Hills is at least 5,000 and may be as high as 10,000–15,150 individuals including populations in Mundandurai and Kalakkadu wildlife sanctuaries, Muthukuzhi Plateau, Papanasam Hills, eastern foothills, Mahendragiri, Kuttalam, western slopes of Agastyamalai; not in immediate danger of extinction	Oates, pers. comm. 1978
	Restricted to Western Ghats between about 8°–12°N and about 76°–77°30′E; abundant in Ashambu (Agastyamalai) Hills: population of at least 5,000; common in Nilgiri Hills; low density in Pandalam Hills; about 150–500 in Palni Hills; apparently increasing in Anaimalai Hills; probably in precarious position or extinct in some of hill ranges (e.g., Brahmagiri Hills); status uncertain in Wynaad	Oates 1979

Table 161 (continued)

Country	Presence and Abundance	Reference
India (cont.)	Plateau, High Wavy Mts., Palghat, Nelliampathi, Cardamom, Peermade, Varushanad, and Andipatti Hills; vulnerable listing is justified	Oates 1979 (cont.)

Table 162. Population parameters of *Presbytis johnii*

Population Density (individuals/km^2)	Group Size $\bar{X}$	Range	No. of Groups	Home Range Area (ha)	Study Site	Reference
		20–30			S India	Webb-Peploe 1947
	15	10–25	5	10	S India	Tanaka 1965
	8.9	3–25	14	65–259	S India	Poirier 1968a, 1969, 1970
		6–12			S India	Krishnan 1971
	18	7–27	3	2.3–3.9	S India	Horwich 1972
21–60	15	9–19	4	25	S India	Oates, pers. comm. 1978
		10–20			S India	Oates 1978

Presbytis obscura Dusky Leaf-Monkey

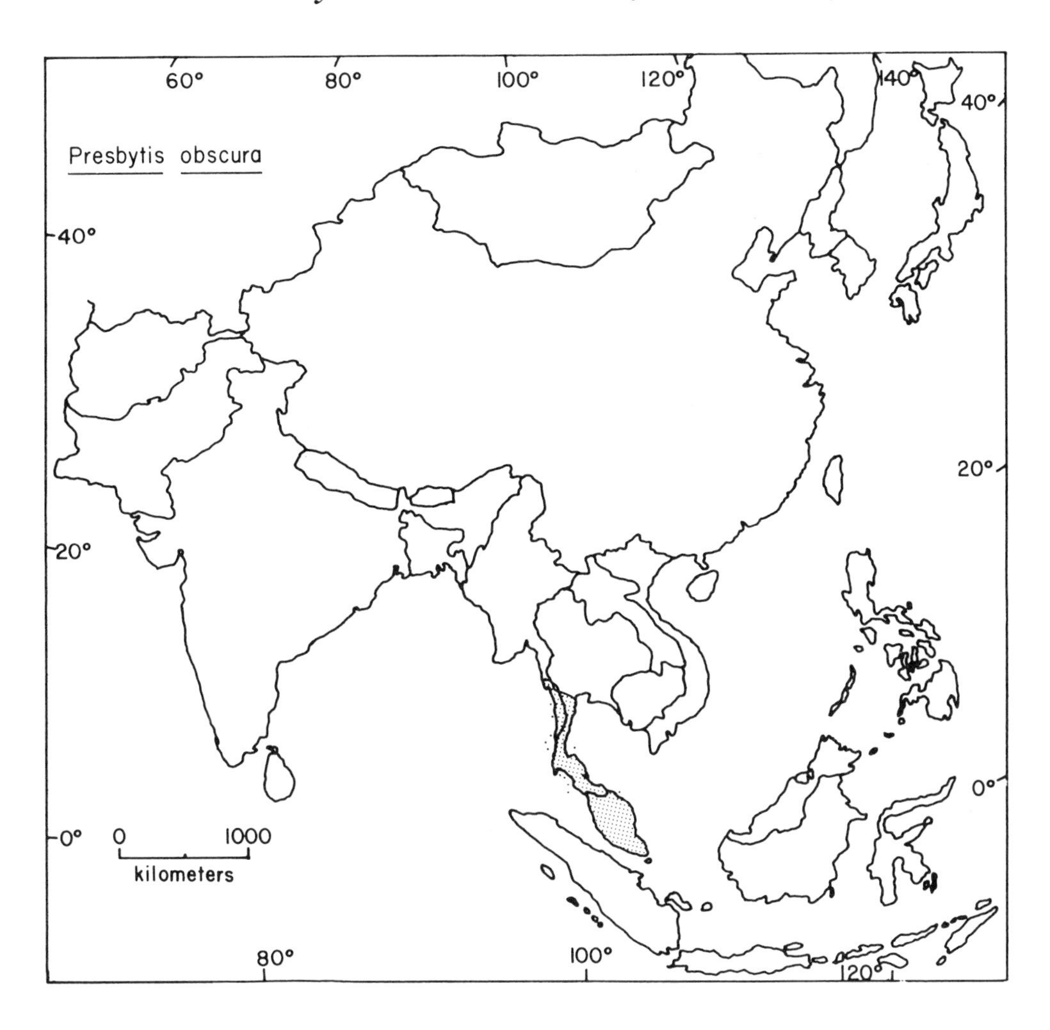

Taxonomic Note

Seven subspecies are recognized (Napier and Napier 1967).

Geographic Range

Dusky leaf-monkeys are found on the Burma-Thailand-Malaysian peninsula and on some of the adjacent small islands (Medway 1970). The northern limits of the range are at Tavoy (14°05′N, 98°12′E) in Tenasserim, Burma, and Phet Buri (13°10′N, 100′E) in Thailand; the southern limit is near the southern tip of the peninsula at about 1°25′N, 104°05′E (Ellerman and Morrison-Scott 1966, Fooden 1976). For details of the distribution, see Table 163.

Abundance and Density

Presbytis obscura has recently been described as abundant in southern peninsular Thailand and locally common but overall declining sharply in peninsular Malaysia (Table 163). Its status in Burma is not known.

This species lives in groups of 3 to 20 individuals, with average group sizes of 5 to 18, and home ranges of 9 to 33 ha (Table 164). Near Kuala Lompat its population density was estimated at 10 per km^2, suggesting that a minimum of 500 km^2 would be necessary as a reserve to protect a population of 5,000 *P. obscura* (Medway and Wells 1971).

Habitat

P. obscura is a forest species found in primary rain forest and old growth secondary forest (Curtin 1976). In one area 62% of groups were in secondary and 25% in primary forest, but population density was slightly higher in primary than in secondary forest (Southwick and Cadigan 1972). Among the kinds of trees *P. obscura* utilizes are dipterocarp, leguminous, fig, bamboo, citrus, jackfruit, and rubber (Fooden 1976).

Dusky leaf-monkeys have been reported in lowland forest (Harrison 1969), submontane dipterocarp forest (McClure 1964), and montane areas up to 1,828 m elevation (Medway 1970). They are more common above 600 m than below (McClure 1964), and their mean elevation of occurrence is 330 m (Chivers 1973).

This species can also utilize rubber plantations adjacent to forest (Medway 1970) and urban forest in close association with man (Southwick and Cadigan 1972).

Factors Affecting Populations

Habitat Alteration

In peninsular Malaysia, where clearcutting is proceeding at the rate of 200 km^2 per year (Southwick and Cadigan 1972), *P. obscura* is being seriously disturbed by logging (Burgess 1971) and by clearing for agriculture and rural development (Khan 1978).

Since *P. obscura* adapts poorly to habitat changes, logging is especially detrimental to it and is responsible for the population decline seen in some areas (Curtin 1976). This species is also being threatened by tin and other kinds of mining, and by illegal forest clearance for cultivation in central Perak, peninsular Malaysia (Weber 1972).

No information is available regarding the effect of habitat alteration on this species in Burma or Thailand.

Human Predation

Hunting. Leaf-monkeys are hunted for meat in peninsular Malaysia (Southwick and Cadigan 1972; Curtin, pers. comm. 1978) and for medicinal purposes in Thailand (Lekagul 1968), but the effect of this hunting on *P. obscura* populations has not been determined.

Pest Control. *P. obscura* damages rubber trees and seedlings (Medway 1969a, Khan 1978), but no mention of control efforts was located.

Collection. This species is seldom kept as a pet or collected for export from peninsular Malaysia (Khan 1978).

Between 1968 and 1973 a total of 51 *P. obscura* were imported into the United States (Appendix), primarily from Thailand. In 1965–75, 62 were imported into the United Kingdom (Burton 1978). None entered the United States in 1977 (Bittner et al. 1978).

Conservation Action

P. obscura occurs in the Krau Game Reserve (Curtin 1976) and is classified as protected in Malaysia (Yong 1973). In Thailand, hunting, capture, and keeping it in captivity is allowed only under license, and export is permitted only by special permission from the Wildlife Advisory Committee (Lekagul and McNeely 1977).

Table 163. Distribution of *Presbytis obscura*

Country	Presence and Abundance	Reference
Burma	In S and on Mergui Archipelago (off W coast, 11–13°N)	Medway 1970
	Mainland Tenasserim and island localities mapped	Fooden 1976
Thailand	In peninsular and on Phuket and Terutao islands (off W coast, 8–9°N)	Medway 1970
	Abundant in S peninsula (localities mapped)	Fooden 1976
	In peninsular (map)	Lekagul and McNeely 1977
Malaysia: Peninsular	Common 35 km E of Kuala Lumpur	McClure 1964
	Widespread throughout	Harrison 1966
	Throughout inland (map)	Lim 1969
	Widespread inland N of Johore; also on islands of Langkawi and Penang (off W coast) and Perhentian Besar (off E coast)	Medway 1969a, 1970
	In N and central (localities mapped)	Chivers 1971
	At Kuala Lompat, Pahang (W central)	Medway and Wells 1971
	Common low on Gunong Benom (W central)	Medway 1972
	Threatened in Limeston Hills area (central Perak)	Weber 1972
	At Kuala Lompat, Pahang (W central)	Chivers 1973
	Common but not abundant, declining slowly in Krau Game Reserve, Pahang (W central)	Curtin 1976
	In W and central, few records in E (localities mapped)	Fooden 1976
	Estimated total population in 1958: 305,000; in 1975: 155,000; this is 49% decrease in 17 years	Khan 1978

Table 164. Population parameters of *Presbytis obscura*

Population Density (individuals/km²)	Group Size			Home Range Area (ha)	Study Site	Reference
	$\bar{X}$	Range	No. of Groups			
		≤15			Malaysia	McClure 1964
	13.2	9–17	5		Malaysia	Bernstein 1967b
		6–20			Malaysia	Medway 1969a
10	5		2		Malaysia	Medway and Wells 1971
	12.5	5–18	4	9–15	Malaysia	Chivers 1973
	7.1	3–20	23		Malaysia	Fooden 1976
	18	17–19		25–33	Malaysia	Curtin 1977

Presbytis phayrei Phayre's Leaf-Monkey

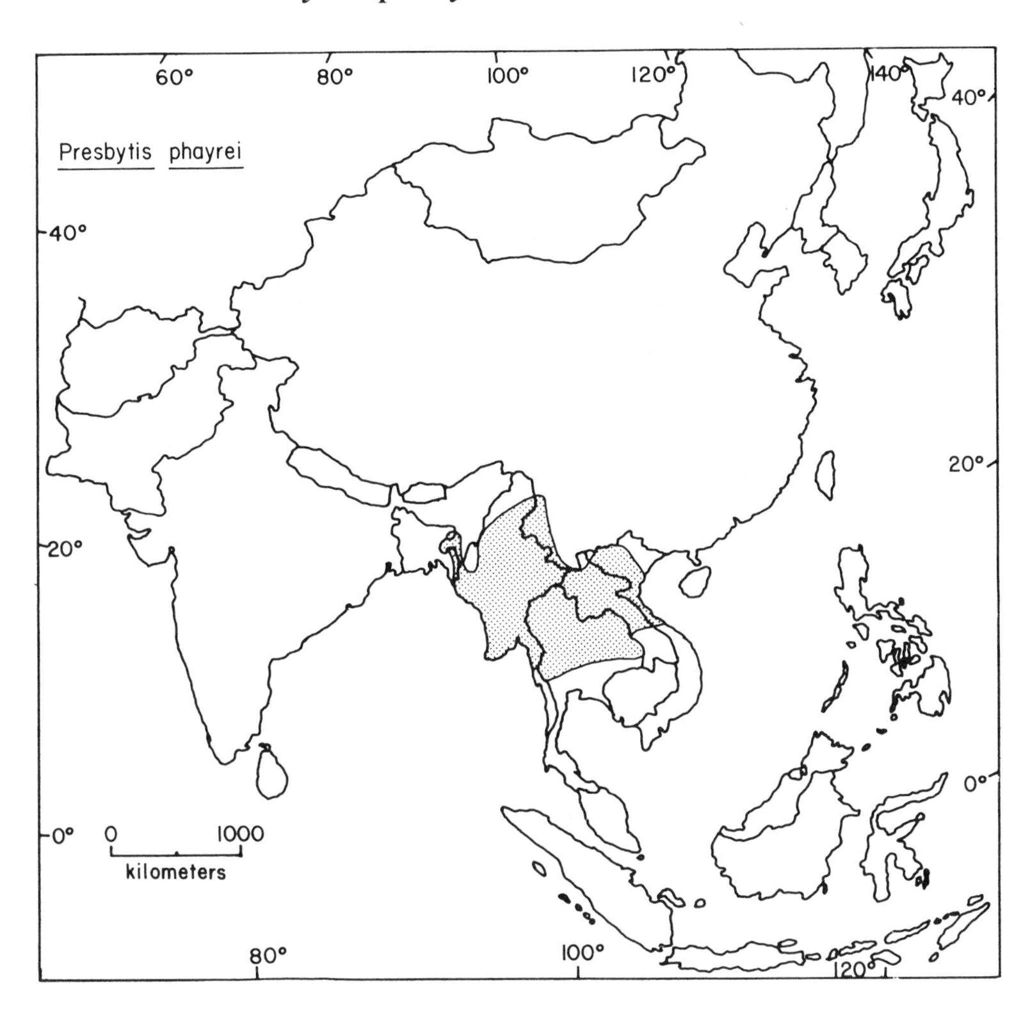

Taxonomic Note

Four subspecies are recognized (Ellerman and Morrison-Scott 1966, Napier and Napier 1967).

Geographic Range

Presbytis phayrei is found in Southeast Asia from Bangladesh to Vietnam. It has been reported from as far west as about 91°18′E (at 23°38′N) in Tripura, India (Agrawal 1974), and as far east as 105°38′E (at 20°19′N) in northern Vietnam (Fooden 1976). It occurs as far north as 25°N, 98°45′E in China and as far south as 14°41′N, 98°52′E in Thailand (Fooden 1976). For details of the distribution, see Table 165.

Abundance and Density

No information is available regarding the abundance of this species in any country. In Bangladesh, 6.8 individuals per km^2 were seen in secondary forest and 2.1 individuals per km^2 in primary forest (Green 1978a). In Thailand 10 groups were sighted, varying in estimated size from 3 to 30 individuals ($\bar{X}$ = 14.4) (Fooden 1971a). No determinations of home range size were located.

Habitat

P. phayrei is found in evergreen forest (Fooden 1971a), secondary forest, primary forest, and tea estates with bamboo thickets (Green 1978a).

Factors Affecting Populations

Habitat Alteration

The destruction of forest, due to the rapidly expanding human population's demand for building materials and firewood, is adversely affecting *P. phayrei* populations in Bangladesh (Green 1978a). Although no direct information is available, military activity in Indochina has undoubtedly been detrimental to its forest habitats there (Orians and Pfeiffer 1970, S.I.P.R.I. 1976).

Human Predation

Hunting. In Burma, Phayre's leaf-monkeys were hunted to extinction in areas around salt springs to obtain their huge lime-induced gallstones (bezoar stones), highly valued as medicine by the Chinese (Lekagul and McNeely 1977).

Pest Control. No information.

Collection. Five *P. phayrei* were imported into the United States between 1968 and 1973 (Appendix). None were imported into the United Kingdom between 1965 and 1975 (Burton 1978) or into the United States in 1977 (Bittner et al. 1978).

Conservation Action

In Thailand, *P. phayrei* may be hunted, captured, or kept in captivity only under license and may be exported only with special permission from the Wildlife Advisory Committee (Lekagul and McNeely 1977). No information is available regarding protection in other countries.

Table 165. Distribution of *Presbytis phayrei* (countries are listed in order from west to east)

Country	Presence and Abundance	Reference
Bangladesh	In E and SE (map)	Fooden 1976
	In S Sylhet (in E) and in E Chittagong (in SE) (localities mapped)	Green 1978a
India	In W and central Tripura (in E) (localities listed)	Agrawal 1974
Burma	In Pidaung Wildlife Sanctuary (in N near 25°N on Irrawaddy R.)	Milton and Estes 1963
	Throughout central and S from Bhamo in NE (24°15′N, 97°15′E) to Tenasserim	Ellerman and Morrison-Scott 1966
	Throughout most, S of about 25°N and N of 16°N (localities mapped)	Fooden 1976
China	In Hsi-Shuan-Pan-Na area (about 22°N, 101°E) extreme SW Yunnan Province	Kao et al. 1962
	In extreme W Yunnan (localities mapped)	Fooden 1976
Thailand	In W as far S as Chong Krong (14°41′N) and as far N as Huai Kwang Pah (about 17°25′N)	Fooden 1971a
	Throughout except in S (localities mapped)	Fooden 1976
	In Huay Kha Khaeng Game Sanctuary (in W, near 15°29′N)	Mittermeier 1977b
Laos	Throughout most of N, N of 17°N (localities mapped)	Fooden 1976
Vietnam	In Phu Quy region (N central)	Dào Van Tien 1963
	In N and W (localities mapped)	Fooden 1976

Presbytis pileata Capped Langur

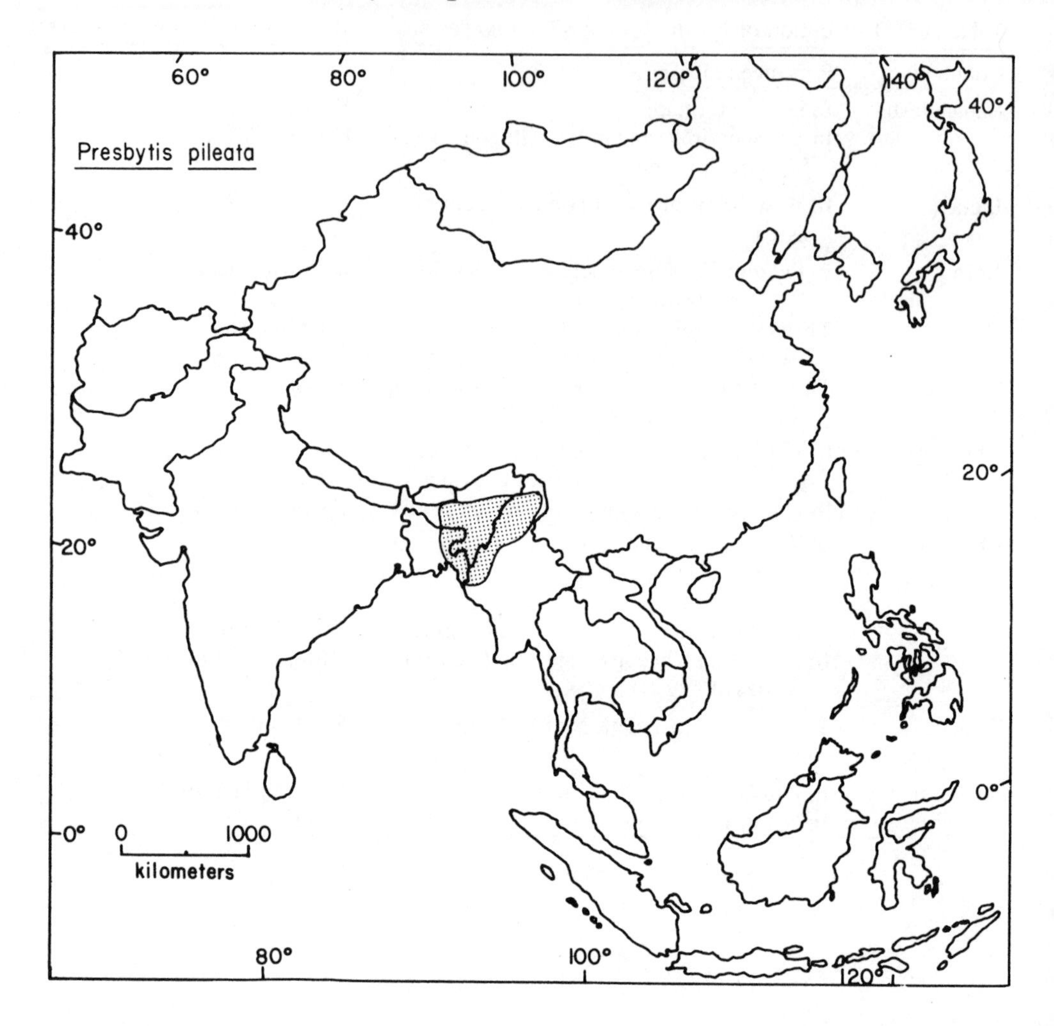

Taxonomic Note

Five subspecies are recognized (Napier and Napier 1967).

Geographic Range

Presbytis pileata is found from eastern Bangladesh and Assam, India, to northern Burma. It has been reported as far west as 90°23′E (at 24°45′N) in Bangladesh (Oppenheimer 1977), as far north and east as about 28°N, 98°E, and as far south as about 21°N, all in Burma (Fooden 1976). For details of the distribution, see Table 166.

Abundance and Density

P. pileata has been described as abundant in one forest in Bangladesh, common in the Garo Hills, and 20 years ago was considered in danger in Assam (Table 166). No recent reports regarding its status throughout Assam, Bangladesh, or Burma are available.

This species has been found at population densities of 1.5 to 31.1 individuals per km^2, and in groups whose average size was 8.7 (Green 1978a). No other population parameters were located.

Habitat

Capped langurs inhabit tropical wet deciduous and tropical evergreen rain forest (Oboussier and Maydell 1959), including dense forest near streams (McCann 1933a) and "wooded areas" (Lahan and Sonowal 1973). In Bangladesh they were observed most frequently in sal (*Shorea robusta*) forest, next most often in primary forest, and least often in secondary forest (Green 1978a).

Factors Affecting Populations

Habitat Alteration

Destruction of forest due to the rapidly expanding human population's demand for building materials and firewood is reducing the range of *P. pileata* in Bangladesh (Green 1978a). Moderate habitat reduction has also occurred in Assam (Tilson, pers. comm. 1975). The situation in Burma has not been investigated.

Human Predation

Hunting. *P. pileata* has been hunted and collected heavily in Assam (Oboussier and Maydell 1959) and is eaten by the local people in the Garo Hills District (Biswas 1970).

Pest Control. No information.

Collection. Two *P. pileata* were imported into the United States between 1968 and 1973 (Appendix). None were imported into the United Kingdom between 1965 and 1975 (Burton 1978), or into the United States in 1977 (Bittner et al. 1978).

Conservation Action

P. pileata is found in two wildlife sanctuaries in India and one reserve in Bangladesh (Table 166). It is listed in Appendix I of the Convention on International Trade in Endangered Species of Wild Fauna and Flora and as endangered under the U.S. Endangered Species Act (U.S. Department of Interior 1977a,b).

Table 166. Distribution of *Presbytis pileata*

Country	Presence and Abundance	Reference
Bangladesh	In E and SE (localities mapped)	Fooden 1976
	In Madhupur Forest S of Mymensingh	Oppenheimer 1977
	In E and SE (localities mapped); abundant in Madhupur Forest; in Kalinga Reserve, Sylhet	Green 1978a
India	In all of Assam including districts of Garo Hills, Jaintia Hills, Mikir Hills (central),	Oboussier and Maydell 1959

Table 166 (continued)

Country	Presence and Abundance	Reference
India (cont.)	Naga Hills (E), Mishmi Hills (NE), Dafla Hills, Khasi and Manipur (SE); in danger	Oboussier and Maydell 1959 (cont.)
	In Tura, Garo Hills District, W Assam	Biswas 1970
	In Kaziranga Wildlife Sanctuary, S of Brahmaputra R., Assam	Lahan and Sonowal 1973
	In Manas Sanctuary W of Manas R., N border of Assam with Bhutan	Ghosh and Biswas 1975
	N and S of Brahmaputra R., E Assam	Tilson, pers. comm. 1975
	Throughout Assam except extreme W (localities mapped)	Fooden 1976
	In N Assam N of Brahmaputra R. in Dafla Hills, N Lakhimpur, Seajulia, Nagaland, Meghalaya, and Tripura	Roonwal and Mohnot 1977
	Common in Garo Hills, status elsewhere uncertain; including Arunachal, Pradesh, and Manipur	Mohnot 1978
Burma	In W and N (localities mapped)	Fooden 1976
	In Upper Chindwin, E of Chindwin R.	Roonwal and Mohnot 1977

Presbytis potenziani Mentawi Island Leaf-Monkey

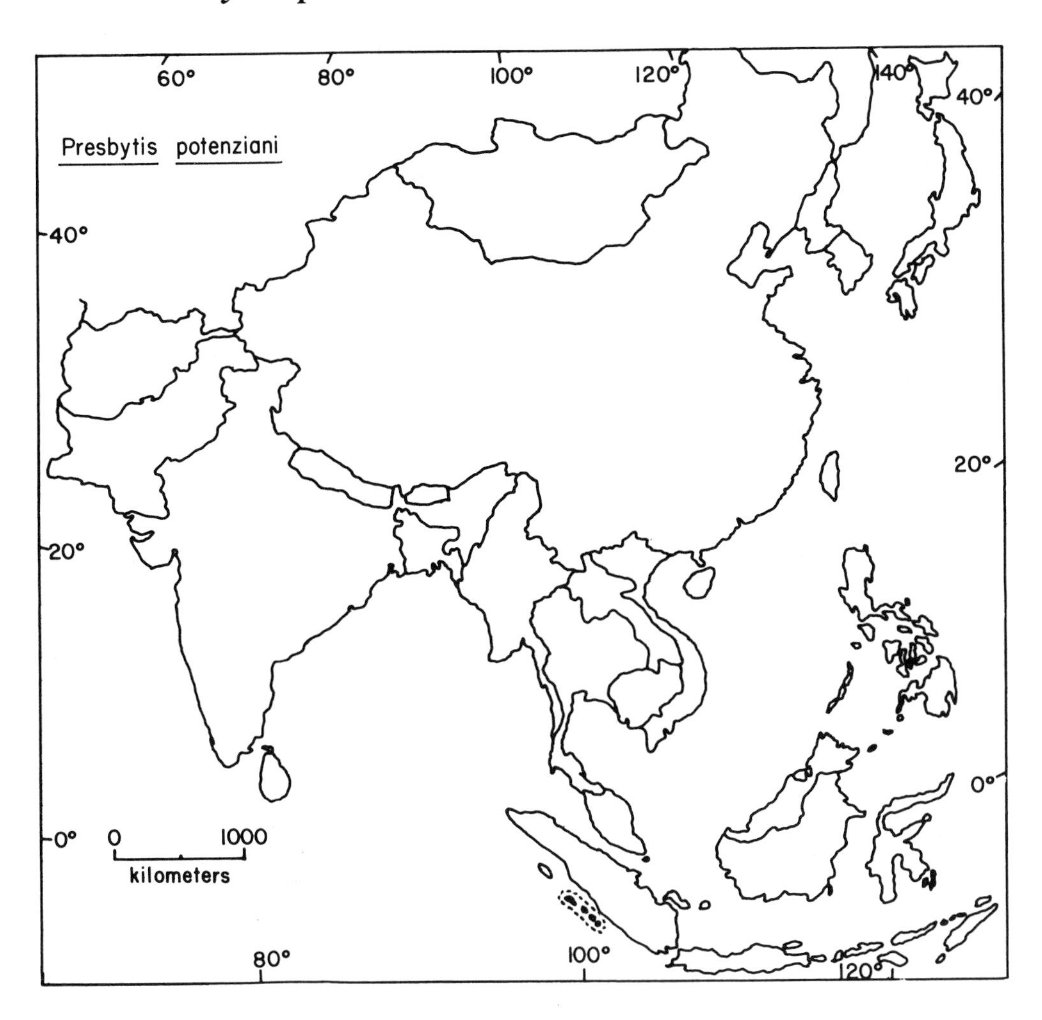

Taxonomic Note

Two subspecies are recognized (Napier and Napier 1967).

Geographic Range

Presbytis potenziani is found on the Mentawi Islands, Indonesia, located between 1°–4°S and 98°–101°E, about 100 km west of Sumatra. It occurs on all four of these islands: Siberut, Sipora, Pagai Utara (north), and Pagai Selatan (south) (Tilson and Tenaza 1976). No further details of its distribution are available.

Abundance and Density

P. potenziani is scarce and declining in all lowland regions and is not abundant anywhere on Sipora, Pagai Utara, or Pagai Selatan, but is "abundant and remaining stable in the hilly interior of Siberut'' (Tilson, pers. comm. 1974). Other workers have described *P. potenziani* as generally endangered or ''status indeterminate'' (I.U.C.N. 1978).

This species has been observed at population densities of 15 individuals per km^2 (Tilson, pers. comm. 1974) and 20 per km^2 (Tenaza, pers. comm. 1974). In central Siberut, nine groups consisted of 2 to 5 members each (Tilson, pers. comm. 1974), with average group sizes of 3.4 to 4 individuals and home ranges of 15 to 22 ha (Tilson and Tenaza 1976).

Habitat

P. potenziani prefers primary forest habitat, and is most abundant in primary hill forest, scarce and declining in primary and secondary lowland forest (Tilson, pers. comm. 1974), and also present in mangrove swamps (Tenaza, pers. comm. 1974).

Factors Affecting Populations

Habitat Alteration

Selective logging, in which trees more than 60 cm in diameter are cut, was begun on Siberut in 1970, at a rate of about 10% of the forest per year (Tenaza, pers. comm. 1974). The effect of this disturbance on primate populations has not yet been determined, but will probably be detrimental not only directly but indirectly, by making remote areas of the forest accessible to settlers and hunters (Tenaza, pers. comm. 1974; Tilson, pers. comm. 1974).

Human Predation

Hunting. The major cause of the decline of *P. potenziani* is extensive hunting for food by the Mentawi Island people (Tilson, pers. comm. 1974). On Siberut, man is a major predator and hunting pressure is intensive (Tilson and Tenaza 1976). On Sipora and the Pagai Islands, human populations are growing rapidly (a fivefold increase during the last 150 years), and pose the threat of excessive hunting (Tenaza, pers. comm. 1974).

Pest Control. P. potenziani does not utilize food from gardens (Tilson, pers. comm. 1974), hence would not be expected to suffer from control efforts.

Collection. Mentawi Island leaf-monkeys seldom survive in captivity for longer than a week, and are rarely captured and kept as pets or sold for research or zoos (Tilson, pers. comm. 1974). None were imported into the United States between 1968 and 1973 (Appendix) or in 1977 (Bittner et al. 1978). Eighteen were imported into the United Kingdom between 1965 and 1975 (Burton 1978).

Conservation Action

In 1974 the Teiteibatti Wildlife Reserve was established on Siberut Island as a sanctuary for *P. potenziani* and other Mentawi Island primates (I.U.C.N. 1978). Reserves are also needed on Sipora and the Pagai Islands, as are well-enforced laws against all hunting of *P. potenziani* on Sipora and against the use of dogs and guns for hunting on Siberut (Tenaza, pers. comm. 1974). Investigation of alternate sources of protein for the local people, and further field studies, have also been recommended (I.U.C.N. 1978).

P. potenziani is listed in Appendix I of the Convention on International Trade in Endangered Species of Wild Fauna and Flora, and as threatened under the U.S. Endangered Species Act (U.S. Department of Interior 1977a,b).

Presbytis rubicunda Maroon Leaf-Monkey

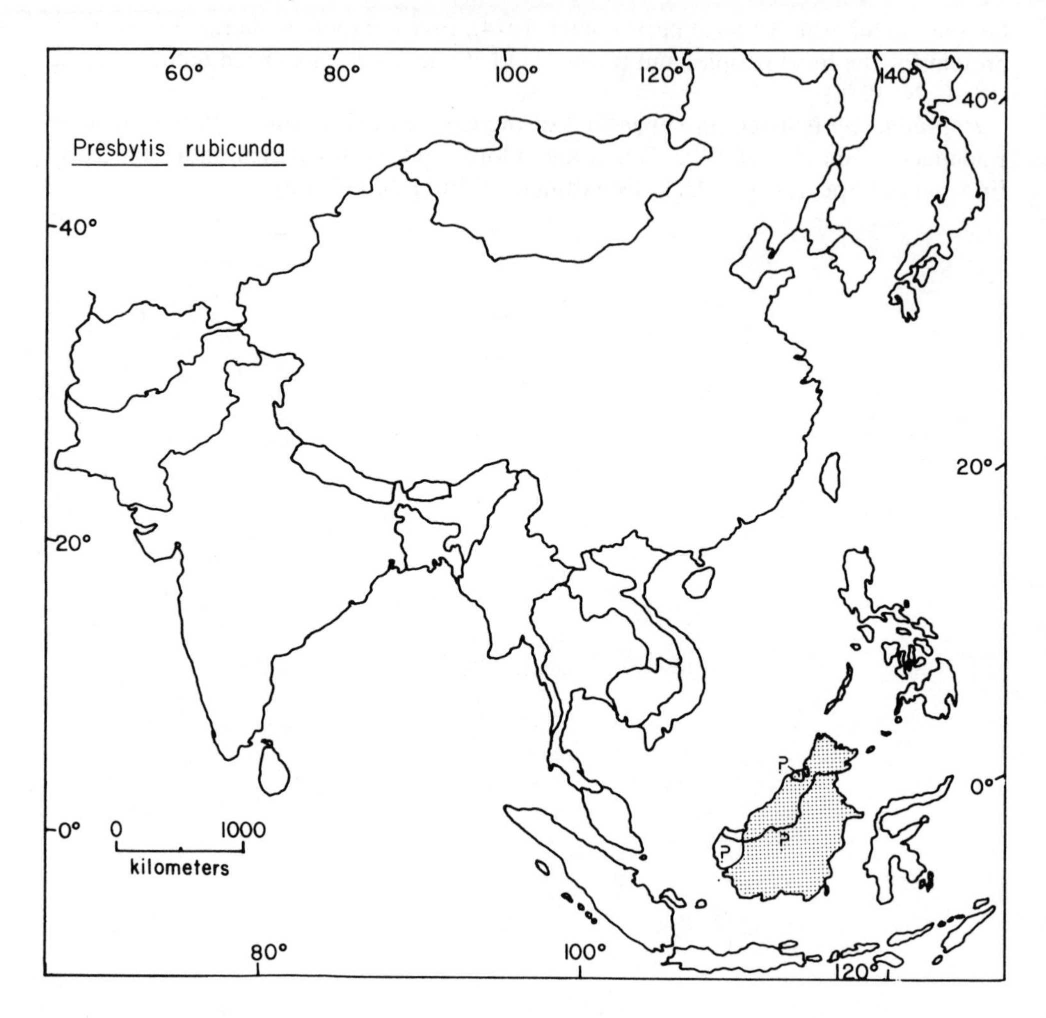

Taxonomic Note

Five subspecies are recognized (Napier and Napier 1967).

Geographic Range

Presbytis rubicunda is found on Borneo and on the Karimata Islands about 150 km west of Borneo (Davis 1962). It has been reported throughout most of Sabah and Sarawak (Malaysian Borneo) in the north, and along the southern coastal areas of Kalimantan (Indonesian Borneo), but not in Brunei or in the interior or the north-western corner of Borneo (Medway 1970). For details of the distribution, see Table 167.

Abundance and Density

P. rubicunda has been described as locally abundant and common in parts of Sabah, and common in one area but rare in another in eastern Kalimantan (Table

167). No information is available regarding its abundance throughout most of Kalimantan or in Sarawak.

At one study site this species was observed at a population density of 12 individuals per km^2 (Wilson and Wilson 1975a). Group sizes of 5 to 8 (Stott and Selsor 1961a), 2 to 8 (Davis 1962), and 5 to 13 (Yoshiba 1964) have been reported. No estimates of home range area have been made.

Habitat

P. rubicunda occupies primary and secondary forest from sea level to 2,000 m elevation (Davis 1962), including lowland rain forest on the inland plains of Sabah (Kawabe and Mano 1972). It is often seen in edge habitats at the margins of forest bordering on cleared land, but also occurs in the interior of primary forest and secondary growth (Stott and Selsor 1961a).

Factors Affecting Populations

Habitat Alteration

Maroon leaf-monkeys frequently utilize the luxuriant second growth that appears in a forest disturbed by selective logging (four trees per hectare) in Sabah (Stott and Selsor 1961a), and occupy selectively logged forest in eastern Kalimantan (Wilson and Wilson 1975a). This species also has been reported in secondary growth bordering railroad track (Stott and Selsor 1961a), and in old logged forest (Davis 1962), and seems to be "indifferent to moderate habitat disturbance" (Wilson and Wilson 1975a).

Human Predation

No information.

Conservation Action

No information.

Table 167. Distribution of *Presbytis rubicunda*

Country	Presence and Abundance	Reference
Malaysia: North Borneo	Present	Burgess 1961
	In NE Sabah, SW of Sandakan; locally abundant	Stott and Selsor 1961a
	In Sabah: near Sandakan Bay, Sapagaya Forest Reserve, Deramakot, Kalabakan (SE), Kinabatangan District (central), Bukit Kretam, Abai; on Karimata Islands (1°25′N, 109°05′E); common	Davis 1962
	Common at Kinabatangan R.; frequent at Labuk Road (W of Sandakan), Sabah	Yoshiba 1964
	In Sabah	de Silva 1968b
	Throughout Sabah and Sarawak (localities mapped)	Medway 1970
	In Ulu Segama Reserve, near junction of Bole and Segama rivers, E Sabah	MacKinnon 1971
	At Kinabatangan R., central Sabah	Kawabe and Mano 1972

Table 167 (continued)

Country	Presence and Abundance	Reference
Brunei	No information	
Indonesia: Kalimantan	In S, SW, and SE (localities mapped); no records from interior or NW	Medway 1970
	In Kutai Reserve (E central, near E coast and equator, Kutai R.)	Rodman 1973
	Rare in Kutai Reserve; common NW of Balikpapan near Sepaku	Wilson and Wilson 1975a

Presbytis senex Purple-faced Langur

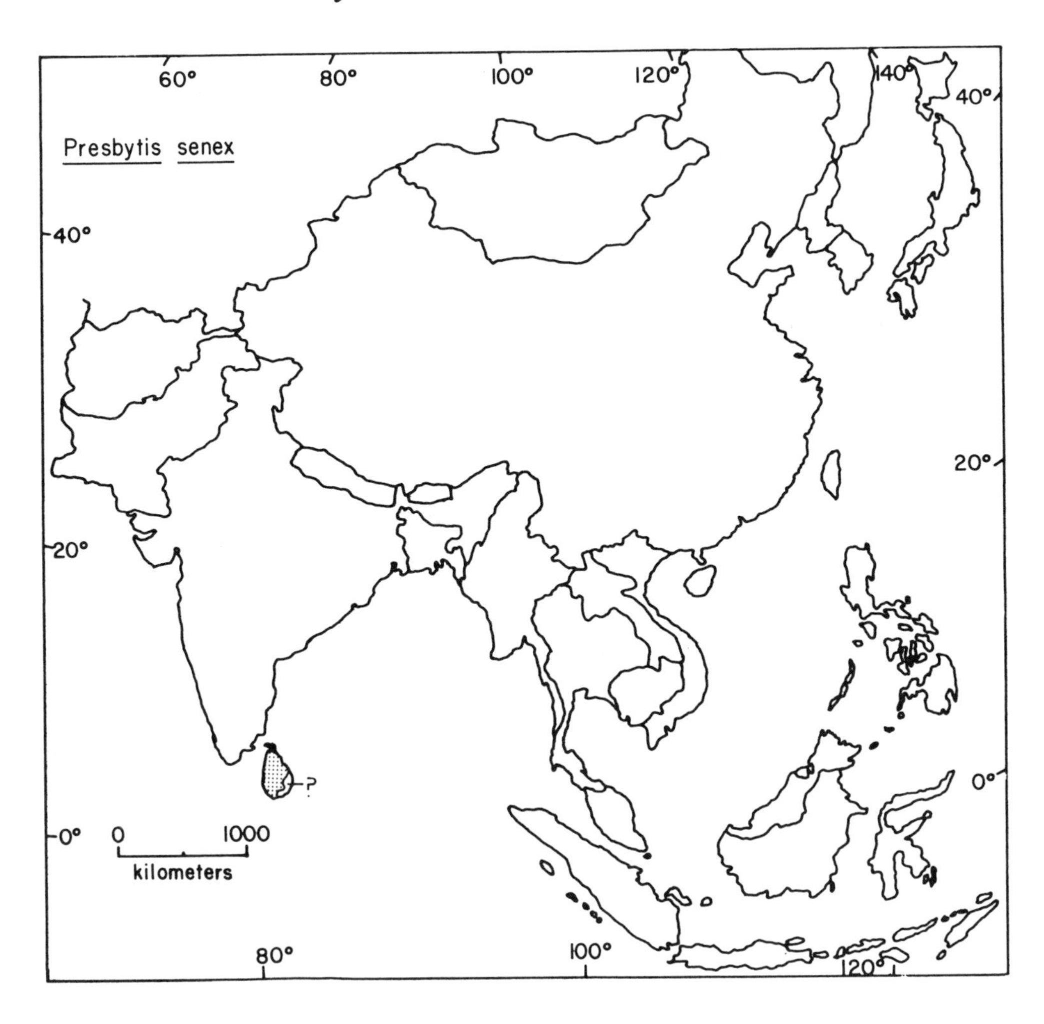

Taxonomic Note

Five subspecies are recognized (Deraniyagala 1955, Napier and Napier 1967).

Geographic Range

Presbytis senex is found on Sri Lanka (Ceylon), located between about 6–10°N and 79°30′–82°E, southeast of the Indian peninsula. It has been reported in almost every part of Sri Lanka except the southeastern region (Muckenhirn, pers. comm. 1974). For details of the distribution, see Table 168.

Abundance and Density

In the northcentral portion of its range *P. senex* is abundant but is suspected to be decreasing, while in the south and southwestern regions it has declined sharply (Table 168). As long ago as 1934 its populations were diminishing rapidly in the

Horton Plains area of southcentral Sri Lanka, and its range there is still shrinking (Table 168).

Purple-faced langurs move in groups of 2 to 20, with average groups of 5.8 to 8.5 members (Table 169). Population densities of 92 to 327 individuals per km^2 and home ranges of 0.9 to 14.9 ha have been reported (Table 169).

Habitat

P. senex is a forest species, found in both wet and dry forests. It occupies monsoon scrub jungle, monsoon forest (Eisenberg and McKay 1970), low altitude humid zone forest (Hladik and Hladik 1972), and semideciduous forest in dry zones (Hladik 1975). In seasonally dry areas it is most common in riverine forest (Eisenberg and Lockhart 1972; Muckenhirn, pers. comm. 1974). At one study site it was observed in mature secondary forest and parkland in the hot, dry, lowland, tropical zone (Rudran 1973a) and can also occupy lowland coastal areas (Rudran, pers. comm. 1974).

The central highland population of *P. senex* is found in undisturbed tropical cloud forest in the cool, wet montane zone (Rudran 1973a) up to 2,195 m elevation (Roonwal and Mohnot 1977).

Factors Affecting Populations

Habitat Alteration

Much of the forested area of Sri Lanka has undergone extensive logging and replacement with rubber, coconut, tea, and teak plantations (Dittus, pers. comm. 1974). Several regions have been completely deforested by government supervised and illegal logging, destroying habitat suitable for *P. senex* (Rudran, pers. comm. 1974). For example, the central highlands habitat has been greatly reduced by the expansion of agriculture, tea plantations, and commercial forestry. Although *P. senex* can utilize the *Eucalyptus* and *Acacia* plantations which have replaced much of the original vegetation, and can occupy small forest patches within tea fields, deforestation there has made large areas unsuitable for it (Rudran, pers. comm. 1974).

In the southwest portion of the range, near Colombo, the habitat has undergone severe disturbance, perhaps eliminating one form of *P. senex* (Muckenhirn, pers. comm. 1974). Farther south, extensive logging has occurred and more is planned (on recommendation of a UN/FAO report of 1969 and with financial aid from Canada), threatening the survival of the last remaining patch of lowland tropical rain forest in Sri Lanka and the southern race of *P. senex*, as well as several other endemic animals (Rudran, pers. comm. 1974).

Some populations of *P. senex* can occupy moderately disturbed habitats such as the one at Polonnaruwa where the underbrush has been cleared to expose archeological ruins (Rudran 1973a). But, this clearance forces the langurs to descend to the ground to cross treeless areas, exposing them to attacks by dogs (Rudran, pers. comm. 1974).

Human Predation

Hunting. No information.

Pest Control. *P. senex* eats cultivated plants such as potatoes and cauliflower (Hladik and Hladik 1972), but no reports of control efforts were located.

Collection. P. senex is sometimes kept as a pet (Rudran, pers. comm. 1974). Five were imported into the United States between 1968 and 1973 (Appendix). None were imported into the United States in 1977 (Bittner et al. 1978) or into the United Kingdom in 1965–75 (Burton 1978).

Conservation Action

P. senex is found in the Wilpattu National Park and the Horton Plains Nature Reserve (Table 168), but elsewhere it and its habitat are in need of protection. Rudran (pers. comm. 1974) has recommended that vigorous efforts be made to persuade the Sri Lanka government to cancel its plans to log the southern forest and conserve the remaining natural habitats there and in the Horton Plains area, where *P. senex* is especially threatened.

Sri Lanka's laws prohibit the killing of all wildlife but are difficult to enforce and do not prevent habitat destruction (Rudran, pers. comm. 1974). *P. senex* is listed as threatened under the U.S. Endangered Species Act (U.S. Department of Interior 1977b).

Table 168. Distribution of *Presbytis senex*

Country	Presence and Abundance	Reference
Sri Lanka	In E (localities listed); in SW and S, N and S of the Kala Ganga R. (about 6°30′N) to Balangda; in central highlands near Hakgala, Bopats, Horton Plains, Moon and Elk Plains: not abundant and declining rapidly	Phillips 1935
	In NW from Thunakai, Northern Province, to Vanativillu (8°8′N) and Tantrimale (80°20′E); in E, E and S of Matale, Kantale, and Polonnaruwa as far W as Paymadu; in central highlands	Deraniyagala 1955
	Rare in Wilpattu N.P. (in NW) but not rare in surrounding area	Eisenberg and Lockhart 1972
	At Polonnaruwa (in E, 7°55′N, 81°E) and Horton Plains (S central, 6°45′N, 81°E)	Rudran 1973a
	In N central (S of 8°30′N), abundant, not in danger; in S near Galle, declining sharply; in SW near Colombo, has declined, may be extinct; also present in NW and Horton Plains; absent in SE Ruhunu Forest (now N.P) (map)	Muckenhirn, pers. comm. 1974
	Abundant but probably declining at Polonnaruwa (NE central, near 8°N, 81°E); range restricted and decreasing, population at low density and probably increasing in Horton Plains Nature Reserve (S central)	Rudran, pers. comm. 1974
	At Polonnaruwa	Hladik 1975

Table 169. Population parameters of *Presbytis senex*

Population Density (individuals/km²)	Group Size			Home Range Area (ha)	Study Site	Reference
	X̄	Range	No. of Groups			
		3–20			Sri Lanka	Deraniyagala 1955
	5.8	4–7	9	1–7.5	Sri Lanka	Hladik and Hladik 1972
92–215	8.5		60	0.9–14.9	Sri Lanka	Rudran 1973b
116–327		2–15		0.9–10.9	Sri Lanka	Manley cited in Roonwal and Mohnot 1977

Pygathrix nemaeus Douc Langur

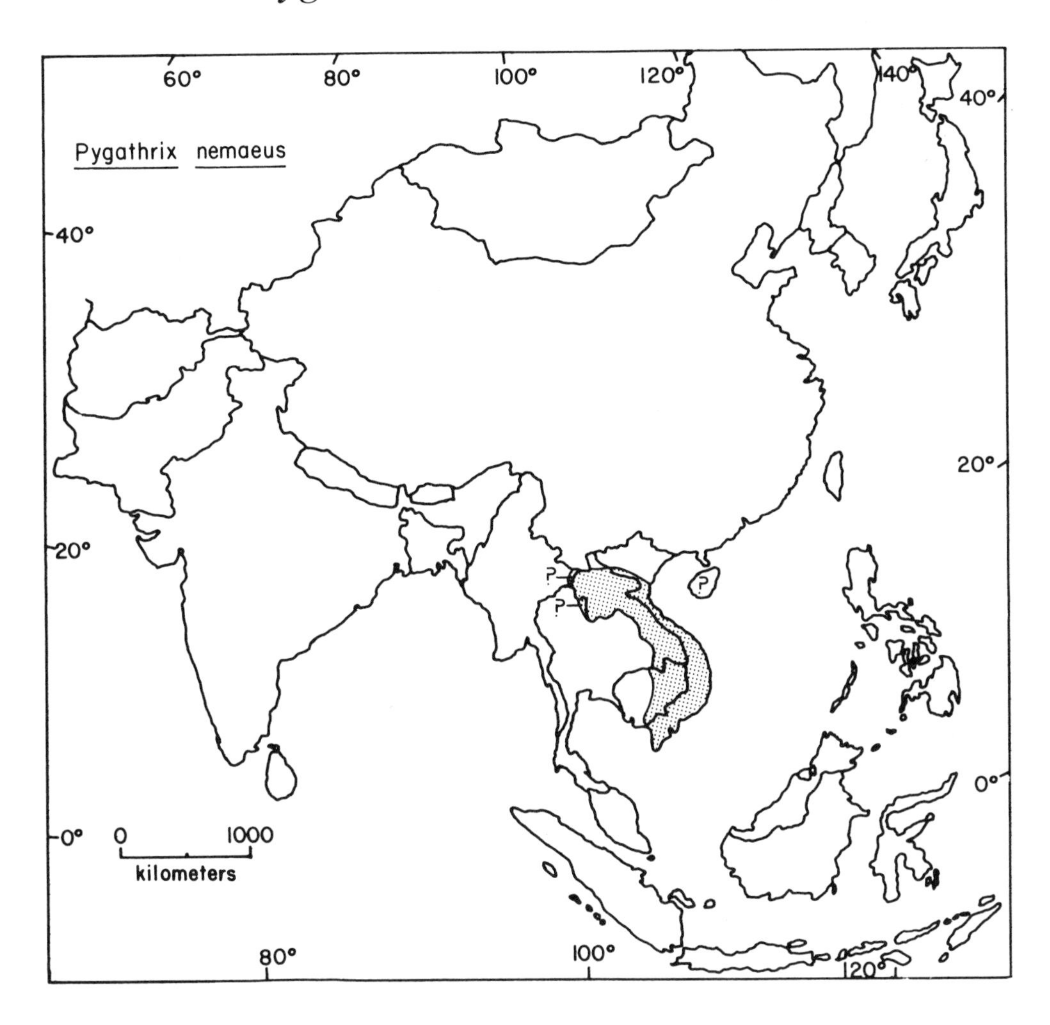

Taxonomic Note

Two subspecies are recognized (Ellerman and Morrison-Scott 1966, Napier and Napier 1967).

Geographic Range

Pygathrix nemaeus is found on the Indochinese peninsula. The only range map available shows the species present throughout Laos, eastern Cambodia, and Vietnam south of about 20°N (Fiedler 1956). This map also shows *P. nemaeus* occurring on the island of Hainan in the Gulf of Tonkin, but Groves (1970d) found no evidence of its presence there. Fiedler's map indicates that the range extends as far north and west as 22°N, 100°E, slightly into northern Thailand and eastern Burma, but no other authors report *P. nemaeus* in these countries. For details of the distribution, see Table 170.

Abundance and Density

P. nemaeus has probably always been rare and is now "in serious danger of extinction" (Southwick et al. 1970). It is classified as endangered by the I.U.C.N. (1972). Its populations were recently described as common in northeastern Cambodia but as reduced and declining in central Vietnam (Table 170). Information is lacking regarding its status throughout most of its range.

This species has been observed in pairs (Pfeffer 1969), groups of 5 and 7 (Gochfeld 1974), and groups of 8, 9, and 11 (Lippold 1977). No data pertaining to population density, average group size, or home range area are available.

Habitat

Douc langurs are forest monkeys that occupy gallery forest (Pfeffer 1969) and rain forest (Van Peenen et al. 1971). On Mount Sontra they are restricted to heavily forested areas of primary and secondary growth (Lippold 1977). They are found from sea level to 2,000 m elevation (Napier and Napier 1967), and on Mount Sontra are usually seen above 300 m (Van Peenen et al. 1969).

Factors Affecting Populations

Habitat Alteration

The recent warfare in Indochina has undoubtedly been detrimental to *P. nemaeus* habitats. In Vietnam especially, heavy bombing and the use of military herbicides have severely damaged many forested areas (Orians and Pfeiffer 1970). "Tens of thousands of acres" that had been sprayed with defoliants still showed no sign of regeneration in 1974 (Lippold 1977). No direct evidence is available, but these activities are believed to be responsible, at least in part, for the decline of *P. nemaeus* populations (I.U.C.N. 1972). The condition of the habitat in Laos and Cambodia is not known.

Human Predation

Hunting. *P. nemaeus* has been heavily hunted for meat by soldiers, guerrillas, and local people (I.U.C.N. 1972, Bassus 1973), and is also shot for target practice (Lippold 1977).

Pest Control. No information.

Collection. Douc langur populations have also been depleted by "waste in non-scientific commercial trade" (Southwick et al. 1970). Between 1968 and 1971 a total of 35 were imported into the United States (Appendix), and in 1970–71, 17 of these came from Thailand, where this species has not recently been recorded. None entered the United States in 1972 or 1973 (Appendix) or in 1977 (Bittner et al. 1978). Five were imported into the United Kingdom in 1965–75 (Burton 1978).

Conservation Action

P. nemaeus is now protected from hunting in Vietnam (Lippold 1977), but the effectiveness of this protection is not known. This species is listed in Appendix I of the Convention on International Trade in Endangered Species of Wild Fauna and

Flora, and as endangered under the U.S. Endangered Species Act (U.S. Department of Interior 1977a,b).

The establishment of six reserves to protect habitats and populations of *P. nemaeus* has been recommended by Lippold (1977). These areas are located (1) near Lao Cai (22°30′N, 104°E) at the Vietnam-China border, (2) near Kim Cnong (18°15′N, 105°E) at the Vietnam-Laos border, (3) at Mount Sontra (16°07′N, 108°18′E), (4) near Ban Me Thuot (12°40′N, 108°03′E) in the central highlands, (5) at Dalat (11°56′N, 108°25′E) in southcentral Vietnam, and (6) near Tinh Bien (10°30′N, 105°E) at the Vietnam-Cambodia border.

Table 170. Distribution of *Pygathrix nemaeus* (countries are listed in order from west to east)

Country	Presence and Abundance	Reference
Burma	In extreme E (map)	Fiedler 1956
Thailand	In extreme N (map)	Fiedler 1956
Laos	At Nape (E central, 18°18′N, 105°07′E)	Delacour 1940
	Throughout (map)	Fiedler 1956
	At Nape and in S	Groves 1970d
Cambodia	Throughout E, E of Mekong R. (map)	Fiedler 1956
	Still common in NE	Pfeffer 1969
Vietnam	Very common at Djiring (S central, 11°35′N, 108°04′E), also at Col des Nuages (central, 16°12′N, 108°08′E)	Delacour 1940
	Throughout except in extreme N (map)	Fiedler 1956
	In central and S (localities listed) in provinces of Bien Hoa, Darlac, Khanh Hoa, Lam Dong, Long Khanh, Quang Nam, Tay Ninh, Thua Thien, and Tuyen Duc	Van Peenen et al. 1969
	In central at Mount Sontra and Col des Nuages; in S at Saigon, Bien Hoa, Trang Bom, Djiring, Dalat, Langbian, and Nha Trang (localities mapped)	Groves 1970d
	Seen often 1965–67 but infrequently 1967–69 near Mount Sontra (central, about 16°N, 108°E)	Van Peenen et al. 1971
	S of Ha Tinh (central) (2 localities mapped)	Bassus 1973
	Have declined on Mount Sontra	Gochfeld 1974
	Population reduced; declining on Mount Sontra; localities listed; may still exist near Tinh Bien, SW of Saigon (10°30′N, 105°E) and in Ban Me Thuot area	Lippold 1977
China	On Hainan Island	Fiedler 1956, Ellerman and Morrison-Scott 1966, Napier and Napier 1967
	"No evidence for presence" on Hainan	Groves 1970d

Rhinopithecus avunculus Tonkin Snub-nosed Langur

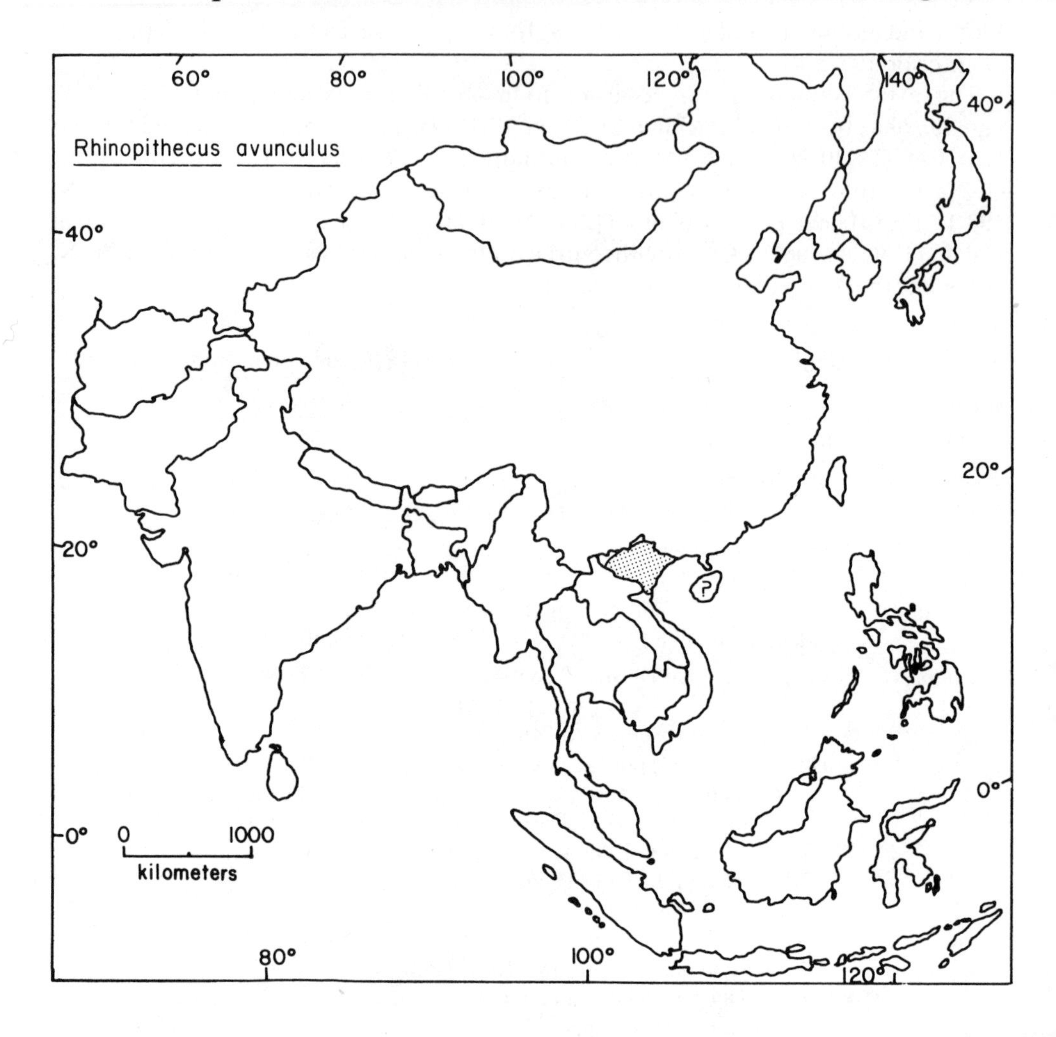

Taxonomic Note

Some recent authors have suggested that *Rhinopithecus* be placed in the genus *Pygathrix* (Groves 1970d, Thorington and Groves 1970). This species is monotypic (Ellerman and Morrison-Scott 1966, Napier and Napier 1967).

Geographic Range

Rhinopithecus avunculus is found in northern Vietnam. The only range map available shows it occurring between about 20°–23°N and 104°–108°E in the northern portion of Vietnam (Fiedler 1956). Bassus (1973) also recorded the "Tonkinlangur" on the island of Hainan, but no other workers report *R. avunculus* there. For details of the distribution, see Table 171.

Abundance and Density

No information.

Habitat

R. avunculus is found in tropical rain forest (Napier and Napier 1967). No other habitat descriptions are available.

Factors Affecting Populations

No information.

Conservation Action

R. avunculus is listed as threatened under the U.S. Endangered Species Act (U.S. Department of Interior 1977b).

Table 171. Distribution of *Rhinopithecus avunculus*

Country	Presence and Abundance	Reference
Vietnam	Near Bac Kan, 22°09′N, 105°50′E (N central)	Delacour 1940
	At Tiuen-Kuang (probably Tuyen Quang, N central, 21°48′N, 105°18′E)	Lapin et al. 1965
	At Bac Kan and at Yen Bay (N central, 21°43′N, 104°54′E)	Groves 1970d
China	On Hainan Island, Gulf of Tonkin (map)	Bassus 1973

Rhinopithecus roxellanae

Snub-nosed Langur or Golden Monkey

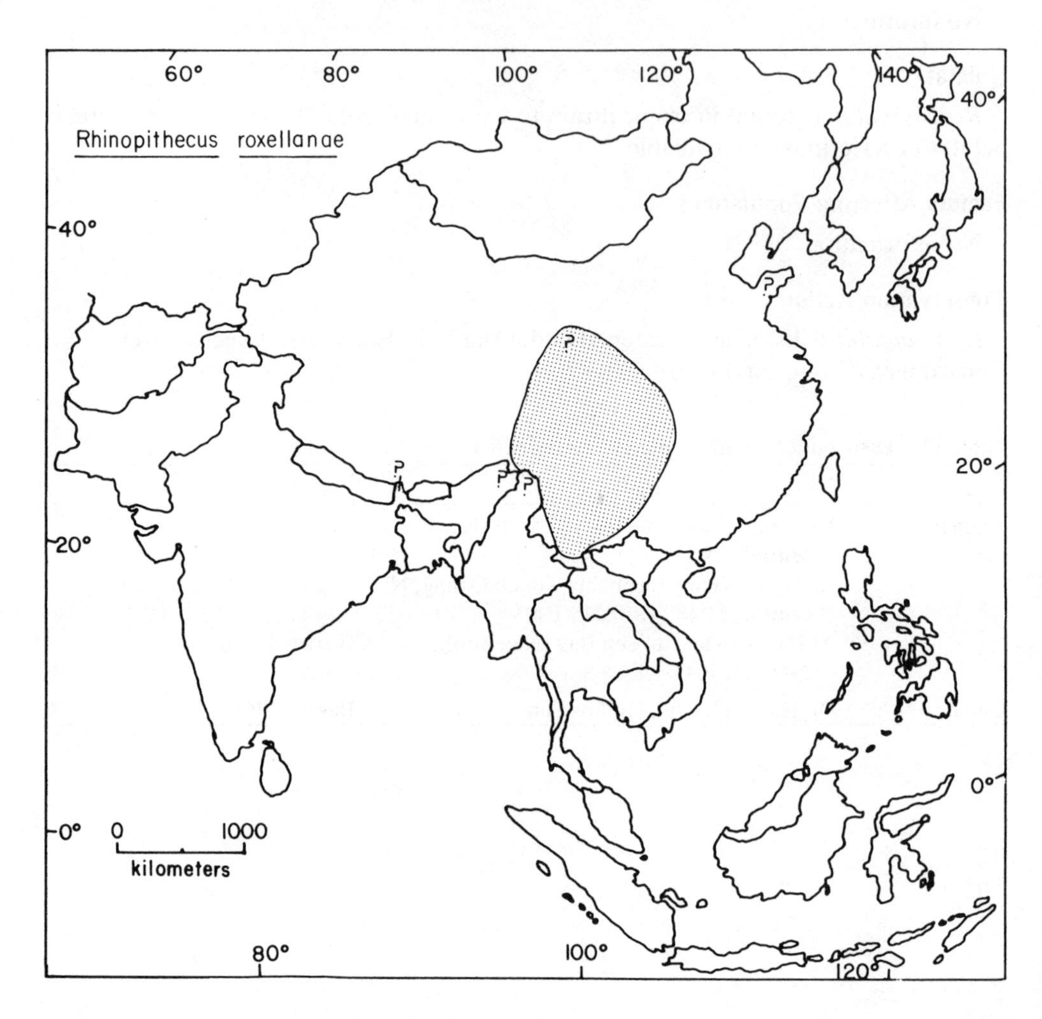

Taxonomic Note

Three subspecies are recognized (Ellerman and Morrison-Scott 1966, Napier and Napier 1967). Two recent authors have suggested placing *Rhinopithecus* within the genus *Pygathrix*, and consider the form *R. roxellanae brelichi* to be a separate species (Groves 1970d, Thorington and Groves 1970).

Geographic Range

Rhinopithecus roxellanae is found in southcentral China, in Kansu, Szechuan, Yunnan, and Kweichow provinces (Ellerman and Morrison-Scott 1966). The only range map available shows the species occurring from about 22°N to 36°N and 98°E to 110°E (Fiedler 1956). The range includes southeastern Tibet (Groves 1970d) and may extend into Assam, eastern India (I.U.C.N. 1976, Roonwal and Mohnot 1977), or even farther west to Sikkim, and east as far as Mou-p'ing (37°23′N, 121°35′E) near the east coast of China (Krumbiegel 1978). For details of the distribution, see Table 172.

Abundance and Density

More than 40 years ago, *R. roxellanae* was "in need of government protection if it [was] to be saved from extermination" (Sowerby 1937). It is now considered rare (I.U.C.N. 1976), but no information about its present status, population density, group size, or home range area is available.

Habitat

R. roxellanae occupies coniferous forest, bamboo jungle, and rhododendron thickets in montane areas up to 3,000 m elevation, and descends to lower cultivated valleys during winter snows (Napier and Napier 1967). No details of its habitat requirements have been reported.

Factors Affecting Populations

Habitat Alteration

No information.

Human Predation

Hunting. R. roxellanae was formerly hunted extensively for its highly prized fur (Napier and Napier 1967) and is still "at risk from constant hunting" (I.U.C.N. 1976).

Pest Control. No information.

Collection. No *R. roxellanae* were imported into the United States between 1968 and 1973 (Appendix) or in 1977 (Bittner et al. 1978), nor into the United Kingdom between 1965 and 1975 (Burton 1978).

Conservation Action

Snub-nosed langurs are found in at least one nature reserve (Table 172). Shooting and capture of this species are both strictly prohibited in China (I.U.C.N. 1976).

Table 172. Distribution of *Rhinopithecus roxellanae*

Country	Presence and Abundance	Reference
China	In E Szechuan Mts.: at Kwanhsien (31°N, 103°37′E), Yangliupa (32°25′N, 109°40′E), Moupin and Wen Chuan Hsien (in Szechuan Province), and Ssigou (Kansu Province); between the Yangtze and Mekong rivers at Atentse and Tsekou (about 29°N, 98°E, Yunnan Province); in Van Gin Shan range, S of middle Yangtze R. (27°–28°N, 106°–108°E, Kweichow Province) (localities mapped)	Groves 1970d

Table 172 (continued)

Country	Presence and Abundance	Reference
China (cont.)	On SE slope of Mount Fang Chin, NE Kweichow (27°40′–28°30′N, 108°30′–109°30′E), in area of 500 km^2; in Wanglang Nature Reserve, Szechuan; overall rare	I.U.C.N. 1976
	Localities listed and mapped include Sikkim (N India), upper Burma, Yunnan, the upper Mekong R., Kansu, Tseku, Szechuan, Kokonoor, S Shensi, Wanglang, Kweichow, and Mou'ping (probably Mou-p'ing, 37°23′N, 121°35′E)	Krumbiegel 1978

Simias concolor

Pig-tailed or Pagai Island Langur or Simakobu Monkey

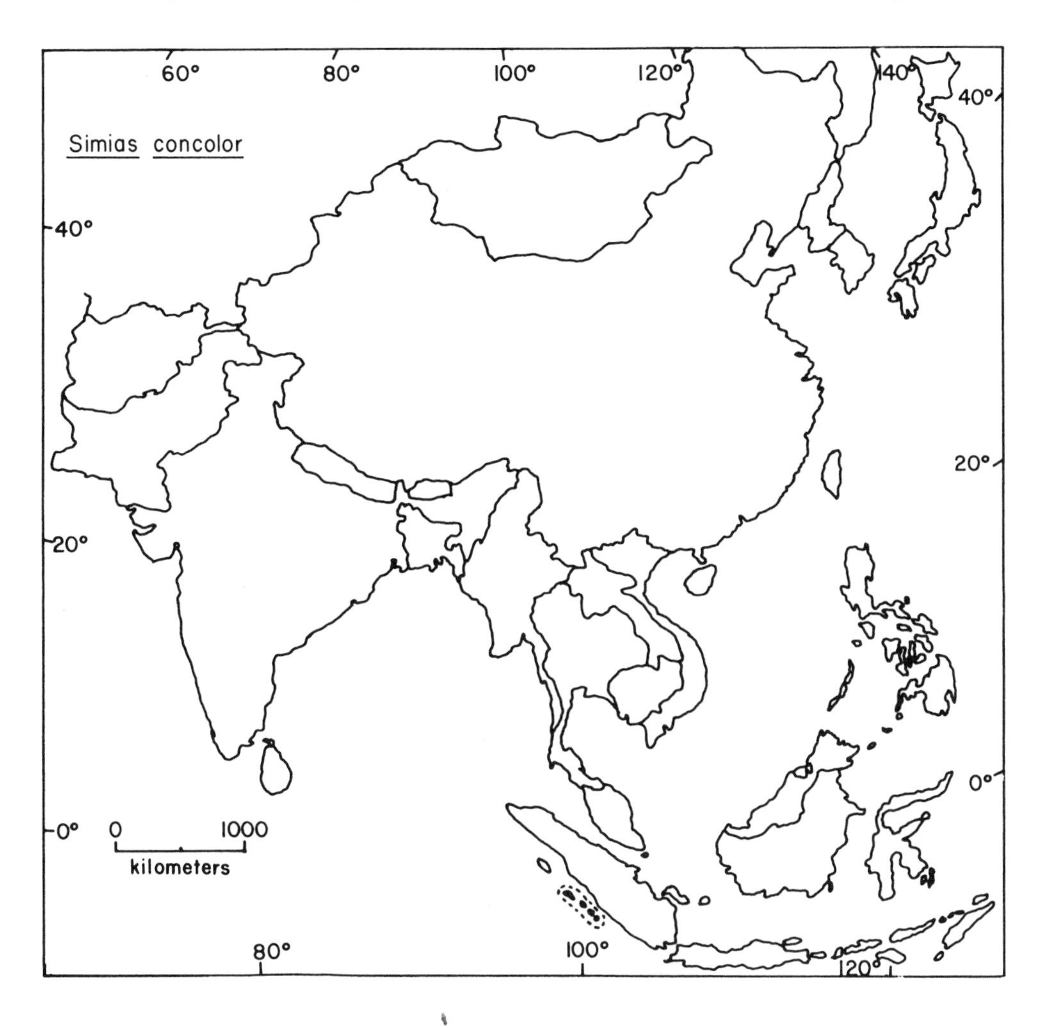

Taxonomic Note

Several recent authors place this species in the genus *Nasalis* (Groves 1970d, Thorington and Groves 1970, Tilson 1977). Two subspecies are recognized (Napier and Napier 1967).

Geographic Range

Simias concolor is found on the Mentawi Islands, Indonesia, between 1°– 4°S and 98°–101°E, about 100 km west of Sumatra. It occurs on all four of the islands: Siberut, Sipora, Pagai Utara (north) and Pagai Selatan (south) (Groves 1970d; Tenaza, pers. comm. 1974; Tilson, pers. comm. 1974). Previously it also occurred on several small adjacent islands, but it has been exterminated on all except one islet southeast of the

Katurei Peninsula, Siberut (Tilson 1977). No other details of the distribution are available.

Abundance and Density

S. concolor populations are declining significantly, and the species is considered endangered (Tenaza, pers. comm. 1974; Tilson, pers. comm. 1974; I.U.C.N. 1978). The maximum total population of *S. concolor* has been calculated at 25,312, and the actual population is thought to be much smaller (Krumbiegel 1978).

In central Siberut, population densities of 5 per km^2 (Tenaza, pers. comm. 1974) and 12 per km^2 (Tilson, pers. comm. 1974) have been reported. Groups there consist of 2 to 5 individuals ($\bar{X} = 3.5$) with home ranges of 25 to 30 ha (Tilson 1977).

Habitat

S. concolor is found mainly in primary forests in hilly areas, and seldom in secondary forest (Tilson 1977). Older sources have described its habitat as flooded and swampy tropical rain forest (Napier and Napier 1967), swampy coastal forest (Groves 1970d), and lowland forest (Medway 1970), but recent studies have shown that *S. concolor* is absent in coastal, mangrove, and lowland forest (Tenaza, pers. comm. 1974; Tilson, pers. comm. 1974; Tilson 1977).

Factors Affecting Populations

Habitat Alteration

Selective logging of trees greater than 60 cm in diameter was started on the Mentawi Islands in 1970, and, if extensive, could be detrimental to *S. concolor* populations (Tenaza, pers. comm. 1974). In addition, logging roads open areas of forest previously inaccessible to the local people, who move in, plant gardens, and hunt the monkeys in the remaining forest (Tilson, pers. comm. 1974).

Human Predation

Hunting. Hunting by the local people is a major cause of the decline of *S. concolor*. The rapidly growing human population and the breakdown in local hunting traditions (Tilson, pers. comm. 1974), as well as the introduction of hunting dogs and guns, have greatly increased the hunting pressure and enabled human hunting seriously to threaten *S. concolor* (Tenaza, pers. comm. 1974). Hunting has already eliminated this species from most of the small islands near the Mentawis (Tilson 1977).

Pest Control. This species is not a crop raider (Tilson, pers. comm. 1974).

Collection. No *S. concolor* were imported into the United States between 1968 and 1973 (Appendix) or in 1977 (Bittner et al. 1978), nor into the United Kingdom in 1965–75 (Burton 1978).

Conservation Action

S. concolor occurs in the 12.5 km^2 Teiteibatti Wildlife Reserve in central Siberut (Tilson, pers. comm. 1975). The species is in need of habitat reserves on the other islands, protection from hunting on Sipora and the Pagai Islands, and banning of the use of dogs and guns for hunting on Siberut (Tenaza, pers. comm. 1974). The introduction of alternate protein sources could help to reduce the hunting pressure on this species (I.U.C.N. 1978).

S. concolor is listed in Appendix I of the Convention on International Trade in Endangered Species of Wild Fauna and Flora, and as endangered under the U.S. Endangered Species Act (U.S. Department of Interior 1977a,b). It is also protected by law in Indonesia (Brotoisworo 1978).

Theropithecus gelada Gelada

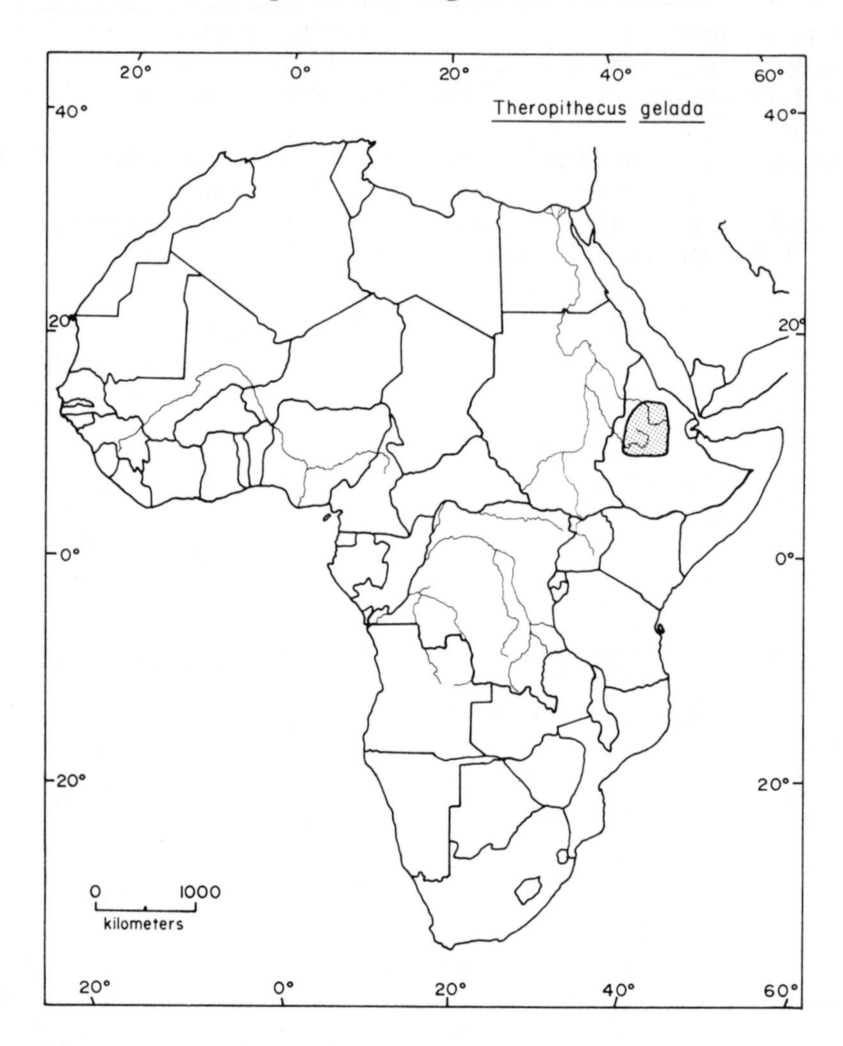

Taxonomic Note

Two subspecies are recognized (Napier and Napier 1967).

Geographic Range

Theropithecus gelada is found in northern and central Ethiopia. Its northernmost record is near Aksum at about 14°N, 39°E (Tappen 1960), and its southernmost locality is in the Upper Cassam Valley, near 9°N, 39°E (Dunbar, pers. comm. 1974). It occurs as far west as about 37°45′E (at 13°N) in the Simien Mountains (Dunbar, pers. comm. 1974) and as far east as 39°47′E (at 9°51′N) at Debre Sina (Bolton 1973). For details of the distribution, see Table 173.

Abundance and Density

T. gelada is locally abundant within its restricted range, and its total population has been estimated at about 600,000 individuals (Table 173). Dunbar (1977a) believes that there is no immediate cause to be concerned about extinction of the species, but that if current rates of habitat alteration and hunting continue (see below), the species could be threatened in the near future.

This species can achieve high population densities in areas of favorable habitat (Table 174). It forms large foraging herds, usually of 30 to 150 individuals but sometimes of up to 600 members (Table 174). These herds consist of aggregations of 20 to 30 one-male units and 3 to 4 all-male units, each with 5 to 10 members (Dunbar and Dunbar 1972). The reproductive units generally average 7 to 14 members each (Dunbar and Dunbar 1975b). The large herds may range over an area of 344 ha (Table 174), but an average one-male group of 10 would require an area of 16 ha for survival (Dunbar and Dunbar 1975b). These authors estimated that the carrying capacity of *T. gelada* habitat was 2.5 animals per ha of grass available during the dry season, and that this grass covers an average of 25% of the land surface.

Habitat

T. gelada is essentially a montane, grassland animal found in cool, seasonally arid, open scrub (Crook and Aldrich-Blake 1968). It occupies areas in which Afro-Alpine forests are interspersed with grassland and dense thickets (Dunbar and Dunbar 1975b), and montane grassland with relict patches of *Acacia*, *Juniperus*, and other types of woodland (Dunbar 1977a). Within these habitats, it is restricted to areas in which large gorges or rocky areas provide sleeping sites and refuge from predators (Dunbar 1977a). It is entirely dependent on these refuges and is rarely found farther than 2 km from them (Crook 1966). Its numbers in an area are also directly related to the quantity of grass forage available (Dunbar and Dunbar 1975b, Dunbar 1977a).

Geladas are found only between 1,400 and 4,500 m elevation (Dunbar 1977b). They prefer steep upper slopes and plateau areas, and only rarely descend to lower elevation valleys (Dunbar and Dunbar 1974a). They visit forest habitats infrequently, and evidently do not feed there (Crook and Aldrich-Blake 1968).

Factors Affecting Populations

Habitat Alteration

The habitats of *T. gelada* have been greatly altered by two activities: the replacement of natural forest with *Eucalyptus*, and intensive cultivation of grass and scrubland. The first process results in deterioration of habitat quality for geladas, because the shallow-rooted *Eucalyptus* trees absorb ground water, thus preventing the growth of herb layers and also promoting erosion due to runoff (Dunbar 1977a).

Cultivation in Ethiopia has greatly increased recently owing to the rapidly expanding human population (Dunbar 1977a), and gelada numbers in the southern part of the range have been sharply reduced because of intensive agriculture and settlement (Crook 1966). *T. gelada* population densities in heavily cultivated areas are one-third to one-half the densities observed in undisturbed habitats (Dunbar 1977a). Already all usable land in the highlands is under cultivation, very steep slopes are being ploughed, and at the present rate of erosion, bedrock will be reached in less than 50 years in some areas (Dunbar 1977a). Overgrazing by cattle and sheep is further depleting the food available to geladas and forcing them onto steeper slopes and poorer habitats. Dunbar (1977a) concluded that since *T. gelada* are highly specialized, they "are probably unable to withstand the kind of encroachment from humans with which many of the baboons and macaques are able to cope."

Human Predation

Hunting. In the southern part of their range — Shoa, Gojjam (Gwejam), and Wollo provinces — adult male geladas are hunted every eight years by the local tribesmen, who use their capes for age-grade change ceremonial headdresses (Dunbar 1977a). Today many of these capes are also sold to tourists (Bernahu 1975). This hunting reduces the percentage of adult males from 9% to 2% of the population, and could lead to deleterious inbreeding (Dunbar 1977a).

Near Debra Libanos, soldiers shoot *T. gelada* for target practice, and tend to shoot more adult males than other age-sex classes because they are larger, bolder, and the last to retreat to cliffs (Crook 1966). This shooting may also change the social structure and genetic composition of a group by reducing the proportion of adult males (Crook and Aldrich-Blake 1968).

Hunting has also removed most of the major predators of *T. gelada*, allowing one population to increase at 15.7% per year (Dunbar and Dunbar 1975b). Such population growth could lead to overexploitation of food resources by this population.

Pest Control. T. gelada does eat crops, especially during harvest time (Crook 1966, Dandelot and Prévost 1972). In densely populated areas where little undisturbed land is available, this behavior is reported to be frequent, and Ethiopian authorities have shot some geladas and permitted farmers to shoot others in order to protect crops (Dunbar 1977a). The number shot by farmers is unknown but probably small (Dunbar 1977a).

Collection. A small number of *T. gelada* are caught and sold as pets, but this probably has an insignificant effect on the population (Dunbar 1977a).

Jolly (1966) discovered scattered reports of use of *T. gelada* as a laboratory animal. He thought that it might be a potentially useful subject for research in adaptation to high altitudes, but added that there was "no particular advantage to using it which would justify an increase in trade in this unique and extremely restricted species."

Between 1968 and 1973 a total of 1,239 *T. gelada* were imported into the United States (Appendix), primarily from Kenya and Tanzania, where this species does not occur. None entered the United States in 1977 (Bittner et al. 1978). In 1965–75, 22 were imported into the United Kingdom (Burton 1978).

Conservation Action

T. gelada is legally protected against hunting, capture, or export from Ethiopia, but native hunting and shooting of crop pests are not controlled, and sales of pets and exporting also occur (Dunbar, pers. comm. 1974).

The population in the Simien National Park is in some of the least disturbed habitat on the Ethiopian plateau, and may remain the only viable population protected in the future (Dunbar 1977a). But, as the pressure builds to put land in cultivation, the park might be in danger of reduction or abandonment. Dunbar (1977a) urged a massive educational effort to counteract public antagonism toward conservation and crop-raiding geladas. He also suggested the translocation of *T. gelada* to another protected national park outside its present range.

T. gelada is listed in Class A of the African Convention, to which Ethiopia is a party (Organization of African Unity 1968).

Table 173. Distribution of *Theropithecus gelada*

Country	Presence and Abundance	Reference
Ethiopia	In N central (map)	Starck and Frick 1958
	In N central including Simien Mts. (in N)	Crook 1966
	At Debra Libanos, NW of Addis Ababa; population in area about 300	Crook and Aldrich-Blake 1968
	At Debra Libanos and in Simien Mts.	Dandelot and Prévost 1972
	Population about 732 in Simien Mts. N.P.	Dunbar and Dunbar 1972
	Plentiful near Debre Sina (9°51′N, 39°47′E) and Debre Berhan (9°40′N, 39°32′E)	Bolton 1973
	Limited to highlands of Amhara Plateau (map); generally abundant, not in danger of extinction; estimate more than 1,500 in Simien Mts. N.P.	Dunbar, pers. comm. 1974
	In N central (localities mapped); population 243 at Bole Valley (9°25′N, 38°00′E)	Dunbar and Dunbar 1974a
	Abundant in Simien N.P.	Dunbar and Dunbar 1974c
	Locally abundant on gorges of N plateau	Bernahu 1975
	Population 762 at Sankaber area, Simien Mts. N.P. and increasing 15.7% annually; 552 in Geech area, Simien N.P.; 243 at Bole Valley (500 km S of Simien Mts. N.P.)	Dunbar and Dunbar 1975b
	At Geech, Simien Mts. N.P.	Ohsawa and Kawai 1975
	Restricted to Amhara Plateau; total population probably about 600,000 (min. 440,000, max. 884,000); not immediately threatened but potentially threatened	Dunbar 1977a

Table 174. Population parameters of *Theropithecus gelada*

Population Density (individuals/km²)	Group Size X̄	Range	No. of Groups	Home Range Area (ha)	Study Site	Reference
		12–70			N Ethiopia	Starck and Frick 1958
	85–156	36–400	70		N Ethiopia	Crook 1966
	30.3	4–169	85		Central Ethiopia	Crook and Aldrich-Blake 1968
		300–400			N Ethiopia	Dandelot and Prévost 1972
		5–400			Ethiopia	Dunbar and Dunbar 1972
50–100		≤600			Ethiopia	Dunbar, pers. comm. 1974
82				78–90	Ethiopia	Dunbar and Dunbar 1974a
46–77.6	127	46–262	6	78–344	Ethiopia	Dunbar and Dunbar 1975b
15–80		30–250			Ethiopia	Dunbar 1977a
		140–260			Ethiopia	Dunbar 1977b

HYLOBATIDAE
Gibbons

Hylobates concolor Black, Crested, or White-cheeked Gibbon

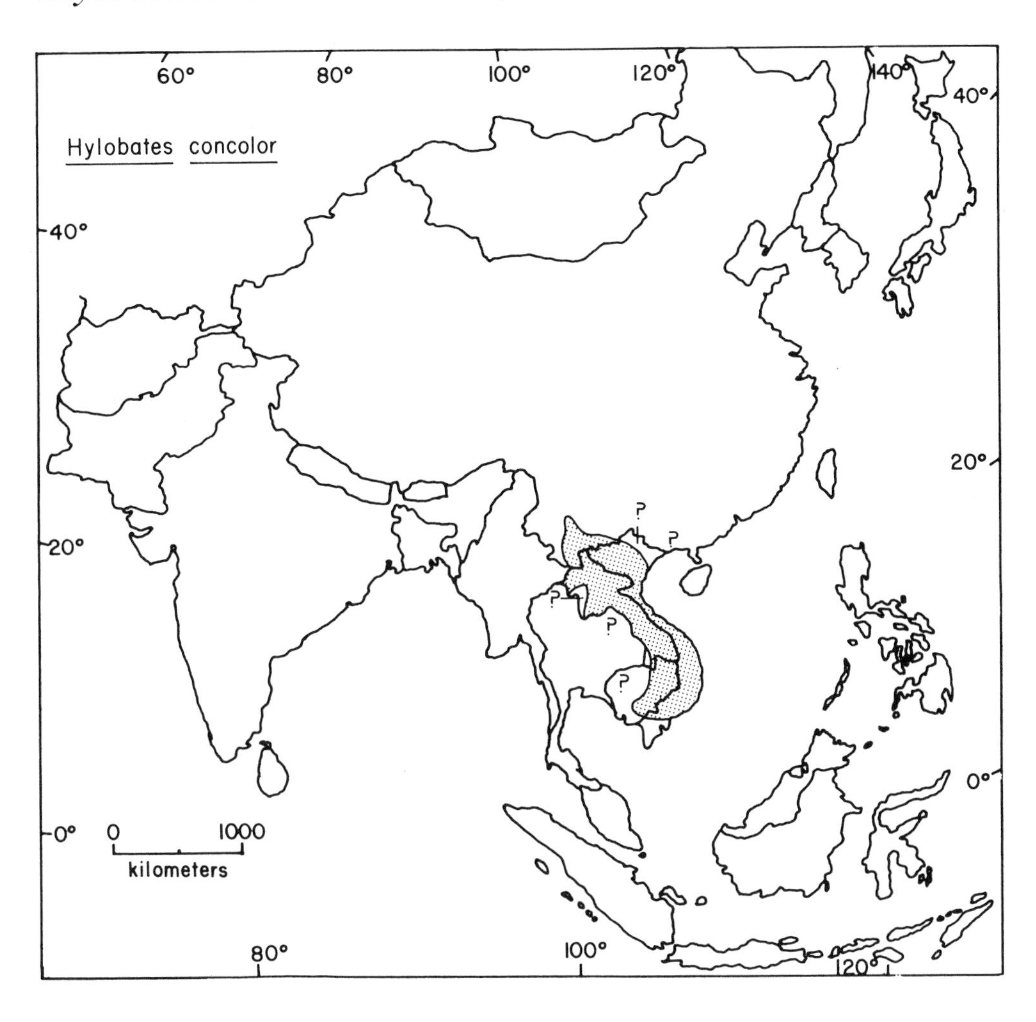

Taxonomic Note

Six subspecies are recognized (Napier and Napier 1967, Groves 1972).

Geographic Range

Hylobates concolor is found from southern China to southern Vietnam. It has been recorded as far north as 24°28′N, 100°54′E in Yunnan, China (Fooden 1969b), and as far south as about 11°N, 107°E in Vietnam (Groves 1972). The species occurs as far east as about 110°E on the island of Hainan, and west in northern Laos at least to the Mekong River and perhaps farther west into northern Thailand (Groves 1972, Brockelman 1975). For details of the distribution, see Table 175.

Abundance and Density

The population of *H. concolor* in Laos is estimated to be about 228,000 individuals and declining rapidly (Chivers 1977a). The status of this species in the other countries of the range is not known (Table 175). No data are available regarding population densities, group sizes, or home range areas of *H. concolor*.

Habitat

H. concolor has been reported in montane forests on Hainan and in forested regions east of the Mekong River (Delacour 1951). No other information was located.

Factors Affecting Populations

Habitat Alteration

The extensive military activities in Vietnam, Cambodia, and Laos have altered large areas of habitat (Committee of Concerned Asian Scholars 1970, Orians and Pfeiffer 1970, Pfeiffer 1973, S.I.P.R.I. 1976), and have undoubtedly been detrimental to populations of *H. concolor*.

Human Predation

Hunting. H. concolor is hunted for its flesh in northern Vietnam (Bassus 1973).

Pest Control. No information.

Collection. H. concolor is sometimes kept as a pet in northeastern Thailand (Brockelman 1975).

In 1968–73 a total of 117 *H. concolor* were imported into the United States (Appendix), primarily from Laos. In 1977 none entered the United States. One individual was imported into the United Kingdom in 1965–75 (Burton 1978).

Conservation Action

H. concolor is legally protected from hunting in Vietnam (Bassus 1973). It is also listed in Appendix I of the Convention on International Trade in Endangered Species of Wild Fauna and Flora, and as endangered under the U.S. Endangered Species Act (U.S. Department of Interior 1977a,b).

Some races of this species are suspected to be in danger of extinction, and habitat reserves are probably badly needed, but very little specific information is available (Chivers 1977a).

Table 175. Distribution of *Hylobates concolor* (countries are listed in clockwise order)

Country	Presence and Abundance	Reference
China	On Hainan Island (maps)	Bourret n.d., Delacour 1951
	In Hsi-Shuan-Pan-Na area (21°59′N, 100°49′E), E of Mekong R., S Yunnan Province	Kao et al. 1962
	Common in SW	Van Gulick 1967
	Northernmost record is Ching-Tung (24°28′N, 100°54′E); between Song Koi R. (Red R.) and Mekong R., S central Yunnan (map)	Fooden 1969b
	In S Yunnan; questionable in Kwangsi (farther E) (map)	Groves 1970b
	Localities mapped in S Yunnan	Groves 1972
	On Hainan and between Song Koi R. (Red R.) and Mekong R.; questionable in Kwangsi (map)	Chivers 1977a
Vietnam	Common throughout N in many areas (listed) including Chapa, Lai Chau, Bac Kan, Thai Nguyen, Hon Gai, Hoi Xuan, and Phu Qui; in central at Da Nang and throughout S except S tip (map)	Bourret n.d.
	Localities listed and mapped throughout S of Red R. and N of Mekong R. delta including Lao Bao, Da Nang, Hue, Quang Tri, Dalat, and Nha Trang	Delacour 1951
	Near Phu Quy (19°20′N, 105°26′E)	Dào Van Tien 1963
	Localities listed in S from provinces of Darlac, Khanh Hoa, Lam Dong, Long Khanh, Quang Nam, Quang Tri, Thua Thien, and Tuyen Duc	Van Peenen et al. 1969
	Throughout S of Song Koi R. (Red R.) except S tip (maps)	Groves 1970b, Fooden 1971b
	Localities mapped as far N as 22°N, 104°E, and throughout as far S as Mekong delta	Groves 1972
	Localities mapped in N as far S as about 18°30′N	Bassus 1973
	Throughout S of Red R. to Mekong delta; questionable N of Red R. (map)	Chivers 1977a
Cambodia	Throughout except extreme W (map)	Bourret n.d.
	Only E of Mekong R. (maps)	Delacour 1951, Groves 1970b, Fooden 1971b, Chivers 1977a
	Localities mapped in SE, questionable in NW (map)	Groves 1972
Thailand	In NE and E (map)	Bourret n.d.
	Questionable W of Mekong R. (maps)	Groves 1970b, 1972
	Reported in NE but unable to confirm	Brockelman 1975

Table 175 (continued)

Country	Presence and Abundance	Reference
Laos	Localities listed in N (Phong Saly), central (Pak Lai to Savannakhet and Tchépone) and S (map)	Bourret n.d.
	Localities listed and mapped in N (except in W corner and questionable in NW), central including Nape, and S including Saravane and Plateau des Bolovens	Delacour 1951
	Throughout E of Mekong R. (maps)	Groves 1970b, Fooden 1971b
	Localities mapped throughout E of Mekong R.; questionable W of Mekong R.	Groves 1972
	Two localities mapped E of Mekong R.	Marshall et al. 1972
	Throughout E of Mekong R. (map); population estimated at 228,000 individuals; to be reduced to 22,800 individuals by 1980	Chivers 1977a

Hylobates hoolock Hoolock Gibbon

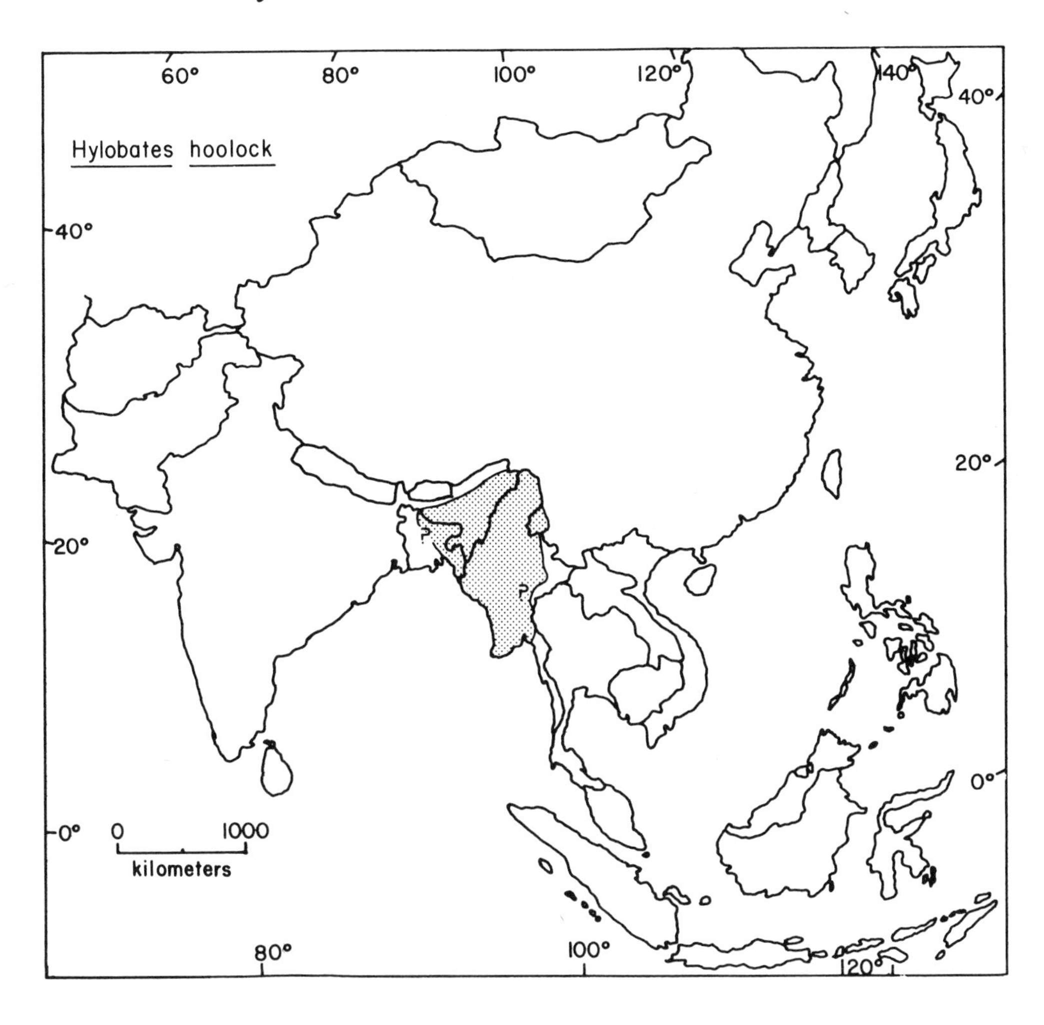

Taxonomic Note

Two subspecies are recognized (Groves 1967a, 1972).

Geographic Range

Hoolock gibbons are found from eastern India and Bangladesh through Burma into southern China. They occur as far west as about 90°E along the Brahmaputra River, Assam, India, and as far east as about 99°E along the Salween River in southwestern Yunnan, China (Groves 1967a). The northern limit of their range is at about 28°N, 96°E north of Sadiya, Assam (Fooden 1969b), and the southern limit at the Mouths of the Irrawaddy (about 16°N) in southern Burma (Fooden 1969b, Chivers 1977a, MacKinnon 1977a). For details of the distribution, see Table 176.

Abundance and Density

In India and Burma, populations of *H. hoolock* were estimated to total 532,000 individuals, which would decline to 91,000 individuals by 1980 (Chivers 1977a). No information could be located regarding their status in Bangladesh or China (Table 176).

This species has been observed at population densities of 14 individuals per km^2 (Chivers 1977a citing Tilson) and 1.6 individuals per km^2 (Green 1978a). Groups of 3 to 6 have been reported (McCann 1933c, Prater 1971), and at one study site 12 groups included 2 to 5 individuals ($\bar{X} = 3.5$) with territories of 18 to 28 ha (Tilson, pers. comm. 1975).

Habitat

Hoolock gibbons are found in thick evergreen forest (McCann 1933c, Mohnot 1978) and hill forest (Prater 1971) in India, and mixed evergreen and scrub forest in Burma (Milton et al. 1964). In Bangladesh they were seen only in primary forest, and seemed to prefer undisturbed areas (Green 1978a). This species is found at elevations from 152 to 1,370 m (Groves 1972).

Factors Affecting Populations

Habitat Alteration

The habitat of *H. hoolock* is being destroyed in Bangladesh, where the rapidly expanding human population is creating a large demand for firewood and building material (Green 1978a), and in Assam, where selective and clearcut logging are both proceeding in remaining forests (Chivers 1977a). No information is available from Burma or China.

Human Predation

Hunting. Hoolock gibbons are eaten and highly desired as food in Assam, India (Biswas 1970), including Nagaland (Chivers 1977a).

Pest Control. No information.

Collection. Two *H. hoolock* were imported into the United States between 1968 and 1973 (Appendix), both from Singapore, where this species does not occur. None entered the United States in 1977 (Bittner et al. 1978). Eight were imported into the United Kingdom in 1965–75 (Burton 1978).

Conservation Action

Hoolock gibbons are found in two reserved forests in Bangladesh and one wildlife sanctuary in Burma (Table 176). They also occur in three sanctuaries or reserves in Assam (Table 176), but although their protected range there includes 235 km^2, these areas are not exempt from habitat disturbance (Chivers 1977a). A reserve where *H. hoolock* and its habitat would receive protection has been proposed for the Sibsagar district of Assam (Chivers 1977a).

The Pidaung Wildlife Sanctuary in Burma is not adequately guarded against agricultural encroachment or poaching and should be upgraded to a national park (Milton et al. 1964). Population surveys are needed to determine the status of the hoolock gibbon (Biswas 1970, Mohnot 1978).

H. hoolock is listed in Appendix I of the Convention on International Trade in Endangered Species of Wild Fauna and Flora, and as endangered under the U.S. Endangered Species Act (U.S. Department of Interior 1977a,b).

Table 176. Distribution of *Hylobates hoolock* (countries are listed from west to east)

Country	Presence and Abundance	Reference
India	In Naga Hills (extreme E)	McCann 1933c
Assam	At Sadiya Frontier Tract, and S of Lohit R. and Brahmaputra R. near Rangdoi and Kenau	Parsons 1942a
	E of Dibang R. but not W of it	Parsons 1942b
	In Kaziranga Wildlife Sanctuary (in E, near 26°30′N, 94°E)	Spillett 1966a
	Localities mapped in E, W, central, and NE	Groves 1967a
	Throughout E and S of Brahmaputra R. (maps)	Fooden 1969b, 1971b; Groves 1970b; MacKinnon 1977a; Roonwal and Mohnot 1977
	In Garo Hills, N Cachar, Mikir Hills (central), and near Gauhati (26°10′N, 91°45′E)	Biswas 1970
	In Garampani Sanctuary	Sankhala 1971
	In Nagaland (in E) and Meghalaya (central) in Garo Hills, Kasi Hills, and near Nongpoh (in N)	Baskaran 1975
	In Hollangapar Reserve (26°30′N, 94°20′E) near Jorhat; and Mikir hills, Sibsagar; Khasi-Jaintia Hills, Meghalaya; and Garo Hills, S of Goalpara	Tilson, pers. comm. 1975
	Throughout E and S of Brahmaputra R. (map); population estimated at 80,000 individuals in area of 31,000 km^2, of which 235 km^2 are protected; population will decline to 1,000 individuals by 1980	Chivers 1977a
	In Meghalaya, Nagaland, and Arunachal Pradesh (N of Brahmaputra R.)	Mohnot 1978
Bangladesh	In E (E of Dacca) (maps)	Fooden 1969b, 1971b; Groves 1970b; Chivers 1977a; MacKinnon 1977a
	In Chittagong (in SE)	Prater 1971
	Localities mapped in S Sylhet in Kalacherra and Lawachera reserved forests and in Chittagong	Green 1978a
Burma	In Shan State (central) and throughout S and E (map)	Bourret n.d.

Table 176 (continued)

Country	Presence and Abundance	Reference
Burma (cont.)	In N on both sides of Chindwin R.	Morris 1943
	In Pidaung Wildlife Sanctuary, Kachin State (in N on Irrawaddy R.); plentiful in Chaukan Pass area (extreme N)	Milton and Estes 1963, Milton et al. 1964
	Localities mapped as far S as Mount Victoria, Chin Hills (about 21°N, 93°E); throughout N to about 27°N along Mali R. and E to Sumprabum (26°38′N, 97°36′E)	Groves 1967a
	Throughout S (maps); northernmost record in Goletu (27°37′N, 97°54′E)	Fooden 1969b
	Questionable SE of Mandalay (map)	Groves 1970b
	Reported along lower Salween R., in Kakhyen Hills on NE border, and on both sides of Irrawaddy R.; S extent of range not known	Groves 1972
	Throughout W of Salween R., S to Mouths of Irrawaddy (map); population estimated at 452,000 individuals, to decline to 90,000 individuals by 1980	Chivers 1977a
China	Localities mapped in SW Yunnan	Groves 1967a
	W of Salween R. as far N as Homu-shu, Yunnan (25°00′N, 98°45′E) (map)	Fooden 1969b
	Reported at several localities W of Salween R., including Ho-Tai and Teng-Yue-chow (25°02′N, 98°28′E)	Groves 1972
	In SW Yunnan W of Salween R. (maps)	Chivers 1977a, MacKinnon 1977a

Hylobates klossii Kloss' Gibbon or Bilou

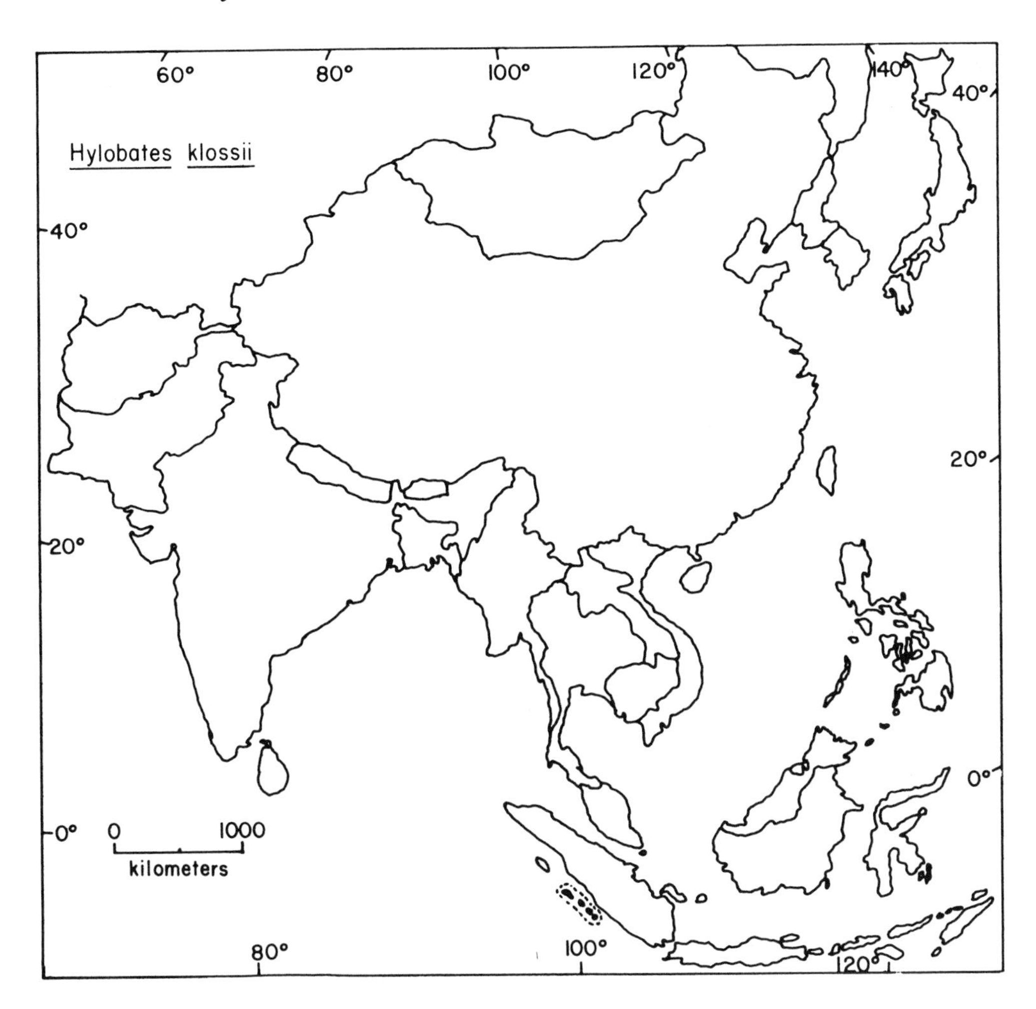

Taxonomic Note

This is a monotypic species (Napier and Napier 1967, Groves 1972).

Geographic Range

Hylobates klossii is found on the Mentawi Islands (Siberut, Sipora, Pagai Utara, and Pagai Selatan) located between about 1°S and 3°30′S, 98°30′E and 101°E, 85 to 135 km southwest of central Sumatra, Indonesia (Chasen 1940, Tenaza and Hamilton 1971, Tenaza 1975). For details of the distribution, see Table 177.

Abundance and Density

H. klossii is locally common but declining in most parts of its range (Table 177). Its total population has been estimated at 84,000 individuals, which would decline to

3,000 by 1980 (Chivers 1977a), but others believe that the population is much smaller than this (I.U.C.N. 1978).

Population densities for this species are 26 (Tilson, pers. comm. 1974) to 30 individuals per km^2 (Tenaza 1975). Kloss' gibbon has been observed in groups of 2 to 6 ($\bar{X}$ = 4, N = 4) (Tenaza and Hamilton 1971), and 2 to 5 ($\bar{X}$ = 3.4, N = 11), in territories of 5.1 to 8.3 ha (Tenaza 1975).

Habitat

H. klossii is found in dense evergreen rain forest (Tenaza 1975, 1976). It utilizes both lowland and hill primary forest, and, infrequently, lowland secondary forest (Tilson, pers. comm. 1974).

Factors Affecting Populations

Habitat Alteration

The local people on the Mentawi Islands have traditionally altered their forest habitat by (1) selective logging for dugout canoes, lumber, and bark, (2) clearing for houses and farms, and (3) trail cutting and rattan harvesting, but *H. klossii* seems to be somewhat tolerant of these limited habitat modifications and able to utilize disturbed or regenerating forest if it is adjacent to relatively undisturbed forest (Tenaza and Hamilton 1971). These authors found that the old established human practices could reduce local *H. klossii* population levels, but they probably had little effect on the total species population compared with the effect of commercial logging by foreign companies. As of 1972 only a small portion of Siberut Island had been cleared by the local people, and this island was not being deforested (Tenaza 1975).

Large-scale logging operations began in the Mentawi Islands in the early 1970s, and the widespread timbering, with its extensive network of wide logging roads, has destroyed some forested areas, fragmented and isolated others, and opened remote regions to colonization and cultivation (Tilson, pers. comm. 1974). Along with the rapidly growing human population, this process could severely affect *H. klossii*'s status (Tilson, pers. comm. 1974; I.U.C.N. 1978).

Human Predation

Hunting. The local people of the Mentawi Islands hunt *H. klossii* for meat with bows and arrows (Tenaza and Hamilton 1971). This hunting has been going on for thousands of years, and the Kloss' gibbon populations seem well able to tolerate the pressure in most parts of Siberut where the human population is stable, but not on the Pagai Islands or Sipora where the human population has increased fivefold in the last 150 years (Tenaza, pers. comm. 1974).

This increasing human population, along with the breakdown of hunting taboos and the introduction of hunting dogs and modern weapons, has led to an intensification of hunting pressure in several areas, and this may pose a serious threat to *H. klossii* (Tilson, pers. comm. 1974).

Pest Control. *H. klossii* does not utilize culvitated foods (Tilson, pers. comm. 1974) and thus is not persecuted as a pest to crops.

Collection. H. klossii mothers are shot to obtain infants for the pet and export trade (Tenaza and Hamilton 1971), and the presence of logging personnel on the islands has increased the market for young gibbons and the number being offered for sale (Tilson, pers. comm. 1974).

No *H. klossii* were imported into the United States between 1968 and 1973 (Appendix) or in 1977 (Bittner et al. 1978), nor into the United Kingdom between 1965 and 1975 (Burton 1978).

Conservation Action

H. klossii is found in one reserve on Siberut, but this reserve contains large farms, villages, and areas of unsuitable mangrove habitat (I.U.C.N. 1978). This reserve should be strictly protected and additional reserves established on the Pagai Islands and Sipora; hunting should be banned on these three islands, and the use of dogs and guns for hunting should be banned on Siberut; and harsh penalties should be levied against persons caught buying or selling primates (Tenaza, pers. comm. 1974).

A survey has been made to determine the distribution and status of *H. klossii* and appropriate sites for reserves, in conjunction with the Indonesian Department of Forestry and Department of Wildlife (Tilson, pers. comm. 1974). Efforts are also under way to provide local villagers with domestically produced protein to reduce hunting pressure, and to limit timber concessions on Siberut (I.U.C.N. 1978).

H. klossii is protected by law in Indonesia (Brotoisworo 1978). It is listed in Appendix I of the Convention on International Trade in Endangered Species of Wild Fauna and Flora, and as endangered under the U.S. Endangered Species Act (U.S. Department of Interior 1977a,b). Its import into the United Kingdom is also restricted to scientific, educational, or propagative purposes (I.U.C.N. 1972).

Table 177. Distribution of *Hylobates klossii*

Country	Presence and Abundance	Reference
Indonesia	On Mentawi Islands (maps)	Groves 1970b, Fooden 1971b, MacKinnon 1977a
	Localities mapped in E Siberut, S Pagai Utara, and N Pagai Selatan	Tenaza and Hamilton 1971
	Endangered; common and widespread but rapidly decreasing	I.U.C.N. 1972, 1976
	Localities mapped in central and S Siberut	Wilson and Wilson 1972–73
	Abundant but declining	Tenaza, pers. comm. 1974
	Abundant and remaining stable only in unpopulated hilly interior of Siberut; less abundant on Sipora, Pagai Utara, and Pagai Selatan; scarce and declining in populated areas of all 4 islands; reduced to scattered pockets in lowland areas; in Tei-Tei-Batti Nature Park, central Siberut	Tilson, pers. comm. 1974

Table 177 (continued)

Country	Presence and Abundance	Reference
Indonesia (cont.)	Study site at Rattan Mountain (Tei-tei Peleigei) at 1°24′S, 99°01′E, between Sirimuri R. and Silabai R., E Siberut	Tenaza 1975, 1976
	Estimated population of 84,000 individuals in area of 4,200 km^2, of which 100 km^2 are protected; by 1980 population to decline to 3,000 individuals	Chivers 1977a
	Vulnerable; widespread and fairly common but rapidly decreasing; in Teiteibatti Wildlife Reserve	I.U.C.N. 1978

Hylobates lar Lar Gibbon

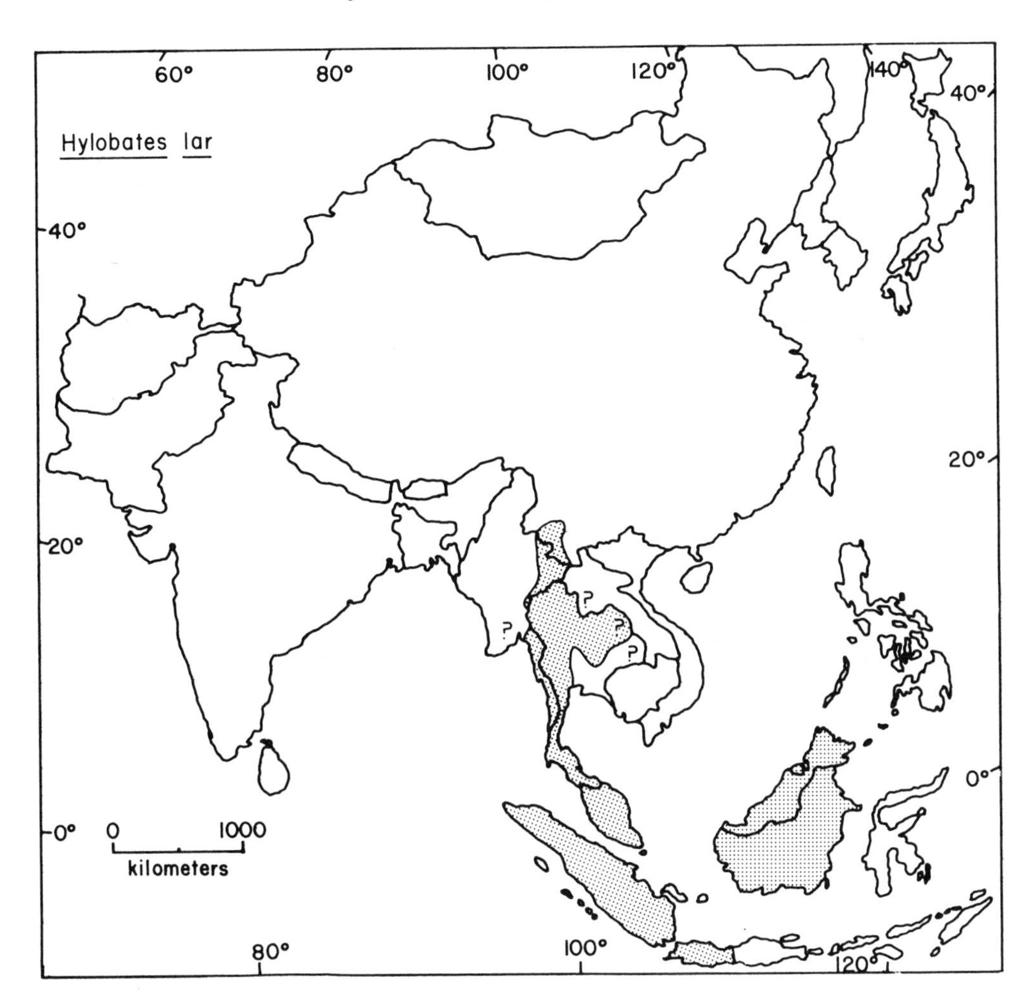

Taxonomic Note

Many authors consider some of the forms included here under *Hylobates lar*, especially *agilis*, *moloch*, and *muelleri*, to be distinct species. In the present report, these three forms, as well as *abbotti*, *vestitus* (=*albimanus*), *entelloides*, *carpenteri*, and *lar*, are all treated as subspecies of *H. lar*, following Groves (1970b, 1971b, 1972). The pileated gibbon, *H. pileatus*, is treated separately.

Geographic Range

Hylobates lar is found from southern China through eastern Burma, Thailand, Malaysia, and Indonesia. The northern limit of its range is at Nam Ka at 23°30′N, 99°00′E on the Namting River in southwestern Yunnan, China (Groves 1972), and the southern limit is at about 7°30′S, 109°E on the southern shore of western Java (Groves 1970b). The species occurs as far west as about 97°15′E (at 4°N) in northern

Sumatra (Fooden 1969b), or even farther west to the northwestern tip of that island (Wilson and Wilson 1972–73, 1977), and as far east as about 119°E in eastern Borneo (Groves 1970b, 1971b). For details of the distribution, see Table 178.

Abundance and Density

Populations of *H. lar* are not considered to be abundant, stable, or likely to increase in any of the countries in which they occur (Table 178). In the three countries for which recent information is available (Thailand, Malaysia, and Indonesia) the total population of lar gibbons is thought to be about 2,839,000, and was predicted to decline to 342,900 by 1980 and to relict populations within 15 to 20 years (Chivers 1977a). The species is decreasing especially rapidly in northern Thailand and Java (Table 178). No recent reports are available from China (where *H. lar* was already rare in 1644), from Burma, or from Laos.

H. lar population densities vary from 0.68 to 21 individuals per km^2, with small family groups of 2 to 12 (usually averaging 3 to 5 members) occupying home ranges of 4.4 to 60 ha (Table 179). Medway and Wells (1971) calculated that a reserve large enough to conserve a population of 5,000 *H. lar* must encompass 1,667 km^2.

Habitat

H. lar is a forest species, found in many different types of lowland and highland, evergreen and deciduous, primary and secondary forest. For example, in Thailand it has been observed in tall, deciduous, monsoon forest of which teak (*Tectona grandis*) is a characteristic species, in subtropical semievergreen forest, and in temperate evergreen forest (Kawamura 1961). It also occupies dry and moist mixed deciduous forest, dry and hill evergreen forest, tropical rain forest, secondary forest (Berkson et al. 1971), and dense evergreen or mixed evergreen-deciduous forest (Marshall et al. 1972, Brockelman 1975). The southern evergreen rain forest seems to be a more suitable habitat for it than the northern mixed dry forest (Chivers 1977a).

In peninsular Malaysia, lar gibbons utilize primary dipterocarp rain forest (McClure 1964, Chivers 1972, Raemaekers 1978) but "evidently exploit a broad variety of forest types and are not restricted to lowland formations dominated by dipterocarps" (Medway 1972). The northern Malaysian form is confined to primary forest, but elsewhere the species also occupies tall secondary forest (Medway 1969a). In one study, 70% of *H. lar* groups were found in secondary forest and 30% in primary forest, but higher densities occurred in the latter habitat (Southwick and Cadigan 1972). Overall, Malaysian lar gibbons occur at almost every inland and coastal locality wherever dry forest still stands (Chivers 1974).

In Borneo, *H. lar* has been recorded in lowland, inland, rain forest (Kawabe and Mano 1972), undisturbed and disturbed primary forest, and secondary forest (Wilson and Wilson 1975a). On Java it occupies mature tropical rain forest and high quality secondary forest (I.U.C.N. 1978). And on Sumatra it utilizes almost every type of primary forest including freshwater swamp, lowland, hill, and submontane, as well as secondary and disturbed forests (Wilson and Wilson 1977), and pockets of forest within savanna and at the edges of fields (Kurt 1973).

In every portion of its range, *H. lar* is found in lowland and highland areas, as high as 2,400 m in Thailand (Groves 1968), 1,400 m in Malaysia (Medway 1972), and 1,980 m in Sumatra (Carpenter 1940). In general, however, it is most common at medium elevations 250 to 500 m above sea level (Chivers 1973, Kurt 1973) and is not often found above 1,000–1,200 m (Chivers 1972, Gittens 1978).

Factors Affecting Populations

Habitat Alteration

The greatest threat to populations of *H. lar* is the reduction of their forest habitats by logging and clearing for agriculture. By the fourteenth century, deforestation had already greatly reduced the range of this species in China (Van Gulick 1967), and the process is continuing in every country of the distribution.

In Thailand, large-scale forest clearance will probably restrict *H. lar* populations to small isolated forest remnants within the next 30 years (Brockelman 1975). The forests of Thailand are being clearcut for lumber (Lekagul 1968, Brockelman and Kobayashi 1971, Cadigan and Lim 1975) and cleared for shifting agriculture and road building (Berkson et al. 1971). Some areas, such as the Korat Plateau, have already been completely deforested (Marshall et al. 1972), and many of the remaining patches in southern Thailand will disappear in the next few years (Brockelman 1978).

Overall, only one-tenth of the original habitat of *H. lar* remains in Thailand; during the last 10 years this habitat decreased an estimated 30%, and since lumbering is continuing, the forests are expected to continue to shrink by several percent each year (Brockelman 1975). In addition, squatters following the loggers prevent regeneration by their slash and burn farming, and the rapidly growing human population is intensifying this pressure (Brockelman 1975, Chivers 1977a).

In peninsular Malaysia, selective logging is practiced, and *H. lar* can survive in the selectively cut forests (Chivers 1972, Ellefson 1974, Cadigan and Lim 1975). But all these authors agree that extensive clearing of the Malaysian forest for agriculture and cutting for wood products is reducing the habitat available to *H. lar* there. About three-fifths of the area of peninsular Malaysia was still forested in the early 1970s (Chivers 1972), but 10% of the forest standing in 1960 was felled by 1970, and the lowland forest habitat of *H. lar* is disappearing quickly (Chivers, pers. comm. 1974; Khan 1978). Some sources have estimated that up to 2,500 km^2 of forest are being cleared each year, and that hydroelectric schemes are also seriously threatening *H. lar* in Malaysia (Chivers 1977a).

Deforestation is also reducing *H. lar*'s habitat on Sumatra (Wilson and Wilson 1972–73). Two studies have shown that population densities are much lower in selectively logged forests than in undisturbed ones there (MacKinnon 1973, Rijksen 1978), although a third survey did not support this conclusion (Wilson and Wilson 1975a). Clearing for agriculture is also destroying large areas of lar gibbon habitat on Sumatra (Wilson and Wilson, pers. comm. 1974).

In Java the forest habitat of *H. lar* has been broken up into many isolated patches, and destruction of these is proceeding rapidly (Brotoisworo 1978). In 1970 one author estimated that only about 33 km^2 of forest were still found in Java, and overexploitation of this remainder for timber is continuing (I.U.C.N. 1978). In Kalimantan, large logging concessions have been granted for selective logging of much of the lowland forest (Chivers 1977a). Although *H. lar* can survive in low intensity selectively logged forest there, it prefers undisturbed forest and would be expected to decline if more than eight trees per hectare were removed from an area (Wilson and Wilson, pers. comm. 1974; 1975a).

Human Predation

Hunting. Lar gibbons are hunted and eaten by the local people throughout Thailand (Berkson et al. 1971), especially in the northwest (Kawamura 1961, Berkson and Ross 1969). Recently, traditional Buddhist restrictions and native taboos against killing gibbons have been losing their force, and gibbon hunting is increasing (Berkson et al. 1971, Brockelman and Kobayashi 1971, Brockelman 1975, Lekagul and McNeely 1977).

Hunting pressure is less severe in peninsular Malaysia (Chivers 1977a), although some aborigines (Khan 1970) and Chinese workmen (Ellefson 1974) consider *H. lar* meat a highly desirable food item. *H. lar* is also hunted on Sumatra (Wilson and Wilson 1972–73), but is not often killed in Kalimantan (Chivers 1977a).

Pest Control. *H. lar* does not feed in plantations or farms and is not killed as a pest to crops (Chivers, pers. comm. 1974; Wilson and Wilson, pers. comm. 1974).

Collection. Lar gibbons are captured for local pets and for export in Thailand (Lekagul 1968), peninsular Malaysia (Khan 1970, Ellefson 1974), North Borneo (Burgess 1961), Sumatra (Wilson and Wilson 1972–73), and Java (I.U.C.N. 1978). In Thailand, for instance, an estimated 1,000 mother *H. lar* have been shot annually to obtain babies for sale at the Bangkok weekend market, where about 20 *H. lar* were sold per week (Berkson et al. 1971). Marshall et al. (1972) did not list specific sighting localities in their report because "publicity for easily observed gibbon populations can lead to their extermination by hunters and scientific collectors." More recently the selling of young *H. lar* in Thailand was still continuing, and the number of pets being kept was probably in the thousands (Brockelman 1975).

In peninsular Malaysia, young lar gibbons are also collected by killing their mothers (Harrison 1966, 1969; Khan 1970, 1978), but the number captured is thought to be low (Khan 1970, Cadigan and Lim 1975).

H. lar is not used extensively in biomedical research in the United States (Comm. Cons. Nonhuman Prim. 1975a), but recent studies have shown it to be susceptible to several human diseases and parasites, and the added pressure of collecting for research could threaten *H. lar* populations, which are "not concentrated enough to enable harvesting in large numbers without depleting populations" (Berkson and Ross 1969).

In 1968–73 a total of 473 *H. lar* were imported into the United States (Appendix), primarily from Laos and Singapore. None entered the United States in 1977 (Bittner et al. 1978). Fifty-eight were imported into the United Kingdom in 1965–75 (Burton 1978).

Conservation Action

H. lar is found in three national parks, twenty reserves, four sanctuaries, and one nature park (Table 178). The total area of its habitat protected in Thailand, Malaysia, and Indonesia is 58,719 km^2 (Chivers 1977a). No information regarding protected areas in the other countries of the range is available.

This species is also protected by law from killing, capture, possession, or export in Thailand (Berkson and Ross 1969, Lekagul and McNeely 1977), but enforcement is difficult or impossible in many areas (Berkson et al. 1971), and hunting and smuggling still occur (Brockelman 1975). Malaysian law also prohibits the hunting, capture, or keeping of gibbons as pets (Harrison 1966, Medway 1969a, Khan 1970, Yong 1973, Lee 1975), although special permits are issued for scientific or display purposes (Khan 1978). In Sabah, *H. lar* may only be killed in defense of human life or crops (de Silva 1971). *H. lar* is also legally protected in Indonesia (Carpenter 1938, Brotoisworo 1978), but the laws are not strictly enforced (Wilson and Wilson 1972–73, I.U.C.N. 1978).

Recommendations for actions to conserve remaining populations of *H. lar* include (1) control and eventual banning of clearcutting in forests, (2) establishment and protection of large reserves in each country, (3) light use of all remaining forests under careful government control, (4) strong enforcement of export restrictions, and (5) education of the people about conservation, the use of natural resources, the importance of forests, and the biology of gibbons (Chivers 1977a).

Efforts to relocate *H. lar* whose habitat is being destroyed should be restricted to areas of low gibbon density to minimize the risk of crowding and the introduction of diseases (Chivers 1977a). A strictly protected reserve is urgently needed for the form in western Java (*H. l. moloch*) (I.U.C.N. 1978). A project has been proposed by Brotoisworo and Ruhiyat to determine the status of this form and to aid in locating and establishing a reserve for it (Mittermeier 1977b).

Captive breeding programs have proved to be costly and slow, and although they may provide *H. lar* for a few research projects, they have not been able to produce large numbers (Berkson and Ross 1969, Brockelman 1975, Cadigan and Lim 1975). Some authors recommend the establishment of forest patches in which *H. lar* groups could breed and produce surplus young for harvesting with live traps or tranquilizer guns (Berkson et al. 1971). Others urge that lar gibbons not be removed for any purpose from areas where forests remain undisturbed (Brockelman and Kobayashi 1971), and add that "the only management strategy that will save adequate populations for the future is vigorous protection of large areas of forest (large enough to support thousands of individuals) from lumbering and hunting" (Brockelman 1975).

H. lar is listed in Appendix I of the Convention on International Trade in Endangered Species of Wild Fauna and Flora, and as endangered under the U.S. Endangered Species Act (U.S. Department of Interior 1977a,b).

Table 178. Distribution of *Hylobates lar* (countries are listed in order from north to south)

Country	Presence and Abundance	Reference
China	Before 1000 A.D.: common as far N as 35°N; by 1644, very rare, confined to SW	Van Gulick 1967
	S of Namting R., 23°30′N, 90°E (*sic*); map shows N limit between Mekong R. and Salween R. at about 29°N, 98°E	Fooden 1969b
	N to Nam Ka on Namting R. (23°30′N, 99°E)	Groves 1972
	In S between Mekong R. and Salween R. to about 25°N, 100°E (map)	Chivers 1977a
Burma	In S around Mouths of the Irrawaddy and in Tenasserim (SE) (map)	Bourret n.d.
	Throughout E of Salween R. and in Tenasserim (map); localities listed and mapped in extreme S	Fooden 1969b
	Localities mapped in SE including Tenasserim	Groves 1970b
	From about 17°N to Isthmus of Kra (about 10°N)	Groves 1972
	In E and SE (map)	Chivers 1977a
	In S from Mouths of Irrawaddy through Tenasserim (map)	Roonwal and Mohnot 1977
Thailand	Throughout W and peninsular (map)	Bourret n.d.
	Plentiful near Chiengmai (about 19°N, 99°E), at Chieng Dao and Doi Angka (all in NW) and along Me Ping R.; not threatened	Carpenter 1940
	Localities listed and mapped in NW in Chieng Mai Province	Kawamura 1961
	In N and part of NE; localities listed in Chiengmai and Loei districts, Khun Tan Mts.; and Siken near Korat	Groves 1968
	In W and S	Lekagul 1968
	In central mountains and Tenasserim Range to S border	Berkson and Ross 1969
	Throughout N and E to Mekong R. and W including peninsular (map)	Fooden 1969b
	Localities mapped in W and peninsular	Groves 1970b
	Localities mapped in W, N central, and central on W and S escarpments of Korat Plateau; plentiful and stable for probably next 10 years	Berkson et al. 1971
	Localities mapped in W in provinces of Mae Hong Song, Tak, Kamphaeng Phet, Nakhon Sawan, Uthai Thani, and Kanchanaburi	Fooden 1971a

Table 178 (continued)

Country	Presence and Abundance	Reference
Thailand (cont.)	Throughout N of Mae Nam Mun R. and E of Bangkok in W and S (map)	Fooden 1971b
	Localities mapped in N (S of 19°30′N), central, W (W of Mae Nam Ping R.), and S	Groves 1972
	Throughout except SE (map); may be extinct in NE; abundant in Khao Yai N.P. (14°25′N, 101°22′E)	Marshall et al. 1972
	Rare and nearing extinction in N	Bruver 1973
	Throughout S, W, and most of N including 3 sanctuaries of Khlong Nakha, Khlong Sang, and Khao Ban That, and Khao Luang N.P.; available habitat about 78,000 km^2, of which 17,000 km^2 are in sanctuaries and park	Brockelman 1975
	In W central and at Ban Thap Plik (near W coast, S peninsular)	Fooden 1976
	Estimated population: 125,000 individuals in 77,500 km^2, of which 17,724 km^2 are protected; by 1980 population to decrease to 60,000 individuals	Chivers 1977a
	In Huay Kha Khaeng Game Sanctuary (near 15°29′N in W)	Eudey, pers. comm. 1977
	In N, NE, W, and S (map); nearly extinct in Doi Angka, Chiengmai	Lekagul and McNeely 1977
	In Phu Khio Reserve, Chaiyaphum Province	McNeely and Laurie 1977
	Abundant in Khao Yai N.P., NE of Bangkok; at Khao Phrik near Takhong R.	Brockelman 1978
	Near Muda Dam (6°08′N, 100°53′E), extreme S on Malaysia border (map)	Gittens 1978
Laos	In W, W of Mekong R. (map)	Fooden 1969b
Malaysia: Peninsular	Throughout (map)	Bourret n.d.
	At Ulu Bernam, Selangor (in W)	Milton 1963
	Near central ridge between Selangor and Pahang E of Kuala Lumpur	McClure 1964
	Common in lowlands; at Kotatinggi (extreme S); in foothills of Perak and Pahang; on Taipang Hills	Harrison 1966
	Present	Bernstein 1967b, Harrison 1969, Groves 1972, Southwick and Cadigan 1972
	Localities listed and mapped in NW, W, central, and SE	Fooden 1969b

Table 178 (continued)

Country	Presence and Abundance	Reference
Malaysia: Peninsular (cont.)	In NW, central, NE, and SE (map)	Lim 1969
	Localities listed in Perlis (extreme NW), Kedah and Perak (in W), Ulu Kelantan (N central), Pehang (E central), Selangor (W), S through Johore (S tip)	Medway 1969a
	Localities mapped throughout W, central, and S	Groves 1970b
	At Fraser's Hill (3°43′N), N of Kuala Lumpur	Kawabe 1970
	Localities mapped in Perak (in W) in N, W, and S including 5 forest reserves of Bubu, Gunong Besout, Segari Melintang, Tanjong Hantu, and Teluk Kopia; population not plentiful but stable	Khan 1970
	Localities mapped throughout	Chivers 1971
	At Kuala Lompat, Pahang, Krau Game Reserve (3°43′N, 102°17′E)	Medway and Wells 1971, Curtin 1977, MacKinnon 1977a, MacKinnon and MacKinnon 1978, Raemaekers 1978
	At Gunong Benom, central, just S of 4°N	Medway 1972
	Localities mapped throughout; total population 211,000–299,000 individuals; range is being fragmented	Chivers 1974
	Localities listed and mapped in Johore, Perak, Kedah, Negri Sembilan, Pahang, Selangor, and Trengganu, including 4 forest reserves of Sedili, Tersang, Klang, and Ranjau Panjang; study sites: Tanjong Triang (2°36′N, 103°46′E), Mersing District, Johore; Jerangau, Trengganu; and Gombak Valley, Selangor	Ellefson 1974
	Estimated population 106,000 individuals in 60,455 km^2, of which 8,420 km^2 are protected; by 1980 population to decline to 50,000 individuals	Chivers 1977a
	Population in 1958: 144,000 individuals; in 1975: 71,000; decreasing	Khan 1978
North Borneo	Not declining	Burgess 1961
	Frequently seen	Stott and Selsor 1961a
	In Lungmanis area	Stott and Selsor 1961b
	Localities listed at Bukit Kretam, Sandakan, Kalabakan, Abai, and Mount Kinabalu (Sabah), and Bellotan and Rayoh (Sarawak)	Davis 1962

Table 178 (continued)

Country	Presence and Abundance	Reference
Malaysia: North Borneo (cont.)	Seen frequently at Labuk Road and less often in Kinabatangan R. area, Sabah	Yoshiba 1964
	In Sabah	de Silva 1968b
	Throughout (maps)	Fooden 1969b, 1971b
	Localities mapped in Sabah and Sarawak	Groves 1970b
	Localities listed and mapped include Kalabakan, Kinabatangan R., Sandakan, Mount Kinabalu, and Madang (Sabah); Bario, Baram R., Marudi, Mount Lambir, Mount Salikan, Mount Dulit, Usan Apau, Belaga, Sut, Balingian, Pelawan R., Sarebas, and Mount Sidong (Sarawak)	Groves 1971b
	In Kinabalu N.P., N Sabah	I.U.C.N. 1971, Phillipps 1973
	Very common in Ulu Segama Reserve, near junction of Bole R. and Segama R., Sabah	MacKinnon 1971
	Along upper Tinjar R., Baram District, Sarawak	Collins et al. 1972
	On inland plain near Kinabatangan R., Sabah	Kawabe and Mano 1972
	In Sepilok Forest, E Sabah	Peters et al. 1976
	Estimated population of 584,000 individuals in 547,700 km^2, of which 9,600 km^2 are protected, including in Kinabalu N.P., Palau Gaya, and Sepilok Reserve (Sabah); by 1980 population to decline to 23,000 individuals	Chivers 1977a
	Localities mapped in Sabah and S and E Sarawak	MacKinnon 1977a
Indonesia	Common on Sumatra	Bourret n.d.
	Abundant throughout Atjeh, N Sumatra, from S and E borders to Koeta Radja	Carpenter 1938
	In Udjung Kulon Nature Park, extreme W Java	Satmoko 1961
	Localities listed and mapped in W, central, and E Sumatra; also in Kalimantan and Java (map)	Fooden 1969b
	Localities mapped throughout Sumatra, in W Java, and in SW, N, and E Kalimantan	Groves 1970b
	Localities listed and mapped in Kalimantan include Pontianak, Landak R., Tajan, Kapaus R., Sukadana, Matan, Belaban, Mankol, Batu Jurong, Klumpang Bay, Pasir R., Balikpapan Bay, Mahakam R., Mount Talisaian, and Birang R.	Groves 1971b
	Localities mapped in Sumatra on N and S central coasts and Musi R.	Groves 1972

Table 178 (continued)

Country	Presence and Abundance	Reference
Indonesia (cont.)	Localities mapped in most areas of Sumatra; numerous but declining	Wilson and Wilson 1972–73
	In Gunung Leuser Reserve, N Sumatra	Kurt 1973
	At Ranun R., W Langkat Reserve, and Sikundur, N Sumatra	MacKinnon 1973
	In Kutai Reserve, E Kalimantan	Rodman 1973
	In Gunung Leuser Reserve, NW of Kutatjane, N Sumatra (3°40′N, 97°40′E)	Rijksen 1974, 1978
	In E Kalimantan	Van Peenen et al. 1974
	Abundant in Kutai Reserve, at Sepaku and at Renggang Road, E Kalimantan	Wilson and Wilson 1975a
	Estimated population of 2,024,000 individuals in 364,300 km^2, of which 22,975 km^2 are protected, including in Tanjong Puting and Kutai reserves (Kalimantan); by 1980 population to decline to 209,900 individuals; Java form is in danger of extinction; others are numerous but will decline to relict populations in 15–20 years	Chivers 1977a
	Localities mapped throughout Sumatra, W Java, and Kalimantan	MacKinnon 1977a
	Localities mapped throughout Sumatra	Wilson and Wilson 1977
	In Java: seriously endangered; in W only, including 2 reserves of Udjung Kulon and Sancang; range shrinking	Brotoisworo 1978
	Form in W Java (*H. l. moloch*) endangered; localities listed include S of Badung, Gunung Honje (estimated population 500–1,000 individuals); Udjung Kulon Nature Reserve; Gunung Halimun and 3 other possible areas; declining	I.U.C.N. 1978

Table 179. Population parameters of *Hylobates lar*

Population Density (individuals /km²)	Group Size X̄	Range	No. of Groups	Home Range Area (ha)	Study Site	Reference
	4.4	2–6	21	12–41	NW Thailand	Carpenter 1940
				4.4–28	Thailand	Berkson et al. 1971
	3.2	2–5	14		W Thailand	Fooden 1971a
2.6–3.8					Thailand	Chivers 1974
4	4			15–>50	Thailand	Brockelman 1975
		3–4	2		SW Thailand	Fooden 1976
					Malaysia	
	4–5		13		Peninsular	Khan 1970
3	3				Peninsular	Medway and Wells 1971
	3.8	2–5	6		Peninsular	Chivers 1972
0.68–1.28		2–3			Peninsular	Southwick and Cadigan 1972
1.9–14.2	2.7–4		15		Peninsular	Chivers 1974
	2.8–3.3	2–6		40	Peninsular	Ellefson 1974
				50–57	Peninsular	MacKinnon 1977a
		5	1	60	Peninsular	Curtin and Chivers 1978
6.1	3.5				Peninsular	MacKinnon and MacKinnon 1978
		4	1	~50	Peninsular	Raemaekers 1978
	5–6	3–9			N Borneo	Stott and Selsor 1961b
	3	3	4		N Borneo	Davis 1962
		2–3			N Borneo	Yoshiba 1964
					Indonesia	
1–4	4.4	2–12	11		N Sumatra	Kurt 1973
8–20					N Sumatra	MacKinnon 1973
4.1–21		3–4			E Kalimantan	Wilson and Wilson 1975a
1.6–14.3	3	2–6			Sumatra	Wilson and Wilson 1977
11	4.5	2–5			N Sumatra	Rijksen 1978

Hylobates pileatus Pileated or Capped Gibbon

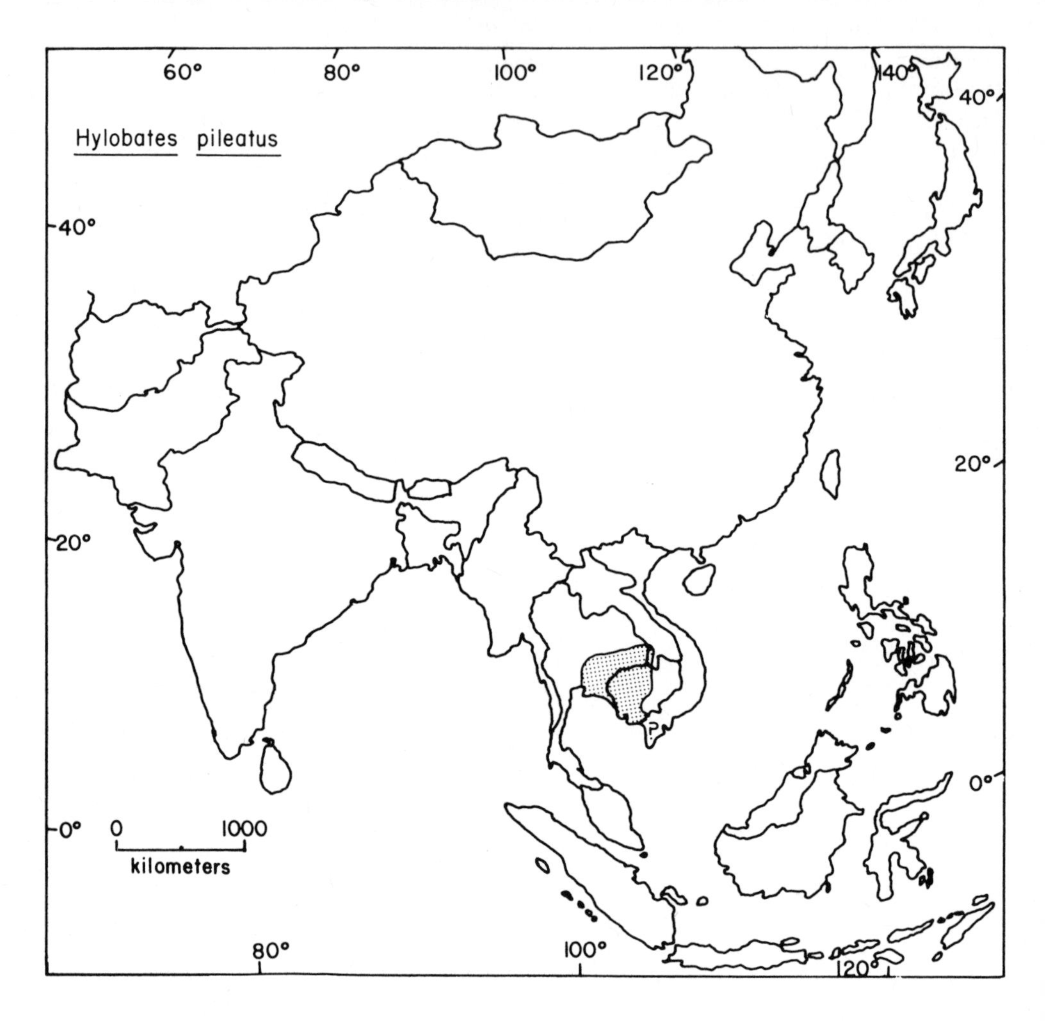

Taxonomic Note

Some authors consider the pileated gibbon to be a subspecies of *Hylobates lar* (Napier and Napier 1967; Lekagul 1968; Fooden 1969b, 1971b), but others regard it as a distinct species (Groves 1970b, 1971b, 1972; Marshall et al. 1972; Brockelman 1975, 1978), and it is treated as such here.

Geographic Range

Hylobates pileatus is found in southern Thailand, southwestern Laos, and western Cambodia. It has also been reported in southern Vietnam, but recent confirmation of its presence there is lacking. The northern and eastern limits of its range are near Pakse, Laos, at about 15°00′N, 105°45′E (Fooden 1969b). *H. pileatus* occurs as far west as Nongkhor, Thailand (13°07′N, 101°02′E), and as far south as the Kiri Rom

Plateau, Cambodia (11°20′N, 104°02′E) (Groves 1972). For details of the distribution, see Table 180.

Abundance and Density

H. pileatus is considered to be endangered and declining rapidly throughout its range (I.U.C.N. 1972, 1976, 1978). In Thailand and Cambodia combined, its population has been estimated at 100,000 individuals in a total area of 42,300 km² of available habitat; these populations were expected to decline to 22,500 individuals by 1980 (Table 180). No recent status information is available from Laos or Vietnam.

Population densities of 2 individuals per km² (Brockelman et al. 1977) and 4 individuals per km² (Brockelman 1975) have been reported. Group sizes vary from 2 to 6 (Lekagul and McNeely 1977), with average sizes of 4 individuals (N=36 groups) (Brockelman et al. 1977), and home ranges of 15 to more than 50 ha (Brockelman 1975). The minimum area in which a group of pileated gibbons could survive is 4 to 6 ha, but the normal dispersal of young adults to form new groups could not occur in such a small area (Lekagul and McNeely 1977).

Habitat

H. pileatus occupies dense diverse evergreen, mixed evergreen-deciduous (Brockelman 1975), and tall, wet montane forests (Brockelman et al. 1977). It has also been observed in dry evergreen forest, tropical rain forest (Berkson et al. 1971), and dry deciduous forest (Rand, pers. comm. 1974). This species requires intact natural forests for survival (Brockelman 1975).

Factors Affecting Populations

Habitat Alteration

The felling of forest for timber and agricultural land has greatly decreased the habitat available to *H. pileatus* in southeastern Thailand (Lekagul 1968; Berkson et al. 1971; Marshall et al. 1972; Rand, pers. comm. 1974; Lekagul and McNeely 1977), and has isolated populations in small pockets of remaining mountain habitats (Berkson and Ross 1969). Deforestation by clearcutting, along with colonization by squatters, which prevents regeneration, poses the greatest threat to the future survival of *H. pileatus* in this region (Brockelman 1975). For example, the area around the Khao Yai National Park is almost completely deforested (Brockelman 1978), and virtually all forested areas in Thailand are shrinking by several percent each year (Brockelman 1975).

In Cambodia, forested areas are also being disturbed and eliminated (Brockelman 1975). Bombing and other military activities have certainly altered natural habitats in Cambodia, Laos, and Vietnam (Orians and Pfeiffer 1970), but no information regarding the effects on pileated gibbon populations is available.

Human Predation

Hunting. H. pileatus is hunted for food in Thailand (Berkson and Ross 1969; Rand, pers. comm. 1974), especially heavily by loggers, settlers, and professional hunters

(Brockelman 1975, Brockelman et al. 1977). No information was located regarding hunting in the other countries of its range.

Pest Control. No information.

Collection. Young *H. pileatus*, obtained by the killing of their mothers, are collected in Thailand for sale as pets (I.U.C.N. 1972, 1976, 1978; Brockelman 1975), but the extent of trade in this species is not known.

No *H. pileatus* were imported into the United States in 1968–73 (Appendix) or in 1977 (Bittner et al. 1978), nor into the United Kingdom in 1965–75 (Burton 1978).

Conservation Action

H. pileatus occurs in three national parks and three sanctuaries and reserves in Thailand (Table 180). A total of 2,632 km^2 of its habitat there are protected (Chivers 1977a) and a population of about 10,000 individuals could be conserved within protected areas (Brockelman 1975). No reports of preserved habitats in the other countries of the range were received.

In Thailand, pileated gibbons may be hunted, captured, or kept in captivity only under license, and exported only by special permit (Lekagul and McNeely 1977), but these regulations are very difficult to enforce (Rand, pers. comm. 1974).

Suggestions for improving protection of this species and its habitat include patrols to prevent loggers and woodcutters from entering valleys leading to the mountains (Brockelman et al. 1977), establishment and effective protection of undisturbed forest reserves (Rand, pers. comm. 1974; Chivers 1977a), field surveys in Laos and Cambodia, and conservation education (I.U.C.N. 1978). Projects have now been funded to survey *H. pileatus* populations in the Khao Soi Dao Sanctuary, determine the extent of poaching there, build a guard station at the sanctuary, and conduct ecological and behavioral studies (Mittermeier 1977b).

H. pileatus is listed in Appendix I of the Convention on International Trade in Endangered Species of Wild Fauna and Flora, and as endangered under the U.S. Endangered Species Act (U.S. Department of Interior 1977a,b).

Table 180. Distribution of *Hylobates pileatus* (countries are listed in order from north to south)

Country	Presence and Abundance	Reference
China	N limit of range was S Kuangsi (misidentification ?)	Van Gulick 1967
Laos	In SW corner (map)	Bourret n.d.
	In S, W of Mekong R.	Delacour 1951
	In SW near Pakse	Fooden 1969b
Thailand	In SE (map)	Bourret n.d.
	In E	Lekagul 1968
	In SE including Saraburi Province	Berkson and Ross 1969
	In SE from Bight of Bangkok to Mekong R. including Khao Yai National Reserve (14°30′N, 101°30′E)	Fooden 1969b
	Reduced in numbers; in danger of extinction	Simon 1969
	Localities mapped in SE	Groves 1970b
	Localities mapped in SE including S escarpment of Korat Plateau; nowhere plentiful	Berkson et al. 1971
	In SE, S of Mae Nam Mun R. (maps)	Fooden 1971b, Chivers 1977a, MacKinnon 1977a
	Localities listed and mapped in SE: Nongnum Kuan, Khao Yai, Klong Yai, Klong Menao, Lem Ngop, Nongkhor, Sakeo, and Khao Soi Dow	Groves 1972
	In SE (map) including Khao Yai N.P.	Marshall et al. 1972
	In SE; declining	Rand, pers. comm. 1974
	Originally about 2–3 million individuals, now about 10,000 individuals could be protected; about 13,500 km^2 of habitat remaining, including Khao Yai N.P. (with 1,000 or more individuals) and Khao Khitchakut N.P. and adjoining Khao Soi Dao Sanctuary; small numbers in Khao Khieo Wildlife Sanctuary and Khao Sabap N.P.; localities listed include Dong Rek Range (in 7 provinces), and other areas in Rayong, Chon Buri, Chachoengsao, and Chanthaburi provinces; range shrinking	Brockelman 1975
	About 500–1,500 individuals in Khao Soi Dao Sanctuary	Brockelman et al. 1977
	Population of about 40,000 individuals in area of 13,500 km^2, of which 2,632 km^2 are protected; population to decrease to 10,500 individuals by 1980	Chivers 1977a

Table 180 (continued)

Country	Presence and Abundance	Reference
Thailand (cont.)	In SE as far W as Siracha, including Khao Yai N.P. and provinces of Prachinburi, Surin, and Buriram (map); rare	Lekagul and McNeely 1977
	Abundant in Khao Yai N.P.; N limit is N slope of Dong Rek Range parallel to Mun R.; other localities listed include Takhong and Phra Phloeng R. reservoir and Sakaerat Biosphere Reserve	Brockelman 1978
Cambodia	Throughout (map)	Bourret n.d.
	In W, W of Mekong R. (map)	Fooden 1969b, 1971b; MacKinnon 1977a
	In SW, questionable in SE (map)	Groves 1970b
	Localities listed and mapped in W include Kiri Rom Plateau and Siemp Reap	Groves 1972
	Abundant and stable in W	Rand, pers. comm. 1974
	Remaining habitat about 28,800 km^2, including areas N of Mongkol R.; Elephant and Cardamom ranges; and in N, W of Mekong R.	Brockelman 1975
	Total population about 60,000 individuals, to decrease to 12,000 individuals by 1980; in W, W of Mekong R. (map)	Chivers 1977a
Vietnam	In S from about 9°–11°N (map)	Bourret n.d.
	Will become extinct there in near future	Ngan 1968
	Reported from Phu Quoc Island	Van Peenen et al. 1969
	In S tip S of Mekong R. (map)	Fooden 1971b

Hylobates syndactylus Siamang

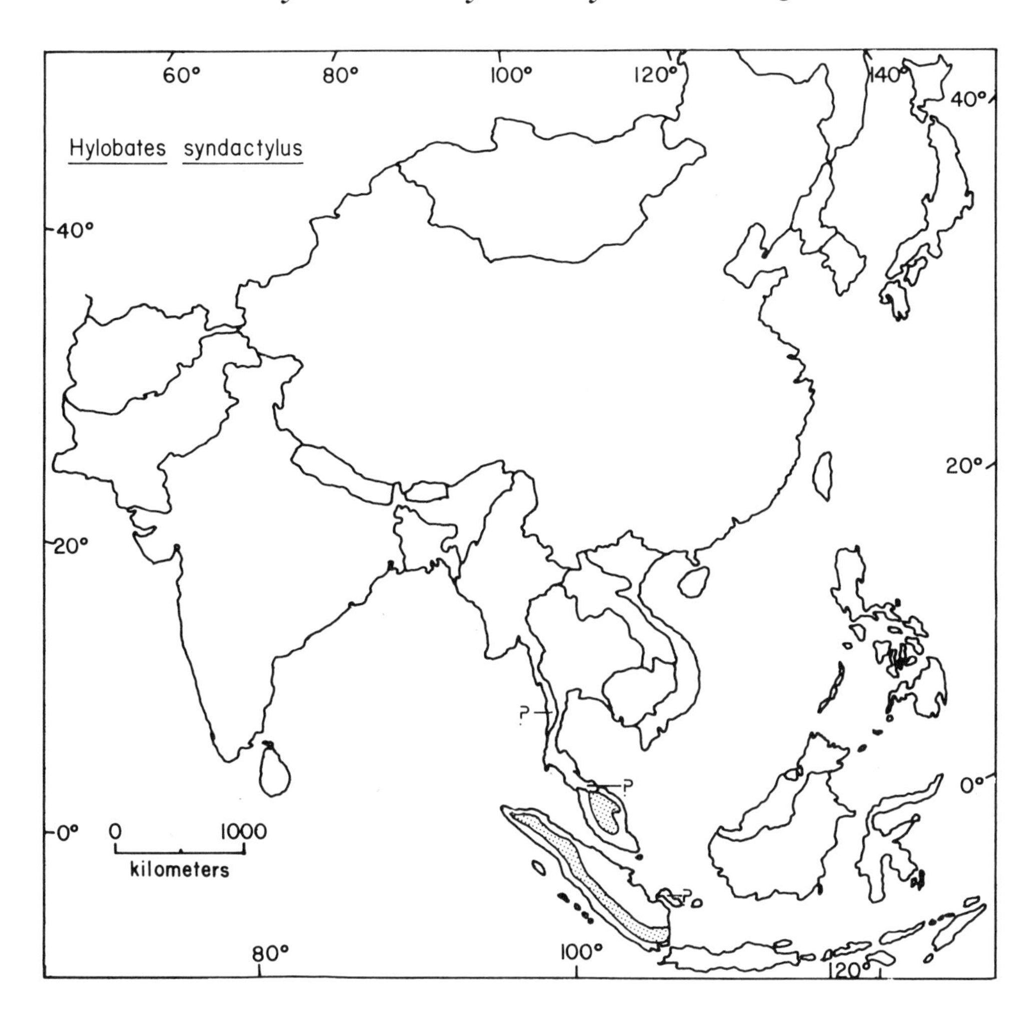

Taxonomic Note

Some authors place the siamang in its own genus, *Symphalangus* (Carpenter 1938; Napier and Napier 1967; Chivers 1975a, 1977b; MacKinnon 1977a), but most regard it as a member of *Hylobates* (Harrison 1966, 1969; Medway 1969a, 1972; Groves 1970b, 1972; Khan 1970, 1978; Rijksen 1978). Two subspecies are recognized (Napier and Napier 1967, Groves 1972).

Geographic Range

Siamangs are found on the Malaysian peninsula and on Sumatra, Indonesia. Their range extends north to about 5°45′N, 101°50′E and west to about 101°E (at 4°45′N) in Malaysia (Chivers 1971, 1974). This species has also been reported farther north into peninsular Thailand (Napier and Napier 1967, Groves 1970b) and in Tenasserim, peninsular Burma (Ellerman and Morrison-Scott 1966), but no modern firsthand confirmations of its presence in either of these two countries were located. On Sumatra, *H. syndactylus* is found as far west as the northwestern end to about 95°30′E, and as far south and east as the southeastern shore, about 5°30′S, 105°50′E (Wilson and Wilson 1977). It may also occur on Bangka, east of Sumatra (Groves 1972). For details of the distribution, see Table 181.

Abundance and Density

Populations of *H. syndactylus* are considered locally numerous but declining and likely to decline further in the near future in both countries of their range (Chivers 1977a). One author estimates the population in Malaysia at about 63,000 individuals, while another calculates that only about 34,000 remain there, that this number would decline to 29,000 by 1980, and that Sumatra has a population of about 133,000, which would decline to 110,000 by 1980 (Table 181). Thus the worldwide population is thought to be about 167,000 to 196,000 individuals. In both countries, populations will decrease to small relicts within 15 to 20 years (Chivers 1977a).

Siamang population densities vary from 0.5 to 26 individuals per km^2 (Table 182). This species lives in small family groups of 2 to 10 individuals, generally averaging 3 to 4 members per group, occupying home ranges of 15 to 50 ha (Table 182). Medway and Wells (1971) calculated that the area necessary to support 5,000 individuals would be 2,000 km^2.

Habitat

H. syndactylus is a tropical rain forest species, found in evergreen lowland and hill dipterocarp, submontane, and montane forest (McClure 1964; Chivers 1972, 1974, 1975b; Ellefson 1974). In Malaysia its main habitat is hill dipterocarp forest (Chivers 1973), but it exploits a broad range of forest types and is not restricted to ones dominated by dipterocarp (Medway 1972). Lim (1969) concludes that it is confined to primary forest, while others have observed it also in secondary and disturbed forests (Chivers 1972). In one study, 57% of groups sighted were in secondary forest, but the population density there was less than half that seen in primary forest (Southwick and Cadigan 1972).

In Sumatra, *H. syndactylus* is found at the highest densities in primary lowland forest and primary hill forest, but also occurs in secondary hill forest, freshwater swamp forest, and submontane forest, 915–1,525 m in elevation (Wilson and Wilson 1977).

Siamangs have been recorded as low as sea level in both parts of their range (Ellefson 1974; Chivers 1974, 1977b; Wilson and Wilson 1977), as high as 3,800 m in Sumatra (Haltenorth 1975), and 1,829 m in Malaysia (Medway 1972), but their average elevation in both areas is 400 to 500 m, and they seem to prefer hill forest to other habitats (Chivers 1974, 1977a).

Factors Affecting Populations

Habitat Alteration

The siamang is an arboreal species that is seriously disturbed by logging (Burgess 1971) and whose habitat is being reduced by timbering and clearing for cultivation in both Malaysia and Sumatra (Chivers 1974). It can adapt well to low intensity selective logging in which only a few trees are removed per unit area (Chivers 1972), but extensive clearing is fragmenting the remaining forest blocks and threatening the rain forest on which this species depends, even in remote areas of Malaysia (Chivers 1974). Other authors who have cited deforestation as a threat to *H. syndactylus* in Malaysia include Khan (1970, 1978) and Cadigan and Lim (1975). An estimated 2,500 km^2 of lowland forest are being cleared annually on the peninsula, and hydroelectric schemes are also reducing forested areas (Chivers 1977a).

In Sumatra, siamangs occupy lightly logged areas, but occur in lower densities there than in undamaged forest (MacKinnon 1973, Wilson and Wilson 1977, Rijksen 1978). Their range in Sumatra is being reduced by lumbering and farming (Wilson and Wilson 1972–73), and additional large logging concessions have been granted (Chivers 1977a).

Human Predation

Hunting. H. syndactylus is eaten by some people in Malaysia (Khan 1970), but hunting was not cited as a major threat to the species there or in Sumatra.

Pest Control. Siamangs do not utilize food from plantations (Chivers, pers. comm. 1974), and would not be expected to suffer from control efforts.

Collection. H. syndactylus mothers are shot to obtain their babies for sale in the local pet trade and for export in both Malaysia (Khan 1970, 1978; Chivers, pers. comm. 1974) and Sumatra (Wilson and Wilson 1972–73).

Between 1968 and 1973, 284 *H. syndactylus* were imported into the United States (Appendix), primarily from Singapore, where this species does not occur. None entered the United States in 1977 (Bittner et al. 1978). Twenty-one were imported into the United Kingdom in 1965–75 (Burton 1978).

Conservation Action

Siamangs are found in four reserves in Malaysia and two interconnected ones in Sumatra (Table 181). The total area of their habitat protected in both regions combined is 23,121 km^2 (Chivers 1977a).

This species is legally protected from killing, capture, trade, and export in Malaysia (Harrison 1966, Medway 1969a, Khan 1970, Yong 1973), except by special permit for scientific or exhibit purposes (Khan 1978), and strenuous efforts are being made to enforce the laws (Chivers, pers. comm. 1974). *H. syndactylus* is also protected by law in Indonesia (Carpenter 1938, Brotoisworo 1978), but enforcement is not adequate there (Wilson and Wilson 1972–73).

Measures which are needed if remaining populations of *H. syndactylus* are to be conserved include (1) the control or cessation of clearcut logging, (2) the establishment of large, protected forest reserves, (3) supervision of use of all remaining forests, (4) enforcement of export restrictions, and (5) widespread conservation education (Chivers 1977a).

Siamangs are listed in Appendix I of the Convention on International Trade in Endangered Species of Wild Fauna and Flora, and as endangered under the U.S. Endangered Species Act (U.S. Department of Interior 1977a,b).

Table 181. Distribution of *Hylobates syndactylus*

Country	Presence and Abundance	Reference
Malaysia: Peninsular	Near ridge between Selangor and Pahang E of Kuala Lumpur (in SW)	McClure 1964
	Throughout except W and N of Perak R. (maps)	Fooden 1969b, 1971b
	In central, including S Kelantan, SE Perak, NE Selangor, and NW Pahang (map)	Lim 1969
	In central from Perak and Kelantan to S Pahang and Negri Sembilan	Medway 1969a
	At Fraser's Hill Reserve, 3°43′N, N of Kuala Lumpur	Kawabe 1970
	Localities mapped in Perak W of Perak R., including Gunong Besout Forest Reserve; population not plentiful but stable	Khan 1970
	Localities mapped in W central and S Kelantan, central Trengganu, N and W Pahang, E Selangor, N and central Negri Sembilan, and E Perak; highest density in Ulu Gombak Jungle Reserve, Selangor	Chivers 1971
	At Kuala Lompat, Krau Game Reserve, Pahang	Medway and Wells 1971, Chivers 1975a, Curtin 1977, MacKinnon 1977a, Raemaekers 1978
	Localities mapped in W central and N; 3 study sites: Ulu Sempam, Pahang (3°46′N, 101°45′E), Kuala Lompat, Pahang (3°43′N, 102°17′E), and Ulu Gombak, Selangor (3°20′N, 101°47′E)	Chivers 1972, 1973
	At Gunong Benom, Pahang	Medway 1972
	Localities mapped in W central and N; mainly in W with an extension E through Taman Negara; total population about 25,000–35,000 individuals	Chivers 1974
	Locally abundant; declining	Chivers, pers. comm. 1974
	Localities mapped at Melaka, Pahang; Chenderoh Dam, Perak; and Gombak Valley, Selangor	Ellefson 1974
	At Ulu Sempam, W Pahang	Chivers 1975b
	In W central and N (map); total population about 34,000 individuals in 13,150 km^2, of which 4,841 km^2 are protected; by 1980 population to decrease to 29,000 individuals; now numerous but will be reduced to relict populations in 15–20 years	Chivers 1977a
	Total population in 1958 about 111,000; in 1975 about 63,000 and decreasing	Khan 1978

Table 181 (continued)

Country	Presence and Abundance	Reference
Indonesia: Sumatra	Abundant throughout Atjeh (W end) from SE border to near Koeta Radja	Carpenter 1938
	Throughout (maps)	Fooden 1969b, 1971b; Groves 1970b
	Localities mapped in W half from N to S ends; range being reduced	Wilson and Wilson 1972–73
	In Gunung Leuser Reserve (in N, NW of Kutatjane)	Kurt 1973; Rijksen 1974, 1978
	Three study sites: Ranun R., West Langkat Reserve, Sikundur (all in N)	MacKinnon 1973, 1974a
	In W, N, and S (maps)	Chivers 1974, 1977a; MacKinnon 1977a; Wilson and Wilson 1977
	Total population about 133,000 individuals in area of 131,350 km^2, of which 18,280 km^2 are protected; by 1980 population to decrease to 110,000 individuals and in 15–20 years will be reduced to relict populations	Chivers 1977a

Table 182. Population parameters of *Hylobates syndactylus*

Population Density (individuals/km²)	Group Size X̄	Group Size Range	Group Size No. of Groups	Home Range Area (ha)	Study Site	Reference
	5				Malaysia	McClure 1964
		2–5			Malaysia	Medway 1969a
	4	3–6	6		Malaysia	Kawabe 1970
		4–5			Malaysia	Khan 1970
3.2	3.3	2–5		20	Malaysia	Chivers 1971
		5	1		Malaysia	Medway and Wells 1971
		4–5	4	23	Malaysia	Chivers 1972
0.5–1		2–3			Malaysia	Southwick and Cadigan 1972
0.8–4.2	3.8	2–5	18	15–34	Malaysia	Chivers 1974
				30–42	Malaysia	Chivers et al. 1975
1–17.8		2–5		15–30	Malaysia	Chivers 1977a
				22–34	Malaysia	MacKinnon 1977a
4.5	3				Malaysia	MacKinnon and MacKinnon 1978
		4	1	50	Malaysia	Raemaekers 1978
20					Indonesia	Carpenter 1938
1–4	3.9	2–10	11		Indonesia	Kurt 1973
6–26					Indonesia	MacKinnon 1973
3.1–11.6	3.8	2–6	16		Indonesia	Wilson and Wilson 1977
15	4.2	2–6		21	Indonesia	Rijksen 1978

PONGIDAE

Great Apes

Gorilla gorilla Gorilla

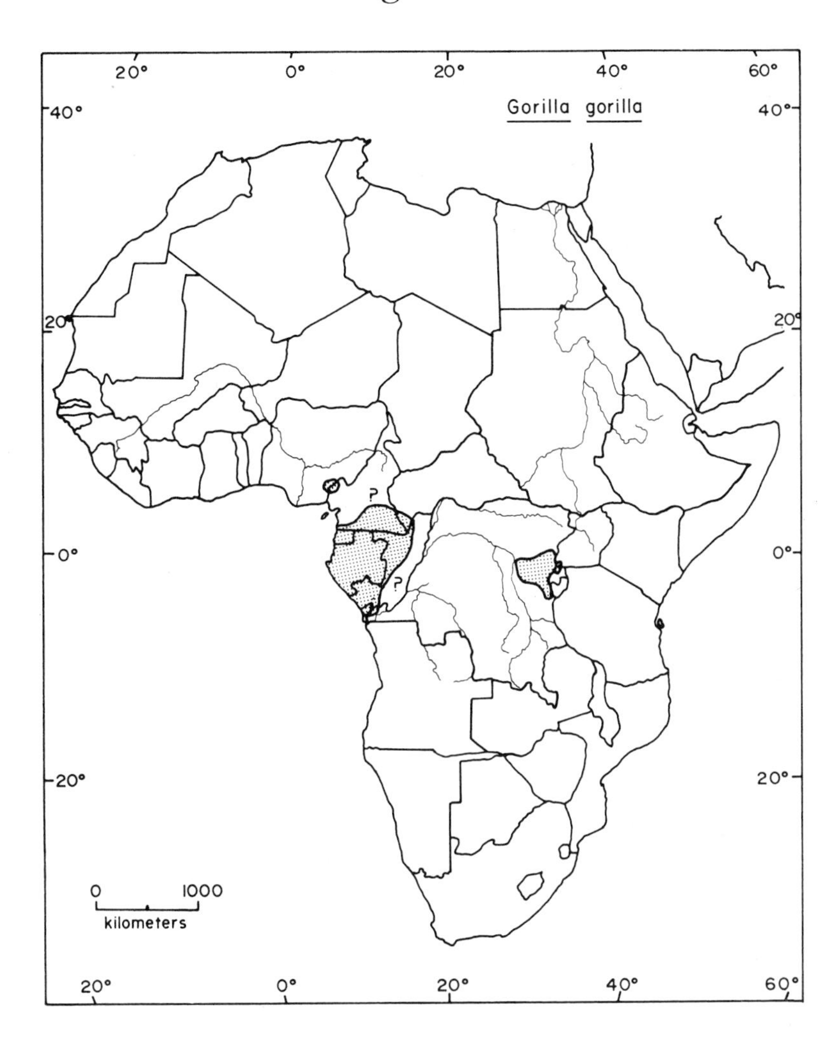

Taxonomic Note

Most earlier workers (e.g., Schaller 1963) discuss two subspecies, but many later ones follow Groves (1967b, 1970a, 1971a) and recognize three: *G. g. gorilla* (western), *G. g. beringei* (true mountain), and *G. g. graueri* (eastern lowland).

Geographic Range

Gorilla gorilla is found in West Africa from southeastern Nigeria to southern Congo, and in East Africa in eastern Zaire, southwestern Uganda, and northwestern Rwanda. The northwestern limit of its range is near 7°N, 9°E north of the Cross River, Nigeria, and the southern limit is near 5°S in Cabinda, Angola (Groves 1971a, Goodall and Groves 1977). In the eastern part of the range, gorillas are found from slightly north of the equator to 4°20′S, and from 26°39′E to 30°E, over an area of

90,650 km^2 (Emlen and Schaller 1960, Schaller 1963, Groves 1971a). No populations occur between northern Congo (about 1°30′N, 16°30′E) and Lubutu, eastern Zaire (0°48′S, 26°39′E), a gap of more than 1,000 km, but evidence indicates that *G. gorilla* may once have occurred in this area (Schaller 1963, Groves 1971a, Goodall and Groves 1977). For details of the distribution, see Table 183.

Abundance and Density

Gorilla populations are reported to be declining in every country for which recent information is available (Table 183). They are described as scarce or facing extinction in Nigeria, Equatorial Guinea, Angola, Uganda, Rwanda, and Burundi, are potentially threatened or vulnerable in Gabon and Cameroon, and have decreased in parts of Congo, Central African Republic, and Zaire. The species is most severely threatened near the edges of its range in the west and in the Virunga Volcanoes area, which straddles the borders of Zaire, Rwanda, and Uganda. The I.U.C.N. (1976) classifies the entire species as vulnerable and the mountain race (*G. g. beringei*) as endangered.

Details concerning the absolute abundance of *G. gorilla* are lacking for many areas, but some rough estimates of population sizes are available. For example, there are thought to be less than 10,000 *G. gorilla* in all of West Africa (Gartlan 1975d), including a few hundred in Equatorial Guinea (Jones and Sabater Pi 1971). The population in Zaire and Uganda was once estimated at 5,000 to 15,000 individuals (Schaller 1963), but more recently Zaire was thought to contain about 2,500 (Verschuren 1975b) and Uganda no more than 200 gorillas (Kingdon 1971). The Rwandan population of *G. g. beringei* is estimated to number 225 to 250 (Cousins 1978c, Harcourt and Curry-Lindahl 1978, Wrangham 1978). The total population of the mountain race is thought to be about 500 individuals (I.U.C.N. 1976).

Population densities in *G. gorilla* vary from 0.38 to 7 individuals per km^2, and groups of 2 to 30 with average sizes of 6 to 17 and home ranges of 200 to 5,000 ha have been reported (Table 184).

Habitat

G. gorilla is a forest species which prefers dense or secondary forest and woodland to other habitats in both West and East Africa. In Nigeria it has been reported in montane rain forest (Petrides 1965), and in Cameroon it utilizes lowland forest, montane forest, and secondary growth (Gartlan, pers. comm. 1974).

In Equatorial Guinea, *G. gorilla* is typical of the thick vegetation of primary dense forest, defective forest, and regenerating forest (Sabater Pi and Jones 1967). There it is confined primarily to humid montane forest up to 3,050 m elevation and lowland regenerating forest, is characterized as an edge species found most frequently near roads, clearings, or fields in various stages of regeneration, and enters mature primary forest, thickets, and meadows relatively infrequently (Jones and Sabater Pi 1971). In Gabon and Congo the preferred habitats are high forest, gallery forest, and forest bordering on cultivated areas (Malbrant and Maclatchy 1949).

In the eastern portion of the range, 75% of *G. gorilla*'s habitat is within moist evergreen lowland forest below 1,500 m elevation, most of the remainder is in humid montane rain forest between 1,500 and 3,500 m, and a minor habitat is the bamboo forest from 2,400 to 3,000 m (Schaller 1963). All of the habitats utilized are moist and lush with a year-round abundance of herbaceous forage near ground level, and *G. gorilla* "favors dense secondary growth where forage in the form of herbs, shrubs, and vines is plentiful, over primary forest which supports only a sparse ground cover" (Schaller 1965). In primary forest, gorillas are found only at edges, near rivers, or where the canopy is broken (Schaller 1965).

In the Virunga Volcanoes area the favored habitat for *G. gorilla* is *Hagenia* woodland; but montane woodland, *Hypericum* woodland, dry colonizing forest, bamboo, giant *Senecio*, tall-grass meadows, and tree heath are also utilized to progressively lesser extents (Schaller 1963). Other workers in this area have observed *G. gorilla* primarily in *Hagenia* and *Hypericum* woodland (Fossey and Harcourt 1977), and in and above the bamboo zone, in moorland, and to the edge of the alpine zone at 3,500 m elevation (MacKinnon 1976). On the Uganda side of the border, *G. gorilla* occupies montane forest, bamboo, and montane savanna from 2,400 to 2,700 m elevation (Harroy 1972).

In the Kahuzi-Biega region, gorillas were seen most often in secondary forest between 1,000 and 2,400 m, dominated by *Neobourtonia*, *Dombeya*, and *Conopharyngia*; they also utilized primary forest, bamboo forest (seasonally), and *Cyperus* marshes (Casimir 1975a,b). At a study site in mixed montane rain forest at 2,000 to 2,500 m, one group visited bamboo forest, swamp, *Myrianthus*, and secondary growth habitat mixtures more often than other vegetation types, spent the majority of its time in secondary growth, and seemed able to utilize a mosaic of biotopes better than either primary or secondary vegetation alone (Goodall 1977, 1978). Other workers have also noted the preference of gorillas in this area for secondary vegetation (Verschuren 1975b, MacKinnon 1976).

In the lowland regions of Zaire, *G. gorilla* also favors secondary forest and open river valleys overgrown with tree ferns, vines, and herbs (Schaller 1963). It occupies several blocks of isolated secondary forest below 1,500 m elevation (Verschuren 1975b), but relatively few reports are available from these lowland areas.

Factors Affecting Populations

Habitat Alteration

Human activities which alter the character of vegetation in an area can have either a positive or negative effect on populations of *G. gorilla*, depending on the nature and intensity of the disturbance. For example, a limited amount of slash and burn rotational clearing for cultivation, in which secondary growth is allowed to regenerate in abandoned fields within a few years of clearing, benefits gorilla populations by opening the forest and allowing the growth of herbs and vines which provide food for the apes (Emlen and Schaller 1960). In fact, these authors found that the optimum habitat for *G. gorilla* is a "mosaic of old fields in various stages of regeneration," and that the distribution of *G. gorilla* in the lowland forest is determined to a great extent by the size and distribution of patches of regenerating forest.

In eastern Zaire, rotational agriculture is very important in creating secondary forest, the favored habitat of gorillas, and the species is concentrated near roads and villages and tends to move into agricultural areas and abandoned settlements, disappearing when the forest becomes mature (Schaller 1963, Goodall 1977). Light logging does disturb gorillas temporarily, but in the long run may improve the habitat by creating new supplies of forage (Schaller 1963).

In West Africa the gorilla also benefits from limited habitat disturbance. In Equatorial Guinea, for instance, "degraded forest" provides much of its food (Sabater Pi 1964), and the distribution of *G. gorilla* coincides with the distribution of regenerating forest, such as that on old fields (Jones and Sabater Pi 1971).

On the other hand, the major threat to gorilla populations in both western and eastern sections of their range is the deforestation of large areas for timber or for permanent agriculture or livestock grazing. Gorillas cannot survive in deforested areas and are even reluctant to cross relatively narrow open areas, such as roads (Lyon 1975a) or meadows without cover (Groom 1973). In Cameroon, for example, clearing of forest has eliminated gorillas from many parts of their former range (Gartlan, pers. comm. 1974). And the large forestry concessions that have been granted for the areas near Lomie, Yokadouma, Abong-Mbang, and Moloundou indicate that most of the primary forest in the southeast will also be destroyed by logging (Cousins 1978a).

Extensive lumbering and clearing for agriculture have also greatly reduced the range of *G. gorilla* in Equatorial Guinea (Sabater Pi and Jones 1967). There *G. gorilla* is found in secondary vegetation only where adjacent to sizable areas of extant primary forest, and lumbering of about one-third of the country has greatly depleted the habitat available to this ape (Jones and Sabater Pi 1971). In Congo and Gabon large tracts of primary forest still remain (Cousins 1978c), but a railroad planned from the coast to the interior of Gabon will open that country for exploitation of these forests and will reduce available gorilla habitat (Cousins 1978a).

In areas of good quality soil in East Africa, such as volcanic regions in the Kayonza Forest and the Lubero district, the range boundaries of *G. gorilla* are being rapidly pushed back by the conversion of land to stabilized agriculture (Emlen and Schaller 1960). In montane areas such as Mount Tshiaberimu, the rich soil can be cultivated for long periods, abandoned fields revert to grassland instead of forest, secondary succession is deflected by grazing or burning, and the forest below 3,000 m is being gradually eliminated and will soon disappear (Schaller 1963).

Habitat alteration has also been detrimental to many other eastern *G. gorilla* populations. For example, the expansion of cultivation in the Ugandan lowlands, resulting in the movement of pastoral people and herds up the mountainsides, has been blamed for the disappearance of *G. gorilla* from Mount Muhavura (Suzuki 1971). Continued encroachment by cultivation, bamboo cutting, and other human activities reduced the area of suitable habitat in the Kigezi Gorilla Sanctuary, Uganda, from 34 to 23.3 km^2 between 1950 and 1970, and population pressure around the Kayonza Forest is expected to make similar inroads on this refuge (Kingdon 1971). Recently the Kayonza Forest still showed no evidence of disturbance from agriculture, lumbering, or livestock grazing (Cousins 1978c).

In Burundi the forest has been fragmented into many tiny areas too small to support gorilla populations (Verschuren 1978b). *G. gorilla* habitats on the slopes of the western rift from Goma to beyond Uvira (the Itombwe area), Zaire, have suffered heavy disturbance from domestic livestock (Vershuren 1975b). And frequent burning of vegetation has prevented secondary forest regeneration in the Kahuzi-Biega region (Goodall and Groves 1977), and may be responsible for the isolation of that population from the one on the Virunga Volcanoes (Schaller 1963, Mallinson 1974).

In the Virunga range, *G. gorilla* is being increasingly confined to the steepest slopes by cattle herds, settlements, honey collectors, wood gatherers, grass cutters (for thatching), and farmers (Campbell 1970, Fossey 1970, Harcourt and Groom 1972, Spinage 1972, Groom 1973, Goodall and Groves 1977). Recently a 10,000 ha portion of the 33,000 ha Rwandan Volcanoes National Park was deforested for a pyrethrum cultivation project, and consequently settlement, road building, and illicit cultivation and cattle grazing have also increased greatly (Harroy 1972, Fitter 1974, Fitter and Fitter 1974, Goodall and Groves 1977). The Rwandan government is planning to excise a further 3,000 to 4,000 ha from the park for cattle pasture (Harcourt and Curry-Lindahl 1978). In addition, human occupation at low elevations has driven large-bodied game to higher altitudes, resulting in severe elephant and buffalo damage in fragile gorilla habitat; a pipeline has been installed to transport water from the Gahinga-Muhavura saddle toward Uganda; another pipeline from the crater lake on Mount Visoke is being considered; and the Rwandan government is under continual pressure to convert land in the park to agricultural projects (Groom 1973).

Human Predation

Hunting. G. gorilla is hunted for its flesh in many parts of its range, including eastern Cameroon (Gartlan, pers. comm. 1974), Gabon (Jouventin, pers. comm. 1974), and Congo (Cousins 1978a). It is hunted intensively in the Mount Alén area of Equatorial Guinea (Jones and Sabater Pi 1971), where its meat is reputed to be delicious, fatty, and tasting of *Afromomum* (Sabater Pi and Groves 1972). It has also been heavily hunted in Zaire using snares, pitfalls, nets, guns, and spears in the Utu and Mount Tshiaberimu regions (Schaller 1963), near Kivu and Maniema (Verschuren 1975b), and near the Kahuzi-Biega National Park (Mallinson 1974, Cousins 1978c), although hunting was stopped in this last area in 1966 (Goodall and Groves 1977).

Gorillas are also killed for witchcraft purposes in both West and East Africa. Populations have been decimated in southern and central Congo by hunters seeking the skulls as fetishes (Bassus 1975), and some tribes in Congo have declared war on *G. gorilla*, seemingly as a substitute for a human enemy (Cousins 1978a). In Rwanda and Uganda, gorilla ears, tongues, genitals, and toes are sought for black magic (Cousins 1978c).

Hunting has had an especially detrimental effect on populations of *G. gorilla* in the Virunga Volcanoes area (Baumgartel 1965, Campbell 1970). Poachers sometimes inadvertently catch gorillas in one of the many snares set for other game, and also kill *G. gorilla* for witchcraft practices (Harcourt and Groom 1972, Groom 1973, Goodall and Groves 1977). Since 1976, hunters using high velocity rifles have also been killing gorillas in the Virunga area for their heads and hands, to be sold as trophies or souvenirs to white residents and tourists (Harcourt and Curry-Lindahl 1978, Wrangham 1978). Overall, man kills gorillas for many reasons, including "sport," and is an important predator of the species (Schaller 1963).

Pest Control. Gorillas are reported to feed on crops in many areas, and local people often cite destructiveness as a justification for shooting these apes. In eastern Cameroon they are frequently killed in retaliation for plantation destruction (Cousins 1978a). In Equatorial Guinea they are said to "devastate the native plantations" (Sabater Pi 1964) and are frequently killed for crop protection, even though they enter native fields only occasionally (Jones and Sabater Pi 1971). This species is also killed to prevent its raids on villages and plantations in Central African Republic and Gabon (Groves 1970c).

G. gorilla is regarded as a culture follower in eastern Zaire, because its principal source of nutrition in some areas is the fruit and pith of cultivated banana plants (Rahm 1966). Local people near Lubero organize gorilla hunts to protect their gardens (Emlen and Schaller 1960), and people in the Walikali region also kill crop-damaging gorillas (Goodall and Groves 1977). Eastern gorillas eat the fruits of peas and maize, the roots of carrots and taro, the tubers of manioc, and the pith of banana plants, whereas western gorillas eat papaya, pineapple, manioc, banana, sugar cane, peanuts, kola nuts, and palm nuts (Schaller 1963). In both sections of its range this species is killed "with the double incentive of eliminating crop raiders and providing meat for the larder" (Schaller 1963), and also for the income from the sale of the meat and the sale of skulls and juveniles to white men (Blancou 1961).

Collection. The capture of live *G. gorilla* for export has had an especially pronounced effect on the western populations. Most of the gorillas in captivity are western ones from Cameroon, Equatorial Guinea, and Gabon (Gartlan 1975d). Cameroon issues export permits for 15 *G. gorilla* per year (Gartlan, pers. comm. 1974). A total of 41 gorillas were exported legally from Equatorial Guinea between 1958 and 1968, but since mothers were killed to obtain infants for export, the number lost to the population was much greater than 41 (Jones and Sabater Pi 1971). The collection of babies for zoos is believed to be a major reason for the decline of *G. gorilla* in Gabon (Jouventin, pers. comm. 1974).

In 1975 the total number of gorillas in zoological collections worldwide was 550 to 560 individuals, of which fewer than 100 were captive born (Cousins 1976). The capture of the other 450 to 460 individuals resulted in an estimated 2,500 deaths of wild gorillas and represents a significant and continuing drain on the remaining natural population (Cousins 1976). The high prices offered for gorillas by purchasers in developed countries encourage illegal capture and export (Spinage 1972, Awunti 1978). The demand for *G. gorilla* by zoos, circuses, and research institutions is thought to be one of the important threats to wild populations (Reynolds 1967b, Fisher et al. 1969).

Between 1940 and 1960 about 150 *G. gorilla* were imported into the United States (Honegger and Menichini 1962). From 1968 to 1973 a total of 32 individuals entered the United States (Appendix), and in 1977 none were imported (Bittner et al. 1978). Thirty-nine were imported into the United Kingdom in 1965–75 (Burton 1978).

Conservation Action

G. gorilla is found in at least six national parks and eight reserves (Table 183) and may also occur in other protected areas (I.U.C.N. 1976). But several of these parks and reserves, especially those in the eastern portion of the range, are not adequately protected against either habitat alteration or hunting. In the Virunga Volcanoes region the Rwandan Volcanoes National Park is suffering from continuing encroachment and human intrusion (Campbell 1970, Groom 1973, Harcourt and Curry-Lindahl 1978), and the Ugandan Gorilla Sanctuary is poorly protected and heavily used (Harcourt and Curry-Lindahl 1978). The Zairean section (Virunga National Park, formerly Albert National Park) is patrolled and guarded much more effectively (Harcourt and Groom 1972), but this improvement was achieved at the cost of the lives of 22 unarmed park staff, who were killed in 1968 attempting to remove settlers, cattle, and poachers, and restore control in the park (Anon. 1969, Groves 1970c). The gorillas in Kahuzi-Biega National Park are well protected, primarily because of the efforts of A. Deschryver, who has recruited former poachers as guides and is taking tourists to see habituated groups of gorillas in an effort to generate a tourism income that will assure continued government protection for the park (Mallinson 1974, Lyon 1975b, Cousins 1978c).

Many suggestions have been made for the conservation of *G. gorilla*. In general, large, undisturbed, continuous forest reserves are needed both east and west of the Rift Escarpment, although shifting agriculture could be allowed to continue in some areas west of the Rift if human population densities were low (Emlen and Schaller 1960, Schaller 1963). These authors also recommended stringent restrictions on the killing of *G. gorilla* and the spread of firearms, as well as on the numbers collected for zoos, medical institutions, and museums, and prohibition of the killing of females to obtain infants.

More specifically, a sanctuary is needed in South East State, Nigeria, to protect any remaining gorilla populations (Happold 1971). In Cameroon, conservators have been posted to the Dja and Campo reserves, and an area of 400 km^2 in the southeast (near Moloundou around Lake Lobeke) has been excised from timber concessions and reclassified for the conservation of gorillas and other wildlife (Awunti 1978). A gorilla sanctuary had been planned for southeastern Equatorial Guinea but was abandoned after independence in 1968 (Cousins 1978a). This last author recommended that an international conservation organization lease large tracts of rain forest from countries such as Gabon, Cameroon, and Congo, to make it profitable for these countries to leave timber intact and to enable international protection of gorilla habitats.

An international agreement is also needed between Rwanda, Zaire, and Uganda to establish effective patrols which can cross national borders in the Virunga Volcanoes region (Groom 1973, Fitter and Fitter 1974, Goodall and Groves 1977). Other measures needed in the Volcanoes National Park are an increase in the number and authority of guards in the park, and improvement in organization, training, equipment, transportation, and housing for these guards, better boundary markings, and increased publicity and organization for tourism, education, and research in the park

(Harcourt and Curry-Lindahl 1978). Rwanda needs financial help if it is to accomplish these goals (Fitter 1974), and in the United Kingdom public subscriptions are being solicited for the Mountain Gorilla Fund to help pay for this improved protection (Harcourt and Curry-Lindahl 1978, Wrangham 1978).

Efforts are being made to increase greatly the size of the Kahuzi-Biega National Park, so that it would include the majority of *G. gorilla* in Zaire (Cousins 1978a). It may be necessary to manage the forest in this park through limited felling to achieve a mosaic of different vegetation types and ages, but any such program must be carefully planned and controlled (Goodall 1977, 1978). Another suggestion that has been made is the translocation of some *G. gorilla* from the Kahuzi-Biega area to the Virunga Volcanoes area to combat the danger of inbreeding in the small population in the latter area (MacKinnon 1976). Such an operation entails many serious risks and should not be undertaken without thorough research, if at all (Harcourt 1977).

Other recommendations for the conservation of *G. gorilla* include standardized census and field surveys to determine both the status of remaining unstudied populations and what conservation measures are needed (Gartlan 1975d), strict enforcement of existing statutes of national parks, development of an overall management plan for eastern gorillas based on a comprehensive census in a variety of habitats, and a gorilla research headquarters to coordinate collection of management information and to train local personnel (Goodall and Groves 1977, Goodall 1978). At least one large area of lowland habitat is needed to insure the survival of the eastern lowland gorilla (MacKinnon 1976).

A census of the gorilla population in the Volcanoes National Park, along with studies of its habitat needs, its conflicts with human activities, and the solutions to these conflicts, is being conducted by A. L. Vedder and A. W. Weber; a study of gorilla populations, management, and conservation programs in Zaire has been proposed by A. G. Goodall (Mittermeier 1977b).

G. gorilla is listed in Class A of the African Convention, giving it protection in all countries of its range except Angola, which was not a party to the convention (Organization of African Unity 1968). The species is also strictly protected by the governments of Cameroon (Awunti 1978, Cousins 1978a), Congo (Bassus 1975), and Gabon (Cousins 1978b,c). In Equatorial Guinea, gorillas were officially protected before independence, but in fact were severely persecuted and openly offered for sale (Sabater Pi 1964, Jones and Sabater Pi 1971). Recent information is not available from that country.

The capture, possession, and export of *G. gorilla* are severely punished in Zaire (Cousins 1978a) and Rwanda (Harcourt and Curry-Lindahl 1978). Previously, gorillas were also well protected in Uganda (Pitman 1954), but this has not been true recently (Cousin 1978a).

G. gorilla is listed in Appendix I of the Convention on International Trade in Endangered Species of Wild Fauna and Flora, and as endangered under the U.S. Endangered Species Act (U.S. Department of Interior 1977a,b). The International Union of Directors of Zoological Gardens and other zoo federations have banned the acquisition of mountain gorillas by their members (I.U.C.N. 1976).

Table 183. Distribution of *Gorilla gorilla* (countries are listed in order from northwest to southeast)

Country	Presence and Abundance	Reference
Nigeria	Small population in extreme SE (map)	Booth 1958
	Along SE border E of Cross R. (map)	Schaller 1963
	In 1955 in Boshi Extension Forest Reserve in Obudu area (NE corner of Eastern Region) (map); uncommon; threatened with extinction	Petrides 1965
	At Cross R. (locality mapped)	Groves 1970a
	Civil war has probably depleted or eliminated them	Groves 1970c
	Only in SE	Happold 1971
	Very rare	Henshaw and Child 1972, Monath and Kemp 1973
	In small area on upper Cross R. from Tinto (5°33′N, 9°35′E) to Obudu plateau (6°38′N, 9°06′E)	Goodall and Groves 1977
	Recently around Ikom and Ogoja (in SE); in danger of extinction as early as 1930; March estimated 200 individuals in Boshi Forest Reserve in 1955; doubtful if any resident populations remain	Cousins 1978b
Cameroon	Throughout S (map); small populations	Booth 1958 [Vandebroek 1958, Schaller 1963]
	Throughout S of Nyong R. and as far N as Batouri (localities listed and mapped)	Groves 1970a
	In S; still numerous near Abong-Mbang and Mendjim Mey but none left near Yaoundé	Groves 1970c
	Isolated population near Akwaya W of Mamfe (in W); in S and SE (map); declining drastically; very vulnerable	Gartlan, pers. comm. 1974
	From Edéa (3°47′N, 10°10′E) to Toukinear Bertoua (4°35′N, 13°30′E) and throughout S	Goodall and Groves 1977
	Depleted in SW, still numerous in SE; may still remain in Assumbo Hills and Mboka, Manyu district, Korup in Ndian, and Bakossi Korup in Meme	Cousins 1978a
	Small number (about 60 individuals) in Takamanda Reserve (along W border); have disappeared from many areas in S including near Kribi, Yaoundé and M'Balmayo; greatly disturbed around Ebolowa-Mbam-Sangmelima; in 2 reserves of Dja and Campo; best populations in SE near Yokadouma, Lomie, Abong-Mbang, and Djaposten	Cousins 1978b
	Outlook is poor in SW but better in SE	Cousins 1978c

Table 183 (continued)

Country	Presence and Abundance	Reference
Equatorial Guinea	Throughout (map)	Vandebroek 1958, Schaller 1963
	At Mbia-Campo R. (NW); near Mobumuom-Monte Mitra (central); at Mokula (SE); and at Nkin (NE) (map); small residual group in Midjumoveng (NE) is decreasing; overall numerous, population probably more than 5,000 individuals but range shrinking	Sabater Pi 1964
	Range mapped; decreasing	Sabater Pi and Jones 1967
	In Monte Raices Territorial Park (in NW, E of Bata)	I.U.C.N. 1971
	Localities listed and mapped in NW, N, central, and SE; total population not more than a few hundred individuals; highly endangered; gravely threatened with extinction	Jones and Sabater Pi 1971
	Recent reports suggest absent in Monte Raices Park; in Mount Alén Partial Reserve	Cousins 1978b
Gabon	Throughout; locally abundant	Malbrant and Maclatchy 1949
	Throughout (maps)	Vandebroek 1958, Schaller 1963
	Localities listed include Sangatanga, Cap Lopez, Libreville, Sette Cama, and Fernan Vaz	Groves 1970a
	In 1964 in NE near Makokou but very rare in S; none recorded in S central near M'bigou since 1957; still numerous	Groves 1970c
	Localities mapped throughout	Groves 1971a
	In Wonga-Wongue N.P. (in W, S of estuary of Gabon R.)	Harroy 1972
	Fair-size population in Wonga-Wongue N.P.	Curry-Lindahl 1974a
	Rare; decreasing; potentially threatened	Jouventin, pers. comm. 1974
	In NE near Mekambo, Makokou, and Belinga; plentiful near Mayumba in SW corner	Cousins 1978a
	In N around Oyem and Bitam on Woleu-N'Tem plateau and around Minvoul; in NE corner; reported to be numerous around Sette Cama and Makokou and decreasing in M'bigou region; in coastal strip between Port Gentil and Mayumba	Cousins 1978b

Table 183 (continued)

Country	Presence and Abundance	Reference
Angola	Locality mapped in Cabinda (on coast N of mouth of Congo R.)	Groves 1967b, 1971a
	Threatened	Bothma 1975
	In Cabinda	Goodall and Groves 1977
	Restricted to N Cabinda near Belize and Buco Zau	Cousins 1978b
Congo	Locally common	Malbrant and Maclatchy 1949
	In W and central (maps)	Vandebroek 1958, Schaller 1963
	Localities listed at Mayombe (in S), and Mambili, Opa, Bade, Zalangoye, and Ouesso (in N)	Groves 1970a
	In interior, e.g., region of Likouala aux Herbes R. (E central); fairly numerous	Groves 1970c
	Localities mapped	Groves 1971a
	About 300 individuals in Niari Valley Faunal Reserve and Mount Fouari Reserve (both in SW) (map); have decreased greatly in S and central	Bassus 1975
	In Odzala N.P. (in NW)	I.U.C.N. 1975
	Localities listed include Sibiti, Komono, Dolisie, and M'Vouti (W of Brazzaville), Mossaka-Likouala, Ewo, Kellé, Boundji, and Makoua (N of Brazzaville), and Quesso (Ouesso?), Imfondo (Impfondo?), and Dongou (farther N)	Cousins 1978a
	Reported to be plentiful W and NW of Brazzaville and E of Dolisie-Kibangou road	Cousins 1978b
Central African Republic	In SW tip (map)	Schaller 1963
	At Nola (3°28′N, 16°08′E) in SW (map)	Groves 1970a
	One captured in SW in 1963	Groves 1970c
	Localities mapped in SW	Groves 1971a
	NE limit of range in W is at Barundu, NE of Nola (3°40′N, 16°15′E)	Goodall and Groves 1977
	Rare and perhaps confined to area W of Sangha R. (SW corner)	Cousins 1978a
	Previously near Nola, Aboghi, and Gounguru; population greatly decreased; not in any reserves; remnant populations only	Cousins 1978b
Zaire	Near Lake Lungwe in Itombwe Mts. NW of Lake Tanganyika and W of Mount Kahuzi	Curry-Lindahl 1956

Table 183 (continued)

Country	Presence and Abundance	Reference
Zaire (cont.)	In E, W of Lake Kivu and Lake Edward and NW of Lake Tanganyika (map)	Vandebroek 1958
	In Virunga Volcanoes, Albert N.P., and at Kabona	Kawai and Mizuhara 1959
	In a triangular area bounded by Lubutu (NW), Lubero (NE), and Fizi (S); population: 3,000–15,000 individuals concentrated in about 60 more or less isolated tracts which together total about 4,900 km^2 (5.5% of range); all in drainage of Zaire-Lualaba R. tributaries: Maiko, Lowa-Osa, Ulindi-Lugulu, Elila, and Luama rivers (localities mapped)	Emlen and Schaller 1960
	In central Virunga Volcanoes	Osborn 1963
	Near Mulungu, Tshibati, Mount Biega	Rahm and Christiaensen 1963
	Population estimated about 5,000–15,000 individuals; decreasing in some areas (e.g., Mount Tshiaberimu); range shrinking; population may be locally increasing in some protected areas (e.g., Virunga Volcanoes); extinction is not imminent but population is highly vulnerable because of small range; area occupied: 19,400–20,700 km^2 (21–23% of range area) (map)	Schaller 1963
	In Kahuzi	Rahm 1965
	Between Rift Valley and Lualaba-Zaire R., Lowa-Osa R., and Elila R. (1°S–3°30′S)	Rahm 1966
	Population estimated at 20,000 individuals	Boné and Haumont 1967
	In E (maps)	Groves 1967b, 1970c
	At Utu, Mwenga-Fizi, Wabembe, Baraka, Itombwe, Tshiaberimu, Lubero, Luofo, Alimbongo, Butemo, Virunga Volcanoes, Mount Kahuzi, Tshibinda, and Mount Nakalongi (map)	Groves 1970a
	Reserve planned for population of 200 individuals near Bukavu; fairly secure in Albert N.P.	Anon. 1971
	Localities mapped in E	Groves 1971a
	On Mount Karisimbi and Mount Visoke, where total population may be as high as 96 individuals; population in Virunga Volcanoes area overall is small, declining, precarious	Groom 1973

Table 183 (continued)

Country	Presence and Abundance	Reference
Zaire (cont.)	Fewer than 1,000 individuals in whole Virunga Volcanoes area, mostly within national parks or reserves	Fitter 1974
	About 250 individuals in Kahuzi-Biega N.P. and 300–350 in Virunga Volcanoes; no interchange	Mallinson 1974
	In Mount Kahuzi area SW of Lake Kivu especially in triangle Lwiro-Mount Biega-Mount Kahuzi; suspects interchange with more E population	Casimir 1975a
	Virunga population isolated from Mwenga-Fizi, Utu, Tshiaberimu one; about 250 individuals in Kahuzi-Biega N.P. (2°25′N, 28°45′E)	Casimir 1975b
	SE of Mambasa and at Uele	Heymans 1975
	In Maiko N.P. (0°–1°S, 27°–29°E, E of Lualaba R.) add Kahuzi-Biega N.P.	I.U.C.N. 1975
	N and W of Lake Kivu (map); total population less than 1,000 individuals, including 300 in Virunga range and 150 in Kahuzi-Biega N.P. where increasing	Lyon 1975a
	Population may be as high as 2,000 individuals, including 250 in Kahuzi-Biega N.P.	Lyon 1975b
	Extinct in SW; in E (map)	Verschuren 1975a
	Contains at least 98% of mountain race; status fairly secure; about 200 individuals in Virunga N.P.; about 200 in Kahuzi-Biega N.P.; about 40 on Tshiaberimu and ridges near Lake Amin on high slopes of W Rift from Goma to beyond Uvira; one race (*G. g. graueri*) numerous over large area of Kivu and Maniema; probably about 2,000 in population; a few in Maiko N.P.; almost certainly extinct in W but may cross border in N of Mayumbe Forest	Verschuren 1975b
	In Kahuzi N.P.; seem to have declined in Virunga N.P.	MacKinnon 1976
	At Tshibinda, Kahuzi-Biega N.P., Kivu Province	Goodall 1977, 1978
	Localities listed and mapped from Lubutu (0°40′S, 26°55′E [actually 0°48′S, 26°39′E]) to Bibugwa (2°30′S, 28°20′E); in Itombwe Mts. between Mwenga (3°00′S,	Goodall and Groves 1977

Table 183 (continued)

Country	Presence and Abundance	Reference
Zaire (cont.)	28°28′E) and Fizi (4°18′S, 28°56′E); Mount Kahuzi (2°10′–2°25′S, 28°40′–28°50′E); Mount Tshiaberimu (0°5′N–0°40′S, 29°00′–29°20′E); declining in Virunga range (1°20′–1°30′S, 29°24′–29°42′E); may be extinct near Djabbir (3°55′N, 23°53′E)	Goodall and Groves 1977 (cont.)
	In Maiko N.P.; about 250 individuals in Kahuzi-Biega N.P. are well protected	Cousins 1978c
	Abundant on Itombwe massif less than 50 km W of Rusizi R. frontier	Verschuren 1978b
Uganda	In Impenetrable (Kayonza) Forest in SW, 32 km from nearest other population	Pitman 1954
	Along SW border in Gorilla Sanctuary N of Mount Mgahinga and Mount Muhavura and E of Mount Sabinio; and in Kayonza Forest (maps)	Kawai and Mizuhara 1959
	In two areas of SW (mapped); population in Kayonza Forest estimated at 120–180 individuals	Schaller 1963
	Near Kisoro (E end Virunga Volcanoes) and N slopes of Mount Muhavura, Mount Mgahinga, and Mount Sabinio	Baumgartel 1965
	Range shrinking; scarce in Virunga Volcanoes; in Kayonza Forest	Reynolds 1967b
	In Impenetrable Forest, Kigezi district	de Vos 1969
	In Kayonza Forest including Kumbi; in Virunga Volcanoes area (maps)	Groves 1970a,c, 1971a
	In Kisoro and Kayonza forests, Kigezi; range shrinking; total population probably does not exceed 200 individuals	Kingdon 1971
	Extinct on Mount Muhavura except for occasional stragglers from Rwanda	Suzuki 1971
	In Gorilla Game Reserve on SW border	Harroy 1972
	Small, declining populations of 13 individuals on Mount Muhavura, 32 on Mount Sabinio; none on Mount Mgahinga; range shrinking; situation precarious	Groom 1973
	Only in Bwindi area of Kayonza Forest	Albrecht, pers. comm. 1974
	Not in Impenetrable Forest	Jahnke 1974
	In Kayonza Forest wildlife sanctuary	I.U.C.N. 1976
	No longer resident in Virunga Volcanoes N of S border; Kayonza (Impenetrable) Forest population is unstudied	Goodall and Groves 1977
	May be extinct on volcanoes near border; may be in Kayonza Forest	Cousins 1978a

Table 183 (continued)

Country	Presence and Abundance	Reference
Rwanda	In Virunga Volcanoes, NW border (maps)	Kawai and Mizuhara 1959, Emlen and Schaller 1960
	Population in Virunga area 400–500 individuals	Schaller 1963
	Situation in Volcanoes N.P. extremely critical: range being drastically reduced	Campbell 1970
	Face extinction within next 20–30 years; studied on Mount Karisimbi, Mount Mikeno, Mount Visoke saddle	Fossey 1970
	Along NW border (maps)	Groves 1970a,c, 1971a
	Situation desperate; habitat drastically reduced	Anon. 1971
	Fossey estimates 375–400 individuals in Volcanoes N.P.; vulnerable	Harcourt and Groom 1972
	On all peaks of Virunga range except Hehu; most common between Karisimbi, Mikeno, and Visoke; rare on W side of Karisimbi; generally rare; declined greatly 1927–60; recently decline has been slower	Spinage 1972
	In a precarious position; only 45 individuals in E end of Virunga range (Sabinio, Gahinga, and Muhavura) and 96 on Visoke; population small, declining	Groom 1973
	Numbers and habitat decreasing	Fitter 1974
	Between Mount Visoke, Mount Karisimbi, and Mount Mikeno	Elliott 1976, Fossey and Harcourt 1977
	Fossey estimated 275 individuals in Virunga region in 1974	I.U.C.N. 1976
	In NW (localities mapped); declining at alarming rate	Goodall and Groves 1977
	About 300 individuals in Virunga area	Harcourt 1977
	Fossey estimates population of 225 individuals; Volcanoes N.P. not secure	Cousins 1978c
	Population in Virunga area 260–290 individuals in 1973; now probably less than 250; primarily in W around Mikeno (Zaire), Karisimbi, and Visoke; about 70 in E on Sabinio, Gahinga, and Muhavura	Harcourt and Curry-Lindahl 1978
	Does not survive S of Gisenyi-Ruhengeri road (extreme NW)	Verschuren 1978b
	Population in Virunga area 300–350 individuals in 1971–73; 200–250 in 1978	Wrangham 1978
Burundi	Previously in SW near Bururi (3°57′S, 29°35′E) but probably extinct there	Verschuren 1978a
	Extinct in 1977 in SW and W and probably in N	Verschuren 1978b

Table 184. Population parameters of *Gorilla gorilla*

Population Density (individuals/km^2)	Group Size $\bar{X}$	Range	No. of Groups	Home Range Area (ha)	Study Site	Reference
1.3					Equatorial Guinea	Sabater Pi 1964
0.58–0.86	6.4–7.1	2–12	13	200–1,200	Equatorial Guinea	Jones and Sabater Pi 1971
		≤25			Congo	Malbrant and Maclatchy 1949
		5–18	6		Zaire and Uganda	Kawai and Mizuhara 1959
0.39–1.9		5–20			Zaire and Uganda	Emlen and Schaller 1960
		3–4	4		Uganda	Osborn 1963
0.38–1.16	6–17	2–30	22	1,036–2,200	Zaire and Uganda	Schaller 1963
		20	1	3,100–5,000	Zaire	Casimir 1975b
		2–21			Zaire	MacKinnon 1976
		20	1	3,400	Zaire	Goodall 1977
2					Zaire	Hendrichs 1977
	13	5–19	9		Rwanda	Fossey 1970
		3–6	1	700–800	Rwanda	Elliott 1976
1.5–2	8		8	400–810	Rwanda	Fossey and Harcourt 1977
2–7			2	3,000	Zaire	Goodall and Groves 1977

Pan paniscus Pygmy Chimpanzee or Bonobo

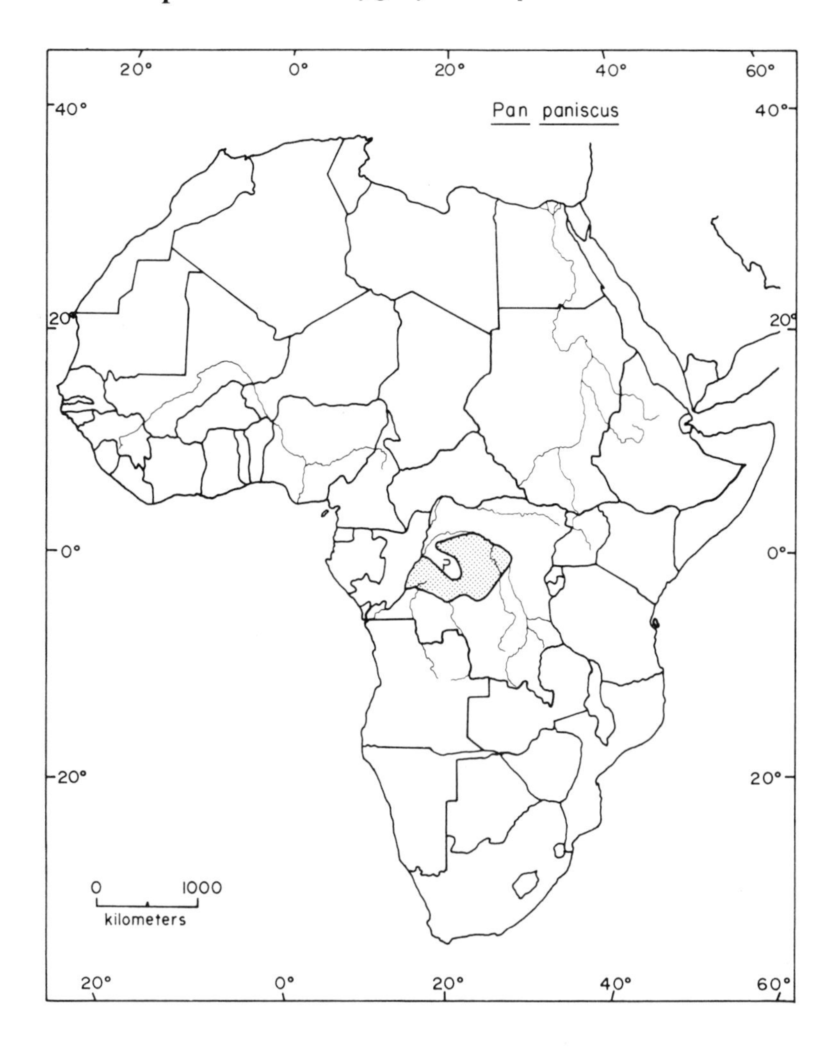

Taxonomic Note

This is a monotypic form (Napier and Napier 1967), which may be a subspecies of *P. troglodytes* (Horn 1979).

Geographic Range

Pan paniscus is found in central and western Zaire, within the area bounded by the Congo River on the west, the Zaire River on the north, the Lualaba River on the east, and the Kasai and Sankuru rivers on the south (Vandebroek 1958). Its range extends from about 16°30′E to 25°30′E, and 1°30′N to 4°30′S, except for the center of this area between the Momboyo River and Busira River (Vandebroek 1958, Kortlandt and Van Zon 1969). The total area of the range is estimated at about 350,000 km^2 (Kortlandt 1976). Isolated reports from north (Hill 1967a, 1969), west (Malbrant and Maclatchy 1949), and south (Reynolds 1967a) of these limits have never been confirmed. For details of the distribution, see Table 185.

Abundance and Density

Very little information was located regarding the abundance of *P. paniscus* anywhere in its range. The few reports available indicate that the total population may be about 100,000 to 200,000 individuals, that the species is rare in the southern portion of the range, and that although it is common in a few parts of the north, it has declined or disappeared in some areas (Table 185).

At one study site west of Lake Tumba, *P. paniscus* population density was estimated at 0.3 to 0.9 individuals per km^2, and groups consisting of about 5 animals each occupied home ranges of 40 to 70 km^2 (Horn, pers. comm. 1978). Elsewhere, groups of 20 to 30 (Vandebroek 1958), 15 to 40 (Nishida 1972b), and 1 to 4 (MacKinnon 1976, Badrian and Badrian 1977) have been reported.

Habitat

Pygmy chimpanzees occupy secondary swamp forest (Nishida 1972b), flood-controlled, low altitude, equatorial rain forest (Hansinger et al. 1974), and lowland secondary rain forest with thick underbrush (Horn, pers. comm. 1974). They have been observed in primary forest dominated by trees of the Leguminosae (e.g., *Albizzia, Dialium, Scrodophloeus, Cynometra*, and *Parkia*), and in seasonally and permanently flooded swamp forest, but seem to prefer the secondary forest characterized by *Musanga smithii*, *Vernonia* spp., *Myrianthus*, and *Afromomum* which regenerates on cleared land (MacKinnon 1976).

Factors Affecting Populations

Habitat Alteration

By 1964 there had been hardly any deforestation within the range of *P. paniscus* except for rare plots of slash and burn agriculture, and wide areas were inaccessible and periodically flooded, thus unlikely to be inhabited or altered in the near future (Kortlandt 1976). More recently, more than 80% of the species' range was found to be still forested (MacKinnon 1976). West of Lake Tumba the rate of habitat destruction was slow, old plantations had regenerated since the departure of the Belgians, and no commercial logging was occurring on a large scale (Horn, pers. comm. 1974).

But habitat destruction is a potential threat to *P. paniscus* (Reynolds 1967b, Fisher et al. 1969) and may have a considerable effect in the future. In the northern part of the range, subsistence agriculture is destroying large areas of primary forest, many trees are being felled to provide wood for houses, canoes, and firewood, and a Canadian-Zairean cooperative venture is now cutting wide transects through undisturbed forest and may soon undertake a major timber exploitation project (Badrian and Badrian 1977).

Human Predation

Hunting. *P. paniscus* has been intensively hunted with bows and arrows and by the use of traps for its meat, which has long been regarded as a delicacy (Nishida

1972 b; Horn, pers. comm. 1974; Verschuren 1975b). Hunting by these traditional methods did little damage to *P. paniscus* populations, but the increasing use of firearms has greatly depleted populations, especially near towns (Badrian and Badrian 1977). Pygmy chimpanzees are also killed to make magical charms, which are believed to confer vigor and strength on the wearer (Nishida 1972b, Badrian and Badrian 1977). Heavy hunting is apparently responsible for the rarity or absence of this species in many areas of former abundance (MacKinnon 1976).

Pest Control. Local people near Lake Tumba report that *P. paniscus* enters plantations and feeds on the roots of manioc, tubers of sweet potato, nuts of oil palm, and fruits of orange, coffee, banana, plantain, and avocado plants (Nishida 1972b). No reports of efforts to control this activity were located.

Collection. Young pygmy chimpanzees are captured by killing the mother and are kept as pets or presented to visiting dignitaries (Badrian and Badrian 1977). This species has also been collected for biomedical research. For example, in 1958–59, 86 *P. paniscus* were captured by the Laboratoire Médicale in Kisangani for American poliomyelitis and arteriosclerosis research programs, but all 86 died within three weeks of capture (Kortlandt 1976). More recently, 5 *P. paniscus* were obtained on a lend-lease arrangement from Zaire for research at the Yerkes Regional Primate Research Center (Bourne 1976).

No *P. paniscus* were imported into the United States in 1968–73 (Appendix) or in 1977 (Bittner et al. 1978). One individual was imported into the United Kingdom in 1965–75 (Burton 1978).

Conservation Action

P. paniscus has been reported in one national park, but recent confirmation of its presence there is lacking (Table 185). The species is fully protected in Zaire, but the ban on hunting it is not enforced (Horn, pers. comm. 1974; Badrian and Badrian 1977).

Forty-seven signers of a petition protested the acquisition of 5 *P. paniscus* by the Yerkes Center, urged the government of Zaire to prohibit further captures, and asked all institutions to desist from receiving any *P. paniscus* until the wild status of the species is clarified (McGrew 1976).

Other workers have concluded that a comprehensive distribution survey for this species is urgently needed, and that a large reserve, in conjunction with a research center, should be established in the area between the Lomako and Yekokora rivers, which is uninhabited, has a good population of *P. paniscus,* is fairly inaccessible, and is surrounded by easily patrolled natural boundaries and buffer zones (MacKinnon 1976, Badrian and Badrian 1977).

P. paniscus is included in Class A of the African Convention (Organization of African Unity 1968). It is also listed in Appendix I of the Convention on International Trade in Endangered Species of Wild Fauna and Flora, and as threatened under the U.S. Endangered Species Act (U.S. Department of Interior 1977a,b).

Table 185. Distribution of *Pan paniscus*

Country	Presence and Abundance	Reference
Zaire	S of great bend of Congo-Zaire R. and N of 5°S (maps)	Yerkes 1943, Reynolds and Reynolds 1965, Reynolds 1967a, Kortlandt and Van Zon 1969
	Confined to left bank of Congo-Zaire R., N of Kasai R. and Sankuru R. (map); rare in S part of range	Vandebroek 1958
	Vulnerable; numbers and habitat decreasing	I.U.C.N. 1972
	Reported W and SW of Lake Tumba, W of Lake Mai Ndombe (both in W), and at high densities in Salonga N.P. (central) (map)	Nishida 1972b
	At Lake Tumba	Hansinger et al. 1974
	Population probably stable W of Lake Tumba; also at Busonjo, S of Lisala and on right bank of Busira R. across from Ingende	Horn, pers. comm. 1974
	Fairly common on left bank of Zaire R.	Heymans 1975
	In Salonga N.P. (1°–3°30′S, 20°–23°E)	I.U.C.N. 1975
	In Salonga N.P. and near Lake Inongo; abundant S of Lisala near Bosondjo	Verschuren 1975b
	Vulnerable; decreasing; no longer in Salonga N.P.	I.U.C.N. 1976
	Not rare; population estimated at about 100,000–200,000 individuals	Kortlandt 1976
	Previously common throughout region S of Zaire R., W of Lualaba R., and N of Kasai R.; now common in only a few scattered localities such as between Lomako R. and Bolombo R., N central portion of range (map); absent or rare in other areas	MacKinnon 1976
	Distribution discontinuous; study site between Lomako R. and Yekokora R. (map); absent in much of Salonga N.P.	Badrian and Badrian 1977

Pan troglodytes Chimpanzee

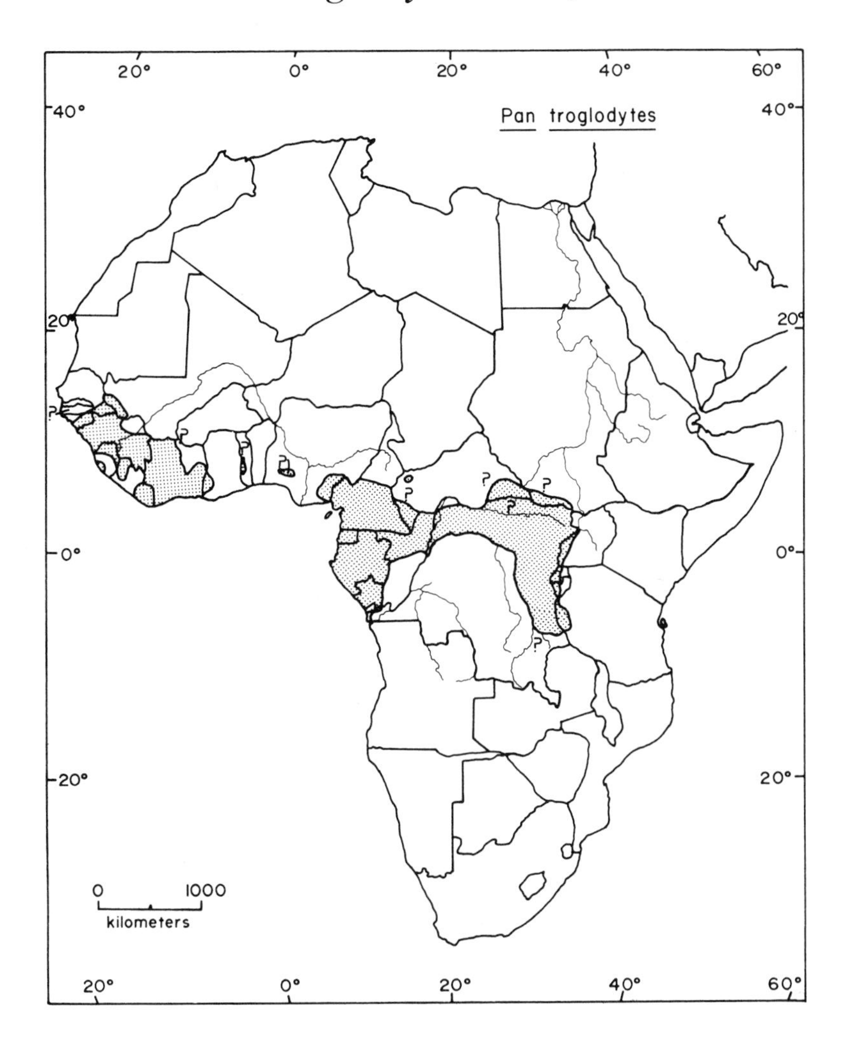

Taxonomic Note

Three races are generally recognized (Yerkes 1943, Vandebroek 1958, Reynolds and Reynolds 1965, Hill 1967b, Napier and Napier 1967). The pygmy chimpanzee, regarded by some authors as a race of *Pan troglodytes*, is treated in this report as a separate species, *Pan paniscus*.

Geographic Range

Pan troglodytes is found in western and central Africa from southeastern Senegal to western Uganda and Tanzania. The range extends north as far as 14°N, 12°W in Mali (Sayer 1977) and south as far as 7°30′S, 30°E in Zaire (Kortlandt and Van Zon 1969). The western limit is at about 15°W (at 11°N) in Guinea-Bissau and the eastern limit at

32°E (at 2°N) in Uganda (Kortlandt and Van Zon 1969). For details of the distribution, see Table 186.

Abundance and Density

P. troglodytes has recently been described as locally common or abundant in parts of Sierra Leone, Ivory Coast, Gabon, Congo, and Zaire but is also known to be decreasing in numbers in the first two countries as well as in Liberia, Cameroon, Sudan, and Uganda (Table 186). It is considered rare in Ghana, Nigeria, and Senegal, but is increasing in this last country as well as in Guinea. It is endangered in Equatorial Guinea, threatened in Angola, and extinct in Benin and parts of Guinea, Nigeria, and Central African Republic (Table 186). To summarize, chimpanzees are thought to be abundant in parts of five countries, decreasing in six, increasing in two, rare in three, endangered or threatened in two, and extinct in all or part of four. No recent firsthand data are available regarding their overall status in Guinea-Bissau, Mali, Upper Volta, Togo, Central African Republic, Rwanda, Burundi, or Tanzania.

The Interagency Primate Steering Committee asked four consultants (J. S. Gartlan, C. Jones, A. Kortlandt, and T. T. Struhsaker) to estimate the numbers of *P. troglodytes* remaining in each country as of 1978 (Blood, pers. comm. 1978). This group concluded that relatively abundant populations exist in Guinea, Cameroon, Gabon, and Zaire, and that an estimated 1,000 to 5,000 individuals each are found in Liberia, Ivory Coast, Central African Republic, and Uganda. They reckoned that populations are very low (few to less than 1,000) in Senegal, Sierra Leone, Mali, Ghana, Nigeria, Sudan, and Tanzania, and are extinct in Upper Volta, Togo, Benin, Rwanda, and Burundi. They list population sizes of zero for Gambia and Guinea-Bissau, and unknown for Congo and Equatorial Guinea (all from Blood, pers. comm. 1978).

P. troglodytes is classified as a vulnerable species (I.U.C.N. 1972), and the westernmost populations (west of the Niger River) are considered to be severely threatened (Kortlandt 1976). According to one authority, the species is "gravely declining almost everywhere in the entire range from Sierra Leone to the Kivu Mountains" (Kortlandt, pers. comm. 1974). In addition, none of the populations in Uganda or Tanzania are large enough to withstand any large-scale interference (Kingdon 1971). Thus *P. troglodytes* seems to be in a vulnerable position throughout its range.

Several studies have shown that chimpanzees live in unstable groups of varying sizes, which coalesce, divide, and wander over wide ranges according to the local abundance of food (Goodall 1965, Reynolds and Reynolds 1965, Nishida 1968, Jones and Sabater Pi 1971, Kano 1971b). Population density estimates vary from 0.05 to 6.7 individuals per km^2, and home ranges are 1,000 to 20,000 ha (Table 187), with the lower densities and larger home ranges in drier, less food-rich savanna regions.

Habitat

Chimpanzees utilize a wide variety of habitats ranging from dry brushy savanna to humid dense rain forest. At the northern limits of their distribution they are found in parklike savanna and small patches of bamboo, gallery forest, and dry forest within this savanna (Kock 1967, Dupuy 1970a, Sayer 1977). Throughout the rest of their range they occur in many different types of woodland and forest from lowland to montane areas up to 3,048 m elevation on Mount Ruwenzori (Reynolds and Reynolds 1965).

In Guinea, *P. troglodytes* has been observed in brushy savanna, scrub forest, park forest, gallery forest, and open forest (Bournonville 1967), and in secondary forest with thick undergrowth and elephant grass (Dunnett et al. 1970). In Sierra Leone this species is confined to remnant strips of riverine forest (Wilkinson 1974), and in Ivory Coast it is found in mature, lowland rain forest (Struhsaker and Hunkeler 1971).

In Ghana, chimpanzees occupy high forest (Jeffrey 1975b), and in Nigeria they have been seen in rain forest and montane forest (Petrides 1965). In Cameroon they have been recorded in young secondary forest, older secondary forest, mature forest, montane forest above 1,000 m elevation (Gartlan and Struhsaker 1972), woodland, and savanna (Gartlan, pers. comm. 1974).

Equatorial Guinean populations are found in stable, mature, primary forest, defective forest, and regenerating forest (Sabater Pi and Jones 1967). These authors concluded that the distribution of *P. troglodytes* in Equatorial Guinea coincided with areas of mature forest, and that the chimpanzees fed, rested, and slept mainly in that habitat, although they also entered secondary forest (Jones and Sabater Pi 1971).

In Zaire, chimpanzees occur in montane forest (Rahm and Christiaensen 1963), middle and low altitude forest (Rahm 1966), rain forest, and secondary forest in all stages of succession (Rahm 1967). They also utilize blocks of riverine forest, some of which are separated by long distances from other populations (Verschuren 1975b). In Uganda, *P. troglodytes* is found in humid forest, montane rain forest, forest-savanna mosaic on the edge of the rain forest (Reynolds 1965), as well as colonizing rain forest, colonizing woodland, mixed forest, swamp forest, and gallery forest (Reynolds and Reynolds 1965). In Rwanda and Burundi this species occupies montane forest (Elbl et al. 1966).

Chimpanzees have been most intensively studied in Tanzania, where they utilize riverine rain forest, *Brachystegia* open forest, deciduous woodlands (*Brachystegia*, *Julbernardia*, and *Isoberlinia*), and savanna woodlands (Azuma and Toyoshima 1961–62; Goodall 1965; Suzuki 1969; Izawa 1970; Kano 1971a,b; Sugiyama 1973; Teleki et al. 1976). In the Mahali Mountains, *P. troglodytes* is found in lowland and riverine forest as well as savanna, woodland, bamboo brush, grassland areas (Nishida 1968), semideciduous forest, and mosaics of gallery forest and open woodlands (Kawanaka and Nishida 1975). At middle elevations it occupies semideciduous gallery forest and *Acacia* woodlands, woodland-forest mosaic (of *Brachystegia*, *Uapaca*, *Combretum*, and *Diplorhynchus*), and bamboo-forest mosaic, while at high altitudes it utilizes montane and alpine-bamboo forest, bamboo brush, and montane grassland (Itani and Nishida, pers. comm. 1974).

Some authors believe that the preferred habitat for *P. troglodytes* is closed canopy forest (Goodall 1965) or tropical rain forest (Reynolds, pers. comm. 1974). Others contend that the optimal habitat consists of a mountainous landscape covered by a variegated mosaic of vegetations including moist and drier types of forests, woodlands, and savannas (Kortlandt 1966) and that the highest population densities of *P. troglodytes* are found not in evergreen primary forest with a closed canopy but only in more open types of semideciduous and secondary forest, and in various types of intermediate and mosaic landscapes (Kortlandt and Van Zon 1969).

Factors Affecting Populations

Habitat Alteration

Because chimpanzees utilize secondary forest, they can survive in areas of limited habitat alteration, such as selective logging (Kortlandt, pers. comm. 1974; Oates 1977a). In some cases, man's activities may even make certain areas more suitable for chimpanzees. For example, the secondary growth in disturbed forests in Sierra Leone provides more food plants for chimpanzees than virgin forests do (Reynolds 1967b). And in western Guinea, irrigation has increased the amount and quality of vegetation available to *P. troglodytes* (Bournonville 1967).

On the other hand, 40% of the land in western Guinea has been drastically altered by man, and much of this land has been turned into a biological desert by pastoral nomadism and agriculture; large areas have also been deforested or devastated by brush fires and road building (Bournonville 1967). Chimpanzees cannot occupy these greatly altered areas.

Deforestation has also reduced the habitat of *P. troglodytes* in many other countries. In Ivory Coast, timber exploitation is the major threat to the last forested area where chimpanzees occur (Struhsaker 1972). In Ghana the high forest cover outside of reserves was reduced by farming from 80% to 20% in 13 years (Jeffrey 1974), and it is still decreasing, owing to clearing for timber and for the cultivation of cash crops such as cacao, oil palm, and rubber (Jeffrey 1975b, Asibey 1978). In Equatorial Guinea about one-third of the country has been logged (Jones and Sabater Pi 1971), and the progressive clearing of forests for agriculture and the degeneration of the soil after a few years of cultivation have transformed many areas into grassy habitats unsuitable for *P. troglodytes* (Sabater Pi and Jones 1967).

The use of arboricides to kill noncommercial trees and speed the regeneration of timber species is altering *P. troglodytes* habitat in Uganda (Reynolds 1967b). In Budongo Forest, for example, arboricides are used to destroy *Ficus* and *Cynometra* trees, whose fruits are important foods for chimpanzees (Suzuki 1971, Albrecht 1976). And the secondary forest which regenerates, often containing planted mahogany seedlings, is an altered ecosystem which can support a less dense population of chimpanzees than the original forest (Albrecht 1976).

Chimpanzees are also threatened by habitat alteration in the other forests of Uganda: in Bwindi by extensive cutting by pit sawyers (Albrecht 1976), in Bwamba by encroachment of cultivation, and in Rabongo by elephant damage (Suzuki 1971). In Tanzania the forest habitat of *P. troglodytes* was dwindling as a result of colonization almost 40 years ago (Moreau 1942), but no recent reports are available.

No direct information was located regarding the effects of habitat alteration on *P. troglodytes* populations in the rest of the countries within its range. But since chimpanzees seem unable to adapt to major changes in their environment anywhere in Africa (Albrecht 1976), it is clear that logging has and will continue to eliminate chimpanzee populations in many areas, both directly, by reducing their habitats, and indirectly, by opening remote areas to settlers and hunters via timber transportation roads (Kortlandt 1966).

Human Predation

Hunting. P. troglodytes is hunted for its flesh in northern Guinea (Bournonville 1967), Sierra Leone (Lowes 1970), Liberia (Curry-Lindahl 1969, Coe 1975, Jeffrey 1977), Ghana (Asibey 1974, 1978), Cameroon (Gartlan, pers. comm. 1974), Equatorial Guinea (Jones and Sabater Pi 1971), Gabon (Hladik 1974), and Zaire (Reynolds 1967b). According to one author this species is eaten over the whole area from western Ivory Coast to the Kivu Mountains, and the killing and eating of tens of thousands of *P. troglodytes* annually is seriously decimating its populations and threatening it with extinction in most of its remaining range (Kortlandt 1966).

In Congo, chimpanzee populations have suffered greatly from hunters seeking skulls to use as fetishes (Bassus 1975).

Pest Control. P. troglodytes feeds on cultivated foods in Ivory Coast (Monfort and Monfort 1973, Bourlière et al. 1974), Cameroon (Gartlan, pers. comm. 1974), Equatorial Guinea (Sabater Pi and Jones 1967, Jones and Sabater Pi 1971), and Zaire (Rahm and Christiaensen 1963). In some areas of Zaire it is considered a "culture follower," because its principal foods are the fruit and pith of cultivated banana plants (Rahm 1966). In Tanzania, chimpanzees consume the stalks of sugar cane and maize, the pith of banana stems, and the nuts of oil palm (Nishida 1972a).

In Uganda one group of chimpanzees was killed when it became a menace to the local people and their crops (Kingdon 1971). And people in Congo and Gabon use crop damage as a pretext for killing chimpanzees for their meat, skulls, or juveniles for sale (Blancou 1961).

At one study site in Guinea, *P. troglodytes* fed in both small subsistence rice and millet farms and in large commercial citrus plantations (Dunnett et al. 1970). Small farmers occasionally shot chimpanzees found raiding their crops but did not make a policy of shooting them on sight or hunting them down. Chimpanzees destroyed an estimated 200 pounds of grapefruit a day at one large plantation, but the manager did not value his crop highly and did not consider the damage serious, thus he took no action against the chimpanzees. A change in manager or attitude, or in the value of grapefruit, could reverse the situation and lead to the extermination of *P. troglodytes* in this area (Dunnett et al. 1970).

Collection. The capture of *P. troglodytes* for export has exerted serious pressure on chimpanzee populations in several countries, especially because young ones are usually collected by shooting their mothers, and many die in the process of collection and transport. Thus it is estimated that 4 to 6 mothers and several infants are lost for each chimpanzee arriving at a laboratory (Kortlandt 1966). This would mean, for example, that the 700 *P. troglodytes* exported from Guinea by the Pasteur Institute between 1917 and 1960 resulted in a loss to the population of 3,000 to 4,000 mothers and all of their potential offspring (Kortlandt 1966).

In Sierra Leone and Liberia, chimpanzees are often captured by driving whole groups into traps and killing unmanageable adults to obtain the young; thus 2,574 and 570 which these two countries respectively exported between 1957 and 1968 represent a large loss to the natural populations (Robinson 1971). Intensive collection in Sierra Leone, if continued, was predicted to lead to the extermination of *P. troglodytes* there by 1976 (Kortlandt 1966).

In 1973–74, 175 *P. troglodytes* were exported from Liberia, the majority to the United States, and more were scheduled for collection in 1975, all by the traditional method of shooting the mother to obtain her baby (Jeffrey 1977). The same method is used to collect chimpanzees for export from Ivory Coast (Aeschlimann 1965) and Ghana (Jeffrey 1974).

Chimpanzees are also collected in Cameroon for export to zoos and research institutions (Gartlan, pers. comm. 1974), and in Equatorial Guinea, where 64 were exported legally between 1958 and 1968 (Jones and Sabater Pi 1971). As early as 1949 Malbrant and Maclatchy cited hunting for commercial selling of the young as a decimating factor of *P. troglodytes* in Gabon and Congo, but no more recent information from these two countries is available.

Collection for export is not as frequent in central Africa as in West Africa because of difficulties in transporting the animals to export points, and is not common in Uganda and Tanzania because of strict export control (Reynolds 1967b).

From 1968 to 1973, a total of 1,305 *P. troglodytes* were imported into the United States (Appendix). During the last four of those years the great majority originated in Sierra Leone and Liberia. In 1977 none were imported into the United States (Bittner et al. 1978). Between 1965 and 1975, 202 *P. troglodytes* were imported into the United Kingdom (Burton 1978).

The United States biomedical research industry has calculated that it needs 300 to 350 new *P. troglodytes* per year from imports or domestic breeding to carry on its vaccine development and research programs (Interagency Primate Steering Committee 1978a). Currently six domestic breeding colonies produce about 75 *P. troglodytes* per year, and some scientists believe that the customary wasteful capture method would certainly be used to collect any chimpanzees imported, since no other method has been perfected (Wade 1978b).

Conservation Action

P. troglodytes is found in 16 to 17 national parks, 17 reserves, and 4 other protected areas throughout its range (Table 186). Additional habitat preserves that have been proposed for its protection include the southern portion of Mount Nimba, Liberia (Coe 1975), the Itama River system in northwestern Bwindi Forest, Uganda (Albrecht 1976), and the Mahali Mountains area south of the Malagarasi River in Tanzania (Kingdon 1971; Itani and Nishida, pers. comm. 1974).

The governments of many countries have given legal protection to *P. troglodytes*. Guinea has prohibited hunting (although the law is difficult to enforce in remote regions), has decreased its rate of exportation, and is attempting to insure that capturing be done only with a tranquilizing gun (Bournonville 1967). But killing of crop-raiding chimpanzees is legal in Guinea, and it is often difficult to determine if claims of crop damage are genuine (Dunnett et al. 1970). These authors also point out that Guinean law prohibits cutting of gallery forests within 5 m of a river, hence conserving riparian habitat.

Sierra Leone has imposed a presidential ban on chimpanzee export and is debating the legality of possession of any young chimpanzees (Anon. 1979). The Liberian government has declared *P. troglodytes* to be fully protected, but enforcement there is difficult (Coe 1975). It is also impossible to exercise police control of hunting of *P. troglodytes* in the forests of Ivory Coast (Aeschlimann 1965). Similarly, chimpanzees are fully protected in Ghana but heavily poached outside of parks and reserves (Asibey 1978).

Chimpanzees are protected from hunting in Gongola State, Nigeria (Hall 1976), fully protected in Congo (Bassus 1975), and previously were well protected by efficient game departments in Uganda and Tanzania (Reynolds 1967b). In addition, they receive full legal protection throughout Nigeria, Gabon, Zaire, and Sudan (I.U.C.N. 1972).

P. troglodytes is listed in Class A (full protection) of the African Convention, to which all of the countries within its range are parties except Angola and Guinea-Bissau (Organization of African Unity 1968). The species is in Appendix I of the Convention on International Trade in Endangered Species of Wild Fauna and Flora, and is listed as threatened under the U.S. Endangered Species Act (U.S. Department of Interior 1977a,b).

Recommendations for the conservation of the habitat of *P. troglodytes* include the cessation of the use of arboricides in the Budongo Forest, establishment of more preserves in Uganda (Suzuki 1971), forestry practices which avoid monoculture and maintain chimpanzee food plants, and additional reserves in the western and central portions of the range (I.U.C.N. 1972). Reynolds (1967b) suggested that the Uganda Forest Department spare 23 *Ficus mucosa* per km^2 of forest, which would reduce timber yield by only 0.5%. The reply expressed doubt that tree poisoning was harming chimpanzee populations but promised to keep a close watch on the situation. More recently, changes in forestry practices—to market rather than poison *Cynometra* and to spare *Ficus mucosa*—are still being urged (Kortlandt 1976).

To lessen the impact of human predation on *P. troglodytes* populations, increased domestic animal meat production, prohibition of firearms (Kortlandt 1966), and the removal of hunting rights in forest reserves (Asibey 1978) would be helpful. Several authors urged strict control of export of chimpanzees (Reynolds 1967b; Kingdon 1971; I.U.C.N. 1972; Kortlandt, pers. comm. 1974). A rehabilitation center at Mount Asserik, Senegal, is attempting to reintroduce small numbers of previously captive chimpanzees into the wild (Brewer 1976, Dupuy and Verschuren 1977, Borner and Gittens 1978).

The task force assigned to determine future United States needs for chimpanzees has recommended (1) government support and coordination of domestic breeding colonies, (2) multiple use and recycling, and (3) additional importation if necessary (Interagency Primate Steering Committee 1978a). Many difficulties have been encountered with captive breeding. For example, in 1974 the Yerkes Center birth rate was only half the rate required to maintain the colony, primarily because of psychological abnormalities caused by removing infants from their mothers (Perry 1976).

The group of consultants on wild chimpanzee populations also suggested that if further importation into the United States is necessary, (1) collection should be done by scientifically qualified personnel, (2) chimpanzees should be captured from areas destined for habitat destruction where possible, (3) if from intact areas, animals should be selected to minimize impact on the wild population and maximize breeding potential, perhaps by the collection of intact matrilineal groups, (4) efficient and humane capture and transport methods should be used, and (5) long-term monitoring of populations should be done to determine the effect of collecting (Blood, pers. comm. 1978).

Table 186. Distribution of *Pan troglodytes* (countries are listed from northwest to southeast)

Country	Presence and Abundance	Reference
Senegal	Relict populations to 13°N	Kortlandt 1966
	Localities mapped in SE	Bournonville 1967
	In SE (map)	Kortlandt and Van Zon 1969
	About 50 in Niokolo-Koba N.P. (in SE); other localities listed; stable or increasing	Dupuy 1970a
	About 100 in Niokolo-Koba N.P.; rare there	Dupuy 1970b
	Not common but increasing; N border mapped near 13°N	Dupuy 1971b
	NW limit of range is Niokolo-Koba N.P.	Dupuy 1971c, Dupuy and Verschuren 1977
	Range expanding eastward	Kortlandt, pers. comm. 1974
	In Basse Casamance N.P. (in SW at mouth of Casamance R.) (this is outside the range limit)	I.U.C.N. 1975
Mali	Relict populations to 12°20′N	Kortlandt 1966
	In extreme SW corner (map)	Kortlandt and Van Zon 1969
	In SW along Bakoye R., Bafing R., and Falémé R. to 14°N (map)	Sayer 1977
Guinea-Bissau	In SE, E of 15°W (map)	Kortlandt and Van Zon 1969
Guinea	Throughout (maps)	Yerkes 1943, Vandebroek 1958
	Became very scarce 1959–60, then recovered; by 1965 not plentiful but increasing	Kortlandt 1966
	Localities listed and mapped throughout W where population is about 12,500 individuals	Bournonville 1967
	Throughout except extinct in NE (map)	Kortlandt and Van Zon 1969
	In S at Bossou and in W at Sougueta; not endangered	Dunnett et al. 1970 [Albrecht and Dunnett 1971]
	In Nimba Mts. Strict Nature Reserve (SE corner)	I.U.C.N. 1971

Table 186 (continued)

Country	Presence and Abundance	Reference
Sierra Leone	Throughout (maps)	Yerkes 1943, Vandebroek 1958
	A few near Freetown (W coast)	Jones 1966
	Rare in N; declining	Kortlandt 1966
	In N and E; isolated populations in central and W (map)	Kortlandt and Van Zon 1969
	Fairly plentiful	Lowes 1970
	Decreasing in most areas	Robinson 1971
	Gravely declining	Kortlandt, pers. comm. 1974
	Locally common in Ducuta area and Sainya Scarp (both in N)	Wilkinson 1974
Liberia	Throughout (maps)	Yerkes 1943, Booth 1958, Vandebroek 1958, Kuhn 1965
	Throughout except central (map)	Kortlandt and Van Zon 1969
	Decreasing in most areas	Robinson 1971
	Rare in Mount Nimba area (in N)	Coe 1975
Ivory Coast	Throughout (maps)	Yerkes 1943, Booth 1958, Vandebroek 1958
	In Banco Reserve	Aeschlimann 1965
	Throughout except NE (map)	Kortlandt and Van Zon 1969
	In Banco N.P. near Abidjan	Happold 1971
	In Tai region between Cavally R. and Sassandra R. (in SW near Troya: 5°41′N, 7°22′W)	Rahm 1971
	Near Troya and Sakré, S of Tai	Struhsaker and Hunkeler 1971
	Abundant in Tai Reserve between Cavally R. and Sassandra R.	Struhsaker 1972
	In Tai N.P.	Monfort and Monfort 1973
	Rare and decreasing near Lamto (S central)	Bourlière et al. 1974
	Abundant in Tai N.P. and Marahoue N.P.; common in Mount Peko N.P. and Asagny Reserve; rare in Banco N.P.	Asibey 1978
	In Tai Forest outside N.P. between Nsé R. and Méno R.	Boesch 1978
Upper Volta	Questionable in SW (map)	Kortlandt and Van Zon 1969
Ghana	Localities mapped in SW; total population not more than 1,000; doomed to extinction soon	Booth 1956a
	Discontinuous distribution even within suitable habitat in SW (map)	Booth 1958
	In SW corner only (map)	Kortlandt and Van Zon 1969
	In Bia Tributaries North Forest Reserves (in SW)	Jeffrey 1970, 1975a,b
	Habitat and numbers declining	Jeffrey 1974

Table 186 (continued)

Country	Presence and Abundance	Reference
Ghana (cont.)	Rare in Bia N.P., Nini-Suhien N.P., and Ankasa Game Production Reserve	Asibey 1978
Togo	S of 10°N (map)	Yerkes 1943
	Along W border in S; questionable farther N (map)	Kortlandt and Van Zon 1969
	In Koué Reserve (central, W of Sokodé)	I.U.C.N. 1971
Benin	S of 10°N (map)	Yerkes 1943
	In extreme S (map)	Vandebroek 1958
	Extinct in S (map)	Kortlandt and Van Zon 1969
Nigeria	Throughout S (map)	Yerkes 1943, Vandebroek 1958
	Isolated populations E and W of Niger R. (map)	Booth 1958
	In Boshi Extension Forest Reserve near Obudu, and Oban Reserve on Cross R. (both in SE); threatened with extinction	Petrides 1965
	In S and SE; extinct in some areas, questionable in others (map)	Kortlandt and Van Zon 1969
	Exterminated there	Struhsaker and Hunkeler 1971
	Very rare	Monath and Kemp 1973
	In Gashaka-Gumti Game Park, Kamatan Game Sanctuary, and Ngel Nyaki Game Sanctuary (all in SE corner of Gongola State, 7°N–8°N and 11°E–12°E) (map)	Hall 1976
Cameroon	Throughout S (maps)	Yerkes 1943, Booth 1958, Vandebroek 1958, Kortlandt and Van Zon 1969
	On Mount Cameroon (in W), at Idenau (4°15′N, 9°E), S Bakundu Reserve (S of Ediki, 4°33′N, 9°28′E), and Campo Reserve and Dipikar (in SW corner) (map)	Gartlan and Struhsaker 1972
	Uncommon in Douala-Edéa Reserve (W central) and Korup Reserve (extreme W)	Struhsaker 1972
	Patchy local distribution S of about 7°N (map) including Douala-Edéa, Korup, and Dipikar Reserves; declining	Gartlan, pers. comm. 1974
	In and near Dja Reserve (S central)	Rowell, pers. comm. 1978
Equatorial Guinea	Throughout (maps)	Yerkes 1943, Vandebroek 1958, Sabater Pi 1966
	Throughout (map); range shrinking	Sabater Pi and Jones 1967
	In Monte Raices Territorial Park (in NW, E of Bata)	I.U.C.N. 1971
	In NW and central (localities mapped); endangered	Jones and Sabater Pi 1971

Table 186 (continued)

Country	Presence and Abundance	Reference
Gabon	Throughout (maps)	Yerkes 1943, Vandebroek 1958
	Common	Malbrant and Maclatchy 1949
	Good stocks in hinterlands	Kortlandt 1966
	Throughout except small areas in W and E (map)	Kortlandt and Van Zon 1969
	Fair-size population in Wonga-Wongue N.P.	Curry-Lindahl 1974a
Congo	Throughout (map)	Yerkes 1943
	Locally abundant	Malbrant and Maclatchy 1949
	In W and N (maps)	Vandebroek 1958, Kortlandt and Van Zon 1969
	Good stocks in hinterlands	Kortlandt 1966
	In N central and in SW including Niari Valley Faunal Reserve (map)	Bassus 1975
	In Odzala N.P. (in W, W of 15°E and N of equator)	I.U.C.N. 1975
Angola	In Cabinda (along Atlantic coast, N of mouth of Congo R.) (maps)	Yerkes 1943, Vandebroek 1958, Kortlandt and Van Zon 1969
	Threatened	Bothma 1975
Central African Republic	In SW corner (maps)	Yerkes 1943, Vandebroek 1958
	As far N as 7°10′N	Kortlandt 1966
	In SW and SE; isolated population in NW; extinct in central, questionable in W and parts of E (map)	Kortlandt and Van Zon 1969
Zaire	Throughout N and NE (maps)	Yerkes 1943, Vandebroek 1958
	In Ituri Forest near Epulu R. (in NE)	Curry-Lindahl 1956
	Very rare in Lemera Forest, W of Lake Kivu (in E)	Rahm and Christiaensen 1963
	In Kahuzi area (in E, W of Lake Kivu)	Rahm 1965
	Fairly numerous near Kabunga (between Lowa R. and Luka R.) and Walikale (between Lake Kivu and Lualaba R.) (in E)	Rahm 1966
	Localities mapped in N near Sudan border	Kock 1967
	Between Rift Valley and Zaire-Lualaba R., 1°30′–1°50′S, 28°0′–28°30′E	Rahm 1967
	Throughout N and NE, N and E of Zaire-Lualaba R.; isolated populations from 4°30′S to 7°30′S, W of Lake Tanganyika; questionable along Luvua R. near 8°S, 28°E and along Uele R. near 4°N, 26°E (map)	Kortlandt and Van Zon 1969

Table 186 (continued)

Country	Presence and Abundance	Reference
Zaire (cont.)	In Albert N.P. (now Virunga N.P.) (along E border N of Lake Kivu) and Garamba N.P. (along N border with Sudan)	I.U.C.N. 1971
	In Basankusu Territory, Equateur Province (in NW)	Ladnyj et al. 1972
	Common in Haut-Zaire Province (in N and NE)	Heymans 1975
	In Kahuzi-Biega N.P. (in E, 2–3°S, 28–29°E, W of Bukavu)	I.U.C.N. 1975
	Relatively common on right bank of Zaire R.	Verschuren 1975b
	In mountains W of Lake Kivu	Hendrichs 1977
Sudan	In extreme S (map); steadily declining	Butler 1966
	Localities mapped in extreme S along Zaire border	Kock 1967, Kortlandt and Van Zon 1969
	In Mbarizunga Game Reserve	I.U.C.N. 1971
Uganda	In SW (map)	Yerkes 1943
	In all W forests from Budongo and Bugoma of Bunyoro District through Toro District to Ankole and Kigezi and Kayonza (Impenetrable) in SW; relatively common	Pitman 1954
	Not in danger in W forests of Budongo, Siba, and Bugoma (Bunyoro), Muhangi (Mubende), Itwara, Semliki, Ruwenzori, and Kibale (Toro), Maramagambo (Ankole), and Kayonza, and forests on Mount Muhavura, Mount Mgahinga, and Mount Sabinio (Kigezi)	Haddow 1958
	In Kibale Forest, Toro	Stott and Selsor 1959
	Population in Budongo Forest: 1,000–2,000 individuals; mapped in 13 forests including Ishasha, Itwara, and Kasyoha	Reynolds and Reynolds 1965
	In Budongo Forest	Sugiyama 1968b, 1973
	In W, S of 2°N (map)	Kortlandt and Van Zon 1969
	In Murchison Falls N.P. (in W) and Queen Elizabeth N.P. (in SW)	I.U.C.N. 1971
	Localities mapped in W, S of Victoria Nile	Kingdon 1971
	In Budongo, Bwamba, and Rabongo forests; threatened	Suzuki 1971
	In Gorilla Game Reserve, S border with Rwanda	Harroy 1972
	Common in Kibale Forest (0°31′–0°41′N, 30°19′–30°32′E)	Struhsaker 1972 [1975, Oates 1977a]
	In Murchison Falls N.P., Queen Elizabeth N.P., Ruwenzori Mts., and Impenetrable Forest (map)	Jahnke 1974

Table 186 (continued)

Country	Presence and Abundance	Reference
Uganda (cont.)	Population of 500–1,000 individuals in Budongo Forest (part is nature reserve); also in Bwindi, Kibale, Kasyoha, Kitomi, Kalinzu, Maramagambo, and Ruwenzori forests	Albrecht 1976
Rwanda	Throughout (map)	Yerkes 1943
	In Rugege Forest, E of Lake Kivu (W central)	Curry-Lindahl 1956, Rahm 1965
	In N and W (map)	Vandebroek 1958
	In Rugege Forest near Uinka (SE of Lake Kivu), at Shangugu (in SW), and at N'doza	Elbl et al. 1966
	In SW, SE of Lake Kivu (map)	Kortlandt and Van Zon 1969
	In Volcanoes N.P. (N border)	I.U.C.N. 1971
	On Mount Hehu, W end of Volcanoes N.P.	Spinage 1972
Burundi	Throughout (map)	Yerkes 1943
	In N'gozi Forest (about 3°S, 30°E), and Rugombo (about 3°S, 29°E) (both in N)	Elbl et al. 1966
	In SW, E of N end of Lake Tanganyika, including Teza forest	Verschuren 1978b
Tanzania	In W, E of Lake Tanganyika (map)	Vandebroek 1958
	At Kabogo Point, N end of Kungwe Bay Forest Reserve; at mouth of Msihezi R.; near Suguna; N of Lugufu R. (all E of Lake Tanganyika)	Azuma and Toyoshima 1961–62
	Semi-isolated population of 60–100 in Gombe Stream Reserve, N of Kigoma, E of Lake Tanganyika	Goodall 1963, 1965
	At Kasoge, S of Kigoma, near Katumbi in Mahali Mts. (map)	Nishida 1968
	In W between Lake Tanganyika and 31°E S to almost 7°S (map)	Kortlandt and Van Zon 1969
	Localities mapped in W, S of Malagarasi R. at Makolongo, Kasakati Basin, Filabanga, Masanguwe, Tumbatumba, and Mulofesi; population 365–530 individuals	Suzuki 1969
	From 5°15′–6°20′S and 29°45′–30°30′E, including Kasakati Basin and Itiye Camp; population there about 110 individuals	Izawa 1970
	Localities mapped from 5°12′–6°38′S and from Lake Tanganyika to Ugalla (31°01′E); including Lilianshimba Hills, Ugalla Hills, Masito Hills, Kururunpeta	Kano 1971a,b

Table 186 (continued)

Country	Presence and Abundance	Reference
Tanzania (cont.)	Range, Mukuyu Hills, Mount Kabogo, Kapalagulu-Igunga-Ipumba Mts., Lugala Hills, and Wansisi Mts.	Kano 1971a,b (cont.)
	Localities mapped in W central	Kingdon 1971
	At Mahali Mts. and Kabogo Head (map)	Sugiyama 1973
	Population about 600 individuals in Mahali Mts. proposed N.P. (6°S, 30°E), E shore Lake Tanganyika; total in country about 2,000 individuals	Itani and Nishida, pers. comm. 1974
	At Kasoge, Mahali Mts.	Kawanaka and Nishida 1975
	At Gombe Stream Research Center	Bryceson 1976, Wrangham 1977
	Population of 100–150 individuals in Gombe N.P. (4°40′S, 29°38′E)	Teleki et al. 1976, Pierce 1978
Malawi	No natural population in Nkata Bay District (W shore of Lake Malawi); individual sighted there in 1959 was an escapee from captivity	Benson 1968

Table 187. Population parameters of *Pan troglodytes*

Population Density (individuals/km^2)	Group Size X̄	Range	No. of Groups	Home Range Area (ha)	Study Site	Reference
				5,000–10,000	Guinea	Kortlandt 1966
0.05–0.50					Guinea	Bournonville 1967
		20–30			Guinea	Albrecht and Dunnett 1971
	4.3	2–10	10		Cameroon	Struhsaker 1969
0.31–1.53	4.7–11.2	2–23	8		Equatorial Guinea	Jones and Sabater Pi 1971
		4–12			E Zaire	Rahm 1966
	6.5	3–16	28		E Zaire	Rahm 1967
2					E Zaire	Hendrichs 1977
3.9				1,554–2,072	Uganda	Reynolds and Reynolds 1965
6.7		2–40			Uganda	Sugiyama 1968b
0.88–1.4	4				Uganda	Struhsaker 1975
		2–30			Tanzania	Azuma and Toyoshima 1961–62
1.27	2–5	2–23	434	1,500–7,800	Tanzania	Goodall 1965
	8.1	2–28		1,040	Tanzania	Nishida 1968
		2–43		1,000–20,000	Tanzania	Suzuki 1969
		3–36		1,230	Tanzania	Izawa 1970
	40	20–60			Tanzania	Kano 1971a
0.2				10,000–20,000	Tanzania	Kano 1971b
1					Tanzania	Sugiyama 1973
	40	20–100	6		Tanzania	Itani and Nishida, pers. comm. 1974
5					Tanzania	Wolfe 1974
		≤ 40		1,295–2,072	Tanzania	Bryceson 1976
3.6–4.6		30–80			Tanzania	Teleki et al. 1976

Pongo pygmaeus Orang-utan

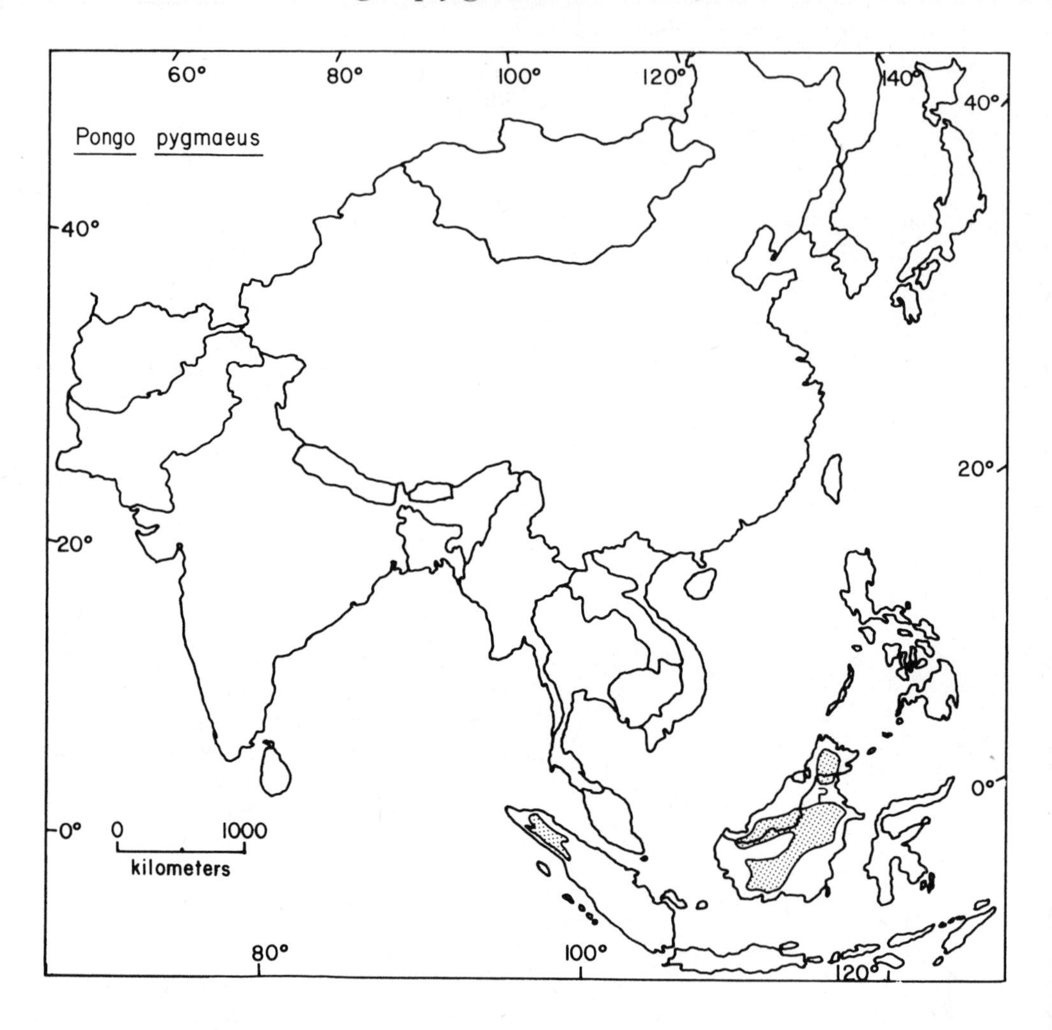

Taxonomic Note

Two subspecies are recognized (Napier and Napier 1967).

Geographic Range

The orang-utan is found in northwestern Sumatra, Indonesia, and on Borneo, in the Indonesian section (Kalimantan), and in the Malaysian states of Sarawak and Sabah. On Sumatra it occurs from the equator to about 5°N, with most of the population concentrated north of 3°N and west of 99°E (Rijksen and Rijksen-Graatsma, pers. comm. 1974; Rijksen 1978). On Borneo its distribution has not been completely delineated but includes at least southcentral, central, and northeastern Kalimantan, southern Sarawak, and eastern and central Sabah (MacKinnon 1974b, Rijksen 1978). For details of the distribution, see Table 188.

Abundance and Density

P. pygmaeus is an endangered species (I.U.C.N. 1976) whose populations are reported to be small and declining in every portion of the range (Table 188). The most recent estimates of its population in Sumatra vary from 3,000 to 15,000 individuals, and even workers who believe the total may be as high as 15,000 consider the species to be severely endangered there (Rijksen 1978). Kalimantan is thought to contain 1,000 to 4,000 *P. pygmaeus* and Sabah 2,000 to 3,000; the population in Sarawak has not been estimated since 1961 and may be extinct (Table 188). Estimates of the total number of orang-utans remaining in the wild vary from 3,000 (Harrison 1970) to 10,000 to 30,000 individuals (I.U.C.N. 1976).

Orang-utan population densities of 0.1 to 7 individuals per km^2 have been reported (Table 189). This species travels in very small groups of 1 to 5 animals each ($\bar{X}$ = 1.45 to 1.91) consisting of a mother and her dependent offspring; subadults and adult males are usually solitary. Individuals occupy home ranges of 42 to 1,000 ha (Table 189).

Habitat

P. pygmaeus is confined to primary tropical rain forest in all areas of its range. In Sumatra it occupies several types of primary forest, from swamp and lowland forests at sea level to hill and mountainous forests at 1,000 to 2,000 m elevation (Rijksen 1978). In lowland areas Dipterocarpaceae and Bombacaceae are common, and other important forest species are members of the genera *Ficus*, *Aglaia*, *Mastixia*, *Walsura*, and *Litsea* (Rijksen 1978). Orang-utans are seen most often in lowland areas (Carpenter 1938, Borner 1976, Wilson and Wilson 1977), and population densities decrease greatly at elevations above 500 m (Kurt 1973).

In Kalimantan, *P. pygmaeus* utilizes primary rain forest, freshwater swamp forest, and nipa-mangrove swamp (Galdikas-Brindamour, pers. comm. 1974). It has been observed in lowland dipterocarp forest characterized by a large proportion of ironwood trees (Lauraceae: *Eusideroxylon zwageri*) (Rodman 1977).

In Sarawak, orang-utans are found in permanently flooded peat swamp forest, riparian forest, lowland and hill dipterocarp forest, and, to minor extents, heath and moss forests, primarily in lowland areas (below 152 m) along the coast and major streams (Schaller 1961). In Sabah they occupy lowland tropical primary forest (Kawabe and Mano 1972, MacKinnon 1974a), lowland dipterocarp rain forest (Horr 1975), and riverine seasonal and permanent swamp forest (Horr 1977), and have been recorded as high as 1,500 m (de Silva 1971) and 1,850 m (Groves 1971d).

Factors Affecting Populations

Habitat Alteration

Since *P. pygmaeus* is dependent on mature primary forest habitats (Davis 1962, Rijksen and Rijksen-Graatsma 1975), a major threat to its survival is deforestation by lumbering and clearing for cultivation. Schaller (1961) reports, "It is doubtful if the orang-utan will ever adapt to habitat other than primary forest and old secondary forest. Although it can survive in very lightly logged forest where the canopy is only partially broken, no sign was ever noted or records obtained from very young secondary forest or from forest so heavily disturbed by logging that large sections of the canopy have been removed."

Some authors have found considerable numbers of orang-utans in areas from which large trees of commercial value had been removed several years before, and believe that orang-utans prefer selectively logged to undisturbed forest (Stott and Selsor 1960). But other evidence contradicts this conclusion (MacKinnon 1971, Rijksen 1978) and indicates that the species is found in smaller numbers in selectively felled areas than in undamaged forest (MacKinnon 1973).

In any case, it is clear that complete deforestation eliminates orang-utans and is reducing their range in both Sumatra and Borneo. As early as the mid 1930s the main threat to *P. pygmaeus* in Sumatra was the clearing of lowland forest for commercial development, cultivation, and rubber plantations (Carpenter 1938). More recently the remainder of orang-utan habitat in Sumatra (aside from the 30% in the Leuser Reserve complex) was under severe pressure from large-scale timber logging, shifting cultivation, and sustained agriculture culminating in eroded, grassy plains (Rijksen 1974). By 1974 more than half of the *P. pygmaeus* habitat in Sumatra had been completely destroyed, and about 10% of the remainder was being destroyed each year (Rijksen and Rijksen-Graatsma, pers. comm. 1974). At least half of the remaining range of the species in Sumatra was expected to be destroyed before the year 2000 (Rijksen 1974–75).

Virtually all the forest habitats in northern Sumatra (except for a few remote regions and reserved areas) have now been leased to timber concessions and will be logged for wood and charcoal for export to Asian, Arabian, American, and European countries, and for locally consumed firewood (Rijksen 1978). The huge demand for wood by developed countries has led to a rapid rate of logging and may soon make it profitable to extract timber from the most inaccessible mountain forests (Borner 1975).

P. pygmaeus living in logged areas move out of their original home ranges and become wanderers under stress from the noise of logging, adjusting to unfamiliar surroundings and increased encounters and competition with nearby orang-utans (Rijksen 1978). In the long run, even selective logging decreases the carrying capacity of orang-utan habitat by reducing the number of food-rich creeping and strangling figs and other climbing and liane-type plants important as food sources and as structures for the orang-utan's arboreal pathways (Rijksen 1978). In addition, road construction for timber extraction causes erosion and facilitates encroachment into the area by migrant people (Rijksen 1978), and oil companies are also building large road networks which are opening up remote areas to timber companies and shifting cultivators (Borner 1975).

Shifting agriculture has already reduced the range of *P. pygmaeus* in northern Sumatra by 20 to 30% since 1935, and it is steadily growing because of the rapid (2.5% per year) increase in human population there (Rijksen 1978). Previously, traditional adat restrictions dictated that certain seedlings and saplings be left on cleared land and that cultivation be shifted to a new plot of forest before the soil was seriously depleted. But the widespread impact of imported ideologies and the huge human population have eliminated these restrictions and allowed cultivation to persist until soil exhaustion and serious erosion prevent further farming or regeneration of forest (Rijksen 1978).

Large areas are also being cleared for plantations of rubber, coffee, pine trees, coconut palms, and *Cannabis*, and together timber logging and agriculture will probably destroy two-thirds of the remaining orang-utan habitat in Sumatra in the near future (Rijksen 1978). The amount of undisturbed rain forest remaining in Sumatra in 1973 appeared to be the critical minimum to support a sufficiently diverse population of *P. pygmaeus*, and this population will suffer greatly if the further predicted habitat decreases occur (Rijksen 1978).

Bornean orang-utan populations face the same pressures. For example, large-scale logging of primary forest is reducing their habitat in southern Kalimantan (Galdikas-Brindamour, pers. comm. 1974), and logging and road building are under way in eastern Kalimantan (Rodman 1977). In western Sarawak, extensive cultivation and almost complete destruction of primary forest eliminated more than 1,500 ha of orang-utan habitat between 1860 and 1960, and isolated several populations in remaining fragments of forest (Schaller 1961). Schaller predicted that "within the next few years . . . all marginal land will be put to use and the remaining patches of forest will be so logged and dissected with fields as to make them unsuitable for orangutan."

In Sabah also, encroachment on the forest by farming and timbering is threatening *P. pygmaeus* (Okano 1965, Davenport 1967), and it is accelerating (de Silva 1970, Phillipps 1973, MacKinnon 1974b). The declared policy of the Sabah government is to exploit its forest resources as quickly as possible in order to raise the standard of living, and all forest reserves in Sabah will probably be cut for timber within the next few years (MacKinnon 1971). At one study site in Sabah the orang-utan population became overcrowded owing to displacement by timber-felling operations in adjoining forests, and its reproductive rate declined markedly (MacKinnon 1974a,b).

In general, this species is extremely sensitive to habitat disturbances and is slow to recolonize logged areas (MacKinnon 1977b). Altogether about 80% of its remaining habitat will probably be destroyed during the next 20 years, eliminating the habitat of 1,000 orang-utans per year (MacKinnon 1977b).

Human Predation

Hunting. The hunting of *P. pygmaeus* for meat was probably the main factor which reduced the range of this species to Borneo and northern Sumatra (Schaller 1961, MacKinnon 1971). Orang-utans are still eaten by one tribe (the Batak) in Sumatra (Rijksen 1974, 1978), by the Chinese and Dyaks of Kalimantan (Galdikas-Brindamour, pers. comm. 1974), by people in Sarawak (Medway 1976), and in Sabah (MacKinnon 1971). They have also been hunted in Sumatra for sport (Rijksen and Rijksen-Graatsma 1975), in Sabah in connection with rituals and the concoction of medicines (de Silva 1970), as substitutes for humans in headhunting rites (MacKinnon 1974b), and because they were considered a dangerous menace (MacKinnon 1971).

Pest Control. P. pygmaeus feeds in plantations of mangoes and mangosteens in

Sabah (Peters et al. 1976). In Sumatra it also strips bark from certain forest tree and liane species and often feeds destructively in fruit trees (Rijksen 1978). It is killed when caught feeding in cultivated fruit trees (*Durio* spp.) (Rijksen and Rijksen-Graatsma, pers. comm. 1974), and in privately claimed, wild-growing fruit trees (Rijksen 1978).

Collection. The collection of young orang-utans (by killing their mothers) for sale as pets or for exhibition is thought to be a major reason for the decline of populations (Boyle 1963; Harrison 1963; Davenport 1967; de Silva 1971; Galdikas-Brindamour, pers. comm. 1974). From 1935 to the early 1960s an increasing demand for live orang-utans in zoological gardens and the high mortality rate of young apes in captivity led to a rapid rate of collection for export (Harrison 1970).

Collection pressure has decreased recently, but has not ceased entirely. For example, about 337 wild-caught *P. pygmaeus* were acquired by zoos in 1964–74 (Perry 1976). One author estimated that in the early 1970s at least 200 orang-utans were removed annually from the Alas Valley, Sumatra (Kurt 1973), while another calculated that about 64 were captured in this area in 1972 (Rijksen and Rijksen-Graatsma 1975). Most of these young orang-utans, which are still collected by killing their mothers, are now kept as high-status pets by Indonesian officials and foreign residents (Kurt 1973, Rijksen 1974, Frey 1975, Borner 1976). Some are smuggled out of Indonesia aboard Japanese timber boats and oil tankers (MacKinnon 1977b).

P. pygmaeus has also been collected for scientific research (Harrison 1970, MacKinnon 1971), but is not often used in research in the United States (Comm. Cons. Nonhuman Prim. 1975a).

In 1968–73, four *P. pygmaeus* were imported into the United States (Appendix). In 1977 none entered the United States (Bittner et al. 1978). Twenty-five were imported into the United Kingdom in 1965–75 (Burton 1978).

Conservation Action

P. pygmaeus has been reported in one national park in Sabah, five reserves (several of them interconnected) in Sumatra, five reserves in Kalimantan, seven in Sarawak, and four in Sabah (Table 188). Of these, the Gunung Leuser Reserve complex in Sumatra is the largest; and recent enlargements, boundary constructions, and management decisions have improved its status (Rijksen 1974–75). But this area is still being encroached upon by timber concessions, and "unless measures are taken to improve the legislative status of the reserve, there is not much hope for the orang-utan's survival" (Rijksen 1978).

The reserves in Kalimantan are also threatened by commercial logging, and the reserved areas in Sarawak and Sabah are too small for effective protection of orang-utan populations (MacKinnon 1971, 1974b; Rijksen 1978). Large areas of primary forest strictly protected from habitat alteration are urgently needed, for example, in Sabah (MacKinnon 1971, Phillipps 1973). Estimates of the minimum size for such reserves include 1,554 km^2 for a population of 300 *P. pygmaeus* (Harrison 1963) and 1 km^2 per individual *P. pygmaeus* (Medway 1976).

Legal protection from killing, capture, and trade has been given to *P. pygmaeus* in both countries of its range. Indonesia granted official protection in 1931 (Carpenter

1938), but enforcement was not achieved until 1963 (de Silva 1968b). Since 1971 the Indonesian army and police force have cooperated in tracing and confiscating captive orang-utans (Rijksen 1974), but enforcement is still not completely effective (Galdikas-Brindamour, pers. comm. 1974; Rijksen and Rijksen-Graatsma, pers. comm. 1974), especially if wealthy or influential violators are involved (Frey 1975, Rijksen and Rijksen-Graatsma 1975).

P. pygmaeus has been legally protected in Malaysia since 1958 (Stott and Selsor 1960, Schaller 1961, Choong 1975), but enforcement there is "variable and difficult" (Davenport 1967) and should be improved (Schaller 1961, Harrison 1968, de Silva 1971, MacKinnon 1971). Laws requiring permits for forest clearance in Sumatra are difficult to enforce, owing to the lack of accurate maps and the authorities' lack of knowledge about the locations of reserve areas (Rijksen 1978).

Rehabilitation centers, designed to serve as headquarters for reintroduction of confiscated orang-utans into natural habitats, have been established in Sarawak (Harrison 1963), later moved to Sabah (de Silva 1970, Mydans 1973), in Kalimantan (Galdikas-Brindamour, pers. comm. 1974; 1975), and in Sumatra (Rijksen 1974, Frey 1975, Borner 1976, Borner and Gittens 1978). Small numbers of *P. pygmaeus* have been reintroduced at each center, but various problems have been encountered, including (1) difficulties in training previously captive orang-utans to feed independently on wild foods (de Silva 1970, Rijksen 1978), (2) the possibility of transmitting serious diseases to healthy wild populations (Rijksen and Rijksen-Graatsma 1975, MacKinnon 1977b), (3) aggravation of local crowded conditions, thus putting stress on wild populations and habitats (MacKinnon 1977b), (4) the misinterpretation of rehabilitation centers as markets for *P. pygmaeus*, thus stimulating further captures (Rijksen and Rijksen-Graatsma 1975), and (5) the diversion of funds needed for habitat conservation and other efforts (MacKinnon 1977b).

Because of these difficulties, the current policies of the rehabilitation centers should be (1) to release orang-utans only into areas that no longer contain local wild populations, (2) to discourage illegal capture by devaluing captive orang-utans and demonstrating national and international concern for *P. pygmaeus*, and (3) to serve as research, education, propaganda, and tourist centers collecting and spreading information about tropical ecology and conservation, and enabling visitors to see free-living orang-utans (Rijksen 1974–75, 1978; MacKinnon 1977b; Borner and Gittens 1978). These authors agree that the function of the rehabilitation centers is not to save *P. pygmaeus* from extinction by reintroducing captive individuals but to increase support for habitat protection and decrease the frequency of captures, by education and persuasion. Thus a better name for the centers would be conservation field stations (MacKinnon 1977b).

Captive breeding has been proposed as a method of reducing collecting pressure, obtaining a secure captive stock (Harrison 1970), and possibly of producing enough surplus animals to release into depleted natural habitats (Brambell 1975). In 1974 there were about 625 *P. pygmaeus* in captivity worldwide, of which more than one-third were captive-born (Perry 1976). But no captive-born female orang-utan has yet given birth, and unless second-generation breeding occurs, the captive population will not sustain itself (Perry 1976, Rock 1977).

Many recommendations have been made for measures to conserve remaining wild populations of *P. pygmaeus*. Those concerning the protection of forested habitats include the upgrading of existing protected areas and the creation of new ones, the logging of only 1% of the forest reserves per year (MacKinnon 1977b), light selective logging in a slow rotation system (Borner and Gittens 1978), the control of growth of the human population in Sumatra, the prohibition of all destructive activities — especially commercial logging — within the range of the species in Sumatra, and a careful policy of sustained yield exploitation, watershed protection, and preservation of large areas for genetic reservoirs in other rain forest areas (Rijksen 1978). It has been suggested that the developed countries should support the protection of needed habitat reserves in Indonesia until that developing country can assume the responsibility itself (Borner 1975).

A large team of qualified rangers and wildlife officers is needed to patrol the boundaries of the Leuser Reserve in Sumatra (Kurt 1973). In addition, Indonesian students should be trained in conservation field work and stationed in the reserve (Rijksen 1974–75).

In Sabah an educational program should aim at instilling in children "a pride in the continued existence of the rare and wonderful orang-utans in their areas," logging roads and deforested areas should be planned so as not to isolate remaining *P. pygmaeus* populations, and orang-utans from logged areas should be transferred to suitable protected habitats (MacKinnon 1971). In Sarawak an appeal to return to traditional taboos governing the killing of *P. pygmaeus*, and an increase in reliance on domestic meat, could help to reduce hunting pressure (Medway 1976).

P. pygmaeus is listed in Appendix I of the Convention on International Trade in Endangered Species of Wild Fauna and Flora, and as endangered under the U.S. Endangered Species Act (U.S. Department of Interior 1977a,b). It is also prohibited from import into the United Kingdom except by special permit (I.U.C.N. 1972). In 1969 all bona fide zoos in the world agreed not to receive any orang-utans that did not have an official clearance document from the I.U.C.N. Survival Service Commission (Rijksen and Rijksen-Graatsma 1975).

Table 188. Distribution of *Pongo pygmaeus*

Country	Presence and Abundance	Reference
Indonesia: Sumatra	Localities mapped in Atjeh (Aceh) (in N, W of 98°E); highest concentrations in W near Lami, Tapa, Toean, and Bakangan, and below fork of Simpang-kanan and Simpang-kiri rivers and near Peureulak R. in E; scarce everywhere	Carpenter 1938
	Population about 1,000 individuals	Reynolds 1967b, Fisher et al. 1969, Basjarudin 1971
	In Rafflesia Serbödjadi Reserve (extreme N end)	I.U.C.N. 1971
	In Mount Löser (Gunung Leuser) Reserve (in N) and Mount Wilhelmina Reserve (in E)	Harroy 1972
	Two localities mapped in S Atjeh	Wilson and Wilson 1972–73
	Population about 3,650 individuals, including about 900 in Gunung Leuser Reserve (localities mapped)	Kurt 1973
	At West Langkat and Ranun R., Sikundur (N Langkat Reserve) and Ketambe (Gunung Leuser Reserve, W of Alas R.) (map)	MacKinnon 1973, 1974a,b
	Population about 4,500 individuals in 1970; becoming rare, threatened; in Gunung Leuser, Langkat, and Kluet reserve complex	Rijksen 1974
	Localities mapped near W coast N of 0° and in N between 3° and 5°N; core area in Gunung Leuser Reserve, Gajo Highlands; strongly declining	Rijksen and Rijksen-Graatsma, pers. comm. 1974
	Population about 15,000 individuals, including about 5,000 in Gunung Leuser Reserve complex	Rijksen 1974–75
	In E, as far N as Krueng Jamboaie R. (Jamboaye) (E of Lhok Seamawe); in W, abundant as far N as Blang Pidi; S to border of Langkat Reserve and S of border of Gunung Leuser Reserve (map); population 3,500–4,500 individuals, of which half outside reserves are severely threatened	Borner 1976
	Only N of Simpang-kiri R. and Wampu R.	Wilson and Wilson 1977
	Population estimates vary from 3,000 to 10,000 individuals	Borner and Gittens 1978
	Concentrated N of Sungei Simpang-kiri (S Alas R.) as far as Benkung and Kluet	Rijksen 1978

Table 188 (continued)

Country	Presence and Abundance	Reference
Indonesia: Sumatra (cont.)	areas and in Kappi plateau N to Serbojadi Mts. and Jambu Aye (Jamboaye) (in E); N boundary: line connecting Tawar and Krueng Woya; W boundary Keudi Pasi to Bakongan and then along coast; E boundary Biruen to Sungei Wampu (maps); population 5,000–15,000 individuals; range shrinking; severely endangered	Rijksen 1978 (cont.)
Kalimantan	Population about 1,000 individuals	Reynolds 1967b
	Population 3,500–4,000 individuals including in Mandor Reserve; in danger of extermination	Basjarudin 1971
	In Kutai Reserve (in E at 0°24′N, 117°16′E)	Rodman 1973, 1977; Horr 1975, 1977; Horr and Ester 1976
	In Tanjung Puting Reserve near Kumai (S central, 2°45′S, 111°44′E); overall rare; in danger of extinction	Galdikas-Brindamour, pers. comm. 1974 [1975]
	In SW and NE	MacKinnon 1974b
	Localities mapped in SW, central, and E	MacKinnon 1977a
	Throughout except SE, extreme W, and 1 region of NW (map), including Bukit Raya Reserve and Kotawaringin Sampit Reserve; questionable in NE, N of about 2°N	Rijksen 1978
Malaysia: Sarawak	Has almost vanished there	Stott and Selsor 1960
	Concentrated in SW between Sadong R. and Batang Lupar R. (110°45′–111°45′E), including in 7 reserves of Sabal, Balai Ringin, Sedilu, Sebayau, Simunjan, Lanjak-Entimau, and Mengiong; otherwise only in local isolated populations, including on upper Baram R., Balui R., and Baleh R. (localities mapped); population 450–700 individuals; declining; range has shrunk greatly; threatened	Schaller 1961
	Threatened	Harrison 1963
	Scattered along border with Kalimantan; virtually extinct	MacKinnon 1971
	In W near Matu (map)	MacKinnon 1974b
	Localities mapped in SW, S, and E	MacKinnon 1977a
	In Ulu Batang–Ai R. system; relatively abundant near headwaters of Ai R., Engkari R., and Lubong Baya R.	Weber 1977
	In SW and along S and E borders (maps)	Rijksen 1978

Table 188 (continued)

Country	Presence and Abundance	Reference
Malaysia: Sabah	Moderate numbers; not in danger	Stott and Selsor 1960
	Fairly common near coast, including along Gaja R. in Bukit Kretam area, in Gomantong Forest Reserve, and along Segaliud R. (all in NE between 5°30′–5°45′N and 117°30′–119°E) (map)	Davis 1962
	At Labuk Road, Bukit Garam, Bakap, and Kinabatangan R. area (all in NE)	Yoshiba 1964
	At Sungei Merak (extreme E)	Okano 1965
	In Sandakan area: Sepilok Forest Reserve, Lokan R. area; increasingly rare	Davenport 1967
	Population about 2,000 individuals	Reynolds 1967b
	Throughout, including Sepilok Forest Reserve near Sandakan; habitat being reduced	de Silva 1970
	Throughout, including Sepilok Reserve, E ridge of Kinabalu, Ulu Dusun Agricultural Station, Lungmanis, Kretam, Sandakan, Kinabatangan, Labuk Road, and Tenegang Besar	de Silva 1971
	In Kinabalu N.P. (in N)	I.U.C.N. 1971, Phillipps 1973
	Rare in Kinabalu N.P.	Jenkins 1971
	In forest reserves along rivers of E coast; in small pockets W of Sandakan; throughout S; in Kinabatangan, Segama, and Tinkayu R. area; in Ulu Segama Game Reserve near junction of Bole R. and Segama R. (map); population of 2,000–3,000 individuals is about one-half total of species	MacKinnon 1971 [1974a]
	In inland plain near Kinabatangan R.	Kawabe and Mano 1972
	In E half (W of Segama R.) and in N (map)	MacKinnon 1974b
	Along Lokan R. (in NE)	Horr 1975, 1977; Horr and Ester 1976
	Localities mapped in E and N	MacKinnon 1977a
	Throughout except coastal areas (maps) including Danum Valley Reserve	Rijksen 1978
Brunei	Has almost vanished there	Stott and Selsor 1960
	No other authors mention it there	

Table 189. Population parameters of *Pongo pygmaeus*

Population Density (individuals /km²)	Group Size			Home Range Area (ha)	Study Site	Reference
	X̄	Range	No. of Groups			
					Indonesia	
0.1–0.4		1–3			Sumatra	Kurt 1973
<1–2	1.8	1–4			Sumatra	MacKinnon 1973, 1974a
3–5				200–1,000	Sumatra	Rijksen and Rijksen-Graatsma, pers. comm. 1974
0.25–1					Sumatra	Borner 1976
2–7					Indonesia	MacKinnon 1977b
3.1					Sumatra	Wilson and Wilson 1977
5	1.91			200–1,000	Sumatra	Rijksen 1978
2–3.7	1.83	1–3			Kalimantan	Rodman 1973
3		1–3			Kalimantan	Galdikas-Brindamour, pers. comm. 1974
				42	Kalimantan	Rodman 1977
					Malaysia	
0.2					Sarawak	Schaller 1961
0.25–0.91					Sabah	Yoshiba 1964
	1.45	1–3	11		Sabah	Davenport 1967
0.97	1.83	1–5	146		Sabah	MacKinnon 1971
<1–2					Sabah	MacKinnon 1974a
		1–4		65–777	Sabah	Horr 1975, 1977
≥1.5					Sabah	MacKinnon 1977a

Discussion

The preceding accounts outline the available information on the status of populations of each species of primate. In this section I compare the data for all these species and discuss the present situation of the order Primates as a whole. Only the most critical evidence is recapitulated, the major conclusions outlined, and some implications and recommendations presented. My goal is to synthesize the pertinent information so as to arrive at an objective assessment of the status of primate species. I shall examine the variation in each of the features critical to the persistence of populations (Table 190), consider the ways these aspects interact to determine species status, and discuss some actions that would help to conserve primate populations.

Geographic Range

Each species of primate is found in a specific area in Africa, Asia, South America, or Madagascar. Natural barriers such as rivers, mountain ranges, or vegetation zones usually determine the limits of these ranges. Geographic range size (Table 190) is a measure of the maximum area in which each species could occur. This measure does not consider the extent of available habitat within the distribution limits, thus often greatly overestimates the actual range size. But it is useful as an approximation of the extent of the distribution of each species, and as a method of comparing the ranges of different forms.

The size of the geographic range is an important determinant of the status of a species. Although widely distributed forms may be severely threatened in other ways, all species with restricted ranges are more vulnerable than species with large ranges, since "where the geographic range of a species is extremely limited, local extermination equals specific extinction" (Hershkovitz 1972). Habitat disturbance or human predation that would have a trivial effect on a wide-ranging form could extinguish a species found only in a limited area.

Eight species have very large ranges: they occur over areas larger than 6 million km^2. Some of these (e.g., *Cercopithecus aethiops*, *Aotus trivirgatus*, and *Ateles paniscus*) are thought by some authorities to consist of several distinct species, and so may not represent truly wide-ranging species. Another 35 species range over large areas (greater than 2 million km^2), and 21 species have medium-size ranges (1 to 2 million km^2).

At the other extreme, 37 species have very small ranges (100,000 to 499,999 km^2). Nine of these are island species; the others are found on the mainland of the three large continents. Another 34 species have tiny ranges (smaller than 100,000 km^2); 25 of these species are island forms.

Altogether, 71 species (47% of the primates) have very small or tiny ranges. These include 12 African monkeys, 15 Asian monkeys, 4 cebids, 9 callitrichids, all 21 lemuroids, 2 lorisids, all 3 tarsiids, 3 hylobatids, and 2 pongids. Also, there are many geographic races of wide-ranging species, such as *Colobus badius*, *C. angolensis*,

Cercocebus galeritus, and *Hylobates lar*, which occur only within limited areas. Because they are so restricted, these species and races are most susceptible to extinction. Whatever their other attributes, they cannot survive extensive pressures from man, and should be the first ones to receive protection.

The number of countries in which a species occurs (Table 190) can serve as an index of range size. It can indicate the vulnerability of species to political policies, since forms whose ranges include several different jurisdictions may experience varying degrees of peril or protection. The number of countries in the range is also useful for comparison with the number of countries in which specific abundance levels or population pressures have been described.

Number of countries is not generally an accurate measure of actual range size because of the great variation in the sizes of countries and the proportion of each country included in the range limits. Similarly, the number of countries in which a phenomenon has been reported is not a precise measure of the prevalence of the phenomenon, because of discrepancies in the size of the areas to which reports refer (e.g., from one study site to whole countries) and the availability of information from different countries, some of which have had no recent field studies. Nevertheless, because other numerical methods of expressing overall abundance and the severity of pressures are not available, the number of countries in the range has been listed for comparison with the columns labeled abundance, habitat alteration, hunting, and pest control (Table 190), which show the number of countries where these phenomena have been reported.

Body Weight

Body size affects conservation status in several ways. Larger primates need more food and larger feeding ranges, and tend to have lower densities than similar small species do (Milton and May 1976, Clutton-Brock and Harvey 1977, Eisenberg 1980). Within one phyletic line, smaller mammals are usually the older and more generalized species and larger ones are more specialized (Hershkovitz 1972). Therefore, larger species tend to be more vulnerable to habitat reduction than smaller forms are.

Larger animals mature and reproduce more slowly than small animals, hence replace lost members of their populations more slowly (see Bonner 1965, Pianka 1974). Large-bodied individuals are shot by hunters in preference to small ones, because they present an easier target and provide more meat per bullet. Larger animals are more likely than small ones to be killed as pests, because they and their damage are more obvious to farmers. Thus larger primates are more severely affected than small forms, by both habitat alteration and human predation, and within one group the largest bodied species will be threatened first.

The primates vary in weight from 0.06 kg (*Microcebus murinus*) to 180 kg (*Gorilla gorilla*) (Table 190). The smallest members of the order generally occur within the Cheirogaleidae, Lorisidae, Tarsiidae, and Callitrichidae, and the largest individuals in the Pongidae and Cercopithecidae. Altogether, 14 species have very large bodies (≥20 kg), and 38 species are either absolutely large (10 to 19 kg) or relatively large: medium size (6 to 10 kg) but among the largest in their families.

Among Old World monkeys, examples of very large forms include *Papio sphinx*,

P. leucophaeus, *Theropithecus gelada*, *Presbytis entellus*, and *Nasalis larvatus*. All the great apes are very large. Among the large or relatively large species are most of the *Colobus*, *Macaca*, and *Presbytis*, as well as *Alouatta*, *Ateles*, and *Lagothricha* (Table 190).

As examples of the correlation between relative size and status, Hershkovitz (1972) compared various members of the New World primates. He found that although marmosets are small, primitive in structure, and "persistent in habitats so drastically altered that no other primate survives there," the most endangered marmoset is the largest one, *Leontopithecus rosalia*. The smallest cebids, *Saimiri* and *Aotus*, are the least threatened, while the largest one, *Brachyteles arachnoides*, is near extinction.

Abundance and Density

There are 14 species of primates for which estimates are available of the total size of the population: *Macaca fuscata*, *M. silenus*, *M. sinica*, *M. sylvanus*, *Presbytis geei*, *Simias concolor*, *Theropithecus gelada*, *Brachyteles arachnoides*, *Leontopithecus rosalia*, *Hylobates klossii*, *H. syndactylus*, *Gorilla gorilla*, *Pan paniscus*, and *Pongo pygmaeus*. Estimates have been made of the size of the population in a portion of the range for a few other species (e.g., *Nasalis larvatus*, *Presbytis entellus*, *Hylobates concolor*, *H. hoolock*, *H. lar*, and *H. pileatus*). For most of the primates there are no estimates of the total population size. In order to evaluate their conditions, one must rely on descriptions of relative abundance or knowledge of population density.

Relative Abundance

A species that is abundant or rare may be so naturally, as a result of its pattern of dispersion, or, secondarily, because of a recent increase or decrease in numbers. In either case, the reports of its relative abundance or rarity can be an indication of its overall status.

The number of countries or parts of countries in which a species is described as abundant (or common), rare (or uncommon, vulnerable, or endangered), and declining is shown in Table 190. One species may have been placed in all three categories even within one country—either in different areas or by different authors. The definitions of abundance and rarity may vary greatly, and no standard criteria have been adopted. Nevertheless, if a species has been described as rare in all or most of the countries in its range, we can probably conclude that it is not a common animal.

A total of 36 species of primates are considered rare in all of the countries in which they occur or in the only country in which they occur (Table 190). These include 3 species of macaques, 5 Asian colobines, 4 cebids, 3 callitrichids, 18 of the lemuroids, 1 gibbon, and 2 of the great apes. Another 9 species were described as rare in 50 to 99% of the countries of their ranges (Table 190). Thus, 45 species (30% of the primates) are thought to be rare in the majority of the countries of their ranges. These species would be expected to be more vulnerable to extinction than abundant species, whether their rarity is the result or the cause of their vulnerability.

Population Density

The natural abundance of a primate species is determined by its reproductive fitness (birth rate, survivorship) and its social structure, group size, intergroup spacing, and home range area. These in turn are influenced by the supply and distribution of critical resources, primarily food, water, cover, and sleeping sites. Dispersion and movements may be determined by the spacing of water holes, sleeping cliffs, or fruiting trees and may vary seasonally. Group size is influenced by the requirements of social interaction (group cohesion, communication, social roles), reproduction, and resource distribution.

The resultant population density also reflects the carrying capacity of the habitat and the genetic diversity necessary to maintain a population. Some species naturally exist at low densities while others achieve high population densities. Above certain limits, density-dependent mechanisms regulate further population growth; below certain minima, detrimental inbreeding and other genetic problems occur and extinction becomes likely (Franklin 1980, Soulé 1980).

Population density, group size, and home range area all affect the probability of extinction through habitat change or human predation. If all other factors are equal, a species with a low population density, small group size, or large home range area will be more vulnerable than species with high densities, large troops, or small home ranges.

For a few forms, the manner in which population structure influences vulnerability is obvious. Monogamous and territorial animals such as *Callicebus* and *Hylobates* live in small groups consisting of one mated pair and their immature offspring, and defend large nonoverlapping areas against other groups of the same species. A forested area must be relatively large to support a resident population. The removal of one adult precludes reproduction in the group until the remaining adult finds a new mate.

In wide-ranging and solitary species, such as *Pongo pygmaeus*, and mother and infant constitute the only long-lasting group. Destruction of their forest habitat prevents these animals from completing their normal circuits of movement, and concentrates populations in remaining patches at such high densities that reproduction may be impaired. Even limited hunting may reduce a population to the point that adults cannot find mates. Other solitary species, such as *Daubentonia madagascariensis*, *Arctocebus calabarensis*, and *Perodicticus potto*, probably face similar problems.

In the species accounts, population parameters are listed where available. It would be desirable to compare these measures to obtain a relative rating for density, group size, and home range area of each species. But these data have been collected in so many different ways and are so variable even within one species, that it is difficult to compare them. For example, densities are often measured at sites selected for their large populations of a species, in optimal habitats, or in very small areas, and thus may not be representative of typical densities for the species. Average group sizes are sometimes calculated using only a few groups, and home range estimates may be based on studies lasting only a few weeks, when longer periods are needed to discover the true extent of the range. A simple analysis of these data would be inadequate, and a complex one is beyond the scope of this book.

Habitat

Specialized species are more vulnerable to habitat alteration than eurytopic forms. Because specialists can utilize only one or a few types of vegetation, or can live and reproduce only within narrow limits of temperature, humidity, or elevation, even limited disturbance may decrease their chances for survival. Thus specialists will suffer first from habitat alteration.

We do not have the detailed information concerning habitat requirements that would allow us to determine the precise degree of specialization of most species of primates. But we do know that 131 species (86% of the primates) live only in forest or woodlands (Table 190). A minimum of 32 species (21% of the primates) are restricted either to one specific type of habitat or to primary forest, and can be considered very specialized or specialized. These include *Macaca silenus, Presbytis johnii, Theropithecus gelada, Colobus satanas, Nasalis larvatus, Presbytis cristata, Alouatta pigra, Ateles paniscus, Hylobates hoolock,* and *Pongo pygmaeus.*

Only 21 species (14% of the primates) are generalists: they utilize both tree-covered and more open habitats (Table 190). Examples are *Cercopithecus aethiops, Macaca fascicularis, M. mulatta, M. radiata, Papio anubis, P. cynocephalus, Presbytis entellus, Callithrix jacchus, Lemur catta, Galago crassicaudatus,* and *Pan troglodytes.* But even these species require at least a few trees or bushes for cover or sleeping sites, and are eliminated by large-scale habitat destruction.

Other Traits

Matters relevant to the status of primate species that are not examined in this volume include (1) characteristics of the location of the geographic range (e.g. coastal, inland, riverine, isolated), (2) dietary habits and requirements, (3) reproductive rates, (4) longevity, (5) adaptability, (6) arboreality, and (7) predator avoidance tactics. These characteristics may have significant effects on the status of certain species and should be investigated.

Factors Affecting Populations

Habitat Alteration

Because almost all primates require trees and some require trees of specific species or shapes, any removal, reduction, or change in the composition of trees in an area will affect the primate fauna. Since many primates have specific niche requirements, any environmental perturbation that alters the plant composition, cover, temperature, humidity, water supply, structural features, food availability, or other variables is likely to change the suitability of a habitat for resident primates.

For most species of primates, all we know is that their habitat is being reduced in certain portions of the range; reports of the rate or extent of disturbance are generally not available. An approximation of the severity of habitat alteration, based on the number of countries in which this pressure has been reported to be harming each species, is provided in Table 190.

Habitat alteration was reported in all countries within the ranges of 42 species. Most of these species have restricted distributions: all but three (*Macaca sylvanus,*

Nasalis larvatus, Pongo pygmaeus) are found in only one country. Those species undergoing severe habitat alteration include many island forms (such as *Macaca cyclopis, M. fuscata, Presbytis senex*, all the Lemuroidea) and several Brazilian species (such as *Alouatta belzebul, Brachyteles arachnoides, Callicebus personatus, Leontopithecus rosalia*). Another 18 species are experiencing deterioration of their habitats in 50 to 99% of the countries in which they occur (Table 190). Some of these species have small ranges (*Cercopithecus diana, Colobus satanas, Alouatta pigra*), but others have extensive distributions (*Macaca mulatta, Cebuella pygmaea, Hylobates lar*).

Altogether, 60 species (40% of the primates) are being affected by habitat alteration in more than 50% of the countries in which they are found, and many other species are experiencing this pressure in smaller portions of their ranges. The effects may be mild, moderate, or severe, but rarely are beneficial to the species concerned.

Destruction of Forest. Deforestation is the most serious threat to primate populations. Tropical moist forests, to which many primate species are restricted, are disappearing at an estimated minimum rate of 11 million hectares or 1.2% per year (Whitmore 1980). Almost every forest species is reported to be suffering loss of habitat in at least part of its range.

The most common method of deforestation is logging and replacement by agriculture. In some areas, only small patches are cut and the agriculture is temporary (slash and burn, swidden, or shifting cultivation). If natural vegetation is allowed to recolonize abandoned fields, certain species, including *Gorilla gorilla, Saguinus oedipus*, and *Miopithecus talapoin*, can occupy the secondary growth if patches of forest remain between the fields. The habitats of these species can be increased by limited clearing for shifting cultivation.

In many places, however, secondary vegetation is soon cut and burned for another planting and the forest cannot regenerate. In Ivory Coast, for example, 28,000 km^2 of forest (30% of the 1956 forest area) were cleared by shifting cultivation between 1956 and 1966. This left only 70,000 km^2 of forest in that country, and the rate of clearing was approaching 5,000 km^2 per year in 1966 (Lanly 1969). Other studies have indicated that about 2.1% of the tropical forest of the world was altered by shifting cultivation during each year before 1969 (Sommer 1976).

In some regions, poor soil or overexploitation precludes both continued cultivation and forest regeneration. Whether grasses overtake these fields or exposure to heavy rains and direct sunlight lead to leaching, laterization, and erosion, these areas are unsuitable as primate habitats (Harrison 1965, Sabater Pi and Jones 1967). This process is occurring in many parts of Africa, Asia, South America, and Madagascar (e.g., see Koechlin 1972; Goodland and Irwin 1975, 1977; Poore 1976, D'Arcy 1977).

Forest is replaced by permanent agriculture in productive areas such as those on volcanic soils or on flood plains. In coastal Cameroon, peninsular Malaysia, and Sumatra, for instance, large portions of primate habitat have been deforested for permanent farms and plantations.

Even where most of the forest remains, primate populations may be harmed by logging operations. Certain species are adapted to mature, primary forest (e.g.,

Cercopithecus diana, *Cercocebus albigena*, and *Ateles paniscus*). These primates are eliminated even by selective logging in which only a few trees per hectare are removed. Other species can tolerate selective logging and may benefit from the undergrowth which increases after limited removal of canopy trees (e.g., *Macaca fascicularis*, *Presbytis rubicunda*, and *Colobus guereza*). But even selective logging can radically change the botanical composition of a forest, eliminating certain tree species entirely (Sternberg 1968). Select cutting damages many more trees than are removed (Whitmore 1980), and the disturbance and hunting which accompany timber operations can be harmful to resident primates (Burgess 1971).

In some areas, herbicides are applied to selectively cut forests to decrease competition from undesirable vegetation and speed the regrowth of commercially valuable trees. The poisoned plants are often important food sources for primates, and their removal reduces the carrying capacity of these habitats. Evidence of the negative effects of arboricide use on primate habitats is available from Uganda (Albrecht 1976, Oates 1977a) and Japan (Azuma 1970, Tamura 1970).

In all, the number of species of primates whose habitat is being disturbed and whose populations are suffering because of logging is far greater than the number that benefit from selective timbering. And no forest primate can survive in areas subjected to extensive clearcutting, as is being practiced in many countries.

Warfare is another major cause of deforestation. Defoliants and bombs have destroyed much of the vegetation in Indochina (Orians and Pfeiffer 1970, Sommer 1976). Although no documentation can yet be obtained, it is very likely that the habitats of 12 species (*Macaca arctoides*, *M. assamensis*, *M. mulatta*, *M. nemestrina*, *Presbytis francoisi*, *P. cristata*, *Pygathrix nemaeus*, *Rhinopithecus avunculus*, *Hylobates concolor*, *H. pileatus*, *Nycticebus coucang*, and *N. pygmaeus*) have been greatly disturbed. Recent warfare in Nigeria, Bangladesh, Angola, Mozambique, Uganda, Afghanistan, and Zimbabwe have probably also reduced primate habitats and populations.

Rapidly growing exotic trees such as pine, eucalyptus, and cedar have become established in previously forested areas of Africa (Curry-Lindahl 1974b; Gartlan 1975a,c), Asia (Southwick and Siddiqi 1973), and South America (Sternberg 1968). Since most primates cannot feed or live in these introduced trees, their spread reduces available primate habitats. Species affected include *Macaca fuscata*, *M. mulatta*, and *Colobus badius*.

Changes in the supply or distribution of surface water, such as those caused by dams, swamp drainage, diversion of rivers, or flood control, can be detrimental to primates adapted to moist forests such as swamp, mangrove, or gallery forest. Species that would suffer if the surface water in their habitats decreased or changed course include *Nasalis larvatus*, *Cercopithecus neglectus*, *Miopithecus talapoin*, and *Chiropotes albinasus*.

Alteration of Nonforest Habitats. Open-country primates have also been affected by habitat disturbances. Most species that inhabit savanna, grassland, or scrub require some trees. Many of these trees, located along rivers and scattered in savanna, have been cut for lumber and fuel. Much open habitat has been planted with crops or converted to pastures. Some of the resident primates may be successful in these areas temporarily by feeding in farms, but subsequently are often exterminated because of their damage to cultivation.

Grazing livestock reduce the food supply for graminivorous primates such as *Papio anubis* and *P. cynocephalus*, and may denude the soil of vegetation by trampling. Domestic stock also compete with the native primates for water supplies. The addition of artificial water sources has improved arid habitats for such primates as *Papio ursinus* and *Erythrocebus patas*. But most human activities reduce the carrying capacity of nonforest habitats for primate populations.

Climatic changes, especially in temperature and rainfall, affect the vegetation of an area and thus its ability to support primates. For example, a severe drought in 1968–73 desiccated much habitat in subsaharan Africa. It is likely that many primate populations decreased because of shortages of water, food, and cover, and that the range limits of some (such as *Papio papio*, *P. anubis*, *Erythrocebus patas*, and *Cercopithecus aethiops*) retreated southward as woodland and savanna became drier. Desertification is a continuing problem in the subsaharn region (Delwaulle 1973, Cloudsley-Thompson 1974).

During drought, livestock and human beings compete with native animals for food, and overgrazing and trampling by stock intensify the effects of drought on the vegetation and soil. Overgrazing can also exacerbate the effect of other natural phenomena. For example, in the Masai-Amboseli Reserve, Kenya, grazing combined with heavy rains, flooding, and a rising water and salt layer caused serious habitat deterioration and a consequent decline in the populations of *Cercopithecus aethiops* and *Papio cynocephalus* (Western and Sindiyo 1972; Altmann, pers. comm. 1973; Struhsaker 1973, 1976a).

To summarize, the main kinds of habitat alteration that are reducing the quantity and quality of primate habitats are deforestation due to clearing for agriculture and timber, selective logging, the use of herbicides, the spread of exotic vegetation, warfare, water control projects, damage by livestock, and climatic changes.

Human Predation

The second major factor affecting populations of primates is the direct killing or removal of individuals by human beings. This predation includes hunting, pest control, and collection.

Hunting. Primates are killed for their flesh, skins, and other parts of their bodies. Meat hunting is especially important because in many parts of the tropics, particularly wet forest areas, domestic livestock do not thrive, sources of animal protein are scarce, and people depend on wildlife for meat (e.g., see Balinga 1977, de Vos 1977). The global limits of the Hunger Belt, within which the average human being consumes less than 2,750 calories per day (*Times Atlas of the World*, 1977), coincide closely with the distribution of primates. In only three areas do primates occur outside the Hunger Belt: northern Argentina, southern South Africa, and Japan. In many parts of the world, primate meat is considered a delicacy, is highly valued, and is obtained with modern firearms as well as primitive weapons and traps.

Hunting for meat severely affects populations of species with large bodies (see above), for example, *Ateles paniscus*, *Lagothrix lagothricha*, *Papio sphinx*, and *Simias concolor*. A few primates are hunted for their fur (such as *Colobus guereza*, *C. polykomos*, *Cercopithecus mitis*), for particular organs used in medicine or witchcraft (e.g., *Presbytis johnii*, *Loris tardigradus*), and for sport or a prestigious trophy (e.g., *Gorilla gorilla*).

Fifty-one species (34% of the primates) are hunted in more than half the countries of their ranges (Table 190). Several other species (e.g., *Pan troglodytes*, *Colobus guereza*, *Cercocebus torquatus*) are heavily hunted in smaller portions of their ranges.

Pest Control. Many primates are subject to human predation because they are considered to be nuisances. Sixteen species are thought to damage agriculture and are the object of extermination efforts in more than half the countries in their ranges (Table 190). Many more (such as *Cercopithecus aethiops*, *Erythrocebus patas*, *Papio anubis*, *P. cynocephalus*, *Macaca mulatta*, *M. nemestrina*, *Cebus albifrons*, and *Pan troglodytes*) are viewed as serious crop pests in certain areas. Primates are also killed for damage to noncrop commodities such as lambs, campsites, and automobiles (all harmed by *Papio ursinus*), and valuable exotic trees (damaged by *Cercopithecus mitis*).

Religious or traditional taboos forbid the killing or eating of certain primates in parts of Africa, Asia, and Madagascar. In the past, these prohibitions protected several species, including *Presbytis entellus*, *Macaca mulatta*, *Pan troglodytes*, and *Daubentonia madagascariensis*, from pest control. But recently food shortages and the declining influence of traditional beliefs have reduced the tolerance of local people to crop damage and removed much of the protection that these species have received.

In some cases, pest control has been used as an excuse for killing primates to obtain their skins (*Colobus guereza*) or their meat (*Pan troglodytes*, *Gorilla gorilla*). A few primates have been falsely persecuted as pests because of their association with destructive species (for example, *Colobus guereza* associating with *Cercopithecus mitis*), or because all monkeys in an area have been killed indiscriminately (*Cercopithecus diana* and *Colobus verus* in Sierra Leone). But in other instances (*Papio anubis*, *P. cynocephalus*, *P. ursinus*, *P. papio*, *M. mulatta*, *M. nemestrina*) substantial damage to crops has been reported, and these species are probably justifiably considered pestiferous.

Collection. Another form of human predation on primates is the collection of live animals for use as pets or as laborers in their countries of origin, and for sale as subjects in biomedical research, drug testing, and zoological exhibits, and as pets and entertainers. Since information about the numbers actually removed from wild populations is difficult to obtain, one must rely on reports of the numbers exported from source countries and imported into user countries.

Export figures are available from only a few countries (e.g., Peru, Colombia, Equatorial Guinea) and include data for only certain species in a few years. Import statistics classified by species are also not consistently available. They have been compiled for the United States for the years 1968–73 and 1977, and for the United Kingdom for 1965–75. Although it is known that other countries, such as Japan,

West Germany, and the Soviet Union, also import primates, I located no reports of the number imported into these countries.

It is evident that since 1968, 17 species of primates have lost more than 1,000 individuals per year to collectors, and another 13 species have been collected by the hundreds (100 to 999) per year in at least two years (Table 190). Thus 30 species, or 20% of the primates, have undergone severe or substantial collecting pressure. And a few of these species (such as *Macaca mulatta* and *Saimiri sciureus*) have lost 10,000 to 47,000 individuals per year to collectors (Appendix).

The number of primates imported into the United States per year decreased by more than 50% from 1968 to 1973 (Table 191). The largest decreases occurred among the Cebidae (66%), Callitrichidae (63%), and Cercopithecidae (29%), and were primarily due to export reductions by source countries, especially Colombia, Peru, and India. Imports into the United States continued to decline steadily after 1974, until by 1978 the total was 25% that of 1968 (Table 191). This decline is at least partly attributable to cessation of the import of primates for pets (see below).

The number of primates imported into the United States from 1968 through 1978 was 691,082 (Table 191). The total imported into the United Kingdom from 1965 through 1975 was 180,463, primarily in the families Cercopithecidae (82%) and Cebidae (13%) (Table 192).

The majority of the primates imported into the United States are used for medical and biological research and testing. For example, in 1972, about 49,000 individuals, or 63%, of the primates that entered the United States were used in biomedical research (Comm. Cons. Nonhuman Prim. 1975a). Recently this proportion has risen, owing to curtailment of the pet trade.

In the future, an estimated 33,912 primates will be needed annually for research programs and the testing of vaccines and drugs. More than 1,000 individuals per year are needed of six primate species: *Macaca mulatta*, *M. fascicularis*, *Aotus trivirgatus*, *Papio anubis*, *Cercopithecus aethiops*, and *Saimiri sciureus*. Smaller numbers are required of each of the following: *Macaca nemestrina*, other macaques, *Saguinus mystax*, *S. oedipus*, *S. fuscicollis*, *S. nigricollis*, *S. labiatus*, *Callithrix jacchus*, *Pan troglodytes*, *Hylobates lar*, *Erythrocebus patas*, *Ateles paniscus*, *Cebus* spp., and *Galago* spp. (Interagency Primate Steering Committee 1978b).

Breeding colonies have been established in the United States for six of these forms: *Macaca mulatta*, *M. fascicularis*, *Saguinus* spp., *Callithrix jacchus*, *Pan troglodytes*, and *Saimiri sciureus*. It was projected that by the year 1980 at least 9,155 primates (27% of the number needed) would be produced annually in these breeding centers (Interagency Primate Steering Committee 1978b). The remaining 24,000 needed animals will have to be obtained by expansion of domestic production, recycling of research subjects, or importation.

Primates are also collected for display in zoological exhibits. I did not gather data concerning the numbers of each species imported for this purpose, but the demand seems to be small in relation to that for research. Nevertheless, birth and survival rates of primates in zoos on the whole seem to be low and death rates relatively high. Thus to maintain their captive populations, zoos must continually import animals from the wild.

This collection can be especially harmful to species, such as the great apes, that are in great demand for zoo exhibits. For example, in 1972 more than 100 individuals each of gorillas, chimpanzees, and orang-utans were on exhibit in 36 major American zoos (Reuler, pers. comm. 1974). The planners of many zoos desire representatives of the apes and of endangered species such as *Macaca silenus*, *Papio sphinx*, and *Cercopithecus diana*. Populations of these species cannot sustain this additional depredation.

In the past, large numbers of primates were imported into the United States for the pet trade. New World monkeys, including *Ateles paniscus*, *Cebus apella*, *C. albifrons*, *C. capucinus*, *Lagothrix lagothricha*, *Saimiri sciureus*, and *Saguinus oedipus*, were especially popular. Macaques, gibbons, and chimpanzees were also purchased as pets. But in 1975 new regulations prohibited the importation of primates for any purpose other than scientific, educational, or exhibitional (U.S. Dept. Health, Education, and Welfare 1975). Thus the importing of primates into the United States for sale or personal use as pets or as entertainers became illegal. Primates are still captured and kept as pets in their native countries, and are exported to other nations for sale as pets, but no information is available on the numbers collected for these purposes.

Regardless of the use to which the captured animals are put, collection can have a profound effect on natural populations of primates. Wasteful capture methods, such as the killing of adult females to obtain clinging young, or of whole groups to obtain juveniles, and high postcapture mortality due to the stresses of handling, shipping, and adjusting to radically different environments, can result in many deaths (estimates range from 4 to 12) for each animal imported. The removal of reproducing females or large numbers of juveniles greatly reduces future reproduction and leads to long-term population declines. Thus the losses to the wild populations due to collection are often much larger than the import statistics indicate. Collection pressure has contributed significantly to the decline of several species, such as *Pan troglodytes*, *Pongo pygmaeus*, *Macaca mulatta*, *Hylobates lar*, *Aotus trivirgatus*, and *Saguinus oedipus*.

Constellations

The status of each animal population is determined by a constellation of ultimate and proximate causes: original geographic range, body size, density, and habitat are the ultimate determinants of population status, while habitat alteration and human predation are the proximate ones (Wolfheim 1976).

Ultimate causes are inherent in the evolutionary history of the species, dependent on geological, ecological, and genetic characteristics. These traits are specific to species or subspecies and determine the degree to which proximate causes affect a species. Proximate causes are those directly responsible for the expansion, reduction, extinction, merging, or isolation of populations. In the absence of human interference, spontaneous events, such as changes in riverbeds, soil, or climate, could lead to modifications in the ranges or habitats of primate groups. But human enterprises are now overshadowing these natural processes. Thus the proximate influences considered here are those resulting from human activities: alteration of the habitat on which the species depend, and exploitation of the animals themselves.

Figure 1 is a model of the way in which the major factors determine the status of a species. They can either help or hurt the chances of a species for survival. If an ultimate factor tends to make the species extinction prone, it may be imagined as a balloon lifting the left end of the balance, while if it tends to increase a species' chances for survival, it will be a weight pulling down the left end. Similarly, proximate factors may act as balloons or weights, buoying or depressing the survival potential of a species. Both balloons and weights can be large or small.

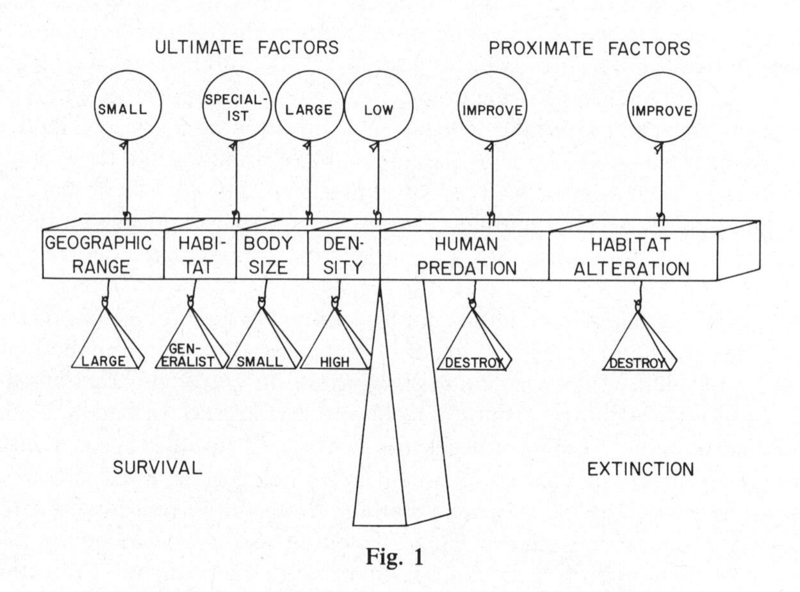

Fig. 1

The species for which the balance will tilt toward survival are those with large weights on the left and not on the right — that is, those with the largest geographic ranges, most generalized niches, smallest bodies, and highest natural densities, which are experiencing the lowest levels of (or most beneficial) habitat alteration and human predation. Species with the opposite characteristics will be the first ones to be threatened.

These factors are neither of equal importance nor independent. They interact in complex fashions, exaggerating or counteracting each other. Geographic range size is the most important one, hence its position at the end of the beam, where it exerts the most leverage. Range size affects the potency of all other factors. Combined with population density and habitat preferences, it dictates overall population size. It is often inversely correlated with degree of specialization, and together these two influence tolerance to habitat changes. Species with very small ranges cannot survive extensive habitat destruction or severe hunting or collection pressure regardless of their other attributes.

Within a phyletic line, body size is often directly correlated with specialization. It influences population density and hence vulnerability to habitat alteration. Body size determines the likelihood that a species will be hunted, the ease with which it is hunted, and the speed with which a population can recover lost members through reproduction.

Preferred habitat, degree of specialization, typical population density, group size, and home range area are all interrelated and influence vulnerability to both habitat alteration and human predation. For example, species that inhabit primary forest are more likely to suffer habitat disturbance than savanna forms, and are less adaptable to habitat changes. They often have smaller groups and are less accustomed to man; thus they are also more vulnerable than open country species to human predation.

Of the proximate causes of change in primate population status, habitat alteration is by far the more crucial. It affects localized species more severely than widely distributed ones, specialists more than generalists, and forms whose populations need large expanses of habitat (those with low densities, small groups, or large home ranges) more than ones that can survive in small patches. A species whose habitat has been extensively disturbed may be exterminated by a degree of hunting or collection that would have little impact on a species whose habitat is unimpaired. Forest clearing and road building open previously inaccessible areas to hunters, trappers, and settlers. Warfare not only destroys habitat but may also lead to food shortages and the breakdown of traditional conservation practices.

Vulnerability to human predation is determined by a species' geographic range, body size, reproductive rate, behavior, and habitat condition. For example, if a species has a restricted distribution, large body, or severely disturbed habitat, hunting or collection can have especially serious effects on its populations. The three types of human predation also interact with each other. For instance, species that are killed as crop pests are also likely to be hunted for meat and trapped for export, because their habits bring them closer to human settlements and make them easier to shoot and trap than primates that do not enter cultivated fields.

Thus, to represent more accurately the functioning of this model, arrows or pipelines should connect every item with every other item on the balance (Figure 1). Instead, one must keep these interactions in mind in considering the status of each species and predicting which ones will be the next to face sharp population declines.

Status Ratings

Each of the eight major factors implicated in the determination of species status has been rated in Table 193 according to its severity. Virtually every entry in this table should contain a question mark, since reports are not available for every variable for every species in each country of its range. Until more studies are completed, we must rely on the present information. Because of the lack of data, the ratings should be viewed as optimistic. When more information is available, some of the situations may be found to be worse. It is unlikely that they are better.

"Threatened" is defined here as "likely to face extinction if current trends continue." "Vulnerable" species are not threatened, but their situation suggests that they will be if conditions worsen. These designations do not coincide with those used by other organizations (see below), but are independent status ratings based on the evidence presented in the preceding accounts.

Figures 2 and 3 show the working of the balance model for two species with different constellations of factors influencing status. One is "threatened substantially" (*Pan troglodytes*), and the other is "threatened severely" (*Colobus satanas*).

A total of 51 species (34%) of the primates are rated as "threatened severely" (++++, Table 193). The greatest number of these (21 species) are Madagascan, and 9 others are Asian island forms. The remainder occur on mainlands but have restricted ranges and greatly disturbed habitats. These species should be the first to receive conservation action.

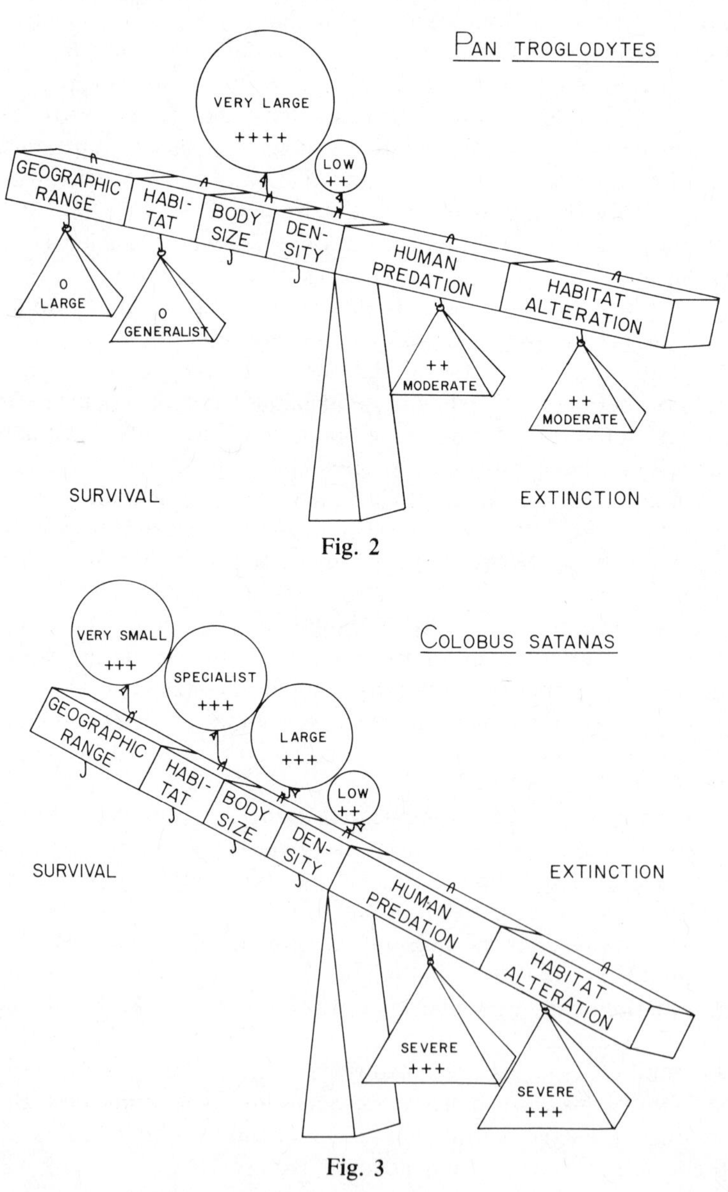

Fig. 2

Fig. 3

Eleven species are "threatened substantially" (+++, Table 193), and 25 are "threatened somewhat" (++, Table 193). Many of the primates in these two categories (such as *Ateles paniscus*, *Colobus badius*, and *Pan troglodytes*) are imperiled in certain parts of their ranges. Only because the species is wide ranging are these forms not considered "threatened severely." According to this analysis, a total of 87 species of primates (58% of the order) are "threatened" to some degree and warrant protection.

Another 34 species are "vulnerable" (+, Table 193). Only 30 species (20% of the primates) are confronting no serious danger and are probably safe at the present time (0, Table 193).

All of the Lemuroidea are menaced, and the Pongidae, Hylobatidae, and Cercopithecidae have high proportions of "threatened" forms (Table 194). A large number of the declining species are leaf eaters: the majority of both the African and Asian colobines are "threatened," as are four species of South American folivores (*Alouatta*).

Comparison with Other Rating Systems

My status ratings (Table 193) do not correspond to those of the United States Endangered Species Act, the Convention on International Trade in Endangered Species of Wild Fauna and Flora (C.I.T.E.S.), or the I.U.C.N. Red Data Book. Table 195 shows a comparison of these four ratings. The U.S. law considers 59 species or races as endangered and 11 as threatened. C.I.T.E.S. lists all or part of 55 species in Appendix I and all other primates in Appendix II. And the I.U.C.N. shows 22 species or races as endangered, 8 as rare, and 19 as vulnerable.

Most of the species classified as "threatened severely" or "substantially" in the present analysis are also listed as endangered or threatened under U.S. law (several are not, including *Colobus polykomos*, *Macaca radiata*, *M. sylvanus*, *Papio papio*, *Presbytis johnii*, *Theropithecus gelada*, *Callicebus personatus*, *Cebus capucinus*, *Saguinus mystax*). Fewer species are in Appendix I of C.I.T.E.S. and in the I.U.C.N.'s categories.

The I.U.C.N. lists several species as rare or vulnerable which do not appear to be "threatened severely" or "substantially" in the present analysis (e.g., *Colobus verus*, *Pygathrix nemaeus*, *Rhinopithecus roxellanae*, *Cacajao melanocephalus*, *Lagothrix lagothricha*, *Pan paniscus*). In all these cases, very little information is available, and the differences in ratings are probably due to my conservative decisions in the absence of supporting data.

Conservation Action

Ultimate Factors

No actions are recommended to influence the ultimate traits affecting the status of primate populations. These phenomena should change only through the slow process of evolution. Some managers have attempted to make such alterations—for example, by translocating populations outside their original range limits or by introducing artificial food sources to increase the carrying capacity of a habitat. These actions may be necessary as a last resort, as in the case of supplying shelter boxes for *Leontopithecus rosalia* (Coimbra-Filho 1978), but most biologists do not advocate manipulation of ultimate traits, because any perturbation of a natural ecosystem is likely to disturb its interdependent components and may result in disaster.

But an understanding of the ultimate determinants of status is necessary for the protection of primate populations. One cannot intelligently advocate management or conservation measures without knowing where a species lives, what type of habitat it requires, or how its populations function. Throughout this volume the many gaps in the body of knowledge about primate populations have been evident. For some species the limits of distribution and levels of abundance in various countries are unknown. Even scarcer are data indicating the actual area that a species inhabits and the amount of suitable habitat available to it. We do not have measurements, in length or weight, for large numbers of both sexes from several populations of each species, needed to assess differences in primate body sizes. Synecological research to reveal the critical components of certain habitats, differences between similar community types, and specific niche requirements would contribute greatly to understanding the degree of specialization of the various primates. Information concerning age structure of populations, ages of individuals at sexual maturity, gestation period, reproductive span, offspring per female per year, and mortality rates is needed to ascertain whether a population is increasing, decreasing, or remaining stable, and to predict future trends.

The study of these ultimate factors should be integrated with investigation of the proximate determinants of population status. For example, comparisons of the area of habitat originally available to a species with the area still intact could reveal the rate of habitat alteration. Data concerning the extent to which various primates are harmed or can adjust to certain disturbances, and methods of ameliorating their effects, would facilitate the formulation of recommendations concerning land use compatible with primate populations. The impact of all types of human predation could be more thoroughly appreciated if we knew the actual numbers killed or removed yearly in each segment of the range, their ages and sexes, the relation of size of groups and population densities to behavior and reproduction, and the rate of replacement of lost individuals.

Field research to increase our understanding of these matters is badly needed. Such work is often difficult, costly, time-consuming, and tedious, but is essential for sound conservation programs. Field biologists must initiate, persuade granting agencies to fund, and carry out the surveys to collect this information. Standardized methods of censusing and of determining and expressing population densities and home ranges would facilitate comparisons between studies. Modern equipment and techniques such as tranquilizer capture guns, radiotelemetry tracking, freeze-brand marking, electrophoretic analysis of blood samples for genetic traits, and dental impressions for population age structure data can be usefully employed along with traditional wildlife study methods.

Life histories are required for all species whose status is in question, especially for those being considered for management programs. Without such information, intelligent decisions regarding the impact of removal of animals for meat or for research are impossible. It is important that exploited populations be monitored to detect changes in their composition or status.

Studies of populations are also needed to indicate which species might be especially vulnerable and to determine the sizes of habitat reserves needed to protect viable populations (see Diamond 1975, 1976; Wilcox 1980). In deciding the smallest effective size for a reserve, several factors deserve consideration: (1) What is the minimum number of individuals necessary to ensure genetic variability and survival of a population? (see Franklin 1980) (2) How much area is necessary to protect a deme of this size? (3) What density can be achieved before self-regulating mechanisms prevent further population growth?

The most urgently needed field studies are those that will locate, document, and help conserve healthy populations of primates dependent on primary forest. Any large areas of forest that contain resident primates, and can be effectively protected from disturbance, should be set aside. Habitat reserves offer the only hope for survival of forest primates. Field biologists are the persons best qualified to find areas suitable as reserves and to collect the information necessary to persuade the public and governments of the value of reserves. Biologists ought to bring a special commitment to such studies, since we have a vested interest in preserving these scientifically valuable animals. If we do not establish reserves for wild forest primates, they will disappear forever.

Some species of primates are already found in habitat reserves. Ten species have been reported in more than 20 reserves each, and 20 species occur in 10 to 20 protected areas (Table 190). But the size, number of resident primates, amount of suitable habitat, quality of protection, and long-term security of these reserves vary greatly and in many cases are not known.

A total of 63 species occur in 3 or fewer protected areas, and 12 of these are found in no reserves (Table 190). Also, for 18 species, occurrence in habitat reserves was not reported. Thus 81 species (54% of the primates) are found in few, no, or an unknown number of reserves. It is imperative that the ranges of these species be carefully scrutinized for reserve sites and that protection be instigated soon.

Establishment and maintenance of habitat reserves is the highest priority for protecting primate populations. To date, only a few large reserves have been established in tropical forest areas such as Tai Forest National Park (N.P.), Ivory Coast; Wonga-Wongue N.P., Gabon; Salonga N.P., Zaire; Manu N.P., Peru; Corcovado National Biological Reserve, Costa Rica; and La Macarena N.P., Colombia, which is not secure (Struhsaker 1976b). Two more reserves (Korup N.P. and Douala-Edéa N.P.) are being established in Cameroon's forests (Gartlan 1976). But many others are needed throughout Africa, Madagascar, Asia, and South America if representatives of every primate species are to be protected. Some suggested forest conservation areas are listed for Central America (D'Arcy 1977), Ecuador (Gentry 1977), and Amazonia (Pires and Prance 1977).

Funds for these parks could be partly provided by visitors using them for tourism, education, and research. But local support in the form of national pride in unique animals and scenic beauty, bolstered by educational programs at every level, will be necessary if the reserves are to endure. Without such local interest, hunting and habitat disturbance inside reserves will continue and regulations will be difficult or impossible to enforce. In the past, wildlife protection has often collapsed in the face of warfare, economic pressure, or changes in political administration. Participation by local residents in reserve affairs, monetary compensation for lands, products, and activities made unavailable by reserves, or international protection for natural areas should be effective in some cases.

Proximate Factors

The most powerful threats to primate populations are direct consequences of human activities and thus can be mitigated by changes in our attitudes, goals, and behavior. The severely depressing processes of habitat alteration and human predation on nonhuman primates are dependent on economic, political, and social phenomena that are complex, interrelated, and difficult to reverse. For example, in many areas deforestation is a prerequisite for land development, with its attendant goals of cultivation, urbanization, industrialization, nutritional advances, improvement in living standards, and the civilization of a region. Natural vegetation and wild animals are seen as obstacles to progress to be eliminated as efficiently as possible.

The conservation of even a fraction of the primate fauna of the world is inextricably interwoven with these problems and cannot be achieved without taking them into consideration. Thus the following suggestions are generalized and are best viewed as possibilities only if they are implemented in a holistic manner accompanied by a widespread acceptance of a conservation ethic by the world's people.

Habitat alteration can only be slowed when the ever-increasing demand for food, land, and products is curtailed and accompanied by control of human population. If the disappearance of tropical rain forest is to slacken, intelligent forest management must reduce the rate of clearcut logging and the intensity of selective logging, switch to reforestation with assorted native tree species, and discontinue the use of herbicides in forest habitats. Large companies from industrialized countries (such as the United States, France, Canada, Germany, and Japan) do much of the extensive logging of tropical forests, and they should aid in efforts to conserve a major portion of these ecosystems. This would greatly improve their public relations.

A decrease in the expansion of agriculture can only be achieved if the demand for food decreases or if food sources increase. Even if these were to occur, the planting of cash crops grown largely for export would probably result in continued clearing. Reduction in consumption of tropical woods, rubber, bananas, coffee, tea, cocoa, and other products grown in tropical forest areas might be effective in removing some of the economic incentives for further deforestation in remote areas. Forest plantations on already degraded lands, and increased production on existing farms, would relieve presssure on remaining wildlands (Budowski 1977).

Tropical forests are not only important as wildlife habitat but are also valuable for watershed protection, soil conservation, temperature control, scenic beauty, genetic reservoirs, and as complex ecosystems whose properties and functioning are not completely understood (Poore 1976). Many governments (for example, in Central America) are so determined to promote development that they encourage habitat alteration by repossessing or taxing undeveloped land (Budowski 1977, D'Arcy 1977). These governments must become convinced of the value of their natural areas, so that they will require that certain regions remain covered by native vegetation. For example, Guinean law prohibits the cutting of gallery forest within 5 meters on each side of a river, hence protecting the riparian habitats of resident primates (Dunnett et al. 1970). Local conservation groups working to save natural habitats in their countries need more support, cooperation, and encouragement (Budowski 1977).

In the long run, habitat protection is most crucial for primate populations. No degree of adaptability or regulation of trade or hunting can save a species if its habitat has been devastated by bombing, paved in concrete, or planted in bananas or cotton. Conversely, a species is unlikely to be extinguished by hunting or collection if its original habitat is left intact. But if a large proportion of the natural habitat of a species has been destroyed, reduction in the rates of human predation may be necessary to save that primate from extinction.

Many countries legally prohibit the hunting of primates but are unable to enforce these laws in remote rural areas. In such cases, control of hunting may be achievable by the provision of alternative protein sources, along with educational programs to teach the people how to use and enjoy this new food and convince them to eat it rather than the native wildlife. More meaningful fines, consistent prosecution of poachers, and gun and ammunition control could also contribute to a reduction in hunting pressure. Methods to reduce the number of primates killed as vermin include capturing pest animals for research (Rowell 1968) or modifying their behavior to eliminate crop raiding (Strum, pers. comm. 1980).

Restrictions of trade through laws regulating export from source countries (such as Colombia and Peru) and import into recipient countries (such as the United States and the United Kingdom) and international agreements such as the African Convention and C.I.T.E.S. have reduced the commerce in and collection of certain primates such as *Saguinus oedipus* and *Pongo pygmaeus*. But exemptions from these laws, such as the legal sale of similar species, crop-raiding animals, captive-born specimens, or animals captured in certain countries, have led to misidentification, transshipment, falsification of papers, bribery, smuggling, and other activities to maintain the flow of the restricted animals or their products. Such loopholes can be closed by enacting consistent and effective laws and enforcing them in all countries involved.

Potential recipients of imported primates (including researchers, pharmaceutical companies, zoos, and pet dealers) should refrain from purchasing any threatened species, large numbers of any primate, and any dependent infants which may have been obtained by killing the mother. It is important that users reduce imports by substituting domestic animals for primates, establish self-sustaining breeding colonies to produce primates, and supervise any necessary collection to reduce deaths during capture and transport.

The use of captive-bred rather than imported primates would benefit both laboratory research and natural populations of primates. Primates born in a laboratory are of known and often controlled age, parentage, nutrition, and medical history, whereas wild individuals arrive with unknown backgrounds and may carry pathogens and parasites. Captive breeding can provide large numbers of animals with specific characteristics to allow for the precise evaluation and accurate analysis of responses to drugs, procedures, or tests. Although they are expensive to produce and maintain primarily because of the labor involved, captive-born primates must be increasingly used in research instead of wild ones.

Wild-caught primates also usually make poor pets. The pet trade has been effectively curtailed in the United States, but it continues in other countries. Prospective pet owners should be informed that primates are difficult to care for, train, tame, and breed. They are often unresponsive, unruly, or morose, and carry diseases such as hepatitis, tuberculosis, and poliomyelitis, which can be transmitted to human beings. Many are short-lived in captivity, or become large, unpredictable, and dangerous as adults. Every purchase of a primate pet creates a monetary incentive for further capture and trade with their concomitant high mortality.

Most of the hardy species of primates that do survive as pets can also thrive in some natural habitats in the temperate zone. Any captive animal is likely to escape, and in some cases escapees have been able to establish a breeding population disruptive to the native fauna and flora of a region (de Vos et al. 1956, Petrides 1968). Feral macaques, baboons, and titi monkeys have been reported living independently in the southern United States (e.g., see Jones et al. 1970, Flor 1973) and could become pests. For all of these reasons, primates are not appropriate as pets, and collection and trade for this purpose can be discouraged by public disapproval and laws including fines and confiscation.

Zoological organizations such as the New York Zoological Society, Smithsonian National Zoological Park, and Freunde des Kölner Zoo have made valuable contributions to primate conservation by their funding of field studies, conferences, and publications. These activities are crucial to the conservation effort and these societies should be supported by donations, membership, and participation.

But some zoos continue to import significant numbers of rare primates, and it is hoped that this will cease. Because so much of the attractiveness of primates is due to their behavior, zoo visitors are better educated and entertained by a few interacting groups of common species such as macaques and baboons than by a great variety of singly housed endangered primates.

Zoos should consolidate their collections and attempt to increase breeding to protect wild populations from further collection. But it is not advisable for them to spend large amounts of money to try to save endangered species of primates by breeding them in captivity for reintroduction to natural habitats. This endeavor has a low probability of success for two reasons. First, it is difficult to breed rare primates, because they are specialized, often delicate animals, and usually do not reproduce well in captivity. Sporadic breeding success has been achieved with some threatened species, such as *Macaca silenus*, *Pygathrix nemaeus*, *Nasalis larvatus*, *Callimico goeldii*, *Pongo pygmaeus*, and *Gorilla gorilla* (M. Jones, pers. comm. 1974). But the captive-born animals have rarely reproduced, thus importation of wild primates must continue to supply breeders. For example, second-generation reproduction of *Pongo pygmaeus* in zoos has been an "almost complete failure" (Conway 1980). And despite intensive efforts at propagation, the captive population of *Leontopithecus rosalia* declined by 30% between 1968 and 1973 (Coimbra-Filho and Mittermeier 1977). This population increased slightly in 1973–75, but the survivorship of both young and adults remained low (Kleiman and Jones 1978).

Second, the feasibility of reintroducing captive-bred primates into their native habitats has not been tested. Because so many of their behavior patterns are acquired through early experience, it is likely that primates raised in captivity could not survive in their natural environments. Even the reintroduction of orang-utans and chimpanzees born in the wild (but reared by human beings) entails a laborious training period, meets with a high rate of failure, and involves the risk of transmitting human diseases to wild ape populations.

Captive breeding of primates is certainly necessary to produce primates for research and display. It can protect species by reducing the demand for wild individuals. But it is unlikely to be able to rescue a primate species that is close to extinction. If zoos have extra money they wish to spend on conservation, they should fund field studies, sponsor wildlife education, or organize expeditions to observe wild animals in nature. They could even lease patches of forest (as do logging companies) and arrange for protection and visits by people wanting to see a "zoo" without bars, windows, or moats: the natural habitat.

Conclusion

The survival of 58% of the species of living primates is being jeopardized, primarily by deforestation. Many other primate species are in vulnerable positions and will probably be threatened soon. Habitat reserves offer the only reasonable hope for the survival of most primates. Our concern, effort, and money should be directed toward establishing more reserves, obtaining a better understanding of primate populations in nature, and educating more people about the beauty, value, irreplaceability, and need for conservation of tropical wildlife.

Table 190. Summary of determinants of population status for all species of primates

Species	Geographic Range (×1,000 km²)	No. of Countries[a]	Max. Body Wt. (kg)[b]	Abundance[a,d] A	R	D	Preferred Habitats[c]	Hab. Alt.[a]	Hunting[a]	Pest Cont.[a]	Collect. (/year)[f]	No. of Reserves
Lorisidae												
Arctocebus calabarensis	960	6	0.5	2	1	?	F	?[g]	1	?	—[h]	3
Euoticus elegantulus	740	7	?	?	2	?	F,R	?	?	?	—	2
Euoticus inustus	110	2	?	?	1	?	F	?	?	?	—	?
Galago alleni	850	6	?	1	1	?	WF,B,Sw	?	?	?	—	1
Galago crassicaudatus	3,810	13	1.2	3	?	?	S,B,Wo,F,R	1	2	?	10	7
Galago demidovii	5,080	26	?	7	?	?	F,B,R,E	?	1	?	10	7
Galago senegalensis	10,650	32	0.3	8	?	?	S,Wo,F,B	1	2	?	100	11
Loris tardigradus	740	2	0.3	1	?	?	F,Wo,Sw	?	?	?	—	1
Nycticebus coucang	4,030	12	1.7	1	?	2	F	?	1	?	10	3
Nycticebus pygmaeus	450	2	?	?	?	?	?	?	?	?	—	0
Perodicticus potto	3,280	19	1.4	3	2	1	F,Sc,R,Wo	2	2	?	10	3
Tarsiidae												
Tarsius bancanus	440	3	0.2	?	?	?	F,E	?	?	?	—	4
Tarsius spectrum	60	1	0.2	?	?	?	?	?	?	?	—	1
Tarsius syrichta	50	1	0.2	?	?	?	F	1	—	?	—	0
Cheirogaleidae												
Allocebus trichotis	tiny	1	?	?	1	?	WF	1	?	?	—	0
Cheirogaleus major	80	1	?	?	?	?	WF,DF	1	?	?	—	5
Cheirogaleus medius	80	1	?	?	1	1	DF	1	?	?	—	7
Microcebus coquereli	40	1	0.06	1	1	1	DF	1	?	?	—	3
Microcebus murinus	330	1	0.06	1	?	?	WF,B	1	?	?	—	15
Phaner furcifer	60	1	?	?	?	1	DF,B,R	1	?	?	—	5
Lemuridae												
Hapalemur griseus	70	1	2.6	1	1	1	WF,Sw	1	1	?	—	7
Hapalemur simus	tiny	1	2.6	?	1	1	?	1	?	?	—	0
Lemur catta	100	1	2.4	?	1	1	WF,DF,B,R	1	1	?	—	6

Lemur coronatus	7	1	?	?	1	1	DF,S,B	1	1	1	—	1
Lemur fulvus	200	1	?	1	1	1	PF	1	1	?	—	10
Lemur macaco	10	1	?	?	1	1	WF	1	—	1	—	2
Lemur mongoz	20	1	?	1	1	1	DF,R	1	?	?	—	1
Lemur rubriventer	40	1	?	?	1	1	WF	1	?	?	—	0
Lepilemur Group	150	1	?	?	1	1	WF,DF,B	1	1	?	—	11
Varecia variegata	50	1	?	?	1	1	WF	1	1	?	?	4
Indriidae												
Avahi laniger	90	1	?	?	1	1	WF,DF	1	1	?	—	3
Indri indri	30	1	?	?	1	1	WF	1	1	?	—	3
Propithecus diadema	70	1	?	?	1	1	WF	1	1	?	—	2
Propithecus verreauxi	140	1	?	?	1	1	WF,DF	1	1	?	—	7
Daubentoniidae												
Daubentonia madagascariensis	?	1	?	?	1	1	WF	1	1	1	—	1
Callitrichidae												
Callimico goeldii	1,070	5	0.5	?	3	?	F	1	1	?	10	1
Callithrix argentata	1,280	2	0.4	2	1	?	F	1	—	?	10	1
Callithrix humeralifer	350	1	0.4	?	1	?	?	?	—	?	—	0
Callithrix jacchus	2,590	1	0.4	1	1	?	F,Wo	1	?	?	100	13
Cebuella pygmaea	1,470	5	?	3	1	?	F,Wo,E,R	3	—	?	1,000	4
Leontopithecus rosalia	50	1	0.6	?	1	1	PWF,R	1	?	?	10	3
Saguinus bicolor	160	1	0.4	1	?	?	SwF,E	?	?	?	—	0
Saguinus fuscicollis	1,820	5	0.4	3	2	?	F,E,FlF	1	—	?	100	5
Saguinus imperator	250	3	0.4	1	1	?	F	?	1	?	—	1
Saguinus inustus	480	2	0.4	?	?	?	WF	?	?	?	?	?
Saguinus labiatus	250	3	0.4	1	?	?	F	?	1	?	10	0
Saguinus leucopus	40	1	0.4	?	?	?	F,E,R	1	—	1	—	0
Saguinus midas	1,570	4	0.4	3	1	?	F,E,R	1	1	?	10	7
Saguinus mystax	990	2	0.4	1	1	?	PF,FlF	1	1	?	1,000	2
Saguinus nigricollis	490	4	0.4	3	1	?	F,E,R	?	?	?	1,000	0
Saguinus oedipus	140	3	0.6	1	1	?	F,E,R	2	1	?	1,000	0

Table 190 (continued)

Species	Geographic Range (×1,000 km^2)	No. of Countries[a]	Max. Body Wt. (kg)[b]	Abundance[a,d] A	R	D	Preferred Habitats[c]	Hab. Alt.[a]	Hunting[a]	Pest Cont.[a]	Collect. (/year)[f]	No. of Reserves
Cebidae												
Alouatta belzebul	1,490	1	7	?	?	?	PF	1	1	?	—	?
Alouatta caraya	2,570	4	7	?	?	?	WF,DF,R	1	2	?	10	2
Alouatta fusca	850	3	7	1	1	?	F,R	1	1	?	—	9
Alouatta palliata	720	10	7	?	1	4	F,R	5	4	?	10	8
Alouatta pigra	270	3	7	?	?	1	PWF	2	1	?	—	1
Alouatta seniculus	5,570	10	7	5	2	1	WF,R	4	7	—	10	26
Aotus trivirgatus	6,840	12	1	2	1	2	F	4	3	?	1,000	8
Ateles paniscus	6,570	17	6.8	6	5	9	PWF	8	10	?	1,000	19
Brachyteles arachnoides	500	1	9.5	?	1	1	PWF	1	1	?	—	17
Cacajao calvus	430	2	0.5	?	1	1	FlF	1	1	?	10	0
Cacajao melanocephalus	640	3	0.5	1	2	?	WF	?	2	?	—	1
Callicebus moloch	4,940	6	0.7	1	1	1	WF,Th,R	2	3	?	10	4
Callicebus personatus	620	1	?	?	1	?	F	1	1	?	—	8
Callicebus torquatus	1,790	4	?	?	2	?	F	1	2	?	—	1
Cebus albifrons	3,660	7	3	?	2	2	PWF	2	5	1	1,000	10
Cebus apella	11,730	11	3	3	2	3	WF,R	3	6	5	1,000	19
Cebus capucinus	450	6	3	2	1	3	WF,Sw	3	?	2	1,000	5
Cebus nigrivittatus	2,020	5	3	?	2	?	PWF	?	2	?	10	10
Chiropotes albinasus	600	1	3	?	1	1	FlF	1	1	?	—	1
Chiropotes satanas	2,090	5	3	?	2	1	WF	2	3	?	—	5
Lagothrix flavicauda	40	1	?	?	1	1	?	1	1	?	?	0
Lagothrix lagothricha	3,580	5	10	?	2	2	PWF	?	3	1	1,000	6
Pithecia monachus	2,350	5	1.6	1	1	1	PWF	?	2	?	100	3

Pithecia pithecia	1,680	5	1.6	?	2	?	WF	?	3	?	10	7
Saimiri sciureus	5,440	11	1.1	6	3	2	WF,R	5	1	?	1,000	14
Cercopithecidae												
Allenopithecus nigroviridis	440	2	?	?	?	?	SwF	?	?	?	—	?
Cercocebus albigena	1,590	7	9[c]	4	1	3	PWF	3	2	2	—	3
Cercocebus aterrimus	1,080	2	?	1	?	?	F	—	1	1	—	?
Cercocebus galeritus	890	7	?	1	1	1	WF	3	2	2	—	2
Cercocebus torquatus	1,400	12	9[c]	5	?	2	PWF	4	5	4	10	10
Cercopithecus aethiops	12,300	36	7[c]	12	5	2	S,Wo,E,R	7	9	11	1,000	47
Cercopithecus ascanius	2,710	9	6.3	3	?	1	WF,Wo	3	3	4	—	4
Cercopithecus cephus	710	7	?	?	?	3	WF	1	3	1	100	3
Cercopithecus diana	330	5	?	?	4	?	PF	2	3	—	10	6
Cercopithecus erythrogaster	40	1	?	?	?	?	F	?	?	?	—	?
Cercopithecus erythrotis	90	3	?	?	1	2	WF	2	2	1	—	4
Cercopithecus hamlyni	420	3	?	?	1	?	F	?	?	?	—	?
Cercopithecus l'hoesti	430	7	?	1	?	1	WF	1	2	1	—	?
Cercopithecus mitis	3,530	16	?	6	4	2	WF	4	7	3	10	40
Cercopithecus mona	2,810	15	?	7	2	?	WF	5	4	5	10	20
Cercopithecus neglectus	2,570	10	11[c]	3	3	?	WF	3	2	—	10	1
Cercopithecus nictitans	1,640	11	?	4	?	4	PWF	4	4	1	10	8
Cercopithecus petaurista	630	9	?	2	2	1	F	3	3	2	—	12
Cercopithecus pogonias	890	6	?	?	?	3	PWF	2	2	—	—	5
Colobus angolensis	1,910	8	?	1	3	?	PWF	1	2	—	—	2
Colobus badius	1,890	18	?	8	6	5	PWF	6	6	—	—	13
Colobus guereza	2,340	13	12[c]	5	1	4	F	5	5	2	10	34
Colobus polykomos	940	10	?	3	3	2	F	3	5	—	10	20

Table 190 (continued)

Species	Geographic Range ×1,000 km²)	No. of Countries[a]	Max. Body Wt. (kg)[b]	Abundance[a,d] A	R	D	Preferred Habitats[c]	Hab. Alt.[a]	Hunting[a]	Pest Cont.[a]	Collect. (/year)[f]	No. of Reserves
Cercopithecidae (cont.)												
Colobus satanas	300	4	?	1	1	2	PWF	2	2	—	—	2
Colobus verus	510	5	4.4	2	2	?	WF	2	2	1	—	6
Erythrocebus patas	6,180	24	12.6	11	6	1	S,Wo	2	7	9	1,000	29
Macaca arctoides	2,950	9	?	?	4	2	F	?	2	2	1,000	?
Macaca assamensis	1,660	9	12.7	1	?	?	F	?	1	?	—	?
Macaca cyclopis	30	1	?	1	?	?	?	1	1	1	100	?
Macaca fascicularis	2,670	11	8.2	3	?	5	FE,Sw,R,CF	4	3	3	1,000	20
Macaca fuscata	180	1	18	?	?	1	F	1	1	1	?	17
Macaca mulatta	6,220	12	10.8	4	1	2	F,T,E	6	?	4	1,000	7
Macaca nemestrina	3,060	10	14.5	1	?	4	WF	3	3	3	100	6
Macaca nigra	90	1	11.2	?	?	?	F	?	?	1	10	1
Macaca radiata	660	1	8.8	1	?	1	F,T,E	1	?	—	—	5
Macaca silenus	40	1	6.7	?	1	1	WF	1	1	—	?	4
Macaca sinica	50	1	8.3	1	1	1	WF,R,Sc	1	—	1	—	5
Macaca sylvanus	20	2	11.1	1	1	2	F	2	1	1	10	2
Miopithecus talapoin	1,110	7	1.3	2	1	?	WF	?	4	2	100	2
Nasalis larvatus	740	3	23.6	1	?	?	RF,CSw	3	?	?	—	7
Papio anubis	7,570	23	30	8	1	5	S,Wo,F	5	5	8	1,000	48
Papio cynocephalus	3,170	9	?	2	?	1	S,Wo,F	1	?	3	100	5
Papio hamadryas	1,300	7	?	4	?	?	De,G	2	1	2	10	3
Papio leucophaeus	150	3	?	1	1	2	WF	1	1	1	10	3
Papio papio	380	8	?	2	?	1	S,Wo,F	2	2	3	100	1
Papio sphinx	410	4	19.5	?	?	2	F	3	3	3	10	3
Papio ursinus	2,890	10	?	4	?	3	Wo,S,G	—	?	5	100	21
Presbytis aygula-melalophus Group	1,220	5	7	2	1	2	WF	2	1	1	—	5
Presbytis cristata	2,000	7	13.6	1	1	1	CF,R,SwF	2	1	1	—	2
Presbytis entellus	3,700	8	20.8	3	1	1	F,R,T	3	3	3	—	16
Presbytis francoisi	100	3	13.6	?	?	?	?	?	?	?	—	?
Presbytis frontata	400	2	7	?	?	?	F	?	?	?	—	?

Presbytis geei	6	2	13.6	?	2	?	F	?	?	?	—	2
Presbytis johnii	20	1	13.1	1	1	1	WF	1	1	1	—	5
Presbytis obscura	220	3	13.6	2	?	1	WF	1	2	?	—	1
Presbytis phayrei	1,200	7	13.6	?	?	?	F	1	1	?	—	?
Presbytis pileata	320	3	13.6	2	?	?	WF	2	1	?	—	3
Presbytis potenziani	8	1	13.6	1	1	1	PF,Sw	?	1	—	—	1
Presbytis rubicunda	680	2	7	2	1	?	F,E	?	?	?	—	?
Presbytis senex	40	1	13.1	1	?	1	WF,DF,R	1	?	?	—	2
Pygathrix nemaeus	540	3	?	1	1	1	F	?	1	?	—	?
Rhinopithecus avunculus	100	1	?	?	?	?	F	?	?	?	—	?
Rhinopithecus roxellanae	1,420	1	?	?	1	?	F	?	1	?	—	1
Simias concolor	8	1	7.1	?	1	1	PF	?	1	—	—	1
Theropithecus gelada	210	1	20.5	1	?	?	G	1	1	1	100	1
Hylobatidae												
Hylobates concolor	660	4	7.9	?	?	1	F	?	1	?	10	0
Hylobates hoolock	820	4	7.9	?	?	2	PF	2	1	?	—	6
Hylobates klossii	8	1	7.9	1	?	1	PF	?	1	?	—	1
Hylobates lar	1,930	6	7.9	?	?	3	WF,DF	4	3	—	10	28
Hylobates pileatus	180	4	7.9	?	4	4	WF,DF	2	1	?	—	6
Hylobates syndactylus	180	2	12.7	2	?	2	PWF	2	—	—	10	6
Pongidae												
Gorilla gorilla	700	11	180	3	8	9	WF,Wo,R	6	7	5	?	14
Pan paniscus	390	1	?	1	1	1	WF,Sw	—	1	?	—	?
Pan troglodytes	2,500	23	48.9	5	9	6	F,Wo,S,R	6	9	9	100	37
Pongo pygmaeus	340	2	69	?	2	2	PWF	2	2	2	10	22

[a] Number of countries.
[b] Napier and Napier 1967.
[c] Rowell, pers. comm. 1978.
[d] A = Abundant. R = Rare. D = Declining.
[e] Habitats: B = Bush. C = Coastal. D = Dry. De = Desert. E = Edge. F = Forest. Fl = Flooded. G = Grassland. P = Primary. R = Riparian. S = Savanna. Sc = Scrub. Sw = Swamp. T = Towns. Th = Thicket. W = Wet. Wo = Woodland.
[f] Collection: 10 = tens (0–99). 100 = hundreds (100–999). 1,000 = thousands (1,000–999,999)/year.
[g] ? = Unknown.
[h] — = Not a factor.

Table 191. Number of primates imported into the United States, 1968–78[a]

Family	Number of Individuals					
	1968	1969	1970	1971	1972	1973
Lorisidae	381	269	154	438	246	185
Tarsiidae	1	0	0	0	0	1
Cheirogaleidae	4	8	4	2	0	0
Lemuridae	3	67	30	15	1	6
Indriidae	4	7	0	0	0	0
Callitrichidae	7,688	6,537	4,882	5,552	5,746	2,861
Cebidae	62,583	66,335	41,544	43,675	39,234	20,799
Cercopithecidae	42,620	35,304	31,467	29,773	31,943	30,188
Hylobatidae	161	137	102	139	214	160
Pongidae	269	294	191	211	236	140
Total	113,714	108,958	78,374	79,805	77,620	54,340
Year	1974[b]	1975[b]	1976[b]	1977[c]	1978[b]	Grand total
Total	46,581	40,814	33,539	28,559	28,778	691,082

[a] 1968–73 figures from Appendix.
[b] Muckenhirn 1975, 1976, 1977, 1979. Statistics from U.S. Department of Commerce reports; not classified by species or family.
[c] Muckenhirn and Cohen 1978. Statistics from U.S. Department of Commerce reports; not classified by species or family.

Table 192. Number of primates imported into the United Kingdom, 1965–75

Family	Number of Individuals
Cheirogaleidae	36
Lemuridae	34
Tarsiidae	64
Callitrichidae	8,560
Cebidae	24,035
Cercopithecidae	147,366
Hylobatidae	101
Pongidae	267
Total	180,463

SOURCE: Burton 1978.

Table 193. Significance of major determinants of population status and status ratings for each species of primate (see Key for criteria)

Species	Small Geographic Range	Large Body Size	Rare	Special-ized	Habitat Alter-ation	Hunting	Pest Control	Collec-tion	Status Rating
Lorisidae									
Arctocebus calabarensis	++	0	+	+	?	+	?	0	+
Euoticus elegantulus	++	0	++	+	?	?	?	0	+
Euoticus inustus	+++	0	+++	+	?	?	?	0	+
Galago alleni	++	0	+	++	?	?	?	0	+
Galago crassicaudatus	0	+	?	0	+	+	?	+	0
Galago demidovii	0	0	?	0	?	+	?	+	0
Galago senegalensis	0	0	?	0	+	+	?	++	0
Loris tardigradus	++	0	?	0	?	?	?	0	+
Nycticebus coucang	0	+	?	+	?	+	?	+	0
Nycticebus pygmaeus	+++	0	?	?	?	?	?	0	+
Perodicticus potto	0	+	+	0	+	+	?	+	0
Tarsiidae									
Tarsius bancanus	+++	0	?	+	?	?	?	0	+
Tarsius spectrum	++++	0	?	?	?	?	?	0	+
Tarsius syrichta	++++	0	?	+	+++	0	?	0	++++
Cheirogaleidae									
Allocebus trichotis	++++	0	++++	++	+++	?	?	0	++++
Cheirogaleus major	++++	0	?	+	+++	?	?	0	++++
Cheirogaleus medius	++++	0	++++	++	+++	?	?	0	++++
Microcebus coquereli	++++	0	++++	++	+++	?	?	0	++++
Microcebus murinus	+++	0	?	+	+++	?	?	0	++++
Phaner furcifer	++++	+	?	+	+++	?	?	0	++++
Lemuridae									
Hapalemur griseus	++++	+	++++	++	+++	+++	?	0	++++
Hapalemur simus	++++	+	++++	?	+++	?	?	0	++++

Table 193 (continued)

Species	Small Geographic Range	Large Body Size	Rare	Special-ized	Habitat Alter-ation	Hunting	Pest Control	Collec-tion	Status Rating
Lemuridae (cont.)									
Lemur catta	+++	+	++++	0	+++	+++	?	0	++++
Lemur coronatus	++++	?	++++	0	+++	+++	+++	0	++++
Lemur fulvus	+++	?	++++	+++	+++	+++	?	0	++++
Lemur macaco	++++	?	++++	++	+++	0	+++	0	++++
Lemur mongoz	++++	?	++++	++	+++	?	?	0	++++
Lemur rubriventer	++++	?	++++	++	+++	?	?	0	++++
Lepilemur Group	+++	+	++++	0	+++	+++	?	0	++++
Varecia variegata	++++	+	++++	++	+++	+++	?	?	++++
Indriidae									
Avahi laniger	++++	?	++++	+	+++	+++	?	0	++++
Indri indri	++++	+	++++	++	+++	+++	?	0	++++
Propithecus diadema	++++	?	++++	++	+++	+++	?	0	++++
Propithecus verreauxi	+++	?	++++	+	+++	+++	?	0	++++
Daubentoniidae									
Daubentonia madagascariensis	++++	0	++++	++	+++	+++	+++	0	++++
Callitrichidae									
Callimico goeldii	+	0	+++	+	+	+	?	+	0
Callithrix argentata	+	0	+++	+	+++	0	?	+	0
Callithrix humeralifer	+++	0	++++	?	?	0	?	0	+
Callithrix jacchus	0	0	++++	0	+++	?	?	++	++
Cebuella pygmaea	+	0	+	0	+++	0	?	+++	0
Leontopithecus rosalia	++++	+	++++	+++	+++	?	?	+	++++
Saguinus bicolor	+++	0	?	++	?	?	?	0	+

Saguinus fuscicollis	+	0	++	++	+	0	?	++	0
Saguinus imperator	+++	0	++	+	?	++	?	0	+
Saguinus inustus	+++	0	?	++	?	?	?	?	+
Saguinus labiatus	+++	0	?	+	?	++	?	+	+
Saguinus leucopus	++++	0	?	+	+++	0	+++	0	++++
Saguinus midas	+	0	++	+	++	++	?	+	0
Saguinus mystax	++	+	+++	+++	+++	+++	?	+++	+++
Saguinus nigricollis	+++	0	++	+	?	?	?	+++	++
Saguinus oedipus	+++	+	++	+	+++	++	?	+++	++++
Cebidae									
Alouatta belzebul	+	+++	?	++	+++	+++	?	0	++
Alouatta caraya	0	+++	?	+	++	+++	?	+	0
Alouatta fusca	++	+++	++	+	++	++	?	0	++
Alouatta palliata	++	+++	+	+	+++	++	?	+	+++
Alouatta pigra	+++	+++	?	+++	+++	++	?	0	++++
Alouatta seniculus	0	+++	+	++	++	+++	0	+	0
Aotus trivirgatus	0	0	+	+	++	++	?	+++	0
Ateles paniscus	0	+++	++	+++	++	+++	?	+++	++
Brachyteles arachnoides	++	+++	++++	+++	+++	+++	?	0	++++
Cacajao calvus	+++	0	++	++++	+++	+++	?	+	++++
Cacajao melanocephalus	++	0	+++	++	?	+++	?	0	+
Callicebus moloch	0	0	+	++	++	+++	?	+	+
Callicebus personatus	++	0	++++	+	+++	+++	?	0	++++
Callicebus torquatus	+	0	+++	+	++	+++	?	0	+
Cebus albifrons	0	0	+	+++	+	+++	+	+++	0
Cebus apella	0	0	+	++	++	+++	++	+++	+
Cebus capucinus	+++	0	+	++	+++	?	++	+++	++++
Cebus nigrivittatus	0	0	++	+++	?	++	?	+	0
Chiropotes albinasus	++	0	++++	++++	+++	+++	?	0	++++
Chiropotes satanas	0	0	++	++	++	+++	?	0	+
Lagothrix flavicauda	++++	+++	++++	?	+++	+++	?	?	++++

Table 193 (continued)

Species	Small Geographic Range	Large Body Size	Rare	Special-ized	Habitat Alter-ation	Hunting	Pest Control	Collec-tion	Status Rating
Cebidae (cont.)									
Lagothrix lagothricha	0	+++	++	+++	?	+++	+	+++	+
Pithecia monachus	0	0	+	+++	?	++	?	++	0
Pithecia pithecia	+	0	++	++	?	+++	?	+	0
Saimiri sciureus	0	0	++	+	++	+	?	+++	0
Cercopithecidae									
Allenopithecus nigroviridis	+++	++	?	++++	?	?	?	0	+
Cercocebus albigena	+	++	+	+++	++	++	++	0	++
Cercocebus aterrimus	+	++	?	+	0	+++	+++	0	+
Cercocebus galeritus	++	++	+	++	++	++	++	0	++
Cercocebus torquatus	+	++	?	+++	++	++	++	+	++
Cercopithecus aethiops	0	++	+	0	+	++	++	+++	0
Cercopithecus ascanius	0	++	?	+	++	++	++	0	0
Cercopithecus cephus	++	++	?	++	+	+	+	++	+
Cercopithecus diana	+++	++	+++	++	+++	+++	0	+	++++
Cercopithecus erythrogaster	++++	++	?	+	?	?	?	0	++++
Cercopithecus erythrotis	++++	++	++	++	+++	+++	++	0	++++
Cercopithecus hamlyni	+++	++	++	+	?	?	?	0	+
Cercopithecus l'hoesti	+++	+++	?	++	+	+	+	0	++
Cercopithecus mitis	0	+++	++	++	++	++	+	+	+
Cercopithecus mona	0	++	+	++	++	++	++	+	0

Cercopithecus neglectus	0	+++	++	++	++	+	0	+	+
Cercopithecus nictitans	+	++	?	+++	++	++	+	+	+
Cercopithecus petaurista	++	++	+	+	++	++	+	0	++
Cercopithecus pogonias	++	++	?	+++	++	++	0	0	++
Colobus angolensis	+	+++	++	+++	+	++	0	0	++
Colobus badius	+	+++	++	+++	++	++	0	0	++
Colobus guereza	0	+++	+	+	++	++	+	+	+
Colobus polykomos	++	+++	++	+	++	+++	0	+	+++
Colobus satanas	+++	+++	++	+++	+++	+++	0	0	++++
Colobus verus	++	0	++	++	++	++	+	0	++
Erythrocebus patas	0	+++	++	+	+	++	++	+++	0
Macaca arctoides	0	++	++	+	?	+	+	+++	+
Macaca assamensis	+	+++	?	+	?	+	?	0	+
Macaca cyclopis	++++	++	?	?	+++	+++	+++	++	++++
Macaca fascicularis	0	++	?	0	++	++	++	+++	+
Macaca fuscata	+++	+++	?	+	+++	+++	+++	?	++++
Macaca mulatta	0	+++	+	0	+++	?	++	+++	++
Macaca nemestrina	0	+++	?	++	++	++	++	++	++
Macaca nigra	++++	+++	?	+	?	?	+++	+	++
Macaca radiata	++	++	?	0	+++	?	0	0	+++
Macaca silenus	++++	++	++++	++++	+++	+++	0	?	++++
Macaca sinica	++++	++	++++	0	+++	0	+++	0	++++
Macaca sylvanus	++++	+++	++++	+	+++	+++	+++	+	++++
Miopithecus talapoin	+	0	+	++	?	+++	++	++	0
Nasalis larvatus	++	++++	?	+++	+++	?	?	0	++++
Papio anubis	0	++++	+	0	+	+	++	+++	0
Papio cynocephalus	0	++++	?	0	+	?	++	++	0
Papio hamadryas	+	++++	?	+++	++	+	++	+	++
Papio leucophaeus	+++	++++	++	++	++	++	++	+	+++
Papio papio	+++	++++	?	0	++	++	++	++	+++
Papio sphinx	+++	++++	?	+	+++	+++	+++	+	++++

Table 193 (continued)

Species	Small Geographic Range	Large Body Size	Rare	Special-ized	Habitat Alter-ation	Hunting	Pest Control	Collec-tion	Status Rating
Cercopithecidae (cont.)									
Papio ursinus	0	++++	?	0	0	?	+++	++	0
Presbytis aygula-melalophus Group	+	++	+	++	++	+	+	0	0
Presbytis cristata	0	+++	+	+++	++	+	+	0	0
Presbytis entellus	0	++++	+	0	++	++	++	0	0
Presbytis francoisi	+++	+++	?	?	?	?	?	0	+
Presbytis frontata	+++	++	?	+	?	?	?	0	+
Presbytis geei	++++	+++	++++	+	?	?	?	0	+++
Presbytis johnii	++++	+++	++++	++++	+++	+++	+++	0	++++
Presbytis obscura	+++	+++	?	++	++	+++	?	0	++
Presbytis phayrei	+	+++	?	+	+	+	?	0	0
Presbytis pileata	+++	+++	?	++	+++	++	?	0	++++
Presbytis potenziani	++++	+++	++++	+++	?	+++	0	0	++++
Presbytis rubicunda	++	++	+++	+	?	?	?	0	+
Presbytis senex	++++	+++	?	+	+++	?	?	0	++++
Pygathrix nemaeus	++	+++	++	+	?	++	?	0	++
Rhinopithecus avunculus	+++	+++	?	+	?	?	?	0	++
Rhinopithecus roxellanae	+	+++	++++	+	?	+++	?	0	++
Simias concolor	++++	++	++++	+++	?	+++	0	0	++++
Theropithecus gelada	+++	++++	?	++++	+++	+++	+++	++	++++
Hylobatidae									
Hylobates concolor	++	++	?	+	?	++	?	+	+
Hylobates hoolock	++	++	?	+++	+++	++	?	0	++
Hylobates klossii	++++	++	?	+++	?	+++	?	0	+++
Hylobates lar	+	++	?	+	+++	+++	0	+	++
Hylobates pileatus	+++	++	++++	+	+++	++	?	0	+++

Hylobates syndactylus	+++	+++	?	+++	+++	0	0	+	+++
Pongidae									
Gorilla gorilla	++	++++	+++	++	+++	+++	+++	?	++++
Pan paniscus	+++	++++	++++	++	0	+++	?	0	++
Pan troglodytes	0	++++	++	0	++	++	++	++	+++
Pongo pygmaeus	+++	++++	++++	+++	+++	+++	+++	+	++++

KEY: In all categories, 0 denotes no detrimental impact, + a low level effect, ++ more significant or severe, and so forth.

Small Geographic Range (size = $\times 10^3$ km^2)

0 = large (≥2,000)
+ = medium (1,000–1,999)
++ = small (500–999)
+++ = very small (100–499)
++++ = tiny (<100)

Large Body Size (max. wt in kg)

0 = small (<6) and not among largest in family
+ = small (<6) but among largest in family
++ = medium (6–9) and not among largest in family
+++ = medium (6–9) but among largest in family, or large (10–19)
++++ = very large (≥20)

Rare (in countries of range)

0 = no
+ = 1–24%
++ = 25–49%
+++ = 50–99%
++++ = all

Specialized (habitats utilized)

0 = forest and others
+ = forest or forest and woodland
++ = wet forest
+++ = primary forest
++++ = 1 vegetation type

Habitat Alteration, Hunting, Pest Control (in countries of range)

0 = no
+ = 1–24%
++ = 25–49%
+++ = ≥50%

Collection (max. no./yr, any 2 yrs)

0 = <10
+ = 10–99
++ = 100–999
+++ = ≥1,000

Status Rating

0 = "Safe"
+ = "Vulnerable"
++ = "Threatened somewhat"
+++ = "Threatened substantially"
++++ = "Threatened severely"

Table 194. Percentage of species in each family in each status rating category (present analysis)

Family	"Threatened"	"Vulnerable"	"Safe"
Lorisidae	—	55	45
Tarsiidae	33	67	—
Cheirogaleidae	100	—	—
Lemuridae	100	—	—
Indriidae	100	—	—
Daubentoniidae	100	—	—
Callitrichidae	38	31	31
Cebidae	44	24	32
Cercopithecidae	60	22	18
Hylobatidae	83	17	—
Pongidae	100	—	—
Order Primates	58	22	20

Table 195. Comparison of status ratings of the present analysis with those of the International Union for Conservation of Nature and Natural Resources (I.U.C.N. 1972–78), Convention on International Trade in Endangered Species of Wild Fauna and Flora (C.I.T.E.S. [U.S. Department of Interior 1977a]), and U.S. Endangered Species Act (U.S. [U.S. Department of Interior 1977 b]). See KEY for explanation of symbols.

Species	I.U.C.N.	C.I.T.E.S.	U.S.	Present Analysis
Lorisidae				
Arctocebus calabarensis	—	II	—	+
Euoticus elegantulus	—	II	—	+
Euoticus inustus	—	II	—	+
Galago alleni	—	II	—	+
Galago crassicaudatus	—	II	—	0
Galago demidovii	—	II	—	0
Galago senegalensis	—	II	—	0
Loris tardigradus	—	II	—	+
Nycticebus coucang	—	II	—	0
Nycticebus pygmaeus	—	II	T	+
Perodicticus potto	—	II	—	0
Tarsiidae				
Tarsius bancanus	—	II	—	+
Tarsius spectrum	—	II	—	+
Tarsius syrichta	—	II	—	++++
Cheirogaleidae				
Allocebus trichotis	R	I	E	++++
Cheirogaleus major	—	I	E	++++
Cheirogaleus medius	V	I	E	++++
Microcebus coquereli	V	I	E	++++
Microcebus murinus	—	I	E	++++
Phaner furcifer	In	I	E	++++

Table 195 (continued)

Species	I.U.C.N.	C.I.T.E.S.	U.S.	Present Analysis
Lemuridae				
Hapalemur griseus	V	I	E	++++
Hapalemur simus	R	I	E	++++
Lemur catta	—	I	E	++++
Lemur coronatus	V	I	E	++++
Lemur fulvus	E,E,—	I	E	++++
Lemur macaco	E,E	I	E	++++
Lemur mongoz	V	I	E	++++
Lemur rubriventer	—	I	E	++++
Lepilemur Group	R,E,E,—	I	E	++++
Varecia variegata	—	I	E	++++
Indriidae				
Avahi laniger	V,—	I	E	++++
Indri indri	E	I	E	++++
Propithecus diadema	R,—	I	E	++++
Propithecus verreauxi	E	I	E	++++
Daubentoniidae				
Daubentonia madagascariensis	E	I	E	++++
Callitrichidae				
Callimico goeldii	In	I	E	0
Callithrix argentata	V,—	II	—	0
Callithrix humeralifer	In	II	—	+
Callithrix jacchus	E,V,—	I,I,II	—	++
Cebuella pygmaea	—	II	—	0
Leontopithecus rosalia	E	I	E	++++
Saguinus bicolor	In	I	E	+
Saguinus fuscicollis	—	II	—	0
Saguinus imperator	In	II	—	+
Saguinus inustus	—	II	—	+
Saguinus labiatus	—	II	—	+
Saguinus leucopus	In	I	T	++++
Saguinus midas	—	II	—	0
Saguinus mystax	—	II	—	+++
Saguinus nigricollis	—	II	—	++
Saguinus oedipus	E,—	I	E	++++
Cebidae				
Alouatta belzebul	—	II	—	++
Alouatta caraya	—	II	—	0
Alouatta fusca	In	II	—	++
Alouatta palliata	—	I	E	+++
Alouatta pigra	In	I?	T	++++
Alouatta seniculus	—	II	—	0
Aotus trivirgatus	—	II	—	0
Ateles paniscus	—	I,I,II	E,E,—	++

Table 195 (continued)

Species	I.U.C.N.	C.I.T.E.S.	U.S.	Present Analysis
Cebidae (cont.)				
Brachyteles arachnoides	E	I	E	++++
Cacajao calvus	In	I	E	++++
Cacajao melanocephalus	V	I	E	+
Callicebus moloch	—	II	—	+
Callicebus personatus	In	II	—	++++
Callicebus torquatus	—	II	—	+
Cebus albifrons	—	II	—	0
Cebus apella	—	II	—	+
Cebus capucinus	—	II	—	++++
Cebus nigrivittatus	—	II	—	0
Chiropotes albinasus	V	I	E	++++
Chiropotes satanas	—	II	—	+
Lagothrix flavicauda	E	II	E	++++
Lagothrix lagothricha	V	II	—	+
Pithecia monachus	—	II	—	0
Pithecia pithecia	—	II	—	0
Saimiri sciureus	E,—	I,II	E,—	0
Cercopithecidae				
Allenopithecus nigroviridis	—	II	—	+
Cercocebus albigena	—	II	—	++
Cercocebus aterrimus	—	II	—	+
Cercocebus galeritus	E,—	I,II	E,—	++
Cercocebus torquatus	—	II	E	++
Cercopithecus aethiops	—	II	—	0
Cercopithecus ascanius	—	II	—	0
Cercopithecus cephus	—	II	—	+
Cercopithecus diana	—	II	E	++++
Cercopithecus erythrogaster	—	II	E	++++
Cercopithecus erythrotis	—	II	E	++++
Cercopithecus hamlyni	—	II	—	+
Cercopithecus l'hoesti	—	II	E	++
Cercopithecus mitis	—	II	—	+
Cercopithecus mona	—	II	—	0
Cercopithecus neglectus	—	II	—	+
Cercopithecus nictitans	—	II	—	+
Cercopithecus petaurista	—	II	—	++
Cercopithecus pogonias	—	II	—	++
Colobus angolensis	—	II	—	++
Colobus badius	E,E,R,R,—	I,I,II	E,E,—	++
Colobus guereza	—	II	—	+
Colobus polykomos	—	II	—	+++
Colobus satanas	V	II	E	++++
Colobus verus	R	II	—	++
Erythrocebus patas	—	II	—	0

Table 195 (continued)

Species	I.U.C.N.	C.I.T.E.S.	U.S.	Present Analysis
Cercopithecidae (cont.)				
Macaca arctoides	—	II	T	+
Macaca assamensis	—	II	—	+
Macaca cyclopis	—	II	T	++++
Macaca fascicularis	—	II	—	+
Macaca fuscata	—	II	T	++++
Macaca mulatta	—	II	—	++
Macaca nemestrina	In,—	II	—	++
Macaca nigra	—	II	—	++
Macaca radiata	—	II	—	+++
Macaca silenus	E	I	E	++++
Macaca sinica	—	II	T	++++
Macaca sylvanus	V	II	—	++++
Miopithecus talapoin	—	II	—	0
Nasalis larvatus	V	I	E	++++
Papio anubis	—	II	—	0
Papio cynocephalus	—	II	—	0
Papio hamadryas	—	II	—	++
Papio leucophaeus	E	II	E	+++
Papio papio	—	II	—	+++
Papio sphinx	—	II	E	++++
Papio ursinus	—	II	—	0
Presbytis aygula-melalophus Group	—	II	—	0
Presbytis cristata	—	II	—	0
Presbytis entellus	—	I	E	0
Presbytis francoisi	—	II	E	+
Presbytis frontata	—	II	—	+
Presbytis geei	R	I	E	+++
Presbytis johnii	V	II	—	++++
Presbytis obscura	—	II	—	++
Presbytis phayrei	—	II	—	0
Presbytis pileata	—	I	E	++++
Presbytis potenziani	In	I	T	++++
Presbytis rubicunda	—	II	—	+
Presbytis senex	—	II	T	++++
Pygathrix nemaeus	E	I	E	++
Rhinopithecus avunculus	—	II	T	++
Rhinopithecus roxellanae	R	II	—	++
Simias concolor	E	I	E	++++
Theropithecus gelada	—	II	—	++++
Hylobatidae				
Hylobates concolor	In	I	E	+
Hylobates hoolock	—	I	E	++
Hylobates klossii	V	I	E	+++

Table 195 (continued)

Species	I.U.C.N.	C.I.T.E.S.	U.S.	Present Analysis
Hylobatidae (cont.)				
Hylobates lar	E,—	I	E	++
Hylobates pileatus	E	I	E	+++
Hylobates syndactylus	—	I	E	+++
Pongidae				
Gorilla gorilla	V,E	I	E	++++
Pan paniscus	V	I	T	++
Pan troglodytes	V	I	T	+++
Pongo pygmaeus	E	I	E	++++

KEY (two entries in one category indicate that the races of a species are rated differently):
I.U.C.N.: E = Endangered. R = Rare. V = Vulnerable. In = Indeterminate. — = Not rated.
C.I.T.E.S.: I = Appendix I. II = Appendix II.
U.S.: E = Endangered. T = Threatened. — = Not rated.
Present Analysis: ++++ = Threatened severely. +++ = Threatened substantially. ++ = Threatened somewhat. + = Vulnerable. 0 = Safe.

References

Adekunle, A. 1971. Progress in wildlife conservation in Western State of Nigeria. Pp. 30–34 in *Wildlife Conservation in West Africa*. I.U.C.N. Publ. New Series, no. 22. D. C. D. Happold, ed. Morges, Switzerland.

Aeschlimann, A. 1965. Notes on the mammals of the Ivory Coast. *Afr. Wildl.* 19(1):37–55.

Agrawal, V. C. 1974. Taxonomic status of Barbe's leaf monkey, *Presbytis barbei* Blyth. *Primates* 15(2–3):235–239.

Aguirre, A. C. 1971. O mono *Brachyteles arachnoides* (E. Geoffroyi). Acac. Bras. Cienc., Rio de Janeiro. 53 pp.

Aiken, S. R., and M. R. Moss. 1975. Man's impact on the tropical rainforest of peninsular Malaysia: A review. *Biol. Conserv.* 8(3):213–229.

Ajayi, S. S. 1974. Nigeria's progress in wildlife management. *Nigerian Field* 39(2):71–76.

Albignac, R., Y. Rumpler, and J.-J. Petter. 1971. L'hybridation des Lémuriens de Madagascar. *Mammalia* 35(3):358–368.

Albrecht, H. 1976. Chimpanzees in Uganda. *Oryx* 13(4):357–361.

Albrecht, H., and S. C. Dunnett. 1971. *Chimpanzees in Western Africa*. R. Piper and Co., Munich. 138 pp.

Aldrich-Blake, F. P. G. 1970a. The ecology and behaviour of the blue monkey, *Cercopithecus mitis stuhlmanni*. Ph.D. diss. University of Bristol.

———. 1970b. Problems of social structure in forest monkeys. Pp. 79–101 in *Social Behaviour in Birds and Mammals*. J. H. Crook, ed. Academic Press, London.

Aldrich-Blake, F. P. G., T. K. Bunn, R. I. M. Dunbar, and P. M. Headley. 1971. Observations on baboons, *Papio anubis*, in an arid region in Ethiopia. *Folia Primatol.* 15:1–35.

Altmann, S. A. 1959. Field observations on a howling monkey society. *J. Mammal.* 40:317–330.

Altmann, S. A., and J. Altmann. 1970. *Baboon Ecology: African Field Research*. Vol. 12 of *Bibl. Primatol.* Pp. 1–220.

Altstatt, L. B. 1971. *S.E.A.T.O. Medical Research Laboratory, Annual Progress Report*.

Alvarez del Toro, M. 1977. *Los Mamíferos de Chiapas*. Universidad Autonoma de Chiapas, Tuxtla Gutierrez, Chiapas, Mexico.

Anderson, J. A. R. 1961. Bako National Park, Sarawak. Nature conservation in Western Malaysia, 1961. *Malay. Nat. J.* 21st anniversary special issue: 128–130.

Andrews, P., C. P. Groves, and J. F. M. Horne. 1975. Ecology of the lower Tana River flood plain (Kenya). *J. East Afr. Nat. Hist. Soc. Nat. Mus.* 151:1–31.

Angst, W. 1973. Pilot experiments to test group tolerance to a stranger in wild *Macaca fascicularis*. *Am. J. Phys. Anthropol.* 38:625–630.

———. 1975. Basic data and concepts on the social organization of *Macaca fascicularis*. Pp. 325–388 in *Primate Behavior,* vol. 4. L. A. Rosenblum, ed. Academic Press, New York.

Annenkov, H. A., A. B. Mirvis, and H. G. Kotrikadze. 1972. Geographical transferrin polymorphism in *Macaca mulatta*. *Primates* 13(3):235–242.

Anonymous. 1969. Disaster for the gorillas. *Oryx* 10(1):7.

———. 1971. Congo's wildlife devastated by war. *New Sci. Sci. J.* 50:138.

———. 1972. The threat to the red colobus. *Oryx* 11(5):304.

———. 1973. Marmoset field station at work. *Oryx* 12(1):9.

———. 1974a. What hope for Zanzibar's *Colobus*? *Oryx* 12(5):522–523.

———. 1974b. A blueprint for conservation in peninsular Malaysia. *Malay. Nat. J.* 27:1–16.

———. 1978. New discoveries on tarsier behaviour in Sulawesi. *Tigerpaper* 5(3):7.

———. 1979. Chimpanzees are seized at Amsterdam airport. *I.U.C.N. Bulletin*. New Series, no. 1:8.

Ansell, W. F. H. 1952. The status of Northern Rhodesian game. Part 1. *Afr. Wildl.* 6(1):21–29.

———. 1957. Northern Rhodesia's protected animals. *Afr. Wildl.* 11(1):19–28.

———. 1959. The type localities of *Colobus angolensis sharpei* Thomas and *Colobus angolensis sandbergi* Lönnberg. *Rev. Zool. Bot. Afr.* 60:168–171.

———. 1960a. *Mammals of Northern Rhodesia*. Government Printer, Lusaka. 155 pp.

———. 1960b. Contributions to the mammalogy of Northern Rhodesia. *Occ. Pap. Nat. Mus. S. Rhod.* 3:351–398.

———. 1969. Addenda and corrigenda to "Mammals of Northern Rhodesia," no. 3. *Puku* 5:1–48.

———. 1973. Addenda and corrigenda to "Mammals of Northern Rhodesia," no. 4. *Puku* 7:1–19.

———. 1974. Some mammals from Zambia and adjacent countries. *Puku* (suppl. 1):1–48.

Ansell, W. F. H., and P. D. H. Ansell. 1973. Mammals of northeastern montane areas of Zambia. *Puku* 7:21–69.

Ansell, W. F. H., C. W. Benson, and B. L. Mitchell. 1962. Notes on some mammals from Nyasaland and adjacent areas. *Nyasaland J.* 15:38–54.

Appell, G. N. 1973. Mammals of Borneo whose survival is threatened. *Borneo Res. Bull.* 5(2):64–66.

Asibey, E. O. A. 1972. Ghana's progress. *Oryx* 11(6):470–475.

———. 1974. Wildlife as a source of protein in Africa south of the Sahara. *Biol. Conserv.* 6(1):32–39.

———. 1978. Primate conservation in Ghana. Pp. 55–74 in *Recent Advances in Primatology*, vol. 2: *Conservation*. D. J. Chivers and W. Lane-Petter, eds. Academic Press, London.

Atapattu, S., and C. S. Wickremasinghe. 1974. Sri Lanka's Gal Oya National Park: Aspects and prospects. *Biol. Conserv.* 6(3):219–222.

Avila-Pires, F. D. de. 1958. Mamíferos colecionados nos arredores de Belém do Pará. *Bol. Mus. Para. Emilio Goeldi*. New Series (Zool.) 19:1–9.

———. 1964. Mamíferos colecionados na reagião do Rio Negro (Amazonas, Brazil). *Bol. Mus. Para. Emilio Goeldi*. New Series (Zool.) 42:1–23.

———. 1966. Observacões gerais sôbre a mastozoologia do Cerrado. *An. Acad. Bras. Cienc.* 38:331–340. Suppl.

———. 1969a. Some problems concerning primates. Pp. 132–135 in *Conservation in Latin America*. I.U.C.N. Publ. New Series, no. 13.

———. 1969b. Taxonomia e zoogeografia do gênero "Callithrix" Erxleben, 1777 (Primates, Callithricidae). *Rev. Bras. Biol.* 29:49–64.

———. 1972. The survival of South American primates. *Int. Zoo. Yearb.* 12:13–14.

———. 1974. Caracterização zoogeográfica da Província Amazônica. II. A Família Callitricidae e a zoogeografia Amazônica. *An. Acad. Bras. Cienc.* 46:159–181.

———. 1976. Nonhuman primates of the American continent. Pp. 3–7 in *First Inter-American Conference on Conservation and Utilization of American Nonhuman Primates in Biomedical Research*. Scientific Publication no. 317. Pan American Health Organization, Washington, D.C.

Awunti, J. 1978. The conservation of primates in the United Republic of Cameroon. Pp. 75–79 in *Recent Advances in Primatology*, vol. 2: *Conservation*. D. J. Chivers and W. Lane-Petter, eds. Academic Press, London.

Ayeni, J. S. O. 1972. Notes taken from a hide overlooking a lick at Yankari Game Reserve, North Eastern State of Nigeria. *West Afr. Sci. Assoc. J.* 17(2):101–112.

Ayres, J. M. 1978. Situação atual da área de ocorrência do cuxiú (*Chiropotes satanas satanas*, Hoffmannsegg–1807). Unpublished manuscript.

Azuma, S. 1970. On the aerial spraying of herbicide over the Japanese monkey habitat on Shimokita Peninsula. *Primates* 11:401–402.

Azuma, S., and A. Toyoshima. 1961–62. Progress report of the survey of chimpanzees in their natural habitat, Kabogo Point Area, Tanganyika. *Primates* 3(2):61–71.

Badrian, A., and N. Badrian. 1977. Pygmy chimpanzees. *Oryx* 13(5):463–468.

Bailey, R. C., R. S. Baker, D. S. Brown, P. Von Hildebrand, R. A. Mittermeier, L. E. Sponsel, and K. E. Wolf. 1974. Progress of a breeding project for non-human primates in Colombia. *Nature* 248:453–455.

Baker, R. 1974. Records of mammals from Ecuador. *Publ. Mus. Michigan State University*, Biol. Series 5(2):129–146.

Baldwin, J. D., and J. I. Baldwin. 1971. Squirrel monkeys (*Saimiri*) in natural habitats in Panama, Colombia, Brazil, and Peru. *Primates* 12:45–61.

———. 1972a. Population density and use of space in howling monkeys (*Alouatta villosa*) in southwestern Panama. *Primates* 13(4):371–381.

———. 1972b. The ecology and behavior of squirrel monkeys (*Saimiri oerstedi*) in a natural forest in western Panama. *Folia Primatol.* 18(3–4):161–185.

———. 1973. Interactions between adult female and infant howling monkeys (*Alouatta palliata*). *Folia Primatol.* 20(1):27–71.

———. 1976. Primate populations in Chiriquí, Panama. Pp. 20–31 in *Neotropical Primates: Field Studies and Conservation*. R. W. Thorington, Jr., and P. G. Heltne, eds. National Academy of Sciences, Washington, D.C.

———. 1977. Observations on *Cebus capucinus* in southwestern Panama. *Primates* 18(4):937–941.

Baldwin, L. A., M. Kavanagh, and G. Teleki. 1975. Field research on langur and proboscis monkeys: An historical, geographical, and bibliographical listing. *Primates* 16(3):351–363.

Baldwin, L. A., T. L. Patterson, and G. Teleki. 1977. Field research on callitrichid and cebid monkeys: An historical, geographical, and bibliographical listing. *Primates* 18(2):485–507.

Baldwin, L. A., and G. Teleki. 1972. Field research on baboons, drills and geledas: An historical, geographical, and bibliographical listing. *Primates* 13:427–432.

———. 1973. Field research on chimpanzees and gorillas: An historical, geographical, and bibilographical listing. *Primates* 14:315–330.

———. 1974. Field research on gibbons, siamangs, and orang-utans: An historical, geographical, and bibliographical listing. *Primates* 15(4):365–376.

———. 1977. Field research on tree shrews and prosimians: An historical, geographical, and bibliographical listing. *Primates* 18(4):985–1007.

Baldwin, L. A., G. Teleki, and M. Kavanagh. 1976. Field research on colobus, guenon, mangabey, and patas monkeys: An historical, geographical, and bibliographical listing. *Primates* 17:233–251.

Balinga, V. S. 1977. Competitive uses of wildlife. *Unasylva* 29(116):22–25.

Barnard, C. 1971. What man owes the baboon. *Int. Wildl.* 1(2):4–9.

Barros Machado, A. de. 1969. Mamíferos de Angola ainda não citados ou pouco conhecidos. *Das Publicaçoes Culturais da Companhia de Diamantes de Angola* 46:93–232.

Basckin, D. R., and P. D. Krige. 1973. Some preliminary observations on the behaviour of an urban troop of vervet monkeys (*Cercopithecus aethiops*) during the birth season. *J. Behav. Sci.* 1:287–296.

Basjarudin, H. 1971. Nature reserves and national parks in Indonesia—Present situation and problems. *I.U.C.N. Publ.* New Series 19:27–33.

Baskaran, S. T. 1975. A note on the hoolock. *J. Bombay Nat. Hist. Soc.* 72(1):194.

Bassus, W. 1973. Wild, Jagd und Naturschutz in der Demokratischen Republik Vietnam. *Arch. Naturschutz Landschaftsforsch.* 13(3):257–267.

———. 1975. Der Wildtierbestand, sein Schutz und seine Bejagung in der Volksrepublik Kongo. *Arch. Naturschutz Landschaftsforsch.* 15(4):247–263.

Baumgartel, M. W. 1965. The gorillas of Virunga. *Afr. Wildl.* 19:16–21.

Bearder, S. 1974. Bushbabies. *Afr. Wildl.* 28(3):20–22.

Bearder, S. K., and G. A. Doyle. 1972. Ecology of bush babies, *Galago senegalensis* and *Galago crassicaudatus*, with some notes on their behavior in the field. Unpublished research seminar on prosimian biology. University of Witwatersrand, Johannesburg.

———. 1974a. Ecology of bushbabies *Galago senegalensis* and *Galago crassicaudatus*, with some notes on their behaviour in the field. Pp. 109–130 in *Prosimian Biology*. R. D. Martin, G. A. Doyle, and A. C. Walker, eds. University of Pittsburgh Press, Pittsburgh.

———. 1974b. Field and laboratory studies of social organization in bushbabies (*Galago senegalensis*). *J. Hum. Evol.* 3:37–50.

Beaton, A. C. 1942. Baboon poisoning. *Sudan Wild Life and Sports* 1(2):19–20.

Beischer, D. E. 1968. The squirrel monkey in aerospace medical research. Pp. 347–364 in *The Squirrel Monkey*. L. A. Rosenblum and R. W. Cooper, eds. Academic Press, New York.

Bennett, C. F. 1968. Human influences on the zoogeography of Panama. *Ibero-Americana* 51:1–112.

Benson, C. W. 1968. An alleged record of the chimpanzee *Pan satyrus* in Malawi. *J. Malawi Soc.* 21:7–12.

Berger, M. E. 1972. Population structure of olive baboons [*Papio anubis* (J. P. Fischer)] in the Laikipia District of Kenya. *East Afr. Wildl. J.* 10:159–164.

Bergner, J. E., Jr., and L.A. Jachowski, Jr. 1968. The filarial parasite, *Macacanema formosana*, from the Taiwan monkey, and its development in various arthropods. *Formosan Sci.* 22:1–68.

Berkson, G., and B. A. Ross. 1969. Factors to consider in a conservation program for gibbons. Pp. 43–47 in *Proc. 2d Int. Congr. Primatol.*, vol. 1. S. Karger, Basel.

Berkson, G., B. A. Ross, and S. Jatinandana. 1971. The social behavior of gibbons in relation to a conservation program. Pp. 226–255 in *Primate Behavior*, vol. 2. L. A. Rosenblum, ed. Academic Press, New York.

Bermant, G., and D. G. Lindburg, eds. 1975. *Primate Utilization and Conservation*. John Wiley and Sons, New York. 196 pp.

Bernahu, L. 1975. Present status of primates in Ethiopia and their conservation. Pp. 525–528 in *Proc. Symp. 5th Congr. Int. Primatol. Soc.* S. Kondo, M. Kawai, A. Ehara, and S. Kawamura, eds. Japan Science Press, Tokyo.

Bernstein, I. S. 1967a. A field study of the pigtail monkey (*Macaca nemestrina*). *Primates* 8:217–228.

———. 1967b. Intertaxa interactions in a Malayan primate community. *Folia Primatol.* 7:198–207.

———. 1968. The lutong of Kuala Selangor. *Behaviour* 33(1-3):1–16.

Bernstein, I. S., P. Balcaen, L. Dresdale, H. Gouzoules, M. Kavanagh, T. Patterson, and P. Neyman-Warner. 1976a. Differential effects of forest degradation on primate populations. *Primates* 17(3):401–411.

———. 1976b. An appeal for the preservation of habitats in the interests of primate conservation. *Primates* 17:413–415.

Bernstein, I. S., and N. B. Guilloud. 1965. The stumptail macaque as a laboratory subject. *Science* 147:824.

Bert, J., H. Ayats, A. Martino, and H. Collomb. 1967a. Le sommeil nocturne chez le babouin *Papio papio*. *Folia Primatol.* 6:28–43.

———. 1967b. Note sur l'organisation de la vigilance sociale chez le babouin *Papio papio* dans l'est Sénégalais. *Folia Primatol.* 6:44–47.

Bertoni, G. T., and J. R. Gorham. 1973. II. The geography of Paraguay. Pp. 9–31 in *Paraguay: Ecological Essays*. J. R. Gorham, ed. Academy of Arts and Sciences of the Americas, Miami, Fla.

Bertrand, M. 1969. *The Behavioral Repertoire of the Stumptail Macaque*. Vol. 11 of *Bibl. Primatol.* Pp. 1–273.

*Bigourdan, J., and R. Prunier. 1973. *Les mammifères sauvages de l'Ouest Africain et leur milieu*. Rudder, Montrouge.

Biswas, S. 1970. White-browed gibbon (*Hylobates hoolock* Harlan) and capped langur (*Presbytis pileatus* Blyth) as human food. *J. Bengal Nat. Hist. Soc.* 36:78–80.

Bittner, S. L., M. R. Boatwright, and R. L. Jachowski. 1978. *Convention on International Trade in Endangered Species of Wild Fauna and Flora, Annual Report for 1977*. Wildlife Permit Office, Department of Interior. U.S. Government Printing Office, Washington, D.C. 77 pp.

Blaffer Hrdy, S. B. 1977. *The Langurs of Abu*. Harvard University Press, Cambridge, Mass. 361 pp.

*Blancou, L. 1958. Notes biogéographiques sur les mammifères de l'A.E.F. *Bull. Inst. Centrafr.* New Series 15–16:7–42.

———. 1961. Destruction and protection of the wild life in French Equatorial and French West Africa. Part V: Primates. *Afr. Wildl.* 15:29–34.

Blower, J. 1973. Rhinos and other problems in Nepal. *Oryx* 12:270–280.

Boesch, C. 1978. Nouvelles observations sur les chimpanzés de la forêt de Tai (Côte d'Ivoire). *Terre Vie* 32:195–201.

Bolton, M. 1970. Conservation in Ethiopia's Rift Valley. *Biol. Conserv.* 3(1):66–68.

———. 1972. Report on a wildlife survey in southeast Ethiopia. *Walia* 4:19–31.

———. 1973. Notes on the current status and distribution of some large mammals in Ethiopia (excluding Eritrea). *Mammalia* 37(4):562–586.

———. 1974. Mount Abaro, Shashemenne District. *Walia* 5:15–16.

Bolwig, N. 1959. A study of the behavior of the chacma baboon (*Papio ursinus*). *Behaviour* 14:136–163.

Bomford, E. 1976a. In search of the aye-aye. *Wildlife* 18:258–263.

———. 1976b. The disappearing home of the lemurs. *Wildlife* 18:309–317.

Boné, E. L., and S. Haumont. 1967. Primatological research in central Africa. Pp. 5–7 in *Progress in Primatology*. D. Starck, R. Schneider, and H.-J. Kuhn, eds. G. Fischer Verlag, Stuttgart.

Bonner, J. T. 1965. *Size and Cycle: An Essay on the Structure of Biology*. Princeton University Press, Princeton, N.J. 219 pp.

Booth, A. H. 1954. A note on the colobus monkeys of the Gold and Ivory Coasts. *Ann. Mag. Nat. Hist.* 7(12):857–860.

———. 1955. Speciation in the mona monkeys. *J. Mammal.* 36:434–449.

———. 1956a. The distribution of primates in the Gold Coast. *J. W. Afr. Sci. Assoc.* 2:122–133.

*Reference not seen but cited by other authors.

———. 1956b. The Cercopithecidae of the Gold and Ivory Coasts: Geographic and systematic observations. *Ann. Mag. Nat. Hist.* 9(12):476–480.

———. 1957. Observations on the natural history of the olive colobus monkey *Procolobus verus* (Van Beneden). *Proc. Zool. Soc. London* 129:421–430.

———. 1958. The zoogeography of west African primates: A review. *Bull. Inst. Fondam. Afr. Noire* 20:587–622.

Booth, C. P. 1962. Some observations on behavior of *Cercopithecus* monkeys. *Ann. N.Y. Acad. Sci.* 102:477–487.

———. 1968. Taxonomic studies of *Cercopithecus mitis* Wolf (East Africa). *Nat. Geogr. Soc. Res. Rep.* 37–51.

Borner, M. 1975. Into the unknown (north Sumatra). *Wildlife* 17(3):157–165.

———. 1976. Sumatra's orang-utans. *Oryx* 13(3):290–293.

Borner, M., and P. Gittens. 1978. Round-table discussion on rehabilitation. Pp. 101–105 in *Recent Advances in Primatology*, vol. 2: *Conservation*. D. J. Chivers and W. Lane-Petter, eds. Academic Press, London.

Boshell, M. J., and G. A. Bevier. 1958. Yellow fever in the lower Motagua Valley, Guatemala. *Am. J. Trop. Med. Hyg.* 7:25–35.

Bothma, J. du P. 1975. Conservation status of the larger mammals of southern Africa. *Biol. Conserv.* 7(2):87–95.

*Bourgoin, P. 1955. *Animaux de Chasse d'Afrique*. La Toison d'Or, Paris.

Bourlière, F., M. Bertrand, and C. Hunkeler. 1969. L'écologie de la mone de Lowe (*Cercopithecus campbelli lowei*) on Côte d'Ivoire. *Terre Vie* 23(2):135–164.

Bourlière, F., C. Hunkeler, and M. Bertrand. 1970. Ecology and behavior of Lowe's guenon (*Cercopithecus campbelli lowei*) in the Ivory Coast. Pp. 297–350 in *Old World Monkeys*. J. R. Napier and P. H. Napier, eds. Academic Press, New York.

Bourlière, F., E. Minner, and R. Vuattoux. 1974. Les grands mammifères de la région de Lamto, Côte d'Ivoire. *Mammalia* 38(3):433–447.

Bourlière, F., G. Morel, and G. Galat. 1976. Les grands mammifères de la basse vallée du Sénégal et leurs saisons de reproduction. *Mammalia* 40(3):401–412.

Bourne, G. H. 1976. Response to pygmy chimpanzee petition. *Lab. Primate Newsl.* 15(1):10–14.

Bournonville, D. de. 1967. Contribution à l'étude du chimpanzé en République de Guinée. *Bull. Inst. Fondam. Afr. Noire* A29:1188–1269.

Bourquin, O., J. Vincent, and P. M. Hitchins. 1971. The vertebrates of the Hluhluwe Game Reserve—corridor (state land)—Umfolozi Game Reserve complex. *Lammergeyer* 14:5–58.

Bourret, R. No date. *Les mammifères de l'Indochine: Les gibbons*. Laboratoire des Sciences Naturelles de l'Universite Indochinoise, Djakarta. 40 pp.

Boyle, C. L. 1963. The orang-utan crisis. *Oryx* 7:106–107.

Brain, C. K. 1965. Observations on the behavior of vervet monkeys. *Cercopithecus aethiops*. *Zool. Afr.* 1(1):13–27.

Brambell, M. R. 1975. Breeding orang-utans. Pp. 235–243 in *Breeding Endangered Species in Captivity*. R. D. Martin, ed. Academic Press, London.

Brewer, S. 1976. Chimpanzee rehabilitation. Special report. International Primate Protection League, Nokomis, Florida. 10 pp.

Bridgewater, D. D. 1972. Introductory remarks with comments on the history and current status of the golden marmoset. Pp. 1–6 in *Saving the Lion Marmoset*. D. D. Bridgewater, ed. Wild Animal Propagation Trust, Wheeling, West Virginia.

Brockelman, W. Y. 1975. Gibbon populations and their conservation in Thailand. *Natur. Hist. Bull. Siam Soc.* 26(1–2):133–157.

———. 1978. Preliminary report on relations between the gibbons *Hylobates lar* and *H. pileatus* in Thailand. Pp. 315–318 in *Recent Advances in Primatology*, vol. 3: *Evolution*. D. J. Chivers and K. A. Joysey, eds. Academic Press, London.

Brockelman, W. Y., D. Damman, P. Thongsuk, and S. Srikosamatara. 1977. Pileated gibbons survey at Khao Soi Dao Sanctuary–Thailand. *Tigerpaper* 4(4):13–15.

Brockelman, W. Y., and N. K. Kobayashi. 1971. Live capture of free-ranging primates with a blow-gun. *J. Wildl. Manage.* 35:852–855.

Brotoisworo, E. 1978. Nature conservation in Indonesia and its problems with special reference to primates. Pp. 31–40 in *Recent Advances in Primatology*, vol. 2: *Conservation*. D. J. Chivers and W. Lane-Petter, eds. Academic Press, London.

Brown, L. H., and E. K. Urban. 1969. DeBrazza's monkey, *Cercopithecus neglectus* Schlegel, in the forests of south west Ethiopia. *East Afr. Wildl. J.* 7:174–175.

———. 1970. Bird and mammal observations from the forests of southwest Ethiopia. *Walia* 2:13–40.

Brumback, R. A. 1973. Two distinctive types of owl monkeys. *J. Med. Primatol.* 2:284–289.

———. 1974. A third species of owl monkey (*Aotus*). *J. Hered.* 65:321–323.

———. 1976. Taxonomy of the owl monkey (*Aotus*). *Lab. Primate Newsl.* 15(1):1–2.

Brumback, R. A., R. D. Staton, S. A. Benjamin, and C. M. Lang. 1971. The chromosomes of *Aotus trivirgatus* Humboldt, 1812. *Folia Primatol.* 15:264–273.

Bruver, W. 1973. Fauna of north Thailand. *Natur. Hist. Bull. Siam Soc.* 24:463–466.

Bryceson, J. G. 1976. A change of pace after the kidnap—but Gombe goes on. *Africana* 6(3):14–16, 31.

Budnitz, N., and K. Dainis. 1975. *Lemur catta:* ecology and behavior. Pp. 219–235 in *Lemur Biology*. I. Tattersall and R. W. Sussman, eds. Plenum Press, New York.

Budowski, G. 1977. A strategy for saving wild plants: Experience from Central America. Pp. 368–373 in *Extinction Is Forever*. G. Prance and T. Elias, eds. New York Botanical Garden, Bronx, New York.

Buettner-Janusch, J. 1966. A problem in evolutionary systematics: Nomenclature and classification of baboons, genus *Papio*. *Folia Primatol.* 4:288–308.

Burgess, P. F. 1961. Wild life conservation in North Borneo. Nature Conservation in Western Malaysia, 1961. *Malay. Nat. J.* 21st anniversary special issue: 143–151.

———. 1971. The effect of logging on hill dipterocarp forest. *Malay. Nat. J.* 24:231–237.

Burkill, H. M. 1961. Protection of wildlife on Singapore Island. Nature Conservation in Western Malaysia, 1961. *Malay. Nat. J.* 21st anniversary special issue: 152–164.

Burton, J. A. 1978. Primate imports into the United Kingdom 1965–1975. Pp. 137–145 in *Recent Advances in Primatology*, vol. 2: *Conservation*. D. J. Chivers and W. Lane-Petter, eds. Academic Press, London.

Buskirk, W. H., R. E. Buskirk, and W. J. Hamilton III. 1974. Troop-mobilizing behavior of adult male chacma baboons. *Folia Primatol.* 22:9–18.

Butler, H. 1966. Some notes on the distribution of primates in the Sudan. *Folia Primatol.* 4:416–423.

Buxton, A. P. 1952. Observations on the diurnal behavior of the redtail monkey (*Cercopithecus ascanius schmidti* Matschie) in a small forest in Uganda. *J. Anim. Ecol.* 21:25–28.

Cabral, J. C. 1967. Mamíferos de Reserva do Luando. *Bol. Inst. Invest. Cient. Angola* 4(2):33–44.

*Cabrera, A. 1929. Catálogo descriptive de los mamíferos de la Guinea Española. *Mem. Soc. Española Hist. Nat. Madrid* 16:121 pp.

*Reference not seen but cited by other authors.

———. 1958. Catálogo de los mamíferos de América del Sur. I. *Rev. Mus. Argentino Cienc. Nat. "Bernardino Rivadavia," Zool.* 4(1):1–307.

Cadigan, F. C., and B. L. Lim. 1975. The future of southeast Asian nonhuman primates. Pp. 83–94 in *Primate Utilization and Conservation.* G. Bermant and D. G. Lindburg, eds. Wiley-Interscience, New York.

Cambefort, Y., and F. Moro. 1978. Cytogenetics and taxonomy of some south Bolivian monkeys. *Folia Primatol.* 29:307–314.

Campbell, R. I. M. 1970. Mountain gorillas in Rwanda. *Oryx* 10:256–257.

Cant, J. H. G. 1978. Population survey of the spider monkey *Ateles geoffroyi* at Tikal, Guatemala. *Primates* 19(3):525–535.

Carpenter, C. R. 1934. A field study of the behavior and social relations of howling monkeys. *Comp. Psych. Monogr.* 10(2):1–168.

———. 1935. Behavior of red spider monkeys in Panama. *J. Mammal.* 16(3):171–180.

———. 1938. A survey of wild life conditions in Atjeh, North Sumatra. Communication no. 12. Netherlands Committee for International Nature Protection, Amsterdam.

———. 1940. A field study in Siam of the behavior and social relations of the gibbon. *Comp. Psych. Monogr.* 16(5):1–212.

———. 1942. Societies of monkeys and apes. *Biol. Symp.* 8:177–204.

Carvalho, C. T. de. 1957. Alguns mamíferos do Acre occidental. *Bol. Mus. Para. Emilio Goeldi,* (Zool.) 6:1–22.

Carvalho, C. T. de, and A. J. Toccheton. 1969. Mamíferos do nordeste do Pará, Brasil. *Rev. Biol. Trop.* 15(2):215–226.

Casimir, M. J. 1975a. Some data on the systematic position of the eastern gorilla population of the Mt. Kahuzi region (République du Zaïre). *Z. Morphol. Anthropol.* 66(2):188–201.

———. 1975b. Feeding ecology and nutrition of an eastern gorilla group in the Mt. Kahuzi region (République du Zaïre). *Folia Primatol.* 24(2–3):81–136.

Casimir, M. J., and E. Butenandt. 1973. Migration and core area shifting in relation to some ecological factors in a mountain gorilla group (*Gorilla gorilla beringei*) in the Mt. Kahuzi region (République du Zaïre). *Z. Tierpsychol.* 33:514–522.

Castro, N. 1976. Guidelines for the conservation of primates in Peru. Pp. 216–234 in *First Inter-American Conference on Conservation and Utilization of American Nonhuman Primates in Biomedical Research.* Scientific Publication no. 317. Pan American Health Organization, Washington, D.C.

Castro R., N. E. 1978. Diagnóstico de la situación actual de los primates no humanos en el Perú y un plan nacional para su utilización racional. Direc. Conservación D.G.F.F., Min. Agric. y Alim. Perú. Mimeographed. 204 pp.

Castro, R., and P. Soini. 1978. Field studies on *Saguinus mystax* and other callitrichids in Amazonian Peru. Pp. 73–78 in *The Biology and Conservation of the Callitrichidae.* D. G. Kleiman, ed. Smithsonian Institution Press, Washington, D.C.

Causey, O. R., H. W. Laemmert, Jr., and G. S. Hayes. 1948. The home range of Brazilian *Cebus* monkeys in a region of small residual forests. *Am. J. Hyg.* 47:304–314.

Chakraborty, S. 1975. On a collection of mammals from Bhutan. *Rec. Zool. Surv. India* 68:1–20.

Chalmers, N. R. 1968. Group composition, ecology, and daily activities of free-living mangabeys in Uganda. *Folia Primatol.* 8:247–262.

———. 1973. Differences in behaviour between some arboreal and terrestrial species of African monkeys. Pp. 69–100 in *Comparative Ecology and Behaviour of Primates.* R. P. Michael and J. H. Crook, eds. Academic Press, London.

Chang, C., and T. Y. Wang. 1963. [Faunistic studies of mammals of the Chinghai Province]. *Acta Zool. Sin.* 15:125–138.
Charles-Dominique, P. 1971. Éco-éthologie des prosimiens du Gabon. *Biol. Gabonica* 7:121–228.
———. 1972. Écologie et vie sociale de *Galago demidovii* (Fischer 1808; Prosimii). *Advances in Ethology: J. Comp. Ethol.*, supp. 9 (Behaviour and ecology of nocturnal prosimians):7–41.
———. 1974a. Ecology and feeding behaviour of five sympatric lorisids in Gabon. Pp. 131–150 in *Prosimian Biology*. R. D. Martin, G. A. Doyle, and A. C. Walker, eds. University of Pittsburgh Press, Pittsburgh.
———. 1974b. Vie sociale de *Perodicticus potto* (Primates, Lorisidés): Étude de terrain en forêt équatoriale de l'Ouest Africain au Gabon. *Mammalia* 38(3):355–379.
———. 1974c. Aggression and territoriality in nocturnal prosimians. Pp. 31–48 in *Primate Aggression, Territoriality and Xenophobia*. R. L. Holloway, ed. Academic Press, New York.
———. 1977a. Urine marking and territoriality in *Galago alleni* (Waterhouse, 1837–Lorisoidea, Primates)—A field study by radio-telemetry. *Z. Tierpsychol.* 43(2):113–138.
———. 1977b. *Ecology and Behaviour of Nocturnal Primates*. Columbia University Press. 277 pp.
———. 1978. Écologie et vie sociale de *Nandinia binotata* (Carnivores, Viverrides): Comparison avec les Prosimiens sympatriques du Gabon. *Terre Vie* 32:477–528.
Charles-Dominique, P., and S. K. Bearder. 1979. Field studies of lorisid behavior: Methodological aspect. Pp. 567–629 in *The Study of Prosimian Behavior*. G. A. Doyle and R. D. Martin, eds. Academic Press, New York.
Charles-Dominique, P., and C. M. Hladik. 1971. Le lepilemur du sud de Madagascar: Écologie, alimentation, et vie sociale. *Terre Vie* 1:3–66.
Charles-Dominique, P., and R. D. Martin. 1970. Evolution of lorises and lemurs. *Nature* 227:257–260.
Chasen, F. N. 1940. A handlist of Malaysian mammals. *Bull. Raffles Mus.* 15:1–209.
Chauvet, B. 1972. The forests of Madagascar. Pp. 191–199 in *Biogeography and Ecology of Madagascar*. R. Battistini and G. Richard-Vindard, eds. W. Junk, The Hague.
Chiarelli, A. B. 1972. *Taxonomic Atlas of Living Primates*. Academic Press, London. 363 pp.
Child, G., and C. R. Savory. 1964. The distribution of large mammal species in southern Rhodesia. *Arnoldia* 1(14):1–15.
Chin, L. 1971. Protected animals in Sarawak. *Sarawak Mus. J.* 19:359–361.
Chivers, D. J. 1969. On the daily behaviour and spacing of the howling monkey groups. *Folia Primatol.* 10(1–2):48–102.
———. 1971. The Malayan siamang. *Malay. Nat. J.* 24:78–86.
———. 1972. The siamang and the gibbon in the Malay peninsula. *Gibbon and Siamang* 1:103–135.
———. 1973. An introduction to the socioecology of Malayan forest primates. Pp. 101–146 in *Comparative Ecology and Behaviour of Primates*. R. P. Michael and J. H. Crook, eds. Academic Press, London.
———. 1974. *The Siamang in Malaya*. Vol. 4 of *Contrib. Primatol.* Pp. 1–335.
———. 1975a. Daily patterns of ranging and feeding in siamang. Pp. 363–372 in *Contemporary Primatology, Proc. 5th Int. Congr. Primatol.* S. Kondo, M. Kawai, and A. Ehara, eds. S. Karger, Basel.
———. 1975b. The behaviour of siamang in the Krau Game Reserve. *Malay. Nat. J.* 29(1):7–22.

———. 1977a. The lesser apes. Pp. 539–598 in *Primate Conservation*. H.S.H. Prince Rainier III of Monaco and G. H. Bourne, eds. Academic Press, New York.

———. 1977b. The feeding behaviour of siamang (*Symphalangus syndactylus*). Pp. 355–382 in *Primate Ecology*. T. H. Clutton-Brock, ed. Academic Press, London.

———. 1977c. The ecology of gibbons: Some preliminary considerations based on observations in the Malay peninsula. Pp. 85–105 in *Use of Non-human Primates in Biomedical Research*. M. R. N. Prasad and T. C. Anand Kumar, eds. Indian National Science Academy, New Delhi, India.

Chivers, D. J., and W. Lane-Petter, eds. 1978. *Recent Advances in Primatology*, vol. 2: *Conservation*. Academic Press, London. 312 pp.

Chivers, D. J., J. J. Raemaekers, and F. P. G. Aldrich-Blake. 1975. Long-term observations of siamang behaviour. *Folia Primatol.* 23:1–49.

Clapp, R. B. 1974. Mammals imported into the U.S. in 1972. Special Scientific Report (Wildlife) no. 181. Bur. Sport Fisheries and Wildlife, Washington, D.C. 46 pp.

Clapp, R. B., and J. L. Paradiso. 1973. Mammals imported into the U.S. in 1971. Special Scientific Report (Wildlife) no. 171. Bur. Sport Fisheries and Wildlife, Washington, D.C. 52 pp.

Cloudsley-Thompson, J. L. 1973. Developments in the Sudan parks. *Oryx* 12(1):49–52.

———. 1974. The expanding Sahara. *Environ. Conserv.* 1(1):5–13.

Clutton-Brock, T. H. 1974. Activity patterns of red colobus (*Colobus badius tephrosceles*). *Folia Primatol.* 21:161–187.

———. 1975a. Feeding behaviour of red colobus and black and white colobus in East Africa. *Folia Primatol.* 23:165–207.

———. 1975b. Ranging behaviour of red colobus (*Colobus badius tephrosceles*) in the Gombe National Park. *Anim. Behav.* 23:706–722.

———, ed. 1977. *Primate Ecology*. Academic Press, London. 631 pp.

Clutton-Brock, T. H., and P. H. Harvey. 1977. Species differences in feeding and ranging behaviour in primates. Pp. 557–584 in *Primate Ecology*. T. H. Clutton-Brock, ed. Academic Press, London.

Coe, M. 1972. The South Turkana expedition, Scientific Papers IX. *Geographic J.* 138(3):316–338.

———. 1975. Mammalian ecological studies on Mt. Nimba, Liberia. *Mammalia* 39(4):523–587.

Coelho, A. M., Jr., C. A. Bramblett, and L. B. Quick. 1977. Social organization and food resource availability in primates: A socio-bioenergetic analysis of diet and disease hypotheses. *Am. J. Phys. Anthropol.* 46:253–264.

Coelho, A. M., Jr., C. A. Bramblett, L. B. Quick, and S. S Bramblett. 1976. Resource availability and population density in primates: A socio-bioenergetic analysis of the energy budgets of Guatemalan howler and spider monkeys. *Primates* 17(1):63–80.

Coelho, A. M., Jr., L. S. Coelho, C. A. Bramblett, S. S. Bramblett, and L. B. Quick. 1977. Ecology, population characteristics and sympatric association in primates: A sociobioenergetic analysis of howler and spider monkeys in Tikal, Guatemala. Pp. 96–135 in *Yearb. Phys. Anthropol.*, vol. 20. J. Buettner-Janusch, ed. Am. Assoc. Phys. Anthropol., Washington, D.C.

Coetzee, C. G. 1969. The distribution of mammals in the Namib Desert and adjoining inland escarpment. *Sci. Papers Namib Desert Res. Station (South West Africa)* 40:23–36.

Coimbra-Filho, A. F. 1969. Mico-leão, *Leontideus rosalia* (Linnaeus, 1766), situação atual da espécie no Brasil (Callithricidae, Primates). *An. Acad. Bras. Cienc.* (suppl.) 41:29–52.

———. 1970. Considerações gerais e situação atual dos micos-leões escuros, *Leontideus*

chrysomelas (Kuhl, 1820) e *Leontideus chrysopygus* (Mikan, 1823) (Callithricidae, Primates). *Rev. Bras. Biol.* 30:249–268.
———. 1971. Os sagüis do gênero *Callithrix* da região oriental Brasileira e um caso de duplohibridism entre três de saus formas (Callithricidae, Primates). *Rev. Bras. Biol.* 31:377–388.
———. 1972a. Conservation and use of South American primates in Brazil. *Int. Zoo Yearb.* 12:14–15.
*———. 1972b. Mamíferos ameaçados de extinção no Brasil. Pp. 13–28 in *Espécies da Fauna Brasileira Ameaçadas de Extinção.* Acad. Bras. Cienc., Rio de Janeiro.
———. 1974. Situação mundial de recursos faunísticos na faixa intertropical. *Brasil Florestal* 5(17):12–37.
———. 1978. Natural shelters of *Leontopithecus rosalia* and some ecological implications (Callitrichidae: Primates). Pp. 79–89 in *The Biology and Conservation of the Callitrichidae.* D. G. Kleiman, ed. Smithsonian Institution Press, Washington, D.C.
Coimbra-Filho, A. F., and A. D. Aldrighi. 1971. A restauracao da fauna do parque nacional da Tijuca. *Publ. Avulsas Museu. Nac.* 57:1–30.
Coimbra-Filho, A. F., and A. Magnanini. 1972. On the present status of *Leontopithecus* and some data about new behavioral aspects and management of *L. rosalia rosalia.* Pp. 59–69 in *Saving the Lion Marmoset.* D. D. Bridgewater, ed. Wild Animal Propagation Trust, Wheeling, West Virginia.
Coimbra-Filho, A. F., and R. A. Mittermeier. 1972. Taxonomy of the genus *Leontopithecus* Lesson, 1840. Pp. 3–22 in *Saving the Lion Marmoset.* D. D. Bridgewater, ed. Wild Animal Propagation Trust, Wheeling, West Virginia.
———. 1973a. Distribution and ecology of the genus *Leontopithecus* in Brazil. *Primates* 14:47–66.
———. 1973b. New data on the taxonomy of the Brazilian marmosets of the genus *Callithrix* Erxleben 1777. *Folia Primatol.* 20:241–264.
———. 1977. Conservation of the Brazilian lion tamarins (*Leontopithecus rosalia*). Pp. 60–94 in *Primate Conservation.* H.S.H. Prince Rainier III of Monaco and G. H. Bourne, eds. Academic Press, New York.
Colillas, O., and J. Coppo. 1978. Breeding *Alouatta caraya* in Centro Argentino de Primates. Pp. 201–214 in *Recent Advances in Primatology*, vol. 2: *Conservation.* D. J. Chivers and W. Lane-Petter, eds. Academic Press, London.
Collias, N., and C. Southwick. 1952. A field study of population density and social organization in howling monkeys. *Proc. Am. Phil. Soc.* 96:143–156.
Collins, W. E., P. G. Contacos, P. C. C. Garnham, M. Warren, and J. C. Skinner. 1972. *Plasmodium hylobati*: A malaria parasite of the gibbon. *J. Parasitol.* 58(1):123–128.
Committee of Concerned Asian Scholars. 1970. *The Indochina Story.* Pantheon Books, New York. 347 pp.
Committee on Conservation of Nonhuman Primates, Institute of Laboratory Animal Resources. 1975a. *Nonhuman Primates: Usage and Availability for Biomedical Programs.* National Academy of Sciences, Washington, D.C. 122 pp.
———. 1975b. Report on the Colombian and Peruvian primate censusing studies. National Academy of Sciences, Washington, D.C. 4 pp.
Conway, W. G. 1980. An overview of captive propagation. Pp. 199–208 in *Conservation Biology.* M. E. Soulé and B. A. Wilcox, eds. Sinauer Assoc., Sunderland, Mass.
Cook, R. S. 1979. Report of the U.S. delegation to the second meeting of the conference of the

*Reference not seen but cited by other authors.

parties to the Convention on International Trade in Endangered Species of Wild Fauna and Flora (C.I.T.E.S.), San Jose, Costa Rica, March 19–30, 1979. Fish and Wildlife Service, U.S. Department of Interior, Washington, D.C. 79 pp.

Cooper, R. W. 1968. Squirrel monkey taxonomy and supply. Pp. 1–29 in *The Squirrel Monkey*. L. A. Rosenblum and R. W. Cooper, eds. Academic Press, New York.

Cooper, R. W., and J. Hernández-Camacho. 1975. A current appraisal of Colombia's primate resources. Pp. 37–66 in *Primate Utilization and Conservation*. G. Bermant and D. G. Lindburg, eds. Wiley-Interscience, New York.

Cousins, D. 1976. Censuses of gorillas in zoological collections with notes on numerical status and conservation. *Int. Zoo News* 23(5):18–20.

———. 1978a. Man's exploitation of the gorilla. *Biol. Conserv.* 13(4):287–297.

———. 1978b. Gorillas: A survey. Part 1. *Oryx* 14(3):254–258.

———. 1978c. Gorillas: A survey. Part 2. *Oryx* 14(4):374–376.

Crespo, J. A. 1950. Notas sobre el habitat del "cai" en el norte de Salta. *Bol. San. Panamer.* 29:711–714.

———. 1954. Presence of the reddish howling monkey (*Alouatta guariba clamitans* Cabrera) in Argentina. *J. Mammal.* 35:117–118.

Crook, J. H. 1966. Gelada baboon herd structure and movement, a comparative report. *Symp. Zool. Soc. London* 18:237–258.

———. 1970. The socio-ecology of primates. Pp. 103–166 in *Social Behaviour in Birds and Mammals*. J. H. Crook, ed. Academic Press, London.

Crook, J. H., and F. P. G. Aldrich-Blake. 1968. Ecological and behavioural contrasts between sympatric ground-dwelling primates in Ethiopia. *Folia Primatol.* 8:192–227.

Curry-Lindahl, K. 1956. Ecological studies on mammals, birds, reptiles, and amphibians in the eastern Belgian Congo. *Ann. Mus. Roy. Congo Belge, Tervuren, Sci. Zool.* 42.

———. 1969. Report to the government of Liberia on conservation, management, and utilization of wildlife resources. *I.U.C.N. Publ.* New Series, Suppl. 24:1–31.

———. 1974a. Conservation problems and progress in equatorial Africa. *Environ. Conserv.* 1(2):111–122.

———. 1974b. Conservation problems and progress in northern and southern Africa. *Environ. Conserv.* 1(4):263–270.

Curtin, S. H. 1976. Niche differentiation and social organization in sympatric Malaysian colobines. Ph.D. diss. University of California, Berkeley.

———. 1977. Niche separation in sympatric Malaysian leaf monkeys (*Presbytis obscura* and *Presbytis melalophus*). Pp. 421–439 in *Yearb. Phys. Anthropol.*, vol. 20. J. Buettner-Janusch, ed. Am. Soc. Phys. Anthropol., Washington, D.C.

Curtin, S. H., and D. J. Chivers. 1978. Leaf-eating primates of peninsular Malaysia: The siamang and the dusky leaf-monkey. Pp. 441–464 in *The Ecology of Arboreal Folivores*. G. G. Montgomery, ed. Smithsonian Institution Press, Washington, D.C.

Dalquest, W. W. 1965. Mammals from Save River, Mozambique, with descriptions of two new bats. *J. Mammal.* 46:254–264.

———. 1968. Additional notes on mammals from Mozambique. *J. Mammal.* 49:117–121.

Dandelot, P. 1965. Distribution de quelques espèces de Cercopithecidae en relation avec les zones de végétation de l'Afrique. *Zool. Afr.* 1(1):167–176.

Dandelot, P., and J. Prévost. 1972. Contribution à l'étude des Primates d'Ethiopie (Simiens). *Mammalia* 36(4):607–633.

Dào Van Tien. 1961. Recherches zoologiques dans la région de Thai-Nguêyn (Nord-Vietnam). *Zool. Anz.* 166:298–308.

———. 1963. Étude preliminaire de la faune des mammifères de la region de Phu-Quy (Province de Nghe-An, Centre Vietnam). *Zool. Anz.* 171:448–456.

———. 1967. Notes sur une collection de mammifères de la region de Yên-Bai (Nord-Vietnam). *Mitt. Zool. Mus. Berlin* 43(1):117–125.

D'Arcy, W. G. 1977. Endangered landscapes in Panama and Central America: The threat to plant species. Pp. 89–104 in *Extinction Is Forever.* G. Prance and T. S. Elias, eds. New York Botanical Garden, Bronx, New York.

Dart, R. A. 1963. Carnivorous propensity of baboons. *Symp. Zool. Soc. London* 10:49–56.

Daugherty, H. E. 1972. The impact of man on the zoogeography of El Salvador. *Biol. Conserv.* 4(4):273–278.

———. 1973. The Montecristo cloud forest of El Salvador: A chance for protection. *Biol. Conserv.* 5(3):227–230.

Davenport, R. K. 1967. The orang-utan in Sabah. *Folia Primatol.* 5(4):247–263.

Davidge, C. 1976. A troop of baboons. *Afr. Wildl.* 30(4):23–24.

Davis, D. D. 1962. Mammals of North Borneo. *Bull. Nat. Mus. Singapore* 31:1–29.

Davis, D. H. S. 1957. Studies on arthropodborne viruses of Tongaland III: The small wild mammals in relation to the virus studies. *S. Afr. J. Med. Sci.* 22:55–61.

Davis, R. 1975. Island of enchantment: Nosy Bé. *Def. Wildl.* 50(2):141–147.

Dawson, G. A. 1976. Behavioral ecology of the Panamanian tamarin, *Saguinus oedipus* (Callitrichidae, Primates). Ph.D. diss. Michigan State University, East Lansing. 164 pp.

———. 1978. Composition and stability of social groups of the tamarin, *Saguinus oedipus geoffroyi*, in Panama: Ecological and behavioral implications. Pp. 23–37 in *The Biology and Conservation of the Callitrichidae.* D. G. Kleiman, ed. Smithsonian Institution Press, Washington, D.C.

Deag, J. M. 1973. Intergroup encounters in the wild barbary macaque. Pp. 316–373 in *Comparative Ecology and Behaviour of Primates.* R. P. Michael and J. H. Crook, eds. Academic Press, London.

———. 1977. The status of the Barbary macaque *Macaca sylvanus* in captivity and factors influencing its distribution in the wild. Pp. 267–287 in *Primate Conservation.* H.S.H. Prince Rainier III of Monaco and G. H. Bourne, eds. Academic Press, New York.

Deag, J. M., and J. H. Crook. 1971. Social behaviour and "agonistic buffering" in the wild Barbary macaque, *Macaca sylvanus* L. *Folia Primatol.* 15:183–200.

Deane, L. M. 1976. Epidemiology of simian malaria in the American continent. Pp. 144–163 in *First Inter-American Conference on Conservation and Utilization of American Nonhuman Primates in Biomedical Research.* Scientific Publication no. 317. Pan American Health Organization, Washington, D.C.

Deane, L. M., J. A. Ferreira-Neto, M. Okumura, and M. O. Ferreira. 1969. Malaria parasites of Brazilian monkeys. *Rev. Inst. Med. Trop. S. Paulo.* 11:71–86.

Deane, L. M., M. P. Deane, J.A. Ferreira-Neto, and F. B. de Almeida. 1971. On the transmission of simian malaria in Brazil. *Rev. Inst. Med. Trop. S. Paulo.* 13:311–319.

DeGraaf, G., K. C. A. Schulz, and P. T. Van der Walt. 1973. Notes on rumen contents of Cape buffalo *Syncerus caffer* in the Addo Elephant National Park. *Koedoe* 16:45–58.

*Dekeyser, P. L. 1950. Contribution à l'étude de l'Aïr. Mammifères. *Mém. Inst. Fr. Afr. Noire* 10:388–455.

*———. 1955. *Les mammifères de l'Afrique noire française.* Inst. Fr. Afr. Noire, Dakar.

Delacour, J. 1940. Liste provisoire de mammifères de l'Indochine francaise. *Mammalia* 4:20–29.

*Reference not seen but cited by other authors.

———. 1951. La systematique des gibbons Indochinois. *Mammalia* 15:118–123.
Delwaulle, J.-C. 1973. Désertification de l'Afrique au sud du Sahara. *Bois et Forêts des Tropiques* 149:3–20.
de Melo Carvalho, J. C. 1971a. The mono in Brazil. *Biol. Conserv.* 3(4):303.
———. 1971b. Three marmosets of the genus *Leontideus* in Brazil. *Biol. Conserv.* 4(1):66.
DeMoor, P. P., and F. E. Steffens. 1972. The movements of vervet monkeys (*Cercopithecus aethiops*) within their ranges as revealed by radio-tracking. *J. Anim. Ecol.* 41:677–687.
Deraniyagala, P. E. 1955. A new race of leaf monkey from Ceylon. *Spolia Zeylan* 27:293–294.
de Rodaniche, E. 1957. Survey of primates captured in Panama, R. P., during the years 1952–1956 for protective antibodies against yellow fever. *Am. J. Trop. Med. Hyg.* 6:835–839.
de Silva, G. S. 1968a. The east coast experiment. Pp. 299–302 in *Conservation in Tropical South East Asia.* I.U.C.N. Publ. New Series, no. 10.
———. 1968b. Wildlife conservation in the State of Sabah. Pp. 144–148 in *Conservation in Tropical South East Asia.* I.U.C.N. Publ. New Series, no. 10.
———. 1970. Training orang-utans for the wild. *Oryx* 10(6):389–393.
———. 1971. Notes on the orang-utan rehabilitation project in Sabah. *Malay. Nat. J.* 24:50–77.
DeVore, I. 1963. Mother-infant relations in free-ranging baboons. Pp. 305–335 in *Maternal Behavior in Mammals.* H. Rheingold, ed. Wiley, New York.
———. 1965. Changes in the population structure of Nairobi Park baboons 1959–1963. Pp. 17–28 in *The Baboon in Medical Research*, vol. 1. H. Vagtborg, ed. University of Texas Press, Austin.
DeVore, I., and K. R. L. Hall. 1965. Baboon ecology. Pp. 20–52 in *Primate Behavior.* I. DeVore, ed. Holt, Rinehart and Winston, New York.
DeVore, I., and S. Washburn. 1963. Baboon ecology and human evolution. Pp. 335–367 in *African Ecology and Human Evolution,* vol. 36. F. C. Howell and F. Bourlière, eds. Viking Fund Publ. Anthropol., Chicago.
de Vos, A. 1969. The need for nature reserves in East Africa. *Biol. Conserv.* 1:130–134.
———. 1977. Game as food. *Unasylva* 29(116):2–12.
de Vos, A., R. H. Manville, and R. G. Van Gelder. 1956. Introduced mammals and their influence on native biota. *Zoologica* 41:163–194.
de Vos, A., and A. Omar. 1971. Territories and movements of Sykes monkeys (*Cercopithecus mitis kolbi* Neuman) in Kenya. *Folia Primatol.* 16:196–205.
De Vos, V., C. A. W. J. Van Nierkerk, and E. E. McConnell. 1973. A survey of selected bacteriological infections of the chacma baboon *Papio ursinus* from the Kruger National Park. *Koedoe* 16:1–10.
Diamond, J. M. 1975. The island dilemma: Lessons of modern biogeographic studies for the design of natural reserves. *Biol. Conserv.* 7:129–146.
———. 1976. Island biogeography and conservation: Strategy and limitations. *Science* 193:1027–1029.
Disney, R. H. L. 1968. Observations on a zoonosis: Leishmaniasis in British Honduras. *J. Appl. Ecol.* 5(1):1–60.
Dittus, W. P. J. 1975. Population dynamics of the toque monkey *Macaca sinica*. Pp. 125–151 in *Socioecology and Psychology of Primates.* R. H. Tuttle, ed. Mouton, The Hague.
———. 1977a. The socioecological basis for the conservation of the toque monkey (*Macaca sinica*) of Sri Lanka (Ceylon). Pp. 237–265 in *Primate Conservation.* H.S.H. Prince Rainier III of Monaco and G. H. Bourne, eds. Academic Press, New York.
———. 1977b. The social regulation of population density and age-sex distribution in the toque monkey. *Behaviour* 63(3–4):281–322.

Dobroruka, L. J. 1967. Über den "Blassgesichtsmakak" nebst einer Übersicht über die Art *Macaca speciosa* (F. Cuvier 1825). *Milu* 2:305–312.

Dorst, J. 1953. A propos de la repartition de quelques mammifères dans l'archipel Malayo-Papou. *Mammalia* 17:306–317.

Dorst, J., and P. Dandelot. 1969. *A Field Guide to the Larger Mammals of Africa*. Houghton Mifflin, Boston. 287 pp.

Downs, W. G., T. H. G. Aitken, C. B. Worth, L. Spence, and A. H. Jonkers. 1968. Arbovirus studies in Bush Bush Forest, Trinidad, W. I., September 1959–December 1964. I. Description of the study area. *Am. J. Trop. Med. Hyg.* 17(2):224–236.

Doyle, G. A. 1974. Behavior of prosimians. Pp. 155–353 in *Behavior of Nonhuman Primates*, vol. 5. A. M. Schrier and F. Stollnitz, eds. Academic Press, New York.

Doyle, G. A., and S. K. Bearder. 1977. The galagines of South Africa. Pp. 1–35 in *Primate Conservation*. H.S.H. Prince Rainier III of Monaco and G. H. Bourne, eds. Academic Press, New York.

Dubost, G., and C.-M. Hladik. 1971. Les gros mammifères de la réserve de Wilpattu à Ceylan II. *Sci. Nat.* 105:3–11.

Dunbar, R. I. M. 1974. Observations on the ecology and social organization of the green monkey, *Cercopithecus sabaeus*, in Senegal. *Primates* 15:341–350.

———. 1977a. The gelada baboon: Status and conservation. Pp. 363–383 in *Primate Conservation*. H.S.H. Prince Rainier III of Monaco and G. H. Bourne, eds. Academic Press, New York.

———. 1977b. Feeding ecology of gelada baboons: A preliminary report. Pp. 251–273 in *Primate Ecology*. T. H. Clutton-Brock, ed. Academic Press, London.

Dunbar, R. I. M., and E. P. Dunbar. 1972. The social life of the gelada baboon. *Walia* 4:4–13.

———. 1974a. Ecological relations and niche separation between sympatric terrestrial primates in Ethiopia. *Folia Primatol.* 21:36–60.

———. 1974b. Ecology and population dynamics of *Colobus guereza* in Ethiopia. *Folia Primatol.* 21:188–208.

———. 1974c. Mammals and birds of the Simien Mountains National Park. *Walia* 5:4–5.

———. 1974d. On hybridization between *Theropithecus gelada* and *Papio anubis*. *J. Hum. Evol.* 3:187–192.

———. 1975a. Guereza monkeys: Will they become extinct in Ethiopia? *Walia* 6:14–15.

———. 1975b. *Social Dynamics of Gelada Baboons*. Vol. 6 of *Contrib. Primatol.* Pp. 1–157.

Dunbar, R. I. M., and M. F. Nathan. 1972. Social organization of the Guinea baboon, *Papio papio*. *Folia Primatol.* 17:321–334.

Dunnett, S., J. Van Orshoven, and H. Albrecht. 1970. Peaceful co-existence between chimpanzee and man in West Africa. *Bijdr. Dierkd.* 40(2):148–153.

Dupuy, A. R. 1969. Première capture d'un colobe bai (*Colobus badius temmincki* Kühl) au Sénégal oriental. *Mammalia* 33(4):733–734.

———. 1970a. Sur la présence du chimpanzé dans les limites du Parc National du Niokolo-Koba (Sénégal). *Bull. Inst. Fondam. Afr. Noire* 32:1090–1099.

———. 1970b. Recensement général de la faune au Parc National du Niokolo-Koba. *Notes Africaines, Inst. Fondam. Afr. Noire* 125:94–96.

———. 1971a. Le Parc National du Niokolo-Koba (République du Sénégal). *Bull. Inst. Fondam. Afr. Noire* 33:253–265.

———. 1971b. Statut actuel des Primates au Sénégal. *Bull. Inst. Fondam. Afr. Noire* 33(1–2):467–478.

———. 1971c. Le Parc National de Niokolo-Koba. *Biol. Conserv.* 3(4):308–310.

———. 1972a. Une nouvelle espèce de Primate pour le Sénégal: Le Cercopithèque hocheur, *Cercopithecus nictitans* (L.). *Mammalia* 36(2):306–307.

———. 1972b. Le Parc National des Oiseaux du Djoudj, République du Sénégal. *Bull. Inst. Fondam. Afr. Noire* 34:774–781.

———. 1973. Premier inventaire des mammifères du Parc National de Basse Casamance (Sénégal). *Bull. Inst. Fondam. Afr. Noire* 35:186–197.

———. 1975. Impressions sur quelques parcs nationaux du Kenya. *Bull. Inst. Fondam. Afr. Noire* 37:242–250.

Dupuy, A. R., and J. Vershuren. 1977. Wildlife and parks in Senegal. *Oryx* 14(1):36–46.

Durham, N. M. 1971. Effects of altitude differences on group organization of wild black spider monkeys (*Ateles paniscus*). Pp. 32–40 in *Proc. 3d Int. Congr. Primatol.*, vol. 3. S. Karger, Basel.

———. 1972. The distribution of nonhuman primates in southeastern Peru. Paper presented at Symposium on Distribution and Abundance of Neotropical Primates, August 1972, Battelle Seattle Research Center and Institute for Laboratory Animal Resources, National Research Council.

———. 1975. Some ecological, distributional and group behavioral features of Atelinae in southern Peru: With comments on interspecific relations. Pp. 87–101 in *Socioecology and Psychology of Primates*. R. H. Tuttle, ed. Mouton, The Hague.

Dutton, P. 1974. Chimanimani Mountains. *Afr. Wildl.* 28(1):27–31.

Editors of *Primates*. 1971. Aftereffects of the incident of aerial spraying of herbicide on Shimokita Peninsula: A continued report. *Primates* 12(1):97–99.

Eisenberg, J. F. 1980. The density and biomass of tropical mammals. Pp. 35–55 in *Conservation Biology*. M. E. Soulé and Bruce A. Wilcox, eds. Sinauer Assoc., Sunderland, Mass.

Eisenberg, J. F., and R. E. Kuehn. 1966. The behavior of *Ateles geoffroyi* and related species. *Smithsonian Misc. Collect.* 151(8):1–63.

Eisenberg, J. F., and M. Lockhart. 1972. An ecological reconnaissance of Wilpattu National Park, Ceylon. *Smithsonian Contrib. Zool.* 101:1–118.

Eisenberg, J. F., and G. M. McKay. 1970. An annotated checklist of the recent mammals of Ceylon with keys to the species. *Ceylon J. Sci.* 8(2):69–99.

Eisenberg, J. F., and R. W. Thorington, Jr. 1973. A preliminary analysis of a neotropical mammal fauna. *Biotropica* 5:150–161.

Eiten, G. 1975. An outline of the vegetation of South America. Pp. 529–545 in *Proc. Symp. 5th Congr. Int. Primatol. Soc.* S. Kondo, M. Kawai, A. Ehara, and S. Kawamura, eds. Japan Science Press, Tokyo.

Elbl, A., U. H. Rahm, and G. Mathys. 1966. Les mammifères et leurs tiques dans la forêt du Ruggege (République Rwandaise). *Acta Trop.* 23:223–263.

Ellefson, J. O. 1974. A natural history of white-handed gibbons in the Malayan peninsula. *Gibbon and Siamang* 3:1–136.

Ellerman, J. R., and T. C. S. Morrison-Scott. 1966. *Checklist of Palaearctic and Indian Mammals, 1758 to 1946*. 2d ed. British Museum (Natural History), London. 810 pp.

Elliot, O., and M. Elliot. 1967. Field notes on the slow loris in Malaya. *J. Mammal.* 48:497–498.

Elliott, R. G. 1976. Observations on a small group of mountain gorillas (*Gorilla gorilla beringei*). *Folia Primatol.* 25(1):12–24.

Emlen, J. T., and G. B. Schaller. 1960. Distribution and status of the mountain gorilla (*Gorilla gorilla beringei*)—1959. *Zoologica* 45(1):41–52.

Encarnación C., F., and N. Castro R. 1978. Informe preliminar sobre censo de primates no humanos en el sur-oriente Peruano: Iberia e Iñapari (Departamento de Madre de Dios) Mayo 15–Junio 14, 1978. Paper presented at Second Inter-American Conference on Conservation and Utilization of American Nonhuman Primates in Biomedical Research, Belém, Brazil. Pan American Health Organization, Washington, D.C.

Encarnación C., F., N. Castro R., and P. de Rham. 1978. Observaciones de primates no humanos en el Río Yavineto (R. Putumayo), Loreto, Peru. Paper presented at Second Inter-American Conference on Conservation and Utilization of American Non-human Primates in Biomedical Research, Belém, Brazil. Pan American Health Organization, Washington, D.C.

Ewel, J. J., and A. Madriz. 1968. Zonas de vida de Venezuela. Republica de Venezuela Ministerio de Agricultura y Cria. Direccion de Investigacion, Caracas.

Farook, S. M. S. 1974. Primates of Ceylon-I. *Loris* 13(3):149, 176.

Fiedler, W. 1956. Übersicht über das System der Primaten. Pp. 1–266 in *Primatologia*, vol. 1. H. Hofer, A. H. Schultz, and D. Starck, eds. S. Karger, Basel.

Fisher, J., N. Simon, and J. Vincent. 1969. *Wildlife in Danger*. Viking Press, New York.

Fitter, R. 1970. Living space for lemurs. *Oryx* 10:365–367.

———. 1974. Most endangered mammals: An action programme. *Oryx* 12(4):436–449.

Fitter, R., and M. Fitter. 1974. Ceylon and Rwanda. *Oryx* 12(4):475–478.

Fleming, T. H. 1973. Numbers of mammal species in North and Central American forest communities. *Ecology* 54:555–563.

Flor, L. 1973. Monkeys live high on the hog. *Washington Star News*, 7 October 1973.

Fogden, M. P. L. 1974. A preliminary field study of the western tarsier *Tarsius bancanus* Horsefield [*sic*]. Pp. 151–165 in *Prosimian Biology*. R. D. Martin, G. A. Doyle, and A. C. Walker, eds. University of Pittsburgh Press, Pittsburgh.

Fontaine, R., and F. V. Du Mond. 1977. The red ouakari in a seminatural environment: Potentials for propagation and study. Pp. 168–236 in *Primate Conservation*. H.S.H. Prince Rainier III of Monaco and G. H. Bourne, eds. Academic Press, New York.

Fooden, J. 1963. A revision of the wooly monkeys (genus *Lagothrix*). *J. Mammal.* 44:213–247.

———. 1964. Stomach contents and gastro-intestinal proportions in wild-shot Guianan Monkeys. *Am. J. Phys. Anthropol.* New Series 22:227–231.

———. 1967. Identification of the stumptailed monkey, *Macaca speciosa* I. Geoffroy 1826. *Folia Primatol.* 5:153–164.

———. 1969a. *Taxonomy and Evolution of the Monkeys of Celebes*. Vol. 10 of *Bibl. Primatol.* Pp. 1–148.

———. 1969b. Color-phase in gibbons. *Evolution* 23:627–644.

———. 1971a. Report on primates collected in western Thailand, January–April 1967. *Fieldiana Zool.* 59:1–62.

———. 1971b. Color and sex in gibbons. *Bull. Field Mus. Nat. Hist.* 42(6):2–7.

———. 1975. Taxonomy and evolution of liontail and pigtail macaques (Primates: Cercopithecidae). *Fieldiana Zool.* 67:1–169.

———. 1976. Primates obtained in peninsular Thailand June–July, 1973, with notes on the distribution of continental southeast Asian leaf-monkeys (*Presbytis*). *Primates* 17:95–118.

Fossey, D. 1970. Making friends with mountain gorillas. *Natl. Geog.* 137(1):48–67.

Fossey, D., and A. H. Harcourt. 1977. Feeding ecology of free-ranging mountain gorilla (*Gorilla gorilla beringei*). Pp. 415–447 in *Primate Ecology*. T. H. Clutton-Brock, ed. Academic Press, London.

Franklin, I. R. 1980. Evolutionary change in small populations. Pp. 135–149 in *Conservation Biology*. M. E. Soulé and B. A. Wilcox, eds. Sinauer Assoc., Sunderland, Mass.

*Freese, C. H. 1974. Progress Report No. 4: Monkey survey along the Samiria River. Pan American Health Organization, Washington, D.C.

———. 1975. A census of non-human primates in Peru. Pp. 17–41 in *Primate Censusing Studies in Peru and Colombia*. Pan American Health Organization, Washington, D.C.

*Reference not seen but cited by other authors.

———. 1976. Censusing *Alouatta palliata*, *Ateles geoffroyi* and *Cebus capucinus* in the Costa Rican dry forest. Pp. 4–9 in *Neotropical Primates: Field Studies and Conservation*. R. W. Thorington, Jr., and P. G. Heltne, eds. National Academy of Sciences, Washington, D.C.

———. 1977. Densities and niche separation in some Amazonian monkey communities. D.Sc. diss. Johns Hopkins University, Baltimore.

———. 1978. The behavior of white-faced capuchins (*Cebus capucinus*) at a dry-season waterhole. *Primates* 19(2):275–286.

Freese, C. H., M. A. Freese, and N. Castro R. 1978. The status of callitrichids in Peru. Pp. 121–130 in *The Biology and Conservation of the Callitrichidae*. D. G. Kleiman, ed. Smithsonian Institution Press, Washington, D.C.

Freese, C. H., P. G. Heltne, N. Castro, and G. Whitesides. In preparation. Patterns and determinants of monkey densities in Peru and Bolivia, with notes on distributions.

Freese, C., M. Neville, R. Castro, and N. Castro. 1976. The conservation status of some Peruvian primates. *Lab. Primate Newsl.* 15(3):1–9.

Frey, R. 1975. Sumatra's red apes return to the wild. *Wildlife* 17(8):356–363.

Frick, F. 1969. Die Hohenstufenverteilung der nepalischen Säugetiere. *Säugertierkd. Mitt.* 17(1):161–173.

Fukuda, K. 1975. A Japanese monkey population in the Boso Mountains: Its distribution and structure. Pp. 384–388 in *Contemporary Primatology. Proc. 5th Int. Congr. Primatol.* S. Kondo, M. I. Kawai, and A. Ehara, eds. S. Karger, Basel.

Furuya, Y. 1961–62a. On the ecological survey of the wild crab-eating monkeys in Malaya. *Primates* 3(1):75–76.

———. 1961–62b. The social life of silvered leaf monkeys (*Trachypithecus cristatus*). *Primates* 3:41–61.

Furuya, Y. 1965. Social organization of the crab-eating monkey. *Primates* 6:285–337.

Galat, G., and A. Galat-Luong. 1976. La colonisation de la mangrove par *Cercopithecus aethiops sabaeus* au Sénégal. *Terre Vie* 30:3–30.

———. 1977. Démographie et régime alimentaire d'une troupe de *Cercopithecus aethiops sabaeus* en habitat marginal au nord Sénégal. *Terre Vie* 31(4):557–577.

Galat-Luong, A. 1975. Notes preliminaires sur l'écologie de *Cercopithecus ascanius schmidti* dans les environs de Bangui (R.C.A.). *Terre Vie* 29:288–297.

Galdikas-Brindamour, B. 1975. Orang-utans, Indonesia's "people of the forest." *Nat. Geog.* 148:444–473.

Galindo, P. and S. Srihongse. 1967. Evidence of recent jungle yellow-fever activity in eastern Panama. *Bull. W.H.O.* 36:151–161.

Gardner, A. L. 1976. The distributional status of some Peruvian mammals. *Occ. Pap. Mus. Zool. Louisiana State University* 48:1–18.

Gartlan, J. S. 1970. Preliminary notes on the ecology and behavior of the drill, *Mandrillus leucophaeus* Ritgen, 1824. Pp. 445–481 in *Old World Monkeys*. J. R. Napier and P. H. Napier, eds. Academic Press, New York.

———. 1973. Influences of phylogeny and ecology on variations in the group organization of primates. Pp. 88–101 in *Precultural Primate Behavior, Symp. 4th Int. Congr. Primatol.*, vol. 1. E. W. Menzel, ed. S. Karger, Basel.

———. 1975a. The African coastal rain forest and its primates—threatened resources. Pp. 67–82 in *Primate Utilization and Conservation*. G. Bermant and D. G. Lindburg, eds. Wiley-Interscience, New York.

———. 1975b. Adaptive aspects of social structure in *Erythrocebus patas*. Pp. 161–171 in *Proc. Symp. 5th Congr. Int. Primatol. Soc.* S. Kondo, M. Kawai, A. Ehara, S. Kawamura, eds. Japan Science Press, Tokyo.

———. 1975c. The African forests and problems of conservation. Pp. 509–524 in *Proc. Symp. 5th Congr. Int. Primatol. Soc.* S. Kondo, M. Kawai, A. Ehara, and S. Kawamura, eds. Japan Science Press, Tokyo.

———. 1975d. A strategy for gorilla conservation efforts. *Lab. Primate Newsl.* 14(1):14–16.

———. 1976. Creation of two rain-forest parks in Cameroon. *Lab. Primate Newsl.* 15(1):4.

Gartlan, J. S., and C. K. Brain. 1968. Ecology and social variability in *Cercopithecus aethiops* and *C. mitis*. Pp. 253–293 in *Primates: Studies in Adaptation and Variability*. P. C. Jay, ed. Holt, Rinehart and Winston, New York.

Gartlan, J. S., and T. T. Struhsaker. 1972. Polyspecific associations and niche separation of rain-forest anthropoids in Cameroon, West Africa. *J. Zool.* 168:221–266.

Gatinot, B. L. 1974. Précisions sur la répartition du Colobe bai (*Colobus badius temmincki* Kuhl, 1820) et de la Mone de Campbell (*Cercopithecus mona campbelli* Waterhouse, 1838) en Sénégambie. *Mammalia* 38(4):711–716.

———. 1975. Écologie d'un Colobe bai (*Colobus badius temmincki* Kühl, 1820) dans un milieu marginal au Sénégal. Thèse de 3° Cycle, Université de Paris VI, Paris. 200 pp.

———. 1976. Les milieux fréquentés par le Colobe bai d'Afrique de l'ouest (*Colobus badius temmincki* Kühl, 1820) en Sénégambie. *Mammalia* 40(1):1–12.

Gautier, J. P., and A. Gautier-Hion. 1969. Les associations polyspécifiques chez les Cercopithecidae du Gabon. *Terre Vie* 23(2):164–202.

Gautier-Hion, A. 1966. L'écologie et l'éthologie du talapoin *Miopithecus talapoin talapoin*. *Biol. Gabonica* 2:311–329.

———. 1970. L'organisation sociale d'une bande de talapoins dans le N.-E. du Gabon. *Folia Primatol.* 12:116–141.

———. 1971. L'écologie du talapoin du Gabon. *Terre Vie* 25:427–490.

———. 1973. Social and ecological features of talapoin monkeys: Comparisons with sympatric cercopithecines. Pp. 148–170 in *Comparative Ecology and Behaviour of Primates*. R. P. Michael and J. H. Crook, eds. Academic Press, London.

Gautier-Hion, A., and J. P. Gautier. 1974. Les associations polyspécifiques de Cercopithèques du plateau de M'passa (Gabon). *Folia Primatol.* 22:134–177.

Gee, E. P. 1961. The distribution and feeding habits of the golden langur, *Presbytis geei* Gee (Khajuria, 1956). *J. Bombay Nat. Hist. Soc.* 58:1–12.

Geerling, C. 1974. Upper Ogun and Opara Game Reserves gazetted in Western State, Nigeria. *Environ. Conserv.* 1(4):304.

Geerling, C., and J. Bokdam. 1973. Fauna of the Comoe National Park, Ivory Coast. *Biol. Conserv.* 5:251–257.

Gengozian, N. 1969. Marmosets: Their potential in experimental medicine. *Ann. N.Y. Acad. Sci.* 162:336–362.

Gentry, A. 1977. Endangered plant species and habitats of Ecuador and Amazonian Peru. Pp. 136–149 in *Extinction Is Forever*. G. Prance and T. S. Elias, eds. New York Botanical Garden, Bronx, New York.

Ghosh, A. K., and S. Biswas. 1975. A note on ecology of the golden langur (*Presbytis geei* Khajuria). *J. Bombay Nat. Hist. Soc.* 72(2):524–528.

Gilbert, C., and J. Gillman. 1951. Pregnancy in the baboon (*Papio ursinus*). *S. Afr. J. Med. Sci.* 16:115–124.

Gittens, S. P. 1978. The species range of the gibbon *Hylobates agilis*. Pp. 319–321 in *Recent Advances in Primatology*, vol. 3: *Evolution*. D. J. Chivers and K. A. Joysey, eds. Academic Press, London.

Glander, K. E. 1971. A report of the adaptations of howling monkeys (*Alouatta palliata*) in a disturbed environment. *Am. J. Phys. Anthropol.* 35:280.

———. 1977. Poison in a monkey's garden of Eden. *Nat. Hist.* 86:34–41.

———. 1978. Drinking from arboreal water sources by mantled howling monkeys (*Alouatta palliata* Gray). *Folia Primatol.* 29:206–217.

Glass, B. P. 1965. The mammals of eastern Ethiopia. *Zool. Afr.* 1(1):177–179.

Gochfeld, M. 1974. Douc langurs. *Nature* 247:167.

Goodall, A. G. 1977. Feeding and ranging behaviour of a mountain gorilla group (*Gorilla gorilla beringei*) in the Tshibinda-Kahuzi Region (Zaïre). Pp. 450–479 in *Primate Ecology* T. H. Clutton-Brock, ed. Academic Press, London.

———. 1978. On habitat and home range in eastern gorillas in relation to conservation. Pp. 81–83 in *Recent Advances in Primatology*, vol. 2: *Conservation*. D. J. Chivers and W. Lane-Petter, eds. Academic Press, London.

Goodall, A. G., and C. P. Groves. 1977. The conservation of eastern gorillas. Pp. 599–637 in *Primate Conservation*. H.S.H. Prince Rainier III of Monaco and G. H. Bourne, eds. Academic Press, New York.

Goodall, J. 1963. Feeding behavior of wild chimpanzees. *Symp. Zool. Soc. London* 10:39–48.

———. 1965. Chimpanzees of the Gombe Stream Reserve. Pp. 425–473 in *Primate Behavior*. I. DeVore, ed. Holt, Rinehart and Winston, New York.

Goodland, R. J. A., and H. S. Irwin. 1975. *Amazon Jungle: Green Hell to Red Desert?* Elsevier, New York. 155 pp.

———. 1977. Amazon forest and cerrado: Development and environmental conservation. Pp. 214–233 in *Extinction Is Forever*. G. Prance and T. Elias, eds. New York Botanical Garden, Bronx, New York.

Goodwin, G. G. 1953. Catalogue of the type specimens of recent mammals in the American Museum of Natural History. *Bull. Am. Mus. Nat. Hist.* 102(3):221–411.

———. 1969. Mammals from the state Oaxaca, Mexico, in the American Museum of Natural History. *Bull. Am. Mus. Nat. Hist.* 141:1–269.

Gorgas, M. 1974. Auf tiersuche in Togo und Dahome. *Z. Köln. Zoo* 17:33–39.

Gray, D. D. 1978. When Boonsong talks, Thailand listens. *Int. Wildl.* 8(6):51–57.

Green, K. M. 1972. Exploitation and regulation of Colombia's primate resources. Paper presented at Third Tropical Symposium on Amazon Biology, Iquitos, Peru, November 21–24, 1972.

———. 1976. The nonhuman primate trade in Colombia. Pp. 85–98 in *Neotropical Primates: Field Studies and Conservation*. R. W. Thorington, Jr., and P. G. Heltne, eds. National Academy of Sciences, Washington, D.C.

———. 1978a. Primates of Bangladesh: A preliminary survey of population and habitat. *Biol. Conserv.* 13(2):141–160.

———. 1978b. Primate censusing in northern Colombia: A comparison of two techniques. *Primates* 19(3):537–550.

Green, S., and K. Minkowski. 1977. The lion-tailed monkey and its south Indian rain forest habitat. Pp. 289–337 in *Primate Conservation*. H.S.H. Prince Rainier III of Monaco and G. H. Bourne, eds. Academic Press, New York.

Greenhall, A. M. In preparation. Mammals imported into the United States in 1973. Unpublished preliminary draft. U.S. Fish and Wildlife Service, Washington, D.C.

Grimwood, I. R. 1969. Notes on the distribution and status of some Peruvian mammals, 1968. *Spec. Publ. Am. Comm. Int. Wildl. Protection* 21:v–86.

Griveaud, P., and R. Albignac. 1972. The problems of nature conservation in Madagascar. Pp. 727–739 in *Biogeography and Ecology of Madagascar*. R. Battistini and G. Richard-Vindard, eds. W. Junk, The Hague.

Groom, A. F. G. 1973. Squeezing out the mountain gorilla. *Oryx* 12:207–215.

Groves, C. P. 1967a. Geographic variation in the hoolock or white-browed gibbon (*Hylobates hoolock* Harlan 1834). *Folia Primatol.* 7:276–283.

———. 1967b. Ecology and taxonomy of the gorilla. *Nature* 213(5079):890–893.

———. 1968. A new subspecies of white-handed gibbon from northern Thailand, *Hylobates lar carpenteri* new subspecies. *Proc. Biol. Soc. Wash.* 81:625–628.

———. 1970a. Population systematics of the gorilla. *J. Zool.* 161:287–300.

———. 1970b. Taxonomic and individual variation in gibbons. *Symp. Zool. Soc. London* 26:127–134.

———. 1970c. *Gorillas: The World of Animals.* Weidenfeld and Nicolson, London.

———. 1970d. The forgotten leaf-eaters and the phylogeny of the Colobinae. Pp. 555–587 in *Old World Monkeys.* J. R. Napier and P. H. Napier, eds. Academic Press, New York.

———. 1971a. Distribution and place of origin of the gorilla. *Man* 6:44–51.

———. 1971b. Geographic and individual variation in Bornean gibbons, with remarks on the systematics of the subgenus *Hylobates.* *Folia Primatol.* 14:139–153.

———. 1971c. Systematics of the genus *Nycticebus.* Pp. 44–53 in *Proc. 3d Int. Congr. Primatol.*, vol. 1. S. Karger, Basel.

———. 1971d. *Pongo pygmaeus.* *Mamm. Species* 4:1–6.

———. 1972. Systematics and phylogeny of gibbons. *Gibbon and Siamang* 1:1–89.

———. 1973. Notes on the ecology and behaviour of the Angola colobus (*Colobus angolensis* P. L. Sclater 1860) in N.E. Tanzania. *Folia Primatol.* 20:12–26.

———. 1974. Taxonomy and phylogeny of prosimians. Pp. 449–473 in *Prosimian Biology.* R. D. Martin, G. A. Doyle, and A. C. Walker, eds. University of Pittsburgh Press, Pittsburgh.

———. 1976. The origin of the mammalian fauna of Sulawesi (Celebes). *Z. Säugetierkd.* 41(4):201–216.

———. 1978. A note on nomenclature and taxonomy in the Lemuridae. *Mammalia* 42(1):131–132.

Groves, C. P., P. Andrews, and J. F. M. Horne. 1974. Tana River colobus and mangabey. *Oryx* 12:565–575.

Grubb, P. 1973. Distribution, divergence, and speciation of the drill and mandrill. *Folia Primatol.* 20:161–177.

Gunderson, V. 1977. Some observations on the ecology of *Colobus badius temmincki*, Abuko Nature Reserve, the Gambia, West Africa. *Primates* 18(2):305–314.

Haddow, A. J. 1952. Field and laboratory studies on an African monkey, *Cercopithecus ascanius schmidti* Matschie. *Proc. Zool. Soc. London* 122:297–394.

———. 1958. Chimpanzees. *Uganda Wildl.* 1(3):18–20.

*Haddow, A. J., G. W. A. Dick, W. H. R. Lumsden, and K. C. Smithburn. 1951. Monkeys in relation to the epidemiology of yellow fever in Uganda. *Trans. R. Soc. Trop. Med. Hyg.* 45:189–224.

Hall, E. R., and K. R. Kelson. 1959. *The Mammals of North America*, vol. I. Ronald Press, New York. 546 pp.

Hall, K. R. L. 1960a. Social vigilance behaviour of the chacma baboon, *Papio ursinus.* *Behaviour* 16:261–294.

———. 1960b. The social behaviour of wild baboons. *New Sci.* 7:601–603.

———. 1962. Numerical data, maintenance activities, and locomotion of the wild chacma baboon, *Papio ursinus.* *Proc. Zool. Soc. London* 139:181–220.

———. 1963. Variations in the ecology of the chacma baboon, *Papio ursinus.* *Symp. Zool. Soc. London* 10:1–28.

*Reference not seen but cited by other authors.

———. 1965a. Behaviour and ecology of the wild patas monkey, *Erythrocebus patas*, in Uganda. *J. Zool.* 148:15–87.

———. 1965b. Ecology and behavior of baboons, patas, and vervet monkeys in Uganda. Pp. 43–61 in *The Baboon in Medical Research*, vol. 1. H. Vagtborg, ed. University of Texas Press, Austin.

———. 1965c. Experiment and quantification in the study of baboon behavior in its natural habitat. Pp. 29–42 in *The Baboon in Medical Research*, vol. 1. H. Vagtborg, ed. University of Texas Press, Austin.

———. 1965d. Social organization of the Old World monkeys and apes. *Symp. Zool. Soc. London* 14:265–289.

———. 1966. Distribution and adaptation of baboons. *Symp. Zool. Soc. London* 17:49–73.

Hall, K. R. L., R. C. Boelkins, and M. J. Goswell. 1965. Behaviour of patas monkeys *Erythrocebus patas* in captivity, with notes on the natural habitat. *Folia Primatol.* 3:22–49.

Hall, K. R. L., and I. DeVore. 1965. Baboon social behavior. Pp. 53–110 in *Primate Behavior*. I. DeVore, ed. Holt, Rinehart and Winston, New York.

Hall, K. R. L., and J. S. Gartlan. 1965. Ecology and behaviour of the vervet monkey, *Cercopithecus aethiops*, Lolui Island, Lake Victoria. *Proc. Zool. Soc. London* 145:37–56.

Hall, P. 1976. Priorities for wildlife conservation in north-eastern Nigeria. *Nigerian Field* 41(3):99–112.

Haltenorth, T. 1975. Gebirgssäugetiere: Eine Ubersicht. *Säugertierkd. Mitt.* 23(1):112–137.

Hamilton, W. J., III, R. E. Buskirk, and W. H. Buskirk. 1976. Defense of space and resources by chacma (*Papio ursinus*) baboon troops in an African desert and swamp. *Ecology* 57:1264–1272.

Hampton, S. H. 1972. Golden lion marmoset conference. *Science* 177(4043):86–87.

Hancock, N. J., ed. 1969. The Oxford and Bristol Expedition to Ethiopia 1968. *Bull. Oxford Univ. Explor. Club* 17(2):iv, 13–24.

Handley, C. O., Jr. 1966. Checklist of the mammals of Panama. Pp. xii, 1–861 in *Ectoparasites of Panama*. R. L. Wenzel and V. J. Tipton, eds. Field Museum Nat. Hist., Chicago.

Hansinger, M. J., E. L. Simons, D. R. Pilbeam, A. D. Horn, and J. S. Gartlan. 1974. The 1972 field study of the pygmy chimpanzee, *Pan paniscus*, in central Africa. *Am. J. Phys. Anthropol.* 40:139.

Hanson, H. M. 1968. Use of the squirrel monkey in pharmacology. Pp. 365–392 in *The Squirrel Monkey*. L. A. Rosenblum and R. W. Cooper, eds. Academic Press, New York.

Happold, D. C. D. 1966. The mammals of Jebel Marra, Sudan. *J. Zool.* 149:126–136.

———. 1971. A Nigerian high forest reserve. Pp. 58–59 in *Wildlife Conservation in West Africa*. I.U.C.N. Publ. New Series, no. 22. D. C. D. Happold, ed. Morges, Switzerland.

———. 1972. Mammals in Nigeria. *Oryx* 11(6):469.

———. 1973. The red crowned mangabey *Cercocebus torquatus*, in western Nigeria. *Folia Primatol.* 20:423–428.

Happold, D. C. D., and B. Philp. 1971. The national parks of northern Dahomey. *Nigerian Field* 36(4):182–188.

———. 1972. The national parks of northern Dahomey. Part II: Appendices. *Nigerian Field* 37(1):39–44.

Harcourt, A. H. 1977. Virunga gorillas: The case against translocations. *Oryx* 13(5):469–472.

Harcourt, A. H., and K. Curry-Lindahl. 1978. The F.P.S. mountain gorilla project: A report from Rwanda. *Oryx* 14(4):316–324.

Harcourt, A. H., and A. F. G. Groom. 1972. Gorilla census. *Oryx* 11(5):355–363.

Harding, R. S. O. 1976. Ranging patterns of a troop of baboons (*Papio anubis*) in Kenya. *Folia Primatol.* 25:143–185.

Harrington, J. E. 1975. Field observations of social behavior of *Lemur fulvus fulvus* E. Geoffroy 1812. Pp. 259–279 in *Lemur Biology*. I. Tattersall and R. W. Sussman, eds. Plenum Press, New York.

———. 1978. Diurnal behavior of *Lemur mongoz* at Ampijoroa, Madagascar. *Folia Primatol.* 29:291–302.

Harrison, B. 1963. The education of young orang-utans to living in the wild. *Oryx* 7:108–127.

———. 1971. Conservation of Non-human Primates in 1970. *Primates in Medicine,* vol. 5. E. I. Goldsmith and J. Moor-Jankowski, eds. S. Karger, Basel. 99 pp.

Harrison, D. L. 1964. *The Mammals of Arabia,* vol. 1. E. Benn, Ltd., London. 192 pp.

———. 1968. The large mammals in Arabia. *Oryx* 9(5):357–363.

Harrison, J. L. 1965. The effect of forest clearance on small mammals. In *Conservation in Tropical Southeast Asia*. I.U.C.N. Publ. New Series, no. 10.

———. 1966. *Mammals of Singapore and Malaya*. Singapore Branch, Malayan Nature Society, Singapore.

———. 1969. The abundance and population density of mammals in Malayan lowland forests. *Malay. Nat. J.* 22:174–178.

Harrison, T. 1970. The orang-utan situation, 1970. *Biol Conserv.* 3(1):45–46.

Harroy, J.-P. 1972. *Addendum-Corrigendum to the Second Edition of the United Nations List of National Parks and Equivalent Reserves*. Hayez, Brussels.

Hassinger, J. D. 1973. A survey of the mammals of Afghanistan. *Fieldiana Zool.* 60:1–195.

Hayashi, K. 1975. Interspecific interaction of the primate groups in Kibale forest, Uganda. *Primates* 16(3):269–283.

Hayflick, L. 1970. The choice of the cell substrate for human virus vaccine production. *Lab. Pract.* 19:58–62.

Hayman, R. W. 1963. Mammals from Angola, mainly from the Lunda District. *Publ. Cult. Comp. Diamantes, Angola* 66:81–139.

Heimpel, H. 1974. Hilfe für den Orangutan. *Kosmos* 70:209–213.

Heltne, P. G. 1978. Census of *Aotus* in the north of Colombia 4 August–30 September 1977. Paper presented at Second Inter-American Conference on Conservation and Utilization of American Nonhuman Primates in Biomedical Research, Belém, Brazil. Pan American Health Organization, Washington, D.C.

Heltne, P. G., C. Freese, and G. Whitesides. 1976. *A Field Survey of Nonhuman Primates in Bolivia*. Pan American Health Organization, Washington, D.C. 36 pp.

Heltne, P. G., and L. M. Kunkel. 1975. Taxonomic notes on the pelage of *Ateles paniscus paniscus, A. p. chamek* (sensu Kellogg and Goldman, 1944) and *A. fusciceps rufiventris* (=*A. f. robustus*, Kellogg and Goldman, 1944). *J. Med. Primatol.* 4(2):83–102.

Heltne, P. G., and R. W. Thorington, Jr. 1976. Problems and potentials for primate biology and conservation in the New World. Pp. 110–124 in *Neotropical Primates: Field Studies and Conservation*. R. W. Thorington, Jr., and P. G. Heltne, eds. National Academy of Sciences, Washington, D.C.

Heltne, P. G., D. C. Turner, and N. J. Scott, Jr. 1976. Comparison of census data on *Alouatta palliata* from Costa Rica and Panama. Pp. 10–19 in *Neotropical Primates: Field Studies and Conservation*. R. W. Thorington, Jr., and P. G. Heltne, eds. National Academy of Sciences, Washington, D.C.

Hendrichs, H. 1970. Schätzungen der Huftierbiomasse in der Dornbuschsavanne nördlich und westlich der Serengetisteppe in Ostafrika nach einem neuen Verfahren und Bemerkungen zur Biomasse der anderen pflanzenfressenden Tierarten. *Säugetierkd. Mitt.* 3:237–255.

———. 1972. Beobachtungen und Untersuchungen zur Ökologie und Ethologie, insbesondere zur sozialen Organisation, ostafrikanischer Säugetiere. *Z. Tierpsychol.* 30:146–189.

———. 1975. The status of the tiger *Panthera tigris* (Linne 1758) in the Sundarbans mangrove forest (Bay of Bengal). *Säugetierkd. Mitt.* 23(3):161–200.

———. 1977. Untersuchung zur Säugetierfauna in einem paläotropischen und einem neotropischen Bergregenwaldgebiet. *Säugetierkd. Mitt.* 25(3):214–224.

Henshaw, J., and J. Ayeni. 1971. Some aspects of big-game utilization of mineral licks in Yankari Game Reserve, Nigeria. *East Afr. Wildl. J.* 9:73–82.

Henshaw, J., and G. S. Child. 1972. New attitudes in Nigeria. *Oryx* 11(4):275–283.

Hernández-Camacho, J., and E. Barriga-Bonilla. 1966. Hallazgo del genero *Callimico* (Mammalia: Primates) en Colombia. *Caldasia* 9(44):365–377.

Hernández-Camacho, J., and R. W. Cooper. 1976. The nonhuman primates of Colombia. Pp. 35–69 in *Neotropical Primates: Field Studies and Conservation*. R. W. Thorington, Jr., and P. G. Heltne, eds. National Academy of Sciences, Washington, D.C.

Hershkovitz, P. 1949. Mammals of northern Colombia, preliminary report no. 4: Monkeys (Primates), with taxonomic revisions of some forms. *Proc. U.S. Nat. Mus.* 98:323–427.

———. 1955. Notes on American monkeys of the genus *Cebus*. *J. Mammal.* 36(3):449–452.

———. 1963. A systematic and zoogeographic account of the monkeys of the genus *Callicebus* (Cebidae) of the Amazonas and Orinoco River basins. *Mammalia* 27(1):1–79.

———. 1966a. Taxonomic notes on tamarins, genus *Saguinus* (Callithricidae, Primates), with descriptions of four new forms. *Folia Primatol.* 4:381–395.

———. 1966b. On the identification of some marmosets, Family Callithricidae (Primates). *Mammalia* 30(2):327–332.

———. 1968. Metachromism or the principle of evolutionary change in mammalian tegumentary colors. *Evolution* 22(3):556–575.

———. 1969. The evolution of mammals of southern continents. VI. The recent mammals of the Neotropical Region: A zoogeographic and ecological review. *Quart. Rev. Biol.* 4(1):1–70.

———. 1972. Notes on New World monkeys. *Int. Zoo Yearb.* 12:3–12.

———. 1975. Comments on the taxonomy of the Brazilian marmosets (*Callithrix*, Callitrichidae). *Folia Primatol.* 24(2–3):137–172.

———. 1977. *Living New World Monkeys (Platyrrhini)*, vol. 1. University of Chicago Press, Chicago. 1,117 pp.

Heymans, J. C. 1975. Contribution à la détermination des Primates de la région du Haut-Zaïres (République de Zaïre): Tableaux synoptiques. *Nat. Belg.* 56:73–82.

Heymans, J. C., and J. C. Meurice. 1973. L'exploitation de la faune sauvage en République du Zaïre. *Nat. Belg.* 54:246–254.

Hickman, R. L. 1969. The use of subhuman primates for experimental studies of human malaria. *Milit. Med.* 134:741–756.

Hill, C. A. 1968. Uakaries . . . white, red, and black-headed. *Zoonooz* 41(6):15–18.

———. 1970. The last of the golden marmosets. *Lab. Primate Newsl.* 9(2):4–8.

———. 1971. Saving the wanderoo. *Zoonooz* 44(12):13.

Hill, W. C. O. 1953. *Primates: Comparative Anatomy and Taxonomy*, vol. 1: *Strepsirhini*. University Press, Edinburgh.

———. 1955. *Primates: Comparative Anatomy and Taxonomy*, vol. 2: *Haplorhini: Tarsoidea*. University Press, Edinburgh. 347 pp.

———. 1957. *Primates: Comparative Anatomy and Taxonomy*, vol. 3: *Pithecoidea: Platyrrhini (Families Hapalidae and Callimiconidae)*. University Press, Edinburgh. 354 pp.

———. 1960. *Primates: Comparative Anatomy and Taxonomy*, vol. 4: *Cebidae*. Part A. University Press, Edinburgh. 523 pp.

———. 1962. *Primates: Comparative Anatomy and Taxonomy*, vol. 5: *Cebidae*. Part B. University Press, Edinburgh. 537 pp.

———. 1963. The Ufiti: The present position. Pp. 57–60 in *The Primates, Symp. Zool. Soc. London*, vol. 10.

———. 1966. *Primates: Comparative Anatomy and Taxonomy*. vol. 6: *Cercopithecinae*. University Press, Edinburgh. 757 pp.

———. 1967a. Taxonomy of the baboon. Pp. 3–12 in *The Baboon in Medical Research*, vol. 2. H. Vagtborg, ed. University of Texas Press, Austin.

———. 1967b. The taxonomy of the genus *Pan*. Pp. 47–54 in *Progress in Primatology*. D. Starck, R. Schneider, and H.-J. Kuhn, eds. G. Fischer Verlag, Stuttgart.

———. 1969. The nomenclature, taxonomy, and distribution of chimpanzees. *Chimpanzee* 1:22–49.

———. 1970. *Primates: Comparative Anatomy and Taxonomy*, vol. 8: *Cynopithecinae: Papio, Mandrillus, Theropithecus*. John Wiley and Sons, New York. 680 pp.

———. 1972. Taxonomic status of the macaques *Macaca mulatta* Zimmerman and *Macaca irus* Cuvier (=*M. fascicularis* Raffles). *J. Hum. Evol.* 1:49–51.

———. 1974. *Primates: Comparative Anatomy and Taxonomy*, vol. 7: *Cynopithecinae: Cercocebus, Macaca, Cynopithecus*. John Wiley and Sons, New York. 934 pp.

Hladik, C.-M. 1974. La vie d'un groupe de chimpanzés dans la forêt du Gabon. *Sci. Nat.* 121:5–14.

———. 1975. Ecology, diet, and social patterning in Old and New World primates. Pp. 3–35 in *Socioecology and Psychology of Primates*. R. H. Tuttle, ed. Mouton, The Hague.

———. 1979. Diet and ecology of prosimians. Pp. 307–357 in *The Study of Prosimian Behavior*. G. A. Doyle and R. D. Martin, eds. Academic Press, New York.

Hladik, C.-M., and P. Charles-Dominique. 1971 Lépilémur et autres lémuriens du sud de Madagascar. *Sci. Nat.* 106:30–38.

———. 1974. The behaviour and ecology of the sportive lemur (*Lepilemur mustelinus*) in relation to its dietary peculiarities. Pp. 23–37 in *Prosimian Biology*. R. D. Martin, G. A. Doyle, and A. C. Walker, eds. University of Pittsburgh Press, Pittsburgh.

Hladik, C. M., and A. Hladik. 1972. Disponibilités alimentaires et domaines vitaux des primates à Ceylan. *Terre Vie* 26:149–215.

*Holdridge, L. R. 1967. *Life Zone Ecology*. Tropical Science Center, San Jose, Costa Rica.

Homewood, K. 1975. Can the Tana mangabey survive? *Oryx* 13(1):53–59.

———. 1977. A well adjusted monkey. *Anim. Kingdom* 80(6):16–22.

Honegger, R. E., and P. Menichini. 1962. A census of captive gorillas with notes on diet and longevity. *Zool. Gart.* 2:203–214.

Hopkins, B. 1970. The Olokemeji Forest Reserve. IV. Check lists. *Nigerian Field* 35:123–144.

Horn, A. D. 1979. The taxonomic status of the bonobo chimpanzee. *Am. J. Phys. Anthropol.* 51(2):273–282.

Horr, D. A. 1975. The Borneo orang-utan: Population structure and dynamics in relationship to ecology and reproductive strategy. Pp. 307–323 in *Primate Behavior*, vol. 4. L. A. Rosenblum, ed. Academic Press, New York.

———. 1977. Orang-utan maturation: Growing up in a female world. Pp. 289–321 in *Primate Bio-Social Development*. S. Chevalier-Skolnikoff and F. E. Poirier, eds. Garland Press, New York.

Horr, D. A., and M. Ester. 1976. Orang-utan social structure: A computer simulation. Pp.

*Reference not seen but cited by other authors.

3–53 in *The Measures of Man: Methodologies in Biological Anthropology*. E. Giles and J. Friedlaender, eds. Peabody Museum Press, Cambridge, Mass.

Horwich, R. H. 1972. Home range and food habits of the Nilgiri langur, *Presbytis johnii*. *J. Bombay Nat. Hist. Soc.* 69:255–267.

Howard, A. N., and G. A. Gresham. 1966. The care and use of baboons in the laboratory. *Symp. Zool. Soc. London* 17:75–89.

Howell, J. H. 1968. The Borgu Game Reserve of Northern Nigeria. Part 2: *Nigerian Field* 33(4):147–165.

———. 1969. The Borgu Game Reserve of Northern Nigeria. Part 3: *Nigerian Field* 34(1):32–35.

Huard, P. 1962. Archéologie et zoologie: Contribution à l'étude des singes au Sahara oriental et central. *Bull. Inst. Fr. Afr. Noire* B24:86–104.

Hufnagl, E. 1972. *Libyan Mammals*. Oleander Press. Stoughton, Wis. 85 pp.

Hunsaker, D., II. 1972. National parks in Colombia. *Oryx* 11(6):441–448.

Husson, A. M. 1957. Notes on the primates of Suriname. *Studies on the Fauna of Suriname and Other Guyanas* 2:13–40.

Hutton, A. F. 1949. Notes on the snakes and mammals of the High Wavy Mountains, Madura District, South India. Part 2: Mammals. *J. Bombay Nat. Hist. Soc.* 48:681–694.

Institute of Laboratory Animal Resources (I.L.A.R.). Newsletter. 1969, 1970. Summary of questionnaire to determine number of animals used for research. National Academy of Sciences, Washington, D.C. 13(1), 14(1).

———. Newsletter. 1971. Survey of nonhuman primates being maintained on 1 Jan. 1971. National Academy of Sciences, Washington, D.C. 15(1).

Interagency Primate Steering Committee. 1978a. Report of the task force on the use of and need for chimpanzees. National Institutes of Health, Bethesda, Maryland. 25 pp.

———. 1978b. *National Primate Plan*. U.S. Department of Health Education and Welfare Publication no. (N.I.H.) 80–1520, Washington, D.C. 81 pp.

International Union for Conservation of Nature (I.U.C.N.). 1971. *United Nations List of National Parks and Equivalent Reserves*. Hayez, Brussels. 601 pp.

International Union for Conservation of Nature and Natural Resources (I.U.C.N.). 1972. *Red Data Book, Mammalia*, vol. 1. I.U.C.N., Morges, Switzerland.

———. 1975. *World Directory of National Parks and Other Protected Areas*, vol. 1. I.U.C.N., Morges, Switzerland.

———. 1976. *Red Data Book: Update Sheets*. I.U.C.N., Morges, Switzerland.

———. 1978. *Red Data Book: Update Sheets*. I.U.C.N., Morges, Switzerland.

Itani, J. 1975. Twenty years with Mount Takasaki monkeys. Pp. 101–125 in *Primate Utilization and Conservation*. G. Bermant and D. G. Lindburg, eds. Wiley-Interscience, New York.

Iwano, T. 1975. Distribution of Japanese monkey (*Macaca fuscata*). Pp. 389–391 in *Contemporary Primatology, Proc. 5th Int. Congr. Primatol.* S. Kondo, M. Kawai, and A. Ehara, eds. S. Karger, Basel.

Izawa, K. 1970. Unit groups of chimpanzees and their nomadism in the savanna woodland. *Primates* 11:1–45.

———. 1971. Japanese monkeys living in the Okoppe Basin of the Shimokita Peninsula: The first report of the winter follow-up survey after the aerial spraying of herbicide. *Primates* 12:191–200.

———. 1972. Japanese monkeys living in the Okoppe Basin of the Shimokita Peninsula: The second report of the winter follow-up survey after the aerial spraying of herbicide. *Primates* 13:201–212.

———. 1976. Group sizes and compositions of monkeys in the upper Amazon basin. *Primates* 17(3):367–399.

———. 1978. A field study of the ecology and behavior of the black-mantle tamarin (*Saguinus nigricollis*). *Primates* 19(2):241–274.

Izawa, K., and A. Mizuno. 1977. Palm-fruit cracking behavior of wild black-capped capuchin (*Cebus apella*). *Primates* 18(4):773–792.

Izawa, K., and T. Nishida. 1963. Monkeys living in the northern limit of their distribution. *Primates* 4(2):67–88.

Jackson, H. D. 1973a. Records of some birds and mammals in the central Chimanimani Mountains of Moçambique and Rhodesia. *Durban Mus. Novit.* 9(20):291–305.

———. 1973b. Faunal notes from the Chimanimani Mountains, based on a collection of birds and mammals from the Mucrera River, Moçambique. *Durban Mus. Novit.* 10(2):23–42.

———. 1975. Further notes on the birds and mammals of the Chimanimani Mountains. *Durban Mus. Novit.* 10:181–187.

Jahnke, H. E. 1974. Conservation and utilization of wildlife in Uganda: A study in environmental economics. *Forschungsberichte der Afrika-Studienstelle, I.F.O.–Instituts für Wirtschaftsforschung*, vol. 54.

Janson, C., and J. Terborgh. In press. Censusing primates in rainforest. *Publ. Mus. Nat. Hist. "Javier Prado."* Lima, Peru.

Jarman, P. J. 1972. Seasonal distribution of large mammal populations in the unflooded middle Zambezi Valley. *J. Appl. Ecol.* 9:283–299.

Jay, P. C. 1965. The common langur of north India. Pp. 197–249 in *Primate Behavior*. I. DeVore, ed. Holt, Rinehart and Winston, New York.

Jay, P. C., and D. G. Lindburg. 1965. Indian Primate Ecology Project: September 1964–June 1965. Unpublished preliminary report.

*Jeannin, A. 1936. *Les mammifères sauvages du Cameroun*. Paul Lechevalier, Paris.

Jeffrey, S. M. 1970. Ghana's forest wildlife in danger. *Oryx* 10(4):240–243.

———. 1974. Primates of the dry high forest of Ghana. *Nigerian Field* 39:117–127.

———. 1975a. Ghana's new forest national park. *Oryx* 13(1):34–36.

———. 1975b. Notes on mammals from the high forest of western Ghana (excluding Insectivora). *Bull. Inst. Fondam. Afr. Noire* 37:950–972.

———. 1977. How Liberia uses wildlife. *Oryx* 14(2):168–173.

Jenkins, D. V. 1971. Animal life of Kinabalu National Park. *Malay. Nat. J.* 24:177–183.

Jewell, P. A., and J. F. Oates. 1969. Ecological observations on the lorisoid primates of African lowland forests. *Zool. Afr.* 4(2):231–248.

Jolly, A. 1966. *Lemur Behavior: A Madagascar Field Study*. University of Chicago Press, Chicago. 187 pp.

———. 1972. Troop continuity and troop spacing in *Propithecus verreauxi* and *Lemur catta* at Berenty (Madagascar). *Folia Primatol.* 17:335–362.

Jolly, C. J. 1966. Introduction to the Cercopithecoidea with notes on their use as laboratory animals. *Symp. Zool. Soc. London* 17:426–457.

———. 1970. The large African monkeys as an adaptive array. Pp. 139–174 in *Old World Monkeys*. J. R. Napier and P. H. Napier, eds. Academic Press, New York.

Jones, C. 1969. Notes on ecological relationships of four species of lorisids in Rio Muni, West Africa. *Folia Primatol.* 11:255–267.

———. 1970. Mammals imported into the U.S. in 1968. Special Scientific Report (Wildlife) no. 137. Bur. Sport Fisheries and Wildlife, Washington, D.C. 30 pp.

*Reference not seen but cited by other authors.

Jones, C., T. Martin, and W. Mason. 1970. Survival of an escaped *Callicebus moloch* in southern Louisiana. *Lab. Primate Newsl.* 9:6–7.

Jones, C., and J. L. Paradiso. 1972. Mammals imported into the U.S. in 1969. Special Scientific Report (Wildlife) no. 147. Bur. Sport Fisheries and Wildlife, Washington, D.C. 33 pp.

Jones, C., and J. Sabater Pi. 1968. Comparative ecology of *Cercocebus albigena* (Gray) and *Cercocebus torquatus* (Kerr) in Rio Muni, West Africa. *Folia Primatol.* 9:99–113.

———. 1971. *Comparative Ecology of* Gorilla gorilla *(Savage and Wyman) and* Pan troglodytes *(Blumenbach) in Rio Muni, West Africa.* Vol. 13 of *Bibl. Primatol.* Pp. 1–96.

Jones, D. M. 1973. Destruction in Niger. *Oryx* 12(2):227–234.

Jones, J. K., Jr., H. H. Genoways, and J. D. Smith. 1974. Annotated checklist of mammals of the Yucutan Peninsula, Mexico. III. Marsupialia, Insectivora, Primates, Edentata, Lagomorpha. *Occ. Pap. Mus. Texas Tech. Univ.* 23:1–12.

Jones, T. S. 1950. Notes on the monkeys of Sierra Leone. *Sierra Leone Agricultural Notes*, no. 22. Department of Agriculture, Sierra Leone.

———. 1966. Notes on the commoner Sierra Leone mammals. *Nigerian Field* 31(1):4–18.

Joubert, E., and P. K. N. Mostert. 1975. Distribution patterns and status of some animals in South West Africa. *Madoqua* 9(1):5–44.

Jouventin, P. 1975. Observations sur la socio-ecology du mandrill. *Terre Vie* 29:493–532.

*Kano, T. 1940. *Zoogeographical Studies of the Tsugitaka Mountains of Formosa.* Shibusawa Inst. Ethnogr. Res., Tokyo. 145 pp.

———. 1971a. Distribution of the primates on the eastern shore of Lake Tanganyika. *Primates* 12:281–304.

———. 1971b. The chimpanzee of Filabanga, western Tanzania. *Primates* 12:229–246.

Kao, Y. T., C. K. Lu, C. Chang, and S. Wang. 1962. Mammals of the Hsi-Shuan-Pan-Na area in southern Yunnan. *Acta Zool. Sin.* 14:180–196.

Karr, J. R. 1973. Ecological and behavioral notes on the lion-tailed macaque (*Macaca silenus*) in south India. *J. Bombay Nat. Hist. Soc.* 70:191–193.

Kavanagh, M., and L. Dresdale. 1975. Observations on the wooly monkey (*Lagothrix lagothricha*) in northern Colombia. *Primates* 16(3):285–294.

Kawabe, M. 1970. A preliminary study of the wild siamang gibbon (*Hylobates syndactylus*) at Fraser's Hill, Malaysia. *Primates* 11:285–292.

Kawabe, M., and T. Mano. 1972. Ecology and behavior of the wild proboscis monkey, *Nasalis larvatus* (Wurmb), in Sabah, Malaysia. *Primates* 13(2):213–227.

Kawai, M., and H. Mizuhara. 1959. An ecological study on the wild mountain gorilla. *Primates* 2:1–43.

Kawamura, S. 1961. A pilot study on the social life of white-handed gibbons in northwestern Thailand. Pp. 159–169 in *Nature and Life in Southeast Asia*, vol. 1. T. Kira and T. Umesao, eds. Fauna and Flora Research Society, Kyoto, Japan.

———. 1975. The present situation of Japanese monkeys and consideration of their conservation. Pp. 501–507 in *Proc. Symp. 5th Congr. Int. Primatol. Soc.* S. Kondo, M. Kawai, A. Ehara, and S. Kawamura, eds. Japan Science Press, Tokyo.

Kawanaka, K. 1973. Intertroop relationships among Japanese monkeys. *Primates* 14:113–159.

Kawanaka, K., and T. Nishida. 1975. Recent advances in the study of inter-unit-group relationships and social structure of wild chimpanzees of the Mahali Mountains. Pp. 173–186 in *Proc. Symp. 5th Congr. Int. Primatol. Soc.* S. Kondo, M. Kawai, A. Ehara, and S. Kawamura, eds. Japan Science Press, Tokyo.

*Reference not seen but cited by other authors.

Kellogg, R., and E. A. Goldman. 1944. Review of the spider monkeys. *Proc. U.S. Nat. Mus.* 3186:1–45.

Kern, J. A. 1964. Observations on the habits of the proboscis monkey, *Nasalis larvatus* (Wurmb), made in the Brunei Bay area, Borneo. *Zoologica* 49:183–192.

Khajuria, H. 1956. A new langur (Primates: Colobidae) from Goalpara District, Assam. *Ann. Mag. Nat. Hist.* 9(12):86–88.

———. 1977. Ecological observations on the golden langur, *Presbytis geei* Khajuria, with remarks on its conservation. Pp. 52–61 in *Use of Non-human Primates in Biomedical Research*. M. R. N. Prasad and T. C. Anand Kumar, eds. Indian National Science Academy, New Delhi, India.

Khajuria, H., and R. K. Ghose. 1970. On a collection of small mammals from Darjeeling District, West Bengal. *J. Bengal Nat. Hist. Soc.* 36:15–36.

Khan, M. K. B. M. 1970. Distribution and population of siamang and gibbons in the State of Perak. *Malay. Nat. J.* 24(1):3–8.

———. 1978. Man's impact on the primates of peninsular Malaysia. Pp. 41–46 in *Recent Advances in Primatology*, vol. 2: *Conservation*. D. J. Chivers and W. Lane-Petter, eds. Academic Press, London.

Kingdon, J. 1971. *East African Mammals: An Atlas of Evolution in Africa*, vol. 1. Academic Press, London. 445 pp.

Kingston, T. J. 1971. Notes on the black and white *Colobus* monkey in Kenya. *East Afr. Wildl. J.* 9:172–175.

Kingston, W. R. 1969. Marmosets and tamarins. *Lab. Anim. Handb.* 4:243–250.

Kinzey, W. G. 1977a. Diet and feeding behaviour of *Callicebus torquatus*. Pp. 127–151 in *Primate Ecology*. T. H. Clutton-Brock, ed. Academic Press, London.

———. 1977b. Positional behavior and ecology in *Callicebus torquatus*. Pp. 468–480 in *Yearb. Phys. Anthropol.*, vol. 20. J. Buettner-Janusch, ed. Am. Assoc. Phys. Anthropol., Washington, D.C.

Kinzey, W. G., A. L. Rosenberger, P. S. Heisler, D. L. Prowse, and J. S. Trilling. 1977. A preliminary field investigation of the yellow-handed titi monkey, *Callicebus torquatus torquatus*, in northern Peru. *Primates* 18(1):159–181.

Kirkpatrick, R. D., and A. M. Cartwright. 1975. List of mammals known to occur in Belize. *Biotropica* 7(2):136–140.

Kittinanda, S. 1975. Country reports: Thailand. Pp. 53–56 in *The Use of Ecological Guidelines for Development in Tropical Forest Areas of South East Asia*. I.U.C.N. Publ. New Series, no. 32. Morges, Switzerland.

Kleiman, D. G. 1972. Recommendations for research priorities for the lion marmoset. Pp. 137–139 in *Saving the Lion Marmoset*. D. D. Bridgewater, ed. Wild Animal Propagation Trust, Wheeling, West Virginia.

———. 1976. International conference on the biology and conservation of the Callitrichidae. *Primates* 17(1):119–123.

———, ed. 1978. *The Biology and Conservation of the Callitrichidae*. Smithsonian Institution Press, Washington, D.C. 348 pp.

Kleiman, D. G., and M. Jones. 1978. The current status of *Leontopithecus rosalia* in captivity with comments on breeding success at the National Zoological Park. Pp. 215–218 in *The Biology and Conservation of the Callitrichidae*. D. G. Kleiman, ed. Smithsonian Institution Press, Washington, D.C.

Klein, L. L., and D. J. Klein. 1975. Social and ecological contrasts between four taxa of neotropical primates. Pp. 59–85 in *Socioecology and Psychology of Primates*. R. H. Tuttle, ed. Mouton, The Hague.

———. 1976. Neotropical primates: Aspects of habitat usage, population density, and regional distribution in La Macarena, Colombia. Pp. 70–78 in *Neotropical Primates: Field Studies and Conservation*. R. W. Thorington, Jr., and P. G. Heltne, eds. National Academy of Sciences, Washington, D.C.

———. 1977. Feeding behaviour of the Colombian spider monkey. Pp. 153–182 in *Primate Ecology*. T. H. Clutton-Brock, ed. Academic Press, London.

Kling, A., and J. Orbach. 1963. The stump-tailed macaque: A promising laboratory primate. *Science* 139:45–46.

Klopfer, P. H., and K. J. Boskoff. 1979. Maternal behavior in prosimians. Pp. 123–156 in *The Study of Prosimian Behavior*. G. A. Doyle and R. D. Martin, eds. Academic Press, New York.

Klopfer, P. H., and A. Jolly. 1970. The stability of territorial boundaries in a lemur troop. *Folia Primatol.* 12:199–208.

Kock, D. 1967. Die Verbreitung des Schimpansen, *Pan troglodytes schweinfurthii* (Giglioli, 1872) in Sudan. *Z. Säugetierkd.* 32:250–255.

———. 1969. Die Verbreitung der Primaten in Sudan. *Z. Säugetierkd.* 34:193–216.

Koechlin, J. 1972. Flora and vegetation of Madagascar. Pp. 145–190 in *Biogeography and Ecology of Madagascar*. R. Battistini and G. Richard-Vindard, eds. W. Junk, The Hague.

Koepcke, M. 1972. Über die Resistenzformen der Vogelnester in einem begrenzten Gebiet des tropischen Regenwaldes in Peru. *J. Ornithol.* 113:138–160.

Koford, C. B. 1969a. Conservation of primates in Latin America. *I.U.C.N. Publ.* New Series 13:142–145.

———. 1969b. Monkeys of the snowy forest. *Animal Kingdom* 72(1):10–14.

Kortlandt, A. 1966. Chimpanzee ecology and laboratory management. *Lab. Primate Newsl.* 5(3):1–11.

———. 1976. Letters: Statements on pygmy chimpanzees. *Lab. Primate Newsl.* 15(1):15–17.

Kortlandt, A., and J. C. J. Van Zon. 1969. The present state of research on the dehumanization hypothesis of African ape evolution. Pp. 10–13 in *Proc. 3d Int. Congr. Primatol.*, vol. 3. S. Karger, Basel.

Koyama, N., K. Norikoshi, and T. Mano. 1975. Population dynamics of Japanese monkeys at Arashiyama. Pp. 411–417 in *Contemporary Primatology, Proc. 5th Int. Congr. Primatol.* S. Kondo, M. Kawai, and A. Ehara, eds. S. Karger, Basel.

*Krieg, H. 1928. Schwarze Brüllaffen (*Alouatta caraya* Humboldt). *Z. Säugetierkd.* 2:119–132.

Krishnamurti, K. 1968. A note on the slow loris and its taxonomy. *Malay. Nat. J.* 21:71–72.

Krishnan, M. 1971. An ecological survey of the larger mammals of peninsular India. *J. Bombay Nat. Hist. Soc.* 68(3):503–555.

Krumbiegel, I. 1978. Die Kurzschwanz-Stumpfnase, *Simias concolor* (Miller, 1903), und die übrigen Nasenaffen. *Säugetierkd. Mitt.* 26(1):59–76.

Kühlhorn, F. 1955. Säugetierkundliche Studien aus Südmattogrosso. 4. Artiodactyla, Primates. *Säugetierkd. Mitt.* 3:156–164.

Kuhn, H.-J. 1965. A provisional checklist of the mammals of Liberia (with notes on the status and distribution of some species). *Senckenb. Biol.* 46:321–340.

———. 1972. Die Geschichte der Säugetiere Madagaskars. *Z. Köln. Zoo* 15(1):28–42.

Kullmann, E. 1970. Die Tierwelt Ostafghanistans in ihren geographischen Beziehungen. *Freunde Kölner Zoo* 13(1):3–25.

*Reference not seen but cited by other authors.

Kumm, H. W., and H. W. Laemmert, Jr. 1950. The geographical distribution of immunity to yellow fever among the primates of Brazil. *Am. J. Trop. Med.* 30:733–748.

Kummer, H. 1968a. *Social Organization of Hamadryas Baboons: A Field Study*. Vol. 6 of *Bibl. Primatol.* Pp. 1–189.

———. 1968b. Two variations in the social organization of baboons. Pp. 293–313 in *Primates: Studies in Adaptation and Variability*. P. C. Jay, ed. Holt, Rinehart and Winston, New York.

———. 1971. *Primate Societies: Group Techniques of Ecological Adaptation*. Aldine-Atherton, Chicago.

Kummer, H., and F. Kurt. 1963. Social units of a free-living population of hamadryas baboons. *Folia Primatol.* 1:4–19.

Kuntz, R. E. 1969. Vertebrates taken for parasitological studies by U.S. Naval-Medical Research Unit expedition to North Borneo (Malaysia). *Quart. J. Taiwan Mus.* 22:191–206.

Kuntz, R. E., and J. A. Moore. 1973. Commensals and parasites of African baboons captured in the Rift Valley Province of central Kenya. *J. Med. Primatol.* 2(3–4):236–241.

Kuntz, R. E., and B. J. Myers. 1969. A checklist of parasites and commensals reported for the Taiwan macaque (*Macaca cyclopis* Swinhoe, 1862). *Primates* 10(1):71–81.

Kurland, J. A. 1973. A natural history of kra macaques (*Macaca fascicularis* Raffles, 1821) at the Kutai Reserve, Kalimantan Timur, Indonesia. *Primates* 14:245–262.

Kurt, F. 1970. Indiens Tierwelt in Gefahr. *Freunde Kölner Zoo* 13:43–56.

———. 1973. Der Gunung Leuser Survey 1970. *Z. Köln. Zoo* 16(2):59–74.

Kurup, G. U. 1975. Status of the Nilgiri langur, *Presbytis johnii*, in the Anamalai, Cardamom, and Nilgiri Hills of the Western Ghats, India. *J. Bombay Nat. Hist. Soc.* 72(1):21–29.

———. 1977. Distribution, habitat, and conservation of the rain forest primates in the Western Ghats, India. Pp. 62–73 in *Use of Non-human Primates in Biomedical Research*. M. R. N. Prasad and T. C. Anand Kumar, eds. Indian National Science Academy, New Delhi, India.

Kutilek, M. J. 1974. The density and biomass of large mammals in Lake Nakuru National Park. *East Afr. Wildl. J.* 12:201–212.

La Bastille, A. 1973. An ecological survey of the proposed Volcan Baru National Park, Republic of Panama. *I.U.C.N. Occasional Paper*, no. 6. Morges, Switzerland. 77 pp.

Ladnyj, I. D., P. Ziegler, and E. Kima. 1972. A human infection caused by monkey pox virus in Basankusu Territory, Democratic Republic of the Congo. *Bull. W.H.O.* 46:593–597.

Laemmert, H. W., Jr., L. de C. Ferreira, and R. M. Taylor. 1946. An epidemiological study of jungle yellow fever in an endemic area in Brazil. Part II. Investigations of vertebrate hosts and arthropod vectors. *Am. J. Trop. Med.* 26(6)Supp:23–60.

Lahan, P., and R. N. Sonowal. 1973. Kaziranga Wildlife Sanctuary, Assam, a brief description and report on the census of large animals (March 1972). *J. Bombay Nat. Hist. Soc.* 70:245–278.

Lamprey, H. 1964. Estimation of the large mammal densities, biomass, and energy exchange in the Tarangire Game Reserve and the Masai Steppe in Tanganyika. *East Afr. Wildl. J.* 2:1–46.

Lancaster, J. B. 1971. Play-mothering: The relations between juvenile females and young infants among free-ranging vervet monkeys (*Cercopithecus aethiops*). *Folia Primatol.* 15:161–182.

Lanly, J. P. 1969. Régression de la forêt dense en Côte-d'Ivoire. *Bois For. Trop.* 127:45–59.

Lapin, B. A. 1965. Reproduction of some human diseases in monkeys, mainly in baboons. Pp. 599–606 in *The Baboon in Medical Research*, vol. 1. H. Vagtborg, ed. University of Texas Press, Austin.

Lapin, B. A., E. K. Dzhikidze, and L. A. Yakovleva. 1965. *Diseases of Primates under Natural Conditions in Vietnam.* Meditsina, Moscow.

Lavieren, L. P. van, and M. L. Bosch. 1977. Évaluation des densités de grands mammifères dans le Parc National de Bouba Ndjida, Cameroun. *Terre Vie* 31:3–32.

Leakey, L. S. B. 1969. *Animals of East Africa.* National Geographic Society, Washington, D.C.

Lee, P. C. 1975. Country report: Malaysia. Pp. 25–32 in *The Use of Ecological Guidelines for Development in Tropical Forest Areas of South East Asia.* I.U.C.N. Publ. New Series, no. 32. Morges, Switzerland.

Lee, V. H. 1972. Ecological aspects of the Jos Plateau, Nigeria. *Bull. W.H.O.* 46:641–644.

Lekagul, B. 1968. Threatened species of fauna of Thailand. Pp. 267–271 in *Conservation in Tropical South East Asia.* I.U.C.N. Publ. New Series, no. 10.

Lekagul, B., and J. A. McNeely. 1977. *Mammals of Thailand.* Association for the Conservation of Wildlife, Bangkok. 758 pp.

———. 1978. Thailand launches extensive reafforestation program. *Tigerpaper* 5(1):9–13.

Leopold, A. S. 1959. *Wildlife of Mexico.* University of California Press, Berkeley.

Leskes, A., and N. Acheson. 1971. Social organization of a free-ranging troop of black and white colobus (*Colobus abyssinicus*). Pp. 22–31 in *Proc. 3d Int. Congr. Primatol.*, vol. 3. S. Karger, Basel.

Leutenegger, W. 1976. Metric variability in the anterior dentition of African colobines. *Am J. Phys. Anthropol.* 45:45–52.

Levine, B. M., J. L. Smith, and C. W. Israol. 1970. Serology of normal primates. *Br. J. Vener. Dis.* 46:307–310.

Lim, B. L. 1969. Distribution of the primates of West Malaysia. Pp. 121–130 in *Proc. 2d Int. Congr. Primatol.*, vol. 2. S. Karger, Basel.

Lim, B. L., and D. Heyneman. 1968. A collection of small mammals from Tuaran and the southwest face of Mt. Kinabalu, Sabah. *Sarawak Mus. J.* 16:257–276.

Lim, B. L., I. Muul, and C. K. Shin. 1977. Zoonotic studies of small mammals in the canopy transect at Bukit Lanjan Forest Reserve, Selangor, Malaysia. *Malay. Nat. J.* 31(2):127–140.

Lindburg, D. G. 1971. The rhesus monkey in North India: An ecological and behavioral study. Pp. 2–106 in *Primate Behavior*, vol. 2. L. A. Rosenblum, ed. Academic Press, New York.

———. 1977. Feeding behaviour and diet of rhesus monkeys (*Macaca mulatta*) in a Siwalik forest in North India. Pp. 223–249 in *Primate Ecology.* T. H. Clutton-Brock, ed. Academic Press, London.

Lippold, L. K. 1977. The douc langur: A time for conservation. Pp. 513–538 in *Primate Conservation.* H.S.H. Prince Rainier III of Monaco and G. H. Bourne, eds. Academic Press, New York.

Lock, J. M. 1972. Baboons feeding on *Euphorbia candelabrum.* *East Afr. Wildl. J.* 10:73–76.

Long, R. C. 1973. A list with notes of the mammals of the Nsanje (Port Herald) District, Malawi. *Soc. Malawi J.* 26(1):60–78.

*Lönnberg, E. 1919. Contributions to the knowledge about the monkeys of Belgian Congo. *Revue Zoologique Africaine* 7:107–154.

Lowes, R. H. G. 1970. Destruction in Sierra Leone. *Oryx* 10:309–310.

Lüling, K. H. 1975. Einiges zur Wiederentdeckung des Gelbschwanz-Wollaffen (*Lagothrix flavicauda*). *Z. Köln. Zoo* 18(4):142–143.

*Reference not seen but cited by other authors.

Lumsden, W. H. R. 1951. The night resting habits of monkeys in a small area on the edge of the Semliki Forest, Uganda. *J. Anim. Ecol.* 20:11–30.
Lynch, C. D. 1975. The distribution of mammals in the Orange Free State, South Africa. *Navorsinge van die Nasionale Museum* 3(6):109–139.
Lynes, H. 1921. Notes on the natural history of Jebel Marra. *Sudan Notes Rec.* 4:119–137.
Lyon, L. 1975a. In the home of the mountain gorillas. *Wildlife* 17(1):16–23.
———. 1975b. The saving of the gorilla. *Africana* 5(9):11–13, 23.
Macedo-Ruíz, H. de, R. A. Mittermeier, and A. Luscombe. 1974. Rediscovery of *Lagothrix flavicauda*, the Peruvian yellow-tailed wooly monkey. Press Conference, Lima, Peru, May 10, 1974.
Mackenzie, A. F. 1952. Notes and exhibitions, the Society for Scientific Business, May 13, 1952. *Proc. Zool. Soc. London* 122(2):541.
MacKinnon, J. R. 1971. The orang-utan in Sabah today. *Oryx* 11(2–3):141–191.
———. 1973. Orang-utans in Sumatra. *Oryx* 12:234–242.
———. 1974a. The behaviour and ecology of wild orang-utans. *Anim. Behav.* 22:3–74.
———. 1974b. *In Search of the Red Ape.* Holt, Rinehart and Winston, New York. 222 pp.
———. 1976. Mountain gorillas and bonobos. *Oryx* 13(4):372–382.
———. 1977a. A comparative ecology of the Asian apes. *Primates* 18(4):747–772.
———. 1977b. The future of orang-utans. *New Sci.* 74(1057):697–699.
MacKinnon, J. R., and K. S. MacKinnon. 1978. Comparative feeding ecology of six sympatric primates in West Malaysia. Pp. 305–321 in *Recent Advances in Primatology*, vol. 1: *Behaviour.* D. J. Chivers and J. Herbert, eds. Academic Press, London.
MacRoberts, B. R., and M. H. MacRoberts. 1971. The apes of Gibraltar. *Nat. Hist.* 80(7):38–47.
Magnanini, A. 1978. Progress in the development of Poço das Antas Biological Reserve for *Leontopithecus rosalia rosalia* in Brazil. Pp. 131–136 in *The Biology and Conservation of the Callitrichidae.* D. G. Kleiman, ed. Smithsonian Institution Press, Washington, D.C.
Magnanini, A., and A. F. Coimbra-Filho. 1972. The establishment of a captive breeding program and a wildlife research center for the lion marmoset *Leontopithecus* in Brazil. Pp. 110–119 in *Saving the Lion Marmoset.* D. D. Bridgewater, ed. Wild Animal Propagation Trust, Wheeling, West Virginia.
Magnanini, A., A. F. Coimbra-Filho, R. A. Mittermeier, and A. Aldrighi. 1975. The Tijuca Bank of lion marmosets *Leontopithecus rosalia*: A progress report. *Int. Zoo. Yearb.* 15:284–287.
Makwana, S. C. 1978. Field ecology and behaviour of the rhesus macaque (*Macaca mulatta*): I. Group composition, home range, roosting sites, and foraging routes in the Asarori Forest. *Primates* 19(3):483–492.
*Malbrant, R. 1952. *Faune du Centre Africain Français.* Lechevalier, Paris.
Malbrant, R., and A. Maclatchy. 1949. Faune de l'Equateur Africain Français. *Encyclopédie Biologique* 36.
Malinow, M. R. 1968. Introduction to *Biology of the Howler Monkey* (Alouatta caraya). M. R. Malinow, ed. Vol. 7 of *Bibl. Primatol.* Pp. 1-12.
Mallinson, J. J. C. 1974. Wildlife studies on the Zaire River expedition with special reference to the mountain gorillas of Kahuzi-Biega. *Ann. Rept. Jersey Wildl. Pres. Trust.* 11:16–23.
Maples, W. R. 1972a. Systematic reconsideration and a revision of the nomenclature of Kenya baboons. *Am. J. Phys. Anthropol.* 36:9–20.
———. 1972b. Survey of the monkey populations of Kenya: A proposal to Division of Sponsored Research, submitted October 27, 1972.

Maples, W. R., M. K. Maples, W. F. Greenhood, and M. L. Walek. 1976. Adaptations of crop-raiding baboons in Kenya. *Am. J. Phys. Anthropol.* 45:309–315.

Maples, W. R., and T. W. McKern. 1967. A preliminary report on classification of the Kenya baboon. Pp. 13–22 in *The Baboon in Medical Research*, vol. 2. H. Vagtborg, ed. University of Texas Press, Austin.

Marler, P. 1969. *Colobus guereza*: Territoriality and group composition. *Science* 163:93–95.

Marsh, C. 1977. Patch of river, ray of hope. *Anim. Kingdom* 80(6):9–15.

Marsh, C., and K. Homewood. 1975. The Tana, a new offbeat safari for Kenya. *Africana* 5(10):17, 21, 24.

Marshall, J. T., B. A. Ross, and S. Chantharojvong. 1972. The species of gibbons in Thailand. *J. Mammal.* 53(3):479–486.

Marshall, J. T., and V. Nongngork. 1970. Mammals of Samui Island, Thailand. *Nat. Hist. Bull. Siam Soc.* 23(4–5):501–507.

Martin, R. D. 1972a. A preliminary field study of the lesser mouse lemur (*Microcebus murinus* J. F. Miller 1777). *Advances in Ethology: J. Comp. Ethol.*, supp. 9 (Behaviour and ecology of nocturnal prosimians):43–89.

———. 1972b. Adaptive radiation and behaviour of the Malagasy lemurs. *Philos. Trans. R. Soc. London, B, Biol. Sci.* 264(862):295–352.

———. 1973. A review of the behaviour and ecology of the lesser mouse lemur (*Microcebus murinus* J. F. Miller 1777). Pp. 1–68 in *Comparative Ecology and Behaviour of Primates*. R. P. Michael and J. H. Crook, eds. Academic Press, London.

Martin, R. D., and S. K. Bearder. 1978. Feeding and ranging behaviour of the lesser bushbaby, studied with radio-tracking. *Primate Eye* 10:9–10.

Mason, W. A. 1966. Social organization of the South American monkey *Callicebus moloch*: A preliminary report. *Tulane Stud. Zool.* 13:23–28.

———. 1968. Use of space by *Callicebus* groups. Pp. 200–216 in *Primates: Studies in Adaptation and Variability*. P. C. Jay, ed. Holt, Rinehart and Winston, New York.

———. 1971. Field and laboratory studies of social organization in *Saimiri* and *Callicebus*. Pp. 107–137 in *Primate Behavior*, vol. 2. L. A. Rosenblum, ed. Academic Press, New York.

Masui, K., Y. Sugiyama, A. Nishimura, and H. Ohsawa. 1975. The life table of Japanese monkeys at Takasakiyama: A preliminary report. Pp. 401–406 in *Contemporary Primatology, Proc. 5th Int. Congr. Primatol.* S. Kondo, M. Kawai, and A. Ehara, eds. S. Karger, Basel.

Matthew, K. M., F. Blasco, and S. Ignacimuthu. 1975. Biological changes at Kodaikanal, 1949–1974. *Trop. Ecol.* 16(2):147–162.

Maxim, P. E., and J. Buettner-Janusch. 1963. A field study of the Kenya baboon. *Am. J. Phys. Anthropol.* 21:165–180.

McCann, C. 1933a. Observations on some of the Indian langurs. *J. Bombay Nat. Hist. Soc.* 36:618–628.

———. 1933b. Notes on some Indian macaques. *J. Bombay Nat. Hist. Soc.* 36:796–810.

———. 1933c. Notes on the coloration and habits of the white-browed gibbon or hoolock (*Hylobates hoolock* Harlan). *J. Bombay Nat. Hist. Soc.* 36:395–405.

McClure, H. E. 1964. Some observations of primates in climax dipterocarp forest near Kuala Lumpur, Malaya. *Primates* 5(3–4):39–58.

McConnell, E. E., P. A. Basson, V. DeVos, B. J. Meyers, and R. E. Kuntz. 1974. A survey of diseases among 100 free-ranging baboons (*Papio ursinus*) from the Kruger National Park. *Onderstepoort J. Vet. Res.* 41(3):97–167.

McCrae, A. W. R., B. E. Henderson, B. G. Kirya, and S. D. K. Sempala. 1971. Chikungunya virus in the Entebbe area of Uganda: Isolations and epidemiology. *Trans. R. Soc. Trop. Med. Hyg.* 65(2):152–167.

McGrew, W. C. 1976. Petition on use of pygmy chimpanzees for research. *Lab. Primate Newsl.* 15(1):7–9.

McNeely, J. A. 1977. Mammals of the Thai mangroves. *Tigerpaper* 4(1):10–15.

McNeely, J. A., and A. Laurie. 1977. Rhinos in Thailand. *Oryx* 13:486–489.

Medway, L. 1969a. *The Wild Mammals of Malaya and Offshore Islands Including Singapore.* Oxford University Press, London. 127 pp.

———. 1969b. The monkeys of Sundaland. Paper prepared in advance for participants in Symposium no. 43, "Systematics of the Old World Monkeys," Wenner-Gren Foundation for Anthropological Research.

———. 1970. The monkeys of Sundaland: Ecology and systematics of the cercopithecids of a humid equatorial environment. Pp. 513–554 in *Old World Monkeys.* J. R. Napier and P. H. Napier, eds. Academic Press, New York.

———. 1972. The Gunong Benom Expedition 1967. VI. The distribution and altitudinal zonation of birds and mammals on Gunong Benom. *Bull. Br. Mus. (Nat. Hist.) Zool.* 23:105–154.

———. 1976. Hunting pressure on orang-utans in Sarawak. *Oryx* 13(4):332–333.

Medway, L., and D. R. Wells. 1971. Diversity and density of birds and mammals at Kuala Lompat, Pahang. *Malay. Nat. J.* 24:238–247.

Meester, J. 1965. The origins of the southern African mammal fauna. *Zool. Afr.* 1(1):87–95.

Mei, D. J. 1977. The prospects of log importation for industrial expansion in peninsular Malaysia. *Malays. For.* 40(2):85–89.

Méndez, E. 1970. *Los principales mamíferos silvestres de Panama.* Privately printed, Panama City. 283 pp.

Middleton, C. C., A. F. Moreland, and R. W. Cooper. 1972. Problems of New World primate supply. *Lab. Primate Newsl.* 11(2):10–17.

Millett, A. R. 1978. *A Short History of the Vietnam War.* Indiana University Press, Bloomington. 169 pp.

Milton, K., and M. L. May. 1976. Body weight, diet and home range area in primates. *Nature* (London) 259:459–462.

Milton, K., and R. A. Mittermeier. 1977. A brief survey of the primates of Coiba Island, Panama. *Primates* 18(4):931–936.

Milton, O. 1963. Field notes on wildlife conservation in Malaya. *I.U.C.N. Special Publ.*, no. 16. Am. Comm. Int. Wildl. Protection, New York. 18 pp.

Milton, O., and R. D. Estes. 1963. Burma wildlife survey, 1959–1960. *I.U.C.N. Special Publ.*, no. 15. Am. Comm. Int. Wildl. Protection, New York. 72 pp.

Milton, O., D. Estes, and H. Z. Kimlai. 1964. Burma wildlife survey report on the Pidaung Wildlife Sanctuary. *Burmese Forester* 14(1–2):54–69.

Mirza, Z. P. 1965. Four new mammal records for West Pakistan. *Mammalia* 29:205–210.

Mitchell, R. M. 1975. A checklist of Nepalese mammals (excluding bats). *Säugetierkd. Mitt.* 23(1):152–157.

Mittermeier, R. A. 1973a. Colobus monkeys and the tourist trade. *Oryx* 12:113–117.

———. 1973b. Group activity and population dynamics of the howler monkey on Barro Colorado Island. *Primates* 14(1):1–19.

———. 1977a. Distribution, synecology, and conservation of Surinam monkeys. Ph.D. diss. Harvard University. 696 pp.

———. 1977b. A global strategy for primate conservation. Unpublished report to International Union for Conservation of Nature and Natural Resources and New York Zoological Society. 325 pp.

———. In preparation. The weird uakaris of Amazonia.

Mittermeier, R. A., R. C. Bailey, and A. F. Coimbra-Filho. 1978. Conservation status of the Callitrichidae in Brazilian Amazonia, Surinam, and French Guiana. Pp. 137–146 in *The Biology and Conservation of the Callitrichidae*. D. G. Kleiman, ed. Smithsonian Institution Press, Washington, D.C.

Mittermeier, R. A., R. C. Bailey, L. E. Sponsel, and K. E. Wolf. 1977. Primate ranching: Results of an experiment. *Oryx* 13(5):449-453.

Mittermeier, R. A., and A. F. Coimbra-Filho. 1977. Primate conservation in Brazilian Amazonia. Pp. 117–166 in *Primate Conservation*. H.S.H. Prince Rainier III of Monaco and G. H. Bourne, eds. Academic Press, New York.

Mittermeier, R. A., A. F. Coimbra-Filho, R. Castro C. 1976. Conference on the biology and conservation of the Callitrichidae. *Lab. Primate Newsl.* 15(3):16–23.

Mittermeier, R. A., A. F. Coimbra-Filho, and M. G. M. van Roosmalen. 1978. Callitrichids in Brazil and the Guianas: Current conservation status and potential for biomedical research. Pp. 20–29 in *Marmosets in Experimental Medicine*. N. Gengozian and F. Dienhardt, eds. Primates in Medicine, vol. 10. S. Karger, Basel.

———. In press. Conservation status of wild callitrichids. In *Proceedings of the Göttingen Marmoset Workshop*. J. Hearn, H. Rothe, and H.-J. Walters, eds.

Mittermeier, R. A., H. de Macedo Ruíz, and A. Luscombe. 1975. A wooly monkey rediscovered in Peru. *Oryx* 13(1):41–46.

Mittermeier, R. A., H. de Macedo-Ruíz, B. A. Luscombe, and J. Cassidy. 1977. Rediscovery and conservation of the Peruvian yellow-tailed wooly monkey (*Lagothrix flavicauda*). Pp. 95–115 in *Primate Conservation*. H.S.H. Prince Rainier III of Monaco and G. H. Bourne, eds. Academic Press, New York.

Mittermeier, R. A., and K. Milton. 1976. Jungle jackpot. *Anim. Kingdom* 79(6):27–31.

Mizuno, A., M. Kawai, and S. Ando. 1976. Ecological studies of forest-living monkeys in the Kibale Forest of Uganda. *Kyoto Univ. Afr. Stud.* 10:1–35.

Mohnot, S. M. 1971. Ecology and behaviour of the Hanuman langur, *Presbytis entellus* (Primates: Cercopithecidae) invading fields, gardens, and orchards around Jodhpur, western India. *Trop. Ecol.* 12(2):237–249.

———. 1978. The conservation of non-human primates in India. Pp. 47–53 in *Recent Advances in Primatology*, vol. 2: *Conservation*. D. J. Chivers and W. Lane-Petter, eds. Academic Press, London.

Molez, N. 1976. Adaptation alimentaire du galago d'Allen aux milieux forestiers secondaires. *Terre Vie* 30:210–228.

Monath, T. P., and G. E. Kemp. 1973. Importance of non-human primates in yellow fever epidemiology in Nigeria. *Trop. Geogr. Med.* 25:28–38.

Monath, T. P., V. H. Lee, D. C. Wilson, A. Fagbami, and O. Tomori. 1974. Arbovirus studies in Nupeko Forest, a possible natural focus of yellow fever virus in Nigeria. I. Description of the area and serological survey of humans and other vertebrate hosts. *Trans. R. Soc. Trop. Med. Hyg.* 68(1):30–38.

Mondolfi, E. 1976. Fauna silvestre de los bosques humedos de Venezuela. Sierra Club-Consejo de Bienestar Rural, Caracas.

Monfort, A., and N. Monfort. 1973. Quelques observations sur les grands mammifères du Park National de Tai (Côte d'Ivoire). *Terre Vie* 27:499–506.

Moreau, R. E. 1942. The distribution of the chimpanzee in Tanganyika territory. *Tanganyika Notes Rec.* 14:52–55.

———. 1969. Climatic changes and the distribution of forest vertebrates in West Africa. *J. Zool.* 158(1):39–61.

Moreno-Black, G., and W. R. Maples. 1977. Differential habitat utilization of four Cercopithecidae in a Kenyan forest. *Folia Primatol.* 27(2):85–107.

Morgan, M. T., and R. H. Tuttle. 1966. Intimate infant-adult male interactions in Rhodesian baboons (*Papio cynocephalus*). *Am. J. Phys. Anthropol.* 25:203.

Moro, M. 1972. Native fauna as a natural resource. Pp. 10–18 in Scientific Publication no. 235. Pan American Health Organization, Washington, D.C.

———. 1976. Nonhuman primates as a natural resource. Pp. 205–215 in *First Inter-American Conference on Conservation and Utilization of American Nonhuman Primates in Biomedical Research*. Scientific Publication no. 317. Pan American Health Organization, Washington, D.C.

———. 1978. Supply and conservation efforts for nonhuman primates. Pp. 37–39 in *Marmosets in Experimental Medicine*. N. Gengozian and F. Dienhardt, eds. Primates in Medicine, vol. 10. S. Karger, Basel.

Morris, R. C. 1943. Rivers as barriers to the distribution of gibbons. *J. Bombay Nat. Hist. Soc.* 43(4):656.

Moynihan, M. 1964. Some behavior patterns of platyrrhine monkeys. I. The night monkey (*Aotus trivirgatus*). *Smithsonian Misc. Collect.* 146(5):1–84.

———. 1967. Comparative aspects of communication in New World primates. Pp. 236–266 in *Primate Ethology*. D. Morris, ed. Aldine, Chicago.

———. 1970. Some behavior patterns of platyrrhine monkeys. II. *Saguinus geoffroyi* and some other tamarins. *Smithsonian Contrib. Zool.* 28.

———. 1976a. *The New World Primates*. Princeton University Press, Princeton, New Jersey. 262 pp.

———. 1976b. Notes on the ecology and behavior of the pygmy marmoset (*Cebuella pygmaea*) in Amazonian Colombia. Pp. 79–84 in *Neotropical Primates: Field Studies and Conservation*. R. W. Thorington, Jr., and P. G. Heltne, eds. National Academy of Sciences, Washington, D.C.

Muckenhirn, N. A. 1975. Trends in primate imports into the United States. *I.L.A.R. News* 18(3):2–3.

———. 1976. Trends in primate imports into the United States. *I.L.A.R. News* 19(3):2–3.

———. 1977. Trends in primate imports into the United States. *I.L.A.R. News* 20(3):5.

———. 1979. Trends in primate imports into the United States. *I.L.A.R. News* 22(3):22–23.

Muckenhirn, N. A., and A. L. Cohen. 1978. Trends in primate imports into the United States. *I.L.A.R. News* 21(2):17–19.

Muckenhirn, N. A., and J. F. Eisenberg. 1978. The status of primates in Guyana and ecological correlations for neotropical primates. Pp. 27–30 in *Recent Advances in Primatology*, vol. 2: *Conservation*. D. J. Chivers and W. Lane-Petter, eds. Academic Press, London.

Muckenhirn, N. A., B. K. Mortensen, S. Vessey, C. E. O. Fraser, and B. Singh. 1975. *Report on a Primate Survey in Guyana*. Pan American Health Organization, Washington, D.C. 49 pp.

Mukherjee, A. K. 1975. The Sundarban of India and its biota. *J. Bombay Nat. Hist. Soc.* 72(1):1–20.

Mukherjee, A. K., and S. Gupta. 1965. Habits of the rhesus macaque *Macaca mulatta* (Zimmermann) in the Sundarbans, 24-Parganas, West Bengal. *J. Bombay Nat. Hist. Soc.* 62:145–146.

Mukherjee, R. P., and G. D. Mukherjee. 1972. Group composition and population density of rhesus monkey [*Macaca mulatta* (Zimmermann)] in northern India. *Primates* 13(1):65–70.

Mukherjee, R. P., and S. S. Saha. 1974. The golden langurs (*Presbytis geei* Khajuria, 1956) of Assam. *Primates* 15:327–340.

Müller, P. 1973. The dispersal centres of terrestrial vertebrates in the neotropical realm. *Biogeographica* 2:1–244.

Murphy, B. L., J. E. Maynard, D. H. Krushak, and K. R. Borquist.1972. Microbial flora of imported marmosets: Viruses and enteric bacteria. *Lab. Anim. Sci.* 22:339–343.

Mydans, C. 1973. Orang-utans can return to the wild with some help. *Smithsonian* 4(8):26–33.

Nagel, U. 1973. A comparison of anubis baboons, hamadryas baboons, and their hybrids at a species border in Ethiopia. *Folia Primatol.* 19:104–166.

Nagel, U., and F. Lohri. 1973. Die Languren der Kanha-Wiesen, Stellung in Ökosystem, Kommensalismus mit Axishirschen und Fress-Soziologie. *Vierteljahrsschr. Naturforsch. Ges. Zür.* 118:71–85.

Napier, J. R., and P. H. Napier. 1967. *A Handbook of Living Primates.* Academic Press, New York. 456 pp.

Napier, P. H. 1976. *Catalogue of Primates in the British Museum (Natural History)*, part 1: *Families Callitrichidae and Cebidae.* British Museum of Natural History, London. 121 pp.

Naumann, C., and G. Nogge. 1973. Die Grossäuger Afghanistans. *Z. Köln. Zoo* 16(3):79–93.

Nel, J. A. J., and J. J. L. Pretorius. 1971. A note on the smaller mammals of the Mountain Zebra National Park. *Koedoe* 14:99–110.

Neville, M. K. 1968. Ecology and activity of Himalayan foothill rhesus monkeys (*Macaca mulatta*). *Ecology* 49:110–123.

———. 1972. The population structure of red howler monkeys (*Alouatta seniculus*) in Trinidad and Venezuela. *Folia Primatol.* 17:56–86.

———. 1975. "Census of primates" in Peru. Pp. 3–15 in *Primate Censusing Studies in Peru and Colombia.* Pan American Health Organization, Washington, D.C.

———. 1976a. The population and conservation of howler monkeys in Venezuela and Trinidad. Pp. 101–109 in *Neotropical Primates: Field Studies and Conservation.* R. W. Thorington, Jr., and P. G. Heltne, eds. National Academy of Sciences, Washington, D.C.

———. 1976b. Census of primates in Peru. Pp. 19–29 in *First Inter-American Conference on Conservation and Utilization of American Nonhuman Primates in Biomedical Research.* Scientific Publication no. 317. Pan American Health Organization, Washington, D.C.

Neville, M. K., N. Castro, A. Mármol, and J. Revilla. 1976. Censusing primate populations in the reserved area of the Pacaya and Samiria Rivers, Department Loreto, Peru. *Primates* 17(2):151–181.

Neyman, P. F. 1977. The protection and management of primates in Sucre and Cordoba (Colombia). Unpublished report to New York Zoological Society. 11 pp.

———. 1978. Aspects of the ecology and social organization of free-ranging cotton-top tamarins (*Saguinus oedipus*) and the conservation status of the species. Pp. 39–71 in *The Biology and Conservation of the Callitrichidae.* D. G. Kleiman, ed. Smithsonian Institution Press, Washington, D.C.

Ng, F. S. P. 1974. A blueprint for conservation in peninsular Malaysia. *Malay. Nat. J.* 27:1–16.

Ngan, P. T. 1968. The status of conservation in South Viet Nam. Pp. 519–525 in *Conservation in Tropical South East Asia.* I.U.C.N. Publ. New Series, no. 10.

Nguyen-Van-Hiep. 1969. Rapport sur la situation de la conservation au Vietnam en 1969. *I.U.C.N. Publ.* New Series 19:54–62.

Niemitz, C. 1973. Field research on the Horsfield's tarsier (*Tarsius bancanus*) at Sarawak Museum. *Borneo Res. Bull.* 5(2):61–63.

———. 1979. Outline of the behavior of *Tarsius bancanus.* Pp. 631–660 in *The Study of Prosimian Behavior.* G. A. Doyle and R. D. Martin, eds. Academic Press, New York.

Nishida, T. 1968. The social group of wild chimpanzees in the Mahali Mountains. *Primates* 9:167–224.

———. 1972a. A note on the ecology of the red colobus monkey (*Colobus badius tephrosceles*) living in the Mahali Mountains. *Primates* 13(1):57–64.

———. 1972b. Preliminary information of the pygmy chimpanzees (*Pan paniscus*) of the Congo basin. *Primates* 13(4):415–425.

Nishimura, A., and K. Izawa. 1975. The group characteristics of wooly monkeys (*Lagothrix lagothrica*) in the upper Amazonian basin. Pp. 351–357 in *Contemporary Primatology, Proc. 5th Int. Congr. Primatol.* S. Kondo, M. Kawai, and A. Ehara, eds. S. Karger, Basel.

Oates, J. F. 1969. The lower primates of eastern Nigeria. *Afr. Wildl.* 23:321–332.

———. 1977a. The guereza and man: How man has affected the distribution and abundance of *Colobus guereza* and other black colobus monkeys. Pp. 420–467 in *Primate Conservation.* H.S.H. Prince Rainier III of Monaco and G. H. Bourne, eds. Academic Press, New York.

———. 1977b. The guereza and its food. Pp. 276–321 in *Primate Ecology.* T. H. Clutton-Brock, ed. Academic Press, London.

———. 1978. The status of the south Indian black leaf-monkey (*Presbytis johnii*) in the Palni Hills. *J. Bombay Nat. Hist. Soc.* 75(1):1–12.

———. 1979. Comments on the geographical distribution and status of the South Indian black leaf-monkey (*Presbytis johnii*). *Mammalia* 43:485–493.

Oates, J. F., and P. A. Jewell. 1967. Westerly extent of the range of three African lorisoid primates. *Nature* 215:778–779.

Oboussier, H., and G. A. v. Maydell. 1959. Zur Kenntnis des indischen Goldlangurs. *Z. Morphol. Ökol. Tiere* 48:102–114.

Ohsawa, H., and M. Kawai. 1975. Social structure of gelada baboons: Studies of the gelada society (I). Pp. 464–469 in *Contemporary Primatology, Proc. 5th Int. Congr. Primatol.* S. Kondo, M. Kawai, and A. Ehara, eds. S. Karger, Basel.

Okano, T. 1965. Preliminary survey of the orang-utan in North Borneo (Sabah). *Primates* 6(1):123–128.

Olivier, T. J., J. Buettner-Janusch, and V. Buettner-Janusch. 1974. Carbonic anhydrase isoenzymes in nine troops of Kenya baboons, *Papio cynocephalus* (Linneaus 1766). *Am. J. Phys. Anthropol.* 41:175–190.

Omar, A., and A. de Vos. 1970. Damage to exotic softwoods by Sykes monkeys (*Cercopithecus mitis kolbi* Neuman). *East Afr. Agric. For. J.* 35(4):323–330.

Oppenheimer, J. R. 1969. Behavior and ecology of the white-faced monkey, *Cebus capucinus*, on Barro Colorado Island, Canal Zone. Ph.D. diss. University of Illinois. *Diss. Abst. Int.* 30B:442–443.

———. 1973. Village dwelling langur monkeys. Pp. 110–117 in *Johns Hopkins Center for Medical Research and Training Annual Report, 1972–73.*

———. 1977. *Presbytis entellus*, the hanuman langur. Pp. 469–512 in *Primate Conservation.* H.S.H. Prince Rainier III of Monaco and G. H. Bourne, eds. Academic Press, New York.

Oppenheimer, J. R., and E. C. Oppenheimer. 1973. Preliminary observations of *Cebus nigrivittatus* (Primates: Cebidae) on the Venezuelan llanos. *Folia Primatol.* 19(6):409–436.

Organization of African Unity. 1968. *African Convention on the Conservation of Nature and Natural Resources.* General Secretariat, O.A.U., Addis Ababa, Ethiopia. 50 pp.

Orians, G. H., and E. W. Pfeiffer. 1970. Ecological effects of the war in Viet Nam. *Science* 168:544–554.

Osborn, R. M. 1963. Observations on the behavior of the mountain gorilla. *Symp. Zool. Soc. London* 10:29–37.

Padua, M. T. J., A. Magnanini, and R. A. Mittermeier. 1974. Brazil's national parks. *Oryx* 12(4):452–464.
Pages, E. 1978. Home range, behaviour and tactile communication in a nocturnal Malagasy lemur *Microcebus coquereli*. Pp. 171–177 in *Recent Advances in Primatology*, vol. 3: *Evolution*. D. J. Chivers and K. A. Joysey, eds. Academic Press, London.
Pan American Sanitary Bureau. 1955. Yellow fever conference, December 21–22, 1954. *Am. J. Trop. Med. Hyg.* 4(4):571–661.
Paradiso, J. L., and R. D. Fisher. 1972. Mammals imported into the U.S. in 1970. Special Scientific Report (Wildlife) no. 161. Bur. Sport Fisheries and Wildlife, Washington, D.C. 62 pp.
Pariente, G. 1974. Influence of light on the activity rhythms of two Malagasy lemurs: *Phaner furcifer* and *Lepilemur mustelinus leucopus*. Pp. 183–198 in *Prosimian Biology*. R. D. Martin, G. A. Doyle, and A. C. Walker, eds. University of Pittsburgh Press, Pittsburgh.
Parsons, R. E. 1942a. Rivers as barriers to the distribution of gibbons. *J. Bombay Nat. Hist. Soc.* 42(2):434.
———. 1942b. Rivers as barriers to the distribution of gibbons. *J. Bombay Nat. Hist. Soc.* 42(3):926.
Parthasarathy, M. D. 1977. Ecology and ethology of the Hanuman langur (*Presbytis entellus*) and the bonnet macaque (*Macaca radiata*). Pp. 46–51 in *Use of Non-human Primates in Biomedical Research*. M. R. N. Prasad and T. C. Anand Kumar, eds. Indian National Science Academy, New Delhi, India.
Paterson, H. E., P. Bronsden, and J. Levitt. 1964. Some culicine mosquitoes (Diptera, Culicidae) at Ndumu, Republic of South Africa. *Medical Proceedings–Mediese Bydraes* 10(1):88–192.
Paterson, J. D. 1973. Ecologically differentiated patterns of aggressive and sexual behavior in two troops of Ugandan baboons, *Papio anubis*. *Am. J. Phys. Anthropol.* 38:641–648.
Peirce, M. A. 1975. Ectoparasites from East African vertebrates. *East Afr. Wildl. J.* 13:153–156.
Peng, M. T., Y. L. Lai, C. S. Yang, and H. S. Chiang. 1973. Formosan monkey (*Macaca cyclopis*): Present situation in Taiwan and its reproductive biology. *Exp. Anim.* (Tokyo), 22(suppl.):447–451.
Perleche M., F. 1974. Después de 50 años hallan mono lanudo en Amazonas. *La Prensa*. Lima, Peru, May 11, 1974.
Perry, J. 1971. The golden lion marmoset. *Oryx* 11(1):22–24.
———. 1976. Orang-utans in captivity. *Oryx* 13(3):262–264.
Peters, W., P. C. C. Garnham, R. Killick-Kendrik, N. Rajapaksa, W. H. Cheong, and F. C. Cadigan. 1976. Malaria of the orangutan (*Pongo pygmaeus*) in Borneo. *Philos. Trans. R. Soc. London, Biol. Sci.* 275(941):439–482.
Petrides, G. A. 1965. Advisory Report on Wildlife and National Parks in Nigeria, 1962. *I.U.C.N. Special Publ.*, no. 18. Am. Comm. Int. Wildl. Protection, New York. 48 pp.
———. 1968. Problems in species' introductions. *I.U.C.N. Bulletin* 2(7):70–71.
Petter, J.-J. 1962a. Ecological and behavioral studies of Madagascar lemurs in the field. *Ann. N.Y. Acad. Sci.* 102:267–281.
———. 1962b. Remarques sur l'écologie et l'éthologie comparées des Lémuriens malgaches. *Terre Vie* 109:394–416.
———. 1965. The lemurs of Madagascar. Pp. 292–319 in *Primate Behavior: Field Studies of Monkeys and Apes*. I. DeVore, ed. Holt, Rinehart and Winston, New York.
———. 1969. Speciation in Madagascan lemurs. *Biol J. Linn. Soc.* 1:77–84.

———. 1972. Order of primates: Suborder of lemurs. Pp. 683–702 in *Biogeography and Ecology of Madagascar.* R. Battistini and G. Richard-Vindard, eds. W. Junk, The Hague.
———. 1975. Breeding of Malagasy lemurs in captivity. Pp. 187–202 in *Breeding Endangered Species in Captivity.* R. D. Martin, ed. Academic Press, London.
———. 1977. The aye-aye. Pp.38–57 in *Primate Conservation.* H.S.H. Prince Rainier III of Monaco and G. H. Bourne, eds. Academic Press, New York.
———. 1978. Ecological and physiological adaptations of five sympatric nocturnal lemurs to seasonal variations in food production. Pp. 211–223 in *Recent Advances in Primatology,* vol. 1: *Behaviour.* D. J. Chivers and J. Herbert, eds. Academic Press, London.
Petter, J.-J., and C. M. Hladik. 1970. Observations sur le domaine vital et la densité de population de *Loris tardigradus* dans les forêts de Ceylan. *Mammalia* 34:394–409.
Petter, J.-J., and G. Pariente. 1971. Les Indridés malgaches. *Sci. Nat.* 106:15–24.
Petter, J.-J., and A. Petter. 1967. The aye-aye of Madagascar. Pp. 195–205 in *Social Communication Among Primates.* S. A. Altmann, ed. University of Chicago Press, Chicago.
Petter, J.-J., and A. Petter-Rousseaux. 1960. Remarques sur la systématique du genre *Lepilemur.* *Mammalia* 24:76–86.
———. 1979. Classification of the prosimians. Pp. 1–44 in *The Study of Prosimian Behavior.* G. A. Doyle and R. D. Martin, eds. Academic Press, New York.
Petter, J.-J., and A. Peyriéras. 1970a. Nouvelle contribution a l'étude d'un lemurien malgache, le aye-aye (*Daubentonia madagascariensis* E. Geoffroy). *Mammalia* 34:167–193.
———. 1970b. Observations éco-éthologiques sur les lémuriens malgaches du genre *Hapalemur.* *Terre Vie* 24(3):356–382.
———. 1974. A study of population density and home ranges of *Indri indri* in Madagascar. Pp. 39–48 in *Prosimian Biology.* R. D. Martin, G. A. Doyle, and A. C. Walker, eds. University of Pittsburgh Press, Pittsburgh.
———. 1975. Preliminary notes on the behavior and ecology of *Hapalemur griseus.* Pp. 281–286 in *Lemur Biology.* I. Tattersall and R. W. Sussman, eds. Plenum Press, New York.
Petter, J.-J., A. Schilling, and G. Pariente. 1971. Observations éco-éthologiques sur deux lémuriens malagaches nocturnes: *Phaner furcifer* et *Microcebus coquereli.* *Terre Vie* 25(3):287–327.
———. 1975. Observations on behavior and ecology of *Phaner furcifer.* Pp. 209–218 in *Lemur Biology.* I. Tattersall and R. W. Sussman, eds. Plenum Press, New York.
Pfeffer, P. 1969. Considerations sur l'écologie des fôrets claires du Cambodge oriental. *Terre Vie* 23:3–24.
Pfeiffer, E. W. 1973. Post-war Vietnam. *Environment* 15(8):29–33.
Phillipps, Q. 1973. Kinabalu, Sabah's national park. *Animals* 15:292–298.
Phillips, W. W. A. 1935. Manual of the mammals of Ceylon. *Ceylon J. of Sci.* Dulau Co., London.
Pianka, E. R. 1974. *Evolutionary Ecology.* Harper and Row, New York. 356 pp.
Pienaar, U. de V. 1967. The small mammals of the Kruger National Park: A systematic list and zoogeography. *Koedoe* (7):1–25.
Pierce, A. H. 1978. Ranging patterns and associations of a small community of chimpanzees in Gombe National Park, Tanzania. Pp. 59–61 in *Recent Advances in Primatology,* vol. 1: *Behaviour.* D. J. Chivers and J. Herbert, eds. Academic Press, London.
Pine, R. H. 1973. Mammals (exclusive of bats) of Belém, Pará, Brazil. *Acta Amazonica* 3(2):47–79.
Pires, J. M. 1974. Tipos de vegetação da Amazônia. *Brasil Florestal* 5(17):48–58.
Pires, J. M., and G. T. Prance. 1977. The Amazon forest: A natural heritage to be preserved.

Pp. 158–194 in *Extinction Is Forever*. G. Prance and T. Elias, eds. New York Botanical Garden, Bronx, New York.
Pirlot, P. 1963. Algunas consideraciones sobre la ecologia de los mamíferos del oeste de Venezuela. *Kasmera* 1:169–214.
Pitchford, R. J., P. S. Visser, J. F. DuToit, U. de V. Pienaar, and E. Young. 1973. Observations on the ecology of *Schistosoma mattheei*, Veglia and Le Roux, 1929, in portion of the Kruger National Park and surrounding area, using a new quantitative technique for egg output. *J. S. Afr. Vet. Assoc.* 44:405–420.
Pitchford, R. J., P. S. Visser, U. de V. Pienaar, and E. Young. 1974. Further observations on *Schistosoma mattheei*, Veglia and Le Roux, 1929, in the Kruger National Park. *J. S. Afr. Vet. Assoc.* 45:211–218.
Pitman, C. R. S. 1954. The influence of Belgian Congo on distribution of Uganda's primates, and some of their characteristics. *Ann. Mus. Congo, Tervuren* 1:47–55.
Poché, R. 1973. Niger's threatened park W. *Oryx* 12:216–222.
———. 1976. Notes on primates in Parc National du W du Niger, West Africa. *Mammalia* 40(2):187–198.
Poirier, F. E. 1968a. Analysis of a Nilgiri langur (*Presbytis johnii*) home range change. *Primates* 9:29–43.
———. 1968b. The ecology and social behavior of the Nilgiri langur (*Presbytis johnii*) of south India. *Diss. Abstr.* 29B:28–29.
———. 1969. Behavioral flexibility and intertroop variation among Nilgiri langurs (*Presbytis johnii*) of south India. *Folia Primatol.* 11:119–133.
———. 1970. The Nilgiri langur (*Presbytis johnii*) of south India. Pp. 251–283 in *Primate Behavior*, vol. 1. L. A. Rosenblum, ed. Academic Press, New York.
———. 1971. The Nilgiri langur, a threatened species. *Zoonooz* 49(7):10–16.
———. 1977. The human influence on genetic and behavioral differentiation among three nonhuman primate populations. Pp. 234–241 in *Yearb. Phys. Anthropol.*, vol. 20. J. Buettner-Janusch, ed. Am. Assoc. Phys. Anthropol., Washington, D.C.
Poirier, F. E., and E. O. Smith. 1974. The crab-eating macaques (*Macaca fascicularis*) of Angaur Island, Palau, Micronesia. *Folia Primatol.* 22:258–306.
Pollock, J. I. 1975. Field observations of *Indri indri:* A preliminary report. Pp. 287–311 in *Lemur Biology*. I. Tattersall and R. W. Sussman, eds. Plenum Press, New York.
———. 1977a. The ecology and sociology of feeding in *Indri indri*. Pp. 38–69 in *Primate Ecology*. T. H. Clutton-Brock, ed. Academic Press, London.
———. 1977b. Madagascar's intriguing indris . . . a mystery no more. *Int. Wildl.* 7(2):12–16.
———. 1979. Spatial distribution and ranging behavior in lemurs. Pp. 359–409 in *The Study of Prosimian Behavior*. G. A. Doyle and R. D. Martin, eds. Academic Press, New York.
Poonai, N. O. 1973. Report for Guiana. *Oryx* 12(2):281–284.
Poore, M. E. D. 1976. The value of tropical moist forest ecosystems and the environmental consequences of their removal. *Unasylva* 28:127–143.
Pope, B. L. 1968. Population characteristics. Pp. 13–20 in *Biology of the Howler Monkey* (*Alouatta caraya*). R. M. Malinow, ed. *Bibl. Primatol.*, vol. 7.
Poulet, A. R. 1972. Recherches écologiques sur une savane sahelienne du Ferlo septentrional, Sénégal: Les mammifères. *Terre Vie* 26:440–472.
Prakash, I. 1960. Breeding of mammals in the Rajasthan desert, India. *J. Mammal.* 41:386–389.
Prance, G. T. 1977. The phytogeographic subdivisions of Amazonia and their influence on the selection of biological reserves. Pp. 195–213 in *Extinction Is Forever*. G. Prance and T. Elias, eds. New York Botanical Garden, Bronx, New York.

Prater, S. H. 1971. *The Book of Indian Animals*. 3d ed. Bombay Nat. Hist. Soc., Bombay, India. 324 pp.
Prijono, H. 1975. Country report: Indonesia. Pp. 24–25 in *The Use of Ecological Guidelines for Development in Tropical Forest Areas of South East Asia*. I.U.C.N. Publ. New Series, no. 32. Morges, Switzerland.
Pringle, J. A. 1974. The distribution of mammals in Natal: Part I. Primates, Hyracoidea, Lagomorpha (except *Lepus*), Pholidota, and Tubulidentata. *Ann. Natal Mus.* 22:173–186.
Puget, A. 1971. Observations sur le macaque rhesus, *Macaca mulatta* (Zimmermann, 1780) en Afghanistan. *Mammalia* 35(2):199–203.
Quevedo, M. M. 1976. Essential policy guidelines for the conservation and utilization of nonhuman primates in Colombia. Pp. 235–241 in *First Inter-American Conference on Conservation and Utilization of American Nonhuman Primates in Biomedical Research*. Scientific Publication no. 317. Pan American Health Organization, Washington, D.C.
Quick, L. B. 1975. Group composition and daily rounds of spider monkeys at Tikal, Guatemala. *Am. J. Phys. Anthropol.* 42:324.
Quris, R. 1975. Écologie et organisation sociale de *Cercocebus galeritus agilis* dans le nord-est du Gabon. *Terre Vie* 29:337–398.
———. 1976. Données comparatives sur la socio-écologie de huit espèces de Cercopithecidae vivant dans une même zone de fôret primitive périodiquement inondée (nord-est du Gabon). *Terre Vie* 30:193–209.
Rabor, D. S. 1955. Notes on mammals and birds of the central northern Luzon highlands, Philippines. *Silliman J.* 2:193–218.
———. 1968. Threatened species of small mammals in tropical south east Asia: The problem in the Philippines. Pp. 272–274 in *Conservation in Tropical South East Asia*. I.U.C.N. Publ. New Series, no. 10.
Rabor, D. S., A. C. Alcala, and R. B. Gonzales. 1970. A list of the land vertebrates of Negros Island, Philippines. *Silliman J.* 17(3):297–316.
Racenis, J. 1952. Some observations on the red howling monkey (*Alouatta seniculus*) in Venezuela. *J. Mammal.* 33:114–115.
Raemaekers, J. J. 1978. The sharing of food sources between two gibbon species in the wild. *Malay. Nat. J.* 31(3):181–188.
Rahaman, H. 1973. The langurs of the Gir Sanctuary (Gujarat): A preliminary survey. *J. Bombay Nat. Hist. Soc.* 70:295–314.
Rahaman, H., and M. D. Parthasarathy. 1967. A population survey of the bonnet monkey (*Macaca radiata* Geoffroy) in Bangalore, south India. *J. Bombay Nat. Hist. Soc.* 64:251–256.
———. 1969. Home range, roosting places, and the day ranges of the bonnet macaque (*Macaca radiata*). *J. Zool.* 157(3):267–276.
———. 1970. Observations on the habits and life span of the slender loris, *Loris tardigradus Lydekkerianus* (L.). *MyForest* 6(4):23–26.
Rahm, U. 1965. Distribution et écologie de quelques mammifères de l'est du Congo. *Zool. Afr.* 1(1):149–166.
———. 1966. Les mammifères de la forêt équatoriale de l'est du Congo. *Ann. Mus. Roy. Afr. Cent., Tervuren, Sci. Zool.* 149:39–121.
———. 1970. Ecology, zoogeography, and systematics of some African forest monkeys. Pp. 591–626 in *Old World Monkeys*. J. R. Napier and P. H. Napier, eds. Academic Press, New York.
———. 1971. L'emploi d'outils par les chimpanzes de l'ouest de la Côte-d'Ivoire. *Terre Vie* 25:506–509.

———. 1972. Zur Verbreitung und Ökologie der Säugetiere des afrikanischen Regenwaldes. *Acta Trop.* 29(4):452–473.

Rahm, U., and A. R. Christiaensen. 1960. Note sur *Colobus polycomos cordieri* (Rahm) du Congo Belge. *Rev. Zool. Bot. Afr.* 41(3–4):215–220.

———. 1963. Les mammifères de la règion occidentale du Lac Kivu. *Ann. Mus. Roy. Afr. Cent., Tervuren, Sci. Zool.* 118:1–83.

———. 1966. Les mammifères de l'île Idjwi (Lac Kivu, Congo). *Ann. Mus. Roy. Afr. Cent., Tervuren, Sci. Zool.* 149:1–35.

Rahm, Ursula. 1967. Observations during chimpanzee capture in the Congo. Pp. 195–206 in *Progress in Primatology*. D. Starck, R. Schneider, and H.-J. Kuhn, ed. G. Fischer Verlag, Stuttgart.

Rainier, H.S.H. Prince III of Monaco, and G. H. Bourne, eds. 1977. *Primate Conservation*. Academic Press, New York. 658 pp.

Rajagopalan, P. K., and C. R. Anderson. 1971. Further studies on ticks of wild monkeys of Kyasanu Forest Disease Area, Shimoga District. *Indian J. Med. Res.* 59:847–860.

Ramírez, M. F., C. H. Freese, and J. Revilla C. 1978. Feeding ecology of the pygmy marmoset, *Cebuella pygmaea*, in northeastern Peru. Pp. 91–104 in *The Biology and Conservation of the Callitrichidae*. D. G. Kleiman, ed. Smithsonian Institution Press, Washington, D.C.

Rathbun, G. B., and M. J. Gache. 1977. The status of *Aotus trivirgatus* in Argentina. Unpublished report. National Institutes of Health. 48 pp.

Rautenbach, I. L. 1976. A survey of the mammals occurring in the Golden Gate Highlands National Park. *Koedoe* 19:133–144.

Reynolds, V. 1965. Some behavioral comparisons between the chimpanzee and the mountain gorilla in the wild. *Am. Anthropol.* 67:691–706.

———. 1967a. On the identity of the ape described by Tulp in 1641. *Folia Primatol.* 5:80–87.

———. 1967b. *The Apes*. E. P. Dutton, New York. 296 pp.

Reynolds, V., and F. Reynolds. 1965. Chimpanzees of the Budongo Forest. Pp. 368–424 in *Primate Behavior*. I. DeVore, ed. Holt, Rinehart and Winston, New York.

Reza Khan, M. A. 1976. Status of the Nilgiri langur *Presbytis johnii* (Fischer) in the Nilgiris. *J. Bombay Nat. Hist. Soc.* 73:517–518.

Richard, A. 1974a. Intra-specific variation in the social organization and ecology of *Propithecus verreauxi*. *Folia Primatol.* 22:178–207.

———. 1974b. Patterns of mating in *Propithecus verreauxi*. Pp. 49–74 in *Prosimian Biology*. R. D. Martin, G. A. Doyle, and A. C. Walker, eds. University of Pittsburgh Press, Pittsburgh.

———. 1978. *Behavioral Variation: Case Study of a Malagasy Lemur*. Bucknell University Press, Lewisburg, Pa. 213 pp.

Richard, A. F., and R. W. Sussman. 1975. Future of the Malagasy lemurs: Conservation or extinction? Pp. 335–350 in *Lemur Biology*. I. Tattersall and R. W. Sussman, eds. Plenum Press, New York.

Richie, T., R. Shrestha, J. Teas, H. Taylor, and C. Southwick. 1978. Rhesus monkeys at high altitudes in northwestern Nepal. *J. Mammal.* 59(2):443–444.

Richter, W. von. 1974. Survey of the adequacy of existing conserved areas in relation to wild animal species. *Koedoe* 17:39–69.

Rijksen, H. D. 1974. Orang-utan conservation and rehabilitation in Sumatra. *Biol. Conserv.* 6(1):20–25.

——— 1974–75. Project 733, Gunung Leuser Reserve, Sumatra: Overall programme. Pp.

167–170 in *World Wildlife Yearbook 1974–75*. P. Jackson, ed. World Wildlife Fund, Morges, Switzerland.

———. 1975. Social structure in a wild orang-utan population in Sumatra. Pp. 373–379 in *Contemporary Primatology, Proc. 5th Int. Congr. Primatol.* S. Kondo, M. Kawai, and A. Ehara, eds. S. Karger, Basel.

———. 1978. *A Field Study on Sumatran Orang Utans (*Pongo pygmaeus abelii *Lesson 1827)*. H. Veenman and Zonen B. V., Wageningen. 420 pp.

Rijksen, H. D., and A. G. Rijksen-Graatsma. 1975. Orang utan rescue work in north Sumatra. *Oryx* 13(1):63–73.

Ringuelet, R. A. 1970. Panorama general de la fauna y sus relaciones ecológicas del N.E. Argentino y del dominio subtropical. *Bol. Soc. Argent. Bot.* 11(supp.):175–183.

Ripley, S. 1967. Intertroop encounters among Ceylon gray langurs (*Presbytis entellus*). Pp. 237–253 in *Social Communication Among Primates*. S. Altmann, ed. University of Chicago Press, Chicago.

———. 1970. Leaves and leaf monkeys: The social organization of foraging in gray langurs, *Presbytis entellus thersites*. Pp. 481–512 in *Old World Monkeys*. J. R. Napier and P. H. Napier, eds. Academic Press, New York.

Roberts, A. 1951. *The Mammals of South Africa*. South Africa News Agency, Johannesburg. 700 pp.

Roberts, T. J. 1977. *The Mammals of Pakistan*. Ernest Benn Ltd., London. 361 pp.

Robinson, G. A. 1976. Notes on mammals encountered in the Tsitsikama National Parks. *Koedoe* 19:145–152.

*Robinson, H. C., and C. B. Kloss. 1915. The zoology of Koh Samui and Koh Pennan. Mammals. *J. Fed. Malay States Mus.* 5:130–139.

Robinson, P. T. 1971. Wildlife trends in Liberia and Sierra Leone. *Oryx* 11(2–3):117–122.

Rock, M. A. 1977. Orang- endangered "man of the forest." *Nat. Parks Conserv. Mag.* 51(8):10–15.

Rodman, P. S. 1973. Population composition and adaptive organisation among orang-utans of the Kutai Reserve. Pp. 171–209 in *Comparative Ecology and Behaviour of Primates*. R. P. Michael and J. H. Crook, eds. Academic Press, London.

———. 1977. Feeding behaviour of orang-utans of the Kutai Nature Reserve, East Kalimantan. Pp. 384–413 in *Primate Ecology*. T. H. Clutton-Brock, ed. Academic Press, London.

Roonwal, M. L., and S. M. Mohnot. 1977. *Primates of South Asia*. Harvard University Press, Cambridge, Mass. 421 pp.

*Rosevear, D. R. 1953. *Checklist and Atlas of Nigerian Mammals*. Nigerian Government, Lagos, Nigeria.

Roth, T. W. 1968. The overkill. *Oryx* 9(5):354–356.

Rothschild, G. 1971. Animals in Bako National Park. *Malay. Nat. J.* 24:163–169.

Rowell, T. E. 1966. Forest living baboons in Uganda. *J. Zool.* 149:344–364.

———. 1968. Primates as a natural resource: A possibility for multiple land use. *East Afr. Ag. For. J.* Special issue. 33:279–280.

———. 1969. Long-term changes in a population of Ugandan baboons. *Folia Primatol.* 11:241–254.

———. 1972. Toward a natural history of the talapoin monkey in Cameroon. *Ann. Fac. Sci. Cameroun* 10:121–134.

*Reference not seen but cited by other authors.

———. 1973. Social organization of wild talapoin monkeys. *Am. J. Phys. Anthropol.* 38:593–597.

Rucks, M. G. 1976. Notes on the problems of primate conservation in Bia National Park. Unpublished mimeograph. Ghana Dept. of Game and Wildlife, Accra, Ghana. 13 pp.

Rudran, R. 1973a. The reproductive cycles of two subspecies of purple faced langurs (*Presbytis senex*) in relation to environmental factors. *Folia Primatol.* 19:41–60.

———. 1973b. Adult male replacement in one-male troops of purple faced langurs (*Presbytis senex senex*) and its effects upon population structure. *Folia Primatol.* 19:166–192.

———. 1978. Socioecology of the blue monkeys (*Cercopithecus mitis stuhlmanni*) of the Kibale Forest, Uganda. *Smithsonian Contrib. Zool.* 249:1–88.

Ruhle, G. C. 1964. Advisory report on a national park system for Thailand, 1959–1960. *I.U.C.N. Special Publ.*, no. 17. Am. Comm. Int. Wildl. Protection, New York. 24 pp.

———. 1966. Advisory report on national parks and reserves for Taiwan, 1965. *I.U.C.N. Special Publ.*, no. 19. Am. Comm. Int. Wildl. Protection, New York. 77 pp.

Rumpler, Y. 1975. The significance of chromosomal studies in the systematics of the Malagasy lemurs. Pp. 25–40 in *Lemur Biology*. I. Tattersall and R. W. Sussman, eds. Plenum Press, New York.

Rumpler, Y., and R. Albignac. 1978. Chromosome studies of the lepilemur, an endemic Malagasy genus of lemurs: Contribution of the cytogenetics to their taxonomy. *J. Hum. Evol.* 7:191–196.

Russell, R. J., and L. W. McGeorge. 1977. Distribution of *Phaner* (Primates, Lemuriformes, Cheirogaleidae, Phanerinae) in southern Madagascar. *J. Biogeography* 4:169–170.

Ryan, D. A. 1978. Recent development of national parks in Nicaragua. *Biol. Conserv.* 13:179–182.

Saayman, G. S. 1971. Behavior of chacma baboons. *Afr. Wildl.* 25:25–29.

Sabater Pi, J. 1964. Distribucion actual de los gorilas de llanura en Rio Muni. *Zoo* 3:26–29.

———. 1966. Los monos de Rio Muni y su distribucion. *Zoo* 6:15–18.

———. 1972a. Contribution to the ecology of *Mandrillus sphinx* (Linnaeus 1758) of Rio Muni (Republic of Equatorial Guinea). *Folia Primatol.* 17:304–319.

———. 1972b. Notes on the ecology of five Lorisiformes of Rio Muni. *Folia Primatol.* 18:140–151.

———. 1973. Contribution to the ecology of *Colobus polykomos satanas* (Waterhouse 1838) of Rio Muni, Republic of Equatorial Guinea. *Folia Primatol.* 19:193–207.

Sabater Pi, J., and C. Groves. 1972. The importance of higher primates in the diet of the Fang of Rio Muni. *Man* 7:239–243.

Sabater Pi, J., and C. Jones. 1967. Notes on the distribution and ecology of the higher primates of Rio Muni, West Africa. *Tulane Stud. Zool.* 14:101–109.

*Sanderson, I. T. 1940. The mammals of the north Cameroons forest area. *Trans. Zool. Soc. London* 24:623–725.

Sankhala, K. S. 1971. National parks of India. *I.U.C.N. Publ.* New Series 19:11–26.

Satmoko, R. K. P. 1961. Udjung-Kulon Nature Park, Java. Nature Conservation in Western Malaysia, 1961. *Malay. Nat. J.* 21st anniversary special issue: 107–124.

Sauer, E. G. F. 1974. Zur Biologie der Zwerg- und Riesengalagos. *Z. Köln. Zoo* 17(2):67–84.

Sayer, J. A. 1977. Conservation of large mammals in the Republic of Mali. *Biol. Conserv.* 12(4):245–263.

Schaller, G. B. 1961. The orang-utan in Sarawak. *Zoologica* 46:73–82.

*Reference not seen but cited by other authors.

———. 1963. *The Mountain Gorilla: Ecology and Behavior*. University of Chicago Press, Chicago. 431 pp.
———. 1965. The behavior of the mountain gorilla. Pp. 324–367 in *Primate Behavior*. I. DeVore, ed. Holt, Rinehart and Winston, New York.
Schenkel, R., and L. Schenkel-Hulliger. 1967. On the sociology of free-ranging colobus (*Colobus guereza caudatus* Thomas 1885). Pp. 185–194 in *Progress in Primatology*. D. Starck, R. Schneider, and H.-J. Kuhn, eds. G. Fischer Verlag, Stuttgart.
Schlichte, H. J. 1975. Nahrungsverhalten von Diademmeerkatzen in National Park Kahuzi-Biega, Kivuhochland, Zaire. *Z. Säugetierkd.* 40:193–214.
Schlitter, D. A., J. Phillips, and G. E. Kemp. 1973. The distribution of the white-collared mangabey, *Cercocebus torquatus*, in Nigeria. *Folia Primatol.* 19(5):380–383.
Schmidt, L. H. 1973. Infections with *Plasmodium falciparum* and *Plasmodium vivax* in the owl monkey-model systems for basic biological and chemotherapeutic studies. *Trans. R. Soc. Trop. Med. Hyg.* 67(4):446–470.
*Schouteden, H. 1947. De Zoogdieren van Belgisch-Congo en van Ruanda-Urundi. *Ann. Mus. Congo Belge* 2(3):576 pp.
Schulz, J. P., R. A. Mittermeier, and H. S. Reichart. 1977. Wildlife in Surinam. *Oryx* 14(2):133–144.
Scott, J. D. 1975. A gentle giant after all. *Int. Wildl.* 5(2):20–25.
Scott, N. J., T. T. Struhsaker, K. Glander, and H. Chirivi. 1976. Primates and their habitats in northern Colombia with recommendations for future management and research. Pp. 30–50 in *First Inter-American Conference on Conservation and Utilization of American Nonhuman Primates in Biomedical Research*. Scientific Publication no. 317. Pan American Health Organization, Washington, D.C.
Service, N. W. 1974. Survey of the relative prevalence of potential yellow fever vectors in north-west Nigeria. *Bull. W.H.O.* 50:487–494.
Setzer, H. W. 1959. Mammals of the Anglo-Egyptian Sudan. *Proc. U.S. Nat. Mus.* 106(3377):447–587.
Sharma, I. K. 1976. The grey langur. *Loris* 14(1):7–11.
Sheppe, W., and P. Haas. 1976. Large mammal populations of the lower Chobe River, Botswana. *Mammalia* 40(2):223–243.
Sheppe, W., and T. Osborne. 1971. Patterns of use of a flood plain by Zambian mammals. *Ecol. Monogr.* 41:179–205.
Shidei, T. 1975. Conservation of woodland for the preservation of mammals population in Japan. Pp. 491–496 in *Proc. Symp. 5th Cong. Int. Primatol. Soc.* S. Kondo, M. Kawai, A. Ehara, S. Kawamura, eds. Japan Science Press, Tokyo.
Siddiqi, M. F., and C. H. Southwick. 1977. Population trends and dynamics of rhesus monkeys in Aligarh District. Pp. 14–23 in *Use of Non-human Primates in Biomedical Research*. M. R. N. Prasad and T. C. Anand Kumar, eds. Indian National Science Academy, New Delhi, India.
Sikes, S. K. 1964a. A game survey of the Yankari Reserve of northern Nigeria. Part 1. *Nigerian Field* 29(2):54–82.
———. 1964b. A game survey of the Yankari Reserve of northern Nigeria. Part 2. *Nigerian Field* 29(3):127–141.
Simon, N. M. 1969. Proposals for field investigations of rare and endangered mammals. *Biol. Conserv.* 1(4):280–290.
Simonds, P. E. 1965. The bonnet macaque in south India. Pp. 175–196 in *Primate Behavior*. I. DeVore, ed. Holt, Rinehart and Winston, New York.

Singh, S. D. 1969. Urban monkeys. *Sci. Am.* 221:108–115.
Smith, C. C. 1977. Feeding behaviour and social organization in howling monkeys. Pp. 97–126 in *Primate Ecology*. T. H. Clutton-Brock, ed. Academic Press, London.
Smith, J. D. 1970. The systematic status of the black howler monkey. *Alouatta pigra* Lawrence. *J. Mammal.* 51:358–369.
Smith, N. J. H. 1976. Utilization of game along Brazil's transamazon highway. *Acta Amazonica* 6(4):455–466.
———. 1978. Human exploitation of terra firme fauna in Amazonia. *Ciênc. Cult.* 30(1):17–23.
Smithers, R. H. N. 1966. *The Mammals of Rhodesia, Zambia, and Malawi.* Collins, London.
———. 1968. *A Checklist and Atlas of the Mammals of Botswana.* Trustees, Nat. Mus. of Rhodesia. 169 pp.
Snyder, P. A. 1972. Behavior of *Leontopithecus rosalia* (the golden lion marmoset) and related species: A review. Pp. 23–49 in *Saving the Lion Marmoset*. D. D. Bridgewater, ed. Wild Animal Propagation Trust, Wheeling, West Virginia.
Soini, P. 1972. The capture and commerce of live monkeys in the Amazonian region of Peru. *Int. Zoo Yearb.* 12:26–36.
———. 1978. Informe sobre estudios primatológicos 1976–1977. III. Ecología y dinámica poblacional de *Cebuella pygmaea* (Primates: Callitrichidae). Paper presented at Second Inter-American Conference on Conservation and Utilization of American Nonhuman Primates in Biomedical Research, Belém, Brazil. Pan American Health Organization, Washington, D.C.
Sommer, A. 1976. Attempt at an assessment of the world's tropical forests. *Unasylva* 28:5–25.
Soria, M. F. 1959. Otros casos de intrusion del vomer en el genero "*Cebus*." *Comun. Mus. Argent. Cienc. Nat. Bernardino Rivadavia (Cienc. Zool.)* 3:131–144.
Soulé, M. E. 1980. Thresholds for survival: Maintaining fitness and evolutionary potential. Pp. 151–169 in *Conservation Biology*. M. E. Soulé and B. A. Wilcox, eds. Sinauer Assoc., Sunderland, Mass.
Sousa, O. E., R. N. Rossan, and D. C. Baerg. 1974. The prevalence of trypanosomes and microfilariae in Panamanian monkeys. *Am. J. Trop. Med. Hyg.* 23:862–868.
Southwick, C. H., M. A. Beg, and M. R. Siddiqi. 1961a. A population survey of rhesus monkeys in villages, towns, and temples of northern India. *Ecology* 42:538–547.
———. 1961b. A population survey of rhesus monkeys in northern India II: Transportation routes and forest areas. *Ecology* 42:698–710.
———. 1965. Rhesus monkeys in north India. Pp. 111–159 in *Primate Behavior*. I. DeVore, ed. Holt, Rinehart and Winston, New York.
Southwick, C. H., and F. C. Cadigan, Jr. 1972. Population studies of Malaysian primates. *Primates* 13:1–18.
Southwick, C. H., A. Ghosh, and C. D. Louch. 1964. A roadside survey of rhesus monkeys in Bengal. *J. Mammal.* 45(3):443–448.
Southwick, C. H., and M. F. Siddiqi. 1970. Primate population trends in Asia with special reference to the rhesus monkey of India. *I.U.C.N. Publ.* New Series 17:135–147.
———. 1973. Population studies of primates in Asia. Paper presented at Joint Meeting of Primate Research Center Directors and the Primate Research Centers Advisory Committee, San Juan, Puerto Rico, May 17, 1973.
———. 1977a. Population dynamics of rhesus monkeys in northern India. Pp. 339–362 in *Primate Conservation*. H.S.H. Prince Rainier III of Monaco and G. H. Bourne, eds. Academic Press, New York.
———. 1977b. Demographic characteristics of semi-protected rhesus groups in India. Pp.

242–252 in *Yearb. Phys. Anthropol.*, vol. 20. J. Buettner-Janusch, ed. Am. Assoc. Phys. Anthropol., Washington, D.C.

Southwick, C. H., and M. R. Siddiqi. 1966. Population changes in rhesus monkeys (*Macaca mulatta*) in India, 1959–1965. *Primates* 7:303–314.

———. 1968. Population trends of rhesus monkeys in villages and towns of northern India, 1959–65. *J. Anim. Ecol.* 37:199–204.

Southwick, C. H., M. R. Siddiqi, and M. F. Siddiqi. 1970. Primate populations and biomedical research. *Science* 170:1051–1054.

Sowerby, A. de C. 1937. Mammals of China, Mongolia, eastern Tibet, and Manchuria requiring protection. *China J.* (Shanghai) 27(5):148–258.

Spencer, C. 1975. Interband relations, leadership behaviour, and the initiation of human-oriented behaviour in bands of semi-wild free ranging *Macaca fascicularis*. *Malay, Nat. J.* 29(2):83–89.

Spillett, J. J. 1966a. A report on wild life surveys in north India and southern Nepal. *J. Bombay Nat. Hist. Soc.* 63:492–612.

———. 1966b. A report on wildlife surveys in south and west India. *J. Bombay Nat. Hist. Soc.* 65:1–46.

———. 1968. A report on wild life surveys in south and west India. November–December 1966. *J. Bombay Nat. Hist. Soc.* 65:633–663.

Spinage, C. A. 1972. The ecology and problems of the Volcano National Park, Rwanda. *Biol. Conserv.* 4(3):194–204.

Starck, D., and H. Frick. 1958. Beobachtungen an aethiopischen Primaten. *Zool. Jahrb., Abt. Syst., Ökol. Geogr. Tiere* 86:41–70.

Starin, E. D. 1978. A preliminary investigation of home range use in the Gir Forest langur. *Primates* 19(3):551–568.

Sternberg, H. O. 1968. Man and environmental change in South America. Pp. 413–445 in *Biogeography and Ecology in South America*, vol. 1. E. Fittkau, J. Illies, H. Kling, G. Schwabe, and H. Sioli, eds. W. Junk, The Hague.

Stockholm International Peace Research Institute (S.I.P.R.I.). 1976. *Ecological Consequences of the Second Indochina War*. Almqvist and Wiksell Int., Stockholm, Sweden. 119 pp.

Stoltz, L. P., and M. E. Keith. 1973. A population survey of chacma baboon in the northern Transvaal. *J. Hum. Evol.* 2:195–212.

Stolz, L. P., and G. S. Saayman. 1970. Ecology and social organization of chacma baboon troops in the northern Transvaal. *Ann. Transvaal Mus.* 26:99–143.

Stott, K., Jr. 1960. Stuhlmann's blue monkey in the Kayonsa Forest, Uganda. *J. Mammal.* 41:400–401.

Stott, K., Jr., and C. J. Selsor. 1959. Chimpanzees in western Uganda. *Oryx* 5:108–115.

———. 1960. The orang-utan in North Borneo. *Oryx* 6:39–42.

———. 1961a. Observations of the maroon leaf monkey in North Borneo. *Mammalia* 25:184–189.

———. 1961b. Color variation in Bornean gray gibbons. *J. Mammal.* 42:99.

Struhsaker, T. T. 1967a. Behavior of vervet monkeys (*Cercopithecus aethiops*). *Univ. Calif. Publ. Zool.* 82:1–64.

———. 1967b. Ecology of vervet monkeys (*Cercopithecus aethiops*) in the Masai-Amboseli Game Reserve, Kenya. *Ecology* 48:891–904.

———. 1969. Correlates of ecology and social organization among African cercopithecines. *Folia Primatol.* 11:80–118.

———. 1970. Notes on *Galagoides demidovii* in Cameroon. *Mammalia* 34:207–211.
———. 1971. Notes on *Cercocebus a. atys* in Senegal, West Africa. *Mammalia* 35:343–344.
———. 1972. Rain-forest conservation in Africa. *Primates* 13(1):103–109.
———. 1973. A recensus of vervet monkeys in the Masai-Amboseli Game Reserve, Kenya. *Ecology* 54(4):930–932.
———. 1974a. Correlates of ranging behavior in a group of red colobus monkeys (*Colobus badius tephrosceles*). *Am. Zool.* 14:177–184.
———. 1974b. Of monkeys and men. *Anim. Kingdom* 77(2):25–30.
———. 1975. *The Red Colobus Monkey*. University of Chicago Press, Chicago. 311 pp.
———. 1976a. A further decline in numbers of Amboseli vervet monkeys. *Biotropica* 8:211–214.
———. 1976b. The dim future of La Macarena. *Oryx* 13(3):298–302.
———. 1977a. Report on a survey of red colobus monkeys in the Magombero Forest Reserve-Lukoga, Msolwa and Pala Ulanga Forest Reserve, Madizini, Tanzania. Unpublished mimeograph. New York Zool. Soc. 8 pp.
———. 1977b. Report on a survey of the Zanzibar red colobus monkeys. Unpublished mimeograph. New York Zool. Soc. 3 pp.
Struhsaker, T. T., and J. S. Gartlan. 1970. Observations on the behaviour and ecology of the patas monkey (*Erythrocebus patas*) in the Waza Reserve, Cameroon. *J. Zool.* 161:49–63.
Struhsaker, T. T., K. Glander, H. Chirivi, and N. J. Scott. 1975. A survey of primates and their habitats in northern Colombia. Pp. 43–78 in *Primate Censusing Studies in Peru and Colombia*. Pan American Health Organization, Washington, D.C.
Struhsaker, T. T., and P. Hunkeler. 1971. Evidence of tool-using by chimpanzees in the Ivory Coast. *Folia Primatol.* 15:212–219.
Struhsaker, T. T., and J. F. Oates. 1975. Comparisons of the behavior and ecology of red colobus and black and white colobus monkeys in Uganda: A summary. Pp. 103–123 in *Socioecology and Psychology of Primates*. R. H. Tuttle, ed. Mouton, The Hague.
Sturm, H., A. Abouchaar L., R. De Bernal, and C. Dettoyos. 1970. Distribución de animales en las capas bajas de un bosque humedo tropical de la region Carare-Opon (Santander, Colombia). *Caldasia* 10:529–578.
Sugiyama, Y. 1964. Group composition, population density, and some sociological observations of hanuman langurs (*Presbytis entellus*). *Primates* 5(3–4):7–37.
———. 1967. Social organization of hanuman langurs. Pp. 221–236 in *Social Communication Among Primates*. S. Altmann, ed. University of Chicago Press, Chicago.
———. 1968a. The ecology of the lion-tailed macaque [*Macaca silenus* (Linnaeus)]: A pilot study. *J. Bombay Nat. Hist. Soc.* 65:283–293.
———. 1968b. Social organization of chimpanzees in the Budongo forest, Uganda. *Primates* 9:225–258.
———. 1971. Characteristics of the social life of bonnet macaques, *Macaca radiata*. *Primates* 12(3–4):247–266.
———. 1973. The social structure of wild chimpanzees: A review of field studies. Pp. 376–410 in *Comparative Ecology and Behaviour of Primates*. R. P. Michael and J. H. Crook, eds. Academic Press, London.
———. 1976. Characteristics of the ecology of the Himalayan langurs. *J. Hum. Evol.* 5:249–277.
Sugiyama, Y., K. Yoshiba, and M. D. Parthasarathy. 1965. Home range, breeding season, male group, and inter-troop relations in hanuman langurs (*Presbytis entellus*). *Primates* 6:73–107.

Sussman, R. W. 1974. Ecological distinctions in sympatric species of *Lemur*. Pp. 75–108 in *Prosimian Biology*. R. D. Martin, G. A. Doyle, and A. C. Walker, eds. University of Pittsburgh Press, Pittsburgh.

———. 1975. A preliminary study of the behavior and ecology of *Lemur fulvus rufus* Audebert 1800. Pp. 237–258 in *Lemur Biology*. I. Tattersall and R. W. Sussman, eds. Plenum Press, New York.

———. 1977a. Feeding behaviour of *Lemur catta* and *Lemur fulvus*. Pp. 1–36 in *Primate Ecology*. T. H. Clutton-Brock, ed. Academic Press, London.

———. 1977b. Socialization, social structure, and ecology of two sympatric species of *Lemur*. Pp. 515–528 in *Primate Bio-Social Development*. S. Chevalier-Skolnikoff and F. E. Poirier, eds. Garland Press, New York.

———. 1977c. Distribution of the Malagasy lemurs. Part 2: *Lemur catta* and *Lemur fulvus* in southern and western Madagascar. *Ann. N.Y. Acad. Sci.* 293:170–184.

Sussman, R. W., and A. Richard. 1974. The role of aggression among diurnal prosimians. Pp. 49–76 in *Primate Aggression, Territoriality, and Xenophobia*. R. L. Holloway, ed. Academic Press, New York.

Sussman, R. W., and I. Tattersall. 1976. Cycles of activity, group composition, and diet of *Lemur mongoz mongoz* Linneaus 1766 in Madagascar. *Folia Primatol.* 26:270–283.

Sutherland, W. J. E. 1972. Development of the Olambwe Valley Game Reserve in south-west Kenya. *Biol. Conserv.* 4(2):148–150.

Suzuki, A. 1969. An ecological study of chimpanzees in a savanna woodland. *Primates* 10:103–148.

———. 1971. On the problems of conservation of the chimpanzees in East Africa and of the preservation of their environment. *Primates* 12(3–4):415–419.

———. 1972. On the problems of the conservation of the Japanese monkey on the Boso Peninsula, Japan. *Primates* 13(3):333–335.

Swanepoel, P. 1975. Small mammals of the Addo Elephant National Park. *Koedoe* 18:103–130.

*Swinhoe, R. 1862. On the mammals of the island of Formosa (China). *Proc. Zool. Soc. London* 42:347–365.

Swynnerton, G. 1958. Fauna of the Serengeti National Park. *Mammalia* 22:435–450.

Tahiri-Zagrët, C. 1976. Le Cercopithecidae de Côte d'Ivoire. *Bull. Inst. Fondam. Afr. Noire* 38(1):206–230.

Tamura, T. 1970. Official statement of the Japan Monkey Centre on the "Shimokita Incident." *Primates* 11:403–405.

Tanaka, J. 1965. Social structure of Nilgiri langurs. *Primates* 6:107–123.

Tappen, N. C. 1960. Problems of distribution and adaptation of the African monkeys. *Curr. Anthropol.* 1:91–120.

———. 1964. Primate studies in Sierra Leone. *Curr. Anthropol.* 5(4):339–340.

———. 1965. Discussion. Pp. 14–15 in *The Baboon in Medical Research*, vol. 1. H. Vagtborg, ed. University of Texas Press, Austin.

Tattersall, I. 1972. Of lemurs and men. *Nat. Hist.* 81(3):32–43.

———. 1976a. Note sur la distribution et sur la situation actuelle des lémuriens des Comores. *Mammalia* 40(3):519–521.

———. 1976b. Group structure and activity rhythm in *Lemur mongoz* (Primates, Lemuriformes) on Anjouan and Moheli Islands, Comoro Archipelago. *Anthropol. Pap. Am. Mus. Nat. Hist.* 53(4):369–380.

*Reference not seen but cited by other authors.

———. 1976c. Notes on the status of *Lemur macaco* and *Lemur fulvus* (Primates, Lemuriformes). *Anthropol. Pap. Am. Mus. Nat. Hist.* 53(2):257–261.
———. 1977a. The lemurs of the Comoro Islands. *Oryx* 13(5):445–448.
———. 1977b. Ecology and behavior of *Lemur fulvus mayottensis* (Primates, Lemuriformes). *Anthropol. Pap. Am. Mus. Nat. Hist.* 54:425–482.
———. 1977c. Distribution of the Malagasy lemurs. Part 1: The lemurs of northern Madagascar. *Ann. N.Y. Acad. Sci.* 293:160–169.
———. 1978. Behavioral variation in *Lemur mongoz* (= *L. m. mongoz*). Pp. 127–132 in *Recent Advances in Primatology*, vol. 3: *Evolution*. D. J. Chivers and K. A. Joysey, eds. Academic Press, London.
Tattersall, I., and J. H. Schwartz. 1974. Craniodental morphology and the systematics of the Malagasy lemurs (Primates, Prosimii). *Anthropol. Pap. Am. Mus. Nat. Hist.* 52(3):139–192.
Tattersall, I., and R. W. Sussman, eds. 1975a. *Lemur Biology*. Plenum Press, New York. 365 pp.
Tattersall, I., and R. W. Sussman. 1975b. Notes on topography, climate, and vegetation of Madagascar. Pp. 13–21 in *Lemur Biology*. I. Tattersall and R. W. Sussman, eds. Plenum Press, New York.
———. 1975c. Observations on the ecology and behavior of the mongoose lemur, *Lemur mongoz mongoz* Linnaeus (Primates, Lemuriformes) at Ampijoroa, Madagascar. *Anthropol. Pap. Am. Mus. Nat. Hist.* 52(4):193–216.
Taub, D. M. 1975. A report on the distribution of the Barbary macaque, *Macaca sylvanus*, in Morocco and Algeria. Unpublished report to New York Zoological Society, Fauna Preservation Society, and I.U.C.N. 71 pp.
———. 1977. Geographic distribution and habitat diversity of the Barbary macaque. *Folia Primatol.* 27:108–133.
Teleki, G., E. E. Hunt, Jr., and J. H. Pfifferling. 1976. Demographic observations (1963–1973) on the chimpanzees of Gombe National Park, Tanzania. *J. Hum. Evol.* 5:559–598.
Telford, S. R., Jr., A. Herrer, and H. A. Christensen. 1972. Enzootic cutaneous leishmaniasis in eastern Panama III: Ecological factors relating to the mammalian hosts. *Ann. Trop. Med. Parasitol.* 66(2):173–179.
Tenaza, R. R. 1975. Territory and monogamy among Kloss' gibbons (*Hylobates klossii*) in Siberut Island, Indonesia. *Folia Primatol.* 24:60–80.
———. 1976. Songs, choruses, and countersinging of Kloss' gibbons (*Hylobates klossii*) in Siberut Island, Indonesia. *Z. Tierpsychol.* 40:37–52.
Tenaza, R. R., and W. J. Hamilton III. 1971. Preliminary observations of the Mentawi Island gibbon *Hylobates klossii*. *Folia Primatol.* 15:201–210.
Thorington, R. W., Jr. 1967. Feeding and activity of *Cebus* and *Saimiri* in a Colombian forest. Pp. 180–184 in *Progress in Primatology*. D. Starck, R. Schneider, and H.-J. Kuhn, eds. G. Fischer Verlag, Stuttgart.
———. 1968a. Observations of squirrel monkeys in a Colombian forest. Pp. 69–85 in *The Squirrel Monkey*. L. A. Rosenblum and R. W. Cooper, eds. Academic Press, New York.
———. 1968b. Observations of the tamarin *Saguinus midas*. *Folia Primatol.* 9:95–98.
———. 1969. The study and conservation of New World Monkeys. *An. Acad. Bras. Cienc.* 41:253–260.
———. 1971. The identification of primates used in viral research. *Lab. Anim. Sci.* 21(6):1074–1077.
———. 1972. Importation, breeding, and mortality of New World primates in the United States. *Int. Zoo Yearb.* 12:18–23.

———. 1975. The relevance of vegetational diversity for primate conservation in South America. Pp. 547–553 in *Proc. Symp. 5th Congr. Int. Primatol. Soc.* S. Kondo, M. Kawai, A. Ehara, and S. Kawamura, eds. Japan Science Press, Tokyo.

———. 1976. The systematics of New World monkeys. Pp. 8–18 in *First Inter-American Conference on Conservation and Utilization of American Nonhuman Primates in Biomedical Research.* Scientific Publication no. 317. Pan American Health Organization, Washington, D.C.

———. 1978. Some problems relevant to the conservation of the Callitrichidae. Pp. 1–11 in *Marmosets in Experimental Medicine.* N. Gengozian and F. Dienhardt, eds. Primates in Medicine, vol. 10. S. Karger, Basel.

Thorington, R. W., Jr., R. Castro, A. Coimbra-Filho, R. Cooper, M. Freese, R. Mittermeier, P. Neyman, and P. Soini. 1978. Conservation of the Callitrichidae: Panel discussion. Pp. 147–158 in *The Biology and Conservation of the Callitrichidae.* D. G. Kleiman, ed. Smithsonian Institution Press, Washington, D.C.

Thorington, R. W., Jr., and C. P. Groves. 1970. An annotated classification of the Cercopithecoidea. Pp. 629–647 in *Old World Monkeys.* J. R. Napier and P. H. Napier, eds. Academic Press, New York.

Thorington, R. W., Jr., and P. G. Heltne, eds. 1976. *Neotropical Primates: Field Studies and Conservation.* National Academy of Sciences, Washington, D.C. 135 pp.

Thorington, R. W., Jr., and R. E. Vorek. 1976. Observations on the geographic variation and skeletal development of *Aotus.* *Lab. Anim. Sci.* 26:1006–1021.

Tilson, R. L. 1977. Social organization of Simakobu monkeys (*Nasalis concolor*) in Siberut Island, Indonesia. *J. Mammal.* 58(2):202–212.

Tilson, R. L., and R. R. Tenaza. 1976. Monogamy and duetting in an Old World monkey. *Nature* 263:320–321.

Times Atlas of the World. 1977. Times Books, London.

Tinley, K. L., A. J. Rosinha, J. L. P. Lobão Tello, and T. P. Dutton. 1976. Wildlife and wild places in Mozambique. *Oryx* 13(4):344–350.

Tinoco V., R. A. 1969. Mono de Colombia los cébidos. *Mensaje Bol. Inform. Fed. Iberoamer. Parques Zool.* 6:13–14.

Trapido, H., and P. Galindo. 1955. The investigation of a sylvan yellow fever epizootic on the north coast of Honduras, 1954. *Am. J. Trop. Med. Hyg.* 4(4):665–674.

Tyler, S. J. 1974. Observations of the mammals and birds of the Chow Bahr or Lake Stephanie area. *Walia* 5:2–3.

Uehara, S. 1975. The importance of the temperate forest elements among woody food plants utilized by Japanese monkeys and its possible historical meaning for the establishment of the monkeys' range: A preliminary report. Pp. 392–400 in *Contemporary Primatology, Proc. 5th Int. Congr. Primatol.* S. Kondo, M. Kawai, and A. Ehara, eds. S. Karger, Basel.

Ullrich, W. 1961. Zur Biologie und Soziologie der Colobusaffen (*Colobus guereza caudatus* Thomas 1885). *Zool. Gart.* 25:305–368.

U.S. Department of Health, Education, and Welfare. 1975. Restrictions on importation of nonhuman primates. *Federal Register* 40(155):part 71 and subpart J-3.

U.S. Department of Interior. 1977a. International trade in endangered species of wild fauna and flora. *Federal Register* 42(35):10462–10488.

———. 1977b. Endangered and threatened wildlife and plants. *Federal Register* 42(135):36420–36431.

Van Citters, R. L., O. A. Smith, Jr., D. L. Franklin, W. S. Kemper, and N. W. Watson. 1967. Radio telemetry of blood flow and blood pressure in feral baboons: A preliminary report.

Pp. 473–492 in *The Baboon in Medical Research*, vol. 2. H. Vagtborg, ed. University of Texas Press, Austin.
Vandebroek, G. 1958. Note écologiques sur les anthropoides africains. *Ann. Soc. R. Zool. Belg*. 89:203–211.
Van Dyke, J. M. 1972. *North Vietnam's Strategy for Survival*. Pacific Books, Palo Alto, Calif. 336 pp.
Van Gulick, R. H. 1967. *The Gibbon in China, an Essay in Chinese Animal Lore*. E. J. Brill, Leiden, Holland. 123 pp.
Van Peenen, P. F. D., M. L. Cunningham, and J. F. Duncan. 1970. A collection of mammals from Con Son Island, Vietnam. *J. Mammal*. 51:419–424.
Van Peenen, P. F. D., S. W. Joseph, A. Saleh, R. H. Light, S. Sukeri, and R. See. 1974. The Indonesian developmental area study: Observations on mammals from south and east Kalimantan (Borneo). *Southeast Asian J. Tropical Med. Pub. Hlth*. 5(3):390–397.
Van Peenen, P. F. D., R. H. Light, and J. F. Duncan. 1971. Observations on mammals of Mt. Sontra, South Vietnam. *Mammalia* 35:126–143.
Van Peenen, P. F. D., P. F. Ryan, and R. H. Light. 1969. *Preliminary Identification Manual for Mammals of South Vietnam*. U.S Nat. Mus., Washington, D.C.
Vanzolini, P. E. 1978. Current problems of primate conservation in Brasil. Pp. 15–25 in *Recent Advances in Primatology*, vol. 2: *Conservation*. D. J. Chivers and W. Lane-Petter, eds. Academic Press, London.
Vargas-Mendez, O., and N. W. Elton. 1953. Naturally acquired yellow fever in wild monkeys of Costa Rica. *Am. J. Trop. Med. Hyg*. 2:850–863.
Veblen, T. T. 1976. The urgent need for forest conservation in highland Guatemala. *Biol. Conserv*. 9(2):141–154.
*Veloso, H. P. 1946. A vegetaçao no municipio de Ilhéus, Estado da Bahia. *Mem. Inst. Oswaldo Cruz* (Rio de Janeiro) 4:13–103, 221–293, 323–341.
Verheyen, W. N. 1963. New data on the geographical distribution of *Cercopithecus* (*Allenopithecus*) *nigroviridis* Pocock 1907. *Rev. Zool. Bot. Afr*. 68(3–4):393–396.
Verschuren, J. 1975a. Wildlife in Zaïre. *Oryx* 13(1):25–33.
———. 1975b. Wildlife in Zaíre. *Oryx*. 13(2):149–163.
———. 1978a. Burundi and wildlife: Problems of an overcrowded country. *Oryx* 14(3): 237–240.
———. 1978b. Les grands mammifères du Burundi. *Mammalia* 42(2):209–224.
Vessey, S. H., B. K. Mortenson, and N. A. Muckenhirn. 1978. Size and characteristics of primate groups in Guyana. Pp. 187–188 in *Recent Advances in Primatology*, vol. 1: *Behaviour*. D. J. Chivers and J. Herbert, eds. Academic Press, London.
*Vieira, C. C. 1944. Os simios do Estado de São Paulo. *Papeis Avul. Dep. de Zool. Sec. de Agric., São Paulo* 4(1):1–31.
*———. 1955. Lista remissiva dos mamíferos do Brasil. *Arq. Zool., São Paulo* 8:341–474.
Vincent, F. 1968. La sociabilité du Galago de Demidoff. *Terre Vie* 22:51–56.
———. 1969. Contribution a l'étude des prosimiens africains: Galago de demidoff, reproduction (biologie, anatomie, physiologie) et comportement. Ph.D. diss. Faculty of Sciences, Paris. Vols. 1 and 2.
———. 1972. Prosimiens africains: V. Répartition géographique de *Euoticus inustus*. *Ann. Fac. Sci. Cameroun* 10:135–141.
———. 1978. Thermoregulation and behaviour in two sympatric galagos: An evolutionary

*Reference not seen but cited by other authors.

factor. Pp. 181–187 in *Recent Advances in Primatology*, vol. 3: *Evolution*. D. J. Chivers and K. A. Joysey, eds. Academic Press, London.

Vogel, C. 1971. Behavioral differences of *Presbytis entellus* in two different habitats. Pp. 41–47 in *Proc. 3rd Int. Congr. Primatol.*, vol. 3. S. Karger, Basel.

———. 1975. Intergroup relations of *Presbytis entellus* in the Kumaon Hills and in Rajasthan (North India). Pp. 450–458 in *Contemporary Primatology, Proc. 5th Int. Congr. Primatol.* S. Kondo, M. Kawai, and A. Ehara, eds. S. Karger, Basel.

———. 1976. Ökologie, Lebensweise, und Sozialverhalten der grauen Languren in verschiedenen Biotopen Indiens. *Fortschr. Verhaltensforsch.* 17:1–159.

———. 1977. Ecology and sociology of *Presbytis entellus*. Pp. 24–45 in *Use of Non-human Primates in Biomedical Research*. M. R. N. Prasad and T. C. Anand Kumar, eds. Indian National Science Academy, New Delhi, India.

Waddell, M. B., and R. M. Taylor. 1946. Studies on cyclic passage of yellow fever virus in South American mammals and mosquitoes. II. Marmosets (*Callithrix penicillata* and *Leontocebus chrysomelas*) in combination with *Aedes aegypti*. *Am. J. Trop. Med.* 26:455–463.

Wade, N. 1978a. India bans monkey export: U.S. may have breached accord. *Science* 199:280–281.

———. 1978b. New vaccine may bring man and chimpanzee into tragic conflict. *Science* 200:1028–1030.

Walker, E. P. 1968. *Mammals of the World*. 2nd ed. Vol. I. Johns Hopkins University Press, Baltimore. 644 pp.

Waller, R. H. 1972. Observations on the wildlife sanctuaries of India. *J. Bombay Nat. Hist. Soc.* 69(3):574–590.

Walsh, J., and R. Gannon. 1967. *Time Is Short and the Water Rises*. Thomas Nelson and Sons, Camden, N.J. 224 pp.

Waltermire, R. G. 1975. A national park in the Bale Mountains. *Walia* 6:20–24.

Waser, P. 1975. Monthly variations in feeding and activity patterns of the mangabey, *Cercocebus albigena* (Lydekker). *East Afr. Wildl. J.* 13:249–263.

———. 1976. *Cercocebus albigena*: Site attachment, avoidance, and intergroup spacing. *Am. Nat.* 110:911–935.

———. 1977. Feeding, ranging, and group size in the mangabey *Cercocebus albigena*. Pp. 183–222 in *Primate Ecology*. T. H. Clutton-Brock, ed. Academic Press, London.

Waser, P. M., and O. Floody. 1974. Ranging patterns of the mangabey, *Cercocebus albigena*, in the Kibale Forest, Uganda. *Z. Tierpsychol.* 35:85–101.

Washburn, S. L., and I. DeVore. 1961a. The social life of baboons. *Sci. Am.* 204:62–71.

———. 1961b. Social behavior of baboons and early man. *Viking Fund Publ. Anthropol.* 31:91–106.

Wayre, P. 1968a. Some observations on the golden langur *Presbytis geei* (Ms. Khajuria) Gee. *J. Bombay Nat. Hist. Soc.* 65(2):473–477.

———. 1968b. The golden langur and the Manas Sanctuaries. *Oryx* 9(5):337–340.

———. 1969. Wildlife in Taiwan. *Oryx* 10(1):46–56.

Webb-Peploe, C. G. 1947. Field notes on the mammals of South Tinnevelly, south India. *J. Bombay Nat. Hist. Soc.* 46(4):629–644.

Weber, B. E. 1972. A parks system for West Malaysia. *Oryx* 11(6):461–469.

———. 1977. Orang-utans. *Tigerpaper* 4(2):24.

Weiss, M. L., and M. Goodman. 1972. Frequency and maintenance of genetic variability in natural populations of *Macaca fascicularis*. *J. Hum. Evol.* 1:41–48.

Wellde, B. T., A. J. Johnson, J. S. Williams, and E. H. Sadun. 1972. Experimental infection with *Plasmodium falciparum* in *Aotus* monkeys. I. Parasitologic, hematologic, and serum biochemical determinations. *Am. J. Trop. Med. Hyg.* 21:260–271.

Western, D., and D. M. Sindiyo. 1972. The status of the Amboseli rhino population. *East Afr. Wildl. J.* 10:43–57.

Wetzel, R. M., and J. W. Lovett. 1974. A collection of mammals from the Chaco of Paraguay. *Univ. Connecticut Occ. Pap.*, Biol. Sci. Ser. 2(13):203–216.

Whiten, A., and T. J. Rumsey. 1973. "Agonistic buffering" in the wild barbary macaque, *Macaca sylvana* L. *Primates* 14(4):421–425.

Whitmore, T. C. 1980. The conservation of tropical rain forest. Pp. 303–318 in *Conservation Biology*. M. E. Soulé and B. A. Wilcox, eds. Sinauer Assoc., Sunderland, Mass.

Whittemore, T. C. 1972. Observations of *Callimico goeldii* and *Saguinus mystax*. M. S. thesis. Harvard University. 12 pp.

Wilcox, B. A. 1980. Insular ecology and conservation. Pp. 95–117 in *Conservation Biology*. M. E. Soulé and B. A. Wilcox, eds. Sinauer Assoc., Sunderland, Mass.

Wilkinson, A. F. 1974. Areas to preserve in Sierra Leone. *Oryx* 12(5):596–597.

Willoughby, D. P. 1978. *All About Gorillas*. Barnes, South Brunswick, N.J. 264 pp.

Wilson, C. C., and W. L. Wilson. 1975a. The influence of selective logging on primates and some other animals in East Kalimantan. *Folia Primatol.* 23:245–274.

———. 1977. Behavioral and morphological variation among primate populations in Sumatra. Pp. 207–233 in *Yearb. Phys. Anthropol.*, vol. 20. J. Buettner-Janusch, ed. Am. Assoc. Phys. Anthropol., Washington, D.C.

Wilson, V. J. 1968. Weights of some mammals from eastern Zambia. *Arnoldia* 32(3):1–20.

———. 1975. *Mammals of the Wankie National Park, Rhodesia*. Museum Memoir no. 5. Trustees of National Museums and Monuments of Rhodesia, Salisbury, Rhodesia.

Wilson, W. L., and C. C. Wilson. 1972–73. Census of Sumatran Primates. Unpublished progress reports 1–4. Washington Regional Primate Research Center, Seattle. 180 pp.

———. 1975b. Species specific vocalizations and the determination of phylogenetic affinities of the *Presbytis aygula-melalophos* group in Sumatra. Pp. 459–463 in *Contemporary Primatology, Proc. 5th Int. Congr. Primatol.* S. Kondo, M. Kawai, and A. Ehara, eds. S. Karger, Basel.

Wolfe, K. A. 1974. Comparative behavioral ecologies of chimpanzees and gorillas. *Collected Papers*, Dept. Anthropol., University of Oregon, 7:53–65.

Wolfheim, J. H. 1975. Activity patterns and sex differences in behavior of talapoin monkeys (*Miopithecus talapoin*). Ph.D. diss. University of California, Berkeley. 167 pp.

———. 1976. The perils of primates. *Nat. Hist.* 85(8):90–99.

Wozniewicz, W. D. 1974. The Amazon Highway system. Pp. 291–314 in *Man in the Amazon*. C. Wagley, ed. University of Florida Press, Gainesville, Fla.

Wrangham, R. W. 1977. Feeding behaviour of chimpanzees in Gombe National Park, Tanzania. Pp. 504–538 in *Primate Ecology*. T. H. Clutton-Brock, ed. Academic Press, London.

———. 1978. Conservation: Digit's death and the future of the Virunga gorillas. *Primate Eye* 10:17–18.

Wright, P. C. 1978. Home range, activity pattern and agonistic encounters of a group of night monkeys (*Aotus trivirgatus*) in Peru. *Folia Primatol.* 29:43–55.

Wycherley, P. R. 1969. *Conservation in Malaysia*. I.U.C.N. Publ. New Series, suppl. no. 22. Morges, Switzerland. 207 pp.

Yamada, M. 1966. Five natural troops of Japanese monkeys in Shodoshima Island. I. Distribution and social organization. *Primates* 7:315–363.

Yerkes, R. M. 1943. *Chimpanzees, a Laboratory Colony*. Yale University Press, New Haven, Conn. 321 pp.

Yong, H.-S. 1973. Totally protected and protected wild mammals of peninsular Malaysia. *Malay. Nat. J.* 26:77–80.

Yoshiba, K. 1964. Report of the preliminary survey on the orang-utan in North Borneo. *Primates* 5(1–2):11–26.

———. 1967. An ecological study of hanuman langurs, *Presbytis entellus*. *Primates* 8:127–155.

———. 1968. Local and intertroop variability in ecology and social behavior of common Indian langurs. Pp. 217–242 in *Primates: Studies in Adaptation and Variability*. P. C. Jay, ed. Holt, Rinehart and Winston, New York.

Yunis, E. J., O. M. Torres de Caballero, C. Ramírez, and E. Ramírez Z. 1976. Chromosomal variations in the primate *Alouatta seniculus seniculus*. *Folia Primatol.* 25:215–224.

Zimmerman, D. A. 1972. The avifauna of the Kakamega Forest, western Kenya, including a bird population study. *Bull. Am. Mus. Nat. Hist.* 149:255–339.

Appendix

Primates imported into the United States, 1968–73 (absence of a species indicates no records of importation for that species in these years).

Species	1968	1969	1970	1971	1972	1973	Total
Lorisidae							
Arctocebus calabarensis	0	0	0	1	0	2	3
Galago alleni	0	0	0	1	0	0	1
Galago crassicaudatus	29	0	0	0	30	0	59
Galago demidovii	47	2	15	151	0	0	215
Galago senegalensis	92	109	61	161	158	99	680
Galago sp. unidentified	121	71	5	0	0	0	197
Loris tardigradus	10	12	0	2	0	0	24
Nycticebus coucang	62	72	73	72	58	84	421
Perodicticus potto	20	3	0	50	0	0	73
Total	381	269	154	438	246	185	1,673
Tarsiidae							
Tarsius sp. unidentified	1	0	0	0	0	1	2
Cheirogaleidae							
Cheirogaleus medius	0	3	0	0	0	0	3
Microcebus murinus	4	5	4	2	0	0	15
Total	4	8	4	2	0	0	18
Lemuridae							
Hapalemur griseus	0	0	3	0	0	0	3
Lemur catta	2	60	8	3	1	0	74
Lemur macaco	1	4	0	0	0	6	11
Lemur mongoz	0	2	14	9	0	0	25
Lepilemur Group	0	1	0	0	0	0	1
Varecia variegata	0	0	5	3	0	0	8
Total	3	67	30	15	1	6	122
Indriidae							
Propithecus verreauxi	4	7	0	0	0	0	11
Callitrichidae							
Callimico goeldii	83	43	49	0	4	0	179
Callithrix argentata	57	40	22	0	0	0	119
Callithrix humeralifer	0	0	3	0	19	5	27
Callithrix jacchus	156	98	233	53	57	14	611
Callithrix sp. unidentified	22	1	44	0	10	0	77
Cebuella pygmaea	197	639	192	166	111	91	1,396
Leontopithecus rosalia	50	149	150	0	0	0	349
Saguinus fuscicollis	289	197	0	293	50	64	893
Saguinus labiatus	92	9	0	0	0	2	103

Appendix (continued)

Species	1968	1969	1970	1971	1972	1973	Total
Callitrichidae (cont.)							
Saguinus leucopus	33	0	0	0	0	0	33
Saguinus midas	0	33	0	1	0	20	54
Saguinus mystax	0	0	1,779	863	1,064	1,244	4,950
Saguinus nigricollis	3,522	1,570	332	1,787	1,933	1,130	10,274
Saguinus oedipus	3,098	3,758	2,077	2,386	2,430	130	13,879
Genus or sp. unidentified	89	0	1	3	68	161	322
Total	7,688	6,537	4,882	5,552	5,746	2,861	33,266
Cebidae							
Alouatta caraya	0	0	0	18	4	33	55
Alouatta palliata	101	93	88	47	33	10	372
Alouatta seniculus	16	12	12	11	7	16	74
Alouatta sp. unidentified	13	36	1	0	0	0	50
Aotus trivirgatus	4,087	5,312	4,209	3,728	3,533	2,636	23,505
Ateles paniscus	2,273	2,659	2,676	1,889	2,070	508	12,075
Brachyteles arachnoides	1	38	0	0	0	25	64
Cacajao calvus	127	69	14	0	0	1	211
Cacajao melanocephalus	4	0	0	0	0	0	4
Callicebus moloch	51	43	40	40	24	14	212
Callicebus sp. unidentified	141	116	121	134	42	34	588
Cebus albifrons	4,913	4,743	3,170	2,221	2,776	970	18,793
Cebus apella	784	1,024	847	2,036	1,975	595	7,261
Cebus capucinus	1,768	1,574	1,764	1,133	1,209	800	8,248
Cebus nigrivittatus	106	78	36	33	0	0	253
Cebus sp. unidentified	106	99	118	196	103	130	752
Chiropotes satanas	4	6	12	0	6	16	44
Lagothrix lagothricha	2,902	3,311	2,244	2,226	2,125	1,275	14,083
Pithecia monachus	162	24	66	83	30	17	382
Pithecia pithecia	10	2	2	1	0	10	25
Saimiri sciureus	45,014	47,096	26,124	29,879	25,297	13,709	187,119
Total	62,583	66,335	41,544	43,675	39,234	20,799	274,170
Cercopithecidae							
Allenopithecus nigroviridis	0	0	1	3	0	0	4
Cercocebus aterrimus	0	0	0	0	0	2	2
Cercocebus torquatus	1	9	7	4	11	13	45
Cercopithecus aethiops	6,352	2,981	3,106	2,823	3,274	3,243	21,779
Cercopithecus ascanius	1	0	0	6	3	0	10
Cercopithecus cephus	4	4	4	0	3	0	15
Cercopithecus diana	24	11	11	0	0	10	56
Cercopithecus erythrotis	0	0	3	0	0	0	3
Cercopithecus hamlyni	0	0	2	0	0	0	2
Cercopithecus mitis	12	5	4	0	0	12	33
Cercopithecus mona	29	12	1	21	14	10	87
Cercopithecus neglectus	23	9	17	3	35	65	152

Appendix (continued)

Species	1968	1969	1970	1971	1972	1973	Total
Cercopithecidae (cont.)							
Cercopithecus nictitans	2	5	6	25	30	45	113
Cercopithecus petaurista	0	0	0	2	0	0	2
Cercopithecus sp. unidentified	0	0	296	15	0	0	311
Colobus guereza	8	23	5	7	9	13	65
Colobus polykomos	9	0	6	0	0	0	15
Colobus sp. unidentified	0	0	0	6	0	0	6
Erythrocebus patas	87	132	156	105	221	263	964
Macaca arctoides	1,043	1,721	1,070	1,207	1,676	878	7,595
Macaca cyclopis	35	0	0	20	15	0	70
Macaca fascicularis	2,137	1,188	1,609	1,727	1,397	2,984	11,042
Macaca fuscata	23	12	0	0	0	0	35
Macaca mulatta	30,933	27,462	23,302	22,097	23,210	20,738	147,742
Macaca nemestrina	572	479	662	436	581	369	3,099
Macaca nigra	47	106	10	11	12	35	221
Macaca radiata	0	51	14	28	0	0	93
Macaca silenus	9	4	2	2	3	0	20
Macaca sylvana	1	13	4	5	0	0	23
Macaca sp. unidentified	0	49	94	34	39	5	221
Miopithecus talapoin	108	0	0	0	5	0	113
Nasalis larvatus	6	4	8	2	0	0	20
Papio anubis	459	189	577	751	1,063	1,119	4,158
Papio cynocephalus	36	20	0	260	20	20	356
Papio hamadryas	159	23	62	22	89	39	394
Papio leucophaeus	3	5	0	0	1	1	10
Papio papio	91	74	0	32	131	274	602
Papio sphinx	10	0	47	0	4	8	69
Papio sp. unidentified	0	142	66	27	20	18	273
Presbytis aygula-melalophus	1	0	0	2	0	0	3
Presbytis cristata	2	0	0	35	34	3	74
Presbytis entellus	3	3	3	20	12	8	49
Presbytis obscura	0	2	13	10	24	2	51
Presbytis phayrei	0	4	1	0	0	0	5
Presbytis pileatus	0	2	0	0	0	0	2
Presbytis senex	0	0	1	0	1	3	5
Presbytis sp. unidentified	0	5	4	0	3	0	12
Pygathrix nemaeus	12	3	15	5	0	0	35
Theropithecus gelada	378	552	278	20	3	8	1,239
Total	42,620	35,304	31,467	29,773	31,943	30,188	201,295
Hylobatidae							
Hylobates concolor	29	34	7	34	11	2	117
Hylobates hoolock	0	0	0	0	0	1	1
Hylobates lar	102	69	81	56	86	79	473
Hylobates syndactylus	30	34	14	43	85	78	284
Hylobates sp. unidentified	0	0	0	6	32	0	38
Total	161	137	102	139	214	160	913

Appendix (continued)

Species	1968	1969	1970	1971	1972	1973	Total
Pongidae							
Gorilla gorilla	13	2	6	5	0	6	32
Pan troglodytes	255	292	185	205	234	134	1,305
Pongo pygmaeus	1	0	0	1	2	0	4
Total	269	294	191	211	236	140	1,341

NOTE: For grand totals see Table 191.

SOURCES: Jones 1970; Jones and Paradiso 1972; Paradiso and Fisher 1972; Clapp and Paradiso 1973; Clapp 1974; Greenhall, in prep.

Presenting a well-organized compilation of the available information on distribution, abundance, and condition of natural populations of each species of monkey, ape, and prosimian, this book is a major reference work based on more than 1,000 sources and about 200 unpublished contributions. Included are 151 succinct yet comprehensive status accounts for all species of nonhuman primates.

Each species account contains an original range map, sections on taxonomy, distribution, abundance and density, habitat, factors affecting populations (habitat alteration and human predation), conservation action, and detailed tables showing particulars of distribution as reported in the literature and in extensive personal communications between fieldworkers and the author.

This book is the first to examine each species in the whole order from the standpoint of its population status. Other recent books on primates contain information on only a few selected species.

In a summary chapter the author synthesizes the salient information regarding all the species of primates and discusses the implications of these data. She rates the severity of each of the six major factors affecting the individual species and draws conclusions about the potential of each species for survival.

JACLYN H. WOLFHEIM became interested in primate biology in 1964 while working on a study of mother-infant behavior of pigtailed macaques at the Regional Primate Research Center in Seattle. She received her Bachelor of Science in zoology (with College Honors) from the University of Washington in 1966. During her graduate work at the University of California at Berkeley, she studied the behavior of talapoin monkeys in the laboratory and in the field in Cameroon, leading to her M.A. (1970) and Ph.D. (1975), both in zoology. Dr. Wolfheim has done research for the U.S. Fish and Wildlife Service at the National Museum of Natural History in Washington, D.C., and has taught biology at George Mason University in Fairfax, Virginia. She is now on the faculty of the Department of Zoology and Entomology, Colorado State University, Fort Collins, Colorado, and prepared the present volume with the financial support of the New York Zoological Society.